Key for world maps inside the front and back covers

A.	Albania	Lie.	Liechtenstein
Afg.	Afghanistan	Lu.	Luxembourg
An.	Andorra	M.	Macedonia
Ar.	Armenia	Mo.	Montenegro
Au.	Austria	Mol.	Moldova
Az.	Azerbaijan	N.	Netherlands
B.	Belgium	Nor.	Norway
Ba.	Bahrain	Nth. Korea	North Korea
Ban.	Bangladesh	Pol.	Poland
Be.	Benin	Q.	Qatar
BH.	Bosnia and Herzegovina	R.	Romania
Bots.	Botswana	S.	Slovakia
Bu.	Burkina Faso	Se.	Senegal
Bul.	Bulgaria	Ser.	Serbia
C.	Croatia	Sl.	Slovenia
CAR.	Central African Republic	S.M.	San Marino
		St.K. & N.	St. Kitts & Nevis
CZ	Czech Republic	St.L.	St. Lucia
Den.	Denmark	St.V.	St. Vincent & the Grenadines
El Salv.	El Salvador		
Eq. Guinea	Equatorial Guinea	Sth. Korea	South Korea
Fin.	Finland	Sw.	Swaziland
G.	Germany	Sw.	Switzerland
G.-B.	Guinea-Bissau	Swe.	Sweden
Gab.	Gabon	Taj.	Tajikistan
Gam.	The Gambia	T.L.	Timor-Leste
Gh.	Ghana	Tu.	Turkmenistan
Gre.	Grenada	U.A.E.	United Arab Emirates
H.	Hungary	U.K.	United Kingdom
Is.	Israel	Ug.	Uganda
Kazak.	Kazakhstan	Ukr.	Ukraine
Kyr.	Kyrgyzstan	Uzb.	Uzbekistan
L.	Lebanon	Zam.	Zambia
Li.	Lithuania	Zimb.	Zimbabwe

Preface

At the end of the first decade of the twenty-first century the world was faced with two overlapping crises. The more immediate was the series of banking failures that cascaded through global financial markets and plunged some of the world's richest economies deep into debt. The more insidious was a creeping environmental disaster, as the legacy of centuries of industrialization and emissions of greenhouse gases continued to threaten irrevocable climate change.

Both crises have demonstrated all too vividly the ramifications of globalization: how mistaken or irresponsible actions in one country can rapidly reverberate across the world. The transmission mechanisms may be different—the electronic ether of the financial markets, or the constantly shifting composition of the upper atmosphere—but both demonstrate the intricate and dynamic interplay within complex global systems.

The implications are certainly global, but the responses have to be national. Global leaders may agree among themselves what to do, but ultimately the buck stops at home. The banks, for all their global reach, have to be bailed out by national treasuries. And carbon emissions cannot be systematically reduced without targets agreed by parliaments and enforced through national courts.

Everyone lives in a sovereign state—where a single authority holds sway over a defined territory. This institution emerged first in the cities of ancient Greece, but started to take a more modern form in the nineteenth century in Europe, to be completed from the 1960s onwards as dozens of new countries laid claim to their patches of land, sectioning the earth's surface into a complex quilt of countries sewn together, somewhat clumsily, along national borders.

Throughout this period, the state has also been deeply enmeshed with the idea of the nation—assumed to be a group of people who share a common culture, language, history and values. The neatest solution would be for each national group to occupy its own territory—giving rise to a quasi-mythical entity, the 'nation-state'.

But that heroic hyphen between nation and state links two very different entities. The state has a precise legal identity—based not least on the power to tax and to enforce its authority, with violence if necessary. But the nation is a much more amorphous notion. It has some intuitive appeal: the French, surely, are different from the Japanese, or the Bantu, or the Australian Aborigines. Travel only a little further back in time, however, and every nation melts away—seen to be the product of centuries of overlapping waves of migration and intermarriage. For all the chauvinistic flag waving, the exercise of defining who is uniquely a member of the French, or Brazilian, or Thai nation is ultimately futile.

No state consists of one national group, and some groups who consider themselves nations, the Chinese perhaps, or the Jews, are scattered across many states. Nevertheless the idea of the 'nation- state' has a persistent power—capable at its most harmless of rallying support for a national sports team or, at its most lethal, stirring visceral racial hatred.

At times, the nation-state has seemed under threat. Right-wing nationalists have, for example, worried that international organizations, from the League of Nations onwards, would absorb proud national institutions into some overarching global administration—a prospect that seems ever more remote. At the other end of the spectrum, are those who fret that the power of governments has been undermined by global corporations, another fear that seems to have receded as some of the mightiest companies have collapsed into the arms of governments.

This is not to deny the value of supranational institutions, or the power and reach of global corporations, but more to reiterate the persistence of national sovereignty— and to justify the organization of this *Guide*, country by country.

The use of 'country' in the title of this *Guide* may appear to evade the disputed notion of the nation-state.

Unfortunately not. According to the Oxford English Dictionary a country is 'a nation with its own government, occupying a particular territory'— ultimately rooted, somewhat depressingly, in the Latin contra, meaning 'against'.

To escape this definitional circle, the choice of entries in this *Guide* might be based on membership of the United Nations. But what of Taiwan, for example, or Kosovo, both of which clearly exist but are not recognized globally as independent states?

Ultimately, the selection of entries is pragmatic and not entirely consistent. The same might be said of the international bodies, whose profiles have been added for this edition.

This *Guide* is meant to offer readers a brief impression of countries with which they may be unfamiliar. In practice, however, I have found that the first country that readers turn to is their own. This is as good a way as any of calibrating what is inevitably a collection of perspectives that are hopefully reasonably objective but inevitably highly personal.

Introduction

Each country has an accompanying box summarizing some of the more significant data. Such information, here and elsewhere, should be treated with caution. Many countries do not collect data in a consistent and comprehensive fashion. The size and distribution of population, for example, can be considered politically sensitive and thus subject to manipulation.

Less reliable still are the estimates on such issues as language, ethnic group structure, and religion. These take little account of the fact, for example, that many people speak a number of languages. People can also profess multiple and overlapping ethnic identities, and while for some people such matters are central to their lives, for others they may be of scant significance. Nevertheless these issues can have political consequences so some indication, however approximate, can be useful. In each case the data are for the latest available year. Where a figure is quoted in dollars, this refers to US dollars.

Sources of information

Economic and social data have been taken from various editions of the UNDP *Human Development Report,* the ILO *World Employment Report*, and the *World Bank World Development Report*. Also useful have been the national *Human Development Reports* prepared locally for individual countries along with their regional equivalents. For consistency, most of the estimates for ethnic structures and language come from the *CIA World Factbook.*

More recent political developments have been traced through a number of printed publications. These include principally: the *Economist*, the *Economist Intelligence Unit* country reports, the *Financial Times*, the *Guardian,* and the *New York Times*.

Of the sources available on the internet, among the most useful ones have been *OneWorld Online* (oneworld.net) and *Oxford Reference Online* (oxfordreference.com) where you will also find the electronic version of this book.

Terms used in the text

This book is intended for the general reader so tries to avoid most jargon. It may be useful, however, to indicate the usage of some of the terms that crop up regularly.

Gross Domestic Product (GDP). This is the total value of all the goods and services produced in a country. This can then be allocated to different forms of activity—agriculture, industry, and services, say—to give an indication of the structure of the economy.

Gross National Product (GNP). This is similar to GDP but adds to it the net income from abroad—from overseas investments, say. The GNP is thus equivalent to a country's total income. Dividing this by the population, to give the GNP per capita, gives an indication of how rich or poor a country is.

Human development. This is an all-encompassing view of human well-being, that takes into account not just income but many other issues such as health and standards of education.

Human development index (HDI). This is an attempt to measure human development. It combines data on income, educational attainment, and life expectancy into a single composite figure. This can then be used to rank countries in order of success. Not all countries can provide sufficient data to do this. The 2009 *Human Development Report* from the United Nations Development Programme was able to rank 182 countries—Norway was at the top; Niger at the bottom.

Poverty rate. This is the proportion of people living on less than a certain sum. For international comparisons of the situation in developing countries, this is set at $1.25 per day. Each country also sets its own national poverty rates these are based on the proportion of people who can afford to consume a certain number of calories daily along with essential non-food items. National poverty rates, being based on national standards, better reflect local circumstances and aspirations, though may not be internationally comparable. In richer countries poverty rates are set differently—typically the proportion of people who earn less than a certain proportion, often half, of the average income. These thus measure 'relative poverty'.

Purchasing power parity (PPP). If you compare the per capita GDP or GNP between different countries using the standard exchange rate to convert the local currency into US dollars you can get a misleading result. This is because the cost of living is lower in some countries than others. Purchasing power parity takes into account the real purchasing power of the local currency. So comparisons of GNP as $PPP give a better indication of relative national incomes.

( ) SEE WEB LINKS

This book has a companion website which offers useful links to official government websites as well as independent national news sources. You will find this at www.oup.com/uk/reference/resources/countriesoftheworld.

Contents

Afghanistan	1	Cuba	81
Albania	3	Cyprus	83
Algeria	5	Czech Republic	84
Angola	7	Denmark	86
Argentina	9	Djibouti	88
Armenia	11	Dominican Republic	89
Australia	13	Ecuador	91
Austria	15	Egypt	93
Azerbaijan	17	El Salvador	95
Bahamas	19	Equatorial Guinea	97
Bahrain	20	Eritrea	98
Bangladesh	21	Estonia	100
Barbados	24	Ethiopia	102
Belarus	25	Fiji	104
Belgium	27	Finland	105
Belize	29	France	107
Benin	30	French Guiana	109
Bhutan	32	French Polynesia	110
Bolivia	33	Gabon	111
Bosnia and Herzegovina	35	The Gambia	113
Botswana	37	Georgia	115
Brazil	39	Germany	117
Brunei	42	Ghana	119
Bulgaria	43	Greece	121
Burkina Faso	45	Guadeloupe	123
Burma (Myanmar)	47	Guam	124
Burundi	49	Guatemala	125
Cambodia	51	Guinea	127
Cameroon	53	Guinea-Bissau	129
Canada	55	Guyana	131
Cape Verde	57	Haiti	132
Central African Republic	58	Honduras	134
Chad	60	Hungary	136
Channel Islands	62	Iceland	138
Chile	63	India	139
China	65	Indonesia	142
Colombia	68	Iran	145
Comoros	70	Iraq	147
Congo	71	Ireland	149
Congo, Dem. Rep.	73	Israel	151
Costa Rica	75	Italy	153
Côte d'Ivoire	77	Jamaica	155
Croatia	79	Japan	157

Jordan	160	Pakistan	242
Kazakhstan	162	Palestine	245
Kenya	164	Panama	247
Korea, North	166	Papua New Guinea	249
Korea, South	168	Paraguay	251
Kosovo	170	Peru	253
Kuwait	172	Philippines	255
Kyrgyzstan	174	Poland	257
Laos	176	Portugal	259
Latvia	178	Puerto Rico	261
Lebanon	180	Qatar	263
Lesotho	182	Réunion	264
Liberia	184	Romania	265
Libya	186	Russia	267
Lithuania	188	Rwanda	270
Luxembourg	190	St Lucia	272
Macedonia	191	St Vincent and Grenadines	273
Madagascar	193	Samoa	274
Malawi	195	São Tomé and Príncipe	275
Malaysia	197	Saudi Arabia	276
Maldives	199	Senegal	278
Mali	200	Serbia	280
Malta	202	Sierra Leone	282
Martinique	203	Singapore	284
Mauritania	204	Slovakia	286
Mauritius	206	Slovenia	288
Mayotte	208	Solomon Islands	290
Mexico	209	Somalia	291
Micronesia	211	South Africa	293
Moldova	212	Spain	295
Mongolia	214	Sri Lanka	297
Montenegro	216	Sudan	299
Morocco	217	Suriname	301
Mozambique	219	Swaziland	302
Namibia	221	Sweden	303
Nepal	223	Switzerland	305
Netherlands	225	Syria	307
Netherlands Antilles	227	Taiwan	309
New Caledonia	228	Tajikistan	311
New Zealand	229	Tanzania	313
Nicaragua	231	Thailand	315
Niger	233	Timor-Leste	317
Nigeria	235	Togo	318
Norway	238	Tonga	320
Oman	240	Trinidad and Tobago	321

Tunisia 323
Turkey 325
Turkmenistan 327
Uganda 329
Ukraine 331
United Arab Emirates 333
United Kingdom 335
United States 337
Uruguay 340
Uzbekistan 342
Vanuatu 344
Venezuela 345
Vietnam 347
Virgin Islands (US) 349
Western Sahara 350
Yemen 351
Zambia 353
Zimbabwe 355

Smaller countries

American Samoa 358
Andorra 358
Anguilla 359
Antigua and Barbuda 359
Aruba 360
Bermuda 360
British Virgin Islands 361
Cayman Islands 361
Christmas Island 362
Cocos Islands 362
Cook Islands 363
Dominica 363
Faroe Islands 364
Falkland Islands 364
Gibraltar 365
Greenland 365
Grenada 366
Isle of Man 366
Kiribati 367
Liechtenstein 367
Marshall Islands 368

Monaco 368
Montserrat 369
Nauru 369
Niue 370
Norfolk Island 370
Northern Mariana Islands 371
Palau 371
St Helena 372
St Kitts and Nevis 372
St Pierre and Miquelon 373
San Marino 373
Seychelles 374
Tokelau 374
Turks and Caicos 375
Tuvalu 375
Vatican 376
Wallis and Futuna 376

International organizations

United Nations 377
Bretton Woods organizations 379
World Trade Organization 380
G8 and G20 381
Organization of American
    States 382
European Union 382
Commonwealth of
    Independent States 384
African Union 384
Arab League 385
Association of South-East
    Asian Nations 386
South Asian Association
    for Regional Cooperation 387
Pacific Islands Forum 387

Indicator tables

Notes on tables 390
1. Income and poverty 392
2. Health and population 402
3. Education and gender 412

Afghanistan

Still a battleground between the Taliban and government and NATO forces.

Land area: *652,000 sq. km.*
Population: *26 million—urban 25%*
Capital city: *Kabul, 3.0 million*
People: *Pashtun 42%, Tajik 27%, Hazara 9%, Uzbek 9%, other 13%*
Language: *Afghan Persian (Dari) 50%, Pashtu 35%, Turkic languages 11%, other 4%*
Religion: *Muslim: Sunni 80%, Shia 19%*
Government: *Islamic republic*
Life expectancy: *44 years*
GDP per capita: *$PPP 1,054*
Currency: *Afghani*
Major exports: *Opium, fruits, carpets*

Afghanistan is a largely mountainous country: more than half the territory is above 2,000 metres. Its central highlands are the western end of the Himalayan chain and include the Hindu Kush mountain range. The richest and most intensively cultivated agricultural land is in the lowland plains of the north. The land to the south and south-west is mostly desert or semi-desert. In recent years the environment has been devastated by conflict and drought.

The main ethnic group in Afghanistan are the Pashtun. They live throughout the country, but particularly in the south and east. Though they are the dominant group, the Pashtun are by no means homogenous and there have been frequent conflicts between different tribes and sub-tribes. The second largest group are Tajiks. Not organized on tribal lines, they are found mainly in the north-east and the west, particularly in the capital, Kabul, where they have made up most of the educated élite. The third main group, and among the poorest, are the Hazara, who live in the central highlands or as labourers or servants in Kabul.

Other groups include Uzbeks and Turkmen, descendants of refugees from the former Soviet Union who live in the northern plain. Almost all Afghans are Muslims, mostly Sunni.

In 2007 the population was estimated at 26 million. Since 1979, at least one million people have been killed. Millions more became refugees; around 4.5 million have returned, but in mid-2006 there were still 2.6 million in Pakistan. Levels of human development in Afghanistan were never high and are now among the worst in the world. Rates of infant and maternal mortality are high and most parts of the country lack access to clean water, adequate sanitation, and sufficient food.

Around one-third of health facilities have been severely damaged and many doctors have left the country. There have been some successful interventions by aid agencies, including a child immunization campaign. But health standards are very low particularly for women—with one of the world's highest rates of maternal mortality. Education too is in a grim state. Most qualified teachers left the country during the Taliban period and girls' schools were closed. Schools have now been reopened but they are struggling to cope with the new influx of pupils. Universities have also reopened. But illiteracy remains the highest in the world.

Women during Taliban rule could not be seen in public unless almost completely covered. But now girls can go to school and a number of women hold national positions.

Most people have traditionally relied on subsistence agriculture, raising livestock or growing crops on irrigated land—chiefly wheat, corn, and rice. But incessant conflict, including the sowing of landmines, has undermined food production and also destroyed most industry and infrastructure. Around half the government's budget has to come from international aid.

The only thriving activity seems to

Opium dominates the economy

be the cultivation of opium. The Taliban had banned production, but since then prices have risen and by 2006 total production was around 6,100 tons—worth around half of GDP. Local warlords make a good income by taxing the trade.

Political power has never been highly centralized in Afghanistan. Allegiances have been based more on region or tribe or on occupational group. The most recent conflicts date from 1978, when Soviet-supported communists seized power and launched an attack on Islam, provoking widespread tribal revolts. The Soviet Union, fearful of an Islamic republic on its doorstep, from 1979 sent in more than 100,000 troops to fight the Mujahideen (holy warriors). Localized rebellions then exploded into a more organized jihad, a holy war, in which the Mujahideen were backed by the USA and Pakistan.

After the Soviet Union withdrew in 1989, warlords representing different militias fought ferocious battles among themselves. Into this chaotic environment came the Taliban, a fundamentalist Islamic group which in 1996 seized Kabul, and imposed their authoritarian rule.

The Taliban soon found themselves in conflict with the United States for harbouring terrorists, notably Osama bin Laden. After the terrorist attack on New York on September 11, 2001, the US launched a sustained bombing campaign that allowed dissidents to oust the Taliban in November 2001.

UN-chaired peace talks led to a presidential election in October 2004 won by Pashtun leader Hamid Karzai. In September 2005, elections were held for the lower house, the Wolesi Jirga, where most people serve as individuals.

From 2005, the Taliban resumed their offensive and caused high casualty rates among NATO troops, primarily US and British. There is also now serious concern among Western governments about the Karzai regime which is considered by many to be corrupt and incompetent. They were not reassured by a fraudulent presidential election in 2009 which gave Karzai more than half the vote.

Albania

One of the poorest of the former communist states, Albania is now moving towards the European Union

Land area: 29,000 sq. km.
Population: 3.1 million—urban 48%
Capital city: Tirana, 0.3 million
People: Albanian
Language: Albanian 95%, Greek 3%, other 2%
Religion: Muslim 70%, Albanian Orthodox 20%, Roman Catholic 10%
Government: Republic
Life expectancy: 77 years
GDP per capita: $PPP 7,740
Currency: Lek
Major exports: Metals, electricity

Though Albania is a coastal country, not much of its territory is at sea level. Mountains make up more than two-thirds of the land area: north, west, and south—highest in the north and centre; lower in the south where they merge with the mountains of northern Greece. Even the coastal plain in the west and centre of the country is scattered with hills.

Most Albanians live in the coastal plain, which has more agriculture and industry. They are one of Europe's more homogenous populations—a product partly of the country's physical isolation. Nevertheless there are some language differences, since there are two dialects of Albanian. Most people in the north speak Gheg while those in the south, along with Albanians in Kosovo, speak Tosk.

Albania previously had very rapid population growth. The communist regime promoted fertility but following the collapse of communism, the birth rate plummeted as many people left the country in search of work. More than one million Albanians are thought to be overseas, chiefly in Italy, former Yugoslavia, and Greece—and most of them unauthorized. Remittances from overseas workers also provide most of the foreign exchange—around $1.5 billion in 2008.

Because of the way Albania's borders were delineated in 1921 many ethnic Albanians also finished up in neighbouring countries: around two million live in Kosovo and hundreds of thousands in Macedonia.

Living standards were low during the communist era but subsequent economic collapse and social chaos saw standards slip further. Health indicators have improved, however, and life expectancy has increased in recent years. Even though education is free only around two-thirds of school-age children are enrolled

Around one-fifth of GDP comes from agriculture, though despite extensive terracing only around one-quarter of the country is suitable for arable farming. By 1994, the collective farms had been dismantled and almost all land had been returned to small farmers. Privatization boosted the output of grains, maize, and vegetables. Albania could become a significant producer of fruit and vegetables for the EU, but irrigation is limited, technology primitive, and productivity low.

Albanian manufacturing industry collapsed when exposed to world markets. Nevertheless, Albania's relatively educated and low-waged workforce has proved attractive to investors for the production of low-tech products like garments and footwear, in companies working as subcontractors for foreign corporations.

Prospects for the mining industry are reasonable. Albania was at one point the world's third largest producer of chrome ore and has significant deposits of other minerals such as copper. The industry has now been fully privatized and foreign companies now exploit the chrome and copper mines.

Economic development has been held back by poor infrastructure. Only a small proportion of the roads are paved, though new highways are now being constructed.

For 40 years until his death in 1985, Albania was in the grip of the repressive, and steadily more isolated, regime of Enver Hoxha. When the communist system collapsed, two main parties emerged. The first was the reformed communist party, now called the Socialist Party of Albania (SP). The second was the right-wing Democratic Party of Albania (DP).

Elections for the People's Assembly in 1992 resulted in a convincing victory for the DP, which used its majority to elect Sali Berisha as president. Berisha pushed through a number of economic reforms but his government became steadily more corrupt and repressive. Berisha and the DP also won the 1996 elections—though the opposition protested that these were fraudulent.

Albania's economic reforms had made the country feel more prosperous and this, combined with remittances, had tempted unsophisticated investors to sink their money into fraudulent financial 'pyramid' schemes. When these

Chaos as pyramid schemes collapse

schemes collapsed spectacularly at the end of 1996 around half the population lost their savings. This provoked widespread riots and an insurrection that required the intervention of an Italian-led peacekeeping force. It also led to the collapse of the DP government, which had been linked to the bogus schemes.

The ensuing 1997 election was won by the SP. The new government was initially led by Fatos Nano who over the next few years had several spells as prime minister. The SP administrations were consistently undermined by internal feuds.

In 2005, the centre-right DP, still led by Sali Berisha, returned after parliamentary elections. In 2007, Bamir Topi of the DP was elected president. The DP also won in 2009 but only narrowly, so had to form a coalition with a smaller party, the Socialist Movement for Integration.

In 2009, Albania applied for EU candidate status, a move which both main parties supported. But membership seems a long way off.

Algeria

Algeria is now much more peaceful, but its democracy is fragile and still vulnerable to occasional terrorism

Land area: 2,382,000 sq. km.
Population: 34 million—urban 66%
Capital city: Algiers, 4.8 million
People: Arab-Berber 99%, other 1%
Language: Arabic, Tamazight (Berber)
Religion: Sunni Muslim 99%, other 1%
Government: Republic
Life expectancy: 72 years
GDP per capita: $PPP 7,740
Currency: Algerian dinar
Major exports: Oil, gas

Algeria has a narrow coastal plain which is regularly broken by parts of the Atlas range of mountains which run across the country from east to west and between which there are high plateaux. To the south, the mountains give way to the arid, sandy expanse of the Sahara that accounts for around 90% of the territory.

Most Algerians, 90% of whom live in the coastal region, are of mixed Arab and Berber descent and are Sunni Muslims. But around one-fifth, particularly in the Kabylia region, consider themselves to be Berber rather than Arab and have protested against official efforts at 'Arabization'. In response, the government declared in 2002 that Tamazight would become an official language.

Although standards of human development have improved in recent years, and primary school enrolment by 2005 had reached 99%, while adult literacy is now around 70%. One of the main problems is unemployment, officially around 12%, but much higher for young people.

Algeria's rapid development after independence in 1962 was based on huge reserves of hydrocarbons in the Sahara. Initially the emphasis was on oil extraction, and Algeria still produces around 1% of the world's oil and has five refineries. But in recent decades the balance has swung in favour of natural gas which now makes up 80% of hydrocarbon production. Hydrocarbons account for more than 98% of exports and 45% of GDP.

Algeria also has extensive reserves of iron ore as well as smaller deposits of other metals. In the past, mining has been relatively neglected, but the government is now keen to develop the mining industry by inviting foreign investment for extracting phosphorus, gold, and diamonds.

In the 1970s, Algeria used its oil revenues to finance industrialization in a number of other sectors, including steel, vehicles, and cement. Around three-quarters of the manufacturing sectors remains in state hands though there have been efforts to attract foreign partners in oil, gas, and other industries. As a result, most people in formal employment work for the state.

Only a small proportion of the land, mostly in the coastal plains and valleys, is suitable for agriculture, which employs around 11% of the workforce. The country does grow cereals, and exports fresh dates, but has to import around half of its food.

Algeria waged a long and bitter war for independence from France. In the decades following independence in 1962, political life in Algeria was dominated by the only legal party, the Front de libération national (FLN). Development initially took the form of state socialism but this started to unravel in the mid-1980s with the fall in oil prices and the government started to liberalize the economy.

From 1988, the younger generation became increasingly resentful of the FLN's grip on power and protested at its austerity measures. The president, Colonel Chadli Benjedid, responded in 1989 with a new constitution that legalized other parties. Radical Islamic groups took full advantage, and in 1990 the Front Islamique du

Radical Islamic groups win an election

Salut (FIS) gained control of local government in most major cities. Then in 1992 they looked like taking power nationally. After the first round of voting in the parliamentary elections, the FIS took 188 of the 231 seats.

Faced with the prospect of an Islamic government, the military leadership stepped in, cancelled the election, suspended parliament, and forced Benjedid to resign. This provoked a ferocious backlash. The armed wing of the FIS, the Islamic Salvation Army (AIS), targeted not just the military but most other secular groups. The government, led by a series of army appointees, responded in kind, and armed village militias unleashed carnage on a horrific scale. The AIS agreed a truce in 1997, but the fighting continued with some of the worst atrocities now committed by other loosely organized groups: the Groupe Islamique Armée (GIA) and the Groupe Salafi pour la predication et le combat (GSPC) as well as by the army. By 1999, the war had cost more than 80,000 lives.

Prospects for peace improved in 1999 with the election of Algeria's first civilian president, Abdelaziz Bouteflika, a former exile whom the military had asked to return. In July 1999, he launched a 'civil accord' and around 3,000 rebels returned to their homes. Life in Algiers is now more normal though there are still attacks by Islamic militants of the GSPC, now called al-Qaida in the Islamic Maghreb, which carried out a series of bombings in 2009.

In 2002 the FLN won almost half the seats in the parliament and in 2004 Bouteflika, with the support of the army, was comfortably re-elected president defeating the FLN candidate Ali Benflis. The FLN won the 2007 parliamentary election, but with a significantly reduced majority.

Bouteflika has achieved a degree of independence from the military but has also entrenched his own power. In 2008 he had the parliament amend the constitution to remove the two-term presidential limit, and in 2009, standing as an 'independent', but with the backing of the three largest parties, he was duly re-elected, with an improbable 90% of the vote.

Angola

Angola's economy is growing thanks to oil exports but the country still faces a long period of post-war reconstruction.

Land area: 1,247,000 sq. km.	
Population: 18 million—urban 59%	
Capital city: Luanda, 4.5 million	
People: Ovimbundu 37%, Kimbundu 25%, Bakongo 13%, other 25%	
Language: Portuguese, Bantu, and others	
Religion: Indigenous beliefs 47%, Roman Catholic 38%, Protestant 15%	
Government: Republic	
Life expectancy: 47 years	
GDP per capita: $PPP 5,385	
Currency: Kwanza	
Major exports: Oil, diamonds, gas	

Angola has a dry coastal plain which is at its broadest in the north where it extends up to 200 kilometres inland. From this plain the territory rises through a number of escarpments to highlands that reach 2,600 metres. Beyond the highlands is a vast plateau that covers around two-thirds of the country. The population is concentrated in and around the highlands.

The largest ethnic group are the Ovimbundu, who are to be found mostly in the central highlands as well as in the coastal towns. The second largest are the Mbundu, who live more along the coast and in the north and north-west; they tend to be the most urbanized and many speak only Portuguese. The third group, the Bakongo, are also to be found in the north, as well as in neighbouring countries.

Decades of civil war until 2002 devastated a potentially rich country. At least half a million people were killed in the fighting.

The health system is still in a poor state and many hospitals lack the most basic equipment. Many of the services are provided by aid organizations. Added to this is the largely unmeasured devastation caused by HIV and AIDS.

The war also destroyed much of the education system, but it is now being rebuilt and around two-thirds of children are now enrolled in school.

Most people struggle to survive from subsistence agriculture. The land in the fertile highlands allows for ample food crops of cassava, beans, maize, millet, and sorghum. The war destroyed agriculture, particularly the commercial farms, and although food production is rising the country remains heavily dependent on food aid. Around one-third of children are malnourished.

The north-west of Angola used also to be a major coffee producer. The output of this and other cash crops like sisal and cotton has yet to recover.

Though agriculture was devastated by the war, two other areas of economic activity were less affected: oil extraction, which funds the government; and diamond mining, which funded the rebel UNITA army.

Oil was first produced offshore in 1955 and Angola has become Sub-Saharan Africa's largest exporter after Nigeria. In 2003, total reserves were over 13 billion barrels. Output has been growing and by 2012 should be around 2.2 million barrels per day. Oil accounts for 60% of GDP and more than 90% of exports. At present most of the associated gas is flared off but there are now a number of projects for liquifying and exporting.

The national oil company Sonangol has joint ventures with many companies, including Chevron. Sonangol's accounts, however, are very opaque, and it is an important part of the system of 'parallel finance' which in recent years has siphoned off around $4.2 billion into private pockets.

Oil riches siphoned off into private pockets

Angola is also the world's fifth largest diamond producer. These alluvial deposits are spread over large areas. Since the war hampered geological surveys, there are probably many more deposits to be discovered. This trade too is largely controlled by the army and government officials.

Though the oil economy is growing, this is not providing much employment. Even on the Chinese-finance reconstruction programme, 60% of the workers are Chinese. More than half the population live below the poverty line.

Angola's civil war started even before independence in 1975. There were three main independence movements. The Movimento Popular de Libertação de Angola (MPLA) was more urban-based and Marxist and had Soviet and Cuban support. The União Nacional para a Independencia Total de Angola (UNITA) was supported by the Ovimbundu in the rural areas and had the backing in particular of South Africa. A third group, the Frente Nacional de Libertação de Angola (FNLA), represented the Bakongo in the north.

The MPLA took control of Luanda when the Portuguese left and gradually achieved international recognition, while the UNITA–FNLA alliance fought a bitter rearguard action in the rural areas. A ceasefire in 1989 was followed by a presidential election in 1992 which was won by the no-longer-Marxist MPLA led by José Eduardo dos Santos. However, the defeated UNITA leader, Jonas Savimbi, refused to accept the result and plunged the country back into civil war.

The war finally ended in 2002 after Savimbi was killed in an ambush. UNITA quickly capitulated and has since transformed itself into a political party, now led by Isaias Samakuva. Legislative elections were held in 2008 and gave the MPLA 80% of the vote and 191 of the 220 seats in the National Assembly.

Meanwhile Mr dos Santos remains all powerful since he controls a dense network of patronage and has the backing of the state-controlled media. In 2010 a new constitution established that the president would no longer be elected but be the head of the party with the most parliamentary seats. This could enable Mr dos Santos to be president until 2022.

Argentina

Painful economic reforms have stimulated rapid economic growth, but economic and social stability remain elusive

Land area: 2,767,000 sq. km.
Population: 40 million—urban 92%
Capital city: Buenos Aires, 17 million
People: White 97%; mestizo, Indian and others 3%
Language: Spanish
Religion: Roman Catholic 92%, other 8%
Government: Republic
Life expectancy: 75 years
GDP per capita: $PPP 13,328
Currency: Peso
Major exports: Oil, cereals, processed food, vegetable oil

To the west, Argentina is bounded by the southernmost section of the Andes, but most of the country is a vast plain that descends gradually eastwards to sea level. The north-eastern part of this plain includes semi-tropical forests and Argentina's section of the arid Gran Chaco region, while to the south lies the semi-desert tableland of Patagonia. But it is the rich agricultural area of the central plain, the pampas, that is home to two-thirds of the population.

Argentines are of European descent, predominantly Italian. The country was originally settled by the Spanish who optimistically named it 'land of silver'. In fact, the territory had few precious metals though good prospects for agriculture. The largest wave of immigration, from the 1880s to the early years of the 20th century, brought people to work on the farms and ranches of the pampas. Around half were Italian, and most of the rest Spanish, though there were also Welsh who settled in Patagonia, as well as English investors and managers who developed many of the roads and railways. There are now few descendants of the original Indian inhabitants.

During its peak years of immigration Argentina was among the ten richest countries in the world, and a land of immense promise. But long decades of political upheaval and social stagnation have stifled development. As a result of a series of economic crises in the 1990s, many people left the country: more than half a million Argentines now live abroad.

In many ways Argentina still has great potential. It has rich agricultural resources. Cattle ranching, the basis of its early wealth, still provides substantial export income and beef remains a mainstay of the diet. Argentines eat 57 kilos each per year, three times as much as Europeans. In recent years the fertile soils of the pampas have also permitted a rapid increase in cereal and oilseed exports. However, agriculture now only employs around 8% of the workforce and has been subject to increasingly heavy taxation.

By the mid-1990s the major export was oil and the country is still a significant producer though largely for local use. Argentina is also an important source of natural gas which it exports to neighbouring countries, notably Brazil and Chile.

Argentina is also a major manufacturer. The opening up of the economy in the 1990s boosted productivity, and output has increased substantially. The food industry, for example, was previously the preserve of family firms but has gradually been penetrated by multinationals such as Nabisco and Cadbury.

This was accompanied by other free-market reforms, including privatizing banks and the national airline. This encouraged foreign investors and delivered growth but also exposed Argentina to international capital markets. By 2001 Argentina was again in crisis and defaulted on some of its international debt. Subsequently, helped by a boom in commodity exports, growth revived until the 2008 global economic crisis.

Historically Argentina's politics has been dominated by 'caudillos', strongmen. One of the most significant was General Juan Peron. Peron was a

The legacy of Juan Peron and Evita

populist, who, with the support of his charismatic wife, Evita, took over as president in 1946. He worked closely with the strong labour unions and nationalized many industries, rising to semi-mythical status. The Peronist party is still a significant force.

Peron's overthrow in 1955 launched Argentina on three decades of economic and political instability

punctuated by military coups. One of the most violent periods followed a coup in 1976 when the military government launched the 'dirty war' against leftist guerrillas. This resulted in widespread torture and murder, with up to 30,000 'disappeared'.

But the military made a fatal miscalculation in its 1982 invasion of the Malvinas Islands ruled by the British as the Falklands. The British responded by sending a task force which retook the islands, a crushing defeat that effectively put paid to the era of military rule. The new presidency of Raul Alfonsín helped to re-establish democracy but he was replaced in 1989 by Carlos Menem.

Menem was a Peronist but rapidly set about dismantling Peron's statist legacy. His free-market policies ushered in a period of rapid economic growth—though at immense social cost.

The Peronists lost the 1999 presidential election to Fernando de la Rua of the Radical Party. He resigned in 2001, following violent demonstrations provoked by the continuing recession, his term being completed by Eduardo Duhalde.

The presidential election of 2003 was won by a Peronist, the governor of Santa Cruz, Néstor Kirchner. He presided over an economic recovery but chose not to stand in the 2007 election, giving way to his wife Cristina Fernández de Kirchner who was duly elected.

They governed as a team until Néstor Kirchner's death in 2010. Cristina Kirchner has retained her husband's supporters and has benefited from a buoyant economy. She is likely to be re-elected in 2011.

Armenia

Armenia is a poor and landlocked country with a worldwide diaspora and a distinctive culture

Land area: 30,000 sq. km.
Population: 3.1 million—urban 64%
Capital city: Yerevan, 1.1 million
People: Armenian 98%, Yezidi (Kurd) 1%, other 1%
Language : Armenian 96%, Russian 2%, other 2%
Religion: Armenian Apostolic 95%, other 5%
Government: Republic
Life expectancy: 74 years
GDP per capita: $PPP 5,693
Currency: Dram
Major exports: Jewellery, manufactured goods

Occupying the north-western part of the Armenian Highlands, Armenia is almost entirely mountainous: the average elevation is 1,800 metres. The landscape is spectacular, with rushing rivers, deep valleys, and many lakes—the largest of which is Lake Sevana. The land can also be violent. The country is dotted with extinct volcanoes, and an earthquake in 1988 killed more than 25,000 people. But the country has few resources in terms of energy or minerals, and much of the land and water is heavily polluted.

Armenia is one of the world's oldest civilizations and its territory once extended from the Mediterranean in the west to the Caspian Sea in the east. Centuries of conquest and occupation have, however, shrunk it to more modest boundaries.

Even within these the population is fairly concentrated. Two-thirds of

Armenians live in towns and cities, with the greatest numbers in the Ararat plain along the south-western border with Turkey. There is also an extensive Armenian diaspora. Around 1.5 million ethnic Armenians live in other parts of the former Soviet Union and 2.5 million more are scattered around the world. The 1988 war with Azerbaijan generated around 250,000 ethnic Armenian refugees from Azerbaijan while ethnic Azeris moved in the other direction. By 2007, there was again a net outflow of people. In 2008 their remittances of over $1 billion were around 10% of GDP.

Armenia's distinctive culture also survived through the Soviet era. Ethnically the country is very homogenous. Most people are Christians—members of the Armenian Apostolic Church, and the Armenian language has retained its own unique alphabet.

During the Soviet era, education and health standards were high, but subsequently deteriorated following sharp cuts in public services. Now around two-thirds of health spending is private.

The Soviet era also transformed Armenia into an industrial economy, but since then heavy industries have suffered a steep decline. One of the more productive areas has been precious stones and gem cutting—based on skilled and low-cost labour, though often using imported diamonds. Other healthier industries in the past few years include energy, telecoms, and chemicals. Gold mining has also benefited from new flows of foreign investment.

Armenia's steep terrain does not make for much productive agriculture and the country has to import a high proportion of its grains and dairy products. Nevertheless, agriculture still employs around one-third of the population, working largely on irrigated land in the Ararat plain where the main crops include potatoes, grapes, and tobacco. Agriculture was boosted by land reform in the 1990s, and most land is now in private hands, though typically in small farms.

The economy shrank dramatically in the early 1990s but subsequently grew quite rapidly which helped reduce formerly high levels of unemployment, though around one-quarter of the population still live below the poverty line.

Following the collapse of the Soviet Union, Armenia became independent in 1991. The new

The dispute over Nagorno-Karabakh

president was the leader of the Pan-Armenian National Movement, Levon Ter-Petrosian, who was elected on a platform of modest reform. Since then political developments have been shaped by the dispute over the Nagorno-Karabakh region of Azerbaijan, which is inhabited by ethnic Armenians. When, in 1992, Nagorno-Karabakh declared its independence, Armenia supplied it with weapons and eventually invaded. Armenia's soldiers had previously been among the élite of the Soviet Army, and in 1993 easily overcame the Azerbaijanis, seizing around 20% of Azerbaijan—including Nagorno-Karabakh. A ceasefire was agreed in 1994 but Armenian troops remain.

Ter-Petrosian was re-elected in 1996, but in 1997 he made the fatal mistake of softening the line on Nagorno-Karabakh. This outraged nationalist sentiment and helped provoke mass defections to the opposition, and in early 1998 forced Ter-Petrosian to resign.

The ensuing presidential election in April 1998 was won by his ex-prime minister Robert Kocharian, also an ex-leader of Nagorno-Karabakh.

On the domestic front, popular discontent with the government took a violent turn in October 1999 when nationalists stormed the parliament, killing the prime minister and seven others. Subsequently, the opposition remained weak and no single leader emerged, which left the field open for Kocharian, to be re-elected in 2003.

In the controversial 2008 election he was replaced by the establishment candidate, Serzh Sarkisian. Ter-Petrosian contested the result and organized protests. Claiming he was organizing a coup, the government clamped down, resulting in ten deaths. Opposition leaders were imprisoned.

The political situation remains tense, partly because of an economic dip but also because Sarkisian has been trying to improve relations with Turkey which maintains a blockade.

Australia

Australia has been establishing stronger links with Asia—but has been unable to shake off the British monarchy

Land area: 7,713,000 sq. km.
Population: 21 million—urban 89%
Capital city: Canberra, 324,000
People: Caucasian 92%, Asian 7%, Aboriginal and other 1%
Language: English
Religion: Roman Catholic 26%, Anglican 21%, other Christian 21%, other 32%
Government: Constitutional monarchy
Life expectancy: 81 years
GDP per capita: $PPP 34,923
Currency: Australian dollar
Major exports: Minerals, agricultural products, manufactures

Australia's landmass—which can be viewed as the world's largest island—is dominated by a vast and largely empty interior of arid plains and plateaux, known as the outback. The most significant mountainous area is the Great Dividing Range, which runs down the east of the country, separating the outback from the coastal plain, re-emerging further south as the island of Tasmania. Most people are concentrated in the more fertile coastal areas, particularly in the east, south-east, and west.

Australia's population has been shaped by a long and continuing process of immigration: one-quarter of the current population were born overseas. Until the mid-1960s the government had a 'white Australia' policy which ensured that immigration was largely from Europe and particularly from the British Isles.

Since then, it has cast its net wider, seeking the people with the highest skills. This, combined with refugee flows, has tilted the balance in favour of Asia. Even so, the largest single source of immigrants is New Zealand. In 2006/07 there were over 140,00 immigrants and a further 373,000 long-term arrivals.

Australia has an increasingly diverse population, but its most distinctive ethnic group are the original inhabitants—a quarter of a million Aborigines. They are the most marginalized part of Australian society: most live in desperate poverty, either in reserves in the outback or in urban slums. In recent decades they have become more assertive and in 2008 the Rudd government officially apologized for the ways in which the state has treated Aborigines.

Australia's economy is dominated by service industries that account for more than 80% of GDP and employment. The largest employer is the retail sector, but many other people work in catering—which also involves feeding more than five million tourists who arrive each year. Manufacturing has declined significantly. Until the 1990s, many Australian producers were protected by high tariffs, but most of these have now been cut, output has fallen, and in 2007 manufacturing employed only 10% of the workforce.

Australia remains a major producer and exporter of raw materials. Its extensive mineral wealth includes iron, aluminium, uranium, zinc, nickel, and lead. It also has gold and diamonds, as well as abundant energy sources in the form of coal, oil, and gas. For many minerals, Australia is the world's leading exporter. Mineral

Aboriginal communities demand their rights

rights are generally vested in the Crown or in individual states which receive royalties from mining companies. But land title is disputed; Aboriginal communities are demanding either a veto over mineral development or a share of the profits.

Much of the territory is arid, but Australia also has rich agricultural land. Agriculture has become less significant, particularly following a series of severe droughts, but it still generates around one-quarter of exports—chiefly from wool, cereals, meat, and sugar. A more recent success has been wine, of which Australia is the world's sixth largest producer.

Australia is a federation of six states, each of which has its own government and governor. The head of state is still the British monarch, though there have been several attempts to despatch her. In a 1999 referendum, 55% of Australians rejected the proposal to become a republic, primarily because they disliked the form of republic on offer,

or because they regarded the whole constitutional process as a waste of time and money.

Australia's political parties have traditionally pitched conservative forces against those of organized labour—though parties of right and left have been moving to the centre ground. The conservative party is the Liberal Party, which draws its support from the urban middle class. The Labor Party has relied more on urban workers. The third major force is the National Party, which draws its support from the rural areas. Two minority parties are the Australian Democrats and the Greens. The racist and anti-immigration One Nation party, which emerged in the 1990s, has now largely disappeared.

Labor's longest period in government stretched from 1983 to 1996, initially headed by former trade union leader Bob Hawke. During this period, Labor continued the process of opening up the economy, floating the Australian dollar, dismantling tariffs, and privatizing utilities.

In the 1996 election, following an economic downturn, Labor was replaced by a Liberal–National coalition led by John Howard. He was re-elected in 1998 and in 2001, in the wake of September 11. He strongly supported the US invasion of Iraq and secured a fourth term in 2004.

In 2007 the Labor Party, led by Kevin Rudd, achieved a clear victory. In 2010, however, the party replaced him as prime minister with Julia Gillard. She then called an election in which Labor lost its majority. Reliant now on independent and Green MPs her government is fragile and may not last very long.

Austria

The resurgence of the far right has abated and Austria has returned to consensus politics

0	Miles	150
0	Km	240

CZECH REP

GERMANY

Linz • **Vienna** ■

• Salzburg

• Innsbruck

ITALY

SLOVENIA

Land area: 84,000 sq. km.
Population: 8 million—urban 68%
Capital city: Vienna, 1.6 million
People: Austrian 91%, others 9%
Language: German
Religion: Roman Catholic 74%, Protestant 5%, Muslim and other 21%
Government: Republic
Life expectancy: 80 years
GDP per capita: $PPP 37,370
Currency: Euro
Major exports: Metal goods, textiles, paper products, chemicals, electronics

---

Austria is one of Western Europe's most mountainous countries. The western two-thirds of the country fall within the Alps, in ranges that run from east to west divided by broad river valleys. Another smaller and less dramatic highland region extends from the north of the country into the Czech Republic. Lowland Austria is largely in the east, including the Vienna basin through which flows the Danube on its way to Slovakia.

In ethnic terms, Austria is fairly homogenous, though there is some representation of neighbouring nationalities. The end of the cold war also saw a further surge of immigration from Eastern Europe, following the war in former Yugoslavia. As a result, between 1989 and 2007 the proportion of foreign-born rose from 4% to 10%—though these flows have eased now as border controls have been tightened.

Immigration did at least temporarily rejuvenate the population. Otherwise, with a fertility rate of only 1.4 children per woman of childbearing age, the population is ageing fast: by 2030 over-60s will be 32% of the population.

Austrians enjoy one of Europe's more generous systems of welfare. Parents are entitled to three years of maternity leave, for example. But many welfare benefits including retirement benefits have recently been cut to reduce future liabilities.

Two-thirds of Austrians are now employed in service industries, from banking to transport. And following the opening up of Eastern Europe, many foreign companies use Vienna as a base. One of the more significant service industries is tourism. In 2005 Austria had around 20 million tourist arrivals which have been growing around 2% per year. People from other European countries are drawn to Austria's rich cultural heritage and to the scenic and winter sports attractions of the Alps.

Austria is also an important centre for manufacturing, which employs around 17% of the workforce and accounts for 21% of GDP. Traditional industries such as textiles and footwear have stagnated, but others have been expanding, including electronics, chemicals, and metals. Much of the output consists of medium-technology intermediate goods for other European countries, particularly Germany.

Agriculture now employs only 1% of the workforce, but Austria is still largely self-sufficient in food, much of it produced using 'organic' methods. Large herds of livestock, particularly in the mountainous west of the country, also allow for the export of some dairy products. And, with around two-fifths of the country forested, there are also exports of timber.

Austria has enjoyed steadier economic progress than many other European countries. Much of this is due to its distinctive system of sozialpartnerschaft (social partnership). This involves regular discussions between employers, trade

Austria has a system of social partnership

unions, academics, and agricultural representatives. Most major economic policies derive from this process of consultation which has contributed to stable labour relations, low inflation, and extensive welfare benefits—though it has also diminished the status of the parliament.

Until recently, consensus was also the major characteristic of Austrian politics. For most of the period since the Second World War, Austria has effectively been run by two parties generally in a coalition. From 1970 onwards, the dominant partner has often been the centre-left Social Democratic Party (SPÖ), which

draws its support from labour and trade unions. In 1997, Viktor Klima took over as party leader and became chancellor—head of the government. The second party was the centre-right Austrian People's Party (ÖVP), led by Wolfgang Schüssel, which is linked to the Roman Catholic Church and draws its support more from the middle class and business.

From the mid-1990s, however, the cosy arrangements between the two parties (which included 'Proporz'—sharing out government jobs between their party card-holders) were challenged from the far right by the Freedom Party and its aggressive former leader Jörg Haider who stood on a populist, nationalist, anti-EU, anti-immigrant platform.

The 1999 elections gave the two major parties a shock. The Social Democratic Party still had the largest share of the vote, but the Freedom Party came second with 29%—narrowly ahead of the People's Party. This resulted in a People's Party–Freedom Party coalition headed by Schüssel, though without Haider.

International opinion was outraged. The EU introduced temporary diplomatic sanctions. Since then the Freedom Party has declined in popularity.

Elections in 2006 produced a grand but bickering coalition of the SPÖ and the ÖVP. The ÖVP resented being the junior partner when it had received almost as many votes. Early elections in 2008 produced a similar result, though the coalition partners are cooperating more. The government is led by Werner Faymann (SPÖ) as chancellor and Michael Spindelegger (ÖVP) as vice-chancellor.

Azerbaijan

Oil and gas-rich Azerbaijan has high levels of poverty and a dynastic political succession

Land area: 87,000 sq. km.
Population: 9 million—urban 52%
Capital city: Baku, 1.9 million
People: Azeri 91%, Dagestani 2%, Russian 2%, Armenian 2%, other 3%
Language: Azeri 90%, Russian 2%, Lezgi 2%, other 6%
Religion: Muslim 93%, Russian Orthodox 3%, Armenian Orthodox 2%, other 2%
Government: Republic
Life expectancy: 70 years
GDP per capita: $PPP 7,851
Currency: Manat
Major exports: Oil, textiles, food

Around half of Azerbaijan comprises the central lowland area through which flow the two main rivers, the Aras and the Kura, which join before draining into the Caspian Sea. These plains are contained by spurs of two mountain ranges, the Greater Caucasus to the north and the Lesser Caucasus to the south-west. In addition, there is a small isolated section of its territory to the south-west, Nakhichevan, sandwiched between Armenia and Iran. Azerbaijan has a varied, and often strikingly beautiful, terrain.

Azerbaijan is ethnically fairly homogenous. Around 90% are Azeris who are Shia Muslims. But there have always been minority groups. This includes a significant number of Armenians, who are Christians, and who live in the Nagorno-Karabakh region in the south, which since 1994 has been occupied by neighbouring

Armenia. There are also many ethnic Azeris elsewhere—indeed more live in Iran than in Azerbaijan itself.

In the past, Azerbaijan's population has achieved high standards of education and health but government services have been undermined by the conflict with Armenia, and by the emigration of many professionals. Although investment in the public healthcare system has increased in recent years and services are officially free, many people make informal payments to receive attention.

Despite the country's oil wealth around half the population live below the poverty line. Over 40% of the workforce are employed in agriculture, largely in the fertile lowlands. Most of the former state and collective farms have now been privatized. But agriculture remains very inefficient, production has been falling and now contributes only 6% of GDP. As a result the country has to import much of its food. Azerbaijan has been a major cotton producer. However, the toxic chemicals used for this have polluted the land and the water. Poor drainage has also led to salination and a threat from the Caspian Sea which is rising by

25 centimetres per year.

The mainstay of the country's economy remains oil, three-quarters of which is extracted offshore in the Caspian Sea. This area has been an oil producer since the 1850s and by the early 1900s it supplied half the world's oil. Baku still has the ornate grand houses built by the first oil millionaires.

Azerbaijan's significance declined with the discovery of larger deposits elsewhere. Nevertheless, the international oil companies have still congregated here looking for new deposits. At the end of 2004 the country had 0.6% of world reserves of oil and 0.8% of reserves of gas, so in global terms is a small player.

In addition to extracting oil and gas, Azerbaijan was a major producer

One of the most polluted places on earth

of equipment for the oil industry— and the city of Sumgait was also one of the main suppliers of chemicals to the Soviet Union; now much of the plant there is rusting in what is probably one of the most polluted places on earth.

Efforts have been made to switch to lighter industries, and the country has some textile factories that use local cotton. But production generally has been held back by the slow pace of economic reform and is weakened by extensive corruption which deters foreign investors.

Given its ethnic fault lines, Azerbaijan's political development was always likely to be painful. Even when it was a Soviet republic, there had been a long-running dispute with Armenia over Nagorno-Karabakh and the difficulties in handling

this crisis cost several leaders their jobs. Independence was declared in 1991 and there were initially some successes in consolidating control over the disputed enclave. But in 1993 the Armenians managed to seize this and neighbouring regions—around 20% of Azerbaijan.

The ensuing political turmoil led in October 1993 to the election as president of Heidar Aliev, a former communist leader. A new constitution in 1995 granted him sweeping new powers. In 1996 Aliev's New Azerbaijan Party (NAP) won most of the seats in the parliamentary elections.

Fighting with Armenia continued until 1994 when a ceasefire was brokered by the Organization for Security and Cooperation in Europe— though Armenia has not withdrawn, and there have been subsequent outbreaks of fighting.

Meanwhile Aliev set about nurturing a personality cult and to no one's surprise was re-elected president in October 1998, after the opposition declared a boycott. The parliamentary elections of November 2000 resulted in a victory for the NAP, but this was largely the result of another blatantly fraudulent process.

In the presidential election of October 2003, the ailing Heidar Aliev was succeeded by his son Ilham (Heidar Aliev died in 2003).

Ilham has followed in his father's footsteps. He rigged the 2005 parliamentary elections. In 2008 he secured re-election, and in 2009 he pushed through a constitutional amendment that removed presidential term limits, so he should be in power for some time.

Bahamas

One of the richest countries in the Caribbean

Land area: 14,000 sq. km.
Population: 304,000—urban 84%
Capital city: Nassau, 212,000
People: Black 85%, white 12%, others 3%
Language: English, Creole
Religion: Baptist 35%, Anglican 15%, Roman Catholic 14%, Pentecostal 8%, Church of God 5%, Methodist 4%, other 19%
Government: Constitutional monarchy
Life expectancy: 73 years
GDP per capita: $PPP 29,723
Currency: Bahamas dollar
Major exports: Lobsters, rum

The Bahamas comprises more than 700 islands in the Caribbean, many with sandy beaches and clear water, although only around 30 are inhabited. The majority of people live on New Providence Island, particularly in the capital, Nassau.

Bahamians generally have good standards of education and, compared with many other Caribbean countries, incomes are high. Health standards too are good but around 3% of 15–49-year-olds are HIV-positive. The islands also attract unauthorized immigrants, particularly from Haiti, to work in hotels and restaurants, and are also a stopover for illegal Chinese migrants being smuggled to the USA.

Tourism accounts for around one-third of GDP and employs around half the workforce. Each year there are more than one million stopover visitors to the country's resorts, and about three million cruise passengers—the majority of visitors coming from the USA. Casino gambling is becoming more important following investment in resort hotels. The largest employer is the South African company Kerzner International.

But not everyone has gained from the tourist boom which still leaves many young people unemployed.

There are no rivers, and the soil is poor, so opportunities for agriculture are limited. But there is a small fishing industry that catches lobsters for local consumption and export.

The Bahamas also has a significant offshore financial sector that contributes about 15% of GDP and employs around 5,000 people. The industry acquired a dubious reputation in the 1980s through association with money laundering and also with the trade in cocaine and marijuana. Following international criticism, the country has taken steps to clean up the financial services industry.

The Bahamas is a parliamentary democracy whose head of state is the British monarch. In the decades after independence in 1973 the government was usually in the hands of Sir Lynden Pindling, leader of the Progressive Liberal Party (PLP). By the end of the 1980s, however, there was an economic downturn along with allegations of corruption.

The elections in 1992 resulted in a victory for the Free National Movement (FNM), which won again in 1997. The elections in 2002, saw the return of the PLP, now led by Perry Christie. But following the 2007 election the FNM, led by Hubert Ingraham, returned with a small majority.

Bahrain

Bahrain's constitutional monarch is faced with widespread protests

The state of Bahrain consists of around 30 islands in the Persian Gulf. Bahrain island itself, which has more than 85% of the territory, is mostly arid. Wells and springs in the north are used to irrigate fruit and vegetables, but the water table is sinking, so future agricultural prospects are poor.

The native population of Bahrain is largely Arab and divided between the majority Shia and minority Sunni sects. Bahrainis enjoy high standards of health and education but, unlike the citizens of most other Gulf oil states, many are not well off, and 15% were unemployed in 2005—with a higher rate among the Sunnis. Even so, Bahrainis employ immigrants, who are 60% of the workforce, primarily from the Indian sub-continent, to do most of the less popular work.

Bahrain's economy is based on oil, but its reserves are small—125 million barrels in 2008. It processes this in its own refineries, along with oil from Saudi Arabia. Since it has less oil than other Gulf countries, Bahrain has had to develop a more diverse economy. Thus, it uses gas reserves to fuel one of the world's largest aluminium smelters and it also has a range of industries that include ship repairing, as well as light engineering and manufacturing.

Bahrain is also a financial, trading,

Land area: 1,000 sq. km.
Population: 0.8 million—urban 89%
Capital city: Manama, 153,000
People: Bahraini 63%, Asian 13%, other Arab 10%, Iranian 8%, other 6%
Language: Arabic, English, Farsi, Urdu
Religion: Muslim 81%, other 19%
Government: Absolute monarchy
Life expectancy: 76 years
GDP per capita: $PPP 29,723
Currency: Bahraini dinar
Major exports: Oil products, aluminium, chemicals

and distribution hub for the other Gulf countries; it is linked by a 25-kilometre causeway to Saudi Arabia. Two-thirds of the workforce are employed in these and other services.

Until 2002, Bahrain was an absolute monarchy run by the al-Khalifa family in cooperation with the Sunni business élite. However, the emir had long faced opposition, both from within the country from the minority Shia, and from the London-based Bharaini Freedom Movement.

In 1999, Sheikh Hamad bin Isa al-Khalifa became emir and started a wave of reforms—transforming Bahrain into a constitutional monarchy with a parliament of two chambers: one elected, one appointed. On a day-to-day basis, however, the country is run by his uncle, the prime minister, Sheikh Khalifa bin Salman al-Khalifa.

In 2002 Sheikh Hamad declared himself king but he also amended the constitution so as to entrench his position. The main opposition is a Shia group, the al-Wefaq National Islamic Society, which in 2006 gained 17 of the 40 seats in the lower house. In 2011, protestors, inspired by the Arab spring, took to the streets and were violently dispersed.

Bangladesh

Bangladesh's democracy has constantly been undermined by bitter political feuding

Land area: 144,000 sq. km.
Population: 159 million—urban 28%
Capital city: Dhaka, 12 million
People: Bengali 98%, some tribal
Language: Bangla
Religion: Muslim 83%, Hindu 16%, other 1%
Government: Republic
Life expectancy: 66 years
GDP per capita: $PPP 1,241
Currency: Taka
Major exports: Garments, jute products

Bangladesh consists of a vast alluvial plain. The waters of the Ganges and the Brahmaputra flow in via India to deposit billions of tons of silt each year across this delta before emerging sluggishly into the Bay of Bengal. The only higher areas are the Chittagong Hill Tracts on the eastern border with Burma.

The annual flooding is a welcome event. Even in a normal rainy season, 20% of the landscape is flooded. This recharges the underground aquifers, deposits the silt that makes for fertile and productive soil, and allows for the spawning and migration of 300 or more species of fish. But flooding can also be more destructive. In 2007 floods covered more than two-thirds of the country driving millions of people from their homes. Bangladesh is also in the frontline of climate change and sea-level rise.

The other major natural hazard is

cyclones. The funnel-shaped coastline regularly sucks in some of the world's most violent storms. In 1991 a tidal surge 25 feet high killed around 140,000 people. In 2009 cyclone Aila claimed hundreds of victims.

Survival in these climatic extremes has made Bangladesh's people remarkably resilient. As a group, they are also relatively homogenous, with a rich language and culture shared with the state of West Bengal in neighbouring India. Bangladesh means 'land of the Bengalis'. Most people are Muslim, though in practice they tend to have a fairly flexible attitude towards religion. The largest distinctive ethnic groups are the tribal peoples who live in the Chittagong Hill Tracts.

Bar some city-states or small islands, Bangladesh has the world's highest population density—over 1,000 per square kilometre. However, a massive, largely aid-financed, family planning programme slowed annual population growth from around 2.8% in the early 1970s to around 1.5% in 2006. The government's goal is for population growth to stop altogether by the year 2045, though by this time the population may well have doubled to around 250 million.

Bangladesh remains very poor. About 40% of the population live below the poverty line. Standards of health are also low. A country where half of the children are malnourished, and where half of the population do not use sanitary latrines, is unlikely to be healthy, and Bangladesh is not. Millions of children have their lives extinguished all too soon—infant mortality is 59 deaths per thousand live births. A further problem is contamination of groundwater by arsenic: around 200,000 people suffer from some kind of arsenic poisoning.

Bangladesh also has low standards of education, though by 2007 the adult literacy rate had increased to 54%. The government has increased investment in education and many more children are now enrolling in school. But most schools are in decrepit buildings, with overcrowded classrooms, and there are shortages of books and equipment. Not surprisingly, attendance is poor and around half of the children do not finish primary school.

Some of this is compensated for by non-formal schools run by

Bangladesh has some of the world's largest NGOs

non-governmental agencies (NGOs). Bangladesh has some of the world's largest and most enterprising NGOs, such as the Bangladesh Rural Advancement Committee (BRAC). They may offer little more than bamboo huts with mud floors, and teachers who themselves have only a basic education, but they have proved very effective at catching children who have missed out on primary school. Other major NGOs include Proshika, and the NGO-like Grameen Bank, which operates in more than half the

country's villages with a system of micro-credit that is being emulated all over the world—and now offers internet and mobile phone businesses.

Most people live in the rural areas and around half the workforce make their living from agriculture. In many respects they have been successful. Over the past 25 years, rice production has more than doubled and the country is now almost self-sufficient in food production. Some estimates suggest that over the next two decades output could increase by 50%.

Nevertheless, harvests are still erratic and heavily dependent on the climate. Plots of land are very small—measured in tenths of a hectare. Yields are still lower than in neighbouring countries, and farmers living on a knife-edge of survival are reluctant to invest. Moreover, even when food is available many people still cannot afford to feed themselves properly.

Another crop that has traditionally been important is jute which employs around three million farm households. Bangladesh is the world's second largest producer. However, production of jute, along with that of another cash crop, tea, has been declining.

Those unable to find work in the rural areas are heading for the cities, or even overseas. There are probably around 3 million Bangladeshis working abroad, especially in the Gulf, sending home around $9 billion per year in remittances.

In terms of industry, Bangladesh's greatest progress has been in light manufacturing, particularly the garment sector, which employs around 1.5 million people, 85% of them women. Until recently Bangladesh enjoyed preferential access to Western markets but these preferences have now been withdrawn so Bangladesh will find it more difficult to compete.

One more optimistic development is the discovery of substantial natural gas reserves with the potential for exports to India.

At independence in 1971, the US Secretary of State famously dismissed the new country as an 'international basket-case'. Bangladeshis have defied this gloomy prognostication. Nevertheless, this remains one of the poorest developing countries. Economic growth has been around 6% annually, though needs to be higher to make a greater dent on poverty.

Bangladesh achieved independence in 1971 after a bloody war of secession from Pakistan. Elections in what was then East Pakistan had produced a victory for the Awami League (AL) led by Sheikh Mujibur Rahman. He demanded greater autonomy and when this was refused organized strikes and demonstrations. The Pakistani army attacked Dhaka and a full-scale civil war erupted in which the East Bengalis were supported by India. By the time the war ended, more than one million people had died.

A bloody war of secession

After independence, 'Sheikh Mujib' and the Awami League (AL) took power, but he became increasingly autocratic and replaced the parliament with a presidential system. The military grew dissatisfied and in 1975 a group of officers assassinated him.

Following this, the de facto leader of the government was the army chief, General Zia ur-Rahman, who took over the presidency in 1977

at the head of his new Bangladesh National Party. In 1981, he in turn was assassinated by rebel army officers led by General Hossain Mohammad Ershad, who took the presidency in 1983. Ershad was deposed in 1990—and subsequently served seven years in prison for corruption and murder.

A general election in 1991 resulted in a victory for the BNP, which was now led by Zia's widow, Begum Khaleda Zia. She defeated the Awami League, now led by Sheikh Mujib's daughter Sheikh Hasina Wajed. In the same year, a referendum altered the constitution to return executive power to the prime minister.

In opposition, the Awami League embarked on a bitter campaign of destabilization, using 'hartals'—general strikes enforced by intimidation. The AL then boycotted the 1996 election demanding that further voting take place under a neutral administration. Eventually they got their way and the AL won the subsequent election. In 1997 the constitution was amended so there would always be a caretaker administration to supervise the elections. In 2001, the BNP, heading a four-party alliance, and following a characteristically violent campaign, won a crushing victory.

In 2006, the AL complained about the BNP-dominated caretaker government. In 2007 the army stepped in, declaring a state of emergency, and by mid-2008 had imprisoned leaders of the three main parties for corruption. But it proved no match for the politicians and eventually had to release them. Elections in 2008 produced a large majority for Sheikh Hasina and the Awami League.

Barbados

A Caribbean success story with a growing economy and a stable system of governance

Land area: 430 sq. km.
Population: 0.3 million—urban 41%
Capital city: Bridgetown, 108,000
People: Black 90%, white 4%, other 6%
Language: English
Religion: Protestant 63%, other 37%
Government: Constitutional monarchy
Life expectancy: 77 years
GDP per capita: $PPP 17,956
Currency: Barbados dollar
Major exports: Sugar, rum, other foods and beverages

Barbados is the furthest east of the Caribbean islands and, except for low, central mountains, is largely flat. It has few natural resources apart from attractive scenery, a good climate, white beaches and coral reefs.

With close to 100% literacy and good health standards, levels of human development here rival those in many industrial countries. Education is compulsory until the age of sixteen and also free to university level. Nevertheless, unemployment remains high and many Barbadians emigrate to work abroad; their remittances come to over $160 million per year.

The country's original wealth was based on sugar cultivation, and sugar cane still provides much of the export income. But agriculture has long since given way to tourism as the main economic activity. Tourism, with around half a million visitors a year, provides half of foreign exchange.

Barbados has also made efforts to diversify into other services and manufacturing. It has, for example, built up a financial services industry and also attracted some of the 'back-office' data processing work for US companies. Like other Caribbean countries, Barbados has also become a transhipment point for drugs, though the government has taken a strong line against this and permits US officials to carry out hot-pursuit searches in Barbadian waters.

Barbados was a British colony. Though it had complete internal autonomy in 1961 it subsequently reverted to being a self-governing colony after the West Indies Federation was dissolved in 1962. In 1966, Barbados achieved full independence and since then has enjoyed a stable parliamentary system. In 1998 a constitutional commission recommended switching to a republic with a ceremonial, elected president.

Two main parties have alternated in power. Between 1986 and 1994 the government was in the hands of the Democratic Labour Party (DLP).

The early 1990s, however, were difficult years. The sugar harvest was poor, unemployment was high, and increasing levels of violence were discouraging tourists. In 1994, power passed to the Barbados Labour Party led by Owen Arthur, who increased his majority in the 1999 elections and won again in 2003.

In the 2008 election, however, after a contentious and divisive campaign, the DLP, led by David Thompson, won 20 of the 30 lower-house seats. Following his death in 2010 he was replaced by Freundel Stuart.

Belarus

Belarus has changed little since the Soviet era and has even cherished ambitions to reunite with Russia

Land area: 208,000 sq. km.
Population: 10 million— urban 74%
Capital city: Minsk, 1.7 million
People: Belarusian 81%, Russian 11%, Polish Ukrainian and others 8%
Language: Belarusian, Russian
Religion: Eastern Orthodox 80%, other 20%
Government: Republic
Life expectancy: 74 years
GDP per capita: $PPP 10,841
Currency: Belarusian rubel
Major exports: Vehicles, oil, minerals

Belarus is largely flat—the highest point in this whole landlocked country is only around 350 metres above sea level. One-third of the territory is forested, including the primeval Belovezhskaya Forest on the western border with Poland. There are also extensive peat bogs and marshes, the largest of which are the Pripet Marshes on the southern border with Ukraine.

Most of the population are ethnic Belarusians, but during the Soviet era Belarus was a significant economic centre, and attracted Russian immigrants. Some have returned, but most remain, and there are also sizeable minorities of Poles and Ukrainians. Meanwhile more than two million Belarusians live in other former Soviet republics.

The Belarusians had achieved good standards of health and education but after 1990 infant mortality rose and life expectancy fell. Since 1993, death rates have exceeded birth rates and the population has been falling. Belarusians are also still suffering the effects of the 1986 Chernobyl disaster; though the explosion was just across the border in Ukraine, most of the fallout was in Belarus.

Belarus industrialized rapidly from the 1950s onwards, becoming an important centre for oil refining and petrochemical production. In addition, the Soviet Union used Belarus to manufacture military equipment and to supply trucks, tractors, and motor cycles. Unfortunately this involved importing oil, gas, and other raw materials from Russia and elsewhere, so when they had to pay market prices industries shrank rapidly.

By 1996, things had started to stabilize and nowadays industry accounts for around 40% of GDP. Nevertheless, the country has been accumulating an energy debt with Russia, which the Russian gas company Gazprom periodically has to write off.

Although the government professes to be following 'market socialism', in practice Belarus still has most of the elements of a command economy. Almost 40% of industrial output still comes from state-owned enterprises and most of the rest from enterprises that are effectively under state control. Russia takes around half of all exports, mostly machinery and foodstuffs, and this leaves the country very vulnerable to fluctuations in Russia's economy.

Agriculture, too, is still mostly a state activity. Small farms have proliferated since independence but most output still comes from state and collective farms, the majority of which are insolvent and need regular injections of state funds.

Prior to 1991, Belarus had never been an independent state. It did have separate UN membership but this, like the membership for Ukraine, was a legal fiction invented by Stalin to give the Soviet Union another UN vote.

Belarus had never been independent

Most of the impetus for Belarus's independence came from the Belarusian Popular Front (BFP) whose nominee, Stanislau Shushkevich, was appointed head of state. But it did not take long for the communists to reassert their authority. In 1994, a commission headed by a former state farm manager, Alexander Lukashenka, ousted Shushkevich on charges of corruption.

Lukashenka was able to lever the exposure this gave him into a victory in the 1994 presidential election, and since then he has proved an effective populist, and an astute political operator. In 1996 he successfully conducted a referendum on a new constitution that extended his term of office and created a new bicameral assembly which the president can dissolve if it 'systematically or seriously violates the constitution'. Most of his support comes from unreformed communists.

Opposition has been heavily repressed: in 1996, protests against his regime were crushed and the BFP leader Zyanon Paznyk sought political asylum in the USA. Opposition radio stations and newspapers were also closed down.

One of Lukashenka's main aims has been to unite Belarus with Russia, preferably with him as leader. In December 1999, this appeared to move forward with the signing of yet another union treaty with Russia. But genuine union between Belarus and Russia now seems unlikely. Some communists in Russia are attracted by the kudos that this expansion would bring them, but others have balked at the economic implications of fully absorbing its backward neighbour, especially with its eccentric leader. Lukashenka's prospects of leading a united country shrank even further when Vladimir Putin was elected president of Russia. He has shown little appetite for unification.

So far, both the autocratic president and his Russophile inclinations are supported by most people—though it is difficult to gauge opposition strength. For the 2000 parliamentary elections, the opposition were denied access to government-controlled media, so boycotted the polls.

There was an atmosphere of fear and intimidation in 2001 when Lukashenka was re-elected president with 76% of the vote. The processes and the results were similar in the 2006 and 2010 presidential elections, both of which were followed by further repression.

Belgium

Belgium has conflicts between its language communities and has been shaken by political scandals

Land area: 31,000 sq. km.
Population: 11 million—urban 97%
Capital city: Brussels, 1.0 million
People: Fleming 58%, Walloon 31%, mixed or other 11%
Language: Dutch 60%, French 40%
Religion: Roman Catholic 75%, Protestant or other 25%
Government: Constitutional monarchy
Life expectancy: 80 years
GDP per capita: $PPP 34,935
Currency: Euro
Major exports: Transportation equipment, chemicals, manufactured goods

Belgium has three main geographical regions running from north-west to south-east. From the sand-dunes and dykes that fringe the North Sea coast the coastal plain extends inland for up to 50 kilometres—with land that is often marshy and intercut with shipping canals. Then the land rises to the rolling hills of the fertile central plateau, and finally to the dense forests of the Ardennes highlands in the south-east.

In addition to Belgium's political boundaries there is also an invisible cultural frontier running across the country from west to east, passing about 20 kilometres south of the capital, Brussels. To the north of this line is Flanders, home to just over half the population who are Flemish, and speak a dialect of Dutch. To the south is Wallonia, home to the one-third of the population who are Walloons

and speak French. There is also a third group, though much smaller, of German speakers on the eastern border.

Superimposed on these divisions is the one million population of Brussels, which, though within Flanders, is a largely French-speaking enclave. It is also home to many of Belgium's immigrants. Belgium has one of Europe's larger immigrant communities—around 9% of the total population. The inflows continue though the proportion of non-nationals has remained fairly stable because it has become easier to obtain Belgian nationality. Around 60% of immigrants are EU citizens attracted by the status of Brussels as unofficial capital of Europe. Brussels is home to the European Commission and to other international organizations such as NATO that generate around 10% of employment. Of the other immigrants, the largest groups are Moroccan and Turkish. The presence of all these people makes Belgium one of Europe's most densely populated, and most urbanized, countries.

Belgium was one of Europe's first countries to industrialize, taking advantage of its formerly extensive coal deposits to process imported raw materials for export. These older industries were in Wallonia, but most coal mines and many of the old factories are now shut. Belgium is still a major steel producer, now using imported coal, but its newer, lighter manufacturing industries, such as chemicals, have been established in Flanders nearer the important ports. The majority of manufacturing companies are foreign owned.

Antwerp is also the world's largest diamond centre—half the world's diamonds pass through the city. This means that Flanders, which used to be a more backward agricultural area, now has a per capita GDP 40% higher than Wallonia—a shift in industrial power that has fuelled resentment between the two communities. The Flemish complain that their taxes are heavily subsidizing the Walloons.

Belgium has run into economic problems in recent years. Though it has many new service industries, unemployment has been high, around 8%, and despite immigration the population is ageing rapidly. Another issue is debt, since governments have frequently attempted to spend their way out of inter-community problems.

Belgium's political landscape is now dominated by the language issue. To deal with this, Belgium has effectively become a federal state, and very heavily governed: in addition to the bicameral federal government there are now separate assemblies not just for Flanders, Wallonia, and Brussels-Capital, but also individual assemblies for the French, Dutch, and German-speaking communities. These various

Belgium is heavily governed

assemblies have complex overlapping memberships. In total there are around 60 ministers or junior ministers.

In the past, the two main unifying factors have been Catholicism and the monarchy: King Albert II acceded to the throne in 1993 and has helped to serve as a mediator. So far, only around one-fifth of the population vote for parties that want to break up the country, but different groups of Belgians are increasingly leading separate lives.

The language split is also matched among the political parties. In the two main regions, the three largest—the Christian Socialists, the Socialists, and the Liberals—each have autonomous parties. In addition there are two right-wing nationalist parties: for Flanders, the Vlaams Blok; for Wallonia, the National Front.

Belgians' faith in their political system was rocked in the late 1990s by a series of scandals. These included the failure to deal with a paedophile ring, allegations of corruption, brutal police treatment of immigrants, and a number of food scares.

These contributed in 1999 to an end of the 40-year dominance of Christian parties. The right-leaning liberals became the biggest party nationally. Guy Verhofstadt, of the Flemish Liberals, formed a coalition and instituted popular reforms and was re-elected in 2003. Elections in 2006 and 2007 failed to produce decisive results. Finally in 2009, a coalition was formed, headed by the Christian Democrats and Liberals.

After the 2010 election the Socialist Party and the New-Flemish Alliance emerged as the largest parties but failed to form a coalition.

Belize

Belize still has a border dispute with Guatemala, but has become steadily more Hispanic

Land area: 23,000 sq. km.
Population: 0.3 million—urban 53%
Capital city: Belmopan, 14,000
People: Mestizo 49%, Creole 25%, Maya 11%, Garifuna 6%, other 9%
Language: English, Spanish, Mayan, Garifuna
Religion: Roman Catholic 50%, Protestant 27%, none 14%, other 9%
Government: Constitutional monarchy
Life expectancy: 76 years
GDP per capita: $PPP 6,734
Currency: Belize dollar
Major exports: Sugar, citrus concentrates, bananas, garments

The northern half of Belize consists largely of lowland swamps, while the main feature in the southern half is the Maya mountain range which rises to 1,112 metres. More than 40% of the territory is covered by rainforests. Just off the coastline there are many small islands, or cays, and one of the world's largest barrier reefs.

As a result of flows of migrants and refugees from El Salvador and Guatemala in the 1980s, probably more than half the population is now Spanish-speaking. Meanwhile English-speaking creoles have emigrated to the USA. In addition, there are smaller numbers of Garifuna in the south.

Around one-quarter of the labour force work in agriculture, mostly on small farms growing maize, beans, and cash crops. The main sources of export income are sugar cane in the north, citrus concentrates in the Stann Creek Valley, and bananas in the south. Forestry is likely to become more important; although there are replanting programmes, the arrival of Asian logging companies raises the prospect of deforestation.

Manufacturing tends to be on a small scale for local consumption. Garment production in export processing zones has fallen as a result of competition from countries with cheaper labour. Tourism is also an increasing source of income; more than 200,000 visitors come each year, many for scuba diving or to visit Mayan archaeological sites. Others visit from cruise ships.

Belize has a border dispute with Guatemala. In 2002, the Organization of American States brokered a new agreement, which neither country has ratified. An increasingly serious problem, however, is the use of Belize for trafficking cocaine from South America, which is leading to rising levels of crime.

Since independence in 1981, Belize has had a stable parliamentary system. The head of state is the British monarch, represented by a governor-general. There are two main parties. The 1998 and 2003 elections were won by the People's United Party (PUP) led by Said Musa. The PUP, previously left-wing, has moved to the centre and tends to be supported by the mestizos.

In 2008, the other main party, the conservative United Democratic Party (UDP), led by Dean Barrow, won a strong parliamentary majority. It is supported more by creoles.

Benin

A busy trading economy, but largely informal so the government gets little revenue

Land area: 113,000 sq. km.
Population: 8 million—urban 42%
Capital city: Porto Novo, 250,000
People: Fon, Adja, Yoruba, Barib, and many others
Language: French, Fon, Yoruba, and others
Religion: Christian 43%, Muslim 24%, Vodoun 17%, other 16%
Government: Republic
Life expectancy: 61 years
GDP per capita: $PPP 1,312
Currency: CFA franc
Major exports: Cotton, oil

Benin has a series of geographical regions that extend northwards from its narrow coastline. Just behind the sandy coast are tidal marshes and lagoons; then there is a flat fertile plateau, the barre. A further series of plateaux lead to the Atakora Mountains in the north-west and to plains that slope down to the Niger River in the north-east.

Benin's people are divided among many ethnic groups, the largest of which are the Fon and the Adja, who make up over half the population. These have an animist religious tradition but there are also significant Muslim influences that come from countries to the north and Christian influences in the south that are a legacy of the French colonial years, so religious beliefs tend to be a mixture of different ideas.

Benin is one of Africa's poorest countries. Around one-third of the population live below the poverty line. Though Benin is self-sufficient in food, more than one-fifth of children under five suffer from malnutrition, and health standards are poor. HIV prevalence, however, is still relatively low at around 1%. Malaria remains a serious problem, along with many cases of gastroenteritis linked with contaminated water supplies.

Education standards are low and only 34% of adults are literate since the country has until recently under-invested in its schools.

Around half the population make their living from agriculture. Farmers grow yams, cassava, millet, sorghum, beans, maize, and rice. Productivity is low, but in a year of reasonable rainfall Benin is self-sufficient in food. Many people also work on cotton and palm-oil plantations. Benin is a major cotton producer and sales of cotton fibre and seeds make up around 45% of foreign exchange earnings. Most industry is linked with processing primary products for export.

Benin is also a bustling centre for trade, which makes up more than half of GDP. Its geographical location makes it a key outlet for the landlocked countries to the north. But most of the trade activity is focused on Nigeria, its giant neighbour to the east. Benin is largely an entrepôt state. Around three-quarters of imports are en route to Nigeria, often to be smuggled.

Most of this business flows through the informal sector, particularly the Yoruba traders who deal with their counterparts in Nigeria. The fairs and markets along the border with Nigeria are always packed. The Beninois sell rice, cigarettes, and spirits, while they buy cars, plastic products, and electrical equipment. But their most important purchase is petrol: every day around 300,000 litres cross the border illegally; the smuggled price is around half the official price.

A frenzy of cross-border trade

Benin's commercial capital, Cotonou, is also a major transit point. Its streets are packed with cars and with motorcycle taxis, the zemidjans. The old bridge that crosses the lagoon and links the country with Togo in the west and Nigeria in the east creaks under the daily weight of traffic.

Much of this frenzy of activity is a response to desperation. Many of those now in the informal sector previously worked in larger enterprises or in government services. But a series of economic crises resulted in heavy job losses. Even government officials need second or third sources of income: many of the zemidjans are owned by civil servants or teachers, who lease them out.

The government largely turns a blind eye to the 'osmosis' in the border region since it provides an important source of funds to so many people. But it does mean that relatively little income is taxed so government finances are weak.

During the early years after independence from France in 1960 political life was punctuated by a series of military coups. In 1972, Major Mathieu Kérékou seized power and from 1974 tried to embark on a Marxist-Leninist course and changed its name from Dahomey to Benin.

But in 1989 in the wake of an economic collapse Kérékou abandoned socialism and in 1990 organized a national conference which led to a new constitution for a multiparty democracy. Nevertheless, he lost the 1991 presidential election to Nicéphore Soglo. Soglo struggled to build a free-market economy and introduced many austerity measures.

These did not make him popular, and the 1996 presidential election was won by Kérékou, who appropriately campaigned under the symbol of the chameleon. He also won the 2001 presidential election, though amid allegations of fraud, and his coalition, the Union pour le Bénin du futur (UBF), followed up with a majority in the 2003 parliamentary elections.

The presidential elections in 2006, however, produced a surprising result. Kérékou was ineligible but neither he nor the UBF could settle on a successor. This left the field open for Boni Yayi, an independent technocrat. He formed an ad hoc alliance with Mr Soglo and others and won in the second round. In the 2007 legislative election, his coalition, Force cauris pour un Bénin émergent, achieved a majority in the National Assembly.

Yayi also won the 2011 presidential election, but the opposition claimed extensive irregularities.

Bhutan

Bhutan's development objective is to maximize 'gross national happiness'

Land area: 47,000 sq. km.
Population: 0.7 million—urban 37%
Capital city: Thimphu, 99,000
People: Ngalop, Sharchop, Nepali
Language: Dzongkha, Tibetan, and Nepalese dialects
Religion: Buddhism 75%, Hinduism 25%
Government: Constitutional monarchy
Life expectancy: 66 years
GDP per capita: $PPP 4,837
Currency: Ngultrum
Major exports: Electricity, timber

Bhutan has three main geographical zones. The north falls within the high Himalayas which reach altitudes of 7,300 metres. To the south are the 'inner' Himalayas which include a number of fertile and well–cultivated valleys—though travel between them can be arduous. To the far south these descend to the narrow, subtropical Duars Plain that runs along the border with India.

The main ethnic group are the Ngalop, who live mostly in the west and centre. Of the other groups, the largest are the Sharchops in the east and ethnic Nepalese in the south.

Concerned at a largely unauthorized influx of Nepalis, the king from the late 1980s enforced a national language and dress. This prompted an exodus of refugees and 107,000 still live in camps in Nepal. Other security problems include the presence of Indian insurgents who use southern Bhutan as a base.

Bhutan has made remarkable progress in human development since it began to open up to the rest of the world after 1959. Between 1960 and 2001, life expectancy increased from 37 to 66 years. This was based on a uniquely Bhutanese model, balancing modernization with Buddhist values and characterized by the king as the pursuit not of gross national product

but of 'gross national happiness'.

Around 75% of people work in agriculture, growing subsistence crops such as rice, maize, and potatoes. The best land is in the fertile valleys and is fairly equally distributed though plots are small. Three-quarters of the territory is forested, and timber is an important export.

Bhutan's recent spurt of economic growth has been based on the development of hydroelectric power for export to India. This is not based on dams but mostly uses 'run-of-river' systems to harness the country's fast-flowing rivers.

Development has largely been financed by aid but now that the Tala power project has been completed the country should need less support. Tourism is also a useful source of income, though careful control of numbers makes this an exclusive and expensive destination.

Until 2008, Bhutan was an absolute monarchy, ruled from 1972 by King Jigme Singye Wangchuk who steered Bhutan to democracy and in 2006 abdicated in favour of his son, Jigme Khesar Namgyel Wangchuk. The first-ever election for the 47-member lower house of parliament in 2008 was won by the Druk Phuensum Tshogpa party with more than 90% of the vote.

Bolivia

Bolivia has taken a sharp left turn with the election of its first indigenous president who wants to 'refound' the country

0	Miles	300
0	Km	480

Land area: 1,099,000 sq. km.
Population: 10 million—urban 67%
Capital city: La Paz, 0.8 million
People: Quechua 30%, mestizo 30%, Aymara 25%, European 15%
Language: Spanish, Quechua, Aymara
Religion: Roman Catholic
Government: Republic
Life expectancy: 65 years
GDP per capita: $PPP 4,206
Currency: Boliviano
Major exports: Soya, zinc, gold, tin, natural gas

Bolivia is one of only two landlocked states in Latin America—having lost its access to the sea in 1883 after a war with Chile. It can be divided into three main regions. First, in the south-west and covering about one-tenth of the country there is the bleak, arid expanse of the Altiplano, a plateau some 3,600 metres above sea level. Second, enclosing this plain, are two branches of the Andes with the highest peak at 6,542 metres. Third, to the east and north, and covering around two-thirds of the country, are the lowlands of the Oriente, which includes grasslands and tropical rainforests.

The Bolivians who occupy this often harsh terrain are mostly of Indian origin. The largest ethnic groups are the Quechua and Aymara, who live mostly in the highland regions. The women have often retained their traditional dress with colourful petticoats, and in some regions distinctive bowler hats. The mestizo and white populations are mostly in the capital, La Paz, and in the richer valleys and lowlands.

This is one of the poorest countries in South America. The rural areas in particular suffer from a lack of safe water and sanitation and poor health services. High levels of infant mortality have depressed population growth. Around two-thirds of Bolivians live below the poverty line. In the face of increasing poverty, many Bolivians, 200,000 or more, have now migrated to work in sweatshops in Argentina.

Agriculture employs around 60% of the labour force, but much of this in the highlands is still primitive cultivation of cereals and potatoes. The more productive land is in the eastern lowlands and is often devoted to commercial farming—including cotton, sugar, and particularly soya which is now the leading official export earner.

Much of Bolivia's wealth, and its potential, lies in its minerals—including tin, silver, zinc, and gold—usually extracted with cheap Indian labour working in harsh conditions.

This reliance on commodities will however continue to leave Bolivia severely exposed to international commodity markets. Other important resources are oil and particularly gas of which there are extensive reserves in the tropical lowlands; 1999 saw the completion of a gas pipeline to Brazil.

Bolivia's largest unofficial export is coca paste. This is derived from coca leaves grown by 35,000 or so peasant farmers in the lowlands. Around 9,000 'laboratories' transform it into coca paste. Bolivians have traditionally chewed the leaf to numb themselves to cold and hunger, but the market for cocaine has transformed this into a lucrative cash crop. Bolivia supplies around 30% of the world's

Source of one-third of the world's coca paste

coca paste, with exports worth around $200 million (3% of GDP). With US aid, the government has made efforts, generally unpopular, to eradicate coca cultivation. The area under cultivation, which is 27,700 hectares, is far less that of a few years ago but is now rising again.

Economic management has often been chaotic. By the mid-1980s, with crashing tin prices, rising international debt, and a world record inflation rate of 20,000%, Bolivia was close to collapse. The subsequent recovery started in 1985 with the 'New Economic Plan', which dramatically reoriented the economy along free-market lines. Foreign firms were allowed to acquire 50% shares in the main state companies. These and other changes enabled Bolivia to enjoy investment-led growth which averaged 4% annually in the period 1990–96. Subsequently, however, the Bolivian

economy, like those of its neighbours, suffered a severe slump. And since many people had derived little benefit, there were frequent protests and demonstrations.

Bolivia's political history has been punctuated by coups and military dictatorships, though since the early 1980s power has generally been transferred constitutionally.

The 1993 presidential election resulted in a coalition led by Gonzalo Sánchez de Lozada, of the Movimiento Nacionalista Revolucionario (MNR). He lost the 1997 presidential election to Hugo Banzer Suarez, formerly a military dictator, but after Banzer stepped down due to ill health, Lozada returned in the 2002 election, just ahead of Evo Morales the radical leader of Bolivia's coca workers.

Sánchez's second term was brief. After violent protests over a proposed gas pipeline he was forced out of office in 2003 to be replaced by a non-party figure, Carlos Mesa. But he too struggled in the face of massive demonstrations and roadblocks.

This opened the path in December 2005 for the election of Evo Morales of the Movimiento al Socialismo (MAS). Morales is Bolivia's first indigenous president. He plans to 'refound' Bolivia and has nationalized the oil and gas industries and launched a programme of land redistribution.

The country's traditional vested interests have mounted vigorous opposition and in 2008 held an unofficial referendum for regional autonomy for Santa Cruz. By 2011, Morales was also losing support because he had been slow to extend indigenous rights.

Bosnia and Herzegovina

Bosnia and Herzegovina is held together by aid and diplomatic pressure

Land area: 51,000 sq. km.
Population: 4 million—urban 49%
Capital city: Sarajevo, 526,000
People: Bosniak (Muslim) 48%, Serb 37%, Croat 14%, other 1%
Language: Serbian, Bosnian, Croatian
Religion: Muslim 40%, Orthodox 31%, Catholic 15%, Protestant 4%, other 10%
Government: Republic
Life expectancy: 75 years
GDP per capita: $PPP 7,764
Currency: Marka

The state of Bosnia and Herzegovina, commonly referred to as BiH, is a loose federation of two entities created in 1995. One is the Serb Republic (RS) with 49% of the territory and approximately 45% of the population. The remainder is itself a federation: the Federation of Bosnia and Herzegovina, which consists of land controlled by the Bosniak (Muslim) and Croat communities—commonly referred to as 'the Federation'.

The region of Bosnia occupies the north and centre of the country while Herzegovina makes up the south and south-west. The political partition gives most of the lower-lying and better agricultural land in the north to the Serb Republic, and the more mountainous remainder, including the Dinaric Alps, and most of the country's industry to the Federation.

The wars since the early 1990s caused rapid changes in population. Apart from those people killed there was considerable migration in and out of the country. The war also affected birth and death rates and the population is rapidly ageing. Education too suffered, not just because of physical damage to schools but also because of a brain drain of teachers as well as the fragmentation of the education system along ethnic lines.

Health also deteriorated as a consequence of poverty, stress, and poor diet, as well as alcohol consumption and smoking. Indeed life expectancy started to fall, though it is now increasing again.

The dissolution of Yugoslavia deprived the country of the markets for its industrial goods and the war destroyed many of its factories. As a result, since the late 1980s industry has declined from 50% to 19% of GDP. This has largely been replaced by services, predominantly public administration and defence. Agriculture accounts for around 10% of GDP and 20% of employment. Many farms were abandoned during the war and the recovery has been very slow.

Most economic growth has come from aid-financed reconstruction. Even so unemployment remains around 40%.

BiH's ethnic strife reflects a long history of occupation and struggles in the Balkans. By the 20th century, this had left the territory with three main communities who are physically identical but have strong cultural and religious differences: the Croats, who are generally Roman Catholics; the Serbs, who are generally Orthodox; and the Muslims, now officially referred to as Bosniaks, who are a legacy of the long occupation by the Ottoman Empire.

The communist government of Marshal Tito managed to keep these groups together, establishing Bosnia and Herzegovina as a republic within Yugoslavia, and giving the Muslims a distinct ethnic status. The distribution of the groups was complex: some areas had majorities of one group but others were ethnically mixed.

When Yugoslavia started to disintegrate the situation in Bosnia was very difficult. At this point, the Serbs wanted to remain part of Yugoslavia, the Croats wanted to unite with Croatia, while the Muslims preferred a multi-ethnic independence. They maintain these ambitions to this day. However, they did all make a choice of sorts in March 1992 when a referendum, which the Serbs boycotted, opted for secession and Bosnia duly declared its independence.

Communities with different ambitions

After independence, the violent conflicts that had accompanied all these events erupted into full-scale civil war as each group sought to seize territory while driving out the members of the other communities in a bitter and savage process of 'ethnic cleansing'. The war was to kill a quarter of a million people, drive one million people out of the country as refugees, and displace another million or so internally.

In 1994, the USA brokered a ceasefire and in 1995 an agreement reached in Dayton, Ohio, brought the war to an end. This established a central government with a rotating three-member presidency and a two-chamber parliament to deal with foreign affairs and monetary policy. But most of the major functions, including economic policy, taxation, defence, and the police forces, would devolve to the two 'entities'.

The Dayton agreement also provided for the appointment of a powerful High Representative, who in 2011 was an Austrian Valentin Inzko. Among other things he keeps the various factions in check, removes officials considered a nuisance, and oversees a 2,000-strong EU peacekeeping force. The office is due to be abolished, to be replaced by a weaker EU envoy, though whether this will actually happen remains uncertain. BiH wants to join the EU but this seems a distant prospect.

Governments at the various levels have been coalitions, often dominated by hardline nationalists.

After the inconclusive 2010 election an already-unstable situation deteriorated further. The main political players are the Party of Democratic Action which represents Bosniaks, the Alliance of Independent Social Democrats which represents most Bosnian Serbs, and the Croatian Democratic Union of BiH, but they have failed to form a government.

Botswana

Botswana's diamonds have financed public services but have yet to lift its people out of poverty

Land area: 582,000 sq. km.
Population: 2 million—urban 61%
Capital city: Gaborone, 225,000
People: Tswana 79%, Kalanga 11%, Basarwa 3%, other 7%
Language: Setswana, English
Religion: Christian 72%, other 8%, none 20%
Government: Republic
Life expectancy: 53 years
GDP per capita: $PPP 13,604
Currency: Pula
Major exports: Diamonds, copper, nickel

Apart from hills in the south-east, Botswana is largely a flat plateau at around 1,000 metres above sea level. In the north-west is the extensive Okavango Swamp and in the south the Kalahari Desert. Most people live in the east, where the climate is cooler and the soil more fertile.

Compared with other countries of Sub-Saharan Africa, Botswana has seen little ethnic conflict. Most people are Tswana: a group of eight ethnic clans. The constitution defines these as the 'majority tribe' and they have official representation in an advisory parliamentary chamber, the House of Chiefs. However, there are also around 50,000 Bushmen, who frequently face discrimination —including removal from the Central Kalahari Game Reserve.

Botswana's income from diamond mining has enabled it to invest in

social services. Education is universal and mostly free, and the primary and curative health services are extensive. Infrastructure has also improved. An increasing proportion of the population are now urbanized—as people are attracted in particular to the capital Gaborone which has expanded ten-fold in the past 30 years.

But the economy has yet to diversify and to distribute economic opportunities more widely. Diamond mining employs only 5% of those in the formal sector. Unemployment is around 20% and many more are underemployed. One-quarter of the population live in poverty.

One of the most alarming recent developments has been the spread of HIV and AIDS. In 2008, 24% of those aged 15–49 were infected—one of the highest rates in the world. This had a dramatic impact on average life expectancy: in the early 1990s this was around 65 years but by 2003 it had slumped to 36, though it has now climbed to 53. As a result population growth has slowed: previously around 3.3% per year, it is now around 1%.

Many people still rely on agriculture—through subsistence farming and particularly raising livestock. But even when rainfall is good, they only grow enough food to meet around one-third of national needs. Agriculture by 2007 had fallen to less than 2% of GDP.

Nowadays, Botswana's economy is dominated by the diamond industry, which accounts for around 40% of GDP. Diamonds were discovered under the sands of the Kalahari Desert shortly after independence and large-scale extraction was started in 1971 by the Debswana Mining Company—a joint enterprise between the government and the South African company De Beers. Botswana is now one of the world's largest producers of diamonds and new diamond 'pipes' are still being discovered.

Diamonds under the desert

There are also significant deposits of copper-nickel and other minerals, including coal, soda-ash, and gold, but the enterprises that are mining these reserves have proved less successful.

Mining employs few people directly. But it does support a large public sector which employs around half of all those working in the formal sector. The government is anxious to build up manufacturing industry and is offering substantial tax concessions for new investment. Recently efforts have been made to build up the textile industry to take advantage of preferential access to the US market but all the inputs have to be imported.

Mining has also enabled the government to build up large savings in anticipation of its diamond reserves eventually being exhausted. In 2007, Botswana had foreign reserves of around $10 billion and has been in the unusual position for an African country of lending funds to the IMF.

Botswana has been one of Sub-Saharan Africa's more democratic countries. Since independence in 1966, it has had a 40-member National Assembly which elects an executive president, in addition to its 15-member advisory House of Chiefs. The first president, Seretse Khama, founder of the Botswana Democratic Party (BDP), was determined to build a multiracial society.

Although the country needed to maintain trading links with South Africa in order to survive, Khama opposed apartheid. On his death in 1980, he was succeeded by his vice-president, Quett Ketumile Masire, who also won elections in 1984, 1989, and 1994. When he retired in 1998, he in turn was replaced by his vice-president Festus Mogae.

Until recently, opposition came from the left-wing Botswana National Front (BNF), but in 1998 eleven of the BNF's assembly members formed the new Botswana Congress Party.

With the opposition split, the BDP achieved another convincing victory in the 1999 National Assembly elections. Mogae chose as vice-president Ian Khama, son of the first president, who started building his own power base, becoming party chairman in 2003.

The BNF, now led by Otsweletse Moupo has continued to suffer from infighting: another faction has left to form the New Democratic Front Party.

The BDP won an easy victory in the 2004 National Assembly elections and in 2008 Ian Khama succeeded Mogae as president. With the opposition still split, the BDP won the 2009 general election, gaining 45 of the 57 seats, though it too is suffering from internal power struggles.

Brazil

Destined by size to be the leading country of South America, Brazil is only just starting to realize its potential

Land area: 8,512,000 sq. km.
Population: 190 million—urban 87%
Capital city: Brasilia, 2.5 million
People: White 54%, mixed white and black 39%, black 6%, other 1%
Language: Portuguese
Religion: Roman Catholic 74%, other 26%
Government: Federal republic
Life expectancy: 72 years
GDP per capita: $PPP 9,567
Currency: Real
Major exports: Manufactures, soya, iron, coffee

Brazil's huge landmass occupies almost half the continent of South America. Geographically, the country's two main features are the Brazilian Highlands and the Amazon Basin. The Brazilian Highlands cover most of the south and east of the country and consist of a vast plateau with an average elevation of 1,000 metres, interspersed, particularly in the east, with rugged mountain ranges, some of which rise above 2,800 metres. Much of this area is forested or opens up to extensive prairies.

The Amazon Basin to the north and west covers more than 40% of the country. This is the world's largest river drainage system and most of it is covered with tropical rainforest. While there are still unexplored areas, many parts of the rainforest have now been penetrated by settlers, ranchers, or mining companies, a process

of deforestation that has alarming environmental implications globally— in terms of climate change and loss of biological diversity. This area contains around one-fifth of world plant species. Most of Brazil has a humid subtropical climate, but the land to the north-east, known as the sertão, suffers from frequent droughts.

Brazil has long been a racial melting pot. There is little overt discrimination, but people of European origin hold the most powerful positions, followed by those of mixed race (who call themselves 'brown') and blacks, with the small and declining Indian population the most marginalized of all. This mixture of races has generated a vibrant and diverse culture. So although Brazil is unified by the Portuguese language and by Roman Catholicism it also has strong African influences.

This social stratification contributes to severe inequalities. The cities in the south, like Rio de Janeiro or São Paulo, are similar in many respects to those in Europe, though also have desperate shanty towns, called favelas. The north-east is almost another country—deep in the Third World.

Brazil has one of the world's most unequal distributions of income. The richest 10% of the population get 47% of the income while the poorest 10% get only 12%. Around 20% of Brazilians live below the poverty line. Inequality has, however, fallen in recent years as a result of faster economic growth and an innovative

Brazil has pioneered conditional cash transfers

programme of 'conditional cash transfers', Bolsa Família, which offers payments to around one-quarter of the population for attending health clinics and sending their children to school.

This should help improve access to education and health services, though this is still very unequal. Brazil has a small proportion of highly educated professionals, but many children drop out of school early. Half of ten-year-olds are illiterate. Health services too are skewed. One estimate suggests that 40% of health expenditure is used for sophisticated curative treatments that benefit only 3% of the population. HIV and AIDS are significant problems but those infected can get free anti-retroviral treatment.

In many respects, Brazil is an advanced industrial economy. Around one-quarter of the labour force work in industry. With such a large population, Brazil was for decades able to direct most of its manufacturing output at its domestic market, and offered local businesses protection from foreign competition.

From 1990, however, the government opened up the markets. This helped stimulate greater efficiency in some areas—though others such as garments or shoes shrank in the face of Asian competition.

Brazil's industry has benefited from its wealth of mineral resources. It has around one-third of the world's reserves of iron ore, as well as large deposits of bauxite, coal, zinc, gold, and tin. It also has extensive oil reserves that have enabled the state oil company, Petrobrás, to supply more than half the country's needs. Vast new oil fields were discovered under the Atlantic in 2007. Most electricity, however, comes from hydroelectric plants—one of the largest in the world being the Itaipú dam shared with Paraguay on the Paraná River.

With such a huge landmass, Brazil might also be expected to be a major agricultural producer. Brazil is indeed largely self-sufficient in basic foods. It also has a large livestock herd, and is an exporter of cash crops like soya and coffee. But output is less than might be expected. This is partly because much of the land area, particularly in the Amazon Basin, is unsuitable for farming. And even the better farming land is often used very inefficiently. In fact much of the land held by the largest landowners is scarcely used at all. Some 58,000 large landowners hold half the country's farmland, while three million small farmers make do on 2%, and millions more have no land at all. Many governments have promised land reform; few have delivered.

Brazil has often been plagued by inflation—usually in at least two digits and sometimes four or five. This pattern appeared finally to have been broken early in 1994 with the introduction of a new currency, the 'real', which was pegged to the US dollar, a move which was strikingly successful at reducing inflation.

Any Brazilian president's task is complicated by the dispersed and fragmented nature of the political system. Brazil is a federal democracy and each of the 26 states has its own legislature and administration. States and their municipalities control over two-fifths of total tax revenues and have considerable freedom.

Federalism has the merit of permitting decentralization, but it also produces confusing overlaps. Universities and hospitals, for example, may be controlled either by the federal or state governments. There are also huge disparities between the states: the average per capita income in the state of São Paulo is around ten times that of the north-eastern state of Piaui or of Amazonas. Governments have tried to narrow these gaps by building infrastructure and offering tax breaks. As a result, many new electronics factories have sprung up in Manaus in the middle of the rainforest.

The federal structure also complicates political manoeuvres in the capital, Brasilia. The president does in theory have considerable power of patronage. But party allegiances are notoriously weak and temporary. Most members of the lower house concentrate on extracting federal largesse for their state or municipality. Party organization is also blurred by powerful cross-party interest groups such as the 'ruralistas', who lobby for large landowners.

For much of the country's history Brazil has been under centralized, authoritarian rule, including a military government from 1964 to

Brazilian states have considerable freedom

1985. In 1988 the country adopted a new constitution that provided for a directly elected executive president in addition to elections for the two houses of Congress. The 1988 election was won by the conservative Fernando Collor de Mellor. He resigned in 1992, having been impeached for corruption.

The 1994 election was won by Fernando Henrique Cardoso who as finance minister had been the architect of the 'real plan'. The success of this also helped him to amend the constitution allowing him a second term in 1988.

By the 2002 elections, however, the situation had deteriorated and the currency came under pressure. Cardoso's chosen successor José Serra proved a poor candidate to face up to Luis Inácio ('Lula') da Silva, the charismatic leader of the left-wing Workers Party (PT), who won in a second round run-off.

Lula had a sure political touch. While pushing through many radical social programmes such as Bolsa Familia, he lifted millions of people out of poverty, he also reassured the international markets by pursuing fairly orthodox economic policies. Hugely popular, Lula was re-elected in 2006 and helped make Brazil a major global force, one of the 'BRIC' countries, bracketed with Russia, India and China.

Lula rejected the idea of a constitutional change that would enable him to seek a third term. The 2010 presidential election was won by his chosen successor Dilma Rousseff of the PT. A less charismatic figure, she has maintaind most of the policies of Lula, who could return in 2014.

Brunei

A high-income sultanate floating on oil, but subjected to autocratic rule

Brunei, officially Brunei Darussalam, occupies a small section of the north coast of Borneo, surrounded by the Malaysian state of Sarawak. A narrow coastal plain rises to low hills in its eastern section and to mountains in the west. Unusually for this part of Asia, forests still cover around 85% of the territory—thanks to its other main natural resource: oil.

Since 1929, oil and gas have been the main sources of wealth. Oil not only gives Brunei's citizens a high per capita income; it also funds extensive free, or highly subsidized, health and education services. The major health concerns are those of affluence: a high-fat diet and a sedentary lifestyle.

Most Brunei citizens are of Malay extraction, and tend to seek work either in the public sector, or in large prestigious companies such as Brunei Shell Petroleum. Most of the entrepreneurial energy comes from the less privileged Chinese minority, who are considered non-citizens, and also from temporary foreign workers who are around 40% of the labour force.

Brunei's oil and gas reserves are expected to last 40 years, and, given the country's extensive foreign investments, should provide an income beyond that. Even so, oil and gas represent a shrinking proportion of GDP—down from 80% in the early 1980s to 70% in 2008, as a result of falling oil prices and an extension of government services. Moreover, there are doubts about the value of the holdings of the Brunei Investment Agency.

As a result the government has concluded that the economy is unsustainable and has proposed a series of measures to streamline expenditure, diversify the economy, and reduce dependence on foreign workers. This includes the construction of an 'agro-technology park' to address Brunei's dependence on food imports.

In theory, Brunei is a constitutional sultanate; in practice it is an absolute monarchy. When the country became independent in 1985, Sultan Hassanal Bokiah (whose family have ruled the country for 600 years) dissolved the existing legislative assembly and has since ruled by decree in a biennially renewed 'state of emergency'.

The sultan occupies numerous government posts, including prime minister, head of the police force, and leader of the Islamic faith. There is only one registered political party, the press is strictly censored, and no public criticism is permitted. Fundamental change seems remote.

Land area: 6,000 sq. km.
Population: 0.4 million—urban 76%
Capital city: Bandar Seri Begawan, 27,000
People: Malay 66%, Chinese 11%, indigenous 3%, other 20%
Language: Malay, English, Chinese
Religion: Muslim 67%, Buddhism 13%, Christian 10%, indigenous and other 10%
Government: Constitutional sultanate
Life expectancy: 77 years
GDP per capita: $PPP 50,200
Currency: Bruneian dollar
Major exports: Oil, gas

Bulgaria

Bulgaria is now politically more stable and is about to join the European Union

Land area: 111,000 sq. km.
Population: 8 million—urban 72%
Capital city: Sofia, 1.2 million
People: Bulgarian 84%, Turk 9%, Roma 5%, other 2%
Language: Bulgarian, Turkish
Religion: Bulgarian Orthodox 83%, Muslim 12%, other 5%
Government: Republic
Life expectancy: 73 years
GDP per capita: $PPP 11,222
Currency: Lev
Major exports: Base metals, chemicals, textiles, machinery, agricultural products

Bulgaria has three main natural regions, each of which extends across the country from west to east. The most northerly, starting from the banks of the Danube, which marks the border with Romania, is a plain with low hills that takes up around one-third of the country. To the south of this plain is the second main region, the Balkan Mountains. Further south still, separated from these mountains by the narrow Thracian Plain, is the third region, the Rhodope Mountains, which form the border with Greece.

Bulgaria has a fairly homogenous population and has so far avoided serious ethnic conflict. The largest minority are around 750,000 ethnic Turks, who live largely in the north-east or in the east of the Rhodope Mountains. The Turks tend to be worse off than most Bulgarians, and in the communist era suffered legal discrimination. They have

their own political party but are not very assertive and many have been emigrating to Turkey. Even worse off are the Roma, or gypsies, who are economically marginalized and frequently the victims of heavy-handed policing. More than 80% live below the poverty line.

Like many other countries of Eastern Europe, Bulgaria now has a shrinking population—a result of a falling birth rate, a higher death rate, and emigration, particularly of ethnic Turks, around 200,000 of whom left during the 1990s. From 1998 to 2007 the total population fell from 9 to 8 million.

Public services deteriorated after the collapse of communism. Education had been one of Bulgaria's strong points but spending has fallen and the quality has declined. Health standards too have suffered.

During the communist era, Bulgaria's economy was heavily industrialized, but production fell steeply during the transition. The industries that held up better were basic metallurgy and chemicals, which along with textiles have been the major exports.

One problem was an erratic and

slow process of privatization, though by 2006 around 80% of formal production was in private hands, added to which Bulgaria has a large black market, estimated at around one-fifth of GDP.

One-quarter of the population still make their living from agriculture, growing basic grains, mostly wheat.

A leading wine exporter to the UK
Most output is now in the private sector. Harvests are vulnerable to an erratic climate with regular droughts.

One successful crop is grapes for the production of wine—with most exports going to the UK.

Bulgaria joined the EU in 2007 but is struggling to meet EU standards in a number of areas, especially in the dairy industry. In 2008 the EU froze some of its $1 billion in transitional aid following allegations of fraud and the padding of contracts. Indeed many argue that Bulgaria was admitted to the EU too early given the scale of corruption, a weak judicial system, and the extent of organized crime. On the other hand the government finances are in reasonable shape with a budget surplus and very little debt.

Bulgaria's post-communist political development has often been conflictive. Indeed no government has been elected for two consecutive terms. One of the most important parties has been the Bulgarian Socialist Party (BSP) formed by former communists, which has support among older people in the rural areas and among those with strong nationalist sentiments.

Another is the Union of Democratic Forces (UDF), a coalition of centre-right organizations that

has greater support in the cities and among the younger and more educated. Other important parties include the party for ethnic Turks, the Movement for Rights and Freedom (MRF).

The BSP won the first post-communist election in 1990. But this government did not last and was followed by a number of short-lived administrations, led either by the BSP or the UDF, which were often undermined by a generally poor performance and corruption.

Such was the disillusionment with the two main parties that both were swept aside in the June 2001 parliamentary elections by a coalition of the MRF and new right-wing party, the Simeon II National Movement (SNM) headed by Simeon Saxe-Coburg, the former king of Bulgaria, who became prime minister. He was successful in promoting stable economic growth, but less so in tackling social issues such as crime.

The 2005 election was complicated by the emergence of a new nationalist coalition, Ataka, hostile to the Turks and the Roma. The outcome was a coalition between the BSP, the SNM, and the MRF, with BSP leader Sergei Stanishev as prime minister.

The 2009 elections gave power to another new centre-right populist party, Citizens for European Development of Bulgaria (GERB) which entered parliament with 40% of the vote on a 60% turnout. GERB is led by Boyko Borisov who had been a successful mayor of Sofia. The BSP came second. Borisov chose to form a minority GERB government with support from Ataka, right-wing nationalist group.

Burkina Faso

Burkina Faso has few natural resources, a fragile environment, and a rapidly growing population

Land area: 274,000 sq. km.
Population: 15 million—urban 20%
Capital city: Ouagadougou, 1.2 million
People: Mossi over 40%, also Gurunsi, Senufo, Lobi, Bobo, Mande, Fulani
Language: French and local languages
Religion: Muslim 50%, indigenous beliefs 40%, Christian 10%
Government: Republic
Life expectancy: 53 years
GDP per capita: $PPP 1,124
Currency: CFA franc
Major exports: Cotton, livestock, gold

Burkina Faso extends over a broad plateau in the Sahel—the fringe of the Sahara Desert. Most of the territory is savannah grassland scattered with small bushes and trees. The climate is generally arid with a short rainy season and a long dry season. But the rains are unpredictable and the land is vulnerable to drought. The country does have major rivers, including the Red, White, and Black Voltas, though these frequently dry up.

Burkina Faso has numerous ethnic groups, of whom the largest, the Mossi, make up more than half the population. Most people live in the rural areas in the east and centre of the country. Around half live below the poverty line. One-third of pre-school children are chronically malnourished and the population is largely uneducated: only around half of children enrol in primary school—one

of the lowest proportions in the world. Health standards are undermined by endemic diseases like malaria and bilharzia, as well as by sleeping sickness transmitted by the tsetse fly, which makes land alongside the rivers virtually uninhabitable. Around 2% of people are infected with HIV/AIDS.

Most people make their living from agriculture and livestock—which provides around 90% of jobs and around 30% of GDP. Farmers grow a number of subsistence food crops such as millet and sorghum. Food production has increased in recent years—though this was more the result of bringing additional land under cultivation than achieving increases in productivity.

The other main crop is cotton. Burkina Faso is one of the region's leading producers, and cotton brings in around 70% of export income. Other cash crops include shea nuts, sesame seeds, and sugar cane. Even so, only around 13% of the land is under cultivation: much of the rest of the territory, particularly in the north, is given over to livestock, largely for export to neighbouring countries such as Ghana and Côte d'Ivoire.

Burkina Faso also has mineral potential. At present, most extraction is of gold by small-scale panners. But the government has attracted a number of foreign mining companies which are looking at the possibilities for zinc, manganese, limestone, phosphates, and diamonds.

The population continues to grow rapidly, by 3% annually, which is putting increasing pressure on the environment. Many parts of the country have already gone beyond what is considered to be a critical limit of 50 inhabitants per square kilometre. Many of Burkina Faso's

A fraying ecological system

traditional systems for managing the fragile ecology are breaking down. Much of the thin topsoil has been lost and in the past 15 years or so a combination of drought, over-grazing, brush fires, and unregulated felling has removed nearly 60% of the trees. Over a similar period the water table has fallen by around 20 metres.

Bringing the country's population and resources into balance will be an increasingly difficult task. Though economic growth and population growth have roughly kept pace, millions of people already migrate seasonally to neighbouring countries in search of work.

As one of the world's poorest countries, Burkina Faso has been a major recipient of foreign aid. Its high levels of poverty and its participation in an IMF-backed structural adjustment programme have also qualified it for debt cancellation. But donors are becoming increasingly concerned about the quality of governance, particularly repression and corruption.

Since independence from France in 1960, governments in Burkina Faso have rarely been free of military influence. One of the most distinctive periods was between 1983 and 1987, when a group of army officers led by Thomas Sankara seized power and embarked on a left-wing agenda determined to redistribute resources to peasant farmers and to free the country from dependence on foreign aid. He also changed the country's name from Upper Volta to Burkina Faso—which means 'land of the dignified'.

This radical experiment ended in 1987 with Sankara's assassination in a violent coup by Captain Blaise Compaoré, who in 1989 founded a new party which later merged with others to become the Congrès pour le démocratie et le progrès (CDP). In 1991, under pressure from donors, he introduced a new constitution permitting opposition parties.

The opposition boycotted the presidential elections in 1991 and again in 1998 allowing Compaoré to be re-elected with a large majority. In 1998, however, there were widespread protests over the murder of a newspaper editor, Norbert Zongo.

For the presidential election in 2005, 16 opposition parties formed a new coalition, Alternance. But they failed to unite around a single candidate and Compaoré was comfortably returned to office. His CDP also won a majority in the 2007 legislative elections.

Compaoré won a clear victory in the 2010 presidential election with more than 80% of the vote. Nevertheless he has faced widespread protests from discontented trade unionists, students and army units.

Burma (Myanmar)

Burma has suffered years of repression from a military regime that has now applied a thin veneer of democracy

Land area: 677,000 sq. km.
Population: 49 million—urban 34%
Capital city: Naypyidaw
People: Burman 68%, Shan 9%, Karen 7%, Rakhine 4%, Chinese 3%, Mon 2%, Indian 2%, other 5%
Language: Burmese and others
Religion: Buddhist 89%, Christian 4%, Muslim 4%, other 3%
Government: Military dictatorship
Life expectancy: 61 years
GDP per capita: $PPP 904
Currency: Kyat
Major exports: Gas, food, timber, prawns

Burma consists largely of the central fertile valley of the Irrawaddy River, encircled by a horseshoe-shaped mountain system running north to south. The highest mountains are to the north, while to the west are two lower ranges, the Chin Hills and the Arakan range. To the east is the extensive Shan Plateau which consists of mountains that intersect with broken ranges of hills and river valleys.

Three-quarters of the population, mostly Burmans, live in the central valley and the coastal strips. Although sparsely settled, Burma has a complex ethnic mix, with 21 major groups and over 100 languages. The second largest group are the Shan, who live on the Shan plateau. The Karen live in the delta, the coastal areas to the south, and the hills bordering on Thailand, while the Rakhine live in the south-west. These groups have engaged in long struggles with the military government. Around 120,000 live in refugee camps in Thailand.

Education standards are low and more than one-third of children are malnourished. Intravenous drug use is growing and 240,000 people are now HIV-positive. Over one-third of public spending is used to finance the army.

Two-thirds of households make their living from agriculture, which accounts for 40% of GDP. But they are short of fertilizer and other inputs so productivity is low. Another source of rural income is forestry. Burma has around 75% of the world's teak reserves, but over-logging, often through government contracts with Thai companies, is rapidly stripping the forests. The most profitable crop is opium poppies, which enable Burma to supply around 60% of the world's heroin. Burma is also a major source of amphetamines.

Burma has made slow progress in industrial development. However, recent offshore discoveries have made natural gas the leading export earner. Foreign investment is limited. A number of Hong Kong and Korean companies have established a garments industry, though in 2003 the US banned imports. Faced with consumer boycotts, Levi-Strauss, Reebok, and British Home Stores have pulled out and Texaco has withdrawn from oil and gas.

The present regime dates to a coup in 1988. Millions of people had taken to the streets to protest against military rule. The response was brutal. Soldiers sprayed bullets indiscriminately into the crowds and during this incident and the subsequent repression 10,000 people may have died. At this point, a new group of senior military officers seized power. They called themselves the State Law and Order Restoration Council (SLORC), and renamed the country Myanmar. In 1989, they placed under house arrest the leading opposition politician Aung San Suu Kyi, daughter of Aung San, a hero of Burma's independence struggle.

Aung San Suu Kyi arrested

In 1990 SLORC, presuming wrongly that their grip on power would intimidate people into voting for military-backed parties, held multiparty elections for representatives to design a new constitution. In the event, 80% of the seats were won by Aung San Suu Kyi's party, the National League for Democracy (NLD). SLORC refused to accept the results. The assembly never met.

This marked the onset of one of the world's most repressive regimes. The UN Human Rights Commission has accused the government of torture, summary executions, and forced displacement and oppression of religious and ethnic minorities. SLORC also overturned the previous regime's socialist model—opening the country up to foreign companies, particularly for timber extraction and the export of natural gas. To build the country's infrastructure it has resorted to forced labour, displaying what an ILO report called 'a total disregard for the human dignity, safety and health of the people'.

Many ethnic groups have stubbornly resisted the regime. But the army, largely staffed by Burmans, has steadily subdued them and most have signed ceasefires. Only the Karen National Liberation Army and two others are still holding out.

In late 1997, SLORC reformed as the State Peace and Development Council, whose chair, General Than Shwe in 2007 moved the capital to a remote jungle location at Naypyidaw.

In 2007, pro-democracy demonstrations, led by monks, were violently repressed. In 2008 the regime under international pressure allowed in foreign aid workers following cyclone Nargis which killed at least 78,000 people.

The most potent focus of non-violent opposition remains Aung San Suu Kyi, who in 1991 won the Nobel Peace Prize but for many years was under house arrest.

In 2010 under a new constitution the SPDC held a tightly-controlled election from which the NLD was barred. The army's puppet organization, the Union Solidarity and Development Party won large majorities in both upper and lower houses, where in addition a quarter of the seats are reserved for military appointees. The SPDC has been dissolved and former-general Thein Sein has been appointed president.

Burundi

After decades of ethnic violence Burundi faces many years of rebuilding

RWANDA

• Ngozi

Bujumbura

• Gitega

DEM.
REP.
OF
CONGO

• Bururi

TANZANIA

| 0 | Miles | 75 |
| 0 | Km | 120 |

Land area: *28,000 sq. km.*
Population: *8 million—urban 11%*
Capital city: *Bujumbura, 400,000*
People: *Hutu 85%, Tutsi 14%, other 1%*
Language: *Kirundi, French, Swahili*
Religion: *Christian 67%, indigenous beliefs 32%, other 1%*
Government: *Republic*
Life expectancy: *50 years*
GDP per capita: *$PPP 341*
Currency: *Burundi franc*
Major exports: *Coffee, tea, cotton*

The west of Burundi lies along the Great Rift Valley, with Lake Tanganyika forming the southern two-thirds of the border. To the east the land rises first to high mountains and then descends across a hilly plateau to the border with Tanzania. Much of this area was originally forested, but most of the land has now been cleared for cultivation. This has resulted in extensive soil erosion.

Burundi's population remains sharply divided between two main ethnic groups. The majority are the Hutu, who are agriculturalists. For centuries they have been dominated by the minority Tutsi, most of whom raise cattle.

In recent decades, there has been a series of Hutu challenges to Tutsi domination, following which the security forces have taken revenge on the Hutu. Waves of killings occurred in 1965, 1969, 1972, 1988, and 1991.

In 1993 it was Hutu militias, however, who killed more than 100,000 Tutsi. And from 1994 onwards, a number of Hutu armed opposition groups fought the Tutsi-dominated armed forces. Half a million people, mainly Hutu, fled into Tanzania and the Democratic Republic of the Congo—and similar numbers, mainly Tutsi, were displaced within Burundi.

Years of civil war largely brought development to a halt. Only around 60% of the population are literate. Schools have frequently been a target of the militias and many teachers and pupils have been killed.

Health standards too are low. Half of the children under five are malnourished and health services have deteriorated as medical staff have been caught up in the fighting.

Around 90% of people live in the rural areas, mostly growing subsistence crops such as cassava, sweet potatoes, bananas, beans, and maize. Rapid population growth in recent decades has intensified pressure on the land. Even so, the soil is relatively fertile and until the mid-1990s the country was mostly self-sufficient in food. Since then, however, Burundi has become reliant on international food aid. The main cash crops are coffee, tea, and cotton, which are also grown on peasant smallholdings.

Until independence in 1962, Belgium administered Burundi and neighbouring Rwanda as one country. In fact, they have always been fairly distinct territories, naturally divided by rivers, and with their people speaking different languages.

Burundi achieved independence as a monarchy but in 1966 the king was deposed by a Tutsi police commander who declared himself president and the country a republic. The decades that followed saw regular purges, massacres, and reprisals, one of which in 1972 resulted in the death of at least 100,000 Hutu.

Another coup by another Tutsi in 1974 brought **Regular purges and massacres** Colonel Jean-Baptiste Bagaza to power. He continued to exert Tutsi dominance but was overthrown in yet another coup by Major Pierre Buyoya. He was a more conciliatory figure and deliberately included Hutu leaders in the government.

In 1993 Burundi had its first free elections. These were won by a new party, the Front pour la démocratie au Burundi (Frodebu). Although ethnically mixed, it had a large Hutu following. A Hutu president, Melchior Ndadaye, took office with a Tutsi prime minister.

National unity was short-lived. Attempts to promote more Hutu within the civil service alarmed the military, who within five months staged another coup—killing Ndadaye and other ministers. An estimated 100,000 people died in the ensuing fighting and up to 700,000 refugees fled to Tanzania.

Attempts to re-establish a legitimate Frodebu government were weakened when the interim president was killed in a plane crash.

Following another coup in 1996, Buyoya returned, though soon entered into a power-sharing agreement with Frodebu. From 1998 to 2001 Burundi was also involved in the war in neighbouring Congo.

External mediation started in 1998, steered initially by former Tanzanian President Julius Nyerere then by Nelson Mandela, resulting in a new constitution and a transitional administration.

According to the new constitution 60% of seats in the National Assembly and the executive must be filled by Hutus, while the two groups share the most senior posts in the army and the Senate.

Elections in 2005 resulted in a victory for Pierre Nkurunziza, who had a Hutu father and a Tutsi mother, and his party the Conseil national pour la défense de la démocratie–Forces pour la défense de la démocratie (CNDD–FDD). Although the CNDD–FDD was formerly a Hutu militia it has in recent years successfully drawn in some Tutsis.

The last group to accept a peace agreement was the Forces nationales pour la libération (FNL) which declared a ceasefire in 2008 transformed itself into a political part and started a process of demobilization.

The 2010 municipal, legislative and presidential elections were boycotted by the opposition, handing victory to the CNDD–FDD, with Nkurunziza returned for a second term.

Cambodia

Decades of war and political upheaval tore Cambodia apart. Democracy has been established but is fragile

Land area: 181,000 sq. km.
Population: 14 million—urban 23%
Capital city: Phnom Penh, 1.0 million
People: Khmer 90%, Vietnamese 5%, Chinese 1%, other 4%
Language: Khmer, French
Religion: Buddhist 96%, other 4%
Government: Constitutional monarchy
Life expectancy: 61 years
GDP per capita: $PPP 1,802
Currency: Riel
Major exports: Logs, timber, rubber, manufactured goods

Three-quarters of Cambodia consists of a large central plain, through which passes the Mekong River on its way from Laos in the north-east to Vietnam in the south. The plain also includes a large lake, the Tonle Sap, which drains into the Mekong—though during the rainy season, while the Mekong is flooding it also sends water back into the lake. Cambodia has a number of mountainous regions—including the Dangrek Mountains on the northern border with Thailand, as well as ranges to the east and south-west.

Cambodians are mostly Khmers, but there are also significant Vietnamese and Chinese minorities. The vast majority live and work in the rural areas. This is one of the poorest countries in Asia. Standards of literacy and health are low; only one-quarter of the population has access to safe drinking water. The health

system is weak, offering care only to about half the population. Around one-third of Cambodia's children are malnourished, more than 40% are working, and many girls are trafficked into the sex trade to Thailand and elsewhere. Cambodia has, however, been successfully addressing HIV and AIDS: adult infection rates are now down to 0.8%.

Three-quarters of the labour force work in agriculture and fishing. The main crop is rice but yields are relatively low. This is partly because production is generally on a very small scale and many farmers lack irrigation systems. Relying on seasonal rains, they can only get one crop per year. Even so, Cambodia is self-sufficient in rice and can even produce a surplus. The government is keen for farmers to diversify into other crops but they are hampered by weak infrastructure and underdeveloped marketing systems. The most important cash crop is rubber which is produced on government-owned plantations.

Forestry is another source of export income. But logging companies are cutting down large numbers of teak trees, often illegally, and deforesting vast areas. Over the past 30 years the proportion of land covered by forests has fallen from 70% to 30%.

Industrial development has been slow. One of the main growth areas has been in garment manufacturing. There are around 200 factories which employed 150,000 workers. Participation in a labour inspection programme has made exports more acceptable to foreign buyers.

Tourism is another important source of foreign exchange. Around two million visitors arrived in 2007, many heading for the extraordinary complex of temples at Ankor Wat.

Cambodia's recent political history has been both tragic and complex. One constant figure was Norodom Sihanouk. Between 1941 and 1970, he ruled either as king or prime minister, attempting to steer a neutral course between right and left. In 1970, he was deposed by more conservative forces in a US-backed coup, though this republic only lasted until 1975 when it fell to the communist Khmer Rouge.

The constant figure of Sihanouk

Headed by Pol Pot, the Khmer Rouge unleashed a regime of extraordinary ruthlessness that killed two million people. Sihanouk returned—to house arrest.

But the Khmer Rouge made the mistake of antagonizing neighbouring Vietnam. In 1979, the Vietnamese responded by invading Cambodia and installing a new regime, forcing the Khmer Rouge back underground. In 1989 the Vietnamese finally withdrew, and in 1991 a UN-brokered peace process led to elections in 1993.

The outcome was evenly balanced between a royalist party, Funcipec, led by Sihanouk's son Prince Ranariddh, and the Cambodian People's Party (CPP), led by a political strongman, Hun Sen, a former Khmer Rouge defector who had been prime minister in the Vietnamese-installed regime. In the same year, a new constitution established a constitutional monarchy with Sihanouk once again as king.

Hun Sen worked with Prince Ranariddh as joint prime minister until 1997 when he ousted him in a violent coup. Ranariddh fled, but returned for new elections in 1998. Again, the CPP came out ahead, but without the two-thirds majority required to rule outright. Funcipec came second, and the 'self-named' party of Sam Rainsy came third. Ranariddh subsequently agreed to join Hun Sen's coalition.

The elections in July 2003 were similar to those in 1998—violent with no clear outcome. In July 2004 Hun Sen and Prince Ranariddh finally agreed to share power with the former as prime minister and the latter as speaker of the national assembly. In October 2004, Sihanouk abdicated and was succeeded by another of his sons, Norodom Sihamoni.

In 2009, the long-awaited tribunal for former Khmer Rouge officials for crimes against humanity eventually started. Five people have been charged, but not Pol Pot who died in 1998, unrepentant to the last. There are, however, concerns about corruption and political manipulation.

In the 2008 general election, the CPP won 90 of the 123 seats. Given Hun Sen's anti-democratic inclinations, and his record of locking up political opponents such as Sam Rainsy, Cambodia is in danger of again becoming an authoritarian one-party state.

Cameroon

Most people have seen little benefit from the exploitation of natural resources

Land area:	475,000 sq. km.
Population:	19 million—urban 58%
Capital city:	Yaoundé, 1.7 million
People:	Cameroon Highlanders 31%, Equatorial Bantu 19%, Kirdi 11%, Fulani 10%, Northwestern Bantu 8%, others 21%
Language:	24 African language groups, English, French
Religion:	Indigenous beliefs 40%, Christian 40%, Muslim 20%
Government:	Republic
Life expectancy:	51 years
GDP per capita:	$PPP 2,118
Currency:	CFA franc
Major exports:	Oil, timber, cocoa, cotton

Cameroon can be divided into four main regions. First, there is a coastal belt with mangrove forests and swamps that stretch up to 60 kilometres inland. These give way to the east to rocky plateaux. Much of the far north is a broad savannah plain with occasional hills leading to the shores of Lake Chad. The highest part of the country is along the border with Nigeria, a region that includes Mount Cameroon, the highest point in West Africa.

Cameroon has a diverse population, fractured along lines of ethnicity and language. There are thought to be around 200 ethnic groups. These differences open up a number of potential divisions: between the Islamic north and the Christian south, for example, and between pastoralists and farmers. But the most significant political differences arise from the colonial experience. The east of Cameroon was colonized by the French and the west by the British. Today the country is around 80% Francophone and 20% Anglophone—with the latter tending to be marginalized.

Although they had been doing well by the standards of Sub-Saharan Africa, most Cameroonians have seen a fall in living standards since the mid-1980s and around 40% live below the poverty line. Primary school enrolment increased to around 95% after school fees were abolished in 2000, but many children drop out. Health has been hit by reductions in public expenditure and around 1% of adults are HIV-positive. One third of the population do not have access to clean drinking water.

Two-thirds of Cameroonians still rely directly or indirectly on agriculture. The land is fertile and the country is usually self-sufficient in basic food crops like cassava, corn, millet, and plantains. The most important cash crops are cocoa, coffee, bananas, and cotton, which are mostly grown on smallholdings, though they are marketed through state corporations. In the south, there are also plantations growing palm oil and rubber. The country's extensive grasslands also offer grazing for livestock, which provide meat both for local consumption and export.

Cameroon's dense forests in the centre and south, as well as in the coastal belt, cover around 40% of the country and have also been an important source of income and the largest source of foreign exchange after oil. Around one-third of the forest area has been exploited, chiefly for the export of mahogany, teak, and ebony. This activity is largely in the hands of multinational companies, but there have been increasing

Rainforests threatened by logging and the oil pipeline

concerns about the over-exploitation of the forests as well as about the effects on the country's oldest inhabitants, the pygmies—hunter-gatherers who live in the southern forests.

Economic prospects brightened with the discovery of offshore oil in 1976. Production started in 1978 and peaked in 1985. Since then, some of the better fields have matured and companies are having to exploit more marginal deposits and production has halved. In response the government has been making greater efforts to exploit deposits of natural gas.

Other important mineral deposits include cobalt, iron ore and bauxite, though these remain underdeveloped.

One of the most contentious development projects has been the 1,100-kilometre oil pipeline from Chad through Cameroon's rainforests to the Atlantic coast. This project, resulted in major ecological damage.

Since 1982, political power has

been in the hands of President Paul Biya. He is from the southern Beti group, but has adroitly manipulated the country's ethnic divides. Initially, he headed a one-party state, but in 1990 he was forced to permit multiparty politics. In a rigged 1992 election Biya was elected president at the head of his own party, the Rassemblement démocratique du peuple Camerounais (RDPC)—narrowly defeating John Fru Ndi of the main Anglophone opposition party, the Social Democratic Front (SDF).

Biya won again in 1997, but this time all three opposition parties boycotted the election. The more radical elements of the SDF, whose power base is among the Bamiléké, were demanding secession for the English-speaking provinces.

For the December 2004 elections the SDF and the other main opposition party, Union démocratique du Cameroun (UDC), ran a joint candidate but he made little impression and Biya was returned for a fifth term with 71% of the vote. The opposition claimed that they were victims of intimidation and fraud.

In 2008 a strike protesting against rising food prices developed into rioting in a number of cities, resulting in the deaths of 100 people. Also in 2008, National Commission on Human Rights and Freedoms also published a report that accused the security forces of unlawful killings, arbitrary arrests and torturing.

A 2008 constitutional amendment removed presidential term limits, but at 76 and in poor health Biya might not get the chance to run again in 2011.

Canada

An increasingly multicultural population largely free from racial tension

Land area: *9,976,000 sq. km.*
Population: *33 million—urban 81%*
Capital city: *Ottawa, 1.2 million*
People: *British origin 28%, French origin 23%, other European 15%, Amerindian 2%, other 32%*
Language: *English, French*
Religion: *Roman Catholic 43%, Protestant 23%, other 34%*
Government: *Constitutional monarchy*
Life expectancy: *81 years*
GDP per capita: *$PPP 35,812*
Currency: *Canadian dollar*
Major exports: *Newsprint, pulp, timber, oil, machinery, vehicles, gas*

Canada is, after Russia, the world's second largest country, though much is barren and sparsely populated. The more remote areas include the icy wastes of the Arctic archipelago in the north, the splendour of the Rocky Mountains in the west, and the stormy Newfoundland coast in the east. The largest region, around Hudson Bay, is the flat, rocky Canadian Shield which is studded with thousands of lakes. To the west are the lowlands of the interior plains, and to the south-east the Great Lakes-St Lawrence region that borders on the USA. Around 85% of Canadians live within 350 kilometres of the border with the US.

Canada's original population, the 'first nations', now make up less than 2% of the population, and are among the poorest people. Many are now claiming land and other rights from provincial governments. In 1997, the Supreme Court ruled that the government has 'a moral, if not legal, duty' to settle their claims.

Everyone else is of immigrant descent, and Canada remains a country of immigration. In 2006, 22% of the population were foreign-born. Immigration is strictly limited to around 250,000 per year, but the pattern of arrivals altered dramatically from the 1960s, following changes that removed the privileges of European immigrants. Now more than half of immigrants come from Asia, many of them of Chinese origin who settled on the Pacific coast. Despite this influx, Canada remains relatively free of racial tension.

The wealth of natural resources and a dynamic immigrant population have made Canada one of the world's richest countries. It also has strong health and education services. Nevertheless there have been some economic problems in recent years, notably unemployment, which in 2009 was around 9%.

Three-quarters of Canadians are employed in service industries, with the most dynamic growth in larger cities like Toronto and Vancouver. Manufacturing, particularly in high-technology goods, has also expanded, boosted by exports to the US.

Mining now employs only around 1% of the workforce but still makes a vital contribution to the economy. Canada is the world's largest producer of uranium and potash and is an important source of many other metals, including nickel, zinc, platinum, copper, and titanium. It also has the world's second largest oil reserves, 95% of which are in the form of 'tar sands' in Alberta, and is a major producer of natural gas.

The Canadian prairies produce vast quantities of wheat, oats, barley, and many other crops, as well as livestock. But agriculture employs only 2% of the workforce. With vast forests Canada is the world's leading exporter of wood pulp and newsprint. And its rich fishing grounds on both Pacific and Atlantic coasts also make it a leading fish exporter.

Primary products account for around one third of export earnings but these are being overtaken by manufactured exports. Around 80% of exports go to the USA, and 70% of imports come from the USA—trade which was boosted by the 1994 North American Free Trade Agreement (NAFTA) with the USA and Mexico. Trade relations with the USA, have however been strained notably over exports of timber, and Canada wants to increase trade with China and the EU.

Trade is mainly with the USA

Canada's head of state is the British monarch—a constitutional anomaly that seems unlikely to change since Canada has a strong federal system with semi-autonomous provincial governments—each of which has a right of veto on constitutional change. In 2005, a former refugee from Haiti, Michaëlle Jean, was appointed governor-general.

For many years, Canadian politics at the federal level was the domain of the centre-left Liberal Party or the centre-right Progressive Conservative Party. The smaller left-wing New Democratic Party has traditionally been stronger in some provincial parliaments. But the 1990s saw dramatic changes, including a collapse in 1993 of the Conservative Party and the rise of the Green Party.

The parliament also has a strong representation for the Bloc Québécois which wants independence for the French-speaking province of Quebec. The province rejected this in a 1995 referendum and support for independence seems to have waned.

The 1997 and 2000 federal elections resulted in overall majorities for the Liberal Party led by Prime Minister Jean Chrétien. In December 2003 he resigned making way for Paul Martin. Then the Canadian Alliance and the Progressive Conservatives merged into a single Conservative Party led by Stephen Harper. The June 2004 election was closely fought and the Liberals only just scraped through to form a minority government.

In 2006, a further election allowed Harper and the Conservatives to form a minority administration. Meanwhile, Michael Ignatieff, a former academic, took over as leader of the Liberals.

In 2010, the opposition provoked yet another election. But this proved a mistake, allowing Harper and the Conservatives finally to win an overall majority. This time the New Democratic Party came second.

Cape Verde

One of Africa's smallest countries—heavily dependent on emigrants' remittances

Land area: 4,000 sq. km.
Population: 0.5 million—urban 61%
Capital city: Praia, 124,000
People: Creole 71%, African 28%, other 1%
Language: Portuguese, Crioulo
Religion: Roman Catholicism and indigenous beliefs
Government: Republic
Life expectancy: 71 years
GDP per capita: $PPP 3,041
Currency: Cape Verde escudo
Major exports: Footwear, garments, fish

 Cape Verde is an archipelago of ten islands off the coast of West Africa. Much of the territory is mountainous with little arable land. Combined with frequent droughts, this has made agriculture difficult, and has resulted in serious environmental degradation.

Two-thirds of Cape Verdeans are creole, of mixed Portuguese and African descent, and half live on the island of Santiago. Because of the harsh terrain and the resultant poverty, they have had a strong tradition of emigration. Today at least half a million Cape Verdeans live abroad, either in the USA or Europe.

Aid-financed public investment in education and health services in the decades following independence in 1975 helped to give Cape Verde one of Sub-Saharan Africa's higher levels of human development.

However this also made the country very dependent on aid—which in 2005 amounted to $320 per capita, mostly in grants. The country has also benefited from emigrants' remittances which in 2008 were equivalent to 10% of GDP.

Agriculture still employs more than one quarter of the workforce but only produces around 15% of the country's food needs. And since it does not offer much prospect of future employment the government has been liberalizing the economy in an attempt to attract foreign investment for light industry, such as textiles, shoe making and food processing. This may help, but seems unlikely to be sufficient to absorb the unemployed, who in 2006 were 21% of the workforce.

Tourism has also become more important and now accounts for around 18% of GDP with around 300,000 visitors a year.

Cape Verde, which was in union with Guinea-Bissau until 1980, was one of the first African one-party states to democratize. The first multiparty elections in 1991 saw a defeat for the long-ruling socialist Partido Africano da Independência de Cabo Verde (PAICV) and victory for the more market-oriented Movimento para a Democracia (MPD)—a victory repeated in 1995.

From 1999 however, the MPD became caught up in faction fighting. In 2001 the PAICV took advantage of this to win both legislative and presidential elections. The strong performance of Prime Minister José Maria Neves and President Pedro Pires gave them another victory in 2006. In the 2011 legislative elections, however, the PAICV won by only one seat.

Central African Republic

A coup-stricken country never far from chaotic violence

Land area: 623,000 sq. km.
Population: 4 million—urban 39%
Capital city: Bangui, 558,000
People: Baya 33%, Banda 27%, Sara 10%, Mandjia 13%, Mboum 7%, M'baka 4%, other 6%
Language: French, Sangho, Arabic, Hunsa, Swahili
Religion: Indigenous beliefs 35%, Protestant 25%, Roman Catholic 25%, Muslim 15%
Government: Republic
Life expectancy: 47 years
GDP per capita: $PPP 713
Currency: CFA franc
Major exports: Diamonds, timber, cotton

The Central African Republic is a landlocked country that sits on a plateau around 700 metres above sea level. The north of the country is savannah with some extensive grasslands. The south has more luxuriant vegetation, particularly along the river valleys and in the dense tropical rainforests.

The country is sparsely populated but ethnically complex. The main division is between savannah dwellers such as the Sara and Mandjia, and the more dominant 'riverines': those such as the M'baka who live in the south along the Ubangi River and who had greater contact with the French colonists. There are many languages, though the national lingua franca is Sangho, and the official language in the education system is French.

The most important transport artery is the Ubangi River along the southern border with the Democratic Republic of Congo.

Years of unrest and ethnic strife have taken their toll on human development. The education system is weak and has been further disrupted by fighting; only around half of the population are literate. Health standards too are low: in the rural areas where more of the people live few have access to safe water or sanitation, and around one-quarter of all children are malnourished. Only half the population have access to basic health care and 6% are HIV-positive. Two-thirds of the population live below the poverty line.

Two-thirds of the population make their living through subsistence agriculture, growing such crops as cassava, peanuts, corn, and millet. They also grow cash crops, including coffee and cotton, but the poor state of communications means that many households effectively live outside the cash economy. Although food production has increased in recent years, and has not been too disrupted by the fighting, the country is still not self-sufficient, and many people have relied on distribution of food aid.

One-fifth of the country is covered by tropical rainforests and there are more than 50 species of commercially viable trees. Transport difficulties have hampered the development of forestry but logging has now made timber the most important export earner. The government has attempted to conserve the rainforest by reducing the proportion of wood exported as timber and encouraging the production of veneers and plywood.

The Central African Republic also has a number of mineral resources, including iron ore, uranium, and gold, but these have not been exploited because of low prices and high transport costs.

More important are alluvial diamonds. Found in the west of the country, these are often mined by individuals and cooperatives and offer a significant source of rural employment. Diamonds are also the second largest export earner, though at least two-thirds of the output is smuggled out of the country. Official export and cutting operations are in the hands of a joint venture between the state and South Korean and Belgian companies.

The Central African Republic achieved independence in 1960, but from 1965 the country's development

One of Africa's most notorious dictators

was overshadowed by the bizarre, tyrannical rule of Jean-Bedel Bokassa, who in 1977 in an extravagant ceremony had himself crowned Emperor Bokassa I. France sustained him in power but in 1979 sent in troops to remove him. However the chosen successor was himself removed in a coup in 1981 by General André Kolingba.

Kolingba offered a degree of stability and attempted to legitimize his rule by creating a single political party, the Rassemblement démocratique centrafricain (RDC) but later had to legalize other parties.

In 1993 Kolingba lost the ensuing presidential election to Ange-Félix Patassé, who had been one of Bokassa's prime ministers. His party, the Mouvement pour la libération du peuple centrafricain (MLPC), gets most of its support from the north-west of the country, where Patassé himself comes from. Kolingba's RDC has its support in the south.

Patassé's government soon proved inept and corrupt and in 1996 the army staged a mutiny, demanding back pay. The French army, which maintained a garrison in the country, came to his rescue.

A further coup attempt in May 2000 was also seen off, this time with the help of Libya. However the sacked commander of the armed forces, General François Bozizé, then attacked from bases in southern Chad. In March 2003, Bozizé finally seized power, dismissing the National Assembly. In 2005 in multiparty elections he defeated Kolingba.

Subsequently, there were two more rebellions. One, in the north-west, from supporters of Mr Patassé, who was again in exile. Another, in the north-east, resulted from ethnic tensions. As a result 300,000 people fled their homes.

By 2011 the groups had joined a peace process and contested the general and presidential elections. But they lost to Bozizé and his party, Kwa Na Kwa. Progress in reintegrating the rebel fighters has been slow, opening up the prospect of more violence.

Chad

Chad has had decades of civil war, and is still subject to many rebellions. Its future could be transformed by oil

Land area: 1,284,000 sq. km.
Population: 11 million—urban 28%
Capital city: Ndjamena, 760,000
People: Sara 28%, Arab 12%, Mayo-Kebbi 12%, Kanem-Bornou 9%, others 39%
Language: French, Arabic, many others
Religion: Muslim 53%, Catholic 20%, Protestant 14%, animist 7%, other 6%
Government: Republic
Life expectancy: 49 years
GDP per capita: $PPP 1,477
Currency: CFA franc
Major exports: Cotton, livestock, textiles

Chad's territory forms a basin centred on Lake Chad on the western border. The terrain rises steadily, eventually reaching mountains in the north and east between which, in the north-east, lies a sandstone plateau. The climate varies from hot and dry in the north to tropical in the south, where the country's small amount of arable land is to be found.

Chad is ethnically very diverse, with more than 200 distinct groups. Conventionally, these are divided into the Islamic north, where most people raise cattle and the common language is Arabic, and the Christian or animist south, where people are farmers and the common language of the élite is French. The dividing line is taken to be the Shari River. Though the two parts have roughly the same population, the south represents less than one-tenth of the territory. This

north–south divide is something of a simplification, since each part has numerous internal divisions, but generally the north has been more traditional, and the south, which was more effectively colonized by the French, is economically more modern.

Throughout the country, poverty is severe and widespread. Infant mortality is high, life expectancy is low, and the literacy rate is only 26%. Over the past 30 years up to 400,000 people have died as a consequence of warfare. Since 2004 Chad has also received 250,000 refugees from Darfur in Sudan and 60,000 from the Central African Republic.

Until very recently Chad's economy has seen little progress since independence, hampered by poor infrastructure and endemic political conflict. Though agriculture now generates only 9% of GDP it employs the majority of the workforce. Sedentary agriculture is almost entirely in the south, on the lake shore, and along the river valleys.

In a good year, production of rice, sorghum, millet, wheat, and other foodstuffs is sufficient for local consumption, but output has frequently been constrained by drought and civil conflict. Cash crops include sugar and peanuts, but the most important, and the major export earner, is cotton—which directly or indirectly is thought to employ about one million people. In the centre and north of the country people rely for income on their livestock—around five million cattle and six million sheep and goats. Most of Chad's industry has also involved processing agricultural products—ginning cotton and producing meat and hides.

But things have changed radically in recent years. Following oil discoveries in 1994, ExxonMobil, Petronas, and Chevron have developed oilfields in Doba in the south and built a 1,100-kilometre pipeline to the port of Kribi in neighbouring Cameroon. The first oil was pumped in July 2003.

Oil income is expected to boost Chad's budget by 50% over the next 25 years. How will this windfall be used? The omens are not good. In 2008, the World Bank, which helped finance the pipeline, subsequently withdrew its support because the government had not abided by its promise to allocate oil revenues for poverty-reducing projects. Local people seem unlikely to gain very little extra direct income. The main beneficiaries seem to be traders and others catering to expatriates.

Oil income going to a select few

The oil companies will be hoping that Chad's multiple conflicts die down. These became evident soon after independence in 1960, and have usually involved direct or indirect participation by neighbouring countries. The first, one-party, regime, supported by the French, came to be seen as a southern dictatorship. This prompted the emergence in the north of a guerrilla movement, the Front de libération du Tchad (Frolinat).

Following a military coup in 1975, the new regime, although again composed mostly of southerners, tried to make peace with Frolinat. This failed, not least because the rebellion was being funded by Libya. In 1982, however, Hissène Habré, a northern leader, seized power, backed by the USA, Sudan, and Egypt, but opposed by Libya. This provoked a rebellion in the south and warfare that was to claim 40,000 victims.

In 1990, Habré's former army head, Idriss Déby, launched the Mouvement patriotique du salut (MPS). With Libyan support, this rapidly ousted Habré and installed Déby as president.

Déby seems to have brought a measure of stability—by Chad's standards. In 1996, a national referendum approved a new constitution that introduced multiparty politics. Déby won the ensuing presidential election, and the MPS obtained a majority of the seats in the National Assembly. Déby was re-elected in 2001 and the MPS also won the 2002 legislative elections. In 2005 Déby introduced a constitutional amendment to remove the two-term limit on the presidency and in 2006 he was duly re-elected.

Déby has had the support of France and of European troops, which in 2008 were transferred to a UN force, MINURCAT, which were able to leave at the end of 2010.

By 2011 the country appeared to be on a steadier course. The MPS won a large majority in elections for the National Assembly and Déby was returned as president.

Channel Islands

Dependencies of the British crown—constitutionally odd and financially rich

Land area: 300 sq. km.
Population: 156,000—urban 31%
Main towns: St Helier, St Peter Port
People: British and Norman French
Language: English, French
Religion: Christian
Government: Dependencies of the British crown
Life expectancy: 80 years
GDP per capita: $PPP 52,000
Currency: Jersey pound, Guernsey pound
Major exports: Vegetables, fruit, flowers

The Channel Islands consist of four main islands and numerous other islets. They are governed as two separate 'bailiwicks' based on the two largest islands: Jersey, with 58% of the population; and Guernsey, with 42%, which also covers the two smaller islands of Alderney and Sark.

Natives of the islands generally speak either English or French, though there are vestiges of a Norman-French dialect. There are also large numbers of settlers: in Jersey only half the resident population were born on the island; most of the rest come from elsewhere in the British Isles, often as tax exiles, and there are also many immigrant workers from Portugal.

The islands have traditionally exported horticultural produce to the UK—fruit, flowers, tomatoes, and potatoes. But agriculture now accounts for less than 5% of GDP. Tourism is more significant, accounting for 24% of GDP in Jersey and 14% in Guernsey. Around 80% of visitors come from the UK, attracted by the mild climate and pleasant scenery, though most come only for short breaks.

More recently, the islands' economies have become dominated by financial services—50% of GDP in Jersey, and 40% in Guernsey. With little or no corporation tax, they became very attractive to British companies in 1979 after the abolition of exchange controls. The business is fairly clean, though a British government report concluded that the scale of tax evasion and fraud was 'unusually hazardous' to assess. The report also discovered that Sark's 575 inhabitants between them held 15,000 company directorships.

All the islands could, however, be affected by increasing intolerance of tax havens following the 2008–09 financial crisis, and tighter EU regulation of hedge funds.

The financial opportunities are created by the islands' unusual constitutional position, with respect to both the UK and the EU. They are neither sovereign states nor part of the UK, but remnants of the duchy of Normandy, a legacy of William the Conqueror, and come under the jurisdiction of the British monarch rather than of the state. They have autonomy in domestic policy, including fiscal policy, and Jersey issues its own coins and notes. The monarch appoints a lieutenant-governor to each of the bailiwicks. Each also has an appointed bailiff who is president of the local Assembly of the States, which has both elected and appointed members.

Chile

Chile has set aside the Pinochet years, and now has a stable democracy and a successful economy

BOLIVIA
BRAZIL
ARGENTINA
Valparaíso
Santiago
Concepción
PACIFIC OCEAN
ATLANTIC OCEAN

| 0 | Miles | 800 |
| 0 | Km | 1280 |

Land area: *757,000 sq. km.*
Population: *17 million—urban 89%*
Capital city: *Santiago, 6 million*
People: *European and mestizo 95%, Mapuche 4%, other 1%*
Language: *Spanish*
Religion: *Roman Catholic 70%, Evangelical 15%, Jehovah's Witness 1%, other 14%*
Government: *Republic*
Life expectancy: *79 years*
GDP per capita: *$PPP 13,880*
Currency: *Peso*
Major exports: *Copper, fruit, cellulose, fishmeal*

Chile is remarkably narrow: 4,329 kilometres long and on average no more than 180 kilometres wide. From north to south run the rolling hills of the coastal highlands in parallel with the massive ranges of the Andes to the east. Between them lies a central valley that is most evident in the middle third of the country. Chile's length also offers striking climatic variations, from the heat of the Atacama Desert in the north, through a temperate centre, then a cool, wet south extending through the lakes and fjords and on to the stormy Straits of Magellan. Chile is vulnerable to earthquakes, the latest of which in 2010 killed over 400 people.

Most of Chile's people are concentrated in the central valley and around the middle of the country. Around one third live in the capital, Santiago. Chileans are almost entirely of mestizo or of European descent. The small Mapuche Indian population is concentrated in an area 700 kilometres south of Santiago.

Chile has at times viewed itself as more European than South American, and has higher standards of education and health than its neighbours. All waged workers contribute to a health insurance system. However the country is steadily becoming more unequal: the income of the richest 10% of the population is 26 times greater than that of the poorest 10%.

Although the economy is now more broadly based it remains very dependent on natural resources. One of the economic mainstays is mining in the northern deserts. Chile is the world's largest producer of copper and has one-third of global reserves. The country is also rich in other minerals, including gold, molybdenum, silver, and iron; and it can count on substantial reserves of oil and gas.

Chile's climatic diversity also permits a diverse agricultural output. Agriculture and fisheries employ around 12% of the workforce. Chile is one of the world's largest fish producers. The climate in the central zone is ideal for multinational companies to grow apples, pears, and grapes for export. Although the industry is profitable, the workforce is often poorly paid. Also doing less well are the 200,000 wheat-growing campesino farmers who in a very liberal trade regime are suffering from cheaper imports. Land ownership is also becoming ever more concentrated into larger farms.

Chilean manufacturing has been diverse. In the past, it was designed for import substitution. But Chile's open economy now means that it tends to operate in areas where it has the greatest competitive advantage. Thus, leading manufactured exports are usually closely linked to agriculture, including cellulose and fishmeal. Chile is also now the world's fifth largest wine exporter.

Chile's economic and political history was transformed in 1973. Previously, the country had enjoyed a long sequence of democratic governments, but this pattern was shattered by a CIA-supported military coup which ousted socialist President Salvador Allende and ushered in the era of General Augusto Pinochet. For the next 17 years, Chile became a byword for torture, repression, and the abuse of human rights: within three years around 130,000 people had been arrested. Ultimately 3,197 people were to die for political reasons, including 1,102 who 'disappeared'.

Pinochet made Chile a byword for torture and repression

Pinochet's political ferocity had been combined with radical economic liberalism—and subsequent governments followed the same economic path. But the 'Chilean model' is actually a mixture of policies. It does espouse open markets, low trade tariffs, deregulation, and privatization, including a privatized pension scheme. But it also has had significant elements of regulation— including controls on capital flows and some indexing of wages and prices.

Democracy was restored only in 1990, and a government was formed by the Concertación de Partidos por la Democracia—a 13-party coalition that included parties from moderate conservative to the far left. Concertación won again in 1993 led by Christian Democrat Eduardo Frei. For most of the 1990s the right-wing opposition had been hampered by its association with Pinochet.

For the 2000 presidential election, however, it seemed that voters were more interested in issues of employment. In a close result, the Concertación candidate, Ricardo Lagos, won a six-year term, defeating Joaquín Lavín of the right-wing coalition, the Alianza por Chile.

Lagos appeared successfully to have balanced the different wings of Concertación, achieved strong economic growth, and worked with the opposition on constitutional reforms, which included reducing the presidential term to four years.

For the 2006 presidential election the Concertación candidate was former health minister Michelle Bachelet of the Partido Socialista who became Chile's first woman president.

By 2008, however, Concertación was running out of steam and as a result of defections lost its majorities in Congress. In 2010 it also lost the presidency, to Sebastián Piñera of the right-wing Coalición por el Cambio.

China

China is reforming into an economic superpower, but the Communist Party retains its political grip

Land area: 9,561,000 sq. km.	
Population: 1,329 million—urban 45%	
Capital city: Beijing, 13.0 million	
People: Han Chinese 92%, Zhuang, Uygur, Hui, Yi, Tibetan, Miao, Manchu, Mongol, Buyi, Korean, and other nationalities 8%	
Language: Mandarin, Yue (Cantonese), Wu (Shanghaiese), Minbei (Fuzhou), Minnan (Hokkien-Taiwanese), Xiang, Gan, Hakka dialects, minority languages	
Religion: Taoism, Buddhism, Muslim, Christian	
Government: Communist state	
Life expectancy: 72 years	
GDP per capita: $PPP 5,383	
Currency: Renminbi (yuan)	
Major exports: Electrical machinery, clothing, footwear, coal	

China's vast landmass incorporates immense topographical and climatic diversity. But it can be divided into three broad areas. First, there is the north-western region, which is predominantly mountainous, including the Tien Shan ranges that average around 4,000 metres—though this area also includes the Tarim and Dzungarian basins and the Takla Makan Desert. A second main region lies to the south-west and includes the Tibetan Plateau and other highlands. The remainder of the country, extending east to the coast, has some mountains but is mostly low-lying and includes extensive river systems flowing west to east: the Huang Ho (Yellow River) to the north and the Yangtze to the south.

This huge territory also has considerable climatic variation— colder and drier to the north and on the mountains and steppes in the interior, and warmer and wetter to the south and east.

Most people live in the eastern part of the country and are ethnically fairly homogenous—more than 90% are Han Chinese. Nevertheless the Han differ regionally, particularly in language. The standardized common language is based on the Mandarin dialect and is used in government as well as in schools and universities, but there are also a number of Han dialects that share most of the same characters though are mutually incomprehensible.

In addition to the Han there are around 100 million Chinese who belong to any of 55 or more different ethnic groups that are settled across more than half the territory. These include the Zhuang, the Hui (Muslims), the Miao, the Mongols, and the Tibetans. Where they are concentrated in specific areas, they theoretically have some autonomy, though this is fairly limited.

One of China's central preoccupations has been population

growth. Population control efforts intensified after 1979 with the introduction of the 'one-child family' policy—with fines for parents who had two or more children. Although subsequently relaxed, this had a dramatic effect. Birth rates are now well below replacement level. This has slowed the growth of the population, which is expected to

China has more men than women

stabilize at around 1.5 billion by 2040. But it has also increased the average age: by 2030 on current trends, one in five people will be over 60. An unusually high proportion will be men. By 2009, there were 106 men for every 100 women.

Poverty has fallen, but 16% of the population still live on less than $1.25 per day. The cities have seen the emergence of a property-owning middle class, but life in the countryside is often harsh. In 2007 household incomes in urban areas were three times those in rural areas. Rural services remain poor. Few people have access to safe sanitation and their health services are inferior to those in the cities. Another pressing health issue for the country as a whole is HIV and AIDS. China currently has around 700,000 people who are HIV-positive.

As a result of economic reforms, China is becoming a steadily more unequal society. The richest 10% of the population earn on average 13 times more than the poorest 10%. The imbalances between rural and urban areas have also generated up to 150 million unregistered migrant workers who have travelled to the cities, many off to work on construction sites.

The first round of market reforms started after 1978, when the government dismantled the collective farms and passed responsibility to households. In response, farm output increased by 50%. But in the past few years agriculture has stagnated. Farmers do not have title to their land, so they are reluctant to make improvements.

China is still able to feed itself—indeed in the late 1990s there were often food surpluses. But the long-term prospects are more worrying. The country already has a relatively small proportion of agricultural land—only around 16% is cultivable—and this continues to shrink as a result of environmental degradation and industrial and housing development. In 2007, agriculture, although it employed 40% of the labour force, only accounted for 11% of GDP.

To feed its industries, China can rely on a wide variety of mineral

The Three Gorges dam displaced one million people

deposits. It has enough iron ore for its steel industry. It also has extensive coal reserves, particularly in the north. In addition, a number of provinces have small onshore oil deposits, and there are offshore possibilities. Hydroelectric power is also important and has benefited from extensive investment, most recently in the controversial $28-billion 'Three Gorges' dam on the Yangtze, a project that became operational in 2008 having displaced more than one million people.

In recent years, China's industry has been transformed. Until the late 1970s, industry was dominated by state-owned industrial enterprises.

Since then, there has been a dramatic shift. Many state-owned enterprises have been reformed, and between 1998 and 2002 laid off more than 26 million workers. Now more than half of industrial output is in the private sector. The government also promoted township and village enterprises which boosted rural incomes, and around one-quarter of the rural labour force are now non-agricultural.

Many foreign businesses have also arrived, tempted by cheap labour and the world's most populous market. Most have invested in coastal areas—originally in Guangdong but increasingly in the Yangtze River Delta area. By 2007 foreign-invested enterprises accounted for more than half of exports. This investment has also changed China's pattern of exports. Much is still in the form of simple manufactured goods: China manufactures one-third of the world's suitcases, and one-quarter of its toys. However there has been a shift away from garments and textiles and towards electronic goods.

Foreign investors flock in

Around one-fifth of exports go to the United States, which has become increasingly concerned about China's economic clout. Exports have also enabled China to build up huge foreign exchange reserves which by 2009 had reached $2.2 trillion.

China's constitution of 1982, as amended in 1988, declares the country to be a socialist dictatorship led by the working class. In practice, it continues to be run by the Communist Party, which with 74 million members, 5% of the population, is the world's largest political party. The party and the government are more or less the same thing and the most important decision-making body is the party's Politburo Standing Committee.

The army has also been a major political and economic force. Its influence has receded, especially since it has been forced to give up all its lucrative commercial interests, but it remains the guarantor of party rule.

China's economic transformation started with Deng Xiaoping who emerged after the death of Chairman Mao Zedong in 1976. In economic policy, Deng was pragmatic: politically, he was more orthodox and less tolerant of dissent. This became chillingly clear in June 1989 when troops opened fire on pro-democracy demonstrators in Beijing's Tiananmen Square, killing hundreds of people.

When Deng died in 1997 he was replaced by his chosen successor Jiang Zemin—and China's leadership has subsequently been more collective. In 2003, China moved on to its fourth generation of leadership when the National People's Congress—a 3,000-member rubber-stamp parliament—appointed a new president, Hu Jintao, and premier, Wen Jiabao.

China remains a repressive society, with little challenge to party rule. This was evident in the crushing of protests in Tibet in 2008 in advance of the Olympic Games. Intellectuals and activists regularly face jail sentences for 'subversion'. The government is also facing unrest in a number of other regions. In 2009, more than 190 people died in the western province of Xinjiang, in ethnic violence that pitted Han Chinese residents against the Uighur minority.

Colombia

Colombia is at last making progress against paramilitary forces and armed insurgents but is still plagued by drugs

Land area: 1,139,000 sq. km.
Population: 44 million—urban 75%
Capital city: Bogotá, 6.6 million
People: Mestizo 58%, white 20%, mulatto 14%, black 4%, mixed black–Indian 3%, Indian 1%
Language: Spanish
Religion: Roman Catholic 90%, other 10%
Government: Republic
Life expectancy: 73 years
GDP per capita: $PPP 8,587
Currency: Peso
Major exports: Oil, coffee, chemicals, coal

The western half of Colombia is dominated by three Andean mountain chains. To the west, they descend to the Pacific and Caribbean coasts. To the east, their foothills lead from plains on to savannah and to rainforests that cover around two-thirds of the country.

Colombia's diverse ethnic make-up changes from one region to another. The highest proportions of blacks and mulattos—mixtures of black and white—are to be found in the coastal regions. The whites and mestizos—mixtures of white and Indian—who make up most of the population, live in the valleys and basins between the mountain ranges. The Indian population are mostly to be found in the isolated lowlands. Though now only 1% of the population they are still very diverse, with more than 180 languages and dialects. Colombia is a very polarized society with wide disparities in income—the richest 10% of the population, predominantly white, get 46% of national income.

Around one-fifth of the workforce are employed in agriculture. They grow a range of food crops such as maize and rice, as well as cash crops for export, of which the most important is coffee; Colombia is the world's leading producer of the mild arabica variety. Colombia also exports large quantities of cut flowers as well as bananas and sugar. However, land distribution is very unequal with around 40% in the hands of large landowners.

Colombia has rich mineral resources. The most important is oil, which accounts for around one-quarter of exports, though production is declining and eventually Colombia could become a net importer. In addition, Colombia has Latin America's largest coal reserves and is a significant producer of gold, emeralds, and nickel. The country also has a diverse manufacturing sector—in areas such as food processing, chemicals, textiles, and clothing, most of which is controlled by large conglomerates.

The great unknown element in

Colombia's economy is the drugs trade which is thought to be worth around $5 billion per year. Colombia processes its own heroin poppies and coca leaf, as well as coca paste imported from Peru and Bolivia. With substantial support from the USA, through 'Plan Colombia' the government is reported to be spending more than $1 billion per year combating drugs, but the traffic continues unabated, and it supplies around 80% of the cocaine that reaches the USA.

Blessed with resources legal and illegal, Colombia has enjoyed steady economic growth. Its economic management has also been fairly steady by Latin American standards, avoiding the cycles of boom and bust as well as the debt crises. Nevertheless, unemployment in 2009 was 12%. The government has had to keep public spending high to pay for the war against the guerrillas.

Colombia is also distinctive in South America for having had a series of democratically elected non-military governments. Even so, the political system has long been steeped in violence. Between 1948 and 1957 this involved warfare between the Liberal and Conservative parties—a period known as 'la violencia' that cost 300,000 lives. Then the 1960s saw the rise of left-wing guerrilla movements, including the Revolutionary Armed Forces of Colombia (FARC), the National Liberation Army (ELN), and M-19.

In recent years there have been a series of ceasefires and amnesties. Eventually, M-19 laid down its arms and entered politics, ultimately unsuccessfully. But political resolution is more difficult nowadays since the guerrillas are less concerned about revolutionary ideals and more preoccupied with making money through drugs, kidnapping, and extortion. Add to these the private armies of drug dealers and right-wing militias and Colombia at times has seemed on the brink of anarchy and civil war. This has taken a terrible human toll.

Colombia on the brink of anarchy

Over the past ten years, around 35,000 people are thought to have died.

When Andres Pastrana of the Conservative Party was elected president in 1998, he promised to make peace with the rebels and reform the army but had little success.

The elections of 2002 reflected public concern about the lack of progress and resulted in the election of an independent candidate, Álvaro Uribe. His policy of 'democratic security' was surprisingly successful at combating the violence. Uribe increased the number of security personnel, ensured that every town had a strong police presence, and launched a military campaign against the guerillas while building a network of one million informants.

As a result, some 30,000 right-wing paramilitaries laid down their arms. The FARC has also become progressively weaker as a result of the death of its leaders and multiple defections. The voters were impressed and in the 2006 elections they awarded Uribe a second term.

Unable to stand again, Uribe was replaced as president in the 2010 elections by another conservative, Juan Manuel Santos, who has continued with similar policies.

Comoros

The Comoros have suffered from endless coups and inter-island tensions

Moroni

INDIAN OCEAN

Land area: 2,000 sq. km.
Population: 0.6 million—urban 28%
Capital city: Moroni, 47,000
People: Antalote, Cafre, Makoa, Oimatsaha, Sakalava
Language: Arabic, French, Comoran
Religion: Sunni Muslim 98%, other 2%
Government: Republic
Life expectancy: 65 years
GDP per capita: $PPP 1,143
Currency: Comoran franc
Major exports: Cloves, vanilla, ylang-ylang

The Comoros consist of three volcanic islands in the Indian Ocean. The largest is Grande Comore (also called Ngazidja), which includes Mount Karthala, an active volcano. To the south-east is the island of Anjouan, and between them is the smallest island, Mohéli. The terrain includes steep hills and some fertile valleys, but the land has been over-exploited and the soil has been eroded.

Comorans are descended from immigrants from Africa, Asia, and beyond, so the population is very diverse. The majority speak Comoran, which is a Bantu language, though the official languages are French and Arabic. Comorans are as poor as the poorest people in other African countries, and their country's limited resources offer a narrow range of opportunities.

Around two-thirds of workers make their living from agriculture which accounts for 40% of GDP. They do grow food but not enough, so much has to be imported. They also grow cash crops, notably vanilla, cloves, and ylang-ylang (a perfume essence), but the country remains reliant on foreign aid. Population density and land shortage are forcing migration to the towns, or overseas; around 150,000 now live abroad. Economic activity is dominated by the government.

The islands were a French colony grouped with the nearby island of Mayotte. Following a referendum in 1974, the three islands voted for independence while Mayotte chose to remain with France. Even so, France has remained a significant influence—and French and other foreign mercenaries have regularly provided the personnel for a series of coups and counter-coups, of which there have been 21 since independence.

In 1997, eyeing the subsidies and aid showered on Mayotte, the islands of Anjouan and Mohéli seceded demanding to be returned to French control. France declined the offer.

In April 1999 the army took over 'to preserve national unity' and subsequently reached an agreement with the separatists in the 'Fomboni Accord' which established a federal structure. In April 2002, the army chief, Colonel Azali Assoumane, was elected federal president.

The 2006 presidential election, which was largely peaceful, was won by Ahmed Abdallah Sambi, leader of an Islamist party, the Front National de la Justice. The 2010 elections were more tense, resulting in a victory for Sambi's candidate, Ikililou Dhoinine.

Congo

The smaller of the two Congos is potentially oil rich, but its recurring civil war has held back development

CAMEROON

GABON

Brazzaville
Loubomo
Pointe Noire

DEM. REP.
OF CONGO

| 0 | Miles | 300 |
| 0 | Km | 480 |

Land area: 342,000 sq. km.
Population: 4 million—urban 62%
Capital city: Brazzaville, 1.1 million
People: Kongo 48%, Sangho 20%, M'Bochi 12%, Teke 17%, other 3%
Language: French, Lingala, Monokutuba, and many local languages
Religion: Christian 50%, animist 48%, other 2%
Government: Republic
Life expectancy: 64 years
GDP per capita: $PPP 3,511
Currency: CFA franc
Major exports: Oil, timber

The country's official name is the Republic of the Congo, though it is also known as Congo (Brazzaville). Its territory consists largely of two river basins separated by mountains and plateaux. The smaller of the two is the Niari basin in the south and south-west. To the north-east of this the land rises to the Batéké and Bembe plateaux, and then descends to the basin of the Congo River, which also serves as the border with the Democratic Republic of the Congo to the east. More than half the land area is covered by dense tropical forests.

Ethnically, Congo is diverse, though around half the population belong to the Kongo ethnic group, who are concentrated in the capital, Brazzaville, and in the south. The Sangho live in the remote forests of the north, the M'Bochi are in the centre-north, while the Teke are in the Batéké plateau. Congo is the most urbanized country in Sub-Saharan Africa and prior to its civil war from 1997 the concentration of people in Brazzaville helped to ensure relatively good educational standards. But the war has undermined the education system and has also taken a toll on health—not only killing more than 10,000 people and destroying clinics but also enabling HIV and AIDS to spread rapidly: up to 4% of the adult population could now be infected.

Since the late 1970s, the Congo's main source of income has been oil. Oil makes up most of export earnings and 70% of GDP. Reserves are largely offshore and serviced by the port of Pointe Noire, with Total and Agip as the leading producers.

Though this income has given the country a relatively high per capita GDP, the wealth has not percolated through to the rest of the economy. There is little other industry and most people in Brazzaville who do not work for the government are employed in the informal sector.

The country's infrastructure is also weak—even Pointe Noire and Brazzaville are not connected by an all-weather road, though there are a number of railways.

People outside the cities generally survive through farming. The country has good soil and ample rainfall, but agriculture is underdeveloped. Most farmers grow crops like cassava, plantains, and groundnuts for their own consumption. Because road links are so poor they find it difficult to market their crops, so most of the food consumed in Brazzaville is actually imported, usually across the Zaire River from Kinshasa.

The country's vast forest resources, which include around 600 species of timber, have also been under-exploited. For loggers too the main handicap is a lack of transport. The only forests that have been felled are near the coast. Environmentalists are now, however, worried about the arrival of Asian logging companies looking for new opportunities.

Congo's current political rivalries have their roots in the 1960s and 1970s when this was the Marxist 'People's Republic of the Congo' run by the Parti congolais du travais (PCT). Colonel Denis Sassou-Nguesso, a northerner from the M'Bochi ethnic group heading the PCT was elected president in 1984, and again in 1989. But by the early 1990s, popular pressure forced the PCT to legalize other parties, and in 1991 they introduced a new constitution and changed the country's name to the Republic of the Congo.

The PCT lost the democratic parliamentary elections in 1992 when the two leading parties were the Union panafricaine pour la démocratie sociale (UPADS), led by Pascal Lissouba whose support came from the centre-south, and the Mouvement congolais pour la démocratie et le developpement intégral (MCDDI), led by Bernard Kolélas who had strong support in Brazzaville. The presidential election in the same year was won by Lissouba.

The country began its slide into chaos in 1993 following parliamentary elections that also gave a victory to Lissouba but were disputed by Kolélas. The three main parties had recruited their own private militias. The PCT had the 'Cobras', the MCDDI had the 'Ninjas', and UPADS the 'Zulus'. Each seized areas of Brazzaville and more than 2,000 died in the ensuing battles. Order was eventually restored when UPADS and the MCDDI agreed to share power while Sassou-Nguesso retired to his northern power base.

Fighting between Cobras, Ninjas, and Zulus

Sassou-Nguesso returned in 1997 launching a six-month civil war that destroyed much of Brazzaville. Sassou-Nguesso regained power and Lissouba fled. Fighting resumed in 1999, but early in 2000 the president of Gabon helped broker a ceasefire.

In 2002, after the introduction of a new constitution, Sassou-Nguesso won a presidential election boycotted by the opposition. Fighting resumed yet again, with a rebel faction led by Pasteur Tata Ntoumi. This was ended by a further peace deal in 2003.

Sassou-Nguesso and the PCT's main support is from minority northern ethnic groups so they have had to rely on independents in parliament. Sassou-Nguesso has done little to reduce ethnic tensions and has regularly manipulated low-turnout elections. He was re-elected in 2009.

Congo, Democratic Republic

Formerly Zaire, the country became a regional battlefield, but has achieved a ceasefire

Land area: 2,345,000 sq. km.
Population: 63 million—urban 35%
Capital city: Kinshasa, 7.8 million
People: Mongo, Luba, Kongo, Mangbetu-Azande, and many others
Language: French, Lingala, Kingwana, Kikongo, Tshiluba
Religion: Roman Catholic 50%, Protestant 20%, Kimbanguist 10%, Muslim 10%, other 10%
Government: Republic
Life expectancy: 48 years
GDP per capita: $PPP 298
Currency: Congolese franc
Major exports: Diamonds, cobalt, coffee, copper, oil

The dominant feature of this vast country, accounting for around 60% of its territory, is the Congo River basin centred on the north-west. This includes some of the world's most dense tropical rainforests. From this depression the land rises to the south and west to plateaux, and then to mountains along the country's borders.

With more than 200 ethnic groups, the population is very diverse. The majority, including the Mongo, Luba, Lunda, and Kongo are Bantu speakers. But in the north there are major Sudanese groups such as the Mangbetu-Azande. There are also pygmy people in the forests. Relations between these groups have often been tense and violent—divisions that politicians have exploited.

Standards of human development have been falling steadily—a result of government neglect, withdrawal of international aid, and civil war. Incomes have declined steeply: GDP per capita in 2009 was lower than at independence.

Most education is provided by Catholic missionaries but fewer children are now attending school and the literacy rate is thought to be falling. Because children are less likely to be immunized, infant mortality is rising, and more than one-third of children are malnourished. AIDS is a growing threat: officially the HIV infection rate is 3% but is probably much higher.

Two-thirds of the workforce are employed in agriculture. Most are subsistence farmers who use fairly primitive techniques, growing plantains, cassava, sugar cane, and maize. Despite ample fertile land, huge quantities of food need to be imported—a result primarily of government indifference to agriculture, and poor transport facilities, as well as the disruption of warfare. There are also some cash crops such as coffee, palm oil, and cotton, but output has declined steeply in recent years. The country's vast forest potential could also be exploited on a sustainable basis.

In the past much of the country's income has come from its lucrative mineral deposits. Traditionally the major earner has been copper in the Katanga region in the south-east. This, along with the cobalt industry, is the responsibility of the state-owned company, Gécamines. But production has fallen steadily as a consequence of low investment and warfare; copper and cobalt now account for less than half of exports.

In 2007 around one-quarter of export earnings came from diamonds, most of which come from Kasaï in the south. Congo is the world's third largest diamond producer, though of industrial rather than gem quality. Diamond production, which is mostly in the hands of the private sector, has been less affected by the recent wars.

The country also has some manufacturing industry, often centred on the processing of minerals. Most foreign investors have stayed away, though Société de Belgique and Unilever have extensive interests.

For most of the period since independence in 1960 the country was under the control of one man, Mobutu Sese Seko. Mobuto came to power in a US-backed military coup in 1965 and established his own political party. In 1991 he changed the

From Congo to Zaire country's name from the Democratic Republic of the Congo to Zaire. Mobuto was subsequently elected unopposed for three consecutive terms of seven years—a period during which he ruthlessly repressed all opposition and looted the economy.

Mobutu's grip weakened in 1996 following a rebellion in Kivu province in the east. This was originally centred on a rebellion by an ethnic group he had persecuted, the Banyamulenge, who by then had seized power in neighbouring Rwanda. With Rwandan help they overcame the Zaïran army and under the leadership of Laurent Kabila, as the Alliance des forces démocratique pour la libération du Congo-Zaire (AFDL), they took control in 1997.

Mobutu fled (and died later that year). Kabila declared himself president and changed the country's name back to the Democratic Republic of the Congo.

Hopes for a more democratic and peaceful era were soon dashed. First it became clear that Kabila was going to be just as heavy-handed. He also sufficiently antagonized his former Tutsi allies that in 1998, with backing from Rwanda and Uganda, they launched another rebellion, but this time against him. Zimbabwe, Angola, and Namibia soon sprang to his defence, leading to a complex and dangerous regional conflict.

Early in 2000 there was a ceasefire following a UN-brokered plan for disengagement of all foreign troops. The prospects then brightened, following the assassination of Kabila in January 2001, and his replacement as president by his son Joseph. In December 2002 there was a final settlement, and in July 2003 a transitional administration was formed, headed by Kabila.

In July 2006 the UN helped organize the first multiparty presidential election for 40 years. Kabila was elected president and has been strengthening his position thanks to a weak opposition.

Costa Rica

Long an island of peace in a troubled region, Costa Rica is struggling to maintain its social achievements

Land area: 51,000 sq. km.
Population: 4.5 million—urban 64%
Capital city: San José, 310,000
People: White or mestizo 94%, black 3%, other 3%
Language: Spanish, English
Religion: Catholic 76%, Protestant 16%, other 8%
Government: Republic
Life expectancy: 79 years
GDP per capita: $PPP 10,842
Currency: Colón
Major exports: Electronics, coffee, bananas, meat

Mountain ranges form a rugged backbone extending down the centre of Costa Rica. On the Pacific side they descend to coastal lowlands and sandy beaches. On the Caribbean side they come down more gradually to heavily forested lowland areas with a rich variety of plant and animal life. Most of the population, however, live in the highlands, in the Meseta Central where the climate is milder and the soil more fertile. Unfortunately, this is also a geologically active zone subject to earthquakes and potentially dangerous volcanoes.

Most people are of European or part-European descent. Few Amerindians remain and the largest minority group consists of black descendants of workers brought from Jamaica at the end of the 19th century to build the railroads and work on plantations. This English-speaking group make up around 30% of the population of the province of Limón on the Caribbean coast.

Costa Rica's development has been relatively equitable and it has created one of the most extensive welfare states in Latin America. Much of this can be traced back to the 1949 constitution which gave the state a strong role in social welfare, established labour rights, and introduced free secondary education. Remarkably, it also abolished the army, to avoid coups and save money—though the country does have a 10,000-strong police force. Although this is one of the most prosperous countries in Central America, with good standards of health and education, around 20% of the population are classified as poor.

Christopher Columbus, who christened Costa Rica (rich coast), had visions of gold and treasure. But the country's wealth has been based on more mundane products—particularly coffee and bananas. Costa Rican arabica coffee is produced largely by 78,000 small farmers in the central highlands—the average grower has ten hectares. Nevertheless, quality and yields are high. In terms of exports, however, coffee has been overtaken by bananas—20% of export value—produced on both the Caribbean and Pacific coasts. The industry employs around 50,000 people, half of them on medium-sized farms. Prices have been lower in the past few years, but Costa Rica, which is the world's second largest exporter, has benefited from a greater opening of European markets to Central American bananas.

Costa Rica has also taken advantage of concerns about global warming to sell carbon offsets created by saving and restoring its rainforests.

Though still relying significantly on agriculture, Costa Rica has nevertheless achieved a very diverse economy. Indeed this soon became primarily an industrial country, initially based on import substitution, in areas like food processing, chemicals, textiles, and metals.

Since then the economy has gone through several stop-go cycles. Lately these cycles have been based more on high-tech industry. Since it has the advantage of a sound democracy, a well-educated workforce, and a good legal system, Costa Rica is now interested in capitalizing on these attributes by attracting high-tech investors. Since Intel built the first **Home to Intel and Microsoft, Costa Rica is becoming a 'chip republic'** of two chip factories in duty-free zones in 1998, high-tech industry now forms the largest share of manufacturing. This includes two major US medical equipment suppliers. Microsoft and other companies also carry out some software development here.

Costa Rica is also now a prime tourist, particularly eco-tourist, destination. More than 1.9 million people come each year to visit the rainforests and volcanoes, though most recent investment has gone into beach resorts. Tourism now employs around one-fifth of the workforce.

Politically, Costa Rica's extensive welfare state and the fact that it has had a stable democracy since 1948 gave it an enviable reputation as the 'Switzerland of Central America'. It avoided the 1980s warfare in neighbouring states; indeed Costa Rica was the driving force of the Central American Peace Agreement of 1987, for which then President Oscar Arias Sánchez was later awarded the Nobel Peace Prize.

Over much of this period, Costa Rica has been ruled by the centre-left Partido Liberación Nacional (PLN), or by the more conservative Partido Unidad Social Cristiana (PUSC).

The 1998 parliamentary and presidential elections were won by the PUSC, with Miguel Ángel Rodríguez elected president. And in 2002, the PUSC again took the presidency, this time for Abel Pacheco.

However in parliamentary elections the two main parties had been losing ground, particularly to a PLN offshoot, the Partido de Acción Ciudadana (PAC), led by Otton Solís.

Nevertheless in the 2006 presidential elections the PLN scraped through, with Arias returning as president, just ahead of Solís. Since then the PLN has been quite successful, achieving strong economic growth and increasing social spending. Laura Chinchilla, the PLN presidential candidate, was elected in 2010 and will follow the same policies.

Côte d'Ivoire

Côte d'Ivoire's former political stability has been shattered by a coup and a civil war

Land area: 322,000 sq. km.
Population: 20 million—urban 50%
Capital city: Yamoussoukro, 227,000
People: Akan 42%, Voltaiques or Gur 18%, Northern Mandé 17%, Krous 11%, Southern Mandé 10%, and many immigrants
Language: French, Dioula, and many local
Religion: Muslim 38%, Christian 32%, Indigenous 12%, other 18%
Government: Republic
Life expectancy: 57 years
GDP per capita: $PPP 10,842
Currency: CFA franc
Major exports: Cocoa, palm oil, wood, coffee

Côte d'Ivoire has a narrow coastal strip of land, which includes in the east a series of deep lagoons. Behind this are two forest areas. The one to the west still has dense rainforest, while the one to the east has now largely been cleared to grow plantation crops. Beyond these to the north the land is mostly savannah, used for grazing cattle.

The population is very diverse with four or five main groups and around 60 sub-groups. Thus, the majority group, the Akan, includes the Baoulé who live mostly in the towns and villages of the south-east. Among the Krous are the Bété, living in the forests of the south-west. In the north-east there are a number of Voltaic groups including the Senoufou.

The country has also attracted many people from Burkina Faso, Guinea, and Mali to work on plantations and in domestic service.

And there are also substantial Lebanese, Syrian, and French communities. First- and second-generation immigrants make up around one-third of the population.

This is one of the most urbanized of African countries and Abidjan is one of Africa's more cosmopolitan cities. By West African standards incomes are reasonable. Nevertheless shrinking government budgets have undermined public services, and education and health standards are low: fewer than half the population are literate, and life expectancy remains quite low, partly as result of HIV/AIDS which now affects around 4% of the adult population.

Around half the workforce are employed in agriculture which accounts for one-quarter of GDP. Farmers grow food crops such as cassava, yams, maize, sorghum, and rice. Peasant farmers also cultivate the two main cash crops, cocoa and coffee, which take up around two-thirds of cultivated land. Côte d'Ivoire is the world's largest cocoa producer and the fifth largest coffee producer. For the first 20 years following independence in 1960, these exports contributed to rapid economic growth. Other important agricultural exports are bananas and pineapples along with tropical hardwood and palm oil.

Côte d'Ivoire also has energy and mineral resources—including deposits of oil and gas that could make the country self-sufficient. In addition there is a significant manufacturing sector. This includes garments and textiles along with some local processing of primary commodities like cocoa—though at present only around one-fifth is transformed into higher value products like chocolate.

Export crops helped to develop Côte d'Ivoire but they also increased its vulnerability to international commodity prices. Dramatic price falls in the late 1970s, exacerbated in 1994 by the devaluation of the CFA franc, led to a severe economic crisis and the country was soon plunged deeply into debt. Commodity prices recovered for a few years and growth resumed, but they then fell again and since the onset of political conflict the economy has suffered a steep decline.

Côte d'Ivoire is vulnerable to changes in cocoa prices

Côte d'Ivoire's previous reputation for political stability rested on the 23-year autocratic reign of the first president, Félix Houphouët-Boigny. For many years, his Parti démocratique de Côte d'Ivoire-Rassemblement démocratique africain was the only legal political party. His liberal economic policies helped to attract investment. But he also promoted some grandiose projects including moving the capital to Yamoussoukro, his home town, where he also built the world's largest Catholic cathedral. In 1990, Houphouët-Boigny introduced a new multiparty constitution and was elected for a seventh term, dying in office in 1993.

He was replaced by Henri Konan-Bédié who was elected president in 1995, though this election had been boycotted by the main opposition parties, who came together in 1997 as the Front populaire ivorienne (FPI) led by Laurent Gbagbo. Konan-Bédié then consolidated his grip and forced his presidential rival, Alassane Ouattara of the Rassemblement des républicains (RDR), to flee into exile.

But Konan-Bédié overestimated the support of the army, and in December 1999 they overthrew him in Côte d'Ivoire's first coup, led by General Robert Guéi, who promised a swift return to democracy.

In 2000 after a new constitution and an election, Guéi claimed victory over Gbagbo, the FPI leader. But the election was clearly fraudulent and popular protest forced Guéi to resign allowing Gbagbo to take over.

All seemed to be going well, but a military uprising in 2002 killed Guéi and triggered a civil war. A number of rebel groups subsequently came together as the 'New Forces' and seized control of the north.

In 2007, Gbagbo and the leader of the New Forces, Guillaume Soro, agreed on terms for power sharing, with Gbagbo as president. The 2010 elections, however, saw renewed violence. Alassane Ouattara won the presidential election, but Gbabo, supported by his guards, refused to step down. In the ensuing fighting more than 2,000 people died. Ouattara finally took office in May 2011.

Croatia

Croatia's fierce nationalism sparked the Balkan wars. Now the country is more democratic and the economy has revived

Land area: 57,000 sq. km.
Population: 4 million—urban 58%
Capital city: Zagreb, 779,000
People: Croat 90%, Serb 5%, other 5%
Language: Croatian
Religion: Catholic 88%, Orthodox 4%, other 8%
Government: Republic
Life expectancy: 76 years
GDP per capita: $PPP 16,027
Currency: Kuna
Major exports: Transport equipment, chemicals, textiles

Croatia's eccentric shape encloses a variety of landscapes and climates. The north-west includes a part of the Dinaric Alps. To the south the dramatic, broken coastline of Istria and Dalmatia consists of mountains that drop steeply into the sea, creating numerous inlets and bays as well as more than 1,100 islands.

When it was a republic within Yugoslavia, this territory had a clear majority of Croatians, as well as around 12% ethnic Serbs. The differences between these two Slav peoples are cultural rather than physical: the Croats are Roman Catholic and Western in outlook; most Serbs are Eastern-oriented Orthodox Christians. As a result of the wars of the 1990s, more than half of the Serbs fled. In addition there are around 800,000 Croatians in Bosnia.

Croatia was, after Slovenia, the second richest of the republics of former Yugoslavia and has an economy similar to that of Western European countries.

The war caused the GDP to shrink dramatically, but in recent years growth has been strong. Some of this is the result of public investment in infrastructure. But industry also played a part and now accounts for 35% of GDP. The break-up of Yugoslavia cost Croatia many of its markets and forced it to re-orient more towards the west. Initially it re-established some industries such as textiles under contract to foreign companies. But since then textile output has fallen and there has been much stronger growth in other sectors, including publishing and printing as well as in the manufacture of metal and electrical products.

Croatia has been a significant oil producer but its reserves are now exhausted and most crude oil has to be imported. Even so, oil refining remains important for both national consumption and exports.

Agriculture is also more significant than in some neighbouring countries. Most agricultural land was already privately owned, though many farms are small and often not very profitable. The best land and most of the larger holdings are to be found in Slavonia in the north-east, which is where most of the capital-intensive production of cereals and other cash crops takes place. Before the war, Croatia was a food exporter but it is now a substantial importer.

As in most industrialized countries, the majority of the workforce—around 60%—are employed in services. Many of these people are employed in the large public sector and most of the rest in trade and transport.

Croatia's spectacular Adriatic coastline had long been a major package-tourist attraction—notably the ancient city of Dubrovnik, which

Dubrovnik took a battering

took a severe battering from the Yugoslav army in 1991. Since then tourism has revived, especially from cruise ships. A high proportion of overnight tourists come from Germany and Slovenia.

Despite Croatia's economic revival, unemployment remains high: in 2009 it was still around 14%. Hopes for economic revival are now placed on accession to the EU for which Croatia is now negotiating.

Croatia's transition from communism was dominated for the first decade by a former general, Franjo Tudjman, who in 1990 created a new party, the Croatian Democratic Union (HDZ). In the first free elections in May 1990 the HDZ won a parliamentary majority and Tudjman became president. Tudjman's hard-line nationalism alarmed Serbs in Krajina and Slavonia who feared breaking with Serbia, and rejected the idea of independence, and in mid-1990 they established 'Serbian Autonomous

Regions' which Tudjman refused to recognize.

When Croatia did declare independence in July 1991, this provoked a full-scale civil war as the Serb paramilitaries united with the Yugoslav army to seize more than one-quarter of Croatia. The UN negotiated a ceasefire in January 1992, but in August 1995 the Croatian forces regained control, causing the Serbs to flee. In December 1995 Tudjman signed the Dayton peace accords.

Croatia is a parliamentary democracy, but in the early years the HDZ had a majority in both houses and Tudjman had a fairly free hand. He was elected president in 1992 and re-elected in 1998, and used his status as a war hero to build a centralized and authoritarian state.

To many people's relief, Tudjman died in December 1999 and the parliamentary elections in 2000 resulted in a surprising turnaround. The main centre-left opposition alliance of Social Democrats (SDP) and Social Liberals (HSLS) won almost half the seats. The SDP's leader, Ivica Racan, became prime minister and a presidential election in February 2000 was won by a political moderate, Stipe Mesic.

But Racan's coalition collapsed and further elections in 2003 returned the HDZ which espoused a more standard conservative approach. Led by Ivo Sanader, the HDZ formed a minority coalition. In 2005, Mesic, was re-elected president. The legislative elections in 2007 produced another HDZ-led coalition, which includes the HSLS and the Croatian Peasants' Party (HSS), led since 2009 by Jadranka Kosor as prime minister.

Cuba

Cuba now has elements of a market economy. Political change has been slower.

Land area: 111,000 sq. km.
Population: 11 million—urban 76%
Capital city: Havana, 2.1 million
People: White 65%, mulatto and mestizo 25%, black 10%
Language: Spanish
Religion: Roman Catholic, Protestant
Government: Communist
Life expectancy: 79 years
GDP per capita: $PPP 6,876
Currency: Peso
Major exports: Sugar, nickel, tobacco, seafood

Around one-quarter of Cuba's territory is mountainous, with three main systems, of which the most extensive is the Sierra Maestra in the south-east. Most of the island, however, consists of plains and gently rolling hills. For a relatively small island, the environment is biologically rich—particularly in its mangrove forests and wetlands.

Cuba's population is racially mixed: most people are white or mulatto. There is little overt discrimination, but those with darker skins tend to have the worst jobs. Very few members of the Communist Party's leadership are black.

Though they are far from rich, Cubans enjoy some of the world's best levels of social development. Health and education services are free. In the mid-1990s the school system lost many teachers to the tourist industry but pay rises have stemmed the flow.

Agriculture accounts for only around 7% of GDP and directly employs only 20% of the workforce, but it remains critically important to the economy. The main crop used to be sugar, but with the disappearance of the main buyer, the Soviet Union, the output of Cuba's inefficient producers collapsed. Between 1990 and 2007 production fell by 90%.

Tobacco was the other main source of income but in this case production has increased. Other crops, including fruit and vegetables, have also done better, but Cuba only grows around one-quarter of its staple food, rice, and has to import the rest.

Cuba's land reform created state farms, but also left around 20% of the land with smallholders and cooperatives. Since then most state farms have been dissolved and transformed into 'Unidades Básicas de Producción Cooperativa' (UBPCs). This involves the government leasing land rent-free to former workers running financially independent cooperatives. This seems to have raised production, though few of the UBPCs are profitable. Farmers can sell surplus production in 'agropecuarias'—farmers' markets.

Manufacturing has suffered from the loss of the Soviet market and the continuing US boycott. To generate more industrial jobs, the government has now created free-trade zones and has invested over $1 billion in research investment in areas such as biotechnology and pharmaceuticals.

Mining is also an important source of foreign exchange. Cuba has large reserves of nickel and cobalt, which are mined in conjunction with the Canadian corporation Sherrit International. High international prices have boosted production.

The most vigorous part of the economy has been tourism. Arrivals have increased dramatically to around 2.2 million in 2005, with Canadians, Italians, and Spanish in the lead—

Tourism to Cuba encourages unofficial enterprise chiefly heading for beach resorts on the northern coast. Tourism brings in around $2 billion annually but three-quarters of this goes straight out again to pay for related imports and as profits to the many foreign companies who run the hotels. Officially tourism employs 100,000 people as well as many more who are working in informal ancillary enterprises, including prostitution.

After 1993, Cubans were able to hold dollars, replaced in 2004 by 'convertible pesos' that they could spend in 'dollar shops'. These are virtually the only source of non-essential consumer goods; prices are around 40 times higher than in the official shops. 'Peso shops' distribute food rations, though Cubans can also buy more expensive produce from farmers' markets.

Despite economic liberalization, political change seems some way off. Cuba remains a 'socialist workers'

state' in which the only legal party is the Partido Comunista de Cuba, which from 1959 for almost 50 years was headed by the revolutionary leader, Fidel Castro. Castro was not initially a communist but, determined to avoid retaliation from the USA, espoused socialism and Cuba came under the patronage of the Soviet Union.

One of the early casualties was democracy, and even today the only opposition comes from human rights organizations and small groups linked to Cuban communities in the USA. According to Amnesty International, Cuba's jails contain 58 prisoners of conscience.

Many of Cuba's difficulties still result from hostility from the USA which, under pressure from a powerful exile lobby in Miami, classifies Cuba as an enemy state and imposes economic sanctions. Relations with the USA are alternately cool and less cool. One of the severest freezes resulted from the 1996 US Helms-Burton Act which imposes sanctions on companies that do business with Cuba. Nevertheless, there have also been some thaws which have enabled flows of emigrants. In 2000, the US Congress adopted bills that allowed sales of food and medicine.

In August 2006, Castro was taken ill and handed the reins of power to his brother Raúl who became president in February 2008. Raúl might try to follow the Chinese example of state-sponsored capitalism but has yet to make any substantial changes.

Under severe economic pressures, he will certainly find it difficult to maintain Cuba's impressive levels of social provision.

Cyprus

A bitterly divided island that has rejected an opportunity for reunification

Land area: 9,000 sq. km.
Population: 0.9 million—urban 70%
Capital city: Nicosia, 307,000
People: Greek 77%, Turkish 18%, other 5%
Language: Greek, Turkish, English
Religion: Greek Orthodox 78%, Muslim 18%, other 4%
Government: Republic
Life expectancy: 80 years
GDP per capita: $PPP 24,789
Currency: Cyprus pound
Major exports: Citrus fruit, potatoes, grapes

The island of Cyprus has two main mountain ranges. The Kyrenian Mountains extend along the northern coast, while the Troodos Mountains are in the south and centre. Between these is the most fertile part of the country, the Mesaorian plain, which includes the capital, Nicosia.

Cyprus is sharply divided between Greek and Turkish Cypriots. Greeks are still in a large majority, virtually all of them living in the south and centre. The Turks are confined to the north; although many have emigrated, others have arrived from Turkey.

The Greek part of the island has shifted decisively away from agriculture and now depends heavily on service industries. The most important is tourism. More than two million tourists arrive each year, half from the UK—accounting for around 12% of GDP. The other main service activity is business and finance, which includes 30 international banks. The economy in the north lags far behind: the main activity is agriculture, growing the traditional export crops of citrus fruit and potatoes.

At times, Greek and Turkish Cypriots have coexisted peacefully, and the constitution adopted at independence in 1960 involved explicit power-sharing. But by 1964 there were violent confrontations

and the UN sent in a peacekeeping force, which is still there, though now down to under 1,000. Matters came to a head after a military coup in 1974. Turkey sent in 30,000 troops to protect Turkish Cypriots and established control over the northern one-third. People moved from one part of the island to the other.

In 1983, the Turkish Cypriot leader Rauf Denktash proclaimed the north to be the Turkish Republic of Northern Cyprus, a claim recognized only by Turkey. The rest of the world recognizes the Republic of Cyprus (ROC), even if it controls only 57% of the territory.

In 2004, the UN proposed a loose federation. While the Turkish part of the island voted in favour, the Greek part voted against. As a result when Cyprus entered the EU in May 2004 the Turkish part was excluded.

Hopes for reunification revived in 2005 when a more conciliatory figure, Mehmet Ali Talat, was elected president in the north. They further improved with the 2008 election as president of the ROC of Demetris Christofias of the Progressive Party of the Working People. But the 2010 elections in the north were won by the hardline Dervish Eroglu so hopes of a solution have receded again.

Czech Republic

A formerly communist country that had a 'velvet revolution' and a fairly painless transition to a market economy

| 0 | Miles | 100 |
| 0 | Km | 160 |

Land area: *79,000 sq. km.*
Population: *10 million—urban 74%*
Capital city: *Prague, 1.2 million*
People: *Czech 90%, Moravian 4%, other 6%*
Language: *Czech*
Religion: *Roman Catholic 27%, Protestant 2%, other 11%, unaffiliated 60%*
Government: *Republic*
Life expectancy: *76 years*
GDP per capita: *$PPP 24,144*
Currency: *Czech crown*
Major exports: *Machinery, manufactures*

The Czech Republic has two main geographical areas. The largest, to the west, is Bohemia, a broad, elevated plateau of plains and low hills centred on Prague and ringed by mountains. Beyond the mountains to the east lies the second main area, the Moravia Valley centred on Brno, whose eastern boundary is formed by the Carpathian Mountains that separate the country from Slovakia.

Ethnically the population is fairly homogenous. But throughout the country there are also small numbers of Slovaks, who moved here when the country was part of Czechoslovakia, and also some Roma, or gypsies, whose lifestyle exposes them to both economic and racial discrimination.

Until the 1980s, the Czech Republic had a young population, but falls in the birth rate are causing the population to age rapidly.

Levels of human development are fairly high for a formerly communist country. In the past, health standards were reasonable, though lifestyles were unhealthy: high fat consumption and smoking reduced life spans. Over the past decade, however, life expectancy has increased.

This is a broadly based industrial economy—though between 1990 and 2006 industry as a proportion of GDP fell from 48% to 28%. Previously it concentrated on heavy industry such as engineering and steel, but more recently some manufacturers have been exporting intermediate goods, particularly to Germany. Lighter industries such as food processing have also been expanding. These changes have however widened regional disparities. Unemployment is now highest in former coal and steel areas such as northern Moravia.

Privatization of state-owned assets took place rapidly. In 2005 the private sector accounted for more than 85% of GDP. Transferring the ownership in this way did little to transform the management of enterprises and many have since stagnated. But some businesses acquired an injection of expertise from foreign companies: the reputation of Skoda cars, for example, was transformed when it was taken over by Volkswagen.

The agricultural sector accounts for only about 4% of GDP and is now entirely in private hands, either as commercial farms or cooperatives. Now that they are in the EU, Czech farmers have greater price support.

Meanwhile, many more people are working in services, particularly in retailing where many of the large stores are now part of foreign-owned chains such as Tesco. Others work in activities catering for 6 million tourists a year. Prague is the main tourist attraction and most new jobs have been created there or in the surrounding regions—heightening the regional disparities.

From 1918 to 1938, and from 1945 to 1992, what is now the Czech Republic was united with Slovakia. As Czechoslovakia, their joint transition from communism in 1989 was so smooth it was termed the 'velvet revolution'. Within the Czech lands, the democracy movement was called the Civic Forum, one of whose leaders was the writer Václav Havel. Civic

From a velvet revolution to a velvet divorce Forum won most of the Czech seats in the election of 1990 while a corresponding movement in Slovakia won the seats there. The new parliament elected Havel as president.

Civic Forum soon reformed into a number of parties, of whom the most important subsequently have been the right-wing Civic Democratic Party (ODS) and the centre-left Czech Social Democratic Party (CSSD).

The union with Slovakia did not last. The disagreements were partly ideological, since the leading Czech party was the right-wing ODS led by

prime minister Václav Klaus, while the leading Slovak party was centre-left.

Eventually, and without much public support, the politicians disbanded the federation. In 1993 they became two countries and Havel was elected president of the new Czech Republic—while Klaus remained prime minister.

The reforms proceeded apace but Klaus lost popularity as the economy slowed and his government became embroiled in corruption. As a result, the 1996 elections reduced the ODS to a minority government which soon collapsed amid economic chaos.

In the ensuing election in 1998 the voters' allegiances swung left. The CSSD became the largest party, the ODS came second, and the unreconstructed Communist Party came third. The new prime minister was Milos Zeman who was obliged to enter into a pact with the ODS.

Prior to the 2002 general election Zeman stepped down as CSSD party leader in favour of Vladimir Spidla who started to distance the party from the ODS. In the election the CSSD strengthened its position and formed a coalition government with two small parties. But divisions within the CSSD forced Spidla to resign in June 2004, to be replaced by Stanislav Gross. In 2003 the parliament had elected Václav Klaus as president.

Parliamentary elections in 2006 led to a dead heat between potential coalitions. The 2010 elections resulted in a new centre-right coalition led by Petr Necas of the ODS supported by two new parties, TOP 09 and Public Affairs. After several corruption scandals this is now looking unstable.

Denmark

A bridge from continental Europe to Scandinavia that is sceptical about the EU

| Land area: | 43,000 sq. km. |

Land area: 43,000 sq. km.
Population: 5 million—urban 87%
Capital city: Copenhagen, 0.5 million
People: Danish, Inuit, Faroese
Language: Danish, Faroese
Religion: Evangelical Lutheran 95%, other Christian 3%, Muslim 2%
Government: Constitutional monarchy
Life expectancy: 78 years
GDP per capita: $PPP 36,130
Currency: Krone
Major exports: Meat products, furniture, pharmaceuticals

Denmark is one of Europe's more physically fragmented countries. The largest part, with around 70% of the territory of Denmark itself, is the Jutland peninsula. To the east of this lie the two largest of Denmark's 400 or so islands: Funen and Zealand; the capital, Copenhagen, is located on the east of Zealand. In addition, Denmark has two distant dependent states— Greenland and the Faroe Islands. Most of Denmark itself is low-lying and fertile, broken occasionally by hills, particularly in the centre and east of Jutland.

Denmark's population is ethnically fairly homogenous, though flows of immigrant workers from the 1970s, and asylum-seekers during the 1980s and 1990s, added variety. Around 5% of the population are foreign-born, with the largest numbers coming from former Yugoslavia and Turkey.

The standard of living is high and the government gives a high priority to education, on which it spends 9% of GDP, one of the highest proportions in Europe. Danes also enjoy free medical care and extensive welfare benefits. Although public provision is not as generous as in other Nordic countries, social welfare spending is equivalent to more than one-quarter of GDP. Income tax rates touch 60% and there is a flat VAT rate of 25%. Poverty levels are among the lowest in Europe.

Even so, it is doubtful that the Danish welfare state will survive in its current form, given the ageing population. The arrival of immigrants rejuvenated the population somewhat but a falling birth rate is still leading to a rising proportion of older people. In the next 30 years the ratio of working people to those over 60 years old will fall from 3.0 to 2.1. A new Welfare Agreement agreed in 2006 incorporates raising the age of retirement

Agriculture accounts for only around 3% of GDP, but is still vital to the economy, since it feeds into industry and exports. Two-thirds of Denmark is used for crops or pasture. Most activity is concerned with livestock—either growing animal feed or raising cattle and pigs. Farms are small and family-owned but technologically very sophisticated. Since they produce around three times Denmark's own requirements, they export most of their output. Denmark also has a major fishing fleet, though over-fishing and controls by the EU have been constraining output.

Agriculture provides important raw materials for Danish industry—some of the cereal crop finishes up, for example, in cans of Carlsberg beer, one of the world's most venerable brands. Other distinctive Danish exports include stylish furniture, the hi-fi equipment of Bang & Olufsen, and the ubiquitous plastic Lego bricks—which have also been used in the construction of Legoland, one of Europe's largest theme parks.

Denmark is the home of Carlsberg and Lego

Apart from its soil, Denmark has limited natural resources. It does not have much coal or many minerals, but oil and gas fields in the North Sea supply the bulk of local needs. With a lot of flat and exposed land, Denmark can also generate a significant amount of wind power.

The Danes joined what is now the EU in 1973, but have not been strong federalists. They originally rejected the Maastricht treaty, for example, though attitudes towards Europe have become more positive.

Denmark is a constitutional monarchy, with Europe's oldest royal family, currently headed by Queen Margarethe II. Most governments have been coalitions or minority administrations. For decades after the Second World War, the main party was the Social Democratic Party, which had promoted the welfare state. This pattern was broken decisively in the 1982 election when the reins of government passed to Poul Schluter of the Conservative Party at the head of a centre-right coalition. As a result, Denmark became the first of the Nordic countries to rein in public spending and deregulate its economy. Schluter resigned in 1993.

He was replaced as prime minister by a Social Democrat, Poul Nyrup Rasmussen, at the head of a coalition that included the Radical Liberals, the Centre Democrats, and the Christian People's Party. The Social Democrats continued with many of the neo-liberal economic policies. They were also returned at the head of coalitions in the 1994 and 1998 elections.

Nevertheless, there were growing signs of a shift to the right, particularly over concerns about immigration which led to street violence and contributed to the rise of the right-wing populist Danish People's Party (DPP) led by Pia Kjaersgaard.

In 2001 Poul Nyrup Rasmussen called an election which he expected to win. But he miscalculated and was replaced by Anders Fogh Rasmussen at the head of a Liberal–Conservative coalition. This coalition was returned after one snap election in 2005 and another in 2007. In 2009 Rasmussen stepped down to become secretary-general of NATO and was replaced by Lars Lokke Rasmussen of the Liberal Party.

In 2006, Denmark found itself in the international spotlight as a result of newspaper cartoons that mocked Islam and provoked violent demonstrations around the world.

Djibouti

A city-state that serves as a transit point for neighbouring countries

Land area: 23,000 sq. km.
Population: 0.8 million—urban 88%
Capital city: Djibouti, 596,000
People: Somali 60%, Afar 35%, other 5%
Language: French, Arabic, Somali, Afar
Religion: Muslim 94%, Christian 6%
Government: Republic
Life expectancy: 55 years
GDP per capita: $PPP 2,061
Currency: Djibouti franc
Major exports: Hides and skins, coffee

Though there are mountains in the north of this small country, most of Djibouti consists of a bare arid plateau—one of the world's hottest places. Most activity centres on the capital, also called Djibouti, and particularly its port, which is crucial to its now landlocked neighbour, Ethiopia.

The country's brief period since independence in 1977 has been soured by conflict between its two main ethnic groups. More than half the population are Somalis of the Issa and other clans, while around one-third are the Afar, who are of Ethiopian origin. However, the precise ethnic division is uncertain, partly because of the flows of refugees, but also because the ethnic population balance is a sensitive issue. Djibouti's population growth rate is around 2%. Three-quarters of these people live in the capital; most of the remainder are pastoral nomads.

Djibouti is a poor country: 20% of people live on less than $1.25 a day and life expectancy is only 55 years. It is also remarkable for its consumption of a mild intoxicant: qat. Every day, Djiboutians together chew their way through a remarkable 11 tons of the drug—which is imported from Ethiopia—devoting a high proportion of household expenditure to the habit. Most of Djibouti's economy depends on its location as a transit point for neighbouring countries: trade and services make up around 80% of GDP. But this makes Djibouti vulnerable to developments around it, such as wars between Ethiopia and Eritrea.

Road and rail links with Addis Ababa are now in a very poor state—and in need of investment. Most food has to be imported and there is little industry.

Since independence in 1977, Djibouti has been ruled by the Rassemblement populaire pour le progrès (RPP), which from 1981 to 1992 was also the only legal party. But the government could not contain the Somali–Afar rivalry and in 1990 an armed rebellion was started by the Afar front pour la restauration de l'unité et la démocratie (FRUD).

Decision-making is centralized around the president. Until April 1999, Djibouti had had only one president, Hassan Gouled Aptidon, who was then replaced by his close confidant, and former security chief, Ismaël Omar Guelleh. He has continued in the same vein, eliminating political opponents and using military force to crush the rebels. In 2005 Guelleh achieved a second six-year term, and in 2011 a third. In the 2008 general election, due to an opposition boycott, his coalition won every seat.

Dominican Republic

The expansion of manufacturing industry and tourism has yet to benefit most of the poor

Land area: 49,000 sq. km.
Population: 10 million—urban 71%
Capital city: Santo Domingo, 2.7 million
People: Mulatto 73%, white 16%, black 11%
Language: Spanish
Religion: Roman Catholic 95%, other 5%
Government: Republic
Life expectancy: 72 years
GDP per capita: $PPP 6,706
Currency: Dominican peso
Major exports: Manufactured goods, ferronickel, sugar, gold, coffee

The Dominican Republic occupies the eastern two-thirds of Hispaniola, an island that it shares with Haiti. Much of the territory is mountainous—the highest peaks in the central highlands rise above 3,000 metres. The best land for agriculture and the most densely populated area, apart from the capital, is the Cibao Valley, which stretches across the north of the country.

Most people are of mixed race, or mulatto—the result of intermarriage between the predominantly Spanish colonists and the slaves brought to work in mines and sugar plantations. Wealth generally correlates with colour of skin—the blacks being the poorest. Unemployment is around 16% and 40% of the population live below the poverty line.

Education standards are low: although most children enrol in primary school around one-quarter

subsequently drop out. Health standards are also low, though access to services such as clean water and sanitation has been improving.

Poor as the Dominican Republic is, neighbouring Haiti is even poorer, and up to 500,000 Haitians, many coming to work on the sugar harvest, are thought to live in the Dominican Republic. Most have no legal status and live in squalid conditions.

Sugar is still important—and one of the largest sources of export earnings. Shrinking demand and under-investment have reduced production, though the industry could benefit from the demand for ethanol.

Other minor export crops, such as cocoa and coffee, have also suffered setbacks due to disease and drought. Exports of tobacco and cigars have done better. Food production too has been slowing. Again this has been the result of low investment—discouraged both by high interest rates and government subsidies for imported food—particularly in election years. By 2007, only around 14% of the population worked in agriculture.

Mining is another source of revenue, though it employs few people. The Canadian-owned Bonau nickel mine has been a major source of export earnings but has been hit by falling world prices.

A more dynamic part of the economy has been manufacturing which in 2006 accounted for 22% of GDP. Apart from more traditional industries like food processing, the fastest growth has been in the free-trade zones. More than 400 companies, the majority US-owned, have established factories in 40 or more zones scattered around the country. Around two-thirds of these factories make garments, while others produce footwear, leather goods, and electronics. The zones do provide employment for a predominantly female workforce, but working conditions are poor and wages are desperately low. These factories are also facing increased competition from low-wage countries in Asia.

Another rapidly growing source of income and foreign exchange is tourism. The Dominican Republic welcomes more than three million visitors each year. More than half are from Europe, though arrivals from North America have been rising as US tourists have become concerned about travelling further afield.

Few tourists venture beyond the beaches

Few visitors venture out onto the streets, preferring the self-contained beach resorts. But even here they are vulnerable to the country's poor infrastructure—particularly to pollution from inadequate sanitation. Most of the resorts are owned by multinational, often Spanish, corporations.

The Dominican Republic is still struggling to build a stable democracy. For 30 years after his first election victory in 1966, the government was dominated by Joaquín Balaguer and his conservative Partido Reformista Social Cristiano (PRSC), whose administrations were corrupt and repressive. These years were marked by instability and violence. There were, however, a couple of interludes of rule by the centre-left Partido Revolucionario Dominicano (PRD).

In 1996, when he was 88 years old and virtually blind, Balaguer's seventh and last term ended prematurely following allegations of electoral fraud. The ensuing election was won by Leonel Fernández Reyna of the Partido de la Liberación Dominicana (PLD), formerly a Marxist split-off from the PRD, though now a party of the centre.

Fernández could not stand again in the 2000 presidential election. This was won by the PRD candidate Hipólito Mejía who won with just under half the vote. Although he campaigned with promises of reform and investment in public services he failed to deliver on most of these, and his administration was tainted by corruption and the collapse of the country's second largest bank.

In the 2004 presidential election Fernández and the PLD returned to power. With loans from the IMF he led the country out of its economic crisis and in 2006 the PLD achieved majorities in both houses of Congress.

He was also re-elected in 2008, despite corruption scandals and a difficult economic situation. He has, however, been facing rising discontent related to poor public services. He will not seek re-election in 2012 but is supporting the candidature of his wife Margarita Cedeño.

Ecuador

Tensions between mestizos and Amerindian groups make for uneasy government

PACIFIC OCEAN

COLOMBIA

■ Quito

Guayaquil

Cuenca

PERU

| 0 | Miles | 200 |
| 0 | Km | 320 |

Land area: 284,000 sq. km.	
Population: 13 million—urban 67%	
Capital city: Quito, 1.4 million	
People: Mestizo 65%, Amerindian 25%, Spanish 7%, black 3%	
Language: Spanish, Amerindian languages	
Religion: Roman Catholic 95%, other 5%	
Government: Republic	
Life expectancy: 75 years	
GDP per capita: $PPP 7,449	
Currency: US dollar	
Major exports: Bananas, coffee, oil	

Ecuador has three main geographical areas. To the west, containing around one-quarter of national territory, are generally low-lying coastal plains. These rise abruptly to two Andean mountain chains that run down the centre of the country from north to south and are separated by high valleys. Beyond the Andes lies the 'Oriente'—the Ecuadorian part of the Amazon basin with its tropical rainforests. In addition, Ecuador has sovereignty over the Galapagos Islands 1,000 kilometres to the west in the Pacific Ocean. As its name indicates, Ecuador sits astride the Equator, and outside its mountain regions has a year-round humid climate.

Most Ecuadorians are mestizo—mixtures of Amerindians and white immigrants who came chiefly from Spain and other European countries, though there are also descendants

of other immigrants including the Lebanese who arrived in the 1900s. The various Amerindian groups are largely Quechua-speaking and can be found throughout the country but particularly in the highlands. They have not in the past exerted much influence but have recently become more assertive. There are also two main black groups—one in Esmeraldas on the north-east coast and the other in the Chota Valley in the northern highlands.

This is a very unequal society: the richest 10% of people receive 43% of national income and more than 40% of the population live below the national poverty line. With few prospects at home around 3 million people have emigrated for work.

Standards of education and health are low, particularly in the highlands. Only about half of all children enrol in secondary school partly because parents have to pay for all books and materials. A quarter of children are malnourished.

Ecuador's distinct geographical areas have tended to impede national integration. People in the mountains, particularly those in the capital, Quito, have generally been more traditional and conservative compared with the more commercially oriented communities on the Pacific coast centred on the port city of Guayaquil.

This divide is evident too in agriculture, which employs 14% of the population. In the mountains, production is chiefly for national consumption—typically corn, potatoes, and beans, as well as livestock. But productivity is low and landholdings are small—tempting some farmers to colonize the sparsely populated Oriente, where they face resistance from indigenous groups. On the coast, much of the land is devoted to large plantations, especially of bananas, of which Ecuador is the world's largest exporter. Fishing is also an important source of both food and export revenue—particularly the cultivation of freshwater shrimps.

Ecuador is the world's leading exporter of bananas

Industry and manufacturing in Ecuador consist largely of goods for local consumption. But participation in the Andean Community and the WTO has stimulated exports particularly of paper and wood.

The largest export industry continues to be oil which is extracted from the Oriente both by the state-owned Petroecuador and by foreign oil companies. Production has risen steadily, but because local refining is limited much fuel still has to be imported.

Governance in Ecuador has been fragile. Since the end of military rule and the subsequent democratic elections in 1979, successive governments have switched between centre-right, centre-left and populist—though none has managed to hold the country together sufficiently to follow a coherent development strategy.

One of the stranger fates was that of Abdalá Bucaram of the Partido Roldista Ecuatoriana, who was elected president in 1996 and dismissed by the Congress in 1997 on the grounds of 'mental incapacity'.

He was replaced by Fabián Alarcón. Then the 1998 presidential election was won by the ex-mayor of Quito, Jamil Mahuad of the centre-right Democrácia Popular. He ended a long-running border dispute with Peru. But the catastrophic economic situation in 1999 forced Ecuador to become the first government in recent history to default on some of its sovereign debt.

Mahuad's downfall came in 2000 as a result of his proposal to replace the sucre with the US dollar. The army stepped in to replace him with his vice-president, Gustavo Noboa, who nevertheless pressed on with replacing the sucre by the dollar.

Lucio Gutiérrez, one of the coup leaders, won the presidential election in 2003 but was removed from office by the Congress, and was later jailed for sedition. He was replaced by former vice-president, Alfredo Palacio who served out that term of office.

The 2006 presidential election was won by Rafael Correa, a left-wing populist who promised political reform and greater control over natural resources. In April 2007, in a referendum he obtained public support for a new constitution that would embody his reforms, and in September 2007 his Alianza País coalition won a majority in Congress. He raised public sector salaries and the minimum wage but by 2009 his popularity was waning.

Egypt

Egypt, the largest Arab country, has made a dramatic breakthrough to democracy

Land area: 1,001,000 sq. km.
Population: 80 million—urban 43%
Capital city: Cairo, 18.4 million
People: Egyptian
Language: Arabic, English, French
Religion: Muslim 90%, Coptic 9%, other Christian 1%
Government: Republic
Life expectancy: 70 years
GDP per capita: $PPP 5,349
Currency: Egyptian pound
Major exports: Petroleum products, cotton products

Egypt's vast territory consists of two main desert areas separated by the fertile Nile Valley. The western desert occupies around two-thirds of the country, a rocky sandy plateau with occasional inhabited oases. The eastern desert, with around one-quarter of the territory, is even less hospitable and more sparsely populated. The country's lifeline, the Nile, provides most of the water, and its valley and delta are home to more than 95% of the population.

Ethnically most Egyptians are the result of generations of intermarriage between Arabs and other groups. Most people are Sunni Muslims, but an important minority are members of the Coptic Orthodox Church, whose religion predates the Arab conquest, and who have a strong influence in both commerce and government. The Copts have also at times been the target of Islamic militancy.

Though some Egyptians have prospered, around 18% the population are below the poverty line. At least one-third of the workforce are unemployed or underemployed. Some four million Egyptians also work abroad, chiefly in the Gulf countries, sending home around $9 billion each year in remittances.

Around 30% of the workforce are still engaged in agriculture, mostly in small plots in the Nile Valley. With intensive applications of fertilizer, and manual irrigation systems, these farms are quite productive—though many are affected by waterlogging and salination.

One of the main crops is Egypt's high-quality, long-staple cotton, but output and export income have been in decline. The main food crop is wheat and production is intensive, but output is insufficient for a fast-growing population, and Egypt is one of the world's largest food importers.

To increase production the government has been reclaiming land from the desert; meanwhile these gains have largely been offset by losses to urbanization and industry. The largest scheme is the Southern Valley development project with an estimated cost of $86 billion.

Egypt's most important industrial enterprise is oil production. Though by Middle Eastern standards the reserves are modest, the oil and petroleum products from Egyptian refineries are still the main export earners, and recently the government has made further investment in gas. These industries, along with others like cotton spinning, are in the hands of state enterprises which employ around one-third of the workforce.

Within the service sector, one of the most important sources of income is the Suez Canal. Around 19,000 vessels pass through each year. Another vital service industry

19,000 vessels per year through the Suez Canal

is tourism. Some 11 million people arrived in 2007, chiefly visiting the historical sights, spending $8 billion and employing two million people. The industry has, however, been vulnerable to the effects of terrorist attacks and political instability elsewhere in the region.

Much recent history has been dominated by the relationship with Israel. Egypt has acted as a peace-broker between other Arab states and Israel for which it has been rewarded with billions of dollars in US aid.

Since it became a republic in 1952, Egypt has yet to have a pluralist democracy. The early years were dominated by Gamel Abdel Nasser, who nationalized many industries, including the Suez Canal. He was replaced on his death in 1970 by Anwar el-Sadat. In 1978, Sadat established the National Democratic Party (NDP). Sadat was assassinated in 1981 by Islamic terrorists and was replaced by Hosni Mubarak.

Mubarak and the NDP maintained a tight grip, ruling the country under an almost permanent state of emergency. The most direct political opposition came from the Islamist movement, the Muslim Brotherhood.

Until 2005 Mubarak guaranteed his own re-election by the People's Assembly which he controlled. For 2005, however, Mubarak allowed multi-candidate presidential elections. When these were held they made very little difference. Following violence and fraud, Mubarak was returned with 89% of the vote on a 23% turnout. Meanwhile hundreds of opposition figures were jailed.

The seeds of Mubarak's downfall were sown in the 2010 election, which was thought to have been Egypt's most fraudulent ever. In January 2011, diverse groups, taking heart from the fall of the regime in neighbouring Tunisia, and organized through social networking sites, embarked on a continuous and largely peaceful protest. They occupied Tahrir Square in Cairo and gripped the world's attention, condemning the corruption of the Mubarak regime and demanding democracy. Crucially, Mubarak could not retain the support of the army, and was removed from office and later arrested.

A military council is administering the country until elections are held towards the end of 2011. A likely presidential candidate is Mohamed ElBaradei, former head of the International Atomic Energy Agency. The Muslim Brotherhood has formed a party, Freedom and Justice, and could enter into a coalition with the nationalist Wafd party. But it will face opposition from other new parties.

El Salvador

El Salvador has moved from a debilitating civil war to an often violent peace

Land area: 21,000 sq. km.	
Population: 6 million—urban 61%	
Capital city: San Salvador, 1.6 million	
People: Mestizo 90%, white 9%, Amerindian 1%	
Language: Spanish	
Religion: Roman Catholic 57%, Protestant, 21%; Jehovah's Witnesses 2%, other 20%	
Government: Republic	
Life expectancy: 70 years	
GDP per capita: $PPP 5,804	
Currency: Colón, US dollar	
Major exports: Coffee, manufactures	

El Salvador is the smallest and most densely populated country in the continental Americas. It has two highland areas: to the north, a rugged mountain range along the border with Honduras, and to the south a range of 20 major volcanoes separated by a series of basins that make up the central plain. The major lowland areas are the valley of the Lempa River that runs from north to south, and a narrow plain along the Pacific coast.

The volcanoes have dominated El Salvador not just physically but also economically. The rich and porous volcanic soils that cover one-quarter of the land area are ideal for coffee cultivation, and the struggle for control over this and other fertile land has shaped the country's history. Land clearing for cultivation has removed most of the forest cover.

Little of El Salvador's Amerindian culture remains. In the 1930s the army launched a vicious military campaign, La Matanza, the 'killing', which not only cost 30,000 lives but also laid waste to the indigenous culture. The civil war of the 1980s also left its mark: a culture of violence particularly from street gangs known as 'maras'.

Levels of human development remain low, though literacy rates and life expectancy have increased. More than one-third of the population live below the national poverty line.

Agriculture employs one-quarter of the workforce and provides 70% of domestic food requirements. Despite post-war gestures at land reform, the best land is still controlled by a few wealthy people: around 1% of landowners control more than 40% of the arable land. Coffee is still one of the main export earners, grown on the slopes of the mountains. Sugar and cotton are cultivated in the lowlands on the coastal plain. The main food crops are corn, beans, and rice, but much of the food has to be imported.

One of the most pressing issues now is environmental degradation. With most of its forest cover removed, El Salvador suffers from extensive soil erosion. Agriculture and rural development generally are also vulnerable to natural hazards such as hurricanes and earthquakes.

El Salvador's economy is now dominated by industry and services. Manufacturing was originally designed for import substitution, but in recent years the emphasis has been on manufacturing for export in 'maquila' factories engaged in offshore assembly, which have overtaken coffee as the leading export earners. Most of this, which has been boosted by the 2007 Central America Free Trade Agreement, involves the production of garments using imported fabrics, by a primarily female workforce.

In an effort to increase foreign investment and trade El Salvador has adopted the US dollar as a legal currency at a fixed exchange rate against the colón. All government transactions are now in dollars. Nowadays the largest source of dollars is actually the two million or more Salvadorans living abroad, most in the USA, who send home around $4 billion annually in remittances. Emigration surged during the war but continues at a steady rate, much of it unauthorized.

Adopting the dollar

El Salvador's long history of injustice and bitterness finally erupted into armed revolt in the 1970s. A series of guerrilla groups united in 1980 as the Farabundo Martí Liberation Front (FMLN), and when Archbishop Oscar Romero was assassinated in 1980 his death triggered an armed insurrection that launched the country into a full-scale civil war.

In 1982, the right-wing Alianza Republicana Nacionalista (Arena) party took over the government and was sustained by massive US aid. Death squads roamed the country, killing anyone promoting peaceful reform. The war lasted until 1992 and cost 80,000 lives. Around one million people were displaced—many ending up in the USA.

A UN-supervised ceasefire was finally achieved in January 1992. The FMLN agreed to end the armed struggle, while the government said it would dismantle the death squads, and redistribute some land.

For the presidential elections in 1994 the FMLN had transformed itself into a political party, but lost to the Arena candidate Calderón Sol. Subsequently politics remained polarized for years between the right-wing Arena and the left-wing FMLN.

In the 1997 legislative elections the FMLN was only one seat behind Arena. Nevertheless in 1999, Arena's Francisco Guillermo Flores Pérez was elected president, albeit on a very low turnout. Then in the 2003 legislative elections the FMLN overtook Arena to become the largest party.

The Arena candidate, media baron Antonio Saca, scored a decisive victory in the 2005 presidential elections. The legislative elections in 2006, however, did not give Arena an overall majority.

The FMLN, finally won the presidency in 2009 after having moved to the centre. It is now headed by a former TV journalist and moderate social democrat, Mauricio Funes. The FMLN is also the largest party in Congress but does not have an overall majority so Funes has to reach out to Arena and another right-wing party, the Partido de Conciliación Nacional.

Equatorial Guinea

Equatorial Guinea has discovered oil but has yet to find democracy

Land area: 28,000 sq. km.
Population: 0.6 million—urban 40%
Capital city: Malabo, 166,000
People: Fang 85%, Bubi 7%; other 8%.
Language: Spanish 68%, other 32% including French, Fang, and Bubi
Religion: Roman Catholic
Government: Republic
Life expectancy: 50 years
GDP per capita: $PPP 30,627
Currency: CFA franc
Major exports: Oil, timber

The mainland of Equatorial Guinea, called Río Muni, consists of a coastal plain and then forested hills and plateaux that extend east to the border with Gabon. The country also includes several islands, the largest of which is Bioko, 200 kilometres to the north in the Gulf of Guinea and home to the capital, Malabo.

The majority ethnic group are the Fang, though these are divided into many subgroups. Those on Bioko, 15% of the population, are mostly Bubi—many of whom are politically and economically marginalized. This was Spain's only colony in Sub-Saharan Africa, and Spanish is the official language. Health standards are poor, but literacy is higher than in neighbouring countries.

During the Spanish period, the country had thriving cocoa plantations, but most of these withered away during the early years of independence. Most of the labour force still work in agriculture growing subsistence crops like cassava and sweet potatoes. The main agricultural export is okoumé wood, which is used to make plywood.

Equatorial Guinea's prospects were transformed in 1991 by the discovery of oil off the coast of Bioko. By 2009, oil represented more than 90% of GDP and government revenue.

However much of the benefit has gone to Western oil companies and the political elite. Less than 2% of its GDP goes on public health and less than 1% on education. Around 20% of children die before the age of five.

Democracy has never had a foothold. At independence the country was ruled by Francisco Macías Nguema, a dictator who dragged Equatorial Guinea into a social and economic abyss. In 1979, after a bloody coup supported by Spain and Morocco, he was overthrown by his nephew, Teodoro Obiang Nguema Mbasogo, who merely delivered a different brand of dictatorship.

In 1991, under pressure from aid donors, Obiang opened the country up to multiparty democracy, but his Partido Democrático de Guinea Ecuatorial (PDGE) remains dominant. In 1996, and in 2002, he was re-elected president, achieving 97% of the vote in elections marked by fraud and intimidation. His repressive reign continues. In 2002, for example, following an alleged coup attempt, hundreds were arrested and tortured.

Obiang has been grooming his son, Teodorín Nguema Obiang, as his successor, but in 2009 he was re-elected for another seven-year term, though with only 95% of the vote.

Eritrea

Eritrea continues a series of debilitating conflicts with its neighbours.

Land area: 125,000 sq. km.	
Population: 5 million—urban 22%	
Capital city: Asmara, 500,000	
People: Ethnic Tigrinya 50%, Tigre and Kunama 40%, Afar 4%, Saho 3%, other 3%	
Language: Afar, Amharic, Arabic, Tigre and Kunama, Tigrinya	
Religion: Muslim, Coptic Christian, Roman Catholic, Protestant	
Government: Republic	
Life expectancy: 59 years	
GDP per capita: $PPP 626	
Currency: Nafka	
Major exports: Livestock, textiles, food	

Eritrea can be divided into three main areas. The eastern part of the country is a long, arid coastal plain inhabited by nomadic herders or fishing communities. The savannah and shrub-covered western lowlands descend to the border with Sudan. The main part of Eritrea is the central highlands which are part of the Ethiopian Plateau. This is the most densely populated part of the country and has the more fertile land, though many of the hills have suffered severe soil erosion.

Eritrea has a number of ethnic groups of which the largest are the Tigrinya, who speak Tigrinya. They form most of the population of the highlands and are also to be found in the adjacent Ethiopian region of Tigray. Another major group are the similarly named Tigre, who speak Tigre, and are to be found in the north of the plateau, as well as in parts of the lowlands. Though their languages have the same root and script, they are mutually unintelligible. Hundreds of thousands of Eritrean refugees who fled during the wars in the early 1990s still live in Sudan and Ethiopia.

Decades of war have kept Eritreans desperately poor. More than 40% of children are malnourished and only 5% use improved sanitation facilities. Education levels are also low. Less than half of children are enrolled in primary school.

More than three-quarters of the population depend on agriculture or fishing. Farmers grow subsistence cereal crops and raise a small number of animals, but output is low. They have little irrigation and have to rely on erratic rainfall. And in the densely populated highlands their land has become increasingly eroded. Given Eritrea's long coastline, the fishing industry has considerable potential for local consumption and export of crab, lobster, and shrimp.

Before its incorporation into Ethiopia in the early 1950s, Eritrea had started to develop a number of industries. But Ethiopian priorities steadily undermined Eritrean enterprises and subsequent conflicts destroyed many of the rest. Even so, there are still light industries processing food and producing textiles and garments, mostly in state-run enterprises.

When it achieved independence from Ethiopia in 1993, Eritrea gained the entire coastline, so in principle Eritrea's ports should be a useful source of income—though this does depend on an abiding peace with Ethiopia. Another important source of income is remittances from Eritreans abroad—worth around $250 million in 2004.

Eritrea got Ethiopia's coastline

The government prefers to manage without external aid, but has accepted funds for post-war reconstruction as well as food aid.

Formerly an Italian colony, Eritrea was federated into Ethiopia in 1952, and in 1962 was formally, and illegally, annexed by Ethiopia. By then an independence struggle had already started but it was weakened by a long conflict between two groups. The Eritrean Liberation Front (ELF), which emerged in 1958, was organized along ethnic and regional lines. But in 1973 a socialist splinter group appeared—what is now the Eritrean People's Liberation Front (EPLF). The two groups fought each other as well as the Ethiopians.

By 1991, the EPLF was dominant and had seized control of the territory. Meanwhile the Ethiopian government in Addis Ababa had been overthrown by the Tigrayan People's Liberation Front, another Marxist-inspired guerilla group which had good relations with the EPLF. In 1993 this led to an amicable parting, and independence for Eritrea.

Early in 1994, the EPLF disbanded and re-emerged as the People's Front for Democracy and Justice (PFDJ), which was to incorporate all political parties, including some former ELF members, with Isaias Afwerki as president.

Eritrea achieved a new and supposedly pluralist constitution in 1997, though in practice opposition parties are banned. The PFDJ imposes this partly through moral force derived from long years of struggle, but also through repression. According to Reporters Without Borders, Eritrea has the world's least-free press.

Eritrea soon entered into conflicts with all its neighbours. The first was with Sudan, each country accusing the other of harbouring its dissidents. Then there were arguments with Yemen and Djibouti.

The most damaging and surprising conflict, however, was with Ethiopia. In May 1998, a border skirmish erupted into a full-scale war that by mid-1999 had cost around 50,000 lives. Ostensibly, the war was fought over a few barren hectares of loosely defined border which Eritrea claims that Ethiopia had been encroaching on, and taxing its people.

In December 2000, a peace agreement was signed and in April 2001 the UN started to monitor the disputed border and in 2007 ruled on where the border should be. But relations with Ethiopia remain tense—a situation referred to as 'no peace–no war'.

There is little sign that the elections postponed since 2001 will take place any time soon, so Afwerki and the PFDJ will retain their iron grip.

Estonia

One of the more successful ex-Soviet republics which has turned towards Europe

0	Miles	100
0	Km	160

Land area: 45,000 sq. km.
Population: 1.3 million—urban 70%
Capital city: Tallinn, 397,000
People: Estonian 68%, Russian 26%, Ukrainian 2%, Byelorussian 1%, other 3%
Language: Estonian, Russian, Ukrainian
Religion: Evangelical Lutheran, Russian Orthodox, Estonian Orthodox
Government: Republic
Life expectancy: 73 years
GDP per capita: $PPP 20,361
Currency: Euro
Major exports: Mechanical equipment, food, textiles, wood products

Estonia's mainland consists largely of a plain that includes many lakes. Around 40% of the territory is covered with forests. In addition, it has more than one thousand islands and islets in the Baltic Sea.

Estonians were disturbed when many Russians moved into the country following the Soviet invasion in 1940. Once Estonians recovered their independence in 1991 they demanded that anyone who wished to become a citizen, or to work in public administration, should speak Estonian—not an easy language.

For the Russians, who made up one-third of the population, this was a significant obstacle. Some have been emigrating but most have stayed. .

Russia has frequently complained about discrimination against Russians, and Estonia's relations with its giant neighbour remain tense. Estonia has, however, been strengthening links with Finland: Estonian is related to Finnish, which helps to sustain cultural and commercial ties.

Education standards are quite high and in response to the increasing demand for educated labour more students are entering tertiary education. Immediately following independence, health standards and life expectancy fell but have since largely recovered.

Of the former Soviet republics, Estonia made one of the most rapid transformations towards a market economy. The government started the privatization programme with land and housing by distributing vouchers to all households, and later extended the scheme to large companies. By 2001, more than 85% of the economy was run by the private sector.

After independence, Estonia's economy contracted suddenly as ties with Russia were cut. But from around 1995 the economy revived, thanks in part to foreign investors impressed by Estonia's enthusiastic liberalization, open trading regime, and low wage rates. This meant that basic industries such as textiles and timber products, along with chemicals, plastics, and metals, revived fairly quickly, finding new markets in Europe. Estonia joined the EU in May 2004 and plans to adopt the euro in 2011.

Finland has been one of the main investors. The electronics company Elcoteq, for example, attracted by wages one-sixth of those in Finland, assembles mobile phones and other items for Nokia and other Scandinavian companies, and is now Estonia's largest exporter.

Estonians assemble mobile phones for Nokia

Estonia has some of its own energy supplies, based on extracting oil from shale, but this is an expensive and polluting process, and the country still relies on Russia for additional, and more expensive, supplies of oil.

Agriculture, which is largely based on livestock, has not performed so well. Between 1989 and 2007, its share of GDP fell from 20% to 3%. The government rapidly dismantled the former collective farms to create more private farms, many of which have now consolidated into larger enterprises. While food output has been declining, forestry has been buoyant as a result of increasing demand for pulp and paper. Fishing has suffered from reductions in EU quotas.

Estonia's market economy has also opened up the potential for a larger and more vigorous service sector that had been stifled during the Soviet years. Tourism has certainly benefited. Estonia abolished visa requirements for visitors from Nordic countries and now welcomes around 1.5 million tourists a year—the majority from Finland arriving by ferry and often to take advantage of low prices for liquor.

Estonia, which had first gained independence in 1918 after centuries of rule by Sweden or Russia, regained its independence in 1991 following the demise of the Soviet Union and adopted a new constitution in 1992. This provides for a single-chamber parliament, the Riigikogu, which elects a president whose role is largely ceremonial.

Since independence Estonia has had a large number of political parties, and a series of coalition governments. The 1992 elections produced a nationalist coalition which pushed through rapid reforms, but its popularity was undermined by economic contraction and political scandals. The 1995 elections resulted in a swing to the left—though the government led by Mart Siimann carried on many economic reforms.

The 1999 elections were contested by more than 30 parties. The outcome was a government formed by a centre-right coalition with Mart Laar as prime minister. However the coalition was fractious and Laar resigned in 2002 with the Reform Party leader Siim Kallas taking over.

The parliamentary elections in March 2003, which were contested by a modest 12 parties, resulted in a centre-right coalition government led by Juhan Parts of Res Publica, a right-wing party. He resigned in 2005 to be replaced by Andrus Ansip, of the Reform Party.

Following an election in 2011, the Res Publica-Reform coalition was returned with a parliamentary majority. Ansip remains prime minister, but given the many personal and policy differences the coalition may not survive.

Ethiopia

A former favourite country of Western aid donors has become increasingly repressive

Land area: 1,221,900 sq. km.
Population: 79 million—urban 18%
Capital city: Addis Ababa, 2.4 million
People: Oromo 40%, Amhara and Tigrean 32%, Sidamo 9%, other 19%
Language: Amharic, Tigrinya, Orominga, Guaraginga, Somali, Arabic, English
Religion: Christian 61%, Muslim 33%, traditional 5%, other 1%
Government: Republic
Life expectancy: 55 years
GDP per capita: $PPP 779
Currency: Birr
Major exports: Coffee, hide and skins, qat

More than half of Ethiopia consists of two high tablelands—the western and eastern highlands which are split, south-west to north-east, by the East African Rift Valley. To the east, much of the territory consists of the lower arid plains of the Danakil depression and the Ogaden.

Ethiopia's population is very diverse, the result of many centuries of linguistic and racial mixing. Officially, there are 64 recognized ethnic groups, but there are probably 200 or more languages. Following a new constitution in 1995, the country became a federal republic and now has nine regions or 'nationalities', though each of these actually contains a mixture of ethnic groups. The largest group are the Oromo, who are to be found in the centre and south, followed by the Amhara and the Tigrayans, who occupy the north and the north-west. Many are Christians, belonging to the previously dominant Ethiopian Orthodox Union Church, but nowadays more are Muslims.

Ethiopia remains one of the world's poorest countries. One third of children are malnourished, infant mortality is high, and life expectancy short. Fewer than 20% of Ethiopians have access to modern healthcare services and their health status is among the worst in the world. Malaria has resurged and AIDS is a major threat: more than 4% of Ethiopians are infected with HIV. Education standards are also very low: only 40% of people are literate. The population grows by around 3% per year and by 2020 it could exceed 100 million.

Some 85% of the population live in the rural areas, and survive mostly through farming—growing a range of subsistence crops, including teff, which is a cereal grass. But agricultural output has been extremely variable. The high population density in the highlands and ensuing deforestation have led to extensive soil erosion. And since almost all agriculture is rain-fed it is very vulnerable to the unreliable climate. Per capita food production has declined sharply. In a good year

Ethiopia is just about self-sufficient, but when the rains fail it has to rely on emergency shipments of food aid. Ethiopia's strategies to improve food security include, on the one hand, stronger early warning systems, and on the other, efforts to increase output through greater use of fertilizer. In 2009 the rains failed again and an estimated 13 million people were depending on food handouts

But some of the problems are political. In the past farmers had greater freedom to migrate temporarily to help harvest cash crops and supplement their incomes. Now they need permission to travel.

The main cash crop is coffee, which is grown mainly in the south by smallholders and which has provided around two-thirds of export earnings—though less recently because of low world prices. Another important agricultural export is qat, a bush whose leaves can be chewed and are a mild stimulant popular in Somalia and Djibouti.

Cultivating qat for its neighbours

Ethiopia has little industry. Most is controlled by the state and is concentrated in food processing. However, some industries, particularly textiles and garments, are being privatized, and there is probably potential for more private-sector activity in the processing of hides and skins, which are also major exports.

In the past three decades, Ethiopia has been through a series of political traumas and civil wars. Following a famine in 1972–74, which cost an estimated 200,000 lives, political unrest led to the overthrow of the monarchy of Haile Selassie and in 1975 to the establishment of a socialist republic. But civilian groups failed to cohere into viable political parties, leaving the way clear for the establishment of a 17-year military regime, the Provisional Military Administrative Council, known in Amharic as the 'Derg'.

From 1977, the Derg was controlled by Colonel Mengistu Haile Mariam, whose repressive regime tried to transform Ethiopia into a Marxist-Leninist state. It survived until 1991 when, having lost the support of the Soviet Union, and following uprisings in Eritrea and Tigray, it was defeated by the Ethiopian People's Revolutionary Democratic Front (EPRDF). This led in 1993 to independence for Eritrea— which left Ethiopia landlocked.

Meles Zenawi, the Tigrayan leader who was head of the transitional administration has remained in charge. The EPRDF, a group which is still dominated by the Tigray People's Liberation Front, regularly wins most of the seats for the legislative assembly, most recently in 2005, when it chose Zenawi as prime minister.

The legislature also elects the largely ceremonial president, currently Girma Wolde-Giorgis.

Strangely, the former allies Ethiopia and Eritrea declared war in 1998 over a border dispute. The fighting stopped in 2000 but relations remain tense.

Zenawi has become increasingly repressive, locking up political opponents and journalists. Prior to the elections in 2010 the government had passed new laws further restricting the space for opposition, and as a result Zenawi and the EPRDF won all but two of the parliamentary seats.

Fiji

Indigenous Fijians and Indians struggle for power

Fiji comprises hundreds of islands and islets in the South Pacific. The two largest islands, Vitu Levu and Vanua Levu, account for more than 80% of the territory.

The population is balanced between Fijians, who are of mixed Melanesian and Polynesian origin, and Indians, who are descendants of 19th-century immigrant sugar workers. Indians were narrowly in the majority until 1987, when many fled after the first coup; now the balance again favours Fijians.

Sugar remains the backbone of agriculture, but drought and worries about land tenure have affected production. Sugar processing also used to dominate industrial production, but manufacturing industry is now more diverse.

Garments now account for one-quarter of export income, though have suffered from a loss of preferential access to Australia and the USA. Earnings from the Vatukoula gold mine are also important. In addition, Fiji depends heavily on tourism but visitors have been discouraged by political instability.

Politics in Fiji has long been afflicted by conflicts between ethnic Fijians and those of Indian origin. A major issue is the ownership of land which has largely been monopolized by indigenous Fijians who lease it to

Land area: *18,000 sq. km.*
Population: *0.8 million—urban 53%*
Capital city: *Suva, 172,000*
People: *Fijian 57%, Indian 38%, other 5%*
Language: *English, Fijian, Hindi*
Religion: *Christian 65%, Hindu 28%, Muslim 6%, other 1%*
Government: *Republic*
Life expectancy: *69 years*
GDP per capita: *$PPP 4,304*
Currency: *Fiji dollar*
Major exports: *Sugar, garments, gold*

Indians—who produce virtually all the sugar. Many leases are now becoming due, and a proposal to offer new leases sparked the latest conflict.

From independence in 1970, indigenous Fijians had controlled the government. The 1987 election, however, was won by two parties that were largely supported by Indians. The Fijians responded with a military coup and a constitutional change to bar Indians from power. This led to international isolation that only ended in 1994 when a new constitution gave the Indians a fairer deal. Eventually, in 1999, with the Fijian vote split among five parties, the Indian-dominated Fiji Labour Party won a majority in parliament and Mahendra Chaudhry became the first Indian prime minister.

But his promise of 30-year land leases to Indians incensed the Fijians. In 2000 businessman George Speight, along with a section of the army, staged a coup. The military installed an interim civilian government led by Laisenia Qarase. He then formed a new indigenous party, the People's Unity Party, which won a majority in the 2001 election and again in 2006. In December 2007, however, claiming widespread corruption, the military, led by Commodore Frank Bainimarama, took over yet again.

Finland

In a short time, thanks to Nokia, Finland has become a leader in telecommunications

Land area: 338,000 sq. km.
Population: 5.3 million—urban 64%
Capital city: Helsinki, 568,000
People: Finn 93%, Swede 6%, other 1%
Language: Finnish, Swedish
Religion: Lutheran National Church 83%, other 3%, none 14%
Government: Republic
Life expectancy: 80 years
GDP per capita: $PPP 34,526
Currency: Euro
Major exports: Metals and machinery, forestry products

Most of Finland is fairly low-lying and the country as a whole is heavily forested. One of its most striking features is the complex of thousands of shallow lakes in the south. Here the climate is fairly temperate but even some of the ports on the Baltic Sea are icebound during the long winter. Most of the higher ground is in the northern one-third of the country, and since this lies within the Arctic Circle the climate is severe.

Finland's population, which is concentrated in the south, is ethnically fairly homogenous. Nevertheless, there are also significant numbers of Swedes, mainly in the southern coastal areas and on the Åland Islands, which lie between the two countries. Swedish is also an official language. Another minority, though now very small in number, are the Sami, or Lapps, who herd their reindeer in the far north. The break-up of the Soviet Union also caused some immigration of ethnic Finns from Russia and Estonia, but the proportion of foreign-born is less than 2% and immigration has often been exceeded by emigration. By European standards, Finland is also fairly rural—more than one-third of people live in the countryside.

Even so, agriculture is not a major source of income, employing only 3.5% of the workforce. The fierce climate and poor soils mean that less than one-tenth of the land is suitable for farming—though the country is largely self-sufficient in basic foods.

The most important rural activity is forestry. Finland's vast forests of pine, spruce, and birch support a number of major industries. The extensive pulp and paper sector, is dominated by three main companies. In addition, Finland is an important producer of building materials and furniture, and has used its forestry experience to specialize in the production of forestry machinery. Finland accounts for around 10% of global exports of forest products.

Finland's other industries have also been expanding fast. Remarkably, most of this development has come from one company, Nokia, which is the world's largest manufacturer of mobile phones and produces other telecommunications equipment. As a result, industry still contributes more than one-quarter of GDP, a higher proportion than in most developed countries.

Finland had made steady economic progress for decades but was rocked by a deep recession in the early 1990s and by the collapse of trade with the Soviet Union, which had taken around one-fifth of exports. The government made deep cuts in public expenditure and this, combined with contributions from the forestry and electronics industry, helped Finland to emerge from the worst of the crisis.

Over the period 2003–2007 economic growth averaged 4% per year. Unemployment has fallen and in 2007 was down to 7%.

However this progress has been quite narrowly focused on telecommunications equipment and is over-reliant on the fortunes of Nokia. Although Oulu in the north is also a technology centre, the industry

Tourists head for the land of the midnight sun

is increasingly concentrated in the south of the country in the region bounded by Helsinki, Turku, and Tampere, to which many people have been migrating.

In addition, Finland has been developing its tourist industry, based on skiing and on visitors wanting to experience Lapland's midnight sun.

Finland joined the EU in 1995 and is an enthusiastic member, being the only Nordic country to switch to the euro. This is due as much to security as to economic concerns. Finland

had a mutual security treaty with the Soviet Union but since 1990 it has tried to follow a neutral course. This means that Finland is reluctant to join NATO but Finns hope that the EU will offer similar security and has been a strong supporter of the EU's expansion to the east.

For the past hundred years or so the government of Finland has been in the hands of coalitions or minority administrations, typically led either by the Social Democratic Party or the rural-based Centre Party.

Since the 1995 election, however, the mosaic has taken on a new pattern. In that year the votes were more evenly shared. The Social Democrats, led by Paavo Lipponen, had the most, though only 25%, and formed a 'rainbow', five-party coalition that included all shades of the political spectrum.

The coalition reformed after the 1999 election, still with Lipponen as prime minister. In 2000, the Finns elected to the largely ceremonial role as president Tarja Halonen, the first woman to hold the office; she was re-elected in 2006.

In the 2003 general election the rainbow's colours changed. This time the Centre Party came out slightly ahead and formed a coalition with the Social Democrats and the Swedish People's Party. Following the 2006 general election. The Centre Party's Matti Vanhanen remained in charge.

The 2011 election was notable for the rise of the right-wing nationalist True Finn party which finished in third place, behind the Social Democratic Party and the National Coalition Party. Negotiations continue over a new coalition.

France

One of Europe's most centralized states, with a distinctive and influential culture

Land area: 552,000 sq. km.
Population: 62 million—urban 78%
Capital city: Paris, 9.6 million
People: French, including ethnic minorities particularly from former colonies
Language: French
Religion: Roman Catholic 85%, Muslim 7%, Protestant 2%, Jewish 1%, other 5%
Government: Republic
Life expectancy: 81 years
GDP per capita: $PPP 33,674
Currency: Euro
Major exports: Machinery, cars, food and agricultural products, chemicals

France has the largest territory in Western Europe. Around two-thirds is lowlands, chiefly to the north and west, including the Paris basin and the Loire Valley. But there are also dramatic mountain ranges. To the south-west the Pyrenees mark the border with Spain. To the south, occupying one-sixth of the territory is the Massif Central, consisting mostly of plateaux at between 600 and 900 metres. And to the south-east, forming a barrier with Italy and Switzerland, are the French Alps, beyond which to the south is the Mediterranean coastline—the Côte d'Azur.

Though France has an increasingly diverse population the government has been determined to sustain a national identity. Thus for more than 250 years, the Académie Française has protected and promoted the French language. The French government has also made strenuous efforts to resist the encroachment of American culture. Nevertheless France does have strong regional identities, and even regional languages such as Breton and Catalan.

The French population continues to be stimulated by immigration. Around 6% of the population have foreign nationalities, the majority coming from Algeria and Morocco. But millions more immigrants have become French citizens. There may also be up to half a million unauthorized immigrants. Rather than encouraging multiculturalism, France has been determined to ensure assimilation, a policy which has led to conflict over girls who wear Muslim headscarves to school.

The education system has also promoted a centralized style of management: most of the country's leadership has been processed through the 'grandes écoles', so they tend to share the same values.

Most French workers are now employed in services and often work for the government. The French state continues to play an important part in the economy—spending around half of GDP and employing more than one in four workers. Many service workers are involved in tourism. France is by far the world's leading tourist destination, with over 160 million visitors per year.

Agriculture nowadays employs only 4% of the labour force—chiefly on smaller farms. Even so, output has increased and France is the EU's leading food producer. The country is largely self-sufficient in food and is a leading exporter of wheat, beef, and other foodstuffs—as well as the world's leading exporter of high quality wine. France is also one of the main supporters of the EU's expensive system of agricultural protection.

France is the world's fourth largest industrial power and has many globally important companies. Danone, for example, is the world's largest dairy products firm, and Peugeot-Citroën is Europe's second largest car maker. One distinctive manufactured export has been the high-speed train, the TGV. France has also been a leading arms exporter. With no oil, France has invested heavily in nuclear electricity, some of which it exports.

Exporting high-speed trains

Like many other European countries, France has been afflicted by high unemployment—around 10% overall, but often above 25% for young people. Many employers blame this on over-regulation of the labour market and a high minimum wage. To reduce unemployment and share work more widely, the government in 2000 introduced an official 35-hour week.

Government in France frequently requires 'cohabitation' between a president of one party and a prime minister of an opposing one. The current constitution, which dates from 1958, strengthened the position of the president, who is directly elected for a seven-year term. The president appoints the prime minister and chairs the weekly Council of Ministers.

François Mitterand of the Parti socialiste (PS) won the presidential election in 1981 but the centre-right party, the Rassemblement pour la république (RPR), subsequently made gains in the National Assembly and obliged Mitterand to appoint a prime minister from its ranks.

Then, after a series of elections, the situation was reversed. The RPR had won the 1993 elections to the National Assembly and in 1995, their candidate, Jacques Chirac, also won the presidential election, defeating Lionel Jospin of the PS. Chirac called a snap National Assembly election in 1997. This was a miscalculation. The PS won, obliging Chirac to appoint Jospin as prime minister.

Chirac won the presidential election in March 2002. Then the mainstream right-wing parties, united as the Union pour la majorité présidentielle (UMP), won the legislative elections, with Jean-Pierre Raffarin as prime minister. Raffarin proved fairly ineffective, however, and was replaced by Dominique de Villepin. He in turn was undermined in late 2005 by rioting in the poverty-stricken suburbs.

This cleared the way for minister of the interior Nicolas Sarkozy to be the UMP's presidential candidate in 2007. He defeated Ségolène Royal of the PS and promised radical economic reform. Although his popularity soon slumped it subsequently revived as a result of his interventions following the global economic crisis.

French Guiana

One of the richest, but most impenetrable, countries of South America

French Guiana's hot and humid territory is largely flat. There are low mountains in the south along the border with Brazil, but most of the country is a vast plateau covered with dense tropical rainforest. The majority of people live along the swampy coastal plain or in the capital, Cayenne.

France used the territory as a penal colony until 1953—most notoriously Devil's Island. Even after they had served their sentences, it was difficult for prisoners to leave. Most of those who stayed succumbed to tropical diseases.

Today, most people are creole or black, but there is also a significant French population. The others include the original Amerindians in the jungles of the south; the Marons, who are descendants of prisoners who escaped to the interior; and immigrants from other French territories. There are even some people from the Hmong ethnic group from Laos.

The country benefits from generous aid from France, particularly since the establishment in 1968 at Kourou, 60 kilometres west of the capital, of the rocket-launching base now used by the French and European space agencies, as well as by a commercial company. There are around a dozen launches each year.

Land area: 90,000 sq. km.
Population: 0.2 million
Capital city: Cayenne, 55,000
People: Black or mulatto 66%, white 12%, Amerindian 12%, other 10%
Language: French, creole
Religion: Roman Catholic
Government: Overseas collectivity of France
Life expectancy: 77 years
GDP per capita: $16,000
Currency: Euro
Major exports: Shellfish, gold, timber

This and other government services employ not just French people but also many local workers. As a result, the average per capita income is one of the highest in South America, though this is distributed somewhat unequally between rocket scientists and Amerindian hunter-gatherers.

The other main economic activities are agriculture, fishing, and forestry. Most farms are small, growing subsistence crops, but there are also a few large commercial fruit plantations that export their produce to France. In addition, fishing for shrimp provides an important source of export income. Because of the European tastes of many inhabitants, the country remains highly dependent on imports from France of food and other goods.

As an overseas territory of France, the country elects two representatives to the French National Assembly and one to the Senate. Locally, there are also consultative general and regional councils.

In the 2010 elections to the General Council, the Socialist Party of Guiana again took the most seats. Few people want independence from France, which for many could result in a catastrophic drop in income. In 2010 voters also rejected an offer of greater autonomy.

French Polynesia

Tahiti and the other islands have become increasingly reliant on tourism

French Polynesia consists of around 130 islands in five groups in the Pacific, of which the largest are Tahiti and nearby Moorea, which have three-quarters of the population. These two islands are volcanic and mountainous, while many others are little more than coral atolls with scant vegetation.

The majority of people are Polynesian, but there is also an influential French minority. Since the islands are a French 'overseas country', EU citizens are free to travel and work there, and many have chosen to do so. But there have been concerns that they may be taking the better jobs and altering the social balance.

The islanders' relatively high standard of living reflects substantial French investment, particularly during the period of nuclear testing that ended in 1996. To compensate for the end of testing France gives an economic restructuring payment of around $170 million per year.

In recent years the economy has shifted decisively towards tourism, which accounts for one-quarter of GDP. Many people also work in government. As a result, more than two-thirds of the population are now employed in services. But many people still work in various forms of agriculture and in fisheries, notably

Land area: 4,000 sq. km.
Population: 0.3 million
Capital city: Papeete
People: Polynesian 78%, Chinese 12%, French 10%
Language: French, Tahitian
Religion: Protestant 54%, Roman Catholic 30%, other 16%
Government: Overseas collectivity of France
Life expectancy: 77 years
GDP per capita: $PPP 18,000
Currency: CFP franc
Major exports: Pearls, coconut products

oyster farming and the production of cultured black pearls, which are the main source of export revenue. The islands continue to enjoy other forms of aid from France, though unemployment is increasing.

Amendments to the French Constitution in 2003 paved the way for more autonomy. After an 'organic law' agreed by the assemblies of France and French Polynesia, the latter in 2004 became an 'overseas country' within the French republic and can pass its own laws.

As citizens of a French overseas country, the islanders elect members to the French parliament, but they also have their own territorial assembly. For 20 years the most seats went to Tahoeraa Huiraatira-Rassemblement pour la république led by Gaston Flosse who served as president. But in recent years the situation has been fluid.

The government in 2008 consisted of a coalition between two unlikely partners who united to unseat the president, Gaston Tong Sang, in a no-confidence vote. Following this the anti-independence Flosse again became president and the pro-independence Oscar Temaru became speaker of the territorial assembly.

Gabon

Oil-rich Gabon is one of Africa's most prosperous countries, but the oil is running out

CAMEROON

Libreville

Port Gentil

Moanda

ATLANTIC OCEAN

CONGO

| 0 | Miles | 200 |
| 0 | Km | 320 |

Land area: 268,000 sq. km.
Population: 1.4 million—urban 86%
Capital city: Libreville, 604,000
People: Four major Bantu groupings: Fang, Eshira, Bapounou, Bateke. Many immigrants
Language: French, Fang, Myene, Bateke, Bapounou/Eschira, Bandjabi
Religion: Christian, animist, Muslim
Government: Republic
Life expectancy: 60 years
GDP per capita: $PPP 15,167
Currency: CFA franc
Major exports: Oil, timber, manganese, uranium

Gabon consists almost entirely of the basin of the Ogooué River and its tributaries, broken by mountains in the centre and the north, with a narrow strip of coastal lowlands. Three-quarters of the country is covered in dense tropical forest—a rainforest second in area only to the Amazon Basin. This has more than 250 species of tree and provides living space for more than four-fifths of the world's chimpanzees and gorillas.

Gabon's rapid development as an oil producer has made it one of the most urbanized countries in Sub-Saharan Africa—more than 80% population are now thought to live in urban areas. This has also left its forest resources relatively unscathed by development. Though there has been some deforestation, particularly along the coast and the river banks, Gabon still has thousands of plant species and thriving wildlife.

The population is very diverse, with more than 40 ethnic groups. But there is also a strong French influence and most of the population is Roman Catholic. The urban population is largely to be found in the coastal strip—in Libreville, the capital, one of the world's most expensive cities for visitors, or in Port Gentil, the centre of oil production.

The interior, on the other hand, is sparsely settled, largely along river banks. Communications generally are poor. Even Port Gentil has no road link with the rest of the country; it must be reached by boat or by air.

Oil has given Gabon one of Africa's highest per capita GDPs but the wealth is unequally distributed. In the rural areas services for health and education are in a poor state. Although water supplies have improved and the infant mortality rate has started to fall, less than half of children are vaccinated. Even in the urban areas 20% of the population live below the poverty line.

There are also many foreign workers—around one-fifth of the workforce—who come from neighbouring countries to perform the more menial tasks.

Agriculture is largely for subsistence, but only 0.6% of the total land area has been cultivated and more than half the country's food has to be imported. Most of this, from vegetables to cheese, comes from France, at considerable cost.

The most important rural product is timber. Around three-quarters of the country is forested and before the discovery of oil, timber was the country's leading source of wealth. Timber is still a major export and is thought to employ around half the labour force. Gabon is the world's leading producer of okoumé and ozigo, softwoods that are used for making plywood. Other traditional cash crops include rubber, coffee, and cocoa. There are also efforts to expand output of palm oil and sugar.

Source of timber for making plywood

Since the late 1960s, Gabon's development has been fuelled by oil, most of which is offshore. Oil accounts for around half of GDP and government income. Some 15 international companies are involved in exploration or production. The most important oilfields are Gamba, which is operated by Shell, and the extensive Rabi Kounga field and its satellites, which are operated by Shell and Elf. Future prospects are uncertain. Most of Gabon's fields are maturing, output is falling, and reserves could be exhausted within the next few decades.

Another major mineral export is manganese. Here the prospects are better. Gabon is the world's third largest producer and its open-cast mines in the Moanda region in the south-east have around one-quarter of global reserves. Mining is in the hands of the Comilog company, which is largely foreign owned, though there is some state and Gabonese participation. The other important mineral is uranium, but the reserves are now almost exhausted.

Gabon embarked on an economic structural adjustment programme in 1995, liberalizing the economy and privatizing a number of state enterprises. But life without oil will be difficult; apart from the possibilities of eco-tourism there do not seem to be many new options on the horizon.

Gabon achieved independence in 1960, and for 42 years was ruled by the corrupt and oppressive Omar Bongo Ondimba. He came to power in 1967, and in 1968 established the Parti démocratique gabonaise (PDG) as the only legitimate political party. Subsequent oil wealth allowed Bongo and his supporters to entrench their position and distribute patronage. Protests were met with violent repression. The regime maintains close relations with France which protected him from its 1,000-strong garrison in Libreville.

In 1991 a financial crisis and the resulting austerity measures led to violent protests and forced Bongo to introduce a multiparty system. He nevertheless continued to be re-elected, for the final time in 2005 with 79% of the vote. In 2007, the PDG won 83 out of 120 seats in the National Assembly.

Bongo died in June 2009 with $183 million in his personal bank account in New York. The ensuing presidential election was won by his son and former defence minister Ali Bongo Ondimba.

The Gambia

A holiday destination with a repressive government

Land area: 11,000 sq. km.
Population: 1.6 million—urban 58%
Capital city: Banjul, 34,000
People: Mandingo 42%, Fula 18%, Wolof 16%, Jola 10%, Serahuli 9%, other 5%
Language: English, Mandinka, Wolof
Religion: Muslim 90%, Christian 9%, other 1%
Government: Republic
Life expectancy: 56 years
GDP per capita: $PPP 1,225
Currency: Dalasi
Major exports: Groundnuts

This long, thin country extends within Senegal for around 300 kilometres inland from the coast, along both banks of the lower Gambia River. The river is fringed with swamps beyond which are flats that give way to low hills and a sandy plateau. The better land is up-river and most people have settled between the river flats and the uplands.

The Gambia has a large number of ethnic groups, but the majority are the Malinke (also called the Mandingo), who are largely to be found higher up the river. More than half the population is urban, and the capital, Banjul, is more than usually crowded since it effectively occupies an island at the mouth of a river and has few opportunities for expansion. There are also a large number of middle-class refugees from Sierra Leone.

The Gambia is one of the world's poorest countries, and poverty has been increasing. According to official estimates, more than half the population is poor and more than one-third are extremely poor. Three-quarters of people make their living from agriculture. They use the lower-lying land that benefits from annual flooding to grow rice, which is the staple food. But yields are low and most of the country's rice now has to be imported. Other subsistence crops include millet, sorghum, maize, and vegetables.

Most farmers get their cash income by growing groundnuts, which are ideally suited to the sandy soil on the higher ground. Groundnuts occupy around half the country's cultivable area and most of the crop is sold through Senegal. The government has made some attempts to diversify by promoting the cultivation of flowers and fruit for air-freighting to Europe, but so far has had little success.

In fact, more than 80% of the country's exports are 're-exports'—manufactured goods that have been imported into the port of Banjul but are en route to neighbouring countries such as Senegal, Mali, Guinea, and Guinea-Bissau.

The other major export earner is tourism. The Gambia's coastline may be short but it has a number of sandy beaches that have attracted package tourism from Europe, particularly from the United Kingdom which provides more than half the 140,000 visitors. Tourism creates hotel jobs for local workers but the benefits do not extend very far beyond the resorts. A military coup in 1994 deterred visitors for a few years, but they have now returned and there has been more investment in new hotels.

Politically and geographically, the Gambia seems an anomalous country that one might expect to be part of Senegal. Its status derives from its history as a colony of the British, whose main preoccupation was to deprive the French of an important navigable waterway. The Gambia's relations with Senegal have sometimes been fraught.

For many years following independence in 1965, the Gambia was held up as a model, albeit imperfect, of African democracy. The first prime minister was the leader of the People's Progressive Party (PPP), Sir Dawda Jawara, who was the first president in 1970 when the country became a republic. Jawara survived a Libyan-backed coup attempt in 1981 with the help of troops from Senegal. This intervention also led in 1982 to the formation of a federation—'Senegambia'—that seemed destined to unite the two countries. Jawara resisted this, however, and the federation was disbanded in 1987.

Gambia gets aid despite human rights abuses

In 1992 Jawara was re-elected as president but in 1994 was overthrown in a military coup led by Lieutenant Yahyah Jammeh. The coup, following which all political parties were banned, was widely condemned internationally. The Gambia found itself isolated and had most of its international aid withdrawn.

International pressure did eventually result in a democratic opening. In 1996, Jammeh formed a new party, the Alliance for Patriotic Reorientation and Construction. Just before the election, however, he took the precaution of banning the three largest opposition parties—and anyone who had held ministerial office for the previous 30 years. This left the United Democratic Party as the main opposition. Jammeh won the presidential election with more than half the vote.

The government has shown scant regard for democratic freedoms. It continues to harass the United Democratic Party, whose leaders it arrests and detains. There have also been regular raids on radio stations and newspapers.

Throughout his regime, Jammeh has been subject to real or supposed conspiracies. In 1998 three lieutenants were executed for plotting against him. And early in 2000 Jammeh claimed to have uncovered another plot hatched by his palace guards, several of whom died while 'resisting arrest'. In the same year troops fired on student demonstrators.

Despite these abuses, aid donors have largely returned, reassured by the façade of democracy, and give around $50 million per year.

Jammeh easily won the 2001 presidential election and he and the APRC then won overwhelmingly in the 2006 presidential, 2007 legislative, and 2008 local elections.

Georgia

A 'rose revolution' was followed by a disastrous fight with Russia.

Land area: 70,000 sq. km.
Population: 4.4 million—urban 53%
Capital city: Tbilisi, 1.1 million
People: Georgian 83%, Azeri 7%, Armenian 6%, Russian 2%, other 2%
Language: Georgian 71%, Armenian 7%, Azeri 6%, Russian 9%, other 7%
Religion: Orthodox Christian 84%, Muslim 10%, Armenian-Gregorian 4%, Catholic 1%, other 1%
Government: Republic
Life expectancy: 72 years
GDP per capita: $PPP 4,662
Currency: Lari
Major exports: Food, chemicals, machinery, tea, wine

Georgia is dominated by the Caucasus mountains. Its northern border is formed by the Greater Caucasus and the southern border by the Lesser Caucasus; more than one-third of the mountains are forested. Between these are the major rivers and their valleys—the Kura which flows east into Azerbaijan and the Rion and others flowing west into the Black Sea. The latter have deposited masses of silt to form the Kolkhidskaya lowlands which make up most of the coastal land. Formerly a sub-tropical swamp, this has been drained to create valuable agricultural land.

Most people live in the lower parts of the country. While ethnic Georgians—who call themselves Kartveli—make up around 70% of the population, they are divided into many regionally based subgroups. Georgians have a distinctive language, alphabet, and culture. Most are Christian, though a subgroup in the south-west are Muslim. Two non-Georgian ethnic groups do not want to be part of Georgia: the South Ossetians on the northern border with Russia; and the Abkhazis in the north-west.

Levels of human development have been high, with good standards of education and health: some Georgians are reputedly very long-lived. But war and economic crisis have devastated the health and education systems and standards are falling. Over half of Georgia's people now live below the poverty line and the population has been declining as a result of emigration and a falling birth rate.

Agriculture has been the most important employer, but has shrunk to only 13% of GDP. The country's geographical diversity permits a wide variety of crops, including tea, citrus fruits, grapes, and tobacco. Around 60% of state-owned land has been privatized; the rest is leased out. Food production is low.

The Soviet Union had promoted heavy industry, but much of this is now obsolete. Georgia has some oil and gas of its own but chiefly relies on imports from neighbouring former-

Soviet republics which have been trying to charge higher prices. Georgia has some important mineral resources, including manganese, iron ore, and gold, but high energy costs and falling demand have affected production.

Georgia has attractive scenery and beach resorts so could have a tourist industry, but political instability and poor infrastructure keep people away.

Many important businesses are still run by the state, but the government has sold majority stakes in some others to foreign companies. The Borjomi mineral water plant, for example, is run by French and Dutch investors. After a catastrophic post-independence decline economic growth revived but has since crashed

Georgia has struggled to build a coherent democratic state. There have been a number of secessionist conflicts. The first has been in South Ossetia which in 1990 hoped to unite with Alania (North Ossetia) in the Russian Federation. Fighting here ended with a ceasefire in 1992 but this did not resolve the issue.

The second secessionist conflict, in Abkhazia, has also proved intractable.

Struggles for secession
The Abkhazis declared independence in 1992, and subsequently expelled more than 250,000 Georgians. Various ceasefires were negotiated, but fighting has persisted between the Abkhazis and the Georgian guerrillas.

The first president, following independence in 1991, was Zviad Gamsakhurdia, who was driven from office in 1992 following accusations of corruption and of human rights violations—sparking a civil war that ended only with his death in 1993.

He was replaced by Eduard Shevardnadze, who had been Soviet foreign minister and was elected president in 1995. His new party, the Union of Georgian Citizens, was returned to power in the 1999 parliamentary elections, and Shevardnadze was re-elected president in April 2000.

Somewhat surprisingly, since Shevardnadze was in little danger of losing, the election was marred by ballot stuffing and other irregularities. After similar activities in the 2003 parliamentary elections, the opposition took to the streets and finally invaded the parliament building and forced Shevardnadze to resign in what was called the 'rose revolution'.

A presidential election in 2004 was won overwhelmingly by Mikhail Saakashvili, a 36-year-old US-educated lawyer and leader of the National Movement. He set about tackling the extensive corruption and was re-elected in January 2008.

In August 2008, however, following an influx of volunteers and mercenaries from Russia to South Ossetia, Georgia responded with a military attack, triggering a five-day conflict with Russia that would cost 350 lives and displace 35,000 civilians. Russia soon came out on top and recognized both South Ossetia and Abkhazia as independent states and thousands of refugees fled into Georgia.

The military defeat, and the subsequent economic crisis might have unseated Saakashvili, as Russia wanted, but so far he has survived. Heading the main opposition group, the Alliance for Georgia, is a former UN envoy, Irakli Alasania.

Germany

Politically powerful in Europe, Germany also has a large and successful economy, though it is becoming less competitive

Land area: *357,000 sq. km.*	

Land area: *357,000 sq. km.*
Population: *82 million—urban 74%*
Capital city: *Berlin , 3.4 million*
People: *German 92%, Turkish 2%, other 6%*
Language: *German*
Religion: *Protestant 34%, Roman Catholic 34%, Muslim 4%, unaffiliated or other 28%*
Government: *Republic*
Life expectancy: *80 years*
GDP per capita: *$PPP 34,401*
Currency: *Euro*
Major exports: *Machinery, cars, chemicals, manufactured goods*

Germany has three main geographical regions. From the North Sea and Baltic coasts southwards, covering roughly one-third of the country, are the lowlands of the North German Plain. This leads to a belt of central uplands running west to east, cut through by major rivers, including the Rhine and the Weser. Further south still are the South German Highlands culminating in the Alps at the borders with Austria and Switzerland.

The 1990 reunification of Germany created by far Europe's largest national population. Today there are 15 million in what was East Germany and 67 million in what was West Germany. Germany also has the largest number of immigrants in Europe—around 9% of the population are foreign nationals, the largest proportion from Turkey. Many of these are former 'guest workers', who

arrived in the 1970s and have chosen to settle. Others are refugees and asylum seekers who came at the end of the 1980s, including more than two million people of German ancestry.

From the 1950s, West Germany developed into Europe's most dynamic industrial nation. East Germany lagged far behind and the merger of the two economies proved difficult and expensive. Germans still have to pay a 'solidarity tax' which adds around 5% to the normal tax bill. This pays for extending social benefits to the East, raising wages, and repairing extensive environmental damage. Even so, wide gaps remain. In 2007, unemployment nationally was 8% but in the East was 15%.

Nevertheless, this remains a powerful industrial economy—the world's third largest after the USA and Japan. Its strength has been highly efficient manufacturing industry, which accounts for around one-fifth of GDP and two-thirds of exports of goods and services. The most important sectors are machinery, cars, and chemicals. Built around this is a large service sector. Agriculture employs few people but the country is still 70% self-sufficient in food.

The economy has been organized

in a fairly coherent way—built around the idea of the 'social market'. Rather than relying on the short-term vagaries of stock markets, investment has been based on long-term partnerships between banks and companies.

Another feature has been cooperation between companies and the powerful trade unions. Though membership has been falling, one-third of workers still belong to trade unions and wages are settled annually in national rounds. In addition, workers are entitled to participate in management through works councils and have seats on the boards of the largest companies.

Germany also has one of the most generous welfare systems with high unemployment benefits and pensions. Employers argue that these costs make the country uncompetitive. Hourly wage costs are higher than in the USA or the UK.

The German model has, as a result, come under increasing pressure. Many years of low growth, combined with the costs of reunification, have been eroding Germany's economic position,

Germany's economic model under fire
and have taken the budget deficit beyond the levels permitted by the EU's growth and stability pact.

In response, early in 2004, the government introduced Agenda 2010, a major reform of the welfare state, which included making it more difficult to get unemployment benefits.

Another notable feature of the German economy and society is a concern for the environment. The country has intensive systems of recycling—and spends around 2% of GDP on environmental protection.

Germany is a federal state with 15 states, or länder (ten from the West, five from the East), plus the federal capital Berlin. Each land has its own parliament and government responsible for such issues as policing and education. The state governments are also represented in the federal parliament's upper house, the Bundesrat. But most power rests in the 662-member lower house, the Bundestag. In 2004, the parliament elected Horst Köhler to the largely ceremonial function of president.

After 1949, most West German governments were coalitions and the tradition continues in the new Germany. From 1982 the head of government, the chancellor, was Helmut Kohl, leader of the centre-right Christian Democratic Union (CDU) who was the architect of German reunification.

This era ended with the 1998 election which resulted in a coalition of the Social Democratic Party (SPD) with the Green Party. The SPD is further left than the CDU.

Gerhard Schröder of the SPD became chancellor and Green politician, Joschka Fischer deputy chancellor. They narrowly won the 2002 election but in 2005, when Schröder called an early election, the government was replaced by a grand coalition of the CDU/CSU and SPD with Angela Merkel of the CDU as chancellor.

Merkel proved an astute politician and has established herself as a dominant figure She was not too happy working with the SPD, but following success in the 2009 election she was able to form a new centre-right coalition with the smaller liberal Free Democratic Party (FDP).

Ghana

Ghana has achieved one of Africa's few transfers of power through the ballot box

Land area: 239,000 sq. km.
Population: 23 million—urban 52%
Capital city: Accra, 1.6 million
People: Akan 45%, Moshi-Dagomba 15%, Ewe 12%, Ga 7%, Guuan 4%, other 17%
Language: English, Twi, and other African languages
Religion: Christian 69%, Muslim 16%, traditional 9%, other 6%
Government: Republic
Life expectancy: 57 years
GDP per capita: $PPP 1,334
Currency: Cedi
Major exports: Gold, cocoa, timber

Apart from scattered hills and plateaux along its eastern and western borders, Ghana consists mostly of lowlands. These include a coastal plain that extends up to 80 kilometres inland and incorporates many lagoons and marshes. Almost two-thirds of the country, however, consists of the basin of the Volta River, which now includes Lake Volta, which was formed by the Akosombo dam—the largest artificial lake in the world. There is a sharp climatic division between the dry north and the wetter and lusher south.

Ghana's population comprises dozens of ethnic groups of whom the largest are the Akan. The population has been growing rapidly—at around 3% per year, though this rate is expected to drop.

By the standards of Sub-Saharan Africa, Ghana is relatively well off. Even so, 28% of the population live below the poverty line—and poverty is generally more severe in the north. Health standards have been low but in 2003 the government introduced a National Health Insurance Scheme which now provides insurance for almost 60% of the population. Education standards have also been low but literacy rates are rising.

Around 50% of people still make their living from agriculture, which accounts for around 42% of GDP. The main food crops are maize, yams, and cassava, while the main cash crop, and one of the leading export earners, is cocoa. Cocoa is produced by more than 1.6 million peasant farmers, mostly on plots of under three hectares and although production collapsed in the mid-1980s it has now bounced back—partly because the government has raised producer prices and farmers have replaced trees.

Even so, overall agricultural productivity has been falling, a consequence of low investment and the removal of subsidies on inputs such as fertilizers, which many farmers can no longer afford. Some are now reverting to subsistence production. Another important agricultural export is timber, though there are concerns about deforestation.

Ghana, whose colonial name was the 'Gold Coast', once again has gold as its leading export. The main producer is the Ashanti Goldfields Corporation. Other important mining activities involve manganese and diamonds. In 2007, oil was discovered off the coast. Production will start in 2010 and is expected to contribute around $1.2 billion annually to the state coffers.

Ghana also has a diverse collection of manufacturing industries, a legacy of earlier efforts at self-sufficiency.

In 1983, Ghana started to follow the market-led structural adjustment recipes of the IMF and the World

Ghana has important historical slavery sites

Donors rewarded Ghana with considerable aid. This stimulated entrepreneurial activity, particularly in financial services and trade, and imports flooded in. Tourists also started to arrive to visit Ghana's ecological reserves as well as historical sites related to slavery.

But development did not reach far into the rural areas, and the loans steadily piled up. By 1999, Ghana's total debt was equivalent to four years of export earnings and the economy was in steep decline. In the past few years, however, the economy has been growing at around 5% annually.

Most of this activity was presided over by Jerry Rawlings. He first entered the political scene in 1979 as 28-year-old Flight-Lieutenant Rawlings, leading a military coup. He briefly handed power over to a civilian administration but returned in a second coup in 1981. His Provisional National Defence Council (PNDC)

government opted for 'party-less' government to be run by technocrats. Initially this was socialist but, Rawlings later turned sharply to the right into the embrace of the IMF, and from 1983 embarked on the sustained programme of structural adjustment that aid donors held up as a model.

Donors were less happy with Rawlings' dictatorial, if popular, rule. In 1992 he responded with a new constitution that allowed multiparty elections and he refashioned the PNDC as a political party—the National Democratic Congress (NDC). Rawlings won the 1992 presidential election, overcoming the candidate of the main opposition party, the business-oriented New Patriotic Party (NPP). In 1996 the NDC again proved victorious, though the NPP did gain one-third of the seats. Rawlings was re-elected with 57% of the vote.

Rawlings' intended successor was his vice-president, John Atta Mills, but he lost the 2000 presidential elections in a run-off with the NPP candidate John Agyekum Kuofor and for the first time Ghana achieved a transfer of power through the ballot box.

Kuofor turned his attention to the previous regime and prosecuted some ex-ministers for corruption. This proved divisive but, with the economy growing, the government remained fairly popular and he won a second term in 2004

For the 2008 election the NPP chose Nana Akufo-Addo as its candidate but, with the economy weak, he narrowly lost in the second round to John Atta Mills of the NDC. The NDC lacks an overall parliamentary majority, so Mills may find it hard to deliver reforms.

Greece

The collapse of the Greek economy and mass protests against austerity measures threaten the euro zone

Land area: 132,000 sq. km.
Population: 11 million—urban 61%
Capital city: Athens, 3.8 million
People: Greek 93%, other 7%
Language: Greek
Religion: Greek Orthodox 98%, other 2%
Government: Republic
Life expectancy: 79 years
GDP per capita: $PPP 28,517
Currency: Euro
Major exports: Manufactures, food

Most of Greece is mountainous. The mainland is dominated by the rugged Pindus Mountains, which extend from the northern border to the hand-shaped peninsula of the Peloponnese. The main lowland areas are the plains in the north-east along the Aegean coast. Greece's coastline is highly indented so the country is penetrated from all directions by the sea. Around one-fifth of the territory consists of two thousand or more islands, the largest of which is Crete, to the south.

The country is subject to regular earthquakes, one of which in 1999 killed 150 people. A more beneficial natural phenomenon is sunshine which allows Greece to use Europe's greatest surface area of solar panels.

Officially, the population is almost entirely ethnic Greek, though it can also be considered a mixture of the many different groups that have

arrived from neighbouring countries, including Turkey, Macedonia, Albania, and Bulgaria. Greeks may not be rich but they have been relatively long-lived—thanks to a healthy diet.

Greece continues to be a country of emigration and immigration. Emigration was still continuing to richer countries of the EU in the 1990s, notably to Germany. But many people are also arriving. The break-up of the Soviet Union encouraged more than 60,000 ethnic Greeks to move to Greece. There are also around 500,000 unauthorized immigrants, mostly from Albania. Around 9% of the population are foreign born.

Around 10% of the population still work in agriculture, but only one-quarter of the land is cultivable, and since the soil is poor and there is not much rain, yields are often low. The coastal plains are used for the cultivation of major cereal crops and sugar beet as well as cash crops such as cotton and tobacco. Greece is also a major producer of olives, tomatoes, grapes, and other fruits.

Although the fishing industry declined in the 1990s because of an ageing fleet and over-fishing, the sea catch now seems to be growing again, as is fish farming in which Greece is Europe's largest producer.

For a European country, industry

is underdeveloped. Greece has few minerals, apart from lignite, and its manufacturing tends to be small-scale. Even so, wages have risen in recent decades and some of the simpler industries such as garments and textiles have migrated to poorer neighbouring countries.

Many of the larger industrial activities are controlled by the state, either owned directly or through state-owned banks.

Historically, one of the most successful industries has been shipping. Greece owns around one-sixth of the global fleet, 3,300 vessels, typically bulk carriers, though most of these sail under other flags using foreign—and cheaper—crews. Now that Greece has joined the euro, more of the earnings from shipping are being invested at home.

Another important source of foreign exchange has been tourism—an industry that employs 18% of the labour force and contributes a similar **More tourists than its population** proportion to GDP. With many historical sites, and spectacular scenery and beaches on its many islands, Greece often attracts around 12 million visitors each year. While these are welcome the sheer numbers have been causing environmental damage.

Internationally, an important issue is the relationship with Turkey, for which Greece is officially supporting EU accession. Another is the reunification of Cyprus for which Greece favours a federal solution.

From 1967 Greece had a military government. But the military were ousted in 1974 following their backing for a failed coup in Cyprus that resulted in a Turkish invasion. Since then Greece has been ruled either by the conservative New Democracy Party (NDP), or the Panhellenic Socialist Movement (Pasok).

The period 1989–90 saw a series of hung parliaments until the NDP, led by Constantine Mitsotakis, took over. But early elections in 1993 were won by Pasok, led by Andreas Papandreou. When the latter fell ill he was replaced in 1996 as prime minister by Costas Simitis. Pasok steadily drifted towards the centre and narrowly won a third term in 2000, defeating the NDP, led by Constantine Karamanlis.

The elections in 2004 marked a turning point. Both parties had a new generation of leaders, albeit from familiar political families: Costas Karamanlis for the NDP and George Papandreou for Pasok. This time the NDP won a clear overall majority.

The NDP and Karamanlis won again in 2007, but in 2009, Karamanlis called a snap election. The result was a large victory for Pasok, allowing George Papandreou to follow his grandfather and father into the prime minister's job.

Greece was the poorest member of the EU when it joined in 1981. Subsequent economic development, often driven by EU subsidies, enabled rapid growth and Greece adopted the euro in 2001. But by 2010 the country was clearly living beyond its means and was deep in debt. The EU, the European Central Bank and the IMF launched a rescue package but this demanded deep cuts in public expenditure and tax rises that led to mass protests. Greece could now default on its debts – threatening the survival of the euro.

Guadeloupe

An overseas department of France that occasionally agitates for independence

Guadeloupe comprises two volcanic islands in the Caribbean, Basse-Terre and Grande-Terre, that together form a butterfly shape, along with several smaller islands. The western wing, Basse-Terre, is mountainous, with dense rainforests as well as mangrove swamps. It also has a smoking volcano, La Soufrière, though this has not erupted for 30 years, along with some spectacular waterfalls. Much of the island has been designated a national park. Grande-Terre has lower hills and plains and is more suitable for agriculture.

The population is mostly black or of mixed race and speaks French, though many also use a creole dialect. The majority of the workforce are employed in the service sector, particularly in tourism. Most tourists come from the USA, many on brief visits from cruise ships, to enjoy the French colonial atmosphere of the largest town, Basse-Terre, or to head for the quiet beaches, the rainforests, or the outer islands.

Agriculture is less significant nowadays and much of the country's food has to be imported from France. Historically, sugar was the major crop but efforts have been made to diversify into bananas and other fruits as well as aubergines and flowers.

Land area: 2,000 sq. km.
Population: 0.5 million
Capital city: Basse-Terre
People: Black or mixed race 90%, other 10%
Language: French
Religion: Roman Catholic
Government: Department of France
Life expectancy: 78 years
GDP per capita: $PPP $7,900
Currency: Euro
Major exports: Sugar, bananas

There is also some light industry, including rum manufacture.

Guadeloupe enjoys substantial subsidies from France—which help to offset high levels of unemployment. Since Guadeloupians are French citizens many young people take the opportunity to emigrate to France.

As an overseas department, Guadeloupe sends representatives to the French parliament—two senators, and four representatives in the National Assembly. But it also has its own elected general council and regional council.

During the 1960s and 1970s, there was agitation for independence. Enthusiasm seems to have waned. In 2007 the islands of Saint-Martin and Saint-Barthélemy seceded to become a seperate collectivity of France.

The local assemblies have usually been dominated by centre and right-wing parties, though the left gained power during the 1980s. In the 1998 elections the majority of seats in the regional council went to the centre-right Rassemblement pour la république. But following a severe drought and a series of strikes, it lost the 2004 elections to the Socialist Party, led by Victorin Lurel. In 2009, he too faced a violent general strike for higher wages. This was eventually resolved but had damaged tourism.

Guam

As a US military outpost, Guam is sensitive to changes in defence priorities

Guam is one of the Mariana Islands in the Pacific Ocean between Hawaii and the Philippines. The northern half of the island is a limestone coral plateau, much of which has been levelled to form airfields. The southern half has a range of volcanic hills. Around one-third of the territory is owned by the US military—and the country has suffered from much indiscriminate dumping of waste, either on the land or in the ocean.

The oldest inhabitants are the Chamorro people who still make up more than one-third of the population. They have their own distinctive language and culture. But years of colonization, by Spain and then by the USA, have taken their toll. Guam has, for example, the world's highest per capita consumption of Spam.

Many other people have arrived. In addition to more than 20,000 US military and other federal employees and their dependants, the bases have drawn immigrants from elsewhere in the Pacific, particularly the Philippines. In a few decades it seems likely that Filipinos could outnumber Chamorros. More recently, Guam has also been attracting unauthorized immigrants.

Guam's military bases, mostly those of the navy, with around $10 billion-worth of infrastructure, are

Land area: 1,000 sq. km.
Population: 0.2 million—urban 93%
Capital city: Hagåtña/Agaña
People: Chamorro 37%, Filipino 26%, other Pacific Islander 11%, white 7%, other 19%
Language: English, Chamorro, Filipino
Religion: Roman Catholic
Government: Unincorporated territory of the USA
Life expectancy: 78 years
GDP per capita: $PPP $21,000
Currency: US dollar
Major exports: Transhipments of oil products, construction materials, fish

one of the mainstays of the economy. Military cutbacks in the 1990s saw some defence-related jobs disappear. This will be offset by the transfer of around 8,000 marines and their dependants from the US base at Okinawa in Japan to be completed by 2014, but their arrival has drawn protests from Chamorro groups worried about the loss of more land.

The second main source of income is tourism. Around one million visitors arrive each year, mostly from Asia and the Pacific, and particularly Japan. At present they usually head for the beaches or massage parlours. Guam, however, wants to attract higher value tourists so is building more luxury spas. Other suggestions for increasing visitor numbers range from promoting Chamorro culture to introducing casino gambling.

Guamanians are US citizens and send a delegate to the US House of Representatives with limited voting rights. In 2008, Democrat Madelein Bordallo was re-elected as the delegate. The 2006 elections for the local governor were again won by a Republican, Felix Camacho. In the 2008 legislative elections the Democrats won a large majority.

Guatemala

Guatemala endured 36 years of civil war. But the injustices that led to the war remain unresolved

Land area: *109,000 sq. km.*
Population: *13 million—urban 50%*
Capital city: *Guatemala, 2.9 million*
People: *Ladino and European 59%, Amerindian 40%, other 1%*
Language: *Spanish, Amerindian*
Religion: *Roman Catholic, Protestant*
Government: *Republic*
Life expectancy: *70 years*
GDP per capita: *$PPP 4,562*
Currency: *Quetzal*
Major exports: *Coffee, sugar, bananas*

Guatemala is one of the most picturesque countries in Central America. The northern, sparsely inhabited part juts out between Mexico and Belize. But around 60% of the population live either on the narrow Pacific coastal plain or in the mountainous areas of the south and west that are studded with spectacular lakes and volcanoes.

Guatemala's population distribution still reflects the legacy of the Spanish conquest. The invaders seized the most fertile soil on the plains and lower slopes, driving the Indian population up to the steeper, less valuable land—where their descendants largely remain.

Today more than half of the population would consider themselves 'ladinos'—though the term is more social and cultural than racial: apart from Spanish descendants it also includes Amerindians who have adopted Spanish language and culture.

Most of the rest of the population consist of 23 Mayan indigenous groups, whose women are particularly noticeable because of their unique and colourful form of dress. The majority are Roman Catholics but Protestant evangelical groups have been making increasing inroads.

Around 40% of the population make their living from agriculture. But this covers vast disparities. Guatemala has probably the most unequal land distribution in Latin America. The most recent data refer to 1979, when 3% of farms had 65% of the land, and 88% of farms had only 16%, and since then the situation has probably deteriorated.

The larger holdings mostly grow export crops such as sugar and bananas on the coastal plantations, while on the lower mountain slopes they grow coffee and cardamom. The poorest farmers concentrate on corn, rice, and beans, but as many as 650,000 have to make the annual pilgrimage to work on the plantations.

Industry accounts for only 20% of employment, mostly concentrated around Guatemala City. Some efforts have been made to expand simple assembly work for export to the

USA, with Korean-owned plants making T-shirts and underwear. But competition from Mexico and elsewhere has been stiff and many factories have closed.

Agriculture also accounts for around one-third of export earnings, led by coffee, sugar, and bananas. But poor prices in the 1990s encouraged diversification into such crops as mange-tout, fruit, and flowers. The production of staple crops, like corn and beans, has also fallen as a result of, among other things, cheap imports and the high cost of inputs.

More than half the population live below the poverty line. Almost half of children are chronically

Half the population live in poverty

malnourished. This is partly because the government seems incapable of taxing the rich. Tax revenues amount to only 11% of GDP.

The lack of prospects at home has caused many people to migrate. Around 10% of Guatemalans are now thought to be living in the United States and many others work in Mexico. In 2007 their remittances were equivalent to 12% of GDP.

Tourism could provide an additional source of income and visitor numbers are now growing.

Guatemala has long history of repression. For 39 years after a US-inspired coup in 1954, the country was dominated by the army and its right-wing supporters. Many of those on the left joined guerrilla groups that in 1982 came together as the Unidad Revolucionária Nacional Guatemalteca (URNG). From the early 1980s in particular, during the presidency of Efraín Ríos Montt, the

government launched a 'dirty war' that cost up to 100,000 lives and a further 40,000 'disappearances'.

Formal democracy was restored only in 1985, and in 1990 the government started negotiations with the URNG. International support for the process included awarding the 1992 Nobel Peace Prize to the Mayan woman activist, Rigoberta Menchú.

The peace agreement was signed in December 1996. This required constitutional amendments to rein in the military, to outlaw discrimination, and to allow the use of Mayan languages in schools. But progress has been slow, particularly in curbing the human rights abuses. In April 1998 a Roman Catholic bishop, Juan Geradi, was murdered after issuing a critical report on the army, and the country became increasingly lawless, beset by violent crime and drug smuggling.

From 1999 Guatemala was ruled by Alfonso Portillo, of the right-wing Frente Republicano Guatemalteco (FRG). His corrupt administration had little economic impact. After a presidential election he was replaced in 2004 by Oscar Berger of the conservative Gran Alianza Nacional.

The presidential election in 2006, then saw a swing to the left with the election of Alvaro Colom Caballeros. He promised to increase spending on education, healthcare and rural development, but has struggled because his Unidad Nacional de la Esperanza government, does not have a working majority in Congress and is also suffering from internal splits.

Guatemala is still beset by violence but one hopeful sign in 2009 was a prosecution for army atrocities committed during the civil war.

Guinea

Despite rich agricultural and mineral potential, Guinea remains poor. But it does now have a fragile democracy.

Land area: 246,000 sq. km.	
Population: 9.6 million—urban 35%	
Capital city: Conakry, 2.1 million	
People: Peuhl 40%, Malinke 30%, Soussou 20%, other 10%	
Language: French, Sousou, and others	
Religion: Muslim 85%, Christian 8%, indigenous beliefs 7%	
Government: Republic	
Life expectancy: 54 years	
GDP per capita: $PPP 1,140	
Currency: Guinean franc	
Major exports: Bauxite, alumina, gold, coffee	

Moving in from the coast, Guinea has four main geographical regions. The coastal plain reaches around 50 kilometres inland, leading to the Fouta-Djallon Highlands which rise to more than 1,500 metres. Then there are the savannah plains of Upper Guinea which slope towards the Sahel in the north-east, and finally the forested highlands of the far south-east. Guinea is also the 'reservoir' of West Africa. The region's major rivers, the Senegal, the Gambia, and the Niger, all have their origins in Guinea.

Ethnic divisions correspond to the four regions. The largest group are the Peuhl (also called the Fulani), who are mostly to be found in the Fouta-Djallon Highlands. Then there are the Malinke in Upper Guinea, the Soussou who occupy the coastal plain, and a number of smaller groups, many

of whom can be found in the forested highlands. The local population was also swollen in the late 1990s by refugees from fighting in neighbouring Liberia and Sierra Leone. In the 1980s, up to two million Guineans also fled the repressive regime in their own country and do not seem to have returned in significant numbers.

Their lack of enthusiasm reflects the country's persistently low achievements in human development. Around 40% of the population live below the poverty line. Life expectancy is short, health services are poor, and the literacy rate is only 30%. Ironically, for a country that has the sources of many major rivers, one-third of the population do not have access to clean drinking water.

Two-thirds of the population depend for survival on agriculture—the main food crops being rice, cassava, maize, plantains, and vegetables. Despite having much fertile agricultural land, only around 20% of the available land is cultivated. Most production is for subsistence, leaving a wide food gap that must be met with imports of rice.

With the help of international donors, the government has been struggling to increase food production and to diversify into cash crops such as coffee, cotton, and fruits. This has required extensive investment in infrastructure, particularly in roads.

Nevertheless, Guinea's fortunes still depend critically on bauxite. Guinea has half of global reserves—which are to be found both in the coastal region and in Upper Guinea. Guinea is the world's second largest producer. In the past, bauxite has been responsible for 90% of the country's exports and 70% of government revenue. These proportions have been falling in recent years as a result of falling world prices and inefficient production at the major smelter that converts part of the output to alumina (aluminium oxide). Much of the industry has been controlled by state-owned enterprises—though there has been greater foreign investment in the past few years, with the participation of companies from Russia, Ukraine, Iran, and Australia.

The world's second largest bauxite producer

Guinea also has significant exports of gold and diamonds. But mining has mostly been in the hands of smaller enterprises and the trade is largely clandestine: only around 15% of diamond production is recorded. In addition, Guinea has extensive reserves of other minerals, particularly iron ore (6% of world reserves), that remain largely unexploited. And it has enormous hydroelectric potential.

Guinea's economic weaknesses and its over-centralized and corrupt bureaucracy are a legacy of the 26-year dictatorship of Sékou Touré. He led the country to independence in 1958 and his Marxist-tinged and ultimately repressive and isolated regime only ended with his death in 1984. At this point the army took over, led by Colonel Lansana Conté. In 1993, Conté, as leader of his newly constituted Parti de l'unité et du progrès (PUP), won the presidential election, and in 1995 the PUP won the majority of seats in the National Assembly. The PUP's power base was mostly in the coastal areas—while other parties, grouped as the Coordinated Democratic Opposition, were more regionally confined.

In 1998, Conté beat four other candidates to win a second presidential term. After the election he cracked down and started arresting opposition leaders.

In June 2002, the PUP duly won a large majority in the legislative elections and in 2003 Conté was re-elected president with 96% of the vote—after a boycott by the leading opposition party the Front républicain pour l'alternance démocratique.

Conté died in December 2008. Within hours there was a military coup led by Captain Moussa Dadis Camara. The junta shocked the country in 2009 when it attacked an opposition rally, killing 150 people and raping dozens of women.

Nevertheless a presidential election was held in December 2010—the first real election since independence. This pitted Alpha Condé, a long-time opposition leader, whose support is largely from the Peuhl ethnic group, against Cellou a former prime minster, Dalein Diallo, who is supported by the Malinke. Condé won and despite some subsequent violence and tension Diallo seems to have accepted the outcome.

Guinea-Bissau

Guinea-Bissau in West Africa is probably the world's most indebted country, and has recently suffered a civil war

Land area: 36,000 sq. km.
Population: 1.5 million—urban 30%
Capital city: Bissau, 397,000
People: Balanta 30%, Fulani 20%, Manjaca 14%, Mandinga 13%, Papel 7%, other 16%
Language: Portuguese, Criolo, African languages
Religion: Muslim 50%, indigenous beliefs 40%, Christian 10%
Government: Republic
Life expectancy: 48 years
GDP per capita: $PPP 477
Currency: CFA franc
Major exports: Cashew nuts

Guinea-Bissau is low-lying and monotonously flat. Much of the territory comprises coastal swamps, and daily tidal flows through the mangrove forests stretch up to 100 kilometres inland. Beyond the swamps and plains are dense forests leading in the north to broad savannah.

The country has a large number of ethnic groups, though around half the population are either Balanta or Fulani. Many follow traditional animist religions, but Islam has been making steady inroads, particularly among the economic and political leadership.

Standards of human development are low. Two-thirds of the population live below the poverty line and do not have access to safe water or sanitation. Life expectancy at 48 years is one of the lowest in the world. The country's relative isolation may initially have protected it from the worst ravages of HIV and AIDS but the infection rate is now around 4%. Education standards are also very poor. Years of economic crisis and structural adjustment have taken their toll on the school system: only around half of children enrol. Girls in particular lose out: only one-quarter of women are literate.

Most economic life centres on agriculture, which accounts for almost 60% of GDP, more than 80% of employment, and more than 90% of exports. The staple food crop is rice, which is grown on one-third of arable land, along with millet, sorghum, cassava, and other root crops. But the country still has to import around 40% of its rice. In the interior, the Fulani and other groups raise cattle.

The main cash crop is cashew nuts, which provide most export income. Cashew nut trees can survive even on very poor land and require little or no cultivation. Production has been rising but Guinea-Bissau has earned less than it could from this. It has few facilities for processing, so most of the nuts are exported to India and then make their way to Europe where they are eaten as a cocktail snack. There are other cash crops such as groundnuts, and mangoes are another with export potential.

One of the obstacles to better agricultural performance is land tenure or, more precisely, deciding who owns what. Most food is produced by the 90,000 small villages, the tabancas. But there are also 1,200 larger farms, the poteiros, that are operated under state concessions. Unfortunately, many of these two types of holding overlap and it has proved difficult to resolve ownership.

Confusing patterns of land-holding

Guinea-Bissau also has considerable fishing potential. Fishing employs only 10% of the workforce but provides additional revenues as a result of licences sold to foreign fleets. The agreement with the EU involves funds to develop local capacity but also allows EU fleets to overexploit local resources.

Until around the mid-1980s, Guinea-Bissau's government attempted to develop along state-socialist lines. But recent decades have seen a decisive shift towards a market economy. This made the country more acceptable to international donors, particularly the World Bank and the IMF, though they have at times suspended their programmes because of chronic political uncertainty.

The country has also received many other forms of aid, particularly from Portugal, the ex-colonial power. But not all aid has been well used and a combination of corruption and economic crisis has made Guinea-Bissau one of the world's most heavily indebted countries.

In an effort to stabilize the economy, the government in 1996 took the unusual step of abandoning its currency, the peso, and entering the currency zone of the former French West African countries.

Guinea-Bissau became independent in 1974, at which point it was federated with Cape Verde. Between 1974 and 1999 the country was ruled by one party—the Partido Africano da Independência de Guiné e Cabo Verde (PAIGC). In 1980, the party leadership was seized by General João Bernardo Vieira in a bloodless coup. He broke the union with Cape Verde and from 1985 he steered the PAIGC away from state socialism, and from the early 1990s opened the political system to other parties.

Faced with a divided opposition he won subsequent presidential elections in 1991, and the PAIGC won the 1994 parliamentary election. By mid-1998, however, a combination of corruption and economic mismanagement had led to an army mutiny and a civil war, though. In 1999 Vieira was finally overthrown and presidential and legislative elections were won by the Partido para a Renovaçao Social (PRS), with Kumba Yala as president.

Yala proved erratic and incompetent. In 2003, with the country in chaos, the military intervened. A bloodless coup followed by elections in 2004 returned the PAIGC to power. But presidential elections in 2005 then brought back Mr Vieira as an independent.

In March 2009, for reasons still unclear, Viera was assassinated by soldiers. The July presidential election was won by Malam Bacai Sanhá of the PAIGC. However, he may be acting as a proxy for the military.

Guyana

Guyana in South America remains split between its Indian and black communities

Guyana has a narrow coastal plain that soon leads into the dense tropical rainforests that cover more than 80% of the territory—though there are mountains in the west along the borders with Venezuela and Brazil.

The country's ethnic mix reflects its colonial history. Most of today's Guyanese, concentrated along the coast, are descendants of slaves from Africa and subsequent indentured labourers from India who came to work on coastal sugar cane plantations. The surviving Amerindians are now in scattered communities along the riverbanks.

The Guyanese have made significant progress in health and education, but 3% of the population are now HIV-positive.

The country has strong agricultural potential. Sugar is the main crop and investment has reinvigorated the industry. Rice too has seen record production for export. As yet, the country's forests, with 1,000 varieties of tree, have been relatively under-exploited but the arrival of Malaysian logging companies should change that—and raise fears of deforestation.

Guyana also has important mineral deposits. The opening of the Canadian-owned Omai gold mine led to a rapid increase in production. Bauxite from two government-owned enterprises used to be a leading export but high production costs have reduced their competitiveness.

Politics in Guyana has usually been sharply divided along racial lines. The two main parties are the People's Progressive Party (PPP), which represents the Indo-Guyanese, and the People's National Congress (PNC), which represents the Afro-Guyanese. Elections have typically been bitterly contested affairs. In 1997, for example, the majority of seats went to a coalition of the PPP and Civic, a small party formed by professionals. The PNC refused to accept the result, leading to street protests and rioting.

In 1998 an intervention by the Caribbean Community (Caricom) produced the 'Herdmanston Accord', which included constitutional reform.

The elections in 2001 again produced a victory for the PPP-C coalition, led by Bharrat Jagdeo, and, although Caricom observers declared the elections fair, again there were PNC protests followed by rioting. The lead-up to the postponed 2006 elections seemed to show an improvement—until the assassination of the agriculture minister. Jagdeo and the PPP won a narrow majority.

Haiti

Already living in the poorest country in the Americas, Haitians suffered a devastating earthquake

Land area:	28,000 sq. km.
Population:	9.7 million—urban 50%
Capital city:	Port-au-Prince, 2.1 million
People:	Black 95%, mulatto plus white 5%
Language:	French, creole
Religion:	Roman Catholic 80%, Protestant 16%, other 4% (including voodoo)
Government:	Republic
Life expectancy:	61 years
GDP per capita:	$PPP 1,155
Currency:	Gourde
Major exports:	Coffee

Haiti occupies the western one-third of the island of Hispaniola, the rest of which is the Dominican Republic. The territory consists of mountain ranges interspersed with fertile lowlands. Around half takes the form of two peninsulas to the north and south of the Golfe de la Gonâve, within which there are a number of offshore islands.

The country sits along the boundary between two plates on the earth's crust. On January 12 2010 these shifted, generating a huge earthquake which killed up to 85,000 people, left one million people homeless, devastated the capital and caused more than $4-billion worth of damage.

Decades of plunder by the dictatorial Duvalier family, followed by years of violent political turmoil, had reduced an already poor population to destitution and even starvation. Now the country faces rebuilding costs equivalent to around half of annual economic output.

Haiti is the only state in the Western Hemisphere classified by the UN as a Least Developed Country. Average income is one-quarter of that in the Dominican Republic. One third of people lack access to safe water, and a similar proportion are illiterate. Around one-quarter of children are undernourished. There is scarcely any tax revenue and social services are weak or non-existent. Only 10% of schools are government run; most of the rest are in the hands of the Catholic church or non-governmental organizations.

Unsurprisingly, many Haitians emigrate—either across the border to work on the sugar plantations of the Dominican Republic, where at least 500,000 Haitians live, or to the USA; around 300,000 live in New York. Migrant remittances are worth around $1.3 billion annually—one of the country's main sources of income.

Two-thirds of the population struggle to make a living from agriculture. Much of this is subsistence farming on tiny plots growing maize, sorghum, millet, and beans. The better, irrigated land is also used to grow rice and beans for sale in the urban areas. The main cash crop is coffee but export earnings have slumped as a result of reduced output and low world prices. Sugar used to be an important export earner but falling prices and foreign competition have discouraged production. Agriculture in Haiti has also suffered from environmental degradation, particularly deforestation. Only around 2% of the land is now forested. Industry is limited, with only a few garment factories.

Around 60% of the workforce are unemployed or underemployed and many rely on feeding programmes and other forms of international aid.

The country's political troubles are the persistent legacy of the years of dictatorship. From 1957 to his death in 1971, 'Papa Doc' Duvalier ruled through fear and repression, whilst draining the country of its meagre resources.

The arrival of Aristide

His son Jean Claude, 'Baby Doc', continued in the same fashion until national and international pressure ousted him in 1986. Several years of upheaval ensued until 1990, when a fiery Catholic priest, Jean-Bertrand Aristide, heading a grass-roots movement called Lavalas ('avalanche'), became president in February 1991. However, his radical policies favouring the poor so upset the army and the business community that he was ousted that September in a military coup by General Raul Cédras.

The coup in turn outraged the international community. Eventually Cédras withdrew, 20,000 US troops arrived and Aristide returned. Constitutionally, he was not eligible to stand in the 1991 elections. In his place the Lavalas candidate,

duly elected, was René Préval. The government was obliged to impose IMF-style spending cuts and privatization programmes.

Aristide carefully distanced himself from these measures and formed a breakaway party called Fanmi ('family') Lavalas. After several postponements an election in May 2000 gave Aristide's Lavalas a resounding victory. But the election was marred by irregularities, and 15 opposition groups united as Convergence Démocratique to challenge the legitimacy of the government. Meanwhile, in an election boycotted by the opposition, Aristide was re-elected president in November 2000.

Aristide unfortunately had developed the autocratic characteristics of previous Haitian leaders and sought absolute control, using street gangs where necessary. In 2004 the opposition and students stepped up demonstrations calling for his resignation. In February 2004 an improvised rebel army ousted Aristide who was flown out in a US plane.

The Supreme Court Chief Justice took over as interim president supported by a UN-authorized peacekeeping force. Elections were finally held in May 2006 and, were won again by René Préval, by then estranged from Aristide, at the head of multiparty coalition Lespwa subsequently called Inite.

Préval was criticized for his handling of the earthquake relief programme. Nevertheless the Inite candidate, Michel Martelly, campaigning on a faster reconstruction effort, was elected president in April 2011.

Honduras

Honduras is slowly recovering from a coup that drew international condemnation

Land area: 112,000 sq. km.		
Population: 7 million—urban 49%		
Capital city: Tegucigalpa, 922,000		
People: Mestizo 90%, Amerindian 7%, black 2%, white 1%		
Language: Spanish, Amerindian dialects		
Religion: Roman Catholic 97%, Protestant 3%		
Government: Republic		
Life expectancy: 72 years		
GDP per capita: $PPP 3,796		
Currency: Lempira		
Major exports: Coffee, bananas		

Honduras occupies the centre of the Central American isthmus. Most of the land is mountainous—around three-quarters has slopes of 20 degrees or more. Lowland areas are confined to the narrow coastal strips and to the river valleys.

The majority of Hondurans are of mixed Indian-European extraction. There are some Indians who live mostly along the Guatemalan border and on the Caribbean coast and there are groups of Garifuna, who are descendants of the Carib Indians and black Africans.

Around half of Hondurans live below the national poverty line and the income distribution is very skewed: the top 20% of the population get 58% of the income, while the bottom 40% get only 11%. The rural population in particular have poor standards of education and health and there are increasing problems of TB and malaria. Around one-quarter of children are malnourished. The northern commercial capital, San Pedro Sula, has also become an epicentre for the AIDS epidemic in Latin America. Nationally, around 1% of adults are HIV-positive.

Agriculture employs around one-third of the workforce though only 14% of the total land area is suitable for arable farming. Half the population are in the rural areas struggling to make a living, cultivating corn and beans on the steep slopes. Agriculture is also responsible for more than half of export earnings. Nowadays the leading export crop is coffee, which is produced by around 105,000 independent producers—though the quality can be low.

The other export crops are grown in the lowland areas which have the most fertile land. The Caribbean coast in particular has one of the world's best climates for bananas.

Since the beginning of this century, the Honduran banana industry has been dominated by two corporations which are subsidiaries of US multinationals: Standard Fruit (selling under the Dole brand name) and the Tela Railroad Company (Chiquita). These companies have had a powerful influence on commercial and political life and employ around 15% of the workforce. There have also been efforts to diversify into other crops such as melons, pineapples, and mangoes.

Recent developments in shrimp farming have offered new export **Perilous diving for lobsters** markets along with lobsters, most of which come from 5,000 Miskito Indian divers who risk their lives for the catch.

Land distribution remains a critical issue. Pressure from peasant farmers in the 1960s and 1970s resulted in some land reform. In 1992, however, an Agrarian Modernization Law removed most of the grounds for automatic confiscation.

In the urban areas, one of the most important new sources of employment has been the 'maquila' factories which assemble goods for export. Around 130,000 people work in a number of export processing zones in the north, chiefly in garment manufacture using imported US cloth and taking advantage of well-developed port facilities around San Pedro Sula. Maquila output has grown strongly, and is the leading source of foreign exchange. Although the industry has diversified into other products such as car parts and furniture it is facing increased competition from Asia.

Tourism is also becoming an important source of foreign exchange. More than one million people come to the beaches, for example, or to the coral reefs or to visit historical Mayan sites. Another important foreign exchange contribution comes from Hondurans abroad whose remittances in 2008 were $2.8 billion.

More worrying types of visitor are the drug cartels who use Honduras as a transit point, adding to the violence from street gangs, the 'maras'.

Honduras has many of the same social divides as its neighbours, but managed to escape the civil wars that devastated other countries in Central America. Nevertheless, it was still deeply involved and heavily militarized, since the country offered military bases both for the 'contra' fighters from Nicaragua and also for the US forces—for which it was rewarded generously with military aid.

Politics in Honduras has long been a two-party struggle between liberals and conservatives—strongly influenced by US commercial and political interests. The centre-right Partido Liberal (PL), headed by Carlos Flores Facussé, won the 1997 presidential elections. But the 2001 elections saw a victory for the more conservative Partido Nacional (PN) with Ricardo Maduro Joest emerging as president.

The presidential election at the end of 2005 resulted in a victory for the PL candidate, Manuel Zelaya Rosales. Zelaya over-reached himself. In 2009 following plans for constitutional reform to enable him to serve another term, Zelaya was overthrown and expelled from the country in a military coup that drew international condemnation.

In November 2009, with Zelaya back, and holed up in the Brazilian embassy, a presidential election was won by the PN candidate Porfirio Lobo Sosa. He has US backing and seems to be a peacemaker who can help restore international recognition of the Honduras government.

Hungary

Hungary is one of the more successful of the former communist countries that have now joined the EU

Land area: 93,000 sq. km.	
Population: 10 million—urban 68%	
Capital city: Budapest, 1.7 million	
People: Hungarian 92%, Roma 2%, other 6%	
Language: Hungarian	
Religion: Roman Catholic 51%, Calvinist 16%, Lutheran 3%, other 30%	
Government: Republic	
Life expectancy: 73 years	
GDP per capita: $PPP 18,755	
Currency: Forint	
Major exports: Machinery, manufactured goods	

Hungary is divided into roughly two halves by the Danube, which cuts through the country from north to south. The area to the west is called Transdanubia. The land is predominantly flat but has two higher areas: the uplands of Transdanubia in the south-west and the loftier Northern Mountains along the border with Slovakia. Hungary's plains also come in two main parts. In the north-west of Transdanubia is the Little Hungarian Plain. To the east of the Danube the Great Hungarian Plain accounts for more than half the country.

Most of Hungary's people are related to Finns and Estonians and are ethnically distinct from their Slavic neighbours. The most significant minority population are 190,000 Roma, or gypsies, who are among the country's poorest people. Most Hungarians suffered a steep drop in living standards after 1990. Since then there has been a steady recovery. Education standards are high, but health is worse than might be expected: middle-aged men tend to have particularly poor health, due to bad diet, smoking, and high alcohol consumption. Hungary also has one of the world's highest suicide rates.

Around one-quarter of ethnic Hungarians live outside Hungary. As a result of the historical redrawing of boundaries there are around 1.6 million in Romania, half a million in Slovakia, and many others in Serbia and Ukraine. This has frequently been a source of friction: Hungary has protested to Slovakia in particular about discrimination against Hungarians, though Hungary tends to have better relations with Romania.

Industry employs around one-third of the workforce and their output of machinery, other equipment, and chemicals accounts for most exports. Production slumped when markets in Eastern Europe collapsed, but thanks to extensive foreign investment these have now been replaced by sales to Germany and Austria—mostly of component parts, particularly for the car industry. Hungary produces around one million of Audi's engines annually, for example. Many of the companies involved come from the EU, the US, and Japan, but one-quarter of sales are by the local firm, Raba. Other important industries include pharmaceuticals and chemicals and more recently electronic goods and computer software. Around 80% of all Hungary's exports go to other countries in the EU.

Even in the communist era, Hungary had been creating elements of a market economy. And the government has since carried out a fairly effective privatization programme. The private sector is now responsible for 80% of GDP.

With fertile soils and a helpful climate, Hungary can grow a wide range of crops. And though agriculture accounts for only 4% of GDP and employs 6% of the workforce, the country is generally self-sufficient in food. The government has converted collective farms to cooperatives.

Workers leaving agriculture and industry have mostly been absorbed into a mushrooming service sector which generates around two-thirds of GDP. Tourism plays an increasingly important part, generating up to 10% of GDP; expenditure per visitor is increasing. Unemployment has come down steadily, to around 8%.

Hungary's political orientation has been steadily westward. In 1999 it joined NATO and, after a favourable referendum in 2003, Hungary joined the European Union in May 2004 and hopes to adopt the euro in a few years.

Over the last decade the political scene has seen regular swings between left and right. By 1989, the Communist Party had itself become a leading advocate of reform. Renamed the Hungarian Socialist Party (MSZP), it had brought together the more liberal intellectuals, social democrats, and trade unions. Even so, it lost the 1990 election to the right-wing Hungarian Democratic Forum, which formed a coalition with two other conservative parties.

In the 1994 election, the MSZP came out ahead. Led by prime minister Gyula Horn, they entered into an alliance with the country's principal liberal party, the Alliance of Free Democrats (SZDSZ). But faced with an economic crisis they were forced to make cuts in welfare spending. And although foreign investment was creating new wealth the benefits were spread unevenly—concentrated around Budapest, leaving much of the east of the country behind.

Wealth is concentrated in Budapest

In the 1998 elections, the voters rejected the socialists. But this time they chose a different alternative—a coalition headed by Fidesz, which has a populist conservative stance.

The elections in 2002 produced yet another switch, bringing back the MSZP and the SZDSZ. The prime minister was Peter Medgyessy, though he was replaced in late 2004 by Ferenc Gyurcsany of the MSZP. The coalition won a second term in 2006.

In 2010, however, Fidesz, led by Victor Orban, returned to power with more than two-thirds of the seats. It was thus able to rewrite the country's rather messy constitution. In 2011 it duly did so, putting a strong emphasis on Christianity and defining marriage as the union of a man and a woman. The opposition, demanding a referendum, boycotted the vote.

Iceland

A rich economy crushed by a banking crisis

Despite its proximity to the Arctic Circle, the climate in most of Iceland is relatively mild. The landscape varies from Mount Hvannadals at 2,119 metres, which is on the edge of a vast glacier, through grassy lowlands to a complex coastline indented with fjords. Iceland is one of the most geologically active places on earth, and has regular minor earthquakes, numerous active volcanoes, and bubbling hot springs.

The country's isolated position has sustained a very homogenous population, but they are far from insular. In addition to Icelandic, most speak English, and education standards are high: Iceland publishes more books per person than any other country. They have also been among the richest people in the world and have a good system of welfare.

Much of the wealth has been based on fish. The fishing industry still employs around 6% of the workforce and accounts for 60% of exports. The main catches are of cod and capelin, though over-fishing has required the introduction of quotas. In 1975 Iceland imposed a 200-mile fishing zone—provoking the 'Cod War' with the UK—and since it also takes 15% of its catch from distant waters it frequently has disputes with other fishing nations.

Iceland's geological activity is also a major resource, providing

Land area: 103,000 sq. km.
Population: 0.3 million—urban 92%
Capital city: Reykjavik, 113,000
People: Icelandic 94%, others 6%
Language: Icelandic
Religion: Lutheran 81%, other 19%
Government: Republic
Life expectancy: 82 years
GDP per capita: $PPP 35,742
Currency: Krona
Major exports: Fish, aluminium

geothermal energy that heats more than 85% of homes. The fractured landscape also has numerous steeply falling rivers that have huge hydroelectric potential. This is being exploited for smelting aluminium and, with a cable to the UK, Iceland could even export electricity.

A major ongoing issue is that of EU membership. At times there has been a majority of the population in favour, but there are worries about the effects on the fishing industry.

Iceland has one of the world's oldest parliaments, the Althing. Since independence from Denmark in 1944, the system of proportional representation has never given one party an absolute majority and politics has usually been consensual.

After the 2007 election the Independence Party joined up with the centre-left Social Democratic Party.

Iceland's economy was plunged into recession from 2008. Three major banks had over-reached themselves with foreign transactions and collapsed. The government had to nationalize them, for which it needed a massive IMF loan.

The 2009 election produced a centre-left coalition of the Social Democratic Alliance and the Left-Green Movement with Johanna Sigurdardottir as prime minister.

India

The world's largest democracy has a rich and diverse culture. Now, it is also achieving more rapid economic growth

Land area: 3,288,000 sq. km.
Population: 1.2 billion—urban 30%
Capital city: New Delhi, 12.8 million
People: Indo-Aryan 72%, Dravidian 25%, Mongoloid and other 3%
Language: Hindi 41%, plus 14 other official languages and English
Religion: Hindu 81%, Muslim 12%, Christian 2%, Sikh 2%, other 3%
Government: Federal republic
Life expectancy: 63 years
GDP per capita: $PPP 2,753
Currency: Rupee
Major exports: Garments, gems and jewellery, cotton textiles, cereals

India's vast territory can be divided into three main regions from north to south. The far north and north-east of India cover part of the Himalayas including Kanchenjunga, the world's third-highest mountain. To the south, these mountains descend to the second region, a broad northern plain formed by the deposits of a number of major river systems, the most important of which is the Ganges, which emerges from the southern slopes of the Himalayas, flowing slowly south-east across the plain before spreading out into a broad delta in the Bay of Bengal.

To the west, the plain becomes a desert that extends into Pakistan. To the east, beyond the Ganges delta, India's territory circles round that of Bangladesh. The third main region, to the south of the plain beyond the Narmada River, is the Deccan plateau, a triangular region of low hills and extensive valleys, bounded along the coasts by two mountain systems: the eastern and western Ghats.

India now has over one billion people and on present trends will be the world's largest nation by 2030. The population can be subdivided in many different ways. The broadest distinction is in terms of religion. India's constitution declares it to be a secular state, but religious intolerance, known here as 'communalism', is never far below the surface. Hindus now account for over 80% of the population and Muslims 12%, and there has always been tension between the two. In 2002, for example, communal riots in Gujarat killed 900 people. The other main source of friction has been with the Sikh community concentrated in the northern state of Punjab. Christians too have come under attack.

Beyond religious divisions, India is also a source of vast cultural and linguistic diversity. The national language, Hindi, is widely spoken in the north but elsewhere it is displaced by dozens of other languages, from Malayalam to Tamil to Bengali, many of which use different scripts.

Another way of dividing India is by caste—a hereditary social system that has Brahmins at the top and the 'untouchables', who perform the most menial social tasks, at the bottom. The latter, who are now referred to as 'scheduled castes', have specific allocations or 'reservations' of government jobs and parliamentary seats. Similar positive discrimination is exercised in favour of 'scheduled tribes'— a smaller disadvantaged group of tribal communities.

In terms of average income, India is still very poor. But the population is so large that the top 10% of the population would on its own constitute one of the world's largest industrial nations. Another significant group are the 16 million or so Indians who now live abroad, whether as permanent residents or contract workers in the Gulf and elsewhere.

At home, around one-third of the population live below the poverty line and around 40% are illiterate. But there are striking regional contrasts.

India has striking regional contrasts Thus, one of the wealthier states, Punjab, has a per capita income twice that of West Bengal. And the state of Kerala, with many enlightened social policies, has a 90% literacy rate.

In India it is also important to consider gender. In most states women fare worse than men. Around half of women are illiterate, and women's and girls' health and nutrition are frequently neglected. As a result of this, and selective abortion, there are only 933 women per 1,000 men.

Two-thirds of India's people are still to be found in the rural areas and rely on agriculture for survival. Land is scarce. By the mid-1980s, the average farm size was down to 1.7 hectares and today one-third of rural households have no land at all. Despite the shortage of land, India's food production, particularly on the larger landholdings, has, thanks to the 'Green Revolution', largely kept pace with population growth. Nevertheless, millions still go hungry because they cannot afford to buy the food; one-fifth of the population are undernourished.

India is one of the world's major industrial powers. For the first 40 years after independence, industrial policy was based on state-owned heavy industry, combined with private production of consumer goods which was protected from foreign competition by high tariffs. At the same time, entrepreneurs had to fight their way through thickets of regulations and licences and negotiate with often corrupt officials.

All this began to change from 1991 when the then finance minister, Manmohan Singh, took the first steps by cutting tariffs, liberalizing imports, and encouraging foreign investment. Liberalization has a long way to go. The state still employs around 70% of formal-sector workers who are deeply resistant to change. But India is now much more open to the outside world. And economic growth, which between 1960 and 1990 had averaged only 4% per year (known as the 'Hindu rate of growth'), has at times touched 8%.

All industries, including textiles, steel, and petrochemicals, have been affected by liberalization, but some of the most visible changes have been in areas such as cars and other consumer goods. There has also been a boom in consumer electronics and computers. The latter can take advantage of India's large pool of computer programmers, especially in Bangalore, India's 'cybercity'.

India is a nursery for software talent

India's English-speaking graduates also staff many call centres for companies in the USA and Europe and perform many other 'outsourced' functions. Software and computer services now account for around one-quarter of exports.

These developments have been transforming the face of India's cities. TV programming has now been opened up to a multitude of satellite channels. Life in the rural areas is slower to change. Villagers may also have satellite TV but little prospect of buying the wares on offer.

India can rightly be proud of having maintained a democratic government for almost the entire period since independence. This is based on a federal constitution with 29 states and 6 union territories. Each state has its own chief minister and regional assembly. Historically, the states have had relatively little power. Nevertheless, Kerala and West Bengal have had communist governments with distinctive policies. In 2009 the government agreed to create a new state of Telangana split off from Andhra Pradesh.

The most important foreign policy issue concerns relations with the old enemy, Pakistan. The main problem is Kashmir, a territory whose status the two countries have disputed since the partition of India in 1947. Up to 100,000 people have died in cross-border fighting. The enmity has often

escalated though the countries have been moving towards reconciliation.

India's federal government consists of a lower house, the Lok Sabha, which elects the prime minister, and the upper house, the Rajya Sabha, whose members are elected by the regional assemblies.

For most of the first 50 years the government was the domain of the Congress Party, which was usually led by the Nehru dynasty; first by the leader of the independence struggle, Jawaharlal Nehru, later by his daughter Indira Gandhi (assassinated in 1984 by Sikh extremists), and then by her son Rajiv (assassinated in 1991 by a Tamil extremist).

Congress's almost uninterrupted rule ended with the election in 1996 that gave the majority of seats to the right-wing Hindu nationalist Bharatiya Janata Party (BJP) led by Atal Behari Vajpayee who headed a series of coalition governments until 2004.

While India's economy seemed to be healthy under the BJP, most progress was in the cities, leaving the rural areas behind. The Congress Party, now led by Rajiv's Italian-born widow, Sonia, capitalized on this discontent and in the elections in 2004 took the largest share of the vote. However, without a majority in the Lok Sabha, Congress had to form a coalition, the United Progressive Alliance, with a number of smaller parties. Sonia Gandhi refused to become prime minister, instead nominating elder statesman Manmohan Singh. The coalition was re-elected in 2009.

Domestically the most violent opposition comes from the Maoist 'Naxalites' in more remote areas.

Indonesia

Indonesia, the world's largest Muslim country and one of the largest democracies, is making steady progress

Land area: 1,905,000 sq. km.		
Population: 224 million—urban 53%		
Capital city: Jakarta, 8.4 million		
People: Javanese 41%, Sundanese 15%, Madurese 3%, other 41%		
Language: Bahasa Indonesia, English, local languages		
Religion: Muslim 88%, Protestant 5%, Roman Catholic 3%, Hindu 2%, Buddhist 1%, other 1%		
Government: Republic		
Life expectancy: 71 years		
GDP per capita: $PPP 3,712		
Currency: Rupiah		
Major exports: Oil, gas, plywood, garments, rubber		

Indonesia is one of the world's largest and geographically most dispersed countries, consisting of more than 17,000 islands. Of the southern chain of islands, the largest are, from west to east: Sumatra, Java, Bali, and Lombok.

To the north of this chain is the island of Borneo, the southern part of which consists of the Indonesian provinces of Kalimantan. To the east of Borneo is the island of Sulawesi, and furthest east are the Molukas and New Guinea, the western half of which is the Indonesian province of Papua. Most of these islands have volcanic mountains covered in dense tropical rainforests that descend to often swampy coastal plains.

Indonesia's people are as diverse as its topography. By some counts, there are more than 300 ethnic groups with almost as many languages. Many of these groups are related and most are of Malay origin. The largest is the Javanese, who live on the centre and east of Java—the western portion of which is occupied by the Sundanese.

Other large groups are the Madurese and the coastal Malays. The many smaller groups include tribal peoples, such as the Dayak, who inhabit the interior of Kalimantan, and people of Melanesian descent in Papua. In addition, Indonesia also has eight million people of Chinese origin who live mostly in the cities and allegedly control half the economy. Ethnic and religious differences have often led to violence particularly in the Malukus.

Indonesia has also suffered from natural disasters. In recent years these have included a tsunami that struck Sumatra in 2004, killing 233,000 people and levelled the capital of Aceh, and major earthquakes—in 2006 in Java and in 2009 in Sumatra.

Despite the large land area, Indonesia's population is fairly concentrated. Half the islands are uninhabited and around two-thirds of the population live on Java, Madura, and Bali. The majority of people profess to be Muslim, though religion has not so far been a dominant force in political life. The official language is Bahasa Indonesia, which is based on Malay. Only around 10% of people use this as their first language; most have it as their second.

Faced with a fast-growing population, the government in the 1970s and 1980s instituted a programme of family planning and the annual growth rate has slowed to around 1.4%. A more controversial demographic initiative was the 'transmigration' programme which moved people from densely settled areas like Java—which has around 60% of the population—to other islands. There has also been extensive labour migration—particularly to Malaysia, where over one million Indonesians work on plantations.

As a result of steady economic progress, Indonesia has seen a reduction in the poverty rate, to around 16%—though unemployment is still a problem with around 30% of the workforce unemployed or underemployed. And although standards of education and health have improved, many schools and health centres, particularly in the rural areas, are still in a poor condition.

Around half the population still make their living from agriculture. Only about one-tenth of the country is suitable for agriculture, but the rich volcanic soil on the major islands is very fertile. Farmers here, most of whom are smallholders, primarily grow rice in the lower areas, and fruit, vegetables, tobacco, or coffee on the higher slopes. By the mid-1980s, with heavy use of 'green revolution' techniques, they had made the country self-sufficient in rice, though imports subsequently resumed. The land on the outer islands is less fertile and is primarily used for tree crops such as oil-palm, rubber, and coconuts, as well as cocoa and coffee.

Forests cover around 60% of the land area and the timber industry has

Choking forest fires grown rapidly, mostly based on hardwoods such as teak and ebony. There has been an increase in illegal logging which has added to deforestation. In addition there are forest fires—a result of slash-and-burn cultivation as well as of companies clearing land for plantations. These fires can send a vast pall of smog across neighbouring countries.

Indonesia's rich mineral deposits have been a major source of income. State companies exploit these through production-sharing agreements with foreign investors. Among the major minerals are tin, bauxite, copper, nickel, gold, and silver.

But the most important mineral is oil, and Indonesia is also the world's largest producer of liquefied natural gas. Oil and gas account for around 20% of exports. The industry is controlled by the state company Pertamina which has production-sharing agreements with transnational oil companies. Nevertheless, oil output is low and falling, and Indonesia could become a net importer.

Indonesia has also developed a highly diversified manufacturing sector. This includes iron and steel, aluminium smelting, oil refining, cement, and, more recently, pulp and paper production. Manufacturing for export has concentrated more on labour-intensive items such as garments and footwear, though this is now facing low-wage competition from countries like China.

Indonesia's political system has in the past been authoritarian with a strong military influence. Between 1950 and 1967, following independence from the Dutch, the country was controlled by President Sukarno. During the mid-1960s his army chief, Mohamed Suharto, suppressed an allegedly communist **Suharto's** coup—and followed up **notorious** with a slaughter that cost **nepotism** up to one million lives. In 1968 Suharto himself was elected president and backed by his political group, Golkar, was subsequently re-elected unopposed for six further terms.

Suharto's 'New Order' administration reinvigorated the economy but seized much of the wealth for itself—a period characterized by 'corruption, collusion and nepotism'. Meanwhile the army brutally suppressed most opposition.

Suharto's reign ended in 1998 as a result of economic collapse. In Indonesia, as elsewhere in Asia, companies had over-borrowed, banks became insolvent, and thousands of companies crashed. Faced with rises in food prices many people rioted, and scapegoated Chinese businesses.

In May 1998, Suharto abruptly resigned, handing over power to his vice-president, B. J. Habibie. He made some crucial decisions. One was to offer the people of East Timor their independence. Another was to decentralize significant political authority to Indonesia's hundreds of districts—a move designed to ward

off further secessionist threats, not just from historically dissident provinces like Aceh and Papua but also from richer provinces like Riau which complained that income from oil was being siphoned off to Jakarta.

Habibie was replaced when the 1999 elections for the People's Consultative Assembly gave most seats to the Democratic Party of Struggle (PDI-P), led by Megawati Sukarnoputri—Sukarno's daughter. Golkar came second, followed by the Islamicist National Awakening Party (PKB). The assembly surprisingly chose as president Abdurrahman Wahid of the PKB—whose chaotic administration ended in 2001 with impeachment for incompetence.

Megawati took his place, but achieved little and was punished in the 2004 assembly elections when Golkar emerged as the largest party.

Over this period Indonesia suffered from Islamic terrorism, notably a 2002 bomb attack in Bali and another on the Jakarta Marriott Hotel in 2003. But the security forces now appear to have the upper hand.

The 2004 presidential election, the first to be based on a direct popular ballot, was won by a former general, Susilo Bambang Yudhoyono. He contributed to significant economic progress, so that Indonesia did not suffer too much in the global recession. He also attacked corruption. But one of his most notable achievements was to end a 29-year rebellion in Aceh by offering autonomy.

'SBY' was rewarded in 2009 when his Democratic Party won the most parliamentary seats and he was convincingly re-elected as president.

Iran

Conservative forces have retained their grip but the regime has been threatened by widespread protests.

Land area: 1,648,000 sq. km.
Population: 72 million—urban 69%
Capital city: Tehran, 7.7 million
People: Persian 51%, Azeri 24%, Gilaki and Mazandarani 8%, Kurd 7%, Arab 3%, Lur 2%, Baloch 2%, other 3%
Language: Persian and dialects 58%, Turkic and dialects 26%, Kurdish 9%, other 7%
Religion: Shia Muslim 89%, Sunni Muslim 10%, other 1%
Government: Islamic republic
Life expectancy: 71 years
GDP per capita: $PPP 10,955
Currency: Rial
Major exports: Oil and gas, carpets, fruit

Much of Iran consists of a high central plateau ringed by mountains, the highest of which are in the volcanic Elburz range in the north. Iran is not very fertile—more than half the territory is barren wasteland, most of which is salt desert and uninhabited. Only around 10% is suitable for arable farming, much of this in the fertile northern plateau. Another 20% can be used for grazing. Iran is vulnerable to severe earthquakes, the most recent of which in 2003 in the city of Bam killed 43,000 people.

Ethnically, Iran is quite diverse. Just over half the people are ethnic Persians, who are to be found throughout the country. But one-quarter are Azeris, who live in the north-west. The least assimilated minority are the Kurds, who make up 7% of the population and have a largely nomadic existence in the

mountains of the west. This diversity is also reflected in language: Persian (Farsi) is the official language but one-quarter of people use Turkic languages. Iran has also received millions of refugees, chiefly from Afghanistan and Iraq. On the other hand, around 200,000 Iranians leave each year, seeking a better life abroad.

Most people are Shia Muslims, but around one-tenth, including the Kurds, belong to the Sunni sect. In theory there is religious freedom, but in practice those who are not Shia Muslims are second-class citizens. Islam also places restrictions on women, particularly on their dress. But by the standards of other Islamic countries, Iran is relatively liberal— women have good access to education and can enter most professions.

Despite Iran's hostile terrain, agriculture accounts for about one-tenth of GDP and one-quarter of the workforce. The main crops are wheat, barley, and potatoes. But only one-quarter of the potential area is being cultivated and productivity is low—hampered by poor management and political uncertainties. Hopes of reaching food self-sufficiency were dashed by a series of droughts. Around one-third of agriculture is based on livestock, chiefly sheep and goats.

At the heart of Iran's industrial economy is oil. Iran has enough reserves for 75 years of extraction at current rates. In 2007 oil accounted for over 90% of export earnings. Reserves are in the south-west, and also offshore in the Persian Gulf. Iran also has the world's second largest reserves of natural gas but these remain under-exploited.

Iranian industry was nationalized after the 1979 revolution. Most is inefficient and runs far below capacity. Foreign investors have been discouraged by official coolness to foreigners, and also by a US ban on trade and foreign investment imposed since 1995. The most promising manufacturing area is petrochemicals, though steel and cement also offer prospects for diversification from oil.

Most Iranians struggle to survive. Unemployment is officially around 13%, but many professional people are on such low salaries that they need to take on two or three extra jobs. Corruption is widespread.

Political life in Iran was transformed by the 1979 revolution that swept aside the Pahlavi dynasty of the Shah and by 1981 had established a strict Islamic Republic which was to be run by 'religious jurists'. At the head of these was the formerly exiled Ayatollah Khomeini, supported by another cleric, Hashemi Rafsanjani, who in 1980 became the speaker of the Majlis, the 207-member parliament.

An Islamic republic run by religious jurists

The early years of the revolution were chaotic and turbulent and included an eight-year war with Iraq, 1980–88, and a persistent enmity with the 'Great Satan', the USA. Khomeini's death in 1989 left no obvious successor. Instead, political and religious functions were re-allocated. The role of rahbar, or religious leader, was taken by Sayed Ali Khamenei, while most political functions were vested in the president. Rafsanjani was elected president.

This set up a tension between the conservative religious leadership and the nominally secular presidency. Rafsanjani tried to open up the economy, with limited success. He was succeeded in 1997 by an even more progressive cleric, Muhammad Khatami. The parliamentary elections in February 2000 altered the picture again, giving a resounding majority to the reformers, but they made little progress against the conservatives.

Since 2004, the conservatives have been firmly in power. Prior to the 2004 parliamentary election the hardline Council of Guardians barred 2,000 reformist candidates, giving the conservative forces a majority. They repeated this in 2005 with the election as president of a more populist figure Mahmoud Ahmadinejad.

The elections in 2009, however, were to prove much more divisive. The opposition candidate Mir Hussein Mousavi claimed the election was stolen, and protestors took to the streets, resulting in dozens of deaths.

Internationally, the most contentious issue has been Iran's potential to build a nuclear bomb—which has led to a confrontation with the EU and the USA.

Iraq

Post-war Iraq descended into chaos but has now achieved a degree of political stability

Land area: *438,000 sq. km.*
Population: *30 million—urban 66%*
Capital city: *Baghdad, 6.3 million*
People: *Arab 75%–80%, Kurdish 15%–20%*
Language: *Arabic, Kurdish*
Religion: *Muslim 97%, other 3%*
Government: *Transitional democracy*
Life expectancy: *68 years*
GDP per capita: *$PPP 3,400*
Currency: *Iraqi dinar*
Major exports: *Oil*

The heart of Iraq, forming around one-third of the territory, is the alluvial and often marshy basin of the Tigris and Euphrates. These rivers flow the length of the country before uniting as the Shatt al-Arab River, which flows into the Persian Gulf. To the north of this basin are uplands, and in the north-east is the mountainous region of Kurdistan. The rest of the country, to the west and south, around two-fifths of the territory, is largely desert.

There are three main population groups. The largest, accounting for around half the population, are Arab Shia Muslims, who live mostly in the centre and south. Next are the Arab Sunni Muslims who are around one-third of the population and are concentrated in the east around Baghdad. Then there are the Kurds, also Sunni Muslims, who make up one-fifth of the population and live in the north and north-east. There are

also around three million exiled Iraqis.

The lives of most Iraqis have been devastated both by the former oppressive political regime, and by the 2003 war and its aftermath. Until the 1990s health standards were good, but then fell steeply, chiefly as a result of a lack of access to clean water and sanitation. Child malnutrition has come down somewhat but remains high. Schools were also in a poor condition though most have now been rebuilt.

Iraq's economy has long depended on the export of oil. Iraq has around 10% of world reserves, the third largest after Saudi Arabia and Iran, along with the world's tenth largest reserves of natural gas. But production has been low. Prior to the war, Iraq was unable to export even its UN-sanctioned quotas and now the industry faces widespread sabotage that deters investment.

Manufacturing remains limited. Before the war efforts had been made to diversify industry away from oil but there has been little progress. Many people are still on the payroll of defunct state-owned enterprises.

Iraq's farmers should have been less affected by the fighting. The rich alluvial soil with ample supplies of water ought to make the country very productive. Around one-quarter of the labour force are engaged in farming. In practice, many peasant farmers have reverted to subsistence production. Production has revived lately but around half of the country's food needs still have to be imported and one-quarter of the population rely on the public food distribution system. Up to half the workforce are probably underemployed.

For 25 years political life in Iraq was dominated by Saddam Hussein who came to power in a coup in 1979. He summarily executed his rivals and subsequently dealt ruthlessly with any opposition. Saddam was politically cunning but he also made disastrous mistakes—notably the war with Iran over the period 1980–88, and the 1990 invasion of Kuwait that provoked intervention and defeat by a US-led coalition in the Gulf War.

Saddam was cunning but made disastrous mistakes

Following that war there were a number of attempts to topple Saddam but uprisings by Kurds in the north and the Shia in the south were easily crushed. The international community made efforts to monitor his weapons production using the UN inspection force UNSCOM. But Saddam denied access to most sites and, after the team was withdrawn in 1998, the USA and the UK responded with bombing raids.

In 2002 US President George Bush referred to Iraq as part of a global 'axis of evil' and made it clear he wanted to get rid of Saddam. Attempts to get UN backing for military action failed so in March 2003 the USA started to bomb Baghdad following which US and British forces invaded

from the south. Within three weeks the regime had collapsed.

The US established a Coalition Provisional Authority subsequently and appointed a 25-member Governing Council. It also had more military successes, notably the killing of Saddam's sons and the capture of Saddam himself in 2003 (he was executed in 2006).

In June 2004, the CPA handed over power to a transitional administration. Despite continuing violence, a general election took place in January 2005. This resulted in a Shia-dominated transitional government headed by prime minister Ibrahim Jaafari.

The first regular parliamentary elections were held in December 2005, again producing a Shia majority. After six months of wrangling a government of national unity finally emerged under a new prime minister, Nuri al-Maliki of the main Shia block.

The violence has subsided somewhat since 2007 and although the security position remains fragile, with frequent suicide bombings, the insurgents probably lack sufficient support to overthrow the government.

The USA, which has been training the Iraqi army, has steadily been withdrawing its combat forces and is aiming for a complete withdrawal at the end of 2011. Most British forces have already left.

Elections in March 2010 eventually resulted in a coalition between Maliki's State of Law coalition, the (mostly Shia) Iraqi National Alliance, the Kurdistan Alliance and the Iraqi National Movement. Maliki remained prime minister. However, the coalition is divided on a number of issues and some members could leave.

Ireland

After decades of prosperity Ireland's economy crashed following the global financial crisis

0	Miles	200
0	Km	320

Land area: 70,000 sq. km.
Population: 4 million—urban 62%
Capital city: Dublin, 506,000
People: Irish
Language: English, Gaelic
Religion: Roman Catholic 88%, other 12%
Government: Republic
Life expectancy: 80 years
GDP per capita: $PPP 44,613
Currency: Euro
Major exports: Machinery, chemicals, food

The Republic of Ireland occupies 80% of the island of Ireland. The remainder, Northern Ireland, is a part of the UK. The interior is largely lowlands interspersed with numerous lakes and bogs. Ringing the lowlands, around most of the coast, are low mountain ranges—the highest peak is only 1,040 metres above sea level.

Ireland's people are largely of Celtic descent and Roman Catholic, though there are also small numbers of Anglo-Irish Protestants, a legacy of the period of English rule. Irish, or Gaelic, is the first official language but only a tiny minority use it; English is spoken almost universally.

Until recently, Ireland was a country of emigration, chiefly to the UK and the USA. More than one million Irish-born people live abroad—around 40 million Americans claim Irish ancestry. But recent years

of rapid development have encouraged some of the diaspora to return—and attracted many other nationalities. In 2007, net immigration was 67,000.

Agriculture remains a significant, though declining, part of the economy. Much of this is based on the rearing of livestock for milk and beef which, along with the export-oriented food processing industry, still employs around one-fifth of the workforce. Ireland's agriculture benefited greatly from EU subsidies, though these have now declined.

Irish workers now have plenty of alternatives to farming. In the past decade manufacturing and service industries have been booming. Tourism is also a major earner. The eight million visitors each year employ 8% of the workforce.

The Irish government can take credit for the industrial transformation. It offered generous grants and tax concessions while restraining wage claims through 'social contracts' with the trade unions. It has also invested heavily in higher education.

Attracted by financial incentives and an inexpensive, well-educated, English-speaking population within the European Union, foreign companies understandably flocked. Ireland now has affiliates of more than 1,000 foreign enterprises, including major computer companies such as IBM, Intel, Fujitsu, and Dell. Computer equipment along with pharmaceuticals, chemicals and electrical equipment now account for the bulk of exports.

This stimulated rapid economic growth—turning Ireland into an 'emerald tiger' with one of Europe's most rapid growth rates. But the wealth was not evenly spread: property prices in Dublin rocketed while the west of the country was left behind. Moreover, this unbalanced growth inflated a credit and property-based bubble which was suddenly deflated by the global financial crisis. In 2008 the economy sank into one of Europe's deepest recessions.

Ireland has also been going through rapid social and political changes. Under the influence of the Roman Catholic Church, Ireland

The Catholic Church's influence has waned

has been very conservative in such matters as abortion, homosexuality, and divorce. But the Church's influence has waned and divorce finally became legal in 1997.

At the same time, the political landscape was transformed by the settlement of the long-running conflict in Northern Ireland. The island was partitioned by the 1921 Anglo-Irish Treaty and the south became independent in 1922. For decades, Irish politics was shaped by attitudes towards the north. The largest party, the nationalist Fianna Fail, was formed in opposition to the treaty, insisting on Irish jurisdiction over the whole island—the name means 'militia of Fál', a stone monument. The second largest is the centre-right

Fine Gael ('Irish nation'), which represented landowners and the business community who accepted the terms of the treaty. Irish politics did not therefore split along right–left lines, but according to attitudes to partition. There is also a Labour Party, but it has always been much smaller.

In recent years, electoral outcomes have usually required coalitions of one of the major parties with one of the smaller ones. After the 1997 election for the parliament, the Dáil, a Fine Gael–Labour coalition gave way to one between Fianna Fail and the Progressive Democratic Party, which was a 1985 right-wing breakaway from Fianna Fail. Bertie Ahern became Taoiseach—prime minister. Mary McAleese took the largely ceremonial post of president.

In 1998 Ahern signed the 'Good Friday' Northern Ireland peace agreement with the UK. This required Ireland to remove from its constitution a claim on Northern Ireland. In 2005 the Provisional Irish Republican Army decommissioned its weapons.

The Fianna Fail and the Progressive Democratic Party coalition retained power after the elections in 2002 and 2007 with Ahern again as Taoiseach, the first time a government had been returned in over three decades. Nevertheless Sinn Fein, the political wing of the Irish Republican Army, and the Green Party, also made gains.

By the 2011 election, however, voters had lost faith in government that was now carrying out painful budget cuts. Instead, Fine Gael was returned as the largest party and entered into a coalition with Labour with Enda Kenny as prime minister.

Israel

After half a century of war and hostility, peace with the Palestinians seems as remote as ever

Land area: 21,000 sq. km.
Population: 7 million—urban 92%
Capital city: Jerusalem, 746,000
People: Jewish 76%, others, mostly Arab, 24%
Language: Hebrew, Arabic, English
Religion: Judaism 76%, Islam 16%, Christian 2%, other 6%
Government: Republic
Life expectancy: 81 years
GDP per capita: $PPP 26,315
Currency: Shekel
Major exports: Industrial goods, cut diamonds

Israel can be considered to have four main geographical regions. To the north is a hilly region that includes the hills of Galilee and extends down through the Israeli-occupied West Bank. Along the Mediterranean coast there is a narrow plain that is home to most of the country's commerce and population. The border in the north-east is formed by the valley of the river Jordan which flows south into the Dead Sea—the lowest point on earth—and the same fissure continues south to the Gulf of Aqaba. Most of the south is the Negev Desert.

The state of Israel was established in what was formerly Palestine in 1948 as a Jewish homeland and has since attracted immigrants from almost every country. Now the population is three-quarters Jewish, of whom more than half are native-born and the remainder immigrants.

Immigration accelerated after 1989 with the arrival of more than 800,000 Jews from the former Soviet Union. Around one-quarter are descendants of the original Arab population—who are typically the poorest people. In addition there are hundreds of thousands of temporary migrant workers from Eastern Europe and Asia.

The Jewish population is usually divided into two groups. First there are those of European and North American origin—the Ashkenazim—who tend to be richer. Second are those from North Africa and the Middle East—the Sephardim or Misrahi—who form most of the working class. But there are many other distinctions, principally between secular and religious Jews.

Israelis have generally good standards of health and are also highly educated. About one-fifth of immigrants are professionals. Israel has the highest proportion of scientists and engineers per capita in the world.

Over its lifetime, Israel has shifted from a simple agricultural society to an export-driven high-tech industrial economy—particularly during the 1990s, with the influx of new professionals and the establishment of communications technology and software enterprises concentrated around Tel Aviv.

Israel also has high-tech agriculture which has to make sophisticated use of limited water resources. As well as being self-sufficient in food, Israel is a major exporter of citrus and other fruits. Most farming,

Israel has high-tech farms that work with little water
reflecting the country's socialist origins, is still organized by kibbutzim (collectives) or moshavim (cooperatives).

Although fairly efficient, their survival still depends on cheap supplies of water and subsidies and many are now heavily in debt.

Israel's Zionist and socialist origins have also influenced its politics, which have largely been dominated by the Labour Party—though now more a liberal party. Labour was for decades the majority ruling party, but was dislodged in 1977 by the conservative Likud. Since the mid-1980s, neither has had a majority and each has had to govern in coalition with a plethora of smaller religious parties.

Throughout its existence, Israel's governments have been embroiled in conflicts with their Arab neighbours, most notably in the six-day war of 1967 as a result of which Israel seized the Gaza Strip from Egypt, the West Bank from Jordan, and the Golan Heights from Syria. The status of these territories has been disputed ever since. Generally, Likud and the orthodox religious parties have wanted to keep them—and encouraged Jewish settlements there—while Labour has been more equivocal.

In the longer term it seems likely that the West Bank and Gaza will constitute an independent Palestine. Major steps towards this were taken in 1993 when, following secret negotiations in Oslo, Labour prime minister Yitzhak Rabin signed a peace agreement with the Palestine Liberation Organization (PLO) that allowed for Palestinian self-rule. Rabin was assassinated in 1995.

The peace process faltered after 1996 when Likud returned to power, though it seemed to come back on track following the 1999 direct election for prime minister, which was won by the moderate Ehud Barak of the One Israel party. Barak met with the PLO at Camp David in July 2000 but the talks stalled and September 2000 saw the start of a second 'intifada'—uprising—by the Palestinians. Hopes of peace faded further with the election in February 2001 of hard-liner Ariel Sharon of Likud as prime minister in coalition with Labour, who started to build a high security fence along the border.

In January 2003, a general election allowed Likud to make many gains. Sharon then secured the withdrawal of settlers from Gaza, and founded a new party, Kadima, but he suffered a massive stroke in January 2006.

He was replaced by Ehud Olmert who led Kadima for the election in 2006, emerging at the head of a new broad coalition. He launched an unsuccessful attack on Hizbullah in southern Lebanon. In 2008, Olmert had to step down following allegations of corruption. This led to a fresh election in 2009 and a new right-wing administration led by Binyamin Netanyahu of Likud.

Italy

Despite regional divisions and a series of unstable governments, Italy remains an important player in Europe

Land area: *301,000 sq. km.*
Population: *59 million—urban 68%*
Capital city: *Rome, 2.7 million*
People: *Italians*
Language: *Italian and some regional*
Religion: *Roman Catholic 90%, other 10%*
Government: *Republic*
Life expectancy: *81 years*
GDP per capita: *$PPP 30,353*
Currency: *Euro*
Major exports: *Machinery, textiles, clothing, cars*

Italy is dominated by mountains. The Alps loom over the north of the country and merge with the Apennine Mountains, which run down the centre. Mountains also make up much of the Italian islands of Sicily and Sardinia. Overall, more than one-third of Italy's territory is covered by ranges higher than 700 metres.

The difficult terrain is one reason why the country has been regionally divided. Italy was united as one country only in 1871, at which point only 3% of the population spoke Italian. Since then there has been a steady process of integration, but strong regional identities persist.

For much of the 20th century, Italy's population grew rapidly, and with limited prospects of work at home, millions of Italians emigrated: even as late as 1968–70 net emigration was over 250,000. Now

the position is reversed. The fertility rate, at 1.3 children per woman, is among the lowest in the world.

Italy has become a country of net immigration: currently it has about two million immigrants, most of whom have come from outside the EU, chiefly from Morocco, former Yugoslavia, Tunisia, and Albania.

Thin soils and steep terrain have always made agriculture difficult, and as a result Italy has been a net importer of most kinds of food, except for fruit and vegetables. It is also poorly endowed with other natural resources: it has no coal and has to import 75% of its energy needs.

In economic terms, the country's great strength has been in manufacturing which accounts for around one-quarter of GDP. It does have some large companies, notably Fiat, Pirelli, and also Fininvest which is controlled by the family of prime minister Silvio Berlusconi. But manufacturing is dominated by networks of thousands of small firms, chiefly in clothing, furniture, kitchen equipment, and 'white goods' such as refrigerators and cookers. These often cluster together regionally: wool textiles in Prato, for example, shoes in Verona, spectacles in Belluno.

These small and medium-sized

enterprises have been the engine for economic growth. But in recent years they have not been able to expand sufficiently and Italy has been falling behind other euro-based economies. Moreover, much of this activity has, been in the underground economy which is thought to control one-quarter of GDP.

Production is concentrated in the north. Here manufacturing accounts for 30% of value-added, compared with 13% in the south. Indeed, the GDP of the south is only 70% of the Italian average.

There are also striking regional differences in unemployment—3% in the north but 11% or more in the south. The impact of unemployment

Many Italian men still live with their mothers

is also cushioned by the extended family which remains stronger in Italy than in most other industrial countries. Around two-thirds of unmarried men under 30 still live with their mothers. And when they marry they do not move very far.

Tourism also makes an important economic contribution, with 41 million visitors per year.

Italy's economic progress has been hampered by its system of governance. Until 2001 the post-war period saw a sequence of weak and transient administrations: between 1947 and mid-2000 it had 58 governments. This weakness was to some extent deliberate. In an effort to avoid repeating the experience of the fascist governments of the 1930s and 1940s, the constitution adopted in 1948 resulted in reduced powers for the prime ministers—who were often beholden to party leaders who

were not themselves in government. Governance was also undermined by widespread corruption and the power of the Mafia.

Between 1947 and 1992, all coalition governments were dominated by the Christian Democrats. After 1993, the political scene shifted dramatically when Italians moved from a system of proportional representation to one in which most seats in both houses of parliament would be determined by straight majorities. This produced a different set of alignments. The 1996 election was won by the centre-left 'Olive Tree' coalition and Romano Prodi took over as prime minister. Prodi's government fell in 1998 following a dispute between the coalition partners, and Massimo D'Alema, a former communist, took over at the head of yet another coalition in 1999, but was replaced by Giuliano Amato.

The 2001 election saw a victory for Silvio Berlusconi of the right-wing Forza Italia who formed a centre-right coalition. Berlusconi is a multimillionaire who owns, among other things, three TV channels. He has been one of Italy's most determined leaders and survived various corruption cases launched against him and his business interests.

In the general election of May 2005, Berlusconi lost to the centre-left Unione coalition, led by a returning Romano Prodi.

But Berlusconi returned to power after elections in 2008 at the head of another centre-right coalition. In 2009 he was assailed by allegations of sexual impropriety, but with a large majority and a divided opposition he seems likely to survive.

Jamaica

Culturally vigorous and a tourist attraction—but prone to gang violence

Land area: 11,000 sq. km.
Population: 3 million—urban 54%
Capital city: Kingston, 658,000
People: Black 91%, mixed 7%, East Indian 1%, other 1%
Language: English, Creole
Religion: Christian 65%, other 35%
Government: Parliamentary, with the British monarchy
Life expectancy: 72 years
GDP per capita: $PPP 6,079
Currency: Jamaican dollar
Major exports: Alumina, bauxite, sugar

Jamaica's mountainous, and often thickly wooded, territory rises to 2,200 metres at the peak of the Blue Mountain in the east—a landscape criss-crossed by rivers and attractive waterfalls. In the centre of the island, and accounting for around half the country, is a limestone plateau that includes extensive systems of underground caves. The most densely cultivated parts of the country, however, are the coastal lowlands of the south and west.

Jamaica's people are predominantly descendants of slaves brought to work on the sugar plantations, along with smaller communities of Asian and European origin. While officially speaking English and mostly professing Christianity, Jamaicans have developed their own characteristic culture, notably their distinctive dialect, the Rastafarian religion, and reggae music.

Jamaica has achieved reasonable levels of human development, but improvements are being hampered by limited budgets. Illiteracy is emerging as a result of a lack of investment in primary schooling. Health standards are good, though a formerly free health service is now charging.

Limited economic opportunities have driven successive generations of young Jamaicans to emigrate—in the 1960s and 1970s to the UK, but more recently to the USA and Canada.

Around one-fifth of the population still make their living from the land. Apart from food crops such as plantains, yams, and corn, Jamaica also has major agricultural exports, notably sugar on plantations in the coastal plain, bananas grown on small farms, and coffee in the highlands. But agriculture in general has been stagnating—hit by droughts, floods, and hurricanes. One of the more consistent crops has been ganja, or marijuana, which provides solace to about half the male population.

Jamaica's most dependable export earner has been bauxite—of which it is the world's fourth largest producer. Most deposits are in the centre of the island and because they are not very deep they are relatively easy to extract. Some is exported as ore, but most is now first converted to alumina (aluminium oxide). With around half of export earnings from this source, Jamaica is vulnerable to world prices, particularly because it is a relatively high-cost producer.

Jamaica has also been trying to develop exports of manufactured goods but has had little success, particularly in garments, as a result of foreign competition.

The other main source of foreign exchange and employment is tourism. Jamaica attracts more than 3 million visitors each year, most of whom come from the USA and head for the

Tourism employs 250,000 Jamaicans

beautiful beaches of the north and west. Directly and indirectly, tourism probably employs around a quarter of a million Jamaicans. The industry grew rapidly in the 1980s and early 1990s, but stalled in the late 1990s, perhaps due to the relatively high rate of exchange—as well as the country's reputation for street violence. But it seems to be reviving with the construction of new Spanish-owned hotels.

Another source of export income is drugs. Some of this is marijuana, but the greatest profits come from transhipping Colombian cocaine. The trade is though to be the equivalent of 7% of GDP.

Politics in Jamaica has frequently been a dangerous business, involving armed confrontations between the two main parties. The violence has its roots in the mid-1970s. Between 1972 and 1982, the prime minister was former union organizer Michael Manley who led the People's National

Party (PNP). Manley tried to pursue socialist policies and gain greater control over the foreign-dominated bauxite industry, and became a leading international figure. Opposing him was the pro-business, anti-communist, Edward Seaga, who led the Jamaican Labour Party (JLP). But their left–right confrontations spread far beyond the parliament to become a kind of tribal warfare with pitched battles between gangs in the streets of Kingston.

Manley could not deliver the economic prosperity he had promised and Seaga won convincing victories in the 1980 and 1983 elections, pursuing more market-friendly policies and opening the country up to foreign investment. But with low prices for bauxite, economic development was slow and Seaga's popularity waned.

In 1989, the PNP and Manley returned to power, though continuing the same liberalization policies as the JLP. Following Manley's retirement in 1992, Percival Patterson took over as prime minister and won convincing majorities in 1993 and 1997. In October 2002 he led the PNP to a fourth successive victory, but by a much narrower margin. When Patterson retired in 2006 he was replaced as prime minister by one of his ministers, Portia Simpson-Miller.

In 2005 Seaga stepped down from the JLP to be replaced by Bruce Golding who led the JLP to victory in the 2007 general election – after 18 years in opposition.

Politics may be less ideologically charged nowadays, but the gangs live on, thriving on income from drugs and extortion. Jamaica still has one of the world's highest murder rates.

Japan

Japan was already looking economically frail. Now it has suffered a devastating earthquake and tsunami

Though the Japanese archipelago has more than 1,000 islands, most of the territory comprises four main islands. From north to south these are: Hokkaido, Honshu, Shikoku, and Kyushu, of which the largest, with more than half Japan's land area, is Honshu. These islands have mountain chains running north to south divided by steep valleys and interspersed with numerous small plains. The mountains include both extinct and active volcanoes, of which the highest, at 3,700 metres, is the distinctively symmetrical and snow-capped Mount Fuji on Honshu.

This is one of the world's most unstable geological zones and Japan experiences more than 1,000 tremors per year. In March 2011 the largest-ever recorded earthquake in Japan triggered a massive tsuami, killing around 18,000 people, and damaging the Fukushima nuclear power plant.

Japan's climate runs from the sub-arctic to the sub-tropical but most of the country enjoys fairly mild temperatures with plentiful rain that supports lush vegetation.

Japan has one of the world's most homogenous populations. This is partly due to isolationism, since for centuries Japan effectively cut itself off from the rest of the world. There is one small indigenous ethnic group, around 24,000 Ainu people on Hokkaido, but the largest group of non-Japanese are the 690,000 Koreans, many of whom are descendants of people brought to Japan during the Second World War to work as forced labourers. During the post-war boom years, however, Japan was the only developed country not to rely extensively on foreign workers to meet labour shortages. Although immigrants arrived from elsewhere in Asia, they often did so unauthorized.

The only group actively welcomed were the 'nikkeijin', the descendants of a previous generation of Japanese emigrants. From the early 1900s, many Japanese had emigrated to the Americas, and in the 1990s thousands of their descendants came back—around 230,000 from Brazil.

Japan's rapid economic development has been accompanied by striking social changes. With a healthy diet and a good health service average life expectancy, at 83 years, is among the highest in the world. Education standards are high—though learning by rote tends to stifle creativity. Japanese houses are well equipped but cramped, typically with less than half the floor area found in other rich countries. Many workers commute long distances.

Japan's population is ageing rapidly. There are now only 1.3 children born per woman of childbearing age—well below the replacement rate. One-fifth of the population is now over sixty-five. With a steady breakdown of the extended family, this has serious implications for the country's pensions system. It has also contributed to a steady rise in inequality in what was formerly a very egalitarian society.

Salarymen replaced by freeters

Social security has already come under pressure as economic crisis has forced many large companies to abandon their former commitment to loyal employees, the 'salarymen', whom they had offered jobs for life. As a result of corporate downsizing and the recession, unemployment in 2009 was around 6%, alarmingly high by Japanese standards. In addition there are now more than four million 'freeters', young people who only work part time.

Japan's development surged after the Second World War—boosted by demand from the Korean War and by high levels of savings that could be channelled into heavy industries

such as steel and chemicals, and later into the provision of consumer goods and services. This process was based on 'keiretsus'—tightly knit groups of manufacturers, suppliers, and distributors that work closely with the banks. The Ministry of Trade and Industry also had a pivotal influence. This, combined with protectionism, helped to trigger two major economic booms: 1965–70 and 1986–91.

Manufacturing still accounts for around one-fifth of GDP, and cars and semiconductors are the leading exports. But rising labour costs, an appreciating currency, and the need to operate closer to overseas markets have long obliged Japanese companies to invest in other countries. By the mid-1990s around 10% of manufacturing was located overseas.

Japan's manufacturing success has been all the more striking since it has few mineral resources and has to import more than 80% of its energy needs. This has meant exploiting nuclear energy—Japan has more than 50 nuclear power plants.

Agriculture employs relatively few people—less than 5% of the labour force. But unlike in many other rich countries, they work predominantly on small family farms, chiefly growing rice. This is the legacy of an extensive post-war land reform imposed by the USA. Since then, farmers have remained a politically powerful lobby and have benefited from protection and high subsidies. In recent years the government has deregulated the system somewhat, but Japan still has highly protected food markets.

Japan's economic expansion began to falter after 1991 when the escalating property prices and booming stock market could not be sustained. This exposed the weaknesses of the keiretsu system. Many banks had continued to lend to unsound companies. Companies were reluctant to restructure to reduce massive over-capacity and the government was slow to force banks into liquidation. By 1998, the economy had slipped into recession. Growth revived somewhat over the next ten years but sank again in 2008.

Japan has found it difficult to take tough decisions because its political system is neither as robust nor transparent as might be expected in one of the world's most advanced countries. For most of the period since 1955 Japan has been ruled by the conservative Liberal Democratic Party (LDP). The LDP has always had close links with both the business community and the farmers—each of whom allied themselves with different LDP factions. For a long time the main opposition party was the centre-left party now called the Social Democratic Party (SDP).

The LDP continued in office despite a series of scandals that

A weak and scandal-hit government

included the prosecution of one prime minister for taking bribes in 1976 from the Lockheed corporation, and the Recruit scandal in 1989, when almost all the LDP factions were found to have received bribes from that Japanese conglomerate.

Throughout this period, it had seemed that the Japanese electorate would continue to forgive a party that had delivered such a long period of prosperity. But the LDP's rule came to an end in 1993. A series of defections

caused it to lose a no-confidence motion in the Diet—the lower house of parliament. It was replaced by a coalition that included the SDP and others. But this too collapsed in 1994 and the LDP returned—this time in a coalition that included the SDP.

For the rest of the 1990s Japanese politics was in a state of flux—delivering 10 prime ministers between 1993 and 2001, though with the LDP as the leading coalition partner.

Meanwhile the opposition parties went through numerous upheavals, changing names and policies with bewildering rapidity. The main opposition party, following the elections of 2003, was the Democratic Party of Japan (DPJ), first formed in 1996 and which in 2003 merged with the Liberal Party.

Japan's next LDP leader, who became prime minister in 2001, was Junichiro Koizumi. A relatively youthful figure he made a number of successful reforms. In 2006, he called an election in which he scored a resounding victory. But left office that year and was replaced by finance minister Shinzo Abe. Abe lasted only about a year and was replaced by Yasuo Fukuda.

Finally in 2009 the LDP was ejected, following a landslide election victory for the DPJ led by Yukio Hatoyama who formed a coalition with the SDP and the People's New Party. Hatoyama proved ineffectual and was ousted in June 2010 to be replaced by Naoto Kan. He did not seem destined to last very long either, until the task of recovering from the 2011 earthquake and the resulting nuclear crisis reduced the appetite for more political change.

Jordan

Jordan's King Abdullah has started to establish a more stable democracy

Land area: *89,000 sq. km.*
Population: *6 million—urban 79%*
Capital city: *Amman, 2.2 million*
People: *Arab*
Language: *Arabic 98%, other 2%*
Religion: *Sunni Muslim 92%, Christian 6%, other 2%*
Government: *Constitutional monarchy*
Life expectancy: *72 years*
GDP per capita: *$PPP 4,901*
Currency: *Dinar*
Major exports: *Phosphates, potash*

Jordan has three main zones. First, there is the Jordan Valley region, which lies to the east of the Jordan river and finishes in the south at the Red Sea, which is the world's lowest body of water. Second, further to the east and running parallel with the valley, are the eastern uplands, whose elevation is mostly 600 to 900 metres. Thirdly, beyond the uplands, and occupying around 80% of the country to the east and north, is the desert.

Most people live in the upland areas and almost everyone is Arab. But there are numerous social divisions based on politics and history. One group consists of the 'East Bankers'—descendants of the people who lived on the East Bank of the Jordan prior to 1948. The second, and now making up around 60% of the population, are descendants of the Palestinian refugees who crossed the river from the West Bank following

the formation of the state of Israel in 1948—who were followed by another surge in 1967 when Jordan lost control of the West Bank to Israel. A third group are the immigrant workers, around 7% of the population—the majority from Egypt, Iraq, and Syria.

The population is young and growing—at around 3% per year. This has put severe pressure on public services. Nevertheless the government does seem determined to invest in a high quality health service, and especially in education on which it spends 20% of the budget.

With relatively little by way of industry or natural resources, Jordan has relied heavily on the service industries that account for more than 70% of GDP. Many of these, including transport and communications, derive from Jordan's close links with its neighbours.

Agriculture is divided between fairly primitive rain-fed cereal production in the uplands and the higher-tech irrigated farms of the Jordan valley that are largely given over to fruit and vegetables, some of which are exported, and rely significantly on immigrant labour, generally from Egypt. Water is a perennial problem. Groundwater is rapidly being exhausted. Planned projects include conveying water from Disi in the south to Amman.

Industry is limited. The most significant enterprises are the state-owned mining operations that extract potash and phosphates. Jordan is the world's third largest exporter of raw phosphates. However the government has also been creating special duty-free economic zones, such as the one at Aqaba, to encourage investors.

In recent years industrial output has been boosted by manufacturing, particularly of garments, in 'qualifying industrial zones' which have duty-free access to the US market.

Tourism is also becoming more important, but one of the most reliable sources of foreign exchange is migrant remittances. Though Jordan

$4 billion in migrant remittances

is home to many immigrants, it also has around 900,000 Jordanians overseas, most of whom work in the Gulf states, and who send home $4 billion in remittances each year.

Since the late 1980s, Jordan has been burdened with a large external debt and has had to turn to the World Bank which required a far-reaching programme of structural adjustment. In addition, however, good relations with many donor countries have enabled Jordan to benefit from substantial flows of aid.

In principle, Jordan is a constitutional monarchy with executive power vested in its 80-member parliament and its prime minister. In practice, the parliament has always been directed by the king. For most of the second half of the 20th century, this was King Hussein, who reigned in an often autocratic

fashion—bolstered by his links with tribal leaders and with the army. Somehow, he managed to cultivate close ties with Israel and keep the USA as an ally while also placating his vociferous Palestinian citizens.

Hussein allowed a democratic opening in 1989—instituting reforms that permitted new elections. To his displeasure, however, this gave a strong representation to radical Islamic groups. As a result he took a series of measures to reduce the number of Islamicists in parliament and introduced strict press laws.

Hussein died in 1999 and was succeeded by his son Abdullah ibn Hussein al-Hashemi. The British-educated Abdullah knew relatively little of Jordanian politics. But he soon had a considerable impact and embarked on a number of social and economic reforms. In 2002 his 'Jordan First' campaign envisaged reforms in the education system. Women's rights have also been a priority for Abdullah and his Palestinian wife, Queen Rania.

On the political front, Abdullah has to walk the same tightrope between keeping the Palestinians on side and maintaining good relations with the USA. He has been wary of the main Islamicist party, the Islamic Action Front, and twice postponed elections that were finally held in 2003 when Islamicists won 10% of the vote.

In 2007, however, the parliamentary arm of the Muslim Brotherhood, the Islamic Action Front (IAF), saw its representation fall from 17 seats to six – partly because of infighting within the IAF.

The king remains generally popular though has been criticized for not cutting diplomatic ties with Israel.

Kazakhstan

Kazakhstan has yet to capitalize on its rich mineral resources, or the opportunity for democracy

Land area: 2,717,000 sq. km.	
Population: 15 million—urban 59%	
Capital city: Astana, 313,000	
People: Kazakh 53%, Russian 30%, Ukrainian 4%, German 2%, Uzbek 2%, Tatar 2%, other 7%	
Language: Kazakh, Russian	
Religion: Muslim 47%, Russian Orthodox 44%, other 9%	
Government: Republic	
Life expectancy: 65 years	
GDP per capita: $PPP 10,863	
Currency: Tenge	
Major exports: Oil, gas, metals	

Kazakhstan is a landlocked country in central Asia that consists of a vast plain stretching from the Caspian Sea in the west to the mountainous border with Kyrgyzstan and China to the east. Most of the territory is grassy steppes or desert.

Kazakhs form around half the population in a country that has more than 100 ethnic groups. Russian settlers moved in from the 18th century and continued to do so during the Soviet era, along with ethnic Germans who were exiled here by Stalin in the 1930s. Most Kazakhs live in the east, while Russians are in the majority in the north.

Since independence, there have been efforts at 'Kazakhization', particularly the promotion of the Kazakh language. When Kazakhstan was effectively a Russian colony, Kazakhs had to learn Russian which then displaced Kazakh.

Demography, too, is starting to swing back in the Kazakhs' favour. This is partly due to emigration. Many emigrants were Russians: between 1989 and 2000, 1.6 million left. Around 500,000 ethnic Germans headed for Germany. The average age of Kazakhs is 20, while that of Russians is 46 and because of emigration and their higher birth rate Kazakhs are now in a majority again. After 2002, the population started to grow again, partly as a result of immigration from Uzbekistan.

In the past Kazakhstan's people had reasonable standards of human development, but health services deteriorated in the 1990s, and there were sharp rises in infectious diseases such as tuberculosis, as well as lifestyle diseases, particularly alcoholism. Services have now started to improve though health expenditure is still low.

Poverty has also started to fall: only around 15% of the population now live below the national poverty line. Nevertheless there are still wide regional gaps: the per capita income in the poorest province is only one-fifth that of the richest.

One-third of the workforce are employed in agriculture, though this now generates a shrinking proportion of GDP, under 6% in 2007. The soil is good but the climate is unpredictable and it swings between hostile extremes—fiercely hot summers and bitterly cold winters.

Agriculture has not developed much since the Soviet era. Privatization has been slow. Most farmers lease their land from the state and are often so short of credit, inputs, and equipment that they cannot complete their harvests.

Kazakhstan's other major source of employment—though more for Russians and Ukrainians—is heavy industry which centres on the production of oil and gas. Oil was first found here in 1899. Kazakhstan still has substantial oil deposits along the Caspian Sea coast. At the end of 2007 the country was estimated to have 3.2% of global reserves. And despite the difficulties of exporting from this landlocked country, these deposits have proved very attractive to

Oil companies have flocked to Kazakhstan companies from all over the world who are keen to engage in joint ventures.

More than 80% of output now comes from foreign companies, the rest from the state-owned company Kazmunaigaz.

The country also has substantial deposits of metal ores, including chromium, lead, copper, zinc, and tungsten.

Kazakhstan's industrial past has bequeathed severe environmental problems. It was also a Soviet nuclear test site—between 1949 and 1989, 470 tests were carried out in the north, which now has to deal with a legacy of cancer and birth defects. In the south, the main environmental problem is the drying up of the Aral Sea, whose water level has been falling as a result of withdrawals from feeder rivers elsewhere in Central Asia. The sea is also heavily polluted by fertilizer run-off—which is killing the fishing industry and is also contaminating water supplies.

When Kazakhstan achieved independence from the Soviet Union in 1991 the sole presidential candidate was Nursultan Nazarbaev—a former head of the Kazakh Communist Party, who initially had opposed independence. Like other leaders in Central Asia, he continued to rule in much the same autocratic fashion. And when it was time for re-election in 1999 he took no chances—suddenly bringing the election forward while the economy was in better shape, and banning his most serious rival. Standing against a communist candidate, he was duly re-elected with 82% of the vote.

Opposition started to surface in 2002 with the formation of the Democratic Choice of Kazakhstan (DVK). But the DVK and other parties have been effectively stifled by repression and electoral fraud and in the 2004 parliamentary elections were allowed to have only one seat between them. The process was similar for the presidential election of 2005. Unsurprisingly, Nazarbaev won with 91% of the vote. Altynbek Sarsenbaev, an opposition figure, was murdered in February 2005.

Nazarbaev shows no signs of stepping down and is now aiming for re-election in 2012.

Kenya

Formerly one of Africa's most stable democracies, Kenya has been shaken by violence

ETHIOPIA

Kisumu
Nakuru
Nairobi

TANZANIA
Mombasa
INDIAN OCEAN

| 0 | Miles | 300 |
| 0 | Km | 480 |

Land area: 580,000 sq. km.
Population: 38 million—urban 22%
Capital city: Nairobi, 1.3 million
People: Kikuyu 22%, Luhya 14%, Luo 13%, Kalenjin 12%, Kamba 11%, other 28%
Language: English, Swahili, and others
Religion: Protestant 45%, Roman Catholic 33%, Muslim 10%, other 12%
Government: Republic
Life expectancy: 54 years
GDP per capita: $PPP 1,542
Currency: Kenyan shilling
Major exports: Tea, coffee, horticultural products

Most of the eastern half of Kenya consists of a broad plateau with scattered hills that slope down to a narrow coastal strip. This is an area of uncertain rainfall, merging into the arid and semi-arid lands of the north and north-east. The western half has several mountain systems split in two by the snaking line of the Rift Valley, while to the south-west the mountains descend to a basin around the shores of Lake Victoria.

The population is concentrated in the western half of the country, particularly on the fertile central highlands and on the land to the west. The land in the north-east is occupied mostly by nomadic herding communities. Kenya has many different ethnic communities, but they can be divided into three main groups. The largest, with about two-thirds of the population, are the Bantu group which includes the Kikuyu and the

Luhya. A second group, the Nilotic, making up around a quarter of the population, includes the Luo and Kalenjin and the Masai nomads. A third, smaller group are the Cushitic. Others are of European, Asian, and Arab descent.

Kenya has had very rapid population growth though the rate has now fallen to 2.8% as a result of more intensive family planning as well as the death toll from AIDS—5% of adults are HIV-positive. Even so, the population density is already high in the areas that have better land, putting pressure on the environment and driving people into the cities, particularly to the slums of Nairobi.

Kenya also suffers from wide income disparities. At independence, much of the best land was taken and kept by the Kikuyu élite. And subsequent liberal free-market development led to rapid economic growth but also heightened inequality. The richest fifth of the population get half of national income. Around half the population live below the national poverty line.

Agriculture accounts for around one-quarter of GDP and for around one-fifth of formal employment.

Kenya's small farmers grow food crops such as cassava and maize for subsistence, and also produce around 60% of two major export items, tea and coffee—the rest being grown on large estates. Kenya is the world's largest exporter of black teas.

Over recent decades Kenyan farmers have increasingly turned to horticulture for export, primarily fruit, vegetables, and cut flowers. These are air-freighted to Europe where they meet around 25% of the demand for off-season produce. Output continues to rise as Kenya finds new horticulture markets in Asia and North America.

Florist to the world

Manufacturing accounts for 10% of GDP, with the emphasis on food processing and small-scale consumer goods. Output has been constrained, however, by poor infrastructure, especially power supplies: the country faces frequent power cuts, following a badly managed liberalization of the electricity industry.

Kenya also has a vast informal sector working in small-scale manufacturing making everything from household goods to car parts—and contributing around one-fifth of GDP.

Tourism has been a further source of foreign exchange, and accounts for around one-fifth of GDP, as half a million visitors each year head for the game reserves and the beaches. However, the industry has suffered from political uncertainty and weak infrastructure as well as tougher competition from neighbouring countries.

For much of the period following independence in 1963, Kenya was effectively a one-party state—run by the Kenya African National Union (KANU). The first president was a Kikuyu, Jomo Kenyatta, who on his death in 1978 was succeeded by a Kalenjin, Daniel Arap Moi. In 1982, Moi made KANU the sole legal party, which it remained until 1992 when he was forced to concede multiparty elections. Moi and KANU won the 1992 elections and won again, though more narrowly, in 1997.

The final years of Moi's reign saw a steady descent into greater corruption, poverty and violence. KANU was originally a Kikuyu-dominated party, but Moi transformed it into one that primarily served his own group, the Kalenjin.

This dismal scene brightened at the end of 2002 when opposition groups united as the National Rainbow Coalition. They achieved a landslide victory in the presidential election—for Emilio Mwai Kibaki—and a corresponding win in the parliamentary election. Kibaki said he was determined to fight corruption and revive the economy and he introduced free primary education.

The presidential elections of December 2007, however, proved traumatic. Kibaki was awarded a narrow victory, beating Raila Odinga of the ODM. But the ODM claimed fraud and soon there was rioting across the country that left 600 dead and 250,000 displaced, in what amounted to ethnic cleansing of tribal areas. Following international mediation the crisis was resolved in 2008 when Kibaki agreed to share power, with Odinga as prime minister.

In 2010 there was a peaceful ballot to approve a new Constitution.

Korea, North

Increasingly unpredictable, North Korea now presents a significant nuclear threat

Land area: 121,000 sq. km.
Population: 24 million—urban 63%
Capital city: Pyongyang, 2.7 million
People: Korean
Language: Korean
Religion: Buddhism and Confucianism
Government: Communist
Life expectancy: 67 years
GDP per capita: $PPP 1,800
Currency: North Korean won
Major exports: Minerals, agricultural products

Officially, this is the Democratic People's Republic of Korea. Around four-fifths of the territory comprises mountains or high plateaux—the highest of which are to be found in the north-east. The main lowland areas, home to most of the population, and where most agriculture takes place, are to be found in the south-west.

The population is ethnically among the most homogenous in the world, a factor that has helped sustain national solidarity in the face of extreme hardship. There are also several million ethnic Koreans living across the border in China—including around 100,000 refugees.

One of the achievements of the communist regime has been education. Children have good access to pre-school activities, partly to allow their mothers to work, but also to expose them early to the ideals of the state. Health standards were also good until undermined by the famines of the 1990s. In 2007, more than one-third of children were thought to be malnourished. Estimates of the death toll from famine since 1995 range from one to three million. Since so many people have fled to China and the birth rate is low, the population is either static or declining.

Agriculture still employs one-third of the workforce, primarily growing rice, corn, and potatoes. Officially, this is produced by cooperatives and collectives—though in recent years much has also been traded through markets. Until the mid-1990s, agriculture was relatively successful, but from 1995 a series of climatic setbacks, including floods, droughts, and tidal waves, combined with mismanagement, crippled production and triggered a sequence of famines. Most people now rely on food aid.

North Korea once had a flourishing industrial sector because the partition of Korea in 1948 left it with most of the minerals and heavy industry. Though it has no oil, the country has plenty of brown coal and considerable hydroelectric potential. Industry, which is largely on the east coast, is dominated by iron and steel, machinery, textiles, and chemicals. But much of the equipment is now antiquated and production has shrunk. There is some light industry that includes textiles exports and TV manufacturing by South Korean firms such as Samsung and LG.

The Democratic People's Republic of Korea is a communist state which has been underpinned by a bizarre personality cult. The state was formed in 1948 from the territory of Korea above the 38th parallel which had been occupied by the Soviet Union. With Soviet help, the communist leader Kim Il-Sung rapidly took control. In 1950, he also attempted to seize South Korea, provoking the 1950–53 Korean War. Although he failed, Kim nevertheless entrenched his position and with the support of China, the Soviet Union, and a highly disciplined party, he managed to rebuild the country.

By the mid-1950s, however, he had drifted away from the Soviet sphere and developed his own philosophy 'Juche', an obscure mixture of communism and self-reliance, woven into a personality cult around Kim himself. The country was never as self-reliant as it proclaimed. When the Soviet Union collapsed and China started to liberalize its economy and build stronger ties with the West, North Korea lost its two main props.

A bizarre personality cult

When Kim Il-Sung died suddenly in 1994 he left a confusing power vacuum. The situation became clearer in 1998 when the late Kim Il-Sung was posthumously elevated to 'president for eternity' while his son Kim Jong-Il became Chairman of the National Defence Committee and head of state. Even so, the real location of

power remains unclear—distributed between the state, the party, and the army.

With persistent famines and a bankrupt government, North Korea seems permanently on the brink of implosion. South Korea does not want the economic burden of having North Korea fall into its lap and has been sending fertilizer and food aid. It has also established a fund of $450 million to help the north. China, nervous of a capitalist Korea on its doorstep, has been allowing hundreds of thousands of starving North Koreans to enter in search of food.

North Korea certainly frightens its neighbours. It has the world's fifth largest armed forces—1.2 million strong—which soak up around one-third of GDP.

Even more alarming are its nuclear threats. In 1994 North Korea and the US signed the General Framework Agreement, through which North Korea would cap its nuclear weapons programme in exchange for US aid for the construction of new light-water reactors. By 2002, however, having been described by the US president as belonging to an 'axis of evil', North Korea claimed it was again developing nuclear weapons. The US has offered economic aid provided North Korea gives up its weapons, while also insisting on six-way talks that include South Korea, Russia, and Japan. But Kim Jong-Il seems determined to provoke the world with missile tests.

With Kim Jong-il looking increasingly unwell, the leadership may now pass to his third son Kim Jong-un, though this could lead to instability.

Korea, South

South Korea has one of Asia's most successful economies but worries about its alarming northern neighbour

Land area: 99,000 sq. km.
Population: 48 million—urban 82%
Capital city: Seoul, 9.8 million
People: Korean
Language: Korean
Religion: Christianity 26%, Buddhism 23%, other or none 51%
Government: Republic
Life expectancy: 79 years
GDP per capita: $PPP 24,801
Currency: Won
Major exports: Electronic goods, textiles, cars, ships

The country's official name is the Republic of Korea. Almost three-quarters of its territory is mountainous—the largest range being the T'aebaek, which occupies most of the eastern half of the country. Most of the population are to be found in the lowlands of the north-west and south-west and in the Naktong River basin in the south-east.

Almost everyone is ethnically Korean—and speaks the Korean language. Traditionally most Koreans have followed Confucian and Buddhist principles, but Christianity has had an increasing impact, particularly the evangelical protestant sects. One of these, though less significant now, is the controversial Unification Church of Reverend Sun Myung Moon, which has been exported around the world.

South Korea is one of the world's most densely populated countries. Until the early 1980s, South Korea sent emigrant workers to other countries, but the flow has now been reversed and South Korea has attracted more than 200,000 foreign workers, legal and illegal, to do the more menial work. Most come from South-East Asia and China and many are of Korean ethnic origin.

Koreans have good standards of health and are well educated. But the curriculum puts much emphasis on rote learning that does not encourage creativity.

Today few Koreans work as farmers—around 10%. Thanks to a ban on rice imports, the remaining farmers, who are mostly smallholders, receive high prices for their crops and are so productive that the country is usually self-sufficient in rice. Meanwhile consumer food prices are heavily subsidized.

Nowadays around one-third of the workforce are employed in manufacturing industry. The partition of Korea left the South worse off—since the North had most of the mineral resources and the heavy industry. Manufacturing in the South initially concentrated therefore on light industry for local consumption and then for export. But from the 1970s South Korea turned its attention to heavy industry, particularly steel and cars—largely in the Seoul-Inchon region—and later to advanced electronic products. The result was a phenomenal rate of growth: from the 1990s until 1997 it averaged around 9% per year. This led to steady rises in income, and in labour costs, so much of the simpler labour-intensive assembly work in footwear and toys has now moved overseas.

Service industries, on the other hand, have been less sophisticated than those of other countries of similar wealth—whether in retail, finance, or entertainment. This has started to change, however, particularly in retailing, with a relaxation in restrictions on foreign participation.

At the heart of South Korea's rapid economic expansion were the 'chaebol'—vast diversified family-owned conglomerates. In 1999, five of these—Hyundai, Samsung, Daewoo, LG, and the SK group—still accounted for one-third of sales and almost half of exports. They benefited from decades of state favours and easy credit but recklessly overstretched and were deep in debt by 1997 when the Asian financial crisis struck. The currency lost half its value. In July 1999 the Daewoo group collapsed in the world's largest ever corporate bankruptcy.

The world's biggest bankruptcy

Meanwhile, foreign companies such as Hyundai Motors, Renault, and DaimlerChrysler took the opportunity to buy stakes in Korea's car industry.

Since then, the economy has bounced back and growth in recent years has been around 3%. Unemployment has come down to around 3%. North Korea, however, presents an economic threat since a collapse followed by a takeover could cost up to $1.2 trillion.

During most of its period of rapid growth, South Korea was in the grip of authoritarian military-backed regimes, which repressed opposition groups and especially the militant labour unions. The democratic opening appeared in 1987, when the government acceded to demands for democratic elections. South Korea now has a strong presidential system in which the president appoints the prime minister.

In 1997 the presidential election was won by a coalition led by the left-of-centre Kim Dae-Jung of the Millennium Democratic Party (MDP). He had to rule in coalition with several smaller parties. The largest party was the Grand National Party (GNP).The 2002 presidential election was also won by the MDP's candidate, Roh Moo-Hyun. In 2003, part of the MDP split off to form a new party, Uri, which subsequently became United Democratic Party.

The December 2007 presidential elections, however, were decisively won by a former Hyundai executive Lee Myung-bak of the conservative GNP which also narrowly won the general election in 2008, with 153 out of the 299 seats.

In May 2009 Roh Moo-Hyun committed suicide while being investigated for corruption charges which his supporters claim were politically motivated. This resulted in violent clashes in the National Assembly.

Kosovo

Kosovo has broken away from Serbia but has yet to achieve full international recognition.

Land area: 11,000 sq. km.
Population: 1.8 million–urban 37%
Capital city: Pristina, 500,000
People: Albanians 88%, Serbs 7%, other 5%
Language: Albanian, Serbian, Bosnian
Religion: Muslim, Serbian Orthodox, Roman Catholic
Government: Republic
Life expectancy: 69
GDP per capita: $PPP 2,300
Currency: Euro
Major exports: Agricultural produce, textiles, metals

Kosovo's claim to statehood is disputed, since Serbia still claims the territory as a province, but for most purposes Kosovo is now independent.

This landlocked territory consists largely of two basins—both around 500 metres above sea level—divided by low mountain ranges running from north to south. To the west, and including the capital Pristina, is the Metohija Basin whose principal river is the Beli Drim which flows first south and then west into Albania. To the east is the Sitnica which flows north into Serbia. The highest mountains, both above 2,000 metres, are the Šar which form the border with Macedonia and the Dinaric Alps in the east. Around two-fifths of the territory is covered by forests.

Kosovo's ethnic and religious balance is a legacy of centuries of occupation and conquest. In the Middle Ages it was part of the Serbian Empire and thus Orthodox Christian. In the fifteenth century it was absorbed into the Ottoman Empire and became largely Islamic. Ethnically, however, the stronger influence was to come from Albania. By the end of World War II the majority of the population spoke Albanian.

Kosovars are among Europe's poorest people. In 2007, 39% of the population were below the poverty line, of whom one-third were living in extreme poverty. This reflects high levels of unemployment—estimated at around 40%, most of which is long-term. There has been some progress but largely confined to the urban areas, so inequality is rising. The Serb minority are growing steadily poorer, partly because richer Serbs have been leaving.

Education levels are similar to those in neighbouring countries. Standards of health, however, are low, and life expectancy at 69 is the lowest in the region. As a result of weak government provision, most people have to pay for their own medicines.

Around one-third of the workforce is employed in agriculture. The soils are relatively fertile and the temperate climate is suitable for agricultural production. Most of the land is used for wheat, maize, and vegetables. But farms are small and productivity is low—so production is largely for subsistence.

Kosovo has modest deposits of a number of minerals, including lignite, lead, zinc, ferro-nickel, and magnesium. Output has fallen since the 1990s but could revive given sufficient investment.

The war and political upheaval over the past 20 years have undermined industrial production and much of the infrastructure, particularly that for power supply, is in a poor state. Government finances have often relied on international donors but aid flows are now set to fall.

With few opportunities at home, many Kosovars work overseas. One in five households has a migrant abroad. In 2007 their total remittances were around $450 million, about 10% of GDP.

Kosovo is the seventh state to have emerged from former Yugoslavia. Unlike the others, however, it was not a constituent republic but a

Formerly a province of Serbia province of Serbia. It had, nevertheless, been granted special autonomy that gave it a status similar to that of the other republics.

This had created some resentment among Serbian politicians so when Slobodan Milošević became Serbian president in 1989 he stripped Kosovo of its autonomy. He also violently repressed protesters and closed Albanian-language schools. As they saw Yugoslavia fall apart the Kosovars too wanted independence. In 1996 some formed the guerilla Kosovan Liberation Army and by 1998 they

were engaged in open warfare with the Serbs. By 1999 Kosovo was feeling the full force of Serbian ethnic cleansing. Hundreds of thousands of refugees fled into neighbouring countries while NATO forces bombed Serbia. Later that year a ceasefire was agreed, calling for Serbian troops to be withdrawn and replaced by UN peacekeepers. Kosovar refugees returned. Now it was the Serbians' turn to flee.

In 2007 the UN put forward a plan for self rule in Kosovo. The Kosovar Albanians accepted this but Serbia refused even though it would not mean independence.

In the event, the Kosovars took matters into their own hands and declared independence in February 2008. By 2010 Kosovo had been recognized by 64 UN member countries, which included most of those in the EU. The most vigorous opposition has come from Russia.

There seems little likelihood that independence will be reversed. Although Kosovo is not a member of the UN, it has joined the IMF and the World Bank. Kosovo still has 15,000 NATO troops who are now largely protecting the Serb enclaves in the north, and a large EU mission with hundreds of policemen, judges and customs officials.

Just before independence, the Kosovo parliament elected as prime minister a former KLA political leader, Hashim Thaçi, of the centre-left Democratic Party of Kosovo (PDK). His government is a coalition with the centre-right Democratic League of Kosovo (DLK). In 2011 Atifte Jahjaga, who has no party affiliation, was elected president.

Kuwait

Kuwait's hereditary monarchy retains power but faces greater demands for democracy

Land area: 18,000 sq. km.	
Population: 3 million—urban 98%	
Capital city: Kuwait City, 499,000	
People: Kuwaiti 45%, other Arab 35%, South Asian and other 20%	
Language: Arabic, English	
Religion: Muslim 85%, Christian, Hindu, Parsi, and other 15%	
Government: Hereditary emirate	
Life expectancy: 78 years	
GDP per capita: $PPP 47,812	
Currency: Kuwaiti dinar	
Major exports: Oil, oil products	

Kuwait is a small state at the north of the Persian Gulf between Iraq and Saudi Arabia. It has some oases and a few small fertile areas, but the country consists largely of a sloping plain that is almost entirely desert.

A couple of generations ago, Kuwaitis were nomadic tribesmen. But since 1946 oil has made Kuwait very wealthy. Kuwaitis are among the richest people in the world—and are also entitled to free education and subsidized health care, and are spared the inconvenience of income tax.

The government spends around 6% of its total budget on health, though it now requires expatriates to pay for services at public hospitals and clinics. Education is compulsory and has given Kuwait one of the higher literacy rates of Arab countries.

Most of the work is done by expatriates. In 2006 Kuwaitis made up only 35% of the population and only 17% of the labour force—the rest being immigrant workers. The largest numbers nowadays are from Egypt, Palestine, and Lebanon, but there are also many from South Asia. Almost all employed Kuwaitis choose to work for the government or for public corporations.

Kuwait is still floating on a sea of oil. It has about 10% of world reserves, mostly in the Burgan field south of Kuwait City. At present rates of extraction, this should last for 100 years or more. Since the 1970s, the oil industry has been nationalized. The government owns local refineries and also markets around 40% of its own oil. Kuwait Petroleum International operates more petrol stations using the 'Q8' brand name in five European countries. In addition, Kuwait has large reserves of gas.

Oil accounts for 52% of GDP and around 95% of export earnings. It also provides 75% of government income with most of the rest coming from overseas investments made with previous oil income.

Kuwait has some manufacturing industry—also linked to oil. In the past this has typically involved such low-value products as ammonia, urea, and fertilizers. But there have been efforts to move to more profitable products such as polypropylene.

With scarcely any arable land or water for irrigation, Kuwait's agriculture is very limited. Some farms grow vegetables and fruits but almost all the country's food has to be imported. There is also a fishing industry whose catches are dedicated to the local market.

Much of the oil infrastructure was restored since the 1990 Gulf War, but Kuwait's economy will never be the same. To pay for the war the government had to sell off around half its $100-billion overseas investments.

The government also responded by introducing some charges for health care, for example, and there have been suggestions for income or consumption taxes. But there is stiff opposition from conservatives who argue that the government should instead attack corruption and reduce defence expenditure.

Disturbed by the thought of taxation

One of the most important external influences has been neighbouring Iraq. In 1961, when Kuwait achieved full independence, Iraq claimed the territory as part of Iraq. Finally in August 1990 Iraq, led by Saddam Hussein, occupied Kuwait until it was expelled by a US-led coalition in Operation Desert Storm. Kuwait was naturally also a strong supporter of the US invasion of Iraq in 2003 and of the fall of Saddam Hussein.

Kuwait is a hereditary emirate and the ruling al-Sabah family holds all the key government posts. When the ailing Sheikh Saad Abdullah al-Salem al-Sabah was forced by the family to step down he was replaced in 2006 by Sheikh Sabah al-Ahmed al-Jabr al-Sabah. He then appointed his half-brother, Sheikh Nawaf al-Ahmed al-Jabr al-Sabah, as crown prince, and his nephew, Sheikh Nasser Mohammed al-Ahmed al-Sabah, as prime minister.

Kuwait does have a parliament which consists of 50 members elected for four-year terms plus 15 cabinet members appointed by the emir. Its powers are limited but it serves as a place to air grievances, since members can question ministers. Kuwait also has a relatively free press.

Formal political parties are banned, but there are more informal groupings both inside and outside the parliament. On the one hand, there are the liberals who broadly support the government and are concentrated in the centre of Kuwait City; on the other, there are the more oppositional conservative tribal-based Islamic groupings.

The 2006 elections represented something of a landmark. First, because for the first time women were allowed to vote and stand for election, though none succeeded. Second, because these elections resulted in a large majority for the opposition Islamist forces.

The election, in May 2009, which had been called three years early, was remarkable, since for the first time women were elected, though only four. The secular and Shia candidates also made gains while the previously dominant Sunni groupings lost ground. Members of parliament frequently quarrel with the government—especially over measures to open up the economy. But they are weakened by their inability to form strong groups.

Kyrgyzstan

Another of the former Soviet republics to experience a popular insurrection

Land area: 199,000 sq. km.		
Population: 5 million—urban 37%		
Capital city: Bishkek, 804,000		
People: Kyrgyz 65%, Uzbek 14%, Russian 13%, Dungan 1%, other 7%		
Language: Kyrgyz, Russian		
Religion: Muslim 75%, Russian Orthodox 20%, other 5%		
Government: Republic		
Life expectancy: 68 years		
GDP per capita: $PPP 2,006		
Currency: Som		
Major exports: Food, metals, electricity		

Kyrgyzstan is a landlocked country with spectacular scenery. It is almost entirely mountainous, lying at the junction of several ranges that run mostly east to west and whose peaks are permanently covered in snow and ice. More than half the territory is above 2,500 metres. Some of the main lowland basins are in the south-west and there is one in the north where the capital Bishkek is to be found. The main river, flowing east to west, and a major source of irrigation, is the Naryn. The country also has many lakes, the largest of which, Ysyk-Köl in the north-east, is 1,500 metres above sea level.

The Kyrgyz people make up around two-thirds of the population—a proportion that has risen as a result of the exodus of Russians and ethnic Germans: half a million departed between 1990 and 1995.

The majority of Kyrgyz live in the rural areas. The second largest minority are the Uzbeks, who are concentrated in the south, around the city of Osh. Ethnic relations have at times been tense.

While standards of health have generally been good, health services are deteriorating and there have been outbreaks of contagious diseases. Education services too have suffered from budget cuts and school enrolment has fallen. Around 40% of the population are poor with those in the south being worst off.

Agriculture is the largest source of income, employing around half the workforce. Livestock has long been a mainstay of the economy—producing wool, hides, and meat. And many people still work as nomadic herders, living in their traditional round felt tents called 'gers'.

Only a small proportion of the land is arable, but more of the grazing land is now being used for crops—benefiting from extensive irrigation from rivers rushing down from mountain peaks. Most of the previous collective farms are now joint-stock companies and relatively efficient. Crops include cotton, wheat, and vegetables and the country is now self-sufficient in grains.

During the Soviet period, Kyrgyzstan underwent a rapid process of industrialization, chiefly based on the exploitation of mineral deposits. The country has extensive coal reserves, as well as non-ferrous metal deposits, including gold, uranium, mercury, and tin. Gold production, for example, has been stepped up following the opening of the Canadian-operated Kumtor mine near Bishkek which now accounts for two-fifths of industrial production. Much of the heavy industrial manufacturing has been fairly inefficient and output has declined sharply, largely as a result of low investment.

Kyrgyzstan was one of the faster reforming countries of the ex-Soviet republics in Central Asia. More than 60% of eligible enterprises were rapidly privatized, though the state retained 'strategic enterprises'. This, combined with tight macroeconomic policies, ensured considerable support from international donors and financial institutions. But the economy has been slow to respond—hampered by the exodus of skilled Russian workers.

Kyrgyzstan has lost many skilled Russians

Growth resumed after 1995 but has been erratic, depending largely on output from Kumtor and to a lesser extent on agriculture. The economic outlook remains problematic—the country's relative isolation has been deterring foreign investment in anything other than mining.

Kyrgyzstan's economic reforms were driven largely by former President Askar Akaev, who was in office between 1990 and 2005. The riots in 1990 had fatally weakened the Communist Party and opened the way for reformists to elect Akaev, a former physicist, as president. He subsequently banned the Communist Party and, following independence in 1991, he was elected unopposed.

At that point, the parliament had three main groups: the communists, who were now legal again but were hostile to reform, Akaev's Union of Democratic Forces (SDS), and a social democratic grouping that tried to resist the increasing presidential powers.

In 1995, Akaev was convincingly re-elected and took the opportunity to push through yet another constitutional amendment to tilt power more in his favour. Akaev and the SDS duly won the elections in 2002. But since Akaev had banned several opposition groups and jailed one of his main challengers, the opposition became more vocal.

Akaev's administration came to a sudden, and rather surprising, end in March 2005 after a blatantly fraudulent election provoked a full-scale insurrection. Thousands of people took to the streets, eventually causing Akaev to flee the country.

One of the leaders of the 'tulip revolution', Kurmanbek Bakiev, won a landslide victory in the presidential election in 2005. But Bakiev became increasingly authoritarian and following violent protests he was ousted in April 2010. This was followed in June by ethnic violence between Kyrgyz and Uzbeks with 470 deaths. In October, a parliamentary election resulted in a three-party coalition. Roza Otunbayeva was appointed temporary president.

Laos

An isolated communist state that is slowly opening up to the rest of the world

Land area:	237,000 sq. km.
Population:	6 million—urban 33%
Capital city:	Vientiane, 698,000
People:	Lao 55%, Khmou 11%, Hmong 8%, other 26%
Language:	Lao, French, English, and various ethnic languages
Religion:	Buddhist 60%, animist and other 40%
Government:	Communist
Life expectancy:	65 years
GDP per capita:	$PPP 2,165
Currency:	Kip
Major exports:	Wood, coffee, electricity

Laos, officially the Lao People's Democratic Republic, is a landlocked and almost entirely mountainous country. The highest mountains are to the north, rising to 2,800 metres above sea level, while another range lies to the east along the border with Vietnam. The only extensive lowland areas are to the south in the floodplains of the Mekong River—where around half the population live. Only around 5% of the territory is suitable for agriculture and most of the rest is covered with forests—broadleaf in the north, but tropical hardwoods in the south.

Laos is sparsely populated but it has, according to the government, around 68 'nationalities' divided into three main groups. The largest, and politically the most dominant, are the Lao. These include the Lao-Loum, who live in the cities and in the lowlands of the Mekong delta. The Lao-Theung, also called the Mon-Khmer, live in the uplands throughout Laos and can also be found in neighbouring countries. Finally, there are the highland Lao-Soung, who include the Hmong (Meo) and the Yao (Mien).

Levels of human development are low. Around one-third of the population live below the poverty line, consuming less than 2,100 calories per day, though with much higher poverty rates in the rural areas: 40% of the population live more than six kilometres from a main road. Laos has more than 80% of children enrolled in primary school, but the quality of education is low. Health standards are also poor and more than one-third of children are malnourished. Many people are also addicted to opium, and the rate of HIV infection, though low, could soon rise.

Around three-quarters of Laotians depend for survival on agriculture, which accounts for almost 40% of GDP. But farming is not very efficient. Upland farmers rely on slash-and-burn techniques and even in the more fertile lowlands rice yields are low—vulnerable to droughts and floods and hampered by lack of credit and inputs, as well as by crumbling infrastructure that has suffered from decades of war and low investment.

More recently, Laos has been trying to produce more cash crops for export—including cotton, groundnuts, sugar cane, and tobacco. But the most notable export success has been coffee. This was originally introduced by the French colonialists and is now being revived by small farmers, who grow good-quality robusta coffee for export to Europe.

Nevertheless, the country's leading export earners remain timber and wood products. Laos has valuable forest resources, including teak, rosewood, and ebony. IKEA, for example, has a contract to buy wood products. A less official export is drugs: Laos is thought to be the world's third-largest producer of opium and heroin. The country also serves as a transit route for heroin from Burma.

Laos has no heavy industry but does have some small-scale manufacturing around Vientiane.

Laos exports hydro-electricity

Some of the more successful products include handicrafts, garments, foodstuffs, and wood products. Another export prospect is hydroelectric power. Laos's rugged terrain is cut through with fast-flowing rivers, some of which are dammed to produce power for export to Thailand, and in future to Vietnam. The largest source is the Nam Theun II project which opened in 2009.

Political life in Laos is controlled by the Lao People's Revolutionary Party (LPRP). Laos has been a communist country since 1975 and

the most recent constitution in 1991 confirmed that power would remain exclusively in the hands of the governing party. But whether Laotian communism amounts to socialism is more questionable. Politics is based less on class and more on regional, family, and ethnic loyalties. From the outset, it looked as though socialism would remain a long-term goal, and it seems more distant than ever.

Since the mid-1980s the government has pursued a more liberal economic path. In 1986 it abandoned collectivization of farms and introduced the 'New Economic Mechanism', which also removed many restrictions on private enterprise. The government has also been privatizing state-owned enterprises: between the early 1990s and 2000 the number of state-owned enterprises fell from 800 to 29—though those surviving tend to be the largest undertakings.

Politically, the situation is less fluid and the party hardliners in Laos retain their grip. Non-communist political parties are proscribed and the government clamps down on dissent. In the National Assembly elections in 2011 the LPRP won 128 of the 132 seats. There seems little prospect of political liberalization.

The most visible opposition comes from insurgent Hmong tribesmen, known as the 'Chao Fa'. They are the remnants of a guerrilla Hmong army that the CIA financed during the Vietnam War. At times they seem to have got more popular support. In 2007 there was fighting around the tourist resort of Vangviang. Since then, however, the opposition has been fairly quiet.

Latvia

Latvia has now joined the EU as its poorest member

Land area: 65,000 sq. km.
Population: 2 million—urban 68%
Capital city: Riga, 717,000
People: Latvian 58%, Russian 30%, Belarusian 4%, Ukrainian 3%, other 5%
Language: Latvian, Lithuanian, Russian
Religion: Lutheran 20%, Orthodox 15%, other Christian 1%, other 64%
Government: Republic
Life expectancy: 72 years
GDP per capita: $PPP 10,109
Currency: Lat
Major exports: Timber, wood products, textiles, metals

Latvia consists of gently rolling plains interspersed with low hills. Around one-quarter of the land is forested and there are numerous lakes, marshes, and peat bogs. The long coastline includes major ports as well as holiday resorts and extensive beaches.

Latvia's people, the Letts, have come close to being a minority in their own country. Before Latvia was annexed by the Soviet Union in 1940 the Latvians were 77% of the population, but a combination of deportation of Latvians and immigration from other Soviet republics, particularly Russia, dramatically altered the ethnic balance. By 1989, the Letts made up only 52% of the population. As a result of emigration of Russians and falling birth rates, the population has been shrinking.

After Latvia regained independence in 1991 the government was determined to reassert Latvian control and in 1994 introduced a rigorous Citizenship Law. Although, under pressure from Russia, this was relaxed somewhat in 1998 it still requires people taking Latvian citizenship to pass stiff tests in language and history. And unless you are a citizen you cannot vote, buy land, or enter a number of professions. Russians who arrived after 1940 are regarded as 'occupiers' and do not get automatic citizenship.

Latvia has been one of the more successful ex-Soviet republics. Living conditions and standards of health deteriorated sharply in the first years after independence, with a rise in diseases of poverty such as TB, though the situation has gradually started to improve.

Poverty tends to be greater in the eastern regions where unemployment can be as high as 20%, compared with 5% in the capital, Riga.

Industry in Latvia had been geared to producing industrial and consumer goods, including processed food, for the Soviet Union. After independence, when the country was opened up to foreign competition, many industries such as machine building went into steep decline, but have since revived. By 2009 as a result of a steady process of privatization most production is now in private hands.

Although agriculture represents a small and declining proportion of the economy it is still a major source of employment. Most of the land was collectivized during the Soviet era. This now been restored to private hands but in many cases the ownership of land remains unclear. The main activities are dairy farming and livestock raising, along with cereal production. It seems likely that accession to the EU, along with increased competition from other countries, will further reduce agricultural output.

Probably the most important rural activity is timber production. Forests cover around 45% of the land area. Since 1996, the output of timber-related industries has more than doubled and timber and wood products make up around one-quarter of exports to the EU.

Forestry is one of Latvia's boom industries

Another striking economic development in recent years has been the rise of the service sector, which now accounts for around two-thirds of GDP. Much of this is connected with trade since Latvia, with three important ports, is a major transit route for goods in and out of Russia.

Following a positive result in a referendum in 2003, Latvia entered the EU in May 2004, where it is one of the poorest members. It is also one of the most corrupt, in a country where political parties are closely aligned with business.

Since independence in 1991, politics has been a shifting and unstable affair, producing a series of weak and transient governments. This is partly because of the diverse and shifting nature of political parties

Among the other largest parties in the parliament are the People's Party which was formed in 1998, largely as a vehicle for Andris Skele who twice became prime minister, but has since resigned from the party. When in government the People's Party has been associated with vested interests and corruption.

The largest right-wing nationalist group is Fatherland and Freedom-Latvian National Independence Movement (TB-LNNK). Another major party is the New Era Party, which was formed in 2002, led by a former physicist and central bank chairman, Einars Repse. Others, formed in 2002, include the eurosceptic Union of Greens and Farmers (ZZS) and the First Party-Latvia's Way (LPP-LC) which aims to promote Christian values.

The president is also elected by the Saeima. Although neutral, the president, has some political authority and appoints the prime minister. Since 2007, the president has been Valdis Zatlers.

These parties have formed and reformed multiple coalitions with great rapidity. In March 2009 a new centre-right, five-party Unity coalition took over. This was headed by Valdis Dombrovskis from New Era as prime minister.

Following the elections in 2010, Dombrovskis remained as prime minister, heading a centre-right coalition consisting of his Unity group and the Union of Greens and Farmers (ZZS).

Lebanon

Decades of warfare appear to have ended but conflicts can always resurface

Land area:	*10,000 sq. km.*
Population:	*4 million—urban 87%*
Capital city:	*Beirut, 391,000*
People:	*Arab 95%, Armenian 4%, other 1%*
Language:	*Arabic, French*
Religion:	*Muslim 60%, Christian 40%*
Government:	*Republic*
Life expectancy:	*72 years*
GDP per capita:	*$PPP 10,109*
Currency:	*Lebanese pound*
Major exports:	*Paper products, foodstuffs, textiles*

Along its Mediterranean coast Lebanon has a narrow, flat coastal plain. Inland there are two parallel mountain ranges running north-east to south-west: the Lebanon Mountains, which ascend abruptly from the plain; and the Anti-Lebanon Mountains along the border with Syria. Between these two ranges is the fertile Bekaa Valley.

Almost all Lebanese are ethnically Arab, but they are sharply divided into 'confessional' groups—Muslim and Christian sects. At independence in 1943, political leaders established an unwritten National Pact that shared political power according to the proportions of each sect in the 1932 census. At that point, the most numerous were the Maronite Christians, followed by Sunni Muslims, Shia Muslims, Greek Orthodox Christians, Druze (Muslims), and Greek Catholics.

Overall, Christians were in a narrow majority and for many years were politically dominant. Since then, waves of immigration, notably of Palestinians, have altered the balance. Of today's four million population, around 60% are Muslim, including around 200,000 Palestinian refugees living in camps as well as 300,000 migrant workers, mostly from Syria.

Prior to the civil war of 1975–90, Lebanon was one of the most developed Arab states—a major centre for trade, finance and tourism. Much of that was wrecked in the civil war which virtually levelled Beirut. Today, the Lebanese population remains highly educated with good standards of health, but many people are unemployed, income disparities are widening.

Until the 2006 war, Lebanon had been recovering slowly. Unsurprisingly, one of the most vibrant sectors has been construction. From 1992, money flooded in for the rebuilding of Beirut, often from Lebanese expatriates.

But as before, the main sources of employment are services of various kinds, many connected with trade, which account for three-quarters of GDP. The financial sector is also being redeveloped and tourists have been coming back—arrivals were rising at 5% per year. Many new manufacturing enterprises have also sprung up, most of which are small, engaged, for example, in food production, furniture-making, and in textiles and clothing.

Agriculture, despite favourable soil and climate, has been less significant: it now accounts for only 5% of GDP and employs 8% of the labour force. Farmers grow fruit and vegetables on the coastal plain, and wheat and barley in the Bekaa Valley, but there has been relatively little investment and many farmers find it difficult to compete with imports.

Lebanon's political troubles date back to the late 1960s when the Palestine Liberation Organization (PLO) started using Lebanon as a base. The Muslims, many of whom supported the PLO, had long resented Christian political domination and from 1975 their disagreements erupted into a full scale civil war. In 1978, as part of its campaign against the PLO, Israel invaded and seized a 'security zone' on Lebanon's southern border.

The PLO used Lebanon as a base

The war stopped in 1990—after 150,000 deaths and $25 billion-worth of damage. The end came when Syrian forces took control, helping the government to disarm the militias. Israel finally withdrew from southern Lebanon during 2000.

Political loyalties in Lebanon align not behind parties but more along personal or religious lines. The largest group now are the Shia, who include the militant Hizbullah. The Sunni are a less coherent group and during the war had weaker militias.

In the deal that ended the civil war, Lebanon's National Pact was amended to increase the powers of the prime minister (always a Sunni Muslim) and the speaker of parliament (a Shia Muslim), and to reduce those of the president (a Maronite Christian).

In 1992 Rafiq Hariri, a Saudi-Lebanese billionaire, emerged as prime minister. He set about the reconstruction of Beirut.

In 1998 Syria supported the election as president of a Christian former army commander, Emile Lahou. Hariri then resigned. Following parliamentary elections in 2000, however, Hariri was reinstated but in 2005, was killed in a bomb blast. One million people took to the streets blaming Syria and demanding that Syrian troops withdraw—which they did in May 2005.

In June 2005, parliamentary elections gave the most seats to a coalition led by Hariri's son Saad. Fouad Siniora became prime minister.

Meanwhile Hizbullah had sustained its attacks on Israel. In July 2006, after the capture of two of its soldiers, Israel attacked southern Lebanon. Before a ceasefire the month-long war had killed 1,100 Lebanese.

In 2008 Lebanon seemed again to be headed for civil war, after a violent conflict between Hizbullah and supporters of Siniora. This was resolved following the election of the former army commander, Michel Suleiman, as president.

After elections in 2009, Saad al-Hariri became prime minister. In 2011, however, his coalition collapsed and was replaced by another, led by Najib Mikati.

Lesotho

Lesotho's democracy is fragile—and dependent on South Africa. Now it faces devastation from HIV and AIDS

Land area: 30,000 sq. km.	
Population: 2 million—urban 27%	
Capital city: Maseru, 250,000	
People: Basotho	
Language: Sesotho, English, Zulu, Xhosa	
Religion: Christian 80%, indigenous 20%	
Government: Constitutional monarchy	
Life expectancy: 45 years	
GDP per capita: $PPP 1,541	
Currency: Loti	
Major exports: Garments, footwear	

The landlocked kingdom of Lesotho is embedded in South Africa. Most of the territory is mountainous with its highest points in the east and north-east. The lowest and most densely populated areas, which have the best land, are in the north-west, particularly in the valley of the Caledon River.

The kingdom was formed in the 19th century when the Basotho ethnic group took refuge in the mountains to avoid being killed by the more aggressive Zulus and Boers. Today, almost all the people are still Basotho, though this includes a number of subgroups.

Lesotho is a poor country but by African standards has relatively high levels of human development. More than 80% of the population are literate, and 86% of children are enrolled in secondary schools. Until

the arrival of HIV and AIDS health standards too had been quite good, partly as a result of the fresh non-tropical climate, but also because of well-distributed health services.

Around 250,000 of Lesotho's people live temporarily or permanently in South Africa, many working on farms, but fewer are employed as miners because South African mines have become less profitable. By 2005 the number of miners had dropped to 47,000.

However, this still means that around 10% of the labour force is in South Africa. This has had a major social impact. Family structures are very disrupted, and most of the agricultural work in Lesotho is done by women.

A large migrant population also exposes Lesotho to HIV and AIDS. In 2007 almost one-quarter of the adult population were HIV-positive. This has led to a steep decline in life expectancy, from 60 years in the 1990s to 45 years.

For more than half the population, however, the main source of income is still farming. Only around one-tenth of the land is suitable for agriculture, and even this has become less productive as a result of over-exploitation and soil erosion, and agriculture contributes only 15%

of GDP. The main crops are maize, wheat, and sorghum, but even in good years around one-quarter of food needs to be imported. Many people raise cattle which they buy with migrant remittances. They also raise sheep and goats which generate some export income from wool and mohair.

In recent years, Lesotho has seen a surge in manufacturing industry. Investors from Taiwan, Hong Kong, and Singapore have established a series of joint ventures with the Lesotho National Development Corporation to manufacture clothing and footwear for export. Thanks to preferential access, more than half of these goods go to the United States. Other companies are now making handicrafts and furniture for export. More companies are also engaged in food processing and making other goods for local consumption.

Making garments and shoes for the USA

Lesotho's political relationship with South Africa has always been close. During the colonial era, South Africa had wanted to incorporate what was then Basutoland. The Basotho successfully resisted and in 1966 Lesotho was born as a constitutional monarchy. Elections the previous year had narrowly been won by the Basotho National Party (BNP), led by Chief Leabua Jonathan who became prime minister. Chief Jonathan was soon in conflict with King Moshoeshoe II, who had tried to extend his authority.

But Jonathan was no democrat. In 1970, when it looked as though the opposition Basotho Congress Party (BCP) had won the election, he suspended the constitution, dispatched the king temporarily into exile, arrested opposition politicians, and continued to rule by force.

In 1974 the BCP, following a failed uprising, formed the Lesotho National Liberation Army. But Jonathan was not overthrown until 1986, when the military took over and re-established King Moshoeshoe. In 1990, the army quarrelled with the king, exiled him again, and installed his son as King Letsie III.

This regime ended after a bloodless coup in 1991, which led in 1993 to new elections. These were convincingly won by the BCP, led by Ntsu Mokhehle. He reinstalled Moshoeshoe—who died in 1996 and was again succeeded by Letsie.

The BCP tried to promote national dialogue but eventually in 1997 fell apart. Mokhehle himself quit to form the Lesotho Congress for Democracy (LCD), though Mokhehle did not stay as leader, being replaced by Pakalitha Mosisili. In the 1998 election, the LCD won only 61% of the vote but gained 78 of the 79 seats.

In September 1998, an electoral commission refused to overturn the result, provoking riots and a revolt by junior officers. The government asked for intervention from South Africa and Botswana. The next election, it was decided, would have an element of proportional representation. This was held in 2002 with virtually the same result, and Mosisili returned as prime minister.

Much the same thing happened in the 2007 election which was won by the LCD still headed by Mosisili. The opposition again protested, resulting in international mediation and several court cases.

Liberia

Liberia has yet another peace agreement that could end decades of warfare

Land area: *98,000 sq. km.*
Population: *4 million—urban 62%*
Capital city: *Monrovia, 1.1 million*
People: *African 95%, Americo-Liberian 3%, other 2%*
Language: *English and about 20 others*
Religion: *Traditional 40%, Christian 40%, Muslim 20%*
Government: *Republic*
Life expectancy: *58 years*
GDP per capita: *$PPP 362*
Currency: *Liberian dollar*
Major exports: *Iron ore, timber, diamonds*

The most developed part of Liberia is the narrow coastal strip that extends around 50 kilometres inland. Behind this rises a belt of low hills and plateaux, and beyond this is the densely forested mountainous interior.

Liberia's 15 or more ethnic groups were left relatively undisturbed by the colonial powers until freed American slaves arrived in the mid-19th century and established the Republic of Liberia. Nowadays, the Americo-Liberians make up only around 3% of the population but until the early 1980s they dominated political life. The wars have partly been a struggle between this élite group and the rest of the population, as well as between other ethnic groups.

The wars are thought to have killed around 200,000 people, and displaced around 700,000, though most of these have now returned home, Even before the wars, this was one of Africa's

most urbanized countries, but the fighting and the flows of refugees have driven many people permanently from the rural areas.

Levels of human development have fallen steeply over the past few years. Infant mortality is high and the health system has lost many of its staff. The education system too collapsed in many parts of the country as more than 40,000 children were recruited into the different militias. In the past agriculture has provided work to around three-quarters of the workforce—the staple foods being rice and cassava. But the fighting drove many subsistence farmers from the land, and the population remains heavily dependent on food imports.

Many small farmers were also responsible for the country's major cash crop, rubber. One-third of this was produced by smallholders and the rest on large plantations now owned by the Japanese Bridgestone Company. The rubber industry now needs to be rebuilt—with investment not just in plantations but also in roads and in employee housing, much of which was destroyed in the war.

Logging has also been an important source of income. About 50% of Liberia is covered with trees and there are an estimated 3.8 million hectares of productive forests. Now that trade sanctions have been lifted, exports should revive.

Liberia is well endowed with minerals, including iron ore, gold, and diamonds. Indeed much of the fighting involved struggles for control over these resources. Iron ore was the most significant in terms of export income. Production had already fallen in the mid-1980s as a result of a global slump in the demand for steel. But the war brought the industry to a rapid halt. Again, heavy investment will be needed to replace the damaged equipment. But with proven reserves of over one billion tons of high-grade ore this should be profitable.

One billion tons of high-grade iron ore

Gold and diamond production will revive more quickly, since these are small-scale operations with about five thousand mining and dealing operations. But since much of the output is smuggled out of the country these make a limited contribution to the national economy.

The seeds for Liberia's civil war were sown in 1980 when a military coup led by Master-Sergeant Samuel Doe ended 150 years of domination by the Americo-Liberians. His regime promoted the interests of his own ethnic group, the Kahn, and rapidly degenerated into ruthless repression. After a rigged election in 1985 and attempted coup by the Gio group, full-scale civil war erupted in 1989 when Charles Taylor (an Americo-Liberian) invaded with rebel forces.

In 1990 a group of neighbouring countries through the Economic Community of West African States sent in a peacekeeping force, but this in turn became involved in offensive operations. Doe was assassinated in November that year. The ensuing years of anarchy saw many failed peace deals and a bewildering succession of alliances and splits between warring factions.

In 1995, another peace deal led to elections in 1997 which were won by Taylor. He ruled in lavish style, with a heavy hand, and with armed guards, as well as with an 'Anti-Terrorist Unit' commanded by his son.

Most opposition politicians fled, some of them to form 'Liberians United for Reconciliation and Democracy' (LURD), a rebel group that entered the country in 1999 launching a new round of civil wars. Faced with US pressure and advances by LURD and another group, Taylor, by then an indicted war criminal, finally resigned in August 2003 and left for Nigeria.

The ensuing Comprehensive Peace Agreement, which was backed by 15,000 UN peacekeepers, led to a two-year provisional administration.

In October 2005 presidential elections were held, and won by a former UN and World Bank civil servant, Ellen Johnson-Sirleaf, who defeated former footballer George Weah. Johnson-Sirleaf is well respected across the continent, has a good relationship with donors and has widespread support in the country. In 2009 she merged her Unity Party with two others, which will strengthen her position in the legislature.

Meanwhile Charles Taylor has been put on trial in The Hague for war crimes committed in Sierra Leone.

Libya

Libya's dictatorship faces a violent insurrection supported by NATO air strikes

Land area: 1,760,000 sq. km.
Population: 6 million—urban 74%
Capital city: Tripoli, 1.1 million
People: Berber and Arab
Language: Arabic, Italian, English
Religion: Sunni Muslim
Government: Military dictatorship
Life expectancy: 74 years
GDP per capita: $PPP 14,364
Currency: Libyan dinar
Major exports: Oil, petroleum products, natural gas

Most of Libya is a vast barren plain of rocks and sand. The majority of the population live in Tripolitania, in the north-west of the country, a region that includes the capital and a number of coastal oases. The other main inhabited area is Cyrenaica in the north-east.

Many of Libya's people descend from Bedouin Arab tribes—and the tribe is still often the basic social unit. The original Berber population has also largely been absorbed into Arab culture. Nowadays, however, most people live in the cities and the population is very young—60% are under 20 years old.

Libya's oil wealth has not been widely distributed. Around 70% of Libyan workers are employed in one way or another for the government on salaries that average only $190 per month. Unemployment is high, at around 30%.

But Libyans do benefit from subsidized housing and largely free education and health care. Notably for an Arab country, women are at least as well educated as men.

In addition to the native population, there are also around two million immigrant workers, chiefly from Egypt, often doing jobs that local workers refuse. Sometimes there have been tensions with the local population, and even riots.

Since 1959, when oil was first discovered, Libya has become one of the world's largest producers. Oil makes up more than 95% of exports and provides over 90% of government income. At current rates of extraction the oil should last 50 years or more. Most oil production is controlled by the National Oil Company but there are many exploration and coproduction contracts with European oil companies. Libya also has huge reserves of natural gas that have yet to be exploited.

Most other industrial operations are run by the state. They do now produce many consumer goods, though companies from Italy and Korea are manufacturing some items like refrigerators and video recorders.

Less than 2% of the land is suitable for farming so agriculture is limited and employs only 7% of the workforce. Around 80% of cereals must be imported. The main problem is water. Along the coast, groundwater has been over-extracted so sea water may seep in to replace it. The government's solution is the $20-billion 'great man-made river' scheme which will pump groundwater from aquifers in the south-east to the north. This project already delivers water, at great expense, to Tripoli and Benghazi, but has hit technical problems and is far from completion.

For the past 35 years, Libyan politics has revolved around Muammar Qaddafi. Following a military coup in 1969, Libya's monarchy was overthrown and a group of military officers led by the then 29-year-old Captain Qaddafi seized control and redirected Libya towards Arab socialism.

In 1977 he changed the country's name to the Socialist People's Libyan Arab Jamahiraya ('state of the masses'). In theory this is popular government, with everyone participating in Basic People's Congresses and choosing representatives to a General People's Congress which selects members of the government. The philosophy is spelled out in Qaddafi's 'Green Book' —a mixture of Islam and socialism.

Libya is a 'socialist state of the masses'

In reality, all power resides with the idiosyncratic Colonel Qaddafi, whose picture is ubiquitous. He is assisted by 'revolutionary committees' who control most institutions, including the media. Citizens also find their activities scrutinized by 'purification committees'. Human rights abuses range from arbitrary arrest and torture to extra-judicial executions and disappearances.

Qaddafi's main fear in the past was an Islamic uprising. In response, he appropriated Islamic symbols and set out an Islamic 'third path' between communism and capitalism.

Qaddafi has also been a thorn in the side of Western governments, particularly the USA. In the past he used Libya's oil wealth to bankroll a range of guerrilla groups from the PLO to the IRA. A series of terrorist incidents culminated in the bombing in 1988 of a Pan Am jet over Lockerbie in Scotland, killing 270 people. Libya originally denied involvement but in 1999 handed over the suspects for trial—one of whom was found guilty—and in 2003 admitted 'civil responsibility' and agreed to pay $2.7 billion to the victims' families.

Relations with the UK and the USA improved when Qaddafi condemned the 9/11 attack and promised to give up Libya's weapons of mass destruction.

In February 2011, however, taking inspiration from the 'Arab spring' in Tunisia and Egypt, protesters took to the streets demanding democratic reforms. Qaddafi responded violently, triggering an insurrection based in the eastern city of Benghazi that has now developed into a civil war. In order to protect the civilian population the UN Security Council declared a no-fly zone over Libya. Since then Qaddafi's forces have faced hundreds of air strikes from NATO. By mid-2011, however, the conflict appeared to have reached a stalemate.

Lithuania

The most ethnically homogenous Baltic state, but with a shrinking population

Land area: 65,000 sq. km.	
Population: 3.4 million—urban 67%	
Capital city: Vilnius, 556,000	
People: Lithuanian 83%, Polish 7%, Russian 6%, other 4%	
Language: Lithuanian, Polish, Russian	
Religion: Roman Catholic	
Government: Republic	
Life expectancy: 72 years	
GDP per capita: $PPP 17,575	
Currency: Litas	
Major exports: Mineral products, textiles, machinery	

Lithuania's terrain is mostly flat—a plain dotted with low hills and around 3,000 small lakes. The highest points are in the Baltic Highlands in the east and south-east. A feature of the sandy coastline is the narrow 100-kilometre sand spit that creates a distinctive lagoon.

Compared with the other Baltic states—Estonia and Latvia—Lithuania's population has a higher proportion of its own national group: more than 80% of people are ethnic Lithuanian. As a result, when it regained independence in 1991 Lithuania was less nervous than the other Baltic states about offering automatic citizenship to all—an offer that most people took up. Lithuanians also differ in that, like the Poles to the south, they maintain a strong Roman Catholic tradition.

The population has shrunk since 1991, initially as a result of emigration

to Russia, Poland, and elsewhere, but more recently because of falling birth rates and rising death rates. Mortality rates, particularly from heart disease and cancers linked to poor diets, rose steeply until 1995 though since then have they levelled off. Many men's health problems are linked to high consumption of alcohol.

The transition also caused an increase in poverty: beggars and street children are becoming part of a new underclass. In 2009 official unemployment rose steeply to 15% but the real figure is certainly much higher. And in any case most workers have to do a second job in order to survive.

Lithuania had industrialized rapidly in the Soviet era, but after 1991 the country lost its major markets and its sources of cheap energy. By 1995, industrial output had halved. Having reoriented more towards the West, some industries, particularly textiles, chemicals, and wood products, have since recovered. By 2007 manufacturing accounted for 19% of GDP and employed 17% of the workforce.

With only small amounts of domestic oil and gas, the country remains very dependent on energy imports. Two Chernobyl-style reactors at Ignalia provide over 80% of electricity—making this probably the most nuclear-dependent country in the world. Under EU pressure Lithuania has promised to close them—though this will mean buying more gas from Russia or building a new reactor.

Independence boosted the service sector, which by 2009 accounted for around two-thirds of GDP. Much

Lithuania embraces an orphaned Russian province

of this is concerned with transport and communications with Russia. A significant proportion of exports are actually re-exports to Russia of imported cars and consumer goods. Moreover, Russians have to cross Lithuania to reach their 'orphaned' province of Kaliningrad.

Agriculture continues to make an important if declining contribution—chiefly livestock and cereals. By 2007 it accounted for only 5% of GDP and 10% of employment. One of the first steps after independence was to break up the collective farms and cooperatives and return land to the descendants of former owners. Since most of this was in fairly small plots, efficiency and output fell.

In May 2001 Lithuania joined the EU which has further unsettled farmers who, although they benefit from subsidies, also have to meet higher standards.

One-quarter of the country is covered by forests which supply raw materials for wood and paper industries. With a strong world market the industry has been growing rapidly.

Lithuania's independence struggle had been led in the late 1980s by Sajudis—the Lithuanian Movement

for Restructuring. But in 1992 in the first free election to the unicameral parliament, the Seimas, the voters reverted to the former communist party, now the Democratic Labour Party (DLP). In 1993, the DLP candidate Algirdas Brazauskas was elected president. The president has considerable powers—including the appointment of ministers.

The DLP government made some progress but was rocked by a banking crisis and following the 1996 election Sajudis, now renamed the Homeland Union Party (TS), formed a coalition government with the Christian Democratic Party (LKD).

The general election in October 2000 produced another coalition government, New Policy. The main partner was the Liberal Union, led by Rolandas Paksas. The other partner was New Union–Social Liberals (NU). The coalition collapsed in June 2001 and the NU formed another with the Social Democratic Party whose leader, former president Algirdas Brazauskas, took over as prime minister.

Since then there have been several more coalition governments. In 2008 the TS merged with the LKD and a general election that year resulted in a centre-right, four-party coalition led by Andrius Kubilius of TS-LKD. But the coalition, which has only a small majority, is coming under greater strain as a result of the global economic recession which in 2009 caused a 12% fall in GDP.

The 2009 presidential election was won by Dalia Grybauskaite, a former diplomat who had been the EU's budget commissioner who so far has been broadly supportive of the government.

Luxembourg

Luxembourg is the world's richest country—but with a lot of help from foreigners

The northern one-third of Luxembourg is mountainous, with dense forests. The more fertile part of the country is in the southern two-thirds—the rolling plains of the 'Bon Pays'.

Luxembourg's people have their own dialect and a strong national identity, but most also speak French and German. They have the world's highest per capita income—and usually Europe's lowest rate of unemployment.

But the birth rate is also low, so immigrants are needed to fill the labour gaps. Foreign-born residents make up 40% of the population, and more than half the labour force. Portugal is the main provider. In addition there are some 60,000 'frontaliers'—cross-border workers who commute daily from France, Belgium, and Germany.

Most immigrants used to come to work in Luxembourg's heavy industry, particularly steel which took advantage of iron-ore deposits along the southern border. Steel is still an important industry, and the main export. And since an international merger in 2006 Luxembourg is the base for the world's largest steel company: Acelor Mittal. There are also other major manufacturing industries, including a Goodyear tyre plant and a DuPont chemical plant.

Land area: 3,000 sq. km.
Population: 0.5 million—urban 82%
Capital city: Luxembourg, 83,000
People: Luxembourgish and immigrants
Language: Luxembourgian, French, German, English
Religion: Roman Catholic 87%, other 13%
Government: Constitutional monarchy
Life expectancy: 79 years
GDP per capita: $PPP 79,485
Currency: Euro
Major exports: Steel, chemicals, rubber products

Today, however, the core of Luxembourg's economy is formed by financial and business services, which contribute around one-third of government revenue. Luxembourg is one of Europe's leading financial centres. Its low tax rates attract funds from Germany, and it controls more than 90% of Europe's offshore investment funds.

Luxembourg is a constitutional monarchy—a Grand Duchy. In September 2000, Grand Duke Jean abdicated in favour of his son Henri.

All governments since 1919 have been coalitions, usually led by the centre-right Christian Socialist Party (CSV). The CSV were the major party after the 1984 election. They governed in coalition with the Socialist Party (LSAP), with Jacques Santer as prime minister, and were re-elected in 1989 and 1994. Santer was replaced in 1995 by Jean-Claude Juncker.

In the 1999 elections the LSAP were pushed to third place by the right-wing Democratic Party which became the coalition's junior partner. The elections in June 2004 also produced a CSV-led coalition with Juncker as prime minister. This coalition was returned in a stronger position in the 2009 election.

Macedonia

Macedonia avoided the worst of the Balkan wars, but has persistent ethnic tensions

Land area: *26,000 sq. km.*
Population: *2 million—urban 68%*
Capital city: *Skopje, 467,000*
People: *Macedonian 64%, Albanian 25%, Turkish 4%, Roma 3%, Serb 2%, other 2%*
Language: *Macedonian 67%, Albanian 25%, Turkish 4%, Serbian 1%, other 3%*
Religion: *Macedonian Orthodox 64%, Muslim 33%, other 3%*
Government: *Republic*
Life expectancy: *74 years*
GDP per capita: *$PPP 9,096*
Currency: *Macedonian denar*
Major exports: *Food, tobacco, machinery*

This landlocked country is largely mountainous. In the past it was also heavily forested, and though forests remain, particularly in the west, the cleared land revealed thin soil, much of which has now been eroded. The best land is to be found in the valley of the Vardar River, which runs through the centre of the country from north to south. Tectonic fault lines run in the same direction and the country remains vulnerable to earthquakes.

Macedonia's ethnic composition is a source of political tension. The 2002 census showed two-thirds of the population as ethnic Macedonian who belong to the Eastern Orthodox Church. There are also minorities of Turks, Romanians, and Serbs.

But the most significant minority were ethnic Albanians, who made up one-quarter of the population. The Albanians, who are Sunni Muslims, live largely in the north-west. They claim they were under-counted and that in fact they represent around one-third of the population. Albanians are often treated as second-class citizens. In 2001 guerrillas from the Albanian National Liberation Army (NLA) attacked the government. After a peace agreement in Ohrid in August 2001, the government made a number of concessions, concerning power-sharing, the use of the Albanian language, Albanian education, and hiring more Albanian policemen.

Macedonia was the poorest of the constituent republics of former Yugoslavia. It had developed heavy industries such as steel, chemicals, and metal-based manufacturing, and was mining its deposits of lead, zinc, copper, chromium, and coal. It also had light industries such as textiles. But the break-up of Yugoslavia and subsequent trade embargoes sent industry into a steep decline. By 1995, output had dropped by more than one-third. There has been limited foreign investment in mining. Manufacturing industries have since recovered, but between 1998 and 2007 employment in manufacturing continued to fall.

Officially unemployment in 2009

was 35% but many people actually work in the black economy which accounts for around one-third of GDP.

Agriculture also suffered in the early years but has recovered more strongly. Even as part of Yugoslavia, Macedonian agriculture depended largely on small private farms which produce most of the food as well as important cash crops such as tobacco. Tobacco remains an important crop since it supports around 40,000 households. Tutunski Kombinat, which is part of the Imperial Tobacco group, is the country's third-highest taxpayer

The country could be self-sufficient in food, and even export vegetables and fruit, but at present has to import much of its requirements.

Macedonia's emergence as an independent republic in 1991 was a tortuous process. Neighbouring Greece immediately objected to the name on the grounds that this and the constitution implied sovereignty over the adjacent Greek province of the same name. Greece's objections for a time blocked international recognition.

Objections to the country's name
The name problem was solved by calling the country the 'Former Yugoslav Republic of Macedonia'—though few people use this. Macedonia joined the UN in 1993 and almost immediately requested a UN peacekeeping force which successfully prevented annexation by Serbia.

The 1992 parliamentary election was won by the ex-communist Social Democratic Alliance of Macedonia (SDSM), which entered into a coalition with the moderate Albanian Party for Democratic Prosperity. The SDSM increased its representation in the 1994 parliamentary elections which were boycotted by the main opposition party, the Internal Macedonian Revolutionary Organization (VMRO-DPMNE). But the government was beset by pyramid scheme scandals and by riots in the Albanian-dominated city of Gostivar.

In the general election in 1998, the VMRO-DPMNE, formed a centre-right coalition government. Its main partner was the new pro-business Democratic Alternative (DA) Party, along with the Democratic Party of Albanians (DPA), a radical separatist group. It did deal successfully with the Albanian insurgent campaign of 2001 but lost the parliamentary election of 2002.

This saw a return for the SDSM which formed a coalition with the ethnic Albanian party Democratic Union for Integration (DUI), which also represented the NLA, with the SDSM's Branko Crvenkovski as prime minister. After the death of the president, the SDSM's Boris Trajkovski, in a plane crash. Crvenkovski was elected as president.

After the elections in 2006 the government consisted, as usual, of a multi-party coalition, this time led by Nikola Gruevski, of VMRO-DPMNE, as prime minister.

In 2008, Gruevski decided to call a snap election which he won with an increased majority. His position was further strengthened in 2009 with the election of the VMRO-DPMNE candidate, Gjorge Ivanov, as president. In 2011 after the opposition boycotted the parliament, he called another election in which VMRO-DPMNE again had the largest share of the vote.

Madagascar

Madagascar remains riven by corruption and political conflict

Madagascar is dominated by a central mountainous plateau that covers around two-thirds of the island—the highest peak is 2,800 metres. To the east, the mountains drop steeply to a narrow coastal strip. To the west the slope is more gentle to broad fertile valleys. Most of the original forest cover on the mountains has been lost and Madagascar now suffers from severe erosion as millions of tons of soil are washed each year into the sea. Deforestation has also stripped the natural habitats of many rare plant and animal species.

Madagascar is frequently battered by cyclones—and in early 2000 was hit by three in fairly rapid succession, with others in 2003 and 2004.

Most Malagasy are descendants of immigrants who arrived more than 2,000 years ago from South-East Asia. Today, these consist of around 20 ethnic groups, of whom the largest are

Land area: 587,000 sq. km.
Population: 19 million—urban 30%
Capital city: Antananarivo, 1.5 million
People: Merina, Betsileo, Betsimisaraka, and others
Language: French, Malagasy
Religion: Indigenous beliefs 52%, Christian 41%, Muslim 7%
Government: Republic
Life expectancy: 60 years
GDP per capita: $932
Currency: Malagasy franc
Major exports: Coffee, vanilla, cloves

the Merina (around one-quarter of the population) and the Betsileo. Some groups such as the Betsimisaraka are of mixed African, Malayo-Indonesian, and Arab ancestry. All speak dialects of Malagasy, though the education system also works in French.

Levels of human development in Madagascar remain low. Two-thirds of the population live in poverty—with an increasing disparity between urban and rural areas. Life expectancy, however, is a little higher than in other countries of Sub-Saharan Africa, and the island's relative isolation offers some protection from the rapid spread of HIV and AIDS. Nevertheless, one-third of children are malnourished and only one-third of the rural population have access to safe water.

Madagascar's poverty is concentrated in the countryside where 80% of people struggle to survive through agriculture. Much of this is at a subsistence level, primarily growing rice, along with cassava, sweet potatoes, maize, and other crops. But farms are generally small, yields are low, and food has to be imported. The main cash crops are coffee, vanilla, sugar, and cloves, but here too output has been stagnant. Some of the problems result from under-investment and there is relatively little irrigation. In fact much more land could be brought under cultivation to grow food that might displace imports. Livestock rearing which occupies more than half the land is also an important source of income—indeed it often takes up land that might better be used for growing crops.

Prices for cash crops have been liberalized in recent years, so farmers' rewards are linked more to world prices. But with poor infrastructure and inadequate systems for quality control they are at a disadvantage against international competitors.

Fishing, especially for prawns, should be an important source of income but the local industry is relatively underdeveloped, so much of the income actually comes from licences sold to European and Japanese fleets that fish for tuna in Madagascan waters.

Selling fishing licences to the EU and Japan

Madagascar has some basic industries producing goods for local consumption, including plastics, pharmaceuticals, and footwear. It has also had some success in developing export-processing zones which account for around 80% of foreign investment. These include a textiles industry which is based on locally-grown cotton.

Madagascar's dominant political figure until 2001 was Didier Ratsiraka. He first came to power at the head of a military government in 1975 and held a referendum that approved a new constitution allowing for only one political organization. This was to be a 'national front' that embraced all the political parties, though it was actually controlled by Ratsiraka's party which was subsequently renamed the Association pour la renaissance de Madagascar (Arema).

Ratsiraka embarked on a socialist transformation, nationalizing banks and other major companies, but by 1980 was seeking the support of the IMF and in1990, he allowed other political groups to operate.

This led to the formation of a new opposition front, the Forces vives (FV), led by Albert Zafy. Zafy and the FV won the elections in 1993. But Zafy's chaotic presidency was fraught with problems and in 1996 he was impeached for corruption and abuse of power. In the ensuing presidential election Ratsiraka returned.

For the 2001 presidential election the opposition united behind businessman Marc Ravalomanana of the Tiako-I-Madagasikara (TIM) party and mayor of Antananarivo. After the first round Ravalomanana was just short of 50% of the vote but said the election had been rigged and took control of the administration while Ratsiraka and his supporters blockaded the capital. A recount confirmed Ravalomanana's victory and Ratsiraka was eventually exiled.

Ravalomanana was comfortably re-elected in 2006 and the TIM was victorious in the 2007 legislative, and regional elections. In March 2009, however, following widespread popular resentment about his autocratic style and favouritism towards his own businesses, he was overthrown in a coup and replaced by a Haute Autorité pour la Transition, led by Andry Rajoelina, another former mayor of Antananarivo. Elections are due in late 2011.

Malawi

Malawians now have a democracy, but they remain desperately poor and millions are dying from AIDS

Land area: 118,000 sq. km.
Population: 14 million—urban 20%
Capital city: Lilongwe, 890,000
People: Chewa, Nyanja, Tumbuko, and others
Language: English, Chichewa, and others
Religion: Christian 80%, Muslim 13%, other 7%
Government: Republic
Life expectancy: 52 years
GDP per capita: $PPP 761
Currency: Kwacha
Major exports: Tobacco, tea, sugar, coffee

Malawi's dominant geographical feature is the East African Rift Valley, which cuts through the country from north to south. This includes Lake Malawi, through which runs most of the eastern border, and the Shire River valley, which drains the lake southwards into Mozambique. To the west of the Rift Valley in the centre of the country, there is an extensive plateau area. There are also high plateaux in the north and intensively cultivated highlands in the south.

The main ethnic groups are the Chewa, who are the majority of people in the centre of the country and whose language is the most widely spoken. The Nyanja are mostly in the south, and the Tumbuko in the north.

By African standards, Malawi is densely populated and, with few people using contraception, the population growth rate prior to the

AIDS epidemic was 3.5% per year, but could now be around 2%.

Malawi is a poor country. Around two-fifths of people live below the poverty line. Its education levels are higher than in some African countries and it even has its own version of an élite English public school, the Kamuzu Academy. But most children are less fortunate and drop-out rates at ordinary government schools are high. Half the children are malnourished and around 10% of children do not live beyond their fifth birthday.

Mortality rates started to climb even more steeply as a result of AIDS. By 2007, 8% of the population were HIV-positive. By the end of 2005 it was estimated that 40% of teachers, for example, had died. AIDS deaths are pushing the survivors even deeper into poverty, since those who will die are the main breadwinners, and the country will also face the heavy cost of caring for AIDS patients.

With 80% of people living in the rural areas, this is one of the world's least urbanized countries. Farmers have cultivated virtually all the arable land, and in the mid-1980s more than half of rural households had less than one hectare each. Most concentrate on subsistence crops like maize, sorghum, or millet.

In recent years fertilizer subsidies for farmers—making it practically free for the poorest—have substantially increased output. Over the four years to 2009 the maize harvest trebled and over half is now exported to Kenya. This is expensive, costing around 4% of GDP, so may not be sustainable in the long term.

Cash crops have also been important, particularly tobacco, which has provided 50% of export income, along with tea, sugar, and cotton. Previously the policy was to concentrate cash crops on large commercial estates (which still occupy around 20% of arable land) but the government is now encouraging smallholders by offering better prices.

Malawi relies heavily on exports of tobacco

Another important source of food is Lake Malawi, whose fish provide more than two-thirds of animal protein consumption—though catches have fallen in recent years as a result of over-fishing and pollution.

Malawi has some mineral deposits but no substantial mining industry. Now the government is trying to encourage foreign companies to develop bauxite and titanium mines.

For the first three decades following independence in 1962, political life in Malawi was overshadowed by the diminutive but dictatorial figure of Dr Hastings Banda at the head of his Malawi Congress Party (MCP). He was first prime minister and then president, and finally in 1970 declared himself 'president for life'. Banda's rule became increasingly despotic. He had an extensive network of spies and secret police to enable him to jail and torture opponents.

Banda's regime started to fall apart in the early 1990s following strikes and riots. Eventually he was forced to concede a 1993 referendum which approved a return to multiparty democracy.

In the ensuing presidential election in 1994, Banda was defeated by a former cabinet minister, Bakili Muluzi, at the head of the United Democratic Front (UDF). The legislative election also gave the UDF the most seats, ahead of the MCP, but without a majority it initially entered into a coalition with the third party, the Alliance for Democracy (Aford).

After his election Muluzi released a number of political prisoners. He also had Banda arrested on a murder charge, though he was acquitted, and died (101 years old) in 1997. In the 1999 elections, Muluzi scraped through to a second five-year term.

In July 2002, the parliament rejected a motion to remove the limit on the number of presidential terms. Muluzi accepted this and backed Bingu wa Mutharika as the UDF candidate for the May 2004 election, which he duly won.

However Mutharika soon fell out with Muluzi who had remained as UDF chairman. Mutharika then formed his own Democratic Progressive Party (DPP) and started an anti-corruption drive, including a case against Muluzi which was still underway in 2010.

In 2009 Mutharika won the presidential election in a landslide and the DPP won a large majority in the presidential election.

Malaysia

Malaysia's ruling party is coming under increasing pressure.

Land area: 330,000 sq. km.
Population: 27million—urban 72%
Capital city: Kuala Lumpur, 1.5 million
People: Malay 54%, Chinese 23%, indigenous 11%, Indian 7%, other 5%
Language: Bahasa Melayu, English, Chinese, Tamil, and tribal dialects
Religion: Muslim, Buddhist, Hindu, Confucian, Christian, tribal religions
Government: Constitutional monarchy
Life expectancy: 74 years
GDP per capita: $PPP 13,518
Currency: Ringgit
Major exports: Manufactured goods, rubber, palm oil

Malaysia has an unusual geographical composition. Its territory is evenly divided between a portion on the Asian mainland and a similar area on the north-west of the island of Borneo. On the mainland, peninsular Malaysia, which has around 80% of the population, is largely mountainous in the north, with coastal lowlands to the west and south. Eastern Malaysia, 600 kilometres away across the South China Sea, is more sparsely populated, with a swampy coastal plain rising to high mountains that form the border with Indonesia.

Malaysia also has a distinctive racial composition. The majority are classified as ethnic Malays, most of whom are Muslims. But around one-third of the population are of Chinese origin, who live chiefly in the urban areas. There are also a number of South Asians, as well as small tribal groups who are found particularly in Eastern Malaysia. Standards of education and health have improved rapidly though there remain wide disparities between urban and rural areas and between richer and poorer states. There are also notable differences in population growth between ethnic groups. In 2007 the growth rate for Malays was 2.2% while for Indians it was 1.4% and for the Chinese 1.1%.

Since the early 1970s, Malaysia has transformed itself into an export-oriented industrial country. More than one-quarter of the population now work in manufacturing industry, which until recently was largely concentrated in states on the west of the peninsula. However the government has now created 14 'free industrial zones' along with 200 other industrial estates across the country. Much of this has been driven by electronics multinationals that have used Malaysia as an assembly base.

The government has recently been making efforts to have more manufacturing take place in Malaysia and use higher levels of technology. The most ambitious project is a 'multimedia super-corridor', carved out of land that previously was either jungle or palm-oil plantations. This was slow to take off but by mid-2005 had attracted 1,112 companies employing 23,000 workers.

Attracted by better-paid jobs in the cities, many people have been leaving the land. As a result, agricultural production has fallen and Malaysia is now a net importer of rice. Production of rubber, long a major export, has also been declining—though partly because plantations have switched to the more lucrative palm oil of which output continues to expand.

Malaysia is also reliant on immigrant labour. In 2009, foreign residents were officially 6% of the population but there may be one million more migrants there illegally, primarily from Indonesia.

Malaysia's distinctive ethnic mix has also had a profound impact on its political processes. The country is a federal constitutional monarchy, and formally it is an Islamic state.

Rulers take turns as king of Malaysia
Each of the 13 states has a state assembly and an executive council that deals with state issues. Nine of the states have hereditary rulers who take turns to serve a five-year term as king.

Since independence in 1957, political power at the federal level has been in the hands of the main Malay political party, the United Malays' National Organization (UMNO), which has ruled in a National Front coalition with parties representing other racial groups.

The political system in Malaysia turned decisively in a new direction following riots in 1969 when Malays protested against the pervasive economic power of the Chinese minority. From 1971, the government embarked on a 'new economic policy' that would deliberately favour Malays and indigenous ethnic minorities, referred to as 'bumiputra' (sons of the soil). This included preferential access to government jobs, to higher education, and to a share of equity in national and foreign investment.

After becoming prime minister in 1981, UMNO leader Dr Mahathir Mohammed built a dominant position within Malaysia and became an outspoken international figure, notably accusing foreign currency speculators of provoking Asia's financial crisis.

Mahathir's position was weakened somewhat in 1998 when he had his obvious successor, Anwar Ibrahim, jailed on charges of corruption and sexual misconduct. And the main opposition group, the Parti Islam sa-Malaysia (PAS), seemed to be gaining ground. Nevertheless, Mahathir remained in control until he stepped down after 22 years in 2003.

His chosen replacement as prime minister was Abdullah Badawi whose position was confirmed with a striking victory in the parliamentary elections in 2004, when his Barisan Nasional (BN) coalition gained 90% of the seats. This was a crushing rejection of PAS fundamentalism.

In March 2008, however, the Barisan Nasional performed badly in the general and state elections, and in 2009 Badawi retired to be replaced by Najib Razak.

Meanwhile Anwar Ibrahim, now out of jail, and de facto leader of the opposition, Parti Keadilan Rakyat, is gaining support, though bizarrely has been again charged with sodomy.

Maldives

Finally having achieved democracy, the country is threatened by climate change

Land area: 300 sq. km.
Population: 0.3 million—urban 41%
Capital city: Male', 62,000
People: Sinhalese, Dravidian, Arab, African
Language: Maldivian Divehi, English
Religion: Muslim
Government: Republic
Life expectancy: 71 years
GDP per capita: $PPP 5,196
Currency: Rufiyaa
Major exports: Marine products

The Maldives consists of 1,190 coral islands grouped in atolls, spread out over a wide area. Of these 200 are inhabited, while 80 others are tourist resorts. Since few rise more than one metre above sea level, global warming could swamp them. In 2004 the Asian tsunami washed across the islands, though only killed 82 people.

Culturally, the Maldives is a blend of influences from India, Sri Lanka, Arabic countries, and elsewhere. All Maldivians speak the national language, Divehi, and are Muslims. In addition migrant workers from Bangladesh, India, and Sri Lanka make up more than one-quarter of the labour force, providing professionals, such as teachers, as well as labourers.

By South Asian standards, Maldivians are healthy and well off, but rapid development has also been socially disruptive. The Maldives has the world's highest divorce rate and many young people are abandoning the atolls for Male' and some have fallen victim to heroin peddlers.

The country's major source of income is tourism, which brings in around half a million visitors per year at the top end of the market—up to $10,000 per night—providing one-third of government revenue and 70% of foreign exchange. The hedonistic tourists are however mostly kept on

their own islands, well away from the conservative Muslim population.

Fishing is also an important source of income and employment. The main catch is tuna, which is processed locally and exported frozen or canned.

From 1978 until 2008, the winner, through a mixture of guile and repression, was Moumoon Abdul Gayoom. In 2003 he gained a sixth successive term. Under international pressure in 2004, he allowed the registration of political parties, and established a special people's Majlis to draw up a new constitution.

The main opposition came from the Maldivian Democratic Party (MDP), which was run for many years from Sri Lanka by Mohamed Nasheed. In 2008, after many years in either exile or jail, Nasheed finally beat Gayoom in a run-off election. Gayoom to his credit dutifully accepted defeat.

Nasheed is also supported by the religiously conservative Adhaalath Party which could lead to conflict with his more liberal MDP supporters. He also has to deal with an opposition-controlled parliament which has blocked many proposals.

In 2009 Nasheed held a cabinet meeting underwater to dramatize the threat of climate change, and has considered buying land in other countries.

Mali

Mali is one of the poorest developing countries, but also one of the more democratic

| Miles | 0 | 400 |
| Km | 0 | 640 |

Land area: 1,240,000 sq. km.
Population: 12 million—urban 33%
Capital city: Bamako, 1.4 million
People: Mandé 50%, Peuhl 17%, Voltaic 12%, Songhai 6%, Tuareg and Moor 10%, other 5%
Language: French, Bambara, other African languages
Religion: Muslim 90%, indigenous beliefs 9%, Christian 1%
Government: Republic
Life expectancy: 48 years
GDP per capita: $PPP 1,083
Currency: CFA franc
Major exports: Cotton, gold, livestock

Mali is almost entirely flat with only occasional broken rocky hills. The northern one-third of the country falls within the Sahara Desert. The centre is the semi-arid Sahel belt, which gives way further south and south-west to grasslands and to the valley of the Niger—a broad river that periodically floods the land, depositing rich alluvial soil.

Mali has a diverse collection of ethnic groups and clans that tend to be concentrated in specific areas. The main distinction is between the Berbers and the black groups. Berbers include the Tuareg and the Moors, who are nomadic herders and occupy the Sahelian zone. Of the black Mandé group, the largest are the Bambara who are farmers along the Niger River. The Peuhl are another nomadic herding group in the Sahel. Most Malians are Muslims, though religious affiliations are also divided along ethnic lines. Islam is not as strong a political force as in other Islamic countries but there are a number of fundamentalist groups. Mali's ethnic diversity has frequently led to conflict between the nomadic and sedentary groups.

The Tuareg in particular have felt marginalized and in 1991 they rebelled, demanding autonomy from the central government. The army responded forcefully, driving more than 80,000 Tuareg into neighbouring countries. A peace agreement was achieved in 2006 though not all Tuareg factions have abided by it.

There are also around three million Malians working as emigrants, primarily in neighbouring countries such as Côte d'Ivoire, but there are also many in France.

Mali is one of the world's poorest countries: two-thirds of people are below the poverty line. Around a quarter of children are malnourished. Life expectancy is short even though HIV infection rates are low by African standards at less that 1%. Two-fifths of children do not attend school and as a result around two-thirds of the population are illiterate.

Most people depend for survival on subsistence agriculture and livestock. They grow food crops in the south on irrigated land around the Niger and its tributaries, their main crops being millet, rice, wheat, and corn. This is also where they grow the principal cash crops: peanuts and particularly cotton which is a major source of export income. Most cotton is grown by small farmers and cooperatives; quality is high and output has been increasing. Although there are some plans to process cotton locally, most is exported raw.

Livestock production accounts for around one-fifth of GDP. This used to be the preserve of nomadic herding communities, but nowadays most of the cattle are on small farms. Fishing in the Niger river is also an important source of income and around one-fifth of the catch is exported to Côte d'Ivoire.

Mali has little industry beyond processing agricultural output. Most consumer goods are imported, or smuggled in, from neighbouring countries. But one increasingly important activity is gold mining. Output in 2007 was around 57 tons, mostly through South African companies working with the government. In 2000 gold overtook cotton as the leading export earner. The country's reserves are equivalent to at least 600 tons. Production costs are low, and world prices have been rising so Mali is likely to rely more on gold exports.

Mali has around 500 tons of gold

For 23 years, Mali was under the autocratic rule of Moussa Traoré, who became president in 1969 following a military coup. In 1979 the country returned to single-party civilian rule. This made little difference since Traoré was elected president at the head of a military-backed party and re-elected in 1985—as the only candidate. Throughout this period there had been a number of protests, with a former minister Alpha Oumar Konaré as one of the leading figures. These, along with several coup attempts, were violently suppressed.

Traoré survived until 1991 when, after mass protests, he was arrested and replaced by an interim administration headed by Lieutenant-Colonel Amadou Toumani Touré.

In 1992, a new constitution led to multiparty elections and a victory in the National Assembly for a coalition led by Konaré and the Alliance pour la démocratie au Mali (Adema). Konaré was elected president.

Konaré was a veteran pro-democracy campaigner and was rewarded with support from donors. The opposition parties, however, accused Konaré of trying to recreate a one-party state and boycotted the 1997 election which Konaré duly won.

The constitution prevented Konaré standing again in the 2002 presidential election, prompting a lot of infighting in his party for the candidature. This opened the way for former transitional president Amadou Toumani Touré, who won as an independent.

Touré has been a popular president, and also has an international reputation as a mediator in other countries. He was re-elected in 2007. In advance of the 2007 legislative elections, the major parties, including Adema, formed a pro-Touré, Alliance pour la démocratie et le progrès, which won a large majority.

Malta

Malta is now a bridge between the EU and North Africa

Malta comprises a small group of islands in the Mediterranean, of which the largest are Malta, Gozo, and Comino. The land is largely low-lying and dry, without permanent rivers or lakes, so around half of water comes from desalination plants.

Malta is a bridge between Europe and North Africa, and its language has both Latin and Arabic components. In the past, young Maltese often chose to emigrate, creating a Maltese diaspora around the world, notably in Australia. Nowadays people tend to stay, and the population continues to grow, if only slowly. The Maltese enjoy an extensive welfare state with free education and health care.

Not many work in agriculture, which is restricted by thin soil and the lack of water, so although farmers do have crops of cereals, fruits, and vegetables, most food has to be imported.

One-fifth of the workforce are in manufacturing industry, either in small-scale factories making goods for local consumption or in the export factories established as a result of foreign investment—typically in clothing and light engineering, and a large SGS-Thomson electronics factory. However the more labour-intensive industries are declining in importance.

Until 1979, Malta had a major

Land area: 316 sq. km.
Population: 0.4 million—urban 95%
Capital city: Valletta, 6,300
People: Maltese
Language: Maltese, English
Religion: Roman Catholic
Government: Republic
Life expectancy: 80 years
GDP per capita: $PPP 23,080
Currency: Euro
Major exports: Clothing, electronics

British naval base and still has a ship-repair industry. More important to the economy now is tourism, which accounts for 17% of GDP. Around 40% of the 1.2 million visitors are from the UK. Many others arrive from elsewhere on cruise liners.

Maltese politics is frequently confrontational and animated, and as a result electoral turnouts are commonly above 90%. For much of the 1970s and 1980s, Malta was governed by the then socialist Maltese Labour Party (MLP) led by Dom Mintoff, who built up a welfare state. Even today, a high proportion of the workforce are employed in the public sector.

Since 1987, however, with a brief 1996–98 Labour interruption, the country has been in the hands of the centre-right Nationalist Party (PN) which has been trying to privatize and to cut down the public sector. The PN was returned to power in the 2003 general election, led by Eddie Fenech Adami, though he was in 2004 appointed president. The PN, now led by Lawrence Gonzi also won, though narrowly, in 2008.

Ideologically, the two parties have moved closer together. The most divisive issue was EU membership, but the issue was finally settled when Malta joined in 2004—and adopted the euro in 2008.

Martinique

A French dependency with hopes for independence

Martinique is a mountainous island in the eastern Caribbean with an active volcano, Mont Pelée, though it does have some flatter parts in the south-west. Abundant rainfall for most of the year sustains lush vegetation: more than one-third of the island is covered with tropical rainforests. There are also many sandy beaches.

The people of Martinique are a rich racial mixture of black, white, and Indian. Thanks to steady flows of French aid, their standard of living is fairly high. The capital, Fort-de-France, is a smart cosmopolitan city, combining French and West Indian cultures, though its attractions have also encouraged many younger people to emigrate onwards to metropolitan France. The descendants of the white settlers, the Béké, still own plantations but most political and economic power nowadays is in the hands of the creole élite.

Historically, Martinique was a major sugar producer and there are still sugar plantations, which supply the raw material for the rum distilleries, but farmers have been diversifying more into tropical fruits including pineapples, avocados, and bananas. Agriculture only contributes around 5% of GDP and most food has to be imported.

Today, the main industry is tourism. More than three-quarters

Land area: 1,000 sq. km.
Population: 0.4 million—urban 97%
Capital city: Fort-de-France, 94,000
People: African and African-white-Indian mixture 90%, white 5%, other 5%
Language: French, creole
Religion: Roman Catholic 85%, other 15%
Government: Overseas department of France
Life expectancy: 79 years
GDP per capita: $PPP 14,400
Currency: Euro
Major exports: Petroleum products, bananas, rum

of the workforce are employed in services, working in hotels and restaurants, as well as in government.

There is also a refinery which produces oil products for local consumption and for other French dependencies in the Caribbean.

As an overseas department of France, Martinique sends two representatives to the French Senate and four to the French National Assembly. For local consultation, it also has a general council and a regional council.

Martinique has typically voted for left-wing parties and since 1992 the president of the general council has been Claude Lise of the Progressive Martinique Party. Of the French dependencies, Martinique has had the most persistent agitation for independence. While some political groups just want greater autonomy, and to attract more investment, others have been elected on pro-independence platforms.

Given the high levels of unemployment, however, and the dependence of many people on welfare payments and other transfers from France, in the short term a vote for independence seems unlikely.

Mauritania

Nomadic communities have settled in the towns, but many traditional practices remain—including slavery

Land area: 1,026,000 sq. km.
Population: 3 million—urban 41%
Capital city: Nouakchott, 760,000
People: Mixed Maur/black 40%, Maur 30%, black 30%
Language: Hasaniya Arabic, Pular, Soninke, Wolof
Religion: Muslim
Government: Military transitional
Life expectancy: 57 years
GDP per capita: $PPP 1,927
Currency: Ouguiya
Major exports: Fish, iron ore

Mauritania is essentially the western section of the Sahara Desert. It does have a strip of arable land alongside the southern border that is formed by the Sénégal river, and the territory to the north of this consists of dry grasslands with occasional bushes, but more than half the country is covered with sand dunes.

The population represents an overlap between ethnic groups from the north and south. Those from the north are lighter coloured and of mixed Arab and Berber descent. Those from the south are black, from a number of groups, including the Fulani, Soninke, and Wolof. But there has also been extensive intermarriage, so most of the population is a mixture of the two groups.

Mauritanian society has traditionally been very hierarchical—with noble families at the top and then a series of different castes extending down to servants and slaves. Slavery was abolished during the French colonial period, and the prohibition was reiterated with new laws in 1980. However, Amnesty International reports that slavery and serfdom continue—even if owners can no longer call on the law to pursue runaway slaves. The slaves may belong to any ethnic group, but particularly the blacks.

Living standards are very low. Around one-quarter of children are malnourished and infant mortality is high—especially for those outside the towns and cities who have few facilities for health care. Around half the population live in poverty.

In the past, most Mauritanians lived in nomadic herding communities, and livestock still contributes around one-sixth of GNP—one million each of cattle and camels in the south and 14 million goats and sheep further north. But the numbers of nomadic herders have been falling. To some extent this is a response to desertification—a result in part of over-grazing. A series of droughts also wiped out herds, driving destitute families to the cities. Moreover, the urban life in general is attractive to younger generations.

Mauritania's little available agricultural land, around 1% of the territory, is used mostly to grow millet and sorghum. Some rice is produced on irrigated land—usually by agricultural labourers working for richer landlords. Nevertheless, most cereals have to be imported. Many rural communities also maintain herds of cattle, camels, and goats.

The Atlantic coastline opens up opportunities for fishing—which provides around half of export

Fishing grounds depleted by foreign trawlers

revenue. This is one of the richest fishing areas in the world but over-fishing by local boats and foreign 'supertrawlers', has depleted the stock. The government has taken measures to combat over-fishing—including tighter surveillance and new agreements for sharing production with fleets from the EU.

With limited agricultural potential, Mauritania's economic survival has rested on the exploitation of mineral resources which have generated 12% of GDP and half of export revenue. The most important is iron ore. The Zouérat mine near the border with Western Sahara has rich deposits of iron ore that have been extracted since the early 1960s. Most of the output goes to the EU.

One recent development is the discovery of offshore oil and gas which is being exploited by the Malaysian company, Petronas.

Mauritania's poverty and vulnerability to shocks, whether from the climate or international commodity markets, have frequently obliged it to turn to the IMF. Popular protests against the ensuing austerity measures have usually been dealt with fairly harshly. Mauritania is, however, a major aid recipient and has also qualified for debt relief—which has allowed it to increase social spending.

Since independence in 1960, political life in Mauritania has been controlled by the Arabic speakers from the north who have periodically made efforts to further 'Arabize' the rest of the country. In 1984 the one-party government that had been in power since independence fell to a military coup led by Colonel Maaouya Ould Sid'Ahmed Taya.

Under pressure from both inside and outside the country, Taya legalized political parties, including his own Parti républicain démocratique et sociale. He won presidential elections in 1992, 1997 and 2002. But Taya was no democrat and frequently harassed, and several times banned, opposition parties. Never very popular, he was overthrown in 2005 in a bloodless coup by army officers led by Colonel Ely Ould Mohamed Vall.

Vall kept his promise to devolve power to a civilian government within two years, and a free and fair presidential election in March 2007 was won by Sidi Mohamed Ould Cheikh Abdallahi, who had been one of Taya's ministers.

In 2008 Abdallah sacked the head of the presidential guard, Mohamed Ould Abdel Aziz, who retaliated by overthrowing him in another military coup. Aziz, too, quickly restored constitutional rule by getting himself elected president in July 2009 at the head of a new party, Union pour la République, whose supporters also control the National Assembly.

Mauritius

A human development success story with a booming economy

Land area: 2,000 sq. km.
Population: 1.3 million—urban 43%
Capital city: Port Louis, 148,000
People: Indo-Mauritian 68%, creole 27%, Sino-Mauritian 3%, Franco-Mauritian 2%
Language: Creole, French, English, Hindi, Urdu
Religion: Hindu 48%, Roman Catholic 24%, Muslim 17%, other 11%
Government: Republic
Life expectancy: 72 years
GDP per capita: $PPP 11,296
Currency: Mauritian rupee
Major exports: Textiles, garments, sugar

Mauritius, an island in the Indian ocean surrounded by coral reefs, is the site of an ancient volcano. The rainiest and most mountainous part of the island is in the south, falling to a plateau in the centre and then to a drier broad plain in the north.

The Mauritian population contains distinct and diverse groups. Two-thirds are of Indian origin—descendants of around half a million Indians who after 1850 were brought to work as indentured labourers in the sugar plantations—of these around one-quarter are Muslim and the rest Hindu. Then there are the creoles, descendants of slaves—the previous sugar workers. There are also smaller numbers of whites, the European colonizers, as well as Chinese and other Asian immigrants who came later as traders.

While there has been relatively little tension, the communities do remain fairly distinct. This is probably the only country where the end of Ramadan, All Saints' Day, and the Chinese New Year are all official holidays.

Each community uses its own language but three have become dominant: English is the official administrative language; French is the language of the major newspapers; creole the unofficial language used in the streets.

Communal divisions used also to correspond to a division of labour—with the Indians as farmers, creoles as artisans, Chinese as traders, and whites as large landowners. Today, these distinctions have been blurred. This is partly the result of increasing levels of education. But it also reflects a rapid modernization of the economy which has offered new opportunities to all. Indeed, Mauritius is one of Africa's rare success stories.

Standards of human development are high. Education is free to university level and health facilities are good.

Up to a few decades ago the major source of wealth was sugar—capitalizing on preferential access to the European market. Sugar still covers most of the arable land and is responsible for around 2% of GNP, but it now employs only around 7% of the workforce and future prospects are poor as a result of trade liberalization that will reduce preferential access to overseas markets. With little land on which to grow rice, most food has to be imported.

Fortunately, Mauritius has invested much of the earnings from sugar into creating a new manufacturing sector based in export-processing zones. Since the early 1980s, the country has attracted more than 500 companies, employing 13% of the workforce—two-thirds of them women. Around half of production is of garments—again taking advantage of preferential access to markets in Europe and North America. Companies were also attracted by Mauritius's stable, healthy, and well-educated society, good infrastructure, and cheap labour. Most investment has come from Hong Kong and from France and other European countries.

500 export-processing companies

The removal of some trade barriers has eroded some of the Mauritian advantage. And there has been a steady rise in local wages: direct labour costs are now around 50% higher than in southern Africa, and one-third higher than in South-East Asia. Employment in garments has already started to fall but this has been offset to some extent by an expansion in the production of seafood.

The government wants to reinvent Mauritius as a 'cyber island' and by 2007 had attracted over 100 companies providing call-centre and other services to foreign companies.

The third main pillar of the economy is tourism, which accounts for around 9% of GDP and, directly and indirectly, employs more than 60,000 people. More than 900,000 visitors arrive each year, with France as the leading source. Most head for the beaches, but there is also eco-tourism. Mauritius sees itself as an 'exclusive' destination.

Surprisingly perhaps for a country with a booming private sector, politics in Mauritius has been dominated by political parties that are avowedly socialist. This has also involved informal power-sharing between Hindus, Muslims, and whites, though excluding the creoles.

In 1991 the country, which until then had the British monarch as head of state, introduced a new constitution and became a republic.

The 1995 election for the National Assembly was won by a coalition of the Labour Party (LP), led by Navin Rangoolam who became prime minister, and the Mouvement militant mauricien (MMM), though the MMM withdrew in 1997 and went into opposition.

The election in 2000, however, was won by the MMM, led by Paul Bérenger, and the Mouvement socialiste militant (MSM), led by Sir Anerood Jugnauth. Under a power-sharing agreement, Sir Anerood initially became prime minister and was replaced by Bérenger in 2003. Sir Anerood then took over as president.

In the 2005 general elections, however, arguments within the MMM–MSM resulted in a defeat by the Alliance de l'avenir coalition headed by the Labour Party whose leader, Navin Rangoolam, returned as prime minister. The coalition was re-elected in 2010.

Mayotte

Mayotte is a part of France and anxious to remain so

Land area:	375 sq. km.
Population:	0.2 million—urban 41%
Capital city:	Dzaoudzi, 15,000
People:	Mahorais
Language:	Shimaore, French
Religion:	Muslim
Government:	Territory of France
Life expectancy:	63 years
GDP per capita:	$PPP 4,900
Currency:	Euro
Major exports:	Ylang-ylang, vanilla, copra

Mayotte is a volcanic island in the Indian Ocean between Madagascar and Mozambique. Its capital and port, Dzaoudzi, is not on the main island, but on a rock linked by a causeway to the nearby islet of Pamandzi.

The people, the Mahorais, are of Malagasy origin and the vast majority are Muslim. In addition there are minorities of Indians, Creoles, and Madagascans. There are also some 'M'Zoungous'—the local name for people who have come from metropolitan France. The official language is French though most people use Shimaore which is based on Swahili.

Here, French law and Muslim law work in tandem—the latter allowing polygamy, for example. Many young children also attend Koranic school before going to a French primary school later in the day. The population is very young (45% of people are under 15 years old) and continues to grow rapidly, a result both of natural increase and immigration from neighbouring Comoros.

Most people are farmers, and grow some food crops as well as cash crops such as vanilla, cloves, and trees for the perfume extract ylang-ylang. Their island has few natural resources, but their link with France entitles them to free education and health services, a guaranteed monthly minimum wage of $400, and French citizenship.

Mayotte was the fourth island of the Comoros when they were a French overseas territory. The split occurred in 1975 when the Comoros as a whole unilaterally declared its independence—despite protests from Mayotte. The Comoros continues to claim Mayotte.

France would be happy to relinquish an expensive colonial memento but will only do so if the local people want this. In 1976, a referendum in Mayotte produced a 99% vote in favour of retaining the link with France.

In a further referendum in 2000, 70% of people voted for greater autonomy while keeping their French status. In 2011, as a result of a 2009 referendum, Mayotte became an overseas department of France. This will require significant changes, notably moving from a traditional religion-based legal system to one based on French law.

The Mahorais also have a 19-member general council usually made up of four or five parties. The current president, elected in 2011, is Daniel Zaïdani of the Mouvement Populaire Mahorais.

Mahorais also elect members to the French National Assembly and to the Senate.

Mexico

Mexico is no longer a one-party state—elections are now fiercely contested

0	Miles		750
0	Km		1200

Land area: 1,958,000 sq. km.
Population: 108 million—urban 78%
Capital city: Mexico City, 18.0 million
People: Mestizo 60%, Amerindian 30%, other 10%
Language: Spanish and some Indian
Religion: Roman Catholic 77%, Protestant 6%, other 17%
Government: Republic
Life expectancy: 76 years
GDP per capita: $PPP 14,104
Currency: Peso
Major exports: Cars, oil

Some have suggested that Mexico's most important geographical feature is the 2,000-mile border it shares with the USA. But Mexico has a striking diversity all of its own—from the snow-capped mountains of the Sierra Madre, to the deserts of Sonora, to the tropical rainforests of Yucatán. Most of the population is to be found, however, in the central region in the high plateau and the surrounding mountains.

Today's Mexicans are largely mestizo, a mixture of Indian and European ancestry. But about one-third are of purer Indian origin, in more than 60 ethnic groups concentrated in the south. They also tend to be the poorest. Overall, 18% of the population live below the national poverty line.

Education levels have improved in the past ten years and only 10% of adults are now illiterate. Even so, one-third of adults have not completed primary education. Standards of health are also better, with a substantial reduction in infant mortality, though rates vary markedly between states.

Population growth and rural–urban migration have contributed to an explosive growth of cities, particularly of Mexico City, which with a population of 18 million is the second largest city in the world. It also has some of the most polluted air and is suffering from rising levels of crime.

Apart from migrating to cities, Mexicans have also been moving to the USA. By 2004, 10.3 million first-generation Mexicans were living in the USA. At least half are there illegally. They send $26 billion annually in remittances. Around 400,000 people emigrate annually.

A major stimulus for emigration has been the poor state of Mexican agriculture. Agriculture only contributes 4% of GDP but is responsible for 16% of employment. Only about one-fifth of the country is suitable for arable farming, and even then the soil is often thin and levels of technology are low. Over half of this land consists of 'ejidos', small collective farms generally growing maize and beans. From the mid-

1980s, the government liberalized agriculture—reducing subsidies and allowing the ejidos to be sold and consolidated. But this brought few benefits to small farmers who found it difficult to get bank loans. Following the 1994 North American Free-Trade Agreement (NAFTA), Mexican maize farmers also find it very hard to compete with imports from the USA. Corn prices have fallen and one-third of corn is imported. The brightest agricultural development has been in the north, where irrigation has permitted a flourishing export trade in fruit and vegetables to the USA.

Aside from the land, one of Mexico's greatest natural assets is oil. Mexico is the world's fifth largest producer and the third largest source of crude oil to the USA. At present rates of extraction, reserves should last until around 2016. Another valuable asset is silver: Mexico has the world's largest silver production.

Most industrial employment is in manufacturing which since 1994 has been transformed by NAFTA. One of the most dynamic parts of the economy, which preceded NAFTA, has been the 'maquiladora', the 3,300 duty-free assembly plants strung along the border with the USA which employ one million people—10% of formal sector jobs.

3,300 assembly plants on the US border

Elsewhere one of the most important industries is production by foreign companies of cars primarily for export to the US. The industry employs half a million people.

Another major source of foreign exchange is tourism, primarily from the US with 12 million visitors

annually, bringing in $13 billion.

Presiding over Mexico's development for 71 years was the Partido Revolucionario Institucional (PRI), which was in power continuously between 1929 and 2000. But the monolith started to crumble in the 1990s. This was partly because of more vigorous opposition, most dramatically from a rebel group, the Ejército Zapatista de Liberación Nacional, which in 1994 led an uprising in the poor southern state of Chiapas. More disruption comes from the drug mafias that control routes from South America into the USA.

The PRI was finally defeated in the 2000 presidential election. The victor was Vicente Fox of the right-wing Partido Acción Nacional (PAN) whose strongest support is in the north and central states and among the middle classes. However, Fox, a former head of Coca-Cola in Mexico had trouble getting support for economic reforms.

The presidential election of 2006 was a closely fought contest between left and right. On the left was Andrés Manuel López Obrador, a populist former mayor of Mexico City, representing a coalition led by the centre-left Partido de la Revolución Democrática (PRD). On the right was the PAN candidate, Felipe Calderón —who won by 100,000 votes out of 42 million. The PRI candidate came a distant third.

Calderón also benefited from PAN's success in the 2006 legislative elections in which the PRD also did well at the expense of the PRI. In the 2009 mid-term elections, however, the PAN lost ground, giving the PRI a working majority, though it is unlikely to use this to obstruct the government.

Micronesia

Heavily dependent on US aid, Micronesia will need other sources of income

The Federated States of Micronesia are Yap, Chuuk, Pohnpei, and Kosrae, and consist of more than 600 islands and coral atolls in the Pacific.

The people of the federation, though few in number, include a great variety of cultures and languages. Chuuk is the largest state, with around half the population, followed by Pohnpei.

Most productive activity centres on subsistence farming and fishing. The main crops include breadfruit, taro, and coconuts, and there is a small export income from copra, black pepper, and handicrafts. The most important fishing catch is tuna though much of this goes to foreign fleets.

The islanders' primary source of cash income, however, is the government which is the main employer, providing 40% of total employment, and its expenditure accounts for over 40% of GDP.

Micronesia relies on transfers of US aid through a 'Compact of Free Association'. Over the period 1986–2001 the country received $1.3 billion in US grants and in 2003 aid was worth $923 per person. This assistance is now being phased out. From 2004 the US government revised the compact and for three years of the subsequent 20-year period it gave $76 million in economic assistance grants,

Land area: 1,000 sq. km.
Population: 0.1 million—urban 23%
Capital city: Palikir
People: Micronesian and Polynesian groups
Language: English, Chukese, Pohnpeian, Yapese, Kosraean
Religion: Roman Catholic 50%, Protestant 47%, other 3%
Government: Republic in free association with the USA
Life expectancy: 68 years
GDP per capita: $PPP 2,200
Currency: US dollar
Major exports: Copra, fish, pepper

to be allocated among education, health, and other sectors. Another $16 million annually will go to a trust fund.

Tourism offers some potential, but so far has been confined to niche markets such as eco-tourism, visits to some ancient ruins, and shipwreck diving. However there are concerns about the impact of increased arrivals on the fragile environment.

There has also been some investment in manufacturing, notably in garment production, but expansion is hampered by poor infrastructure and the distance from markets.

Until 1986, the states were part of a UN Trust Territory. Then they became an independent federation, self-governing but in free association with the USA. Each of the four states has its own elected governor and legislative assembly. There are no political parties. In addition, there is an elected national congress, which chooses the president. In 2011, it re-selected Manni Mori from Chuuk.

In many respects Micronesia functions more as four separate states. There are, however, signs of greater cooperation on common issues such as the environment and tourism.

Moldova

Moldova has effectively been split in two by a long fight for secession

Land area: 34,000 sq. km.
Population: 4 million—urban 41%
Capital city: Chisinau, 780,000
People: Moldovan/Romanian 78%, Ukrainian 8%, Russian 6%, other 8%
Language: Moldovan, Russian, Gagauz
Religion: Eastern Orthodox
Government: Republic
Life expectancy: 68 years
GDP per capita: $PPP 2,551
Currency: Moldovan leu
Major exports: Food products, beverages, tobacco

Most of Moldova lies between two rivers that flow into the Black Sea: the Prut, which forms the western border with Romania; and the Dniester which flows roughly in parallel with the eastern border with Ukraine. The land is largely a hilly plain, though also cut through with many steep ravines. The soil—the rich, black 'chernozem'—is very fertile, and two-thirds of the country is forested.

Moldova's difficulties in building a new nation since 1991 reflect its fairly recent assembly into one territory. In 1940, the Soviet Union united the land from the Prut to the Dniester, Bessarabiya, which had been part of Romania, with the strip of land from the Dniester to the eastern border, Transdniestria. The two were merged as the Moldavian Soviet Socialist Republic.

After the Second World War, the Soviet Union made strong efforts to weaken the republic's Romanian past. This included changing the name of the language from Romanian to Moldovan, and switching from the Roman to the Cyrillic alphabet, while encouraging immigration from Russia and Ukraine. The break-up of the Soviet Union then provoked a crisis of national identity.

During the late 1980s, two separatist movements had appeared. And when the Moldavian Republic declared its independence in 1991, as 'Moldova', both groups feared that independence would lead to unification with Romania and both declared themselves independent. The smaller group were the Gagauz in the south-east who subsequently laid down their arms in exchange for a measure of autonomy.

The larger and more debilitating rebellion came from the Russians in Transdniestria, and in 1992 this escalated into a civil war, in which Russia backed the separatists while Romania backed Moldova. A peace treaty was signed in 1992 and has been enforced with Russian troops.

This has left the country effectively partitioned in two. Transdniestria, which has 11% of the territory, and 17% of the people, has its own government in Tiraspol headed by Igor Smirnov, who was returned

Transdniestria has its own government

as president for the fourth time in 2006. He is an unreconstructed communist whose heavy-handed and repressive rule is attempting to sustain the Soviet economic model. The leaders of Transdniestria now accept, however, that the best they are likely to achieve is a confederation of equal states.

Moldova has the highest population density of the countries of the former Soviet Union, and it would be higher still if around 10% of the population did not work abroad, chiefly in Russia and Italy. It is also one of the poorest successor states and has the worst health standards.

Economically Moldova has suffered, since a high proportion of heavy industrial activity is in Transdniestria. In the 1990s the government privatized many industries and the private sector now accounts for around 80% of GDP. Most manufacturing consists of food processing and beverages, including a wine industry most of whose output goes to former communist countries.

With limited immediate prospects for industrial development, Moldova remains highly dependent on agriculture which accounts for around one-fifth of GDP and 40% of employment, and is the main source of export income. Following the break-up of collective farms, most production now comes from private farmers or household plots. The most important crops are cereals, sugar beet, tobacco, grapes, and other fruit. Although the land is

fertile, it has suffered from over-intensive cultivation and the heavy use of fertilizers, and production is vulnerable to the weather: harvests are regularly hit by droughts and floods.

In 1994, Moldovans voted for a new constitution to maintain the country's current borders but grant extensive autonomy to Gagauz and Transdniestria.

The 1996 presidential election was won by Petru Lucinschi, a leading official from the Soviet era. And the 1998 parliamentary elections made the Communist Party of Moldova (PCRM) the largest party but unable to form a government.

The situation was only resolved after the 2001 legislative election gave the PCRM 70% of the seats. Meanwhile the constitution had been amended to allow the parliament to elect the president and they chose a former Soviet-era politician, Vladimir Voronin.

The PCRM government did not reverse market reforms but took some controversial measures such as boosting the Russian language and playing down links with Romania. Nevertheless, in 2005 the PCRM again won parliamentary elections and with the cooperation of the opposition re-elected Voronin for a second presidential term.

In 2009, however, following two parliamentary elections, the PCRM finally lost to a four-party non-communist Alliance for European Integration. But the Alliance did not have enough members to elect the president. This resulted in another election in 2010 which again failed to give the Alliance a majority. By mid-2011 Moldova still lacked a president.

Mongolia

Long a Soviet satellite state, Mongolia has steadily been building a market economy

RUSSIA

Erdenet • • Darhan   • Choybalsan

Ulaanbaatar ■

CHINA

| 0 | Miles | 300 |
| 0 | Km | 480 |

Land area: 1,567,000 sq. km.
Population: 3 million—urban 58%
Capital city: Ulaanbaatar, 994,000
People: Mongol 95%, Kazakh 4%, other 1%
Language: Khalkha Mongol 90%, Russian or Turkic 10%
Religion: Tibetan Buddhist
Government: Republic
Life expectancy: 66 years
GDP per capita: $PPP 3,236
Currency: Togrog
Major exports: Fuels, minerals, metals, garments

Mongolia's territory, which is around the size of Western Europe, consists largely of a vast plateau between 900 and 1,500 metres above sea level. It has a harsh semi-arid climate, with bitterly cold winters, but has extensive pasture land. The greatest mountain ranges are to the north and west, while the land to the east and south stretches out into rolling steppes and the Gobi Desert.

Most Mongolians now live in cities—one-third in the capital, Ulaanbaatar. In ethnic terms the country is fairly homogenous, the largest minority being Kazakhs in the south. In the 1930s, a Stalinist government purged much of the country's Tibetan Buddhist heritage, though with a democratic government and freedom of religion Buddhism is now staging a revival.

In 1991, following seven decades of socialism and the sudden withdrawal of Soviet aid, the government embarked on a course of 'shock therapy', shutting some industries, privatizing others, and making severe cuts in social services. Despite more rapid economic growth in recent years, more than one-third of the population are still poor and one-fifth of the workforce are unemployed or underemployed, with the most severe problems for urban youth.

Education services also came under pressure and the government reduced the number of dormitory places that are essential for herder children, though in recent years it has increased expenditure on education. One of the most disturbing developments has been the emergence of thousands of street children in Ulaanbaatar. Because winter temperatures can drop to 30 degrees below zero the children have to live in the sewers.

Around 45% of the workforce work in agriculture. Nomadic Mongol tribes, with their circular felt tents or 'gers', and their huge herds of sheep, goats, horses, and yaks, offer a distinctive picture of Mongolian life. Their herds—40 million animals— had been collectivized during the communist period and then privatized again after 1990. The land remains in state ownership. Privatization revived production but it has also increased inequality and larger herds are harming the pastures—two-thirds of which are now degraded.

Mongolia's small farms can only produce around one-third of the country's cereal requirement so the rest has to be imported.

Mongolia also has considerable mineral wealth—with rich deposits of coal, iron ore, copper, molybdenum, fluorspar, and tungsten. Fuels, minerals, and metals are responsible for more than two-thirds of exports. Revenues have fallen with the decline in world prices, but mineral extraction is one of the country's main priorities. Efforts have also been made to increase manufacturing production, and the next largest export sector is garments and textiles, including cashmere from goat hair.

Mongolia's main trade disadvantage is its isolation. Getting goods out to a seaport overland involves a three-day railway journey

Landlocked Mongolia's shipping fleet

through Russia or China. Surprisingly, however, Mongolia is now getting extra income through a shipping register managed by an agency in Singapore; the 'national' fleet now has 300 vessels.

Mongolia's economy shrank rapidly following the withdrawal of Soviet aid, which had been equivalent to 30% of GDP. It also lost supplies of cheap oil. Since then, the economy has largely recovered.

Mongolia became the world's second communist state in 1924, ruled by the Mongolian People's Revolutionary Party (MPRP). For 70

years it was bankrolled by the Soviet Union as a buffer state against China. When communism collapsed in the Soviet Union it duly collapsed in Mongolia too, and in 1990 and 1992 Mongolia's constitution was amended to allow other parties to compete for seats in a single-chamber legislature, the Great Hural. In addition there was to be a directly elected president.

The MPRP, which by then was espousing economic liberalism, won the first two general elections. In 1996, however, it was defeated in the Great Hural election by the centre-right Democratic Union, consisting of the Mongolian National Democratic Party and the Mongolian Social Democratic Party. But matters were complicated later that year when an MPRP candidate, Natsagiin Bagabandi, was elected president.

In 1998 there was a political crisis when the MPRP started to boycott the Great Hural. Things seemed to settle down subsequently with a new administration. But rising poverty made the government increasingly unpopular. In the July 2000 parliamentary elections the MPRP was returned to power with a substantial majority, and Bagabandi was returned as president in May 2001.

At that point the prime minister was Namburiin Enkhbayar, a modernizer who dubbed himself the 'Tony Blair of the steppes'. The MPRP also gained the most seats in the 2004 elections to the Great Hural.

In May 2005, Enkhbayar was elected president. The MPRP chose as prime minister, first Miyeegombo Enkhbold, then in 2007 Sanjaagiin Bayar, then in 2009 when he resigned due to ill health, Sukhbaatar Batbold.

Montenegro

The newest member of the United Nations

Montenegro has a narrow coastal strip along the Adriatic, behind which rise the spectacular Dinaric Alps. The rest of the country consists of a rugged mountainous massif cut through by deep valleys and gorges.

Around half the population are ethnic Montenegrin (the country had been independent between the mid-1800s and 1918). There is also a substantial proportion of Serbs, particularly in areas bordering on Serbia. Although standards of health and education dipped following the break-up of former Yugoslavia they have now recovered.

Although not a very poor country, average salaries here are low and 12% of the workforce is unemployed.

Only around 3% of workers work in agriculture, and because of the mountainous terrain they have been able to cultivate less than 10% of the country; most settled agriculture is largely in the Zeta valley and around Lake Scutari.

Around 30% work in industry of which the most important is aluminium. The country's largest enterprise is the aluminium smelter KAP, which has been bought by the Russian company Rusal and produces almost half of industrial output.

Nowadays most workers, more than 60%, are employed in services. Many of these are linked to the tourist industry which takes advantage of

Land area: 14,026 sq. km.
Population: 0.6 million—urban 60%
Capital city: Podgorica, 180,000
People: Montenegrin 43%, Serbian 32%, Bosniak 8%, Albanian 5%, other 12%
Language: Serbian
Religion: Orthodox Christian, Muslim
Government: Republic
Life expectancy: 74 years
GDP per capita: $PPP 11,699
Currency: Euro
Major exports: Aluminium

beautiful beaches on the Adriatic coast as well as the mountains and lakes. However the industry will need considerable investment to bring it up to international standards.

Montenegro is another independent fragment of the disintegration of Yugoslavia. From 1992 the country had been yoked with republic Serbia as the rump Federal Republic of Yugoslavia. But this uneasy alliance was cast into doubt from 1997 with the election of pro-western Milo Djukanovic, of the Democratic Party of Socialists (DPS), as president of Montenegro. After years of friction, the two created a loose union in 2002 called Serbia and Montenegro. However, Djukanovic had reserved the right to hold a referendum on independence and in 2006 55.5% of Montenegrins voted to break free,

The first parliamentary election was won by the ruling Coalition for a European Montenegro which consisted of the DPS, the Social Democratic Party, and the Croat Civic Initiative. Djukanovic initially chose not to become prime minister but did so eventually in 2008.

In April 2008 the DPS candidate Filip Vujanovic was re-elected president, and in 2009 the DPS won the parliamentary election.

Morocco

Morocco now has a more liberal monarch but continues its illegal occupation of Western Sahara

Land area: 447,000 sq. km.
Population: 31 million—urban 57%
Capital city: Rabat (with Salé) , 1.4 million
People: Arab-Berber 99%, other 1%
Language: Arabic, Amazigh, French, Spanish
Religion: Muslim
Government: Constitutional monarchy
Life expectancy: 71years
GDP per capita: $PPP 4,108
Currency: Dirham
Major exports: Phosphate rock, phosphoric acid, textiles

Morocco is the most mountainous country of North Africa. The two main chains are the Er Rif, along the Mediterranean coast, and the Atlas Mountains which dominate the country from north-east to south-west. Most people live in the lowlands that lie between these two chains and the Atlantic coast. Morocco also controls the desert area of Western Sahara to the south.

Morocco's people are largely Arabized Berbers, but around one-third are less-assimilated Berbers, most of whom live in the mountains and whose language, Amazigh, was given official recognition in 1994.

Morocco remains one of the poorest Arab countries. Around half the population are illiterate and more than 10% of children do not enrol in primary school. In the rural areas more than one-third of the population lacks access to safe water. Even in urban areas there are high levels of poverty, and unemployment in 2009 was 9%. As a result, many Moroccans have chosen to emigrate clandestinely, making the perilous 15-kilometre sea-crossing to Spain. Around 1.7 million Moroccans now live overseas, chiefly in France and Spain; their remittances in 2008 were $6.7 billion—the largest source of foreign exchange.

Agriculture and fishing employs about half the labour force. Most of the land is in the hands of smaller farmers growing cereals, potatoes, and other staples. Since they lack irrigation they have to rely on fairly erratic rainfall. The tenth of the arable land that is irrigated is mostly in larger farms which grow citrus fruits, grapes, and other export crops. Meanwhile, especially during drought years, much of the country's cereal needs have to be imported.

One of Morocco's main priorities is to improve the irrigation network—taking water from the areas of good rainfall to those regularly hit by drought. A new canal was opened in 1999 to take water from Guerdane in the east to the south of the country to irrigate citrus crops.

Many of the smaller farmers in

the mountains also raise cattle, sheep, and goats. Morocco's fishing industry is a major supplier of sardines to the EU, though catches have fallen due to over-fishing.

Morocco's main industrial enterprises are state-owned and linked to phosphates. With Western Sahara, Morocco has three-quarters of the world's reserves, and is the third largest producer. The other major export industry is textiles, which expanded rapidly in the 1980s, though is now facing intense foreign competition. Many other smaller enterprises produce high-quality leather goods, rugs, and carpets.

Of the service industries, one of the most important is tourism. Around four million visitors arrive each year and provide employment to 6% of the labour force. Earnings have been falling, however, and many of Morocco's tourist facilities are in need of a facelift. The government is aiming to build new resorts and increase hotel capacity to around ten million beds.

Morocco is a constitutional monarchy. There is an elected legislative assembly. But it only has limited powers. The king acts like an

Morocco's king really rules

executive president, appointing both the government and prime minister and presiding over the cabinet. From 1961 Morocco was ruled by King Hassan II—whose autocratic rule resulted in thousands of arbitrary arrests and disappearances.

On his death in 1999 Hassan was succeeded by his son who became King Mohammad VI. King Mohammad promised to be somewhat different and did at the beginning

make some important changes. He released political prisoners, allowed exiles to return, and dismantled repressive security forces. He also supported women's rights by rewriting the 'moudawana', the personal and family law, to give women equal rights in marriage. One major concern is pressure from radical groups such as Al Sunna wal Jamaa which try to enforce 'Islamic' behaviour, often with violence. They opposed the new moudawana.

Another issue is Western Sahara which Morocco seized in 1975—despite armed resistance and international protests. The UN has in the past proposed a referendum on independence but in 2008 the UN envoy for Western Sahara, Peter van Walsum, concluded that an independent Western Sahara was "not a realistic goal".

The country's first really free ballot in 2002 to the House of Representatives distributed the 325 seats among 22 parties. The next elections in 2007, probably reflecting popular doubts about the value of the parliament, had a low turnout and many spoilt ballot papers. This time there were 24 parties. The ruling coalition consists of four of the parties that have just under half the seats. The prime minister is Abbas el-Fassi of the Parti Istiqlal.

Inspired by the examples of Tunisia and Egypt, Moroccans started to protest in the streets from February 2011, demanding greater democracy. King Mohammad responded with wage increases and a promise of constitutional reforms which might remove his power to appoint ministers.

Mozambique

Mozambique has now had a long period of peace and stability while Frelimo retains its grip on power

Land area: 802,000 sq. km.
Population: 22 million—urban 38%
Capital city: Maputo, 1.8 million
People: Makua-Lomwe, Tsonga, Yao, Sena-Nyanja
Language: Portuguese, indigenous dialects
Religion: Catholic 24%, Muslim 18%, Zionist Christian 18%, other 18%, none 32%
Government: Republic
Life expectancy: 48 years
GDP per capita: $PPP 802
Currency: Metical
Major exports: Prawns, cotton, cashew nuts

Mozambique can be divided by the Zambezi River into two main geographical regions. The southern part of the country is mostly flat and low-lying, apart from highlands on the eastern border. North of the river the land rises to plateaux and to mountains along the border with Malawi.

The Zambezi also divides the country's many different ethnic groups. Most of the population is concentrated in the better land north of the river, particularly in the north-east. Groups here include the Makua, who are farmers, and the Muslim Yao. South of the river, where people live more along the coast, are groups such as the Tsonga. None of their languages is widely spoken around the country so the official language is Portuguese.

Mozambique's civil war, which ended in 1992, devastated much of the country's social infrastructure, destroying schools, hospitals, and clinics—in what was already one of the world's poorest countries. Since then, the government has managed to rebuild most of the schools. School attendance has been rising and 60% of children are now enrolled in primary schools. Even so, adult literacy is still only 48%. The health system also needs more investment. The HIV/AIDS infection rate in 2007 was 5%.

Most people are very poor. Although the capital, Maputo, is now a lot smarter, more than half the population languish below the national poverty line. Poverty is severe in the rural areas and particularly in the north. Those who have had few employment opportunities at home have traditionally migrated to neighbouring South Africa which currently has around one million immigrant Mozambicans.

Around 80% of the workforce make their living from agriculture. Given good weather, Mozambique should be self-sufficient in basic foodstuffs such as maize and cassava. The main cash crops are cotton and especially cashew nuts. Production of both food and cash crops has increased, but farmers are still hampered by poor infrastructure; many parts of the country are cut off during the rainy season.

Mozambique also exports prawns, along with some shellfish. The fishing industry was less affected than others by the war. Even so, the local fleet is relatively small and the government gains extra income by selling licences to boats from Europe, South Africa, and Japan.

Mozambique's industrial sector was badly damaged by the fighting. Since the mid-1990s, however, and following an extensive programme of privatization along with foreign investment, industrial output has revived. One of the country's largest enterprises is the Cahora Bassa hydroelectric dam on the Zambezi which exports electricity to South Africa and also supplies local demand, including the new $1.3-billion Mozal aluminium smelter, a joint venture led by BHP-Billeton, which has turned Mozambique into one of the world's largest aluminium producers. There are also plans to export natural gas to South Africa.

Much of this activity, however, has been in the southern half of

Development is widening the north–south divide

the country, which also tends to be economically more integrated with neighbouring countries, thus sharpening the north–south divide.

Mozambique remains heavily dependent on aid which accounts for around 20% of GNP—mostly in the form of grants. Mozambique, as one of the world's poorest countries, has also benefited from debt relief under the Heavily Indebted Poor Countries

initiative and debt servicing is now more manageable.

Since independence in 1975 the government has been in the hands of the Frente de Libertação de Moçambique (Frelimo). Frelimo's Marxist policies alarmed the then white governments of Rhodesia and South Africa, who armed and funded a guerrilla group, Resistência National de Moçambique (Renamo). Renamo became one of the world's most vicious guerrilla armies and Mozambique sank into a civil war that killed around 900,000 people.

The war ended with a stalemate and a peace accord in 1992, followed by very effective UN-organized programmes which demobilized the soldiers and repatriated more than one million refugees. In the elections in 1994, Frelimo, by then headed by Joaquím Chissano, won a surprisingly narrow victory over Renamo.

By the mid-1980s Frelimo had renounced Marxism and continued economic liberalization. In 1999, Chissano won another narrow victory in the presidential election and Frelimo gained a small majority in the legislative assembly.

After legislative and presidential elections in 2004, and despite widespread accusations of fraud, Frelimo were returned to power, but with Armando Guebuza as president. Guebuza has steadily concentrated power into his own hands, but both he and Frelimo were convincingly re-elected in 2009. Meanwhile Renamo is fading away and is being replaced as the leading opposition group by the Movimento Democrático de Moçambique, headed by Daviz Simango.

Namibia

Namibia is one of the world's most unequal societies, and now faces an AIDS crisis

Land area: 824,000 sq. km.
Population: 2 million—urban 38%
Capital city: Windhoek, 234,000
People: Ovambo 50%, Kavangos 9%, Herero 7%, Damara 7%, mixed 7%, white 6%, Nama 5%, other 9%
Language: English, Afrikaans, German, Oshivambo, Herero, Nama
Religion: Christian 80%, indigenous religions 20%
Government: Republic
Life expectancy: 60 years
GDP per capita: $PPP 5,155
Currency: Namibia dollar
Major exports: Diamonds, uranium, gold

Namibia has three main geographical regions. Its long, distinctive Atlantic coastline consists of the dunes and rocks of the Namib desert. This rises to a broad plateau that includes mountains rising to 2,500 metres. To the east, the plateau descends to the sands of the Kalahari desert. The climate is hot and dry and the rainfall erratic—the only year-round rivers flow along the northern and southern borders.

Namibia is sparsely populated. Around half the population live in the far north. The other area of concentration is in and around the capital, Windhoek. Namibia has a number of different ethnic groups. The largest are the Ovambo, who are also politically dominant. Linguistically, the country is very diverse. Only a small minority speak English, which is the official language; most others use languages of the Ovambo or their

own ethnic group, though many also understand Afrikaans or German.

Of the minority groups, the most discontented have been the 100,000 or so Lozi who live in the Caprivi strip, an odd sliver of land in the far north-east. In 1999 a Lozi uprising was brutally repressed by Namibian security forces, though since then the unrest seems to have subsided.

Another fundamental division is between the whites and the rest. Namibia's per capita income is fairly high but it has one of the world's most unequal distributions of income—the richest 10% of households get 65% of the national income. Efforts to overcome this imbalance have included heavy investment in education, which in 2007 took up 21% of the budget.

Health too is a major problem; most facilities are concentrated in the urban areas. Though Namibians suffer from diseases like malaria and tuberculosis, the leading cause of death is now HIV and AIDS—7% of the population are infected.

Agriculture represents only around 5% of GDP, but around 70% of the population rely on farming for at least part of their livelihood. The staple crop is millet, which is grown by subsistence farmers in the north—though output is vulnerable to erratic rainfall and Namibia normally has to import almost half its cereal needs.

Commercial farming, which is mostly the prerogative of white settlers in the centre and south, is devoted primarily to livestock—raising cattle and sheep for export to South Africa and the EU. One of the most contentious political issues is land reform, since whites own a high proportion of the land area. So far, the process has been slow and expensive because the constitution requires the government to provide full compensation for any land taken. As a result, the government has started appropriating land.

Whites own around 40% of farmland in Namibia

A larger component of Namibia's GDP comes from mining which contributes around 13% of GDP. Namibia has high-quality diamonds that are extracted by the Namdeb Diamond Corporation, a joint venture between the government and the South African De Beers Centenary company. An increasingly high proportion of these are now coming from offshore fields using sea-bed crawler mining equipment.

Another important mineral is uranium. Namibia's Rössing mine is the world's largest open-pit uranium mine. Namibia also produces gold, silver, and copper and in 2003 opened a new zinc mine and refinery.

Other industrial activity is more limited. Manufacturing has been concentrated in food processing, though the government has attracted other industries to work in an export-processing zone.

After agriculture, the main employer in Namibia is the government. In 2003, 78,000 people worked in the civil service.

Namibia was previously occupied by South Africa. But the South-West Africa People's Organization (SWAPO) fought a long and successful guerrilla war and since independence in 1988 it has retained political power.

SWAPO won the first election for the national assembly in 1989 and its leader Sam Nujoma was also elected president. The main opposition came from the Democratic Turnhalle Alliance—which had formed the government during the South African occupation.

SWAPO consolidated its position in the 1994 National Assembly and presidential elections. But Nujoma's rule grew increasingly autocratic—handing out most of his favours to the Ovambo people, while clamping down on the Lozi. He also changed the constitution to let him run for president a third time in 1999.

Nujoma won that election convincingly, though against a new opposition party, the Congress of Democrats, whose leader Ben Ulenga is a former SWAPO dissident.

For the 2004 presidential election, the SWAPO candidate, who was duly elected, was the party's vice-president, Hifikepunye Pohamba.

He won again in 2009, this time beating Hidipo Hamutenya, head of another splinter from SWAPO, the Rally for Democracy and Progress which is now the main opposition party.

Nepal

Nepal has deposed its king and declared a republic but is struggling to maintain peace

Land area: 141,000 sq. km.
Population: 28 million—urban 18%
Capital city: Kathmandu, 672,000
People: Chhettri 16%, Brahman-Hill 13%,
Magar 7%, Tharu 7%, other 57%
Language: Nepali and 30 tribal languages
Religion: Hindu 81%, Buddhist 11%,
Muslim 4%, Kirant 4%
Government: Republic
Life expectancy: 66 years
GDP per capita: $PPP 1,049
Currency: Nepalese rupee
Major exports: Carpets, garments

Nepal's territory can be divided roughly into three bands descending from north to south. The northernmost band includes the Great Himalayas dominated by Mount Everest at 8,848 metres. The central band has the lower Mahabharat range and a number of major river systems and valleys, including the densely-settled Kathmandu Valley. The band along the southern border with India has forested lower slopes that descend to the fertile Terai plain.

Nepal's people comprise more than 70 ethnic groups that can be considered in two sets. The first and smaller, the Tibeto-Nepalese, are the result of immigration from Tibet. They are found in the bleak high mountain areas as well as in the middle band, and are usually Buddhist. The larger set, of Indo-Aryan ancestry, have immigrated from India and elsewhere, and are largely Hindu and rigidly stratified into higher and lower castes.

In recent years, the Nepalese have become even poorer. Despite economic growth, 30% of the population live below the poverty line. Standards of education and health are very low: only 51% of the population are literate, 40% have no access to safe sanitation, and around 40% of children are malnourished.

The situation is particularly severe for women. While in most countries women's biological advantage enables them to live longer than men, in Nepal women's lifespan is somewhat shorter. This is the outcome of many kinds of discrimination, especially in health care, which results in high rates of maternal mortality—830 mothers die in childbirth per 100,000 live births.

Survival in Nepal depends primarily on agriculture, which generates about 33% of the country's GDP and involves 80% of the workforce—primarily cultivating rice, maize, and wheat, as well as raising livestock. But productivity is low and output is erratic. Cultivation is particularly arduous in the terraced farms of the hilly regions. Land holdings are small, irrigation is difficult, and the situation is being aggravated by soil erosion and deforestation. The position is somewhat better in the land of the Terai plain, which accounts for about half the cultivable land and where there are better prospects for irrigation. But land ownership here is highly concentrated so the benefits are unevenly spread. The government in 2001 announced fresh plans for land reform and there are officially new ceilings on landholdings, but so far nothing has happened.

Most of Nepal's industry is on a small scale for local consumption. One of the main export industries is carpet weaving, but carpet sales in Europe have been affected by accusations of the exploitation of child labour. Some Indian garment manufacturers have also established factories to take advantage of Nepal's quotas. The service sector has been growing too but this is primarily the result of government development expenditure, 70% of which is financed by foreign aid.

Another important source of income is tourism. However, the industry has been hard hit by worries about law and order. In 2005 there were 277,000 arrivals. At the best of times it is difficult to travel around Nepal, and the vast majority of tourists get little further than Kathmandu. The government has been trying to encourage more arrivals, but even with peace it is doubtful that the infrastructure could cope with many more people.

Few tourists get much further than Kathmandu

Until 1990, Nepal was an absolute monarchy. Inspired by democratic changes elsewhere, many political groupings organized a series of protests that culminated in a mass march on the royal palace. Eventually, King Birendra gave way and in 1990 he introduced a new constitution based on multiparty democracy.

The first election in 1991 was won by the Nepal Congress Party, led by Girija Prasad Koirala – the first in a rapid sequence of coalitions.

After 1996 the political situation was further destabilized by a Maoist insurgency demanding land reform and a republican state. This conflict cost 13,000 lives and led to abuses of human rights on both sides.

In June 2001 the drama moved to the royal palace where Crown Prince Dipendra shot and killed King Birendra, after which Birendra's brother Gyanendra was crowned king.

In 2005 Gyanendra assumed absolute power but proved an inept ruler, provoking widespread opposition and demonstrations. Eventually the army intervened and called on the political parties to form a new government. Girija Prasad Koirala became prime minister for the fifth time.

In 2007, the Maoists signed a ceasefire and joined the government. In 2008, as the Communist Party of Nepal (Maoist), they decisively won an election for a constituent assembly to write a new constitution. This assembly abolished the 239-year-old monarchy and declared a republic.

The Maoists, led by Pushpa Kamal Dahal, formed an interim administration but subsequently withdrew. In 2011, the constituent assembly elected Jhala Nath Khanal as prime minister heading a coalition led by the Communist Party of Nepal (Unified Marxist-Leninist).

Netherlands

One of Europe's most tolerant, and most prosperous, societies

Land area: 37,000 sq. km.
Population: 17 million—urban 83%
Capital city: Amsterdam, 0.7 million; jointly with The Hague, 0.5 million
People: Dutch 80%, EU 5%, other 15%
Language: Dutch
Religion: Roman Catholic 30%, Dutch Reformed 11%, other Protestant 9%, Muslim 6%, other 2%, none 42%
Government: Constitutional monarchy
Life expectancy: 80 years
GDP per capita: $PPP 38,694
Currency: Euro
Major exports: Manufactured goods, chemicals, food, agricultural products

True to its name, most of the Netherlands is 'low lands': more than one-quarter of the country lies below sea level, protected from flooding by coastal dunes and by specially constructed dykes. For centuries, the Dutch have been reclaiming land from the sea and keeping it drained, using pumps that initially were powered by windmills. But even the higher and drier parts of the country seldom rise much above 60 metres.

This is one of Europe's most densely populated countries—456 per square kilometre; most people are concentrated in the west and centre. This makes land expensive and its use is controlled and regulated more closely than in most other countries.

The Dutch enjoy one of the world's most advanced systems of social welfare. They also have very liberal social attitudes: in 1994 this was the first country to permit euthanasia, and in 1998 the first to register homosexual partnerships. The Dutch have also gone a long way towards decriminalizing the use of soft drugs—with marijuana readily available in 'coffee' houses.

Like other European countries, the Netherlands is facing a falling birth rate; its fertility rate is only 1.5 children per woman of childbearing age. This has been offset to some extent by immigration. The foreign born make up 20% of the population; previously they came from former colonies such as Indonesia or Suriname, though now the largest national groups are from Turkey and Morocco.

Reclaiming land from the sea helped make the Netherlands one of Europe's leading agricultural nations, even though only around 2% of the labour force now work on the land. Thanks to very productive small farms, it has steadily increased agricultural output. The country is largely self-sufficient in food. Around 60% of production is exported—primarily dairy products, meat, flowers, and bulbs.

Because it had few raw materials, the Netherlands did not develop as much heavy industry as other advanced economies. Instead it based its economy more on processing imported materials and on international trade. Thus Rotterdam is the world's largest port; it contributes around 10% of GDP, and is a transhipment point for imported oil that is destined for other European countries. Amsterdam's Schipol airport is Europe's second largest hub for air freight. The Netherlands' external orientation has also given rise to important multinational companies such as Unilever, Royal Dutch/Shell, Philips, and Heineken. Even so, two-thirds of the workforce are employed in service industries.

One of the most important economic developments in the 20th century was the discovery in 1959 of huge deposits of natural gas in the north of the country. This rapidly

Acquired and cured the 'Dutch disease'

turned the Netherlands into a major gas exporter. Unfortunately, this also distorted other aspects of

development, driving up the exchange rate and wages, and making industry less competitive—a phenomenon subsequently dubbed the 'Dutch disease'. The disease now seems to have been cured. Following decades of wage moderation, the Netherlands enjoyed rapid growth, low inflation, and low unemployment.

This success is the result of a unique social model that allows many different groups to participate in setting social and economic policy. Thus there is a Social Economic Council, with representation from employers and trade unions, that considers such issues as collective

agreements and welfare provision and has helped keep wage claims within manageable limits. However, the pact has come under strain as the government has made cuts in welfare.

Until recently, Dutch politics had been very consensual. The ceremonial head of state is Queen Beatrix, but almost all power rests in the parliament. In the past, political parties have been divided by ideology or by religion—Catholic or Protestant. But these distinctions have steadily been eroded. The religion-based parties merged in 1977 to form Christian Democratic Appeal (CDA) and ruled until the election of 1994, following which the centre-left Labour Party (PvdA) governed in a 'purple coalition' with the centre-right Liberals and the smaller libertarian party, D-66. This coalition retained power in the 1998 election.

The political landscape was reshaped in 2001 by the rise of an anti-immigration populist Pim Fortuyn. He was murdered in 2002, though his party still joined a short-lived right-wing coalition government.

Following an election in 2003 a new centre-right coalition emerged. This too collapsed, in 2006, and after an election the CDA again came out ahead. It was soon succeeded by a new centre-left coalition in 2007, consisting of the CDA, the PvdA, and a small Protestant party, ChristenUnie. This coalition collapsed in 2010 over disagreements on Afghanistan. It was replaced after an election by a minority coalition of the People's Party for Freedom and Democracy (VVD) and the Christian Democratic Appeal, with Mark Rutte of the VVD as prime minster.

Netherlands Antilles

A dispersed Dutch federation that has just been disbanded

Land area: 800 sq. km.
Population: 0.2 million—urban 70%
Capital city: Willemstad, 94,000
People: Mixed black 85%, Carib Amerindian, white, and East Asian 15%
Language: Papiamento, English, Dutch
Religion: Roman Catholic, Protestant, Jewish, Seventh-Day Adventist
Government: Dependency of the Netherlands
Life expectancy: 77 years
GDP per capita: $PPP 16,000
Currency: Netherlands Antilles guilder
Major exports: Petroleum products

The Netherlands Antilles comprised five Caribbean islands. The northern group had the three smaller and greener islands of Sint Eustatius, Saba, and Sint Maarten (the southern part of St Martin). The southern group, off the coast of Venezuela, was formed by the two larger and drier islands of Curaçao and Bonaire.

The people are racially mixed—of combined African, European, and Amerindian ancestry. Those in the northern islands have a stronger black component; those in the south have a stronger Latin component. Though the official language is Dutch, on the northern islands the usual spoken language is English, while in the south it is the creole language Papiamento. Around three-quarters of the total population live on Curaçao. Standards of human development are quite high, boosted by aid from the Netherlands, but there have been high levels of unemployment.

The islands have few natural resources and rely mostly on oil refining, tourism, and financial services. The proximity of Curaçao to Venezuela, and its location on important shipping lanes, made the island a major centre for oil storage and transhipment, cargo handling, and ship repair. This activity has been boosted by investment from Petróleos de Venezuela. While the oil business has brought many economic benefits it has also caused some environmental damage.

The expansion of tourism helped to offer alternative employment, particularly on Sint Maarten. Bonaire is another tourist centre, with some spectacular scuba diving.

The islands are also an important stopping-off point for money. This is an offshore financial centre, with 42 registered banks which, the USA complains, are used for laundering drug money. Agriculture is very limited, employing only 3% of the workforce, though farmers grow oranges to make the liqueur Curaçao.

Until 2010 the islands were a federation with a Dutch-appointed governor, as well as an autonomous government for internal affairs. The legislature, the Staten, had representatives from each island.

Following referenda on all the islands, however, the federation has now been disbanded. In 2010 Curaçao and Sint Maarten became autonomous associated states of the Netherlands, while Bonaire, Saba, and Sint Eustatius became direct parts of the Netherlands as special municipalities.

New Caledonia

Divided over the question of independence from France

Nouméa

SOUTH PACIFIC OCEAN

Most of this Pacific country is formed by Grande Terre, a long island dominated by a mountain range that runs its entire length and descends steeply to hills and a narrow coastal plain. The country also includes many other islands, notably the Loyalty Islands and the Isle of Pines.

The original Melanesian inhabitants, known as the Kanaks, are now in a minority. France established New Caledonia as a country of settlement with immigrants from France, the 'Caldoches', and indentured workers from elsewhere in the Pacific and Asia.

New Caledonia's main resources are its mineral deposits. The country is the world's fourth largest producer of nickel and has around one-third of world reserves. This, combined with French aid of around $900 million per year, has given New Caledonians a relatively high standard of living, but low prices for nickel have reduced their income, increasing the need to develop other resources such as fishing and tourism to combat rising levels of unemployment.

New Caledonia is an overseas territory of France and sends deputies to the French parliament, as well as electing members to local assemblies. Since the 1970s, there has been a running conflict between the Kanaks, who favour independence, and the

Land area: 19,000 sq. km.
Population: 227,000—urban 65%
Capital city: Nouméa , 91,000
People: Melanesian 44%, European 34%, Wallisian 9%, Tahitian 3%, Indonesian 3%, Vietnamese 1%, other 6%
Language: French, various Melanesian languages
Religion: Roman Catholic 60%, Protestant 30%, other 10%
Government: Overseas territory of France
Life expectancy: 75 years
GDP per capita: $PPP 15,000
Currency: CCP franc
Major exports: Ferronickel, nickel ore

Caldoches, who favour continuing as a French territory.

Although a referendum on independence was planned for 1998, this was ultimately considered too divisive, and instead the two sides reached the 'Nouméa Accords' through which France would allow considerable autonomy. Sometime between 2013 and 2018 there will be a referendum on independence.

Implementation of the accord has been slow and the outcome remains uncertain. The Kanaks have a higher birth rate than the Caldoches, and if migration from France declines they could be in the majority by 2013. On the other hand, the next generation of Kanaks will have had ten more years of French aid, so may depend on it.

The main pro-independence political group is the Front de libération nationale kanak socialiste (FLNKS). There are three main anti-independence parties: the Avenir Ensemble (AE), Calédonie ensemble, and the Rassemblement-union pour un mouvement populaire. In February 2011, the government collapsed and administration largely came to a halt. An early election may be needed.

New Zealand

New Zealand now has one of the world's least regulated economies

Land area: 271,000 sq. km.
Population: 4 million—urban 87%
Capital city: Wellington, 449,000
People: European 70%, Maori 8%, Asian 6%, Pacific Islander 4%, other 12%
Language: English, Maori
Religion: Anglican 15%, Roman Catholic 13%, other Christian 26%, other 46%
Government: Constitutional monarchy
Life expectancy: 80 years
GDP per capita: $PPP 27,336
Currency: New Zealand dollar
Major exports: Wool, meat, dairy products, chemicals, forestry products

New Zealand consists of two main islands. The larger, South Island, is also the more mountainous, dominated by the snow-capped Southern Alps which run down its western half. In North Island the mountains are on the eastern side and are somewhat lower. Parts of New Zealand are of volcanic origin and there are a number of hot springs and geysers.

Three-quarters of the population live in the North Island. Most are of European, and particularly British, extraction—the product of more than a century of immigration that gave preference to 'traditional source countries', a policy that ended officially only in 1986. New Zealand has also been a country of net emigration, primarily to Australia. Worried about this brain drain, the government has made greater efforts to attract skilled immigrants and investors. This has increased immigration, primarily from Asia, and created a political reaction.

The earlier arrivals came around the 14th century. These are the Maori who are now a minority, almost all of whom live on the North Island. 'Maori' is Maori for 'normal people', to distinguish themselves from the whites, the 'Pakeha'. The Maori name for New Zealand is Aotearoa, 'land of the long white cloud'. For a hundred years, the Maori were bitter that the 1840 Treaty of Waitangi that had guaranteed their land rights had not been honoured. They got some recognition in 1994 and 1995 in the form of compensation treaties with the government—and a formal apology from the British queen Elizabeth II.

The third largest group of New Zealanders come from various Polynesian islands, many of whom arrived from the 1960s onwards to meet the demand for unskilled labour.

Agriculture still employs around 10% of the population. New Zealand's soil is not particularly fertile, but the temperate climate and its grassy hills and meadows sheltered by the mountains make it ideal for pastoral farming. The vast sheep herd, the fourth largest in the world, enables

New Zealand to be one of the world's leading producers of lamb, mutton, and wool. It also exports beef and dairy products. The main arable crop is barley for animal feed. Other crops include kiwifruit and apples for export, along with the grapes that have enabled New Zealand to become a leading wine producer.

Manufacturing industry is dominated by the processing of meat and dairy products for export, but New Zealand is now also a major producer of pulp and paper.

New Zealand has the service industries common to most industrial countries. But its tourist industry is increasingly important. More than two million visitors come each year to see the country's attractive scenery—some of which posed as 'Middle Earth'

New Zealand poses as Middle Earth

in the movie trilogy, *The Lord of the Rings*.

New Zealand is a constitutional monarchy, headed by the British queen. Because the population is of overwhelmingly British origin there seems to be little pressure to become a republic.

Until the mid-1980s, New Zealand had a highly regulated welfare state. This changed dramatically from 1984 following an electoral victory by the Labour Party which embarked on a radical programme of economic liberalization, removing agricultural subsidies and import controls, and in its next term from 1987 privatizing many public enterprises.

By the end of the second term, splits in the Labour Party contributed to its downfall. In the 1990 elections the right-wing National Party returned to power and, with Jim Bolger as prime minister, continued and extended Labour's liberalization programme while cutting many welfare benefits. In 1993, the National Party was narrowly re-elected.

Meanwhile a new right-wing nationalist party, New Zealand First, was formed to campaign against immigration, particularly from Asia.

The 1996 election was the first under a new system of proportional representation and resulted in a coalition government that paired the National Party with New Zealand First. This soon started to unravel. Bolger was ousted as National Party leader and Jenny Shipley became New Zealand's first woman prime minister.

The 1999 election brought Labour, led by Helen Clark, back to power in a coalition with the centre-left Alliance Party. After years of neo-liberal economic policies, Labour shifted back somewhat to the left, increasing spending on health and education and taking initiatives for Maoris and Pacific Islanders. However the Alliance party fell apart and its founder Jim Anderton created what is now the Progressive Party.

After elections in 2002 and 2005, Clark also formed coalition governments, though in 2005 as a minority administration.

The 2008 election was more decisive. The National Party, led by investment banker John Key won 59 of 122 seats. It can, however, count on six votes of two small right-wing parties. Key has presented National as a business-friendly party and retains strong support. In 2009, Clark became Administrator of the United Nations Development Programme.

Nicaragua

Nicaraguans of left and right now seem more prepared to work together

| Land area: | 130,000 sq. km. |

Land area: 130,000 sq. km.
Population: 6 million—urban 57%
Capital city: Managua, 1.2 million
People: Mestizo 69%, white 17%, black 9%, Amerindian 5%
Language: Spanish, English, Amerindian
Religion: Roman Catholic 59%, Evangelical 22%, other 19%.
Government: Republic
Life expectancy: 73 years
GDP per capita: $PPP 2,570
Currency: Córdoba
Major exports: Coffee, seafood, meat, sugar

Nicaragua is the largest country in Central America—and the least densely populated. The emptiest area consists of tropical forests in the east that stretch down to the marshy area called the Miskito Coast, named after the Miskito Indians who inhabited it when British buccaneers arrived in the 17th century. Even today, this region, which is also home to most of Nicaragua's black English-speaking population, remains relatively isolated from the rest of the country.

The majority of the population, mestizo and white, are to be found in the plains on the Pacific coast or in the central volcanic highlands, though in the north even these highlands are sparsely populated.

Nicaragua is one of the world's more disaster-prone countries, vulnerable to earthquakes and flooding as well as to hurricanes. In October 1998 Hurricane Mitch killed 3,000

people and caused $1.5 billion-worth of damage.

Standards of human development are low. Basic education levels were boosted by a literacy drive by the Sandinistas in the 1980s, but since then progress has been slow and 10% of children are still not enrolled in primary school. The Sandinistas also started mass health campaigns. Now, however, spending on health is inadequate and health standards, particularly in the rural areas, are among the lowest in the Americas.

This is also one of the poorest countries in the region, with around half the population below the poverty line. Meanwhile inequality is increasing: the richest 20% of the population get almost half the income.

Around 40% of the workforce make their living from agriculture. Many grow basic food crops in the central regions and along the Pacific coast, while in the uplands the main crop is coffee. Agricultural yields for food-crops are very low. This is partly because more than one-third of landholders lack secure title over their land so are reluctant to invest in irrigation or other improvements. Coffee output on the other hand has increased, but farmers have suffered from low world prices.

Nicaragua also has some light industry. The most dynamic production is in the 15 'maquila' or free-trade zones. These employ around 70,000 people making a range of products including garments, shoes, and electrical equipment. But jobs elsewhere are hard to find. In 2007 around one-third of the workforce were unemployed or underemployed; the situation is particularly bad on the Atlantic coast. Almost two-thirds of all jobs are in the informal sector.

Nicaragua found itself the focus of world attention after the 1979 overthrow of corrupt dictator Anastasio Somoza and the arrival of the government of the Frente Sandinista de Liberación National (FSLN). The 'Sandinistas' had strong socialist principles, and distributed land and property to the peasants. But they faced resolute opposition from the Reagan administration in the USA which accused it of fostering communism and for most of the 1980s funded 40,000 'contra' rebels operating from bases in Honduras.

Reagan versus the Sandinistas

The debilitating civil war that ensued, combined with poor economic management, brought the country to its knees.

The FSLN survived but in 1990 they were voted out. As well as protesting against poverty, Nicaraguans had become disillusioned with a party that seemed to have become less democratic. They opted instead for a right-wing coalition led by Violeta Chamorro. She set the country out on a new course, reducing public expenditure, and privatizing

state enterprises. She also adopted a conciliatory line with the FSLN, accepting their land reforms.

The 1997 election was violent and again saw a victory for the right wing, this time a coalition led by Arnoldo Alemán Lacayo of the Partido Liberal Constitutionalista (PLC) which won both the presidential election and a majority in the Legislative Assembly. The FSLN was, and continues to be, led by one of its original personalities, Daniel Ortega.

Alemán was a populist and conservative president, though was prepared to deal with the Sandinistas. The 1997 Property Law agreement, for example, protected beneficiaries of the Sandinista land redistribution.

Alemán and Ortega moved even closer together early in 2000 when the government made constitutional changes that entrenched the position of the PLC and the FSLN. Important institutions, such as the Supreme Court, would now have members nominated by the two parties.

Alemán could not stand for re-election in 2001. The victorious PLC candidate this time was Enrique Bolaños Geyer. Bolaños, however, soon turned on Alemán, eventually having him prosecuted for fraud and money-laundering for which he received a 20-year jail sentence, though the case is still being appealed.

In 2006 Ortega won the presidential election. But the FSLN does not have a majority in the Assembly so often has to rely on support from the PLC. Ortega has successfully persuaded the Supreme Court to issue decrees that have enable him to stand for an unconstituional third term in 2011.

Niger

Niger's underground wealth has been of little benefit to most of its people

0	Miles	400
0	Km	640

Land area: 1,267,000 sq. km.
Population: 14 million—urban 17%
Capital city: Niamey, 675,000
People: Hausa 56%, Djerma 22%, Fulani 9%, Tuareg 8%, Beri Beri 4%, other 1%
Language: French, Hausa, Djerma
Religion: Muslim 80%, indigenous beliefs and Christian 20%
Government: Republic
Life expectancy: 51 years
GDP per capita: $PPP 627
Currency: CFA franc
Major exports: Uranium, livestock, cotton

Niger has three main geographical zones. The zone to the north, approximately half the country, falls within the Sahara Desert. This arid territory includes the Air Mountains with sandy desert on either side. The central zone falls within the Sahel, with thin soil, scrubland, and sparse vegetation. The most fertile zone is the south, which benefits from annual rains, as well as, in the far south-west, flooding from the Niger River which flows south into Nigeria.

Most people live in the fertile south where sedentary farming groups include the Hausa and Djerma. The nomadic herding communities, including the Fulani, who raise cattle, and the Tuareg, who also have extensive herds of goats and camels, are likely to be found in the Sahel area, and sometimes in the desert to the north. But creeping desertification and over-use of the land by a growing population are steadily undermining the nomadic lifestyle—and many nomads are settling as farmers or moving to the cities. The annual population growth rate over the period 2001 to 2015 is expected to be 3.6%.

Levels of human development are desperately low. Niger is one of the few countries that even in peacetime has seen no improvement in social indicators over the past decade. In 2009, Niger came last in UNDP's human development index, a consequence not just of low income, with more than 60% of the population below the poverty line, but also of dismal standards of health and education. Around 18% of children die before their fifth birthday and 40% are malnourished. Only 38% of children enrol in primary school and around 70% work.

Most people make their living from agriculture which accounts for around 40% of GNP and employs 80% of the labour force. Their food crops include millet, sorghum, rice, and cassava, and their cash crops include cotton and groundnuts. Most of this cultivation is in the south-west of the country and since it is rain-fed it remains very vulnerable to the often erratic climate, which in recent years has led to severe shortages. In a good year the country is more or less self-sufficient in food. Even then, during the dry season many Nigeriens migrate to neighbouring countries in search of work. The main agricultural export is livestock, much of which walks to Nigeria to escape taxation.

Niger's main export and an important source of government

Niger's uranium fuels French reactors

revenue has been uranium ore. Niger has extensive deposits of high-grade uranium ore in the desert to the east of the Air Mountains. Niger is the world's fourth largest producer, sending a lot of its output to nuclear-power reactors in France and Spain. The mines are run by two independent companies though the government has shareholdings in both.

A number of companies have also been searching for oil though there have been no large finds.

Dependence on the uranium price has often landed the country in debt. Debt service payments were reduced, however, in 2004 as a result of the Heavily Indebted Poor Countries initiative which cancelled about half the country's debt.

Niger's democratic performance since independence in 1960 has not been impressive. Brief periods of multiparty governance have been interspersed with bouts of military rule. There have also been armed rebellions by the Tuareg who, with the support of Libya, have at times pressed for independence though they signed a ceasefire in 1995.

The last-but-one coup was in 1996, when Colonel Ibrahim Baré Maïnassara seized power and arrested both the president and the prime minister and banned political activity. Later that year he did, however, introduce a new constitution that put power much more firmly in the hands of the president. Unsurprisingly, Maïnassara presented himself for election and took most of the vote in an election boycotted by the opposition. In 1997 he launched his own party, the Rassemblement pour la démocratie et le progrès.

Maïnassara's rule became steadily more repressive and any group protesting against government policy was subject to harassment or arrest. In April 1999, however, he was shot dead by soldiers.

The military seized power until fresh elections were held in November 1999 when Mamadou Tandja, a retired army colonel, won the presidential election and his party, the Mouvement national pour la société de développement (MNSD), won 38 of the 83 seats in the National Assembly. Its coalition partner the Convention démocratique et sociale (CDS) took 17. They were later joined by the Alliance nigérienne pour la démocratie et le progrès.

Tandja and the MNSD were returned to power in the elections in 2004. In 2009, however, Tandja introduced a new constitution and dissolved parliament. As a result, in February 2010 he was overthrown in a military coup.

True to their word, the military did restore democracy. In 2011 Mahamadou Issoufou of the Parti nigérien pour la démocratie et le socialisme (PNDS) was elected president while his party gained a majority in the National Assembly.

Nigeria

Nigeria's democratic government is struggling to reverse decades of economic and social decline

Land area: 924,000 sq. km.
Population: 148 million—urban 50%
Capital city: Abuja, 2 million
People: Hausa-Fulani 29%, Yoruba 21%, Igbo 18%, Ijaw 10%, Kanuri 4%, other 18%
Language: English, Hausa, Yoruba, Igbo, Fulani
Religion: Muslim 50%, Christian 40%, indigenous beliefs 10%
Government: Republic
Life expectancy: 48 years
GDP per capita: $PPP 1,969
Currency: Naira
Major exports: Oil, cocoa, rubber, cotton

Nigeria can be divided into four main regions. The humid coastal belt includes extensive swamplands and lagoons that can extend 15 kilometres inland, and further still in the area around the Niger River delta. Further inland the swamps give way first to hilly tropical rainforests and then the land rises to several plateaux—the Jos plateau in the centre and the Biu plateau in the north-east. Further north, these descend to savannah grasslands and eventually to semi-desert areas. The principal feature of the east-central border area with Cameroon is the Adamawa plateau, which includes the highest point in the country, Mount Dimlang.

Nigeria has Africa's largest population, though the actual size is a matter of some dispute. The government has claimed that it is smaller than UN estimates. As with much else, this is a highly political issue since it has a critical bearing on the distribution of funds between the federal government and the states. The population is likely to grow at an average of 2.5% per year over the next decade or so—and could reach 240 million by 2025.

Nigeria is a federation of 36 states with more than 400 ethnic groups, though more than half belong to the three main groups: the Hausa-Fulani, who are Muslims and live in the north; the Yoruba, who are followers of both Christian and Islamic faiths and live in the south-west; and the Igbo, many of whom are Christians and live in the south-east. They largely live together peacefully, but there are at times outbreaks of violence. In 1999, for example, there were clashes between Yoruba and Hausa and hundreds of people died in various outbreaks of ethnic violence. Even more alarming were the deaths in early 2000 of hundreds of Christians and Muslims in disputes over the latter's determination to impose Shariah law in northern states in which Muslims are in the majority.

Despite its potential wealth, Nigeria has been slipping backwards.

Around one-third of the population live below the poverty line. The education system is in poor shape and the literacy rate is only 68%. Those who can afford private education make progress but most children are

Nigerian children packed into crowded classrooms

packed into crowded and dilapidated classrooms. The public health system too is increasingly over-burdened.

Maternal mortality is three times the average for developing countries. HIV and AIDS are also taking a heavy toll: in 2007 the official estimate was that 1.5% of adults were HIV-positive, though the real figure could be higher.

Despite rapid urbanization, most Nigerians still earn their living from agriculture, which makes up around 40% of GDP. The vast majority are subsistence farmers who have small plots and use primitive tools. In the south they tend to grow root crops like yams and cassava, while in the north the main crops are sorghum, millet, and maize.

With little investment, and over-exploitation of the land, output has often lagged behind population growth. As a result Nigeria, which used to export food in large quantities, is now a major importer. Small farmers also grow some cash crops, but of these only cocoa now provides any significant export income—and output has halved in the last 30 years.

Agriculture, as with much else in Nigeria, has been pushed into the background by the oil industry. Oil was discovered in 1956 and output grew rapidly in the 1960s and 1970s, helped by the high quality and low sulphur content and also

by the location—most of the fields are onshore in the Niger delta and convenient for export. Around half goes to the USA. In 2004, proven reserves were around 31 billion barrels—sufficient for another 35 years. Nigeria also has some of the world's largest reserves of natural gas.

In 2008, oil accounted for 90% of export earnings and more than two-thirds of federal government revenue. So when the oil price falls Nigeria is in serious trouble, and since the mid-1980s Nigerians have watched their infrastructure decay and their public services decline.

The government-owned Nigerian National Petroleum Company has joint venture agreements with most of the oil companies, including Elf, Agip, Mobil, and Chevron. But around half the oil output comes from one company, Royal Dutch/Shell, which has come under heavy local criticism. People in the Niger delta have seen little benefit from oil production that has often caused environmental damage.

The Ogoni people in particular have been demanding greater control over the oil, and in 1995 Ken Saro-Wiwa and eight other Ogoni were executed. Shell was accused of collusion with the government

Shell accused of collusion in human rights abuses

and although it denies this, the company has now accepted greater responsibility for

community development in the area.

In 2009, the prospects for peace in Niger Delta increased when, following the offer of an amnesty, thousands of insurgents surrendered their arms in exchange for a pardon, a little cash and the president's promise to channel more oil income directly to villages.

Most of the hundreds of billions of dollars received in oil wealth since the early 1970s have been squandered, either in wasteful development projects, or in graft or theft. Corruption started at the top. Military leaders simply stole much of the money. But scarcely any service is available without the payment of what is known as 'chop'.

Oil wealth squandered

One of the most ironic effects of corruption is that Nigeria has suffered from chronic shortages of petrol. Local refineries have been deliberately run down so that government officials can get their chop from the sale of lucrative licences to import petrol, much of which finishes up on the black market.

From the early 1970s, the government invested some of the oil money in nationalized heavy industry—particularly petrochemicals, steel, and fertilizers—though such industries have been poorly managed and Nigeria still depends on imports. Most productive industrial activity is based on small-scale manufacturing, and a high proportion of the urban populations struggle in the informal sector.

Nigeria's political history since independence in 1960 has been fraught with ethnic conflicts and demands for greater independence for the states, which wanted to seize greater shares of the oil revenue. In 1967 the military governor of the Eastern Region attempted to secede, provoking a civil war that lasted two-and-a-half years and killed one million people. Most of the years since then have involved a series of coups with brief interludes of civilian rule.

The most recent military regime, that of General Sani Abacha starting from 1993, was by general consent the most brutal and rapacious. Abacha had come to power following the annulment of a presidential election that gave victory to a Yoruba, Chief Abiola. Abacha subsequently imprisoned Abiola, launched ferocious attacks on opposition figures, and hanged Saro-Wiwa— drawing widespread international condemnation. To widespread relief Abacha died in 1998—as did Abiola in prison (of a heart attack).

This led to a presidential election in 1999 and a victory for Olesegun Obasanjo, and his People's Democratic Party (PDP). Although a former military ruler, he had left office voluntarily in 1979.

Obasanjo came to power with the support of the northern elite, but tried to be even-handed in distributing state appointments across regional and ethnic lines. Despite promising much, he made little progress in tackling corruption or ethnic violence, and human rights abuses continued.

Obasanjo and the PDP were re-elected in May 2003. He tried to have the constitution altered to allow him to run again in 2007 but the bill was defeated in the Senate. Instead the 2007 elections saw a victory for his little-known, hand-picked successor, Umaru Yar'Adua. Following health problems in 2010, however, he had to hand power to vice-president Goodluck Jonathan.

In April 2011, following reasonably fair elections, Jonathan and the PDP swept the board in the presidential, legislative and state elections.

Norway

Norway is one of the world's richest countries, but is trying to wean itself from over-reliance on oil

Land area: 324,000 sq. km.		

Land area: 324,000 sq. km.
Population: 5 million—urban 78%
Capital city: Oslo, 560,000
People: Norwegian 94%, other 6%
Language: Norwegian (Bokmål, Nynorsk)
Religion: Church of Norway 86%, other 14%
Government: Constitutional monarchy
Life expectancy: 81years
GDP per capita: $PPP 53,433
Currency: Norwegian krone
Major exports: Oil and gas, metals, machinery

Norway is a mountainous country with a distinctively complex coastline. Much of the landscape consists of rolling plateaux, called 'vidder', interspersed with high peaks. To the west the mountains descend steeply to the coast which, during the Ice Age, was deeply cut by glaciers creating long, narrow inlets, the 'fjords', beyond which lie hundreds of islands. Though one-third of the country is above the Arctic Circle, the warm waters of the gulf stream usually keep the fjords above freezing point.

Norway is ethnically fairly homogenous. Almost everyone speaks the same language, though this exists in two mutually intelligible forms. The older of the two, Bokmål, is more widely used—in the national newspapers and in most schools—and is the product of previous centuries of Danish occupation of Norway. 'New Norwegian', Nynorsk, is both newer and older—the result of efforts to revive earlier dialects of Norse.

The oldest minority are the country's original inhabitants, the Sami, who herd their reindeer in the far north. But from the 1980s Norway started to receive significant flows of asylum seekers, first from Asia and Africa and later from former Yugoslavia. By 2008, 9% of the population were foreign born. Norway occupies first place in the UNDP human development index—a reflection of the country's wealth and its comprehensive system of welfare.

Norway's economic prospects were transformed in 1962 by the discovery of the Ekofisk oilfield in the North Sea. This and numerous other fields discovered subsequently have made Norway the world's fifth-largest oil exporter. Oil production is now declining though this is partly offset by exports of gas.

Oil and gas have fuelled Norway's economic growth and provided more than half of export income, and in around one-quarter of government revenue. Unlike many other countries, Norway has been setting much of this windfall aside, into the 'Government Pension Fund–Global', which by 2005 had reached $165 billion. One disadvantage of the oil bonanza, however, is that it has stifled investment in other activities. Manufacturing, for example, accounts for only 19% of GDP. Meanwhile, some older industries have been stagnating.

Norway's dramatic, serrated coastline has ensured a close relationship with the sea, and the country has established a strong shipping industry that has given it around 10% of the world's commercial fleet—including one-quarter of all cruise vessels and 20% of gas and chemical tankers.

Agriculture has long been in decline, though the government has subsidized it heavily and agriculture still employs 4% of the labour force. Fishing too—for herring, cod, and mackerel—has also declined as a result of over-fishing.

Despite weakness outside the oil sector, Norway's unemployment rate in 2009 was only around 3%.

Norway is a constitutional monarchy, currently ruled by King Harald, who acceded to the throne in 1991. For most of the 20th century

Norwegians reject EU membership

Norway was governed by the social-democratic Labour Party. One of the leading figures in the 1980s and 1990s was Gro Harlem Brundtland, who was a Labour prime minister in the periods 1986–89 and 1990–99. Brundtland applied for Norwegian membership of the EU and achieved significant concessions. But the electorate did not share her enthusiasm and in a referendum in 1994 declined to join—52% voted against. Brundtland resigned in 1996. She had already

been a major international figure, having chaired the World Commission on Environment and Development, and later served as head of the World Health Organization.

Brundtland was replaced as Labour leader by Thorbjorn Jagland. Although Labour received 35% of the vote in the 1997 election, Jagland gave way to a minority centrist government led by Lutheran priest Kjell Magne Bondevik of the Christian Democrat Party (KrF) in coalition with the rural Centre Party and the free-market Liberals. Bondevik took the unusual step in late 1998 of disappearing from view as a result of 'a depressive reaction to overwork'—an act which, if anything, seemed to enhance his popularity.

In 2000 Labour replaced Jagland as their leader with Jens Stoltenberg. He set his sights on the government and in March brought it down through a no-confidence vote, after which the king asked him to form a new coalition government.

At the next general election in 2001 there was no clear mandate and after extensive negotiations Bondevik and the KrF returned at the head of a 'cooperation government' that included the Conservatives and the Liberals.

People were becoming increasingly concerned however that Norway's wealth was not being translated into even better public services—with staff shortages in hospitals and other services. As a result Labour and Jens Stoltenberg returned to office in 2005 with a comfortable majority, forming a coalition with the Socialist Left Party and the Centre Party. In 2009, the same coalition was returned, though with a smaller majority.

Oman

Oman is less well endowed with oil than its neighbours, and is keen to diversify

Land area:	*212,000 sq. km.*
Population:	*3 million—urban 72%*
Capital city:	*Muscat, 719,000*
People:	*Arab, Baluchi, South Asian*
Language:	*Arabic, English, Baluchi, Urdu, Indian dialects*
Religion:	*Ibadhi Muslim 75%, Sunni Muslim, Shia Muslim, Hindu, or other 25%*
Government:	*Sultanate*
Life expectancy:	*76 years*
GDP per capita:	*$PPP 22,816*
Currency:	*Omani rial*
Major exports:	*Oil, animal products, textiles*

Oman has three main geographical areas. The Al-Batinah plain extends for about 270 kilometres along the northern coast but only about 10 kilometres inland. From here, the land rises to the Al-Hajar mountain range, whose highest peak is around 3,000 metres. Beyond the mountains to the south, the rest of the country, around three-quarters of the land area, is largely desert until it reaches the mountains of Dhofar in the far south-west.

Oman's people are culturally diverse. Most Omani citizens are Arab and Muslim, though around three-quarters belong to the Ibadhi sect, an early breakaway from the Shia. The rest are Sunni or Shia. But this is not a strict Islamic state.

One-quarter of the population are immigrants, and make up a high proportion of the labour force—80% of the private sector and 20% of the

public sector. Most come from the Indian subcontinent, though there are also expatriate Europeans. Immigrants do all the lowliest jobs—two-thirds earn less than the minimum wage.

From around 1970, thanks to oil income, Oman's people made swift improvements in human development—some of the most rapid progress ever recorded. Between 1970 and 2008 life expectancy increased from 40 to 76 years, and over the same period infant mortality fell from 126 per thousand live births to 10. Education too expanded rapidly, though it has not produced an optimum balance of skills.

Most Omani graduates expect managerial or professional jobs, preferably with the government, and are less willing to acquire vocational skills. But with the population of Omanis growing at around 3% per year there are not enough of these jobs to go around.

Oil has fuelled much of Oman's progress, accounting for 25% of GDP, but it is a dwindling asset and on present trends will last less than 20 years. Since it must be extracted from over 100 small and complex fields in the interior of the country and in Dhofar, production costs are higher than in neighbouring countries. There are other reserves but these would probably be too expensive to extract. The main producer is Petroleum Development Oman in which the government has a majority stake.

The other main source of income is agriculture. Though accounting for less than 2% of GDP, it employs around one-third of the workforce. On the northern Al-Batinah plain their main crop is dates, while on a smaller scale farmers on the southern coastal plain grow coconuts and bananas. There is also some subsistence agriculture in the mountains as well as livestock-rearing in the interior. But Omani agriculture is running short of water. Most has to be pumped from underground and the advent of diesel pumps has encouraged over-extraction. Coastal areas also suffer from intrusion of saline water.

Farmers are running out of water

Oman's first attempts to diversify involved creating new industrial estates for manufacturing. These factories are now working, producing textiles and other goods for export. But they remain over-reliant on foreign workers.

Better prospects seem likely to emerge from recent discoveries of natural gas which have encouraged three major industrial projects. The most important is a large liquefaction plant to enable gas exports—primarily to South Korea, Japan, and India. The other possible uses are more problematic. Plans to develop an ammonia-urea fertilizer plant have been set back by low world prices for fertilizers. And plans to use the gas to fuel aluminium smelting have suffered a number of setbacks.

As well as diversification Oman is also pursuing 'Omanization' of private industry, setting quotas for different industries and establishing fines and other penalties for enterprises that fall short. But with so few trained or willing recruits most companies will be hard pressed to meet their targets.

Presiding over these developments is an absolute monarch. Since the 18th century Oman has been controlled by the Al bu Said tribe, and since 1970 the sultan has been Qaboos bin Said. In his early years he was faced with a communist insurrection in Dhofar. He finally crushed this in 1975 with the help of the USA, which still has a military presence.

The Basic Law also establishes that the sultan's successor must come from the Al-Turki branch of his family. It says too that if the family cannot agree on a suitable successor within three days of the sultan's death, they will have to accept his choice—which will be left in a letter.

The sultan has been edging towards democracy. In 1997 he promulgated a new 'Basic Law' (based on Islamic law) that serves as a constitution and establishes some limited democratic rights. In addition, he has set up two consultative bodies. The first is the Majlis al-Shura for which elections were held most recently in 2007 with a 60% turnout. The second is the State Council whose members are appointed from tribal leaders and former government officials.

Nevertheless, in 2011, inspired by the Arab spring in Egypt and Tunisia, people started taking to the streets demanding democracy. Faster reform seems likely.

Pakistan

Asia's great underachiever. Democracy has yet to take hold and now the country is under attack from terrorists.

Land area: 796,000 sq. km.
Population: 173 million—urban 37%
Capital city: Islamabad, 901,000
People: Punjabi 45%, Pathan 15%, Sindhi 14%, Sariaki 9%, Muhagirs 8%, other 9%
Language: Punjabi 48%, Sindhi 12%, Siraiki 10%, Pashtu 8%, Urdu 8%, other 14%
Religion: Sunni Muslim 77%, Shia Muslim 20%, other 3%
Government: Republic
Life expectancy: 66 years
GDP per capita: $PPP 2,496
Currency: Pakistani rupee
Major exports: Cotton yarn, garments, cotton textiles

Pakistan has four main geographical regions. First, in the far north is the Hindu Kush and Pakistan's section of the Himalayas, including K2, the world's second highest mountain. Second, in the west and south-west, are a number of other mountain ranges and the Baluchistan plateau. Third, to the east are the desert areas which join with the Thar Desert across the border in India. Fourth, in the centre of the country is the rich agricultural plain of the Indus River, which traverses the entire length of Pakistan, emerging from the foothills of the Himalayas and flowing down to the Arabian Sea.

Pakistan is also divided politically into four provinces: Baluchistan in the west; North-West Frontier Province, on the border with Afghanistan; Sindh in the south; and Punjab in the heartland of the Indus Valley.

The population of Pakistan has been formed by waves of migration from many different directions and is a mixture of many influences with relatively few ethnic divisions. The Punjabis account for around two-thirds of the population, and the Sindhi for 13%. The other main groups are the Baluchs and the Pathans in North-West Frontier. Added to these are the Muhajir ('refugees'), descendants of the eight million Muslims who fled from India in 1947 at the time of partition—most of whom are in the urban areas of Sindh and particularly Karachi.

Faced with limited opportunities at home, millions of Pakistanis have headed overseas, chiefly to the Gulf states. They send home remittances worth around $7 billion per year.

Almost all Pakistanis are Muslims. Pakistan was created out of the partition of British India specifically to provide a state for the Muslim community. Even so, it was not initially envisaged as an Islamic state. Only in recent years has Islam become a dominant political force, having frequently been exploited by authoritarian governments seeking to rally support. There have been frequent clashes between Sunni and Shia Muslims.

Economically, Pakistan appears at times to have been fairly successful But it has failed to translate economic growth into real improvements in human well-being. Around one-third of the population live below the poverty line. Education levels are low: one-third of children do not enrol for primary education and over half of the adult population are illiterate. There are similar problems in health. Health services are poor and one-third of the population do not have safe sanitation.

Particularly disturbing is the slow advance of women. Thus while boys on average have 2.9 years of schooling, girls have only 0.7 years, and the literacy rate for women is only half that of men. Women also have fewer employment opportunities—particularly in the socially more conservative provinces of Baluchistan and North-West Frontier.

Pakistani women have been making very slow progress

Women's generally low status is also reflected in poor standards of reproductive health: a consequence of official neglect, religious intransigence and considerable violence against women. Contraceptive use is low, so the birth rate is high and thousands of women die each year in childbirth. This also contributes to Pakistan's high annual population growth rate of 2%. By 2050 this will be the world's third most populous nation—after China and India.

Agriculture still employs around two-thirds of the workforce and makes up 20% of GDP. But more than 40% of the arable land is controlled by rural élites in farms of 50 acres or more where it is often used poorly— growing the most profitable crops like cotton but using inefficient farming methods that have made the land waterlogged or saline. Since 1959, there have been half-hearted efforts at land reform, but little has happened since the main landowners are also usually politicians. Pakistan used to be self-sufficient in wheat and rice but production has not kept pace with population growth and millions of tons of wheat have to be imported.

Industry has been fairly slow to develop. Pakistan is one of the world's major cotton producers but still exports most of its output as yarn rather than cloth. Immediately after independence, manufacturing was aimed at import-substitution.

More recently there have been efforts to open up the economy— removing some bureaucratic restrictions and lowering tariffs. But manufacturing industry is still unimpressive—dominated by low-tech enterprises. Industrialists claim they cannot invest because unlike the politically powerful landowners they have to pay so much tax.

Pakistan has been equally slow to develop its mineral deposits. The country does have some oil and gas. But oil production is only sufficient for about 20% of the country's needs, while reserves are falling. And exploitation of gas in remote areas of Baluchistan is hindered by opposition from belligerent tribesmen. Pakistan also has large deposits of rock salt, limestone, and other minerals but these have been little exploited.

Pakistan has frequently been embroiled in serious disputes with its neighbours, particularly India. In 1998, Pakistan responded to Indian nuclear tests with six tests of its own (Pakistan is the world's only nuclear-armed Islamic state). The most dangerous issue is Kashmir, a territory whose sovereignty Pakistan disputes with India. For many years, this has involved just sporadic gunfire, but it took an alarming turn from 1999 when Pakistan encouraged rebel fighters to cross into Indian-held territory. Early in 2004, relations seemed to improve, however—a thaw that included an historic series of Pakistan–Indian cricket matches.

The only nuclear-armed Islamic state

Democracy in Pakistan has at best been fragile. For more than half its history since independence in 1947 the country has been under military rule. From 1978, the military ruler was General Zia ul-Haq, who had been steering Pakistan towards becoming a full Islamic state.

Following Zia's death in 1988, Pakistan had a series of short-lived administrations that terminated with their dismissal by the president. Thus the 1988 election resulted in a victory for Benazir Bhutto and the Pakistan People's Party (PPP), the party founded by her father—a former prime minister whom Zia had executed. Incompetent and corrupt, she was dismissed in 1990.

The ensuing election resulted in a victory for a coalition headed by Nawaz Sharif, whose party was the Pakistan Muslim League (PML). He too soon fell foul of the president and was dismissed in 1993. Bhutto returned only to be dismissed again in 1996. The 1997 election saw Sharif

back but his downfall came in October 1999; after he tried to fire the army chief, General Pervez Musharraf, who deposed him in a bloodless coup.

Initially the international community condemned the coup, but following Musharraf's support for the US-led war in Afghanistan and the struggle against al-Qaida this criticism subsided.

Musharraf, survived several assassination attempts, developed a hybrid form of military and quasi-democratic rule. In 2002 he held a referendum to have himself appointed as president for five years.

By 2007, Sharif and Bhutto had returned to Pakistan ahead of legislative elections. Musharraf achieved a second term as president, though then stepped down as head of the army. In December 2007, Bhutto was assassinated while campaigning and was replaced as head of the PPP by her 19-year-old son Bilawal, though in practice his father Asif Ali Zardari is now in charge.

The legislative elections were finally held in February 2008. The PPP came out ahead with 121 seats, followed by the PML with 91. Musharraf's Muslim League came a distant third with 54. The PPP formed a minority administration with Rusuf Raza Gilani as prime minister.

Musharraf eventually resigned in August 2008 and Asif Ali Zardari won the presidential election convincingly. The government has plans to reduce the power of the presidency.

In 2011 the government was strengthened when the ruling coalition was joined by the Muttahida Qaumi Movement and the Pakistan Muslim League (Quaid-i-Azam).

Palestine

The chances of a peaceful settlement with Israel look more remote than ever

Land area: 6,000 sq. km.
Population: 4 million— urban 72%
Capital city: East Jerusalem, 364,000
People: Palestinian Arab and other 83%, Jewish 17%
Language: Arabic, Hebrew, English
Religion: Muslim 75%, Jewish 17%, Christian and other 8%
Government: Republic
Life expectancy: 73 years
GDP per capita: $PPP 2,900
Currency: Israeli shekel, Jordanian dinar
Major exports: Industrial goods, food

Palestine comprises two territories linked by land corridors through Israel. The larger of the two is the West Bank, most of which consists of hills of up to 900 metres that descend eastwards from the border with Israel down to the Jordan river and the Dead Sea. The land in the north-west is relatively well watered, but rainfall is much sparser in the south. The remainder of Palestine is the Gaza Strip, which makes up only 6% of the territory—a flat and sandy region on the Mediterranean coast between Israel and Egypt.

The people are predominantly Arab, but the territory is dotted with Israeli settlements. In 2007, the population was 2.5 million in the West Bank and 1.5 million in the more densely settled Gaza Strip. Of these, the UN classifies 1.7 million as refugees, who live in 19 camps in the West Bank and eight in Gaza. There

are also 170,000 Jewish settlers. The population is young—47% are under 15 years.

Thanks to international aid, standards of education and health are relatively high, though services have been disrupted by violence and unrest. Nevertheless around one quarter of people in the West Bank, and half of those in Gaza are below the poverty line. Many Palestinians have sought work or refugee status overseas. In 2006 there were 1.9 million refugees in Jordan, 0.8 million in other Arab states and 0.5 million in other countries.

Until fairly recently, both the West Bank and Gaza were predominantly agricultural, and agriculture still accounts for around 8% of GDP. Most of this is on small family farms, chiefly growing olives, citrus fruits, and vegetables in sufficient quantities for export, as well as raising livestock, but farmers' incomes have been dramatically reduced by Israel's closing of the border.

Industry remains largely undeveloped, accounting for 15% of GDP. This consists mostly of small enterprises engaged in food processing, textiles, clothing, leather, and metal working. But political uncertainty has hampered investment so productivity is low.

Israel's invasion of 1967 effectively incorporated what is now Palestine into Israel—and made it dependent on Israel for most of its markets; 85% of exports still go there. Just as important, Israel became a major source of work: by 1992 one-third of the workforce were commuting to Israel. At times, however, Israel has effectively closed the border to workers and has now cancelled most Palestinian work permits.

Israel has crippled the West Bank's economy

Were it not for international aid of around $1 billion per year Palestinians would be even worse off. Most of this is the form of emergency grants rather than development funds and an increasing proportion has come from the Arab League.

Until the Israeli invasion of 1967, Jordan controlled the West Bank and continued to claim it until 1987, when it relinquished it in favour of the Palestine Liberation Organization (PLO). The PLO had been formed in 1964 and was recognized by the Arab League as the representative of the Palestinians.

From the outset, the PLO had been engaged in armed struggle and in 1987 it embarked on a more general uprising called the 'intifada'. Subsequently, however, it shifted towards political activity and negotiations took a dramatic turn in 1993 when secret talks in Oslo produced a peace agreement. This involved mutual recognition between Israel and the PLO, as well as a staged Israeli withdrawal.

As a result, by 1996 much of Palestine was being administered by a Palestinian National Authority run by the PLO. Elections in 1996 gave 87% of the vote to PLO chairman Yasir Arafat as president and also gave a majority to Arafat's Fatah faction of the PLO in the Legislative Council. Arafat's administration was autocratic, with no real opposition and tainted by corruption.

A provocative visit in 2000 by Israeli leader Ariel Sharon to the Temple Mount in Jerusalem triggered a second intifada and the creation of a special Fatah unit, the al-Aqsa Martyrs Brigades which, together with other militant groups such as Hamas and Islamic Jihad, carried out suicide bombings in Israel. Israel responded by bombing the PLO headquarters and building a 'security fence' that cuts deep into Palestinian territory.

Arafat died in 2004 and presidential elections in January 2005 resulted in a victory for former prime minister Mahmoud Abbas of Fatah (Abu Mazen) who negotiated an end to the intifada.

But a victory for Hamas in the 2006 parliamentary election created an even more complex situation. Hamas does not recognize Israel, and many Western governments classify it as a terrorist group and withdrew aid.

Since then, Palestine has effectively been divided in two. The West Bank is controlled by President Mahmoud Abbas and a caretaker administration, while Gaza is controlled by Hamas.

Meanwhile, Israel continues to expand settlements in the West Bank and maintains air strikes on Gaza while also jailing many Hamas members of parliament.

Panama

Panama has a new master plan to develop and widen its greatest asset

Land area: 76,000 sq. km.
Population: 3 million—urban 75%
Capital city: Panama City, 0.8 million
People: Mestizo 70%, West Indian 14%, white 10%, Amerindian 6%
Language: Spanish, English
Religion: Roman Catholic 85%, Protestant 15%
Government: Republic
Life expectancy: 76 years
GDP per capita: $PPP 11,391
Currency: Balboa (US dollar)
Major exports: Bananas, seafood

Panama does have mountain ranges, including an extinct volcano, and a temperate zone above 2,000 feet. But 85% of the country is lower and flat, and one-third is forested. Its most significant geographical feature, however, is that at its narrowest it is only 80 kilometres wide—an ideal location for a canal to link the eastern and western coasts of the USA.

The USA has long been the dominant influence here. In 1903, it leased from Colombia the strip of land on either side of a canal that had been started and abandoned by a French company. When the Colombian senate refused to ratify the agreement the USA organized a rebellion that resulted in the secession of Panama from Colombia. A treaty with the new country gave the USA rights to the land in perpetuity and the canal was completed in 1914.

Panama's location as a crossroads for communications has given it a very diverse population, most of whom would be considered mestizo. The largest single group are the black descendants of the West Indians who were brought in to build first a railway and then the canal itself. But there are also significant Amerindian communities, including the Guaymí who live in the west of the country.

Compared with their neighbours, Panamanians are better off, with good standards of health and education, though there are often striking contrasts between the relatively prosperous people linked with the canal and the modern sector and the people in the rural areas.

Most Panamanians occupy the land on either side of the canal—although population pressure is forcing people to clear more of the rainforests. Around one-quarter are engaged in agriculture, growing subsistence crops on small farms or producing for export—chiefly bananas and sugar—or fishing for shrimps.

In the past most other employment has been derived from canal services. The canal itself has a labour force of around 9,000. Some 14,000 vessels go through the canal each year creating annual revenues of around $2 billion.

In 2007, the government embarked on a $5-billion development of the canal with the construction of new locks to take ships currently too large for the system.

But Panama has also diversified into other industries. At the Caribbean end of the canal, Colón has the second largest free-trade zone in the world, though it employs only 1% of the national workforce. On the other side of the country, Panama City has also become a major financial centre with more than 70 offshore banks. In addition, Panama's shipping registry gives it the largest 'fleet' in the world with 10,500 registered vessels.

Panama's turbulent political history has been shaped by its relationship with the USA. The US bases in the Canal Zone were home to the 'Southern Command'. But years of pressure and occasional riots provoked a nationalist backlash and Panamanians demanded sovereignty over the canal. In 1978, the military government of General Omar Torrijos, which had introduced popular measures such as agricultural reform, also negotiated a new treaty with US president Jimmy Carter, who agreed a handover for the end of 1999.

This did not signal an immediate

The USA invades to overthrow Noriega

end to US involvement in Panamanian political life. The most dramatic intervention in recent years was in 1989 when the USA invaded to overthrow the government of General Manuel Noriega—later tried and jailed for 40 years in Miami for corruption and drug smuggling. Several hundred people were killed.

The 1994 presidential election was won by Ernesto Pérez Balladares of the populist Partido Revolucionario Democrático (PRD). But he was not to supervise the handover of the canal. Although his party won a large majority in the 1999 legislative elections, he was defeated in the presidential election by Mireya Moscosa of the conservative-populist Arnulfista party, who became the country's first woman president.

The handover of the canal went ahead as planned. But the Stars and Stripes was lowered in a rather grudging manner, as no senior US official chose to attend. As well as providing more direct income for the government, the handover of the canal also created a property bonanza as ex-US buildings were converted.

Moscosa had also achieved control of the Legislative Assembly at the head of a coalition, but high levels of unemployment and concerns about extensive corruption reduced public support for her party—now called the Partido Panameñista (PP).

The 2004 presidential elections resulted in a victory for Martín Torrijos of the Partido Revolucionario Democrático (PRD). He is the illegitimate son of the former president, Omar Torrijos. Although he managed to start the expansion of the canal, voters became concerned about the rising level of crime.

In the 2009 elections he lost to a newcomer, Ricardo Martinelli of the small Cambio Democrático party in a centre-right coalition with the Partido Panameñista, Unión Patriótica and the Movimiento Liberal Repúblicano Nacionalista. With relatively little experience, he may struggle to hold the coalition together.

Papua New Guinea

More than 800 groups live in the rainforests—above some of the world's richest deposits of minerals

Land area: *463,000 sq. km.*
Population: *6 million—urban 13%*
Capital city: *Port Moresby, 254,000*
People: *Melanesian, Polynesian*
Language: *Pidgin English, English, and more than 700 indigenous languages*
Religion: *Roman Catholic 27%, Evangelical Lutheran 20%, United Church 12%, other Protestant, other 3%*
Government: *Constitutional monarchy*
Life expectancy: *61 years*
GDP per capita: *$PPP 2,084*
Currency: *Kina*
Major exports: *Gold, copper ore, oil, logs, palm oil, coffee, cocoa*

Papua New Guinea consists of the eastern half of the island of New Guinea, along with around 600 other islands. The mainland, with around 85% of the territory, includes extensive swampy plains to the north and south and a rugged chain of central highlands. But there are a number of other substantial islands, mostly of volcanic origin, including New Britain, New Ireland, and Bougainville. More than 70% of the country is dense tropical rainforest.

Communications are difficult and much internal travel is only feasible by air, which makes it hard to help isolated communities at times of natural disaster, as when a tidal wave swept into the north-west coast of the mainland in July 1998, killing up to 10,000 people.

Papua New Guinea has one of the world's most complex societies, with 800 or more ethnic groups or languages. The lingua franca is Tok Pisin—Pidgin English. For most Papua New Guineans, the primary allegiance is to their tribe or clan linked by a common language, called in Pidgin a wontok ('one talk').

Papua New Guineans have a rich culture, but low levels of human development. Around one-third of children under five years old are malnourished, and only two-fifths have access to safe water. Nevertheless, almost everyone benefits from basic health services, either from the government, the churches, or non-governmental organizations. And since people have access to commonly held land few are likely to starve unless there is a severe drought.

The country is also making some progress in education: primary school enrolment is now 73% and the literacy rate is 57%, though secondary education is still weak.

More than 75% of the population depend on agriculture. They grow mainly subsistence food crops such as sweet potatoes, bananas, sugar cane, and maize. But smallholders also grow cash crops for export including coffee, cocoa, and copra. In addition, there are large estates producing rubber, palm oil, and tea. Forestry is another source of export income. But commercial agricultural development generally has been slow—hampered by difficult communications.

Modernization is drawing more people to the towns and particularly to the capital, Port Moresby, where 40% of people live in shanty towns with no sanitation or basic services. Here and in other urban areas there has been considerable civil unrest, along with high levels of crime—both of which have served to slow social and economic development.

Port Moresby has very high levels of crime

Papua New Guinea's greatest cash wealth lies underground in extensive mineral deposits that generate more than half of exports. The earliest to be exploited was gold in the 19th century, then copper, notably on the island of Bougainville in the Panguna mine. The only copper mine now, however, is the Ok Tedi mine near the western border, owned by an Australian company, BHP. Since 1989 a series of major gold mines have also opened up. More recently, there have also been discoveries of oil which now accounts for one-quarter of exports. Gas could also be important, with an undersea pipeline to Australia.

Distributing the country's mineral wealth has long been a vexed issue—complicated by communal land ownership. There is also severe environmental damage. BHP has admitted that the pollution from the Ok Tedi mine is so severe that the only solution will be to close down. The government is pressing for this to happen in 2010.

A perennial problem has been the threat of secession by Bougainville. Local people felt they were getting little benefit from the Panguna copper mine. An armed insurrection in 1989 closed the mine permanently and the next ten years of fighting cost around 15,000 lives. Following mediation, from Australia and particularly from New Zealand, the fighting ended in 1999, and in 2000, under the 'Loloate Understanding', a Bougainville Interim Provincial government was established to offer autonomy. Later there could be a referendum on independence.

Papua New Guinea's political structure matches the complexity of its society. The head of state is the British monarch. But since independence from Australia in 1975 its single-chamber parliamentary system has yet to develop a stable party basis. Voters choose MPs largely on personality or tribe. Party allegiances are weak and, once elected, MPs feel free to 'cross the floor'.

Many governments have been unstable coalitions. The 1997 election produced a four-party coalition which chose as prime minister the somewhat erratic Bill Skate. He resigned in 1999 to be replaced by Sir Mekere Morauta of the People's Democratic Movement.

The elections in 2002 were fairly chaotic with more than 3,000 candidates. The party emerging with the largest number of seats was the National Alliance led by Sir Michael Somare who formed a 13-party coalition government. They were successful again in 2007 with Somare still in charge.

Paraguay

Paraguay produces more electricity per person than any other country, but remains very poor

Land area: 407,000 sq. km.
Population: 6 million—urban 62%
Capital city: Asunción, 0.5 million
People: Mestizo 95%, other 5%
Language: Guaraní, Spanish
Religion: Catholic 90%, Protestant 10%
Government: Republic
Life expectancy: 72 years
GDP per capita: $PPP 4,433
Currency: Guaraní
Major exports: Electricity, soya, timber, cotton

The Paraguay River, which flows through the country from north to south, divides Paraguay into two very different geographical zones. To the west, making up three-fifths of the country is the flat, featureless scrubland of the Gran Chaco, which extends into Bolivia, Argentina, and Brazil. To the east, the remaining two-fifths, where most of the population live, is more fertile and humid and is an extension of the Brazilian Highlands. Much of the eastern and southern border is formed by another major river, the Paraná.

The majority of Paraguayans are mestizo, a mixture of European and Guaraní Indian. But the Guaraní influence remains strong. For about one-third of the population Guaraní is the first language, and most people understand it as well as Spanish. More recent immigrants have been Brazilian farmers who have settled on land in the east. There are also small numbers of Indians—in 17 ethnic groups who live both in the Chaco and the east.

Paraguayans are among the poorest people in South America: one-fifth of them live in poverty, and only half have electricity. Around one-fifth are infected with the debilitating insect-transmitted Chagas disease.

Agriculture remains an important source of income, accounting for one-fifth of GDP and employing one-third of the workforce. But land is very unevenly distributed, with the top 1% of landowners controlling around two-thirds of the land. The main export crop used to be cotton, which is grown primarily by 200,000 or so peasant farmers on small plots of land, but of late they have been trying to diversify into other crops like wheat, rice, and sunflowers. Nowadays, the major export crop is soya, which is typically grown on the more efficient farms. Both cotton and soya, however, remain vulnerable to the vagaries of the weather and international prices.

Livestock is also important, with cattle-raising in the Chaco and in the south. Again, ownership is highly concentrated—1% of producers own around three-fifths of the 10 million total herd.

Paraguay has little manufacturing industry. Most is in the hands of small firms producing for local consumption. Foreign investors have shown little interest—deterred by the poor state of infrastructure and the low levels of education and skill.

Although only around 7% of the workforce is officially unemployed, around 30% more are thought to be underemployed.

One important and expanding industry is electricity production. Paraguay's rivers have enormous hydroelectric potential. The world's largest single source of hydroelectricity is the Itaipú dam on the Paraná River most of the funding for which came from Brazil. Since Paraguay's share of the output is more than 20 times its total consumption, it sells most of the electricity to Brazil for which the treaty was renegotiated in 2009.

A second, more controversial joint enterprise is with Argentina further down the Paraná at Yacyretá. This started production early in 1998,

Protests over a new dam

and provides 40% of Argentina's electricity, but has been the subject of international protests against environmental damage and the displacement of 50,000 local people.

Paraguay's political system has yet to develop any coherent shape following the end of the dictatorship of General Alfredo Stroessner. He seized power in 1954, along with the leadership of Paraguay's dominant Partido Colorado (PC), and his dictatorship continued until he was ousted in a coup in 1989 by General Andrés Rodríguez.

Rodríguez subsequently won

the 1989 election as the Colorado candidate.

The Colorado candidate for the 1993 presidential election was a businessman, Juan Carlos Wasnoy. Wasnoy tried to liberalize the economy further. He crossed swords with General Lino Oviedo, forcing him to resign. In 1996, Wasnoy imprisoned Oviedo for an attempted coup, but Raul Cubas Grau won the 1998 election and pardoned him.

Grau had been opposed in this by his vice-president, Luis Maria Argaña, who threatened to impeach him. In 1999 Argaña was murdered and Grau resigned to avoid impeachment. Oviedo fled to Argentina. An Argaña supporter, Luis Gonzalez Macchi, was sworn in as president.

For the elections in 2003, Oviedo formed a new party from exile in Brazil. The PC presidential candidate was Nicanor Duarte Frutos who duly won while the PC also won a majority in the national assembly.

Finally, in 2008, the PC, which, with 61 years in power was the world's longest-ruling political party, lost its grip on Paraguay. For the 2008 presidential elections its candidate was a former teacher Blanca Ovelar. She faced Oviedo, who was now the candidate of the National Union of Ethical Citizens.

But the strongest opposition came from a former Catholic bishop, Fernando Lugo, who was backed by the main opposition parties, the Partido Liberal Radical Auténtico, the Partido Patria Querida and the Partido Unión Nacional de Ciudadanos Éticos. Lugo won easily, but his Alianza Patriótica para el Cambio does not have a majority in the assembly.

Peru

Peru has returned to democracy but now needs to find ways to cut poverty

Land area: 1,285,000 sq. km.	
Population: 29 million—urban 72%	
Capital city: Lima, 8.1 million	
People: Amerindian 45%, mestizo 37%, white 15%, other 3%	
Language: Spanish, Quechua, Aymara	
Religion: Roman Catholic 81%, other 19%	
Government: Republic	
Life expectancy: 73 years	
GDP per capita: $PPP 7,836	
Currency: Nuevo sol	
Major exports: Copper, fish products, gold	

Peru's striking terrain falls into three well-defined zones. The Pacific coastal region is a long strip of desert, broken only by 60 or so rivers crossing to the sea. Further inland rise the dramatic Andean mountains, whose plateaux and valleys nurtured the Inca civilization. Then to the east, the mountains descend more gently through densely forested foothills to the vast sparsely populated tropical rainforest of the Amazon Basin.

The Quechua and Aymara Indian descendants of the Incas still make up close to half the population. Millions of them are to be found in peasant farming communities in the highlands. But many others, driven by violence or poverty, have now made their way to the towns and cities, particularly to those on the coast.

The warmer tropical land to the east, though it makes up more than 60% of the country has only around 5% of the population, mostly native Indians living in scattered settlements.

Peru remains very poor. Based on the national poverty line, more than half the population are classified as poor. In recent years, the government has increased spending on basic social services but investment still lags far behind what is required.

There is also a wide gap between the urban and rural areas: child malnutrition and under-five mortality are twice as high in rural areas as in the towns and cities. But even in the sprawling capital, Lima, half the population struggle to make a living in the informal sector.

Agriculture employs around one-third of the workforce. The most productive areas are around the rivers in the coastal areas where farmers grow cotton, sugar, and food for the cities. Agriculture in the highlands, the altiplano, remains fairly primitive. Peasant farmers grow maize and potatoes and raise alpacas and llamas. However, the most lucrative crop in the highlands is still coca—worth perhaps $1 billion per year. After Colombia, Peru is the world's second largest source of cocaine.

On the coast one of Peru's most important activities is fishing. Peru is one of the world's largest fishing nations, catching around 10 million tons per year, mostly sardines and anchovies that are processed locally into fishmeal. Peru is the world's largest fishmeal exporter, providing one-third of the total supply.

Peru's most important official foreign exchange earner, however, is still mining. Copper, gold, zinc, and lead provide around half of export earnings. Privatization of many of the mines has encouraged new foreign owners to step up investment. Peru also has reserves of oil, both in the jungle regions and offshore, as well as large natural gas fields.

Tourism to the spectacular Inca sites has also provided important income, and has revived now that guerrilla warfare has subsided, with around 1.9 million visitors per year who provide the second largest source of export earning.

In the 1970s Peru had a series of military or autocratic governments. But even the restoration of democracy in the 1980s did little to improve

Fighting Sendero Luminoso

the lot of the poor. On one side the populist government of Alan García of the Alianza Popular Revolucionaria American (Apra) had by the late 1980s contributed to rampant inflation. On the other side, two guerrilla movements had emerged: the Cuban-inspired Movimiento Revolucionario Tupac Amaru (MRTA); and the Maoist Sendero Luminoso. The ensuing struggles killed 28,000 people and forced more than 700,000 to flee.

It was in this chaotic environment that Peruvians in 1990 elected as president Alberto Fujimori, a former university rector and descendant of

Japanese immigrants. Fujimori proved an effective antidote to many of these problems—though at a democratic cost. In 1992, in an army-backed 'autocoup', he closed the Congress and suspended the judiciary. In 1995, however, he won a free election.

His successes included the 1992 arrest of Sendero Luminoso leader Abimael Guzmán. He also embarked on free-market reforms and opened Peru up to foreign owners.

For the election in 2000 Fujimori narrowly failed to win the first round against the candidate of the Perú Posible party, who withdrew from the second round, accusing Fujimori of fraud. Fujimori duly claimed victory.

Fujimori's downfall came in 2000 when his security chief, Vladimiro Montesinos, was videotaped bribing television stations for their support. Fujimori fled to Japan from where he resigned.

Fresh elections in 2001 gave Toledo victory over former president Alan García of Apra. Toledo struggled to live up to his election promises to increase employment and improve public services. His government also faced several scandals.

The 2006 elections concluded with a run-off between the centre-left former president, Alan García of Apra, and Ollanta Humala of the recently formed Partido Nacionalista Peruano. García won and this time practiced more prudent economic management.

In 2011, Humala, having moved from statist nationalism to the political centre, defeated Fujimori's daughter Keiko in the presidential election. Ambitiously, he now seems to be modelling himself on Brazil's former president Lula.

Philippines

The Philippines has lagged behind the 'tigers' of Southeast Asia—always on the brink of economic take-off

Land area: 300,000 sq. km.
Population: 89 million—urban 66%
Capital city: Manila, 1.6 million
People: Tagalog 28%, Cebuano 13%, Ilocano 9%, Bisaya/Binisaya 8%, other 60%
Language: Filipino (Tagalog), English
Religion: Roman Catholic 81%, Protestant 9%, Muslim 5%, other 5%
Government: Republic
Life expectancy: 72 years
GDP per capita: $PPP 3,406
Currency: Peso
Major exports: Electronics, machinery and transport, garments

The Philippines is an archipelago of more than 7,000 islands, though many are tiny and less than half are named. The two largest are Luzon in the north and Mindanao in the south. In between lie the group of islands known collectively as the Visayas. Most of the islands are mountainous, with the ranges running from north to south, and they include 20 active volcanoes. The Philippines is environmentally rich and diverse but since the 1970s much of the forest land has been cleared and tropical forests now cover only around one-fifth of the territory.

Filipinos are ethnically relatively homogenous, though there are a number of small indigenous groups. The main distinction is between Christians, who make up the majority of the population, and Muslims, most of whom are to be found on Mindanao. The country had two major periods of colonization, first by Spain and later by the USA, and it bears strong traces of both. Though the main language is Filipino, many people also speak English, which is the principal language for higher education. There are also smaller ethnic groups on different islands.

Many Filipinos are well educated, and the literacy rate is high, but nutrition and health standards are less impressive. Around one-quarter of children are malnourished and health services are very unequally distributed: half the doctors work in the National Capital Region.

Around one-quarter of people are living below the poverty line—of whom most are in the rural areas. Emigration is one solution to poverty, and there are thought to be around 11 million Filipinos abroad, with two million in the USA alone. One-third of Filipino children live in households where at least one parent has gone overseas. Their remittances, which in 2008 amounted to $18 billion, not only sustain families, they also prop up the economy.

The largest source of employment at home remains agriculture, which employs around one-third of the population. But the Philippines is a very unequal society, especially in landholding. In 1988, 2% of landowners had 36% of the land. The main food crop is rice, most of which is grown in Luzon. The second largest crop is coconuts, with half the production in Mindanao. Sugar used to be a major cash crop, particularly on the island of Negros, but following a collapse in the world price output has declined steeply. Fishing is another source of livelihood, though over-fishing by commercial fleets has been hitting the catches of inshore subsistence fishing communities.

The Philippines was slower than many other South-East Asian countries to modernize its industries. Like land, industry is often in the hands of powerful families. But its educated, and often English-speaking, population is proving increasingly attractive to foreign investors. Many companies have now established

Assets controlled by rich families

factories in dozens of export processing zones, particularly for assembly of electronic products, which now account for around two-thirds of exports. The Philippines also has some mineral potential: though it has little oil, it does have deposits of copper, nickel, gold, and silver.

The Philippines has had to deal with two main guerrilla movements. The first was the communist New People's Army which was very active in the 1980s, but after a number of ceasefires this is now less important. Of greater concern are various Islamic groups struggling for the independence of Mindanao.

Here too there have been ceasefires, most recently with the Moro Islamic Liberation Front. But other more radical and violent groups, such as Abu Sayyaf, have emerged which have links with al-Qaeda.

Politics in the Philippines tends to be based more on personality than ideology or principle. This was particularly evident during the authoritarian rule of Ferdinand Marcos from 1965 to 1984. Although the economy grew during this period, he and his high-profile wife, Imelda, siphoned off much of the wealth. Marcos was eventually ousted by a 'people-power' revolution and Corazon Aquino, wife of a murdered opposition leader, was elected president in 1987. She was succeeded in 1992 by former army leader Fidel Ramos, who embarked upon a period of economic reform.

In 1998, Ramos was succeeded as president by a former movie actor, Joseph Estrada. By early 2000, however, his government had sunk into cronyism and corruption. People-power surfaced yet again and eventually he was forced to resign.

In 2001 he was replaced by his vice-president, Gloria Macapagal Arroyo. In the 2004 elections she defeated another populist film star, Fernando Poe Junior. Arroyo's government was often unpopular and fragile. She had to survive impeachment in 2005, and a supposed coup attempt in 2006.

In the 2010 presidential election, Estrada stood again, but was decisively beaten by Benigno Aquino, son of Corazon—a victory due as to the electorate's fondness for his mother as to his own merits.

Poland

Poland has been economically one of the most successful countries in Eastern Europe, but politically very volatile

> *Land area:* 313,000 sq. km.
> *Population:* 38 million—urban 61%
> *Capital city:* Warsaw, 1.7 million
> *People:* Polish
> *Language:* Polish
> *Religion:* Roman Catholic 90%, other 10%
> *Government:* Republic
> *Life expectancy:* 76 years
> *GDP per capita:* $PPP 15,987
> *Currency:* Zloty
> *Major exports:* Manufactures, machinery, mineral fuels

Poland has swamps and sand dunes along its northern coast. But the main part of the country comprises the central plains and lowlands; 'Poland' derives from a Slavic word for 'plain'. The mountainous one-third of the country includes the Carpathians along the southern border with Slovakia.

Poland's population has become ethnically more homogenous—largely as a result of the deaths and expulsions of Jews and other minorities during the Second World War. At the same time, many ethnic Poles have also emigrated, a trend which continued in the early 1990s as many people moved to Germany and elsewhere in search of work. Around one-third of ethnic Poles are thought to live abroad. An important unifying factor has been the Catholic Church, which played an important part in the

struggle against communism—and also provided the previous Pope, Karol Wojtyla, John Paul II. However, the Church's political influence has diminished and it now restricts itself more to social teaching.

Poles have not just become wealthier, they have also become healthier. Since the early 1990s, life expectancy has been increasing and infant mortality falling. This was helped by better food and by cleaner air, though Poles still suffer from high levels of smoking and alcohol abuse. Education levels had also been high during the communist period, though low salaries have dissuaded potential teachers, particularly in the rural areas, and standards have fallen.

Poland is considered a successful example of economic 'shock therapy'. From 1989, the government rapidly liberalized prices, imposed wage controls, and reduced subsidies to state-owned enterprises. This ushered in a deep recession with high unemployment, but by the mid-1990s Poland had embarked on rapid economic growth. Although the rate has slowed in recent years in 2010 it is expected to be 2%. Manufacturing led the way, fuelled by foreign investment. Some of the older industries have faded but there has been strong growth in lighter industry, notably food and drink and some heavier manufacturing such as automobiles. Between 1989 and 2006, the private sector's share of GDP rose from 18% to 88%. This has gradually reduced unemployment which in 2009 was below 10%.

Poland is one of the world's largest coal producers. But in recent years output has been falling while wage costs have risen and the industry is losing money.

Agriculture remains important. Although it accounts for only 4% of GDP, it employs 15% of the labour force. Polish farmers had managed to resist collectivization during the communist period. All the 1.8 million farms are privately owned but most are small—averaging only 8.6 hectares—and very labour intensive.

Poland has two million small farms

In May 2004 Poland joined the EU where it has established itself as a tough negotiator. Poland's farmers will benefit from the EU's common agricultural policy but will only gradually get the same level of subsidies as those in other states.

EU membership has also enabled many people to find work in other countries. In 2007 there were thought to be 2.3 million Poles working temporarily abroad.

Poland's political transformation out of communism was famously the work of the Solidarity trade union based in the Gdansk shipyard. This momentum also swept Solidarity into political power and its leader, Lech Walesa, was elected president in 1990. Solidarity achieved many economic reforms but later fragmented into numerous centre-right groups.

In 1993, after a series of short-lived Solidarity governments, parliamentary elections resulted in a victory for the former communists of the Left Democratic Alliance (SLD) and in 1995 its founder, Aleksander Kwasniewski, was elected president. Even so, the SLD continued Solidarity's economic reforms.

Nevertheless, Solidarity had been regrouping and under the leadership of Marian Krzaklewski came back together as a 37-party coalition, Solidarity Electoral Action (AWS) which in 1997 form a government in coalition with Freedom Union (UW).

Kwasniewski won the November 2000 presidential election and in 2001 SLD, headed by Leszek Miller, returned to government. When Miller resigned in 2004 he was replaced by Marek Belka.

In 2005, however, the conservative Law and Justice (PiS) party came out ahead. In addition, the PiS candidate Lech Kaczynski won the presidential election, and in July 2006 his twin brother Jaroslaw took over as prime minister. Their rather primitive approach to politics and diplomacy made them very unpopular in the EU, and eventually at home too.

In 2007, in an election with a high turnout, particularly among younger voters, the 'terrible twins' were defeated by a pro-European party, Civic Platform, led by Donald Tusk. It won 209 seats in the 460-seat assembly and formed a centre-right coalition with the Polish Peasants' Party. In 2010, Civic Platform strenthened its position with the election of Bronislaw Komorowski as president.

Portugal

Political stability and EU membership have stimulated development

Land area: 92,000 sq. km. **Population:** 11 million—urban 61% **Capital city:** Lisbon, 2.1 million **People:** Portuguese **Language:** Portuguese **Religion:** Roman Catholic 85%, other 6% **Government:** Republic **Life expectancy:** 79 years **GDP per capita:** $PPP 22,765 **Currency:** Euro **Major exports:** Textiles, garments, cars, footwear, wood products

Portugal's major river, the Tagus, flows across the country from east to west, reaching the Atlantic ocean at the capital, Lisbon. To the north of the Tagus, the land is mountainous and the climate cooler and wetter. The southern part of the country, the Alentejo ('beyond the Tagus') consists of rolling and generally arid lowlands. Portugal also includes the Madeira Islands and the Azores.

Portugal is one of the least industrialized countries in Western Europe and two-thirds of the population live in rural areas, with greater concentrations in the north.

By the standards of most European countries, the population is fairly homogenous, with relatively small numbers of immigrants from its former African colonies. Indeed until the 1980s Portugal was one of Europe's leading sources of emigrants who were drawn away by higher wages in France and Germany. Some 4.5 million Portuguese still live overseas. However, many former emigrants have been returning and over the period 2002–07 Portugal had net immigration.

Though unemployment is reasonable, at under 10%, wages have also been low, but have recently started rising. Portugal certainly benefited from membership of the EU. Between 1986 when it joined and 1996, per capita income grew from 54% to 74% of the EU15 average, but by 2007 had fallen back to 67%.

Portugal's slow development has been linked to the relatively poor state of agriculture, which employs one-fifth of the workforce. Most farms, particularly in the north, are small and relatively unproductive. Thin soil and unreliable rainfall, combined with low levels of investment, lead to yields one-third lower than the EU average. More than half the country's food has to be imported and without EU subsidies many farms would fail.

Nevertheless, Portugal does have agriculturally based exports, notably tomato paste, port, and other wines. Forestry also generates important exports: more than one-third of the country is forested, much of this with cork-oak trees. Portugal has around half the world's cork output. The country's long coastline also supports a significant fishing industry, largely catching cod for local consumption and sardines for export, though more recently over-fishing has caused catches to fall steeply.

Manufacturing in Portugal is still fairly low-tech—much of it concentrated in textiles, clothing, and footwear—but the government has gone out of its way to attract higher-tech investment. One of the most striking examples was the record $1-billion subsidy for the joint Ford–Volkswagen Auto Europa car factory just outside Lisbon which is responsible for around one-tenth of merchandise exports. Portugal also has some mineral resources, particularly in the Alentejo which has the Neves Corvo copper mine as well as centres of marble production.

Portugal offers $1 billion for a car plant

Within the service sector, one of the most important industries is tourism. Portugal attracts more than 14 million visitors each year, mostly from Europe's chillier countries, heading for Madeira and the southern Algarve coastline. Tourism employs around 10% of the workforce and makes a major contribution to the balance of payments.

The years after 1976, which marked the end of the 36-year dictatorship of Antonio Salazar and the introduction of a democratic constitution, were a period of rapid change. In the first 15 years of democracy, there were 12 changes of government. In the mid-1980s, however, the situation settled down. In 1985 the largely ceremonial post of president was taken by Mario Soares, leader of the centre-left Socialist Party (PS). And in 1987 the Social Democratic Party (PSD) achieved a parliamentary majority and Anibal Cabaço Silva became prime minister. This pairing ushered in a period of relative stability—due to Portugal's 1986 entry into the EU.

Discontent with the PSD led to their defeat in the 1995 parliamentary elections, following which the PS took over again, this time with the urbane Antonio Guterres as prime minister. In 1996 Cabaço Silva ran for president but was defeated by the socialist candidate Jorge Sampaio.

The socialist administration, which moved towards the centre, benefited from several years of rapid economic growth. Guterres remained popular and in the 1999 parliamentary elections the socialists increased their representation. After the election in 2002 the PSD, now led by José Manuel Durão Barroso, formed a coalition with the Popular Party.

In 2004 Barroso resigned to become president of the European Commission, making way for Pedro Santana Lopes, who proved so inept that the president dissolved parliament.

Elections resulted in an outright victory for the the PS led by José Sócrates. After the 2009 election, however, he had to form a minority administration.

By 2011, with the economy in dire straits and needing an IMF bail-out the goverment fell. In the ensuing election the PSD, led by Pedro Passos Coelho, just fell short of a majority and formed a coalition with the conservative Popular Party.

Puerto Rico

Puerto Rico has decided not to become a US state, yet. Now it wants greater economic independence

Land area: 9,000 sq. km.
Population: 4 million—urban 98%
Capital city: San Juan, 428,000
People: White 77%, black 7%, other 16%
Language: Spanish, English
Religion: Roman Catholic 85%, other 15%
Government: Commonwealth (US)
Life expectancy: 79 years
GDP per capita: $PPP 17,800
Currency: US dollar
Major exports: Chemicals, manufactures, food

Puerto Rico is largely mountainous. The Central Cordillera range extends east–west along most of the length of the island, descending steeply to the narrow southern coastal plain, and more gently to the broader plains of the northern coast. There is also a north–south distinction in climate: humid and tropical in the north; drier in the south.

The population is racially fairly homogenous, and there is little overt discrimination though, as elsewhere in Latin America, the people in positions of power tend to be those with lighter skin colour. Puerto Ricans are US citizens, but their standard of living is far lower than that in the USA: the per capita income is around half that of the poorest US state, Mississippi; the unemployment rate is usually far higher than in the USA; and half the population live below the US poverty line. Not surprisingly, many people driven by poverty and unemployment have headed for the USA, which now has around 3.4 million people of Puerto Rican descent, most of whom are in New York.

Almost everyone speaks Spanish, which is the main language of instruction in schools. English is also an official language, but far fewer speak it fluently and many people want to resist any further imposition.

Puerto Rico's main economic activity is manufacturing, which accounts for more than 40% of GDP. Many US firms have been attracted to Puerto Rico by low wages. Although Puerto Rico is subject to US minimum wage legislation, wages tend to be one-quarter less than in the USA. In the past there have also been significant tax concessions. These are now being phased out though local politicians are lobbying for them to be retained.

Initially this drew in firms engaged in labour-intensive industries like garments manufacture. But in recent decades companies have invested in higher levels of technology—around half of current manufacturing is in chemicals, and another quarter is in metal products and machinery. Most of this output is for export, primarily to the USA.

Though some garment production remains, and employs around 30,000 people, many labour-intensive industries, faced with increases in the US minimum wage, have migrated in search of cheaper labour elsewhere—especially when the North American Free-Trade Agreement heightened competition from Mexico.

Little is left of Puerto Rican agriculture, which now contributes less than 1% of GDP.

Another important source of employment and income is tourism. San Juan, the capital, is a favourite port of call for cruise liners. Around 1.3 million cruise visitors disembark each year along with four million stopover tourists, primarily from the USA.

The USA acquired Puerto Rico in 1899 at the end of the Spanish American War. Since then, its status has remained **Puerto Rico's** unresolved. The **strange status** US Supreme Court has asserted that Puerto Rico is an 'unincorporated territory of the United States'—a possession of, but not a part of, the USA. A new constitution in 1952 established a system for local government, making Puerto Rico a 'Commonwealth'.

So although Puerto Ricans are US citizens, they cannot vote in presidential elections and they do not pay federal income tax. They elect their own governor and Congress which have autonomy in many areas such as tax, education, and criminal justice. The US government deals with defence, monetary issues, and trade and also extends some federal social programmes to the island. Around 30% of government expenditure comes from federal grants.

There have been a number of attempts to change the relationship, since Puerto Rico's position is anomalous in international law; unless it is a US state or an independent country it is effectively a colony. On this view, 'commonwealth' is a transitional status and Puerto Ricans should periodically vote to select their ultimate status—to become independent, or become a US state. Non-binding plebiscites in 1993 and 1998, with more than 70% turnouts, voted narrowly in favour of retaining commonwealth status.

The argument for statehood is chiefly financial. As part of the USA, the island would get full representation in Washington; people would pay federal income tax but they would also qualify for more federal funds. The argument against is chiefly cultural—Puerto Rico would lose its distinctive identity.

Given the significance of this issue, politics in Puerto Rico tends to be organized around it. The two main parties are the Partido Nuevo Progresista (PNP), which is pro-statehood, and the Partido Popular Democrático (PDP), which is pro-commonwealth. Since 1968 power has tended to alternate between the two.

In the elections in 2000, for example, after eight years of PNP rule the PDP returned with Sila María Calderón elected as Puerto Rico's first woman governor. She was replaced after the 2004 election by another PDP figure, Aníbal Acevedo Vilá. Then after the 2008 election the PNP was back, with Luis Fortuño as governor, and majorities in both assemblies.

Qatar

With a huge new gas field, Qatar could become the world's richest nation

Though it has some wells and sand depressions that provide water for limited agriculture, the peninsula of Qatar is mostly a flat and arid desert.

Native Qataris are in the minority—only around one-third of the population. Though they are entitled to free education, this is not compulsory and 11% of adults are illiterate. The health service is also free, though some low charges now have been introduced.

Most of the manual work is done by an immigrant workforce, drawn from the Indian subcontinent and Iran.

Oil was discovered in 1939 and has fuelled this small state's rapid development, providing two-thirds of government revenue. The country has also made efforts to diversify. Qatar has had a steel industry since 1978 and many foreign companies have invested in the production of fertilizers and other petro-chemicals.

Even so, the future lies with gas. Qatar's Dukhan field has been exploited since 1980, but most attention is now focused on the North field, which is the world's largest gas field not associated with oil—with reserves that could last 300 years. To pay for the investment in gas Qatar has been over-pumping oil at a rate that could exhaust reserves in about 20 years, as well as borrowing from

abroad. But this investment should pay off over the next few decades, doubling the country's income.

Qatar is ruled by the emir, Sheikh Hamad al-Thani, an absolute monarch who took over in 1995 when he ousted his father in a bloodless coup. As crown prince, Sheikh Hamad had been working to open up Qatar economically and politically, and when he became emir he immediately ended press censorship and established al-Jazira as one of the region's most outspoken TV stations—a key source of information during the war in Iraq. Qatar also has close relations with the US and provided military facilities during the 2003 war.

In 2003 the Sheikh introduced the country's first constitution, agreed in a referendum in 2004, which sets up a new 45-member parliament, the Shura Council, two-thirds of which will be elected and which will have legislative powers—although the emir will retain a power of veto. The first elections for the council have yet to happen. Given the current unrest elsewhere in the region, the emir might consider it wise to hold the election. However there seems little likelihood of popular demands for genuine democracy.

Land area: 11,000 sq. km.
Population: 1 million—urban 96%
Capital city: Doha, 340,000
People: Arab 40%, Pakistani 18%, Indian 18%, Iranian 10%, other 14%
Language: Arabic, English
Religion: Muslim 78%, other 22%
Government: Absolute monarchy
Life expectancy: 76 years
GDP per capita: $PPP 74,882
Currency: Qatari riyal
Major exports: Petroleum products, fertilizers, steel

Réunion

A French dependency in the Indian ocean with high levels of unemployment

Saint-Denis

ATLANTIC OCEAN

Land area: 2,500 sq. km.
Population: 783,000—urban 68%
Capital city: Saint-Denis, 122,000
People: African, French, Malagasy, Asian
Language: French, creole
Religion: Roman Catholic 86%, Hindu, Muslim, or Buddhist 14%
Government: Republic
Life expectancy: 75 years
GDP per capita: not available
Currency: Euro
Major exports: Sugar, rum, molasses, perfume essence

Réunion is a volcanic island with rugged mountains surrounded by basins and plateaux that lead down to the tropical coastal lowlands. The island is often exposed to violent cyclones, and there is still a live volcano.

Most people are of African or creole (mixed) descent and are far poorer than the French and Asian minorities, who are generally the richest. The population is growing fast and unemployment remains high: 33% in 2003. This has at times led to violent protests as people have demanded social security payments equivalent to those in France. Many have also emigrated to France in search of work.

The island was colonized as a potential sugar producer, and sugar still dominates the economy, providing 80% of exports as well as the raw material for rum and molasses. Other crops include vanilla beans and geraniums, which are grown for perfume essences. Another source of income is tourism, from 400,000 visitors per year, of whom 80% come from France. But the country remains heavily dependent on French aid.

Réunion is the largest overseas department of France and sends five deputies and three senators to the French National Assembly. Locally, it is administered by a French-appointed prefect and has a general council and a regional council.

Political parties on the island include one of the main French parties, the right-wing Union pour la démocratie française (UDF). This also included the Rassemblement Pour la République until 2002 when it merged with part of the UDF and another party to form Union pour un Mouvement Populaire (UMP).

But two local parties have also held sway. One is the Parti communiste réunionnais (PCR); the other arose out of a pirate television station Télé-Free-DOM that the authorities had tried to suppress. In the 1992 regional council election the most seats went to candidates supporting the station's right to broadcast. In another election in 1993 Free-DOM won again and the station-owner's wife Marguerite Sudre was elected president of the regional council.

Since then, more conventional parties have been in charge. Following the elections of April 2004 the UMP emerged as the largest party and Nassimah Dindar-Mangrolia, a Muslim, became president of the general council. In 2010 Didier Robert, also of the UMP, became president of the regional council.

Romania

Romania has suffered from poverty and political confusion but has now joined the EU

Land area: 238,000 sq. km.
Population: 22 million—urban 55%
Capital city: Bucharest, 1.9 million
People: Romanian 90%, Hungarian 7%, Roma 2%, other 1%
Language: Romanian, Hungarian
Religion: Romanian Orthodox 87%, Roman Catholic 6%, Protestant 7%
Government: Republic
Life expectancy: 73 years
GDP per capita: $PPP 12,369
Currency: Leu
Major exports: Textiles, footwear, basic metals

Western Romania is dominated by the Carpathian Mountains which take up around one-third of the land area and run from the northern border with Ukraine down the centre of the country before veering west and exiting south-west into Serbia. East of the mountains are broad plains that extend to the south before reaching the coast of the Black Sea.

Most people are ethnically Romanian but there are significant minorities that have given rise to ethnic tensions. The largest minority are the two million Hungarians who are concentrated in the north-west of the country and have been pressing for greater rights. These efforts have born fruit in recent years with greater language rights and the creation of a Hungarian university. The other important minority are the Roma, or gypsies. The official figure is around 500,000 but there are probably nearer

two million. They tend to live apart from other Romanians and have been subject to widespread discrimination and abuse. Many have fled to other countries, notably Germany. Around 1.8 million Romanians work abroad, especially in Italy, Spain, Germany, Israel, and Hungary.

Romania is one of the poorest countries in Eastern Europe. More than one-quarter of the people live in poverty—with particularly high poverty rates in the north-east. Health standards are desperately low: the rates for infant and maternal mortality are among the highest in Europe.

As a result of economic uncertainty and family planning, the birth rate is very low. In addition, many people are leaving to work abroad. Around two million migrants have moved to other EU states mainly to Spain and Italy. Combined with a rising death rate this has led to a steady fall in population— by 7% between 1992 and 2007.

Around one-third of the workforce are employed in agriculture. Almost all the agricultural land is now in private hands, but this is mostly in the form of small inefficient plots in which there has been very little investment. EU funds could help change this, though farmers may not have the capacity to use such funds efficiently.

Communist governments had given a high priority to heavy industry. Romania's manufacturing potential now lies in more labour-intensive light industries, though traditional industries such as textiles struggle to compete with developing countries. There has been some foreign investment in the car industry, notably in the Dacia plant which is owned by Renault, as well as in pharmaceuticals, but manufacturing overall has been slow to take off.

Romania's economic reforms have been sluggish—resisted by the bureaucracy, the trade unions, and by politicians milking the system. State enterprises continue to play a major role in utilities, banking, and manufacturing.

Given the country's scenic beauty and rich culture, there is potential for tourism but this remains largely underdeveloped.

Romania joined the European Union in 2007 but is still being

Monitored by Brussels

monitored by the European Commission to ensure that it fulfils the commitments made during accession. There are still considerable doubts about the judicial system and concerns about corruption.

Romania had a traumatic transition from communism. From 1965 it had largely been the personal fiefdom of a brutal dictator, Nicolae Ceausescu, who was finally overthrown and executed in 1989. But the government remained in the hands of former communists, and their new party, the National Salvation Front (NSF),

won the 1990 parliamentary and presidential elections. A new constitution approved in 1991 was followed by further elections in 1992 which again chose Ion Iliescu as the president.

In 1993 the NSF, having split and merged with other parties, was renamed the Social Democratic Party (PDSR).

This era ended in 1996 with the election of an anti-communist government, led by a centre-right coalition: Democratic Convention (DC). Emil Constantinescu of the DC also defeated Iliescu in the presidential election and appointed Victor Ciorbia as prime minister. The DC governed in conjunction with the Social Democrat Union (SDU).

In the elections in 2000 the most seats went to the PSDR whose candidate, Ion Iliescu, was also elected president. The PSDR subsequently merged with a smaller party as the Social Democratic Party (SDP) and formed a minority government.

The 2004 election resulted in a coalition led by the National Liberal Party (NLP) and the Democratic Party (DP). Traian Basescu was elected president. In 2007, the NLP formed a new, minority coalition with the Hungarian Union of Democrats in Romania (HUDR). In 2008, the NLP and the DP merged to form the Democratic Liberal Party (DLP).

After the parliamentary elections in 2008 the SDP formed a coalition with the DLP, replaced in 2009 by a coalition between HUDR and the DLP, with Emil Bloc of the DLP as prime minister. Meanwhile Basescu was re-elected president.

Russia

Russia's president is determined to modernize this vast country but is also crushing its democracy

Land area: 17,075,000 sq. km.
Population: 142 million—urban 73%
Capital city: Moscow, 10.1 million
People: Russian 80%, Tartar 4%, Ukrainian 2%, Chuvash 1%, other 13%
Language: Russian
Religion: Russian Orthodox, Muslim
Government: Republic
Life expectancy: 66 years
GDP per capita: $PPP 14,690
Currency: Rouble
Major exports: Fuels, metals, machinery

Russia, which is officially the Russian Federation, is by far the world's largest country, spanning two continents—more than 10,000 kilometres across. Russia's territory is huge and diverse but can be considered as a series of regions west to east. First, to the west are the rolling plains and uplands of the European plain, through which flow major rivers including the Don and the Volga. These plains terminate in the east at the Ural mountains, which, running north to south, form a dividing line between Europe and Asia.

Beyond the Urals lies the vastness of the West Siberian plain, a mostly featureless and often marshy landscape that stretches east to the Yenisey River. This leads on to the Central Siberian Plateau and then to Russia's Far East, which ends in a number of mountain ranges before reaching the Pacific coast. Russia's climate is often harsh, with bitterly cold winters and short summers, especially in the north and east.

The majority of people, more than four-fifths, are ethnic Russians, but the country also has over 70 other nationalities, of which the largest are Tartars, Ukrainians, and Chuvash. The ethnic spectrum corresponds to some extent to the country's complex administrative structure. Russia has 89 different administrative units, including 21 'minority' republics, such as Tatarstan, Chechnya, and Karelia. However, these account for only around 20% of the total population, and even they have majority Russian populations.

Russia's people have endured a series of shocks since the demise of communism. Living standards have fallen. In 2008, 19% lived below the official poverty line. There has also been a steep increase in inequality. By 2007 officially the top 20% of the population got 47% of national income, but given the extent of tax evasion they probably got much more.

Russians have also suffered a fall in health standards. Between 1989 and 2005, Russia's death rate rose steeply and life expectancy for men dropped from 65 years to 59, though for women the drop was less steep—74 to 72. This has been ascribed to a combination of stresses and

Bribes for health and education uncertainties that have contributed to heart problems, strokes, and alcoholism. The decline in the health system will also have played a part. Russia still has enough doctors, but they often work in poorly equipped hospitals and anyone who needs urgent treatment has to bribe the staff. Around one million Russians are thought to be HIV-positive.

The education system too has come under strain. School attendance is compulsory and free, but standards have fallen because of low investment. Given the rising demand for qualifications, in practice around half of students now pay their own fees along with bribes required for university entrance.

In the first half of the 20th century Russia, as the core of the Soviet Union, had made striking economic progress—particularly in heavy industry. But by the 1980s its centrally planned economy was unable to produce the goods evident in the rest of the world. Following the collapse of the Soviet Union, Russia from 1992 undertook radical reforms—privatizing industries, liberalizing prices, and reducing tariffs.

The first phase of privatization involved selling off small enterprises, such as shops or small restaurants, primarily to their workers. More difficult was the disposal of large-scale enterprises. Initially this was done in 1992 by giving every Russian citizen a voucher to buy shares.

After 1995, the government started selling enterprises for cash, often at low prices, to large industrial groups, typically those with good political contacts. This created vast fortunes for a new group of 'oligarchs' who now control much of the country's industry and its financial sector.

Russian manufacturing was antiquated and once exposed to international competition went into steep decline. Between 1990 and 1998, output halved before starting to grow again. Compared with other industrial countries, where small and medium enterprises account for around half of GDP, in Russia they still contribute less than 15%. Around two-thirds of Russia's official GDP is controlled by 20 huge conglomerates. However, much commercial activity, perhaps 40%, probably takes place unrecorded in the informal sector.

One of the largest industrial sectors, and the one with the greatest potential, remains energy. Russia has about 5% of global oil reserves and is now the world's second largest exporter. Privatization and subsequent consolidation created major oil

Vast mineral reserves companies—though the government now wants to regain control: one of the largest, Yukos, is effectively now state-owned following the imprisonment of its head, Mikhail Khodorkovsky.

Russia also has around one-third of global gas reserves, most of which are in the hands of the state controlled by Gazprom, Russia's largest company.

Russia is also a major coal producer and has large deposits of many other minerals, including diamonds, nickel, and platinum.

Agriculture, which still employs

around 13% of the workforce, has also suffered a sharp decline and now accounts for only 5% of GDP. Between 1992 and 1998 the grain harvest fell by half—though it has since recovered somewhat. Most of the land is cultivated by partnerships or by companies created by reorganizing former state or collective farms—which still rely heavily on state subsidies. Only since 2002 has it been possible to sell agricultural land.

Although Russia's economy has started to grow again the country still lags way behind most countries in Europe, even if the recent boost in global oil prices is changing the picture.

Privatization and the accumulation of all this new wealth have been accompanied by a proliferation of organized crime. Commerce is now dominated by the 'mafia' who can also rely on police cooperation.

The new era in Russia can be traced from 1985 when Mikhael Gorbachev started to open up the Soviet Union. This unleashed tensions between conservative and reformist forces that culminated in August 1991 in a failed coup by hard-line conservatives which triggered the break-up of the Soviet Union.

Until the end of 1999, political life in Russia was dominated by Boris Yeltsin who had been elected president of the Russian Federation in 1991. Yeltsin strengthened his position in 1993 after winning an overwhelming victory in a referendum that gave the president strong executive powers. The upper house is the Federation Council, consisting of regional governors and heads of regional assemblies.

Yeltsin's grip was constantly shaken by a number of rebellions, the most serious of which was in the southern republic of Chechnya where he unleashed a fierce civil war. Yeltsin was re-elected in 1996 and appointed and sacked a sequence of prime ministers, the final one being, in 1999, a former KGB colonel, Vladimir Putin. In December 1999 Putin and

Brutal war in Chechnya his supporters won a striking victory in the elections for the state parliament, the Duma.

Yeltsin resigned on 1 January, 2000—handing over presidential power to Putin who easily won the March 2000 presidential election. Following the parliamentary elections in 2003, parties supporting Putin then dominated the Duma.

Putin consolidated his position as an autocratic leader, He repressed the media, jailing the owner of the only national independent TV station. He has also manipulated elections and harassed or expelled human rights groups while attacking oligarchs, such as the head of Yukos, who might be an alternative focus of political power.

In March 2004, with the overwhelming and uncritical support of Russia's state-owned media, Putin was re-elected president. In 2007, the pro-Kremlin party, United Russia, won a majority in the Duma election.

Putin was unable to stand for a third term in 2008, but did the next best thing by hand-picking his successor, Dmitry Medvedev, who won with 70% of the vote and duly appointed Putin as prime minister. Medvedev is assumed to be holding the position for Putin to return as president.

Rwanda

Rwanda has emerged from the period of genocide but now has an autocratic government

Land area: 26,000 sq. km.
Population: 10 million—urban 19%
Capital city: Kigali, 750,000
People: Hutu 84%, Tutsi 15%, Twa 1%
Language: Kinyarwanda, French, English, Kiswahili
Religion: Roman Catholic 57%, Protestant 26%, Muslim 5%, indigenous beliefs and other 12%
Government: Republic
Life expectancy: 50 years
GDP per capita: $PPP 866
Currency: Rwandan franc
Major exports: Coffee, tea, hides

in 1994. Since then, a traumatized population has staged a remarkable recovery. The economy is back to pre-war levels and with the help of international aid Rwanda has rebuilt its health and education systems.

Around 80% of Rwandans live in rural areas and depend on subsistence agriculture. Since population density is high most households survive on less than one hectare. For decades, their intensive cultivation of steep slopes has eroded the soil and the violence of 1994 resulted in part from competition for scarce land.

Farmers primarily grow staple crops like plantains, sweet potatoes, cassava, and maize. Food production has recovered from the war but has not kept pace with population growth. Livestock numbers are back to pre-war levels.

Rwandan farmers also grow cash crops, notably high-quality coffee in the north-west. And there is also a good export market for locally grown tea. Most industrial activity is devoted to processing coffee, tea, and other crops, along with production in a few factories manufacturing textiles and other simple consumer goods.

Rwanda also has some mineral

From Lake Kivu on the western border, Rwanda's territory rises to a steep mountain range, with the volcanic Virunga Mountains in the north. To the east of these is the broad hilly plateau that covers most of the country before descending to the marshy land along the border with Tanzania.

The people of Rwanda are sharply divided between two main groups: the Hutu and the Tutsi, though the demarcation is more political and economic than cultural, since they speak the same language and follow the same religions. Rwanda has Africa's highest population density: 350 people per square kilometre.

Traditionally, the minority Tutsi, who were primarily cattle herders, had formed the ruling élite; the majority Hutu worked for the Tutsi while raising their own crops. Conflict between them culminated in genocide resources including columbo-tantalite (coltan) which is used for making mobile phones.

For the longer term, there is also potential for high-value niche tourism, particularly to see the gorillas.

The seeds for modern conflict were sown by the Belgian colonists who ruled through their favoured group, the Tutsi. As independence approached, the Hutu, anticipating majority rule, had become more assertive. In 1959, the Tutsi king fled into exile along with hundreds of thousands of refugees.

Rwanda achieved independence in 1962—with a Hutu government led by the first president, Grégoire

Early indications of conflict

Kayibanda. The ethnic strife continued. The exiled Tutsi made several coup attempts and thousands of people died in reprisals.

Kayibanda's government was overthrown in 1973 by the Hutu army chief, General Juvénal Habyarimana. More violence followed, with massacres of both Hutu and Tutsi. By 1990 more than 600,000 Tutsi had fled into exile. In a one-party state, Habyarimana was subsequently elected at the head of the Mouvement révolutionaire national pour le développement (MRND)—and re-elected in 1989.

Violence surged again in 1990 when the exiled Tutsi united as the Rwandan Patriotic Front (RPF) headed by Paul Kagame. Meanwhile, under donor pressure, the Hutu government had allowed other political parties to organize, and in 1991 Habyarimana agreed to a transitional government which, in 1993, made a peace agreement with the RPF, though many in the MRND opposed this and started training Hutu militias: the 'Interahamwe'.

In 1994, President Habyarimana died in a plane crash. MRND dissidents took over and together with the Rwandan army they launched a genocidal massacre, mostly of the Tutsi. In response, the RPF intensified its campaign and by July 1994 had taken over the country. This ended the genocide, but by then up to 800,000 people had died. The RPF also pursued the Hutu into what is now Congo, provoking another horrific civil war in that country.

In principle the RPF was ruling as part of a coalition and until April 2000 the president was Pasteur Bizimunguy, an ethnic Hutu. When he resigned, however, he was replaced by Tutsi vice-president Paul Kagame, who had really been running the country. Kagame's position and that of the RPF was confirmed in presidential and legislative elections in 2003.

The RPF has remained firmly in control. It claims that it is not a tribal party: indeed it has taken great pains to remove the significance of ethnic identity, having abolished the collection of any statistics based on ethnic distinctions. Returning Hutu rebels go through re-education camps, and all young Rwandans are obliged to imbibe similar non-tribal values in 'solidarity camps'.

This preoccupation has, however, created an authoritarian regime. There is no free press, any potential opposition groupings are harassed or jailed, and a number of opposition figures have disappeared. Kagame was re-elected in 2010.

St Lucia

A Caribbean island nation building a diverse economy

Land area: 620 sq. km.
Population: 0.2 million—urban 28%
Capital city: Castries
People: Black 83%, mixed 12%, East Indian 2%, other 3%
Language: English, French creole
Religion: Roman Catholic 68%, other 32%
Government: Constitutional monarchy
Life expectancy: 74 years
GDP per capita: $PPP 9,786
Currency: East Caribbean dollar
Major exports: Bananas, clothing, cocoa, vegetables, fruits, coconut oil

St Lucia is a volcanic island in the Caribbean, and subterranean activity surfaces via a number of hot springs. The mountain range that runs from north to south still has some dense forest cover though most of the land lower down has been cleared for agriculture.

Most St Lucians are black and English-speaking. But there is also a French influence: the majority are Roman Catholic and many also speak a French creole language.

Around 40% of the labour force work in agriculture on small farms that take advantage of generally fertile soil. They raise subsistence crops such as breadfruit and cassava. In addition, many have traditionally grown bananas which are the country's leading export.

But in recent years prices for bananas have been low and the EU import regime is now less favourable. As a result, almost half of St Lucia's banana farmers are thought to have quit since the 1990s and their numbers are still falling.

The other main source of income is tourism. The island has a number of beach resorts that provide all-inclusive holidays. More than a quarter of a million visitors stay each year, and there are a similar number of short visits by cruise-ship passengers. Most tourists come from Europe and the USA. Poorer St Lucians argue

that relatively little of the income percolates down to them.

Manufacturing output is more diverse than in other Caribbean islands. As well as processing agricultural goods, local factories produce clothing and assemble electronic items. There is also a small offshore financial sector.

St Lucia is a parliamentary democracy with the British monarch as its head of state, represented locally by a governor-general. The first elections after independence in 1979 were won by the St Lucia Labour Party (SLP), which established links with Cuba and North Korea. In 1982, power passed to the more conservative United Workers' Party (UWP), which opened up the economy and embarked on a series of IMF structural adjustment programmes.

These measures failed to reduce persistently high unemployment—of 30% or more—and this, combined with allegations of corruption, contributed to the UWP's downfall.

In the 1997 and 2001 elections, the SLP won. But in 2006, the UWP returned and Sir John Compton again became prime minister. On his death in 2007 he was replaced by Stephenson King.

St Vincent & the Grenadines

Rural poverty alongside hedonistic luxury

The country comprises one main island, St Vincent, which has around 90% of the territory and population, and the Grenadines, a group of 30 or more smaller islands, or cays, to the south. St Vincent is rugged and mountainous, with an active volcano, Soufrière, which last erupted in 1979.

Much of St Vincent is covered with dense tropical rainforests. The Grenadines have white sandy beaches and coral reefs.

There are still a few of the original Carib Amerindians, but most people on St Vincent, who are concentrated on the coast, are black or mulatto. They depend for survival primarily on agriculture, growing subsistence crops such as sweet potatoes and plantains.

The main cash crop, and the core of the country's economy, is bananas. These are grown both on small farms and large estates. New irrigation projects have boosted production and quality. But changes to the EU import regime have dramatically reduced the income and add to unemployment.

Efforts to diversify have had limited success, though marijuana could now be the second-largest export. US marines arrived in 1998 to destroy an estimated $1 million-worth of the crop, prompting local farmers to demand compensation.

The Grenadines are famous for secluded luxury tourism on islands

Land area: 390 sq. km.
Population: 0.1 million—urban 48%
Capital city: Kingstown, 16,000
People: Black 66%, mixed 19%, East Indian 6%, Carib Amerindian 2%, other 7%
Language: English, French creole
Religion: Protestant, Roman Catholic
Government: Constitutional monarchy
Life expectancy: 71 years
GDP per capita: $PPP 7,691
Currency: East Caribbean dollar
Major exports: Bananas

like Mustique, and are occasionally home to celebrities, but the islands also attract more than 100,000 other tourists each year, as well as many visitors from cruise liners.

The Grenadines have a drugs trade too, though the drug is cocaine, which is transhipped to the USA.

A small offshore financial sector provides another source of income. Around 13,000 companies are registered, including internet gambling companies. This used to be sustained by strict legislation to preserve secrecy, though this has had to be relaxed to comply with international regulations.

The country is a constitutional monarchy whose head of state is the British monarch, represented by a governor-general. Government is based on a two-chamber legislature, with a 15-member house of assembly and a six-member appointed senate.

Between 1984 and 2001, the government was in the hands of Sir James Mitchell and the New Democratic Party.

The two main opposition parties had merged in 1994 as the United Labour Party. This finally achieved power in 2001 with Ralph Gonsalves as prime minister, and was returned in the 2005 and 2010 elections.

Samoa

Samoa is forging closer links with its American neighbour

Land area: 3,000 sq. km.
Population: 0.2 million—urban 23%
Capital city: Apia, 38,000
People: Samoan
Language: Samoan
Religion: Christian
Government: Constitutional monarchy
Life expectancy: 71 years
GDP per capita: $PPP 4,467
Currency: Tala
Major exports: Fish, copra, car parts

Samoa, which in 1998 changed its name from Western Samoa, consists of two principal Pacific islands: Upolu and Savai'i—volcanic islands surrounded by coral reefs. Less than one-quarter of the territory is covered by tropical forests. The islands are vulnerable to tropical storms and in 2009 were hit by a tsunami that killed 170 people.

Two-thirds of Samoans live on Upolu. Standards of human development are fairly high. The literacy rate is close to 100% for both men and women, and the leading diseases are those associated with a rich lifestyle, including hypertension, diabetes, and coronary heart disease. Although fertility rates are high, population growth has been slowed by emigration.

Two-thirds of the workforce are engaged in agriculture, mostly working on communally owned land. As well as growing subsistence crops, such as breadfruit and taro, they also grow the cash crops that produce the country's major exports—including cocoa and coconuts.

Samoa also has substantial forest reserves, though these are being depleted by illegal logging and land clearance. Both agriculture and forestry are often affected by the cyclones that regularly hit the islands. Fishing has also expanded in recent years, primarily for tuna, and accounts for more than half of export earnings.

Manufacturing primarily involves agricultural processing and garments, though the Japanese company Yazaki also exports car parts to Australia.

Samoa has been strengthening its links with neighbouring American Samoa—though the country's change of name to Samoa caused some friction. Cooperation helps promote tourism; the islands between them welcome more than 90,000 visitors each year.

Closer integration should also help boost trade; Samoa sends tuna to be canned in American Samoa, along with large quantities of beer.

Since independence from New Zealand in 1962, Samoa has been a constitutional monarchy, based on traditional systems of authority. The head of state is the high chief, who in 2006 was Malietoa Tanumafili II. When he dies his successor will be elected by the legislature, the Fono.

Since 1982 the government has been formed by the conservative Human Rights Protection Party. Its latest victory in elections to the Fono was in 2011 when Tuila'epa Sa'ilele Malielegaoi was returned as prime minister. However, the Tautua Samoa Party also did well this time, winning 13 of the 49 seats and could provide serious opposition.

São Tomé and Príncipe

Bankrupted by political rows and now cursed by oil

Land area: 1,000 sq. km.
Population: 0.2 million—urban 62%
Capital city: São Tomé, 54,000
People: Mestiço, African
Language: Portuguese, creole languages
Religion: Roman Catholic, Protestant
Government: Republic
Life expectancy: 65 years
GDP per capita: $PPP 1,638
Currency: Dobra
Major exports: Cocoa

This small country consists of two main volcanic islands as well as a number of smaller islets in the Gulf of Guinea off the coast of West Africa. São Tomé has around 85% of the land area and almost all the people.

The islands were largely uninhabited until the Portuguese established plantations here—producing first sugar, then coffee and now cocoa. Today, most people are of mixed race, though of distinctly defined origins.

The forros, the ruling élite, are of Portuguese-African origin. The angolares are descendants of slaves from Angola. The serviçais are descendants of former contract labourers from Angola, Cape Verde, and Mozambique. Half the population still live in the rural areas, mostly on plantations. Although education and health standards are respectable by African standards, unemployment is very high and poverty has been increasing—to over 50%.

Since independence in 1975 production of the main crop, cocoa, has dropped by more than half, though still accounts for 85% of exports. There is also limited production of coffee. The second largest source of foreign exchange is licences from foreign vessels, mostly from the EU, fishing for tuna.

Prospects are likely to be transformed however by the discovery of offshore oil, production of which should start in 2010. Given the current widespread corruption and the dismal example of nearby Nigeria, the arrival of new funds that will swamp the existing budget is unlikely to improve the quality of governance.

Since independence, the government has largely been controlled by the Movimento de Libertação de São Tomé e Príncipe (now the MLSTP-PSD). Initially this was a Marxist party and until a new constitution in 1990 it was the only legal one. Today the political scene has less to do with ideology than with personal power struggles.

The presidential election in 2001 was won by Fradique de Menezes supported by the main opposition group, the Movimento Democrático das Forças da Mudança (MDFM). The elections in 2002 were indecisive, and after a short-lived coup a new coalition was formed in 2004 consisting of MLSTP-PSD and Acção Democrático Independente (ADI). In 2006, Menezes was re-elected president. In 2007, Joaquim Rafeal Branco became the MLSTP-PSD prime minister. The 2010 elections were indecisive. With 26 of the 55 seats the ADI's Patrice Trovoada formed a minority government.

Saudi Arabia

Even Saudi Arabia has to tighten its belt when expenditure outpaces income

| Land area: 2,150,000 sq. km. |
| Population: 25 million—urban 82% |
| Capital city: Riyadh, 4.7 million |
| People: Arab 90%, Afro-Asian 10% |
| Language: Arabic |
| Religion: Muslim |
| Government: Monarchy |
| Life expectancy: 73 years |
| GDP per capita: $PPP 22,935 |
| Currency: Riyal |
| Major exports: Oil |

Along the Red Sea, Saudi Arabia has a narrow coastal plain. From here the land rises sharply to highlands that vary from 1,500 metres in the north to 3,000 metres in the south. Beyond these western highlands is a vast plateau that descends gently to the eastern coast along the Persian Gulf. In the centre of this plateau is the rocky expanse of the Najd, around which circles an arc of desert, which includes in the south the world's largest area of sand, the Rub' al-Khali—the 'empty quarter'. Saudi Arabia has no rivers and must take all its water from underground sources.

The native population is almost entirely Arab, though along the Red Sea coast there is also a black population. This is the original home of Islam. Most people are Sunni Muslim with a strict interpretation known as Wahhabism, but there are also up to two million Shia Muslims who live largely in the east and have suffered repression. Many live abroad in exile. The population is growing fast—2.7% per year. Saudi Arabian citizens generally have benefited from the oil wealth. All receive free education and health care, and enjoy a range of subsidies. Those less wealthy are also entitled to a plot of land and a loan to build a house.

Women have traditionally been kept behind closed doors, and are still not allowed to drive, but their education levels are rising and many are now entering businesses. They can also take full advantage of the internet—two-thirds of Saudi users are women.

Although only one-quarter of the resident population consists of foreigners they make up 80% of the labour force. This includes immigrant workers from other Arab countries and from South and South-East Asia, who do the more menial jobs, as well as Europeans and Americans who do more specialized or technical work.

Nevertheless the boom years now seem to be over. Since the population has been growing faster than the economy Saudi Arabia's per capita income has been falling—it has more than halved in the past 20 years. As a result, Saudi nationals are having to accept more routine jobs such as security guards and taxi drivers.

Most economic activity revolves around oil, which is responsible for more than one-third of GDP, 80% of government revenue, and 90% of exports. Saudi Arabia has the world's largest oil reserves—one-fifth of the global total. This is extracted from huge oilfields—most of which are in the east, though there have also been discoveries elsewhere. At current rates of extraction, reserves should last 80 years or more.

The business is almost entirely in the hands of the government through Saudi Aramco. As well as having local refineries, the company also owns extensive refining and marketing operations in other countries.

Saudi Arabia has other important natural resources. Thus it has 4% of global gas reserves and is well endowed with many other minerals, including gold, iron, copper, and phosphates. Having concentrated on oil, it has barely exploited these but is planning to do so in the future.

The government has been making efforts to diversify and to encourage and build industries that would offer more employment—including petrochemicals, fertilizers, and steel. It has also been making greater efforts to attract foreign investment.

Despite its lack of water, and its underdeveloped agriculture, the country in the 1980s aimed to be self-sufficient in food, even in wheat. This meant extracting huge quantities of water from underground aquifers at rates that would exhaust the resource in around 30 years. Wheat production costs are four times the world price. In

Exhausting underground water

recent years, however, the government has been withdrawing subsidies and encouraging farmers to grow vegetables.

Another source of income is tourism. The annual 'haj' pilgrimage to Mecca draws in around three million foreigners a year, many of whom also take advantage of the opportunity to stock up on consumer goods. The government is now more interested in promoting tourism and has set up a tourist board.

Since the 1930s Saudi Arabia has been governed by the Al-Saud family, which in its various branches has 30,000 members, many of whom pursue a profligate lifestyle. The current leader is 85-year old King Abdullah bin Abdel-Aziz al-Saud who succeeded his brother, King Fahd, in 2005. Sultan bin Abdel-Aziz al-Saud was appointed as crown prince.

Successive kings have usually governed with some degree of consultation and in 1992 Fahd introduced a 'Basic Law' which serves as a kind of constitution. There is also the Majlis al-shura, a consultative council with 120 appointed representatives, chiefly from former officials and tribal leaders.

Since 2003, Saudi Arabia has suffered a series of terrorist attacks. Osama bin Laden, who came from a rich Saudi family, and al-Qaeda are bitterly opposed to Saudi Arabia's close relationship with the USA.

There appeared to be a democratic opening in 2005, when for the first time there were open elections for seats on town councils. The country has yet to be affected by the demands for democracy now being voiced in other countries in the region.

Senegal

One of Africa's most stable democracies, though it still faces armed insurgents

Land area: 197,000 sq. km.
Population: 12 million—urban 43%
Capital city: Dakar, 2.4 million
People: Wolof 43%, Peuhl 24%, Serer 15%, Diola 4%, Mandinka 3%, Soninke 1%, other 10%
Language: French, Wolof, Pulaar, Diola, Mandingo
Religion: Muslim 94%, Christian 5%, indigenous beliefs 1%
Government: Republic
Life expectancy: 55 years
GDP per capita: $PPP 1,666
Currency: CFA franc
Major exports: Fish, chemicals, groundnuts, cotton, phosphates

Senegal's northern border is formed by the Sénégal River whose valley provides a fertile strip of land. This soon gives way in the south and west to the flat, dry savannah of the Sahel, suitable only for raising livestock. The centre of the country consists largely of grasslands that are used more intensively for agriculture. Further south, beyond the 'finger' of the Gambia, a country that Senegal completely encloses, the Casamance region is greener, with more dependable rainfall and some tropical forests.

Senegal has more than a dozen ethnic groups of whom the largest are the Wolof, who are found particularly in the west. They make up only around two-fifths of the population and their language is also spoken by many others—though the official language remains French. Another large group, found in the west, are the Serer. Senegal also has many Peuhl (or Fulani). They are found throughout the country, and although traditionally they raise livestock many are now settled farmers. One of the politically most powerful groups is the Islamic Sufi sect, the Mourides.

The most significant dissident group are the Diola people in the Casamance region in the south who feel they have been exploited by northerners. Their armed organization, the Mouvement des forces démocratiques de la Casamance (MFDC), signed a peace agreement in 2004 though there is fighting by some factions within the MFDC.

Senegal is more developed than its immediate neighbours and Dakar, formerly the capital of French West Africa, is one of the region's more cosmopolitan cities, but the Senegalese are still very poor. More than one-third of the population are below the poverty line, only 40% are literate, and average life expectancy is 55 years. Health services are weak and concentrated in the cities.

Most people still depend directly or indirectly on agriculture, typically in small farms using basic production methods. The main food crops are sorghum and millet, grown in the north and centre, and rice in the Sénégal river valley and in the Casamance region in the south. Even so, Senegal still needs to import three-quarters of its rice consumption.

Most farmers also grow cash crops, of which the most important traditionally has been groundnuts. Groundnuts were also the main export earner but harvests over the last decade have been affected by erratic rainfall and more recently by privatization of the marketing system. Another important cash crop is cotton, much of which is bought by the local textile industry. Livestock production too has been affected by drought and the country also has to import meat.

The largest source of foreign exchange is fish. Senegal's Atlantic coastal waters are a rich source

Catching too much tuna

particularly of tuna. Most is caught from small boats by fishermen, who make up 15% of the workforce, but there is also industrial fishing. Tuna is canned and exported to the EU. The industry has expanded recently, making fish the largest source of export income, but over-fishing and increasing international competition have been reducing Senegal's tuna sales.

Senegal is one of the more industrialized countries in West Africa though productivity is not very high. Most manufacturing involves processing local raw materials. Thus, in addition to the fish canneries, there are groundnut-crushing mills, and four cotton-ginning plants, as well as textiles factories.

There is also a significant

chemicals industry, based on the deposits of phosphates in the west of the country. This produces phosphoric acid and fertilizer for export.

Senegal has a reputation for being more democratic than most African countries, though for the first four decades after independence in 1960 it was ruled by the same party, now called the Parti socialiste (PS). The PS tended to win because of its power of patronage, and the whole system had become corrupt.

From 1981 to 2000 the president was Abdou Diouf. He and the PS won elections in 1983 and 1988. The main opposition party, the Parti démocratique sénégalais (PDS), and its leader Abdoulaye Wade alleged that the elections were fraudulent. In 1991, under donor pressure, Diouf brought Wade into the government.

Change came at last in February 2000. Wade had been gathering greater support from the younger urban population, and from the Mourides, and won the presidential election with 60% of the vote. To his credit, Diouf accepted defeat with good grace and stepped aside—one of the few elected African presidents to have done so.

Wade, who finally arrived in office at the age of 74, strengthened his position in 2001 when the PDS won the most seats in the legislative elections. Idrissa Seck was later appointed as prime minister but, seen as a rival to Wade, he was removed in 2004 and charged with corruption.

Wade has been a popular. He was re-elected in 2007 is standing for a third term in 2012. The PDS also swept the board in legislative elections boycotted by the opposition.

Serbia

This remnant of former Yugoslavia has steadily been losing territory

| Land area: *88,361 sq. km.* |
| Population: *10 million—urban 52%* |
| Capital city: *Belgrade, 1.6 million* |
| People: *Serb 66%, Albanian 17%, Hungarian 4%, other 13%* |
| Language: *Serbian, Albanian* |
| Religion: *Serbian Orthodox, Muslim, Roman Catholic* |
| Government: *Republic* |
| Life expectancy: *74 years* |
| GDP per capita: *$PPP 10,248* |
| Currency: *Yugoslav dinar* |
| Major exports: *Manufactures, food, livestock* |

Serbia was one of the core republics of the former Federal Republic of Yugoslavia. It claims two provinces, though one has now declared independence. The first in the north is Vojvodina, which has two-thirds of Serbia's territory, and consists of low-lying fertile plains, through which flows the River Danube, and which rise to forested mountains in the centre and south. Further south is recently independent Kosovo, which borders on Albania and Macedonia and has two inter-mountain basins.

The country is ethnically diverse, but the three main groups have been concentrated geographically—and even more so following the Kosovo War of 1999.

Two-thirds are Serbians who are Eastern Orthodox Christians and are now almost entirely in Serbia. The next largest are the ethnic Albanians, most of whom are Sunni Muslims who now form almost all the population of Kosovo. The third main group are Hungarians, who are in Vojvodina.

Most people are well educated and minority groups are entitled to education in their own language. However the education system suffered as a result of the wars and international isolation.

The health system too deteriorated after 1991. Much of the hospital equipment is antiquated and health expenditure is low.

Before the wars of the 1990s, Serbia formed the core of Yugoslavia and had a thriving economy based on diverse manufacturing industries— including chemicals, vehicles, furniture, and food processing. The NATO bombing campaign of 1999 destroyed much of this and the recovery has been slow. The chemicals industry has made some progress but others such as computers, telecommunications equipment, and textiles are still struggling.

Agriculture was also doing fairly well and Serbia has regularly had a food surplus. The chief crops are wheat and maize, along with a wide range of fruit and vegetables. But a lack of investment is holding back development and yields are around half of what they could be.

Throughout the country, warfare and economic mismanagement have savagely cut living standards. In 2009, inflation was around 10% and 20% of the workforce were still officially unemployed, though many of these actually work in the black economy.

The country's political history from the mid-1980s to 2000 was dominated by Slobodan Milosevic—a former communist leader who in 1989 was elected president of Serbia. He and the Serbs had resisted the break-up of Yugoslavia. By 1992, Serbia and Montenegro were the only republics remaining and declared themselves to be the Federal Republic of Yugoslavia.

Dominated by Slobodan Milosevic

Milosevic was re-elected president of Serbia in 1992 and continued to support the Serbs in Bosnia who wanted to create a 'Greater Serbia'. In 1995, however, he took part in the negotiations in Dayton in the USA which ended the Bosnian War. Having had two terms as Serbian president, Milosevic stood instead for the presidency of the federation, to which he was elected in 1997.

Meanwhile, matters were coming to a head in Kosovo. During the communist era, both Kosovo and Vojvodina had enjoyed considerable autonomy, but to suppress the Albanian community Milosevic had suspended Kosovo's autonomy and imposed rule from Belgrade. Opposition to this turned violent in 1997–98 when a guerrilla movement, the Kosovo Liberation Army, stepped up its attacks on Serbian police.

Milosevic responded ferociously. After a failed peace conference, NATO started to bomb Serbia, while Serbia drove hundreds of thousands of Albanians out of Kosovo. The war ended after 72 days in June 1999 when Milosevic suddenly accepted a peace plan. Most of the Albanians returned, but then it was the Serbians who started to leave—180,000 to Serbia, 30,000 to Montenegro.

The political situation changed dramatically following the presidential election for the federation in October 2000 when Milosevic was defeated by Vojislav Kostunica of the Democratic Party of Serbia (DSS), standing as the candidate of a coalition: the Democratic Opposition of Serbia (DOS). The DOS also won the general election in Serbia in December 2000. In 2001 the government delivered Milosevic to a war crimes tribunal in The Hague where he died in 2007.

In 2003 under a new constitution the federal republic of Yugoslavia was replaced by a looser union: Serbia and Montenegro, with Svetozar Marovic as president. In Serbia, following the legislative election in 2003, Kostunica and the DSS formed a coalition government that included G17 Plus and the Serbian Renewal Movement–New Serbia alliance. Boris Tadic of the DSS was elected president in 2004 and 2008.

Montenegro had in 2006 declared its independence, and in February 2008 was followed by Kosovo, despite strong opposition in Serbia.

The elections in 2008, produced an unstable situation. The final outcome was a 10-party coalition with Mirko Cvetkovic of the DSS as prime minister.

Sierra Leone

Sierra Leone was devastated by years of warfare. After maintaining a fragile peace, it has now started to rebuild

Land area: 72,000 sq. km.	
Population: 5 million—urban 38%	
Capital city: Freetown, 470,000	
People: Mende 30%, Temne 30%, and many smaller groups	
Language: Mende, Temne, Krio, English	
Religion: Muslim 60%, indigenous beliefs 30%, Christian 10%	
Government: Republic	
Life expectancy: 47 years	
GDP per capita: $PPP 679	
Currency: Leone	
Major exports: Rutile, diamonds, bauxite, cocoa	

Sierra Leone takes its name ('lion mountain') from the distinctive shape of the mountainous peninsula that is the site of the capital, Freetown, and north of which there is a fine natural harbour. Most of the coastal area however is swampy for about 60 kilometres inland. Beyond this for around 100 kilometres is a broad plain of grasslands and woods, rising to a plateau broken by mountains that covers the eastern half of the country.

Sierra Leone's largest ethnic groups are the Temne and the Mende, each of which incorporates many subgroups. There is also a small population of Krios—descendants of liberated slaves who settled here from the end of the 18th century. Though few in number, they have had a disproportionate influence in the professions and the civil service.

The level of human development, already low, sank further in recent years when the country was torn apart by violence. Sierra Leone's people have one of the world's shortest life expectancies—47 years. Around one-quarter of children die before their fifth birthday and 2% of all births result in the death of the mother. HIV and AIDS are also a growing threat.

The war destroyed many of the health facilities, though with donor support the government is now reconstructing and resupplying them.

Many parents are now rebuilding schools and employing teachers and around half have reopened. Less than one-third of the population are literate.

During the fighting many people fled to the relative safety of the cities, where they added to already critical problems of overcrowding, poor sanitation, and poverty.

Around two-thirds of people make their living from agriculture, principally growing rice along with cash crops such as food and coffee. Indeed, Sierra Leone was a major rice exporter. For most of the 1990s its people had to rely largely on food aid. Now that the fighting is over, most farmers have returned to the land, and output has recovered strongly and should continue to do so.

Sierra Leone should have been able to invest in much higher standards of human development, given its rich mineral resources. In the south it has extensive deposits of rutile (titanium ore) and bauxite (aluminium ore) that in the past provided the bulk of the country's export earnings. Mining operations have now resumed.

Sierra Leone's other main mineral, diamonds, was less affected by the war. Because the diamonds are alluvial, larger mining companies can use mechanical systems but individuals can also dig and pan. The main problem is that most diamonds are smuggled out of the country.

Surveys have also indicated offshore oil and gas deposits though so far there have been no significant discoveries.

Since independence in 1961, Sierra Leone has suffered from economic mismanagement and corruption on a grand scale. For two decades after 1970 this was effectively a one-party state. That government was overthrown partly as a result of the spillover from the fighting in neighbouring Liberia when in 1991 a local rebel group, the Revolutionary United Front (RUF), with Liberian help, launched an uprising in the south-east.

Corruption on a grand scale

Then in 1992, a group of army officers who had been fighting with the rebels staged a coup. This intensified the ferocious civil war that killed around 15,000 people, and drove millions of people from their homes. The RUF were particularly brutal, recruiting thousands of child soldiers.

It was not until 1996, that Sierra Leoneans had the opportunity to vote in an election. They chose a former UN bureaucrat, Ahmed Tejan Kabbah, of the Sierra Leone People's Party (SLPP), as president. Hopes of a lasting peace were dashed, however, in 1997 when the same army officers staged another coup and invited the RUF, led by Foday Sankoh, to share power, until the Economic Community of West African States, and its peacekeeping force ECOMOG, largely with Nigerian troops, helped to restore Kabbah as president.

The fighting continued until a UN-sponsored peace deal in July 1999. This kept Kabbah as president and put Sankoh in charge of diamond mining. In fact the RUF did not disarm and in April 2000, when Nigerian peacekeepers handed over to UN troops, the RUF captured hundreds of UN soldiers and started to advance towards Freetown.

This alarmed the international community, and particularly the former colonial power, the UK, which sent in troops and managed to protect Freetown and restore order. In May 2000 Sankoh was arrested.

The UN was then able to adopt a more positive peace-building role. By 2001, the majority of the RUF fighters had been demobilized. Some, however, are now being recruited for militias in neighbouring countries.

In 2002, the SLPP gained an absolute majority in the legislative elections and Kabbah was returned as president.

In 2007, he was defeated by Ernest Bai Koroma of the All People's Congress (APC), and the APC became the leading party in parliament.

Singapore

An autocratic city-state that has become a leading manufacturing and financial centre

Land area: 643 sq. km.
Population: 5 million—urban 100%
Capital city: Singapore, 3.6 million
People: Chinese 77%, Malay 14%, Indian 8%, other 1%
Language: Chinese, Malay, Tamil, English
Religion: Buddhist, Muslim, Christian, Hindu
Government: Republic
Life expectancy: 80 years
GDP per capita: $PPP 49,704
Currency: Singapore dollar
Major exports: Petroleum products, machinery, chemicals

Singapore lies at the tip of the Malay peninsula, occupying the island of Singapore, along with 60 other adjacent islets. The city covers the southern part of the main island, but has extended itself further so that there is now little distinction between the city and the countryside. There is a small, highly productive agricultural sector but this is essentially an urban environment.

Singapore has a diverse immigrant-based population. The majority are of Chinese origin, though since they come from a range of provinces they often speak different dialects. The Indian population is also disparate, including Tamils, Malayalis, and Sikhs. The Malay population is linguistically more uniform. But most Singaporeans are at least bilingual and also speak English, the language of government and business. Racial

tensions, although muted, persist.

Singapore's rapid development has also created labour shortages—at both the bottom and the top of the market. More than one-quarter of the resident population are migrants. This includes 140,000 foreign maids as well as 300,000 foreign professionals. The government has a well-organized system for controlling this workforce, involving special levies on employers.

Singapore has now attained first-world status. Standards of education and health are good. The education system is highly competitive, with a strong emphasis on the literacy and numeracy requirements of a modern economy—though there are worries that regimented instruction is producing workers who are not sufficiently creative.

Most of these workers are employed in manufacturing and services. Singapore has for a century or more been an important regional centre for trade and business, and since the early 1960s the government has relentlessly propelled the economy along a path of export-led growth.

Manufacturing accounts for one quarter of GDP, around half of which comprises electronics: Singapore was the world's leading producer of computer disk drives but is now moving more into integrated circuits.

Since the early 1970s, Singapore has become one of the world's largest oil-refining centres—based on imported crude oil—and is a significant producer of chemicals and pharmaceuticals. It is also a centre for high-quality printing and publishing. Most of this activity is in the hands of 5,000 or so international companies along with a few large local enterprises.

A further one-quarter of GDP comprises finance and business services. Singapore's strong financial services sector combines a protected local banking system, dominated by four domestic banks, with a more open offshore banking sector. All are closely regulated.

Singapore's development has demanded a hectic process of construction. Until the mid-1980s this largely meant razing slums and replacing them with soulless apartment blocks and skyscrapers. More recently there have been

Hectic construction of soulless apartments

attempts to preserve older buildings. High urban density has also created serious environmental problems, particularly from the automobile. In response, the government applies a sophisticated electronically monitored system of road pricing, along with a good network of public transport. One persistent problem is a shortage of water, which has to be piped from Malaysia.

This frantic expansion has been directed by an autocratic government. For two years following independence in 1963, Singapore was part of the Malaysian Federation, but it broke away in 1965 to become an independent country.

From the outset, Singapore has been governed by the multiracial People's Action Party (PAP), which for three decades was presided over by Lee Kuan Yew, who became a major international figure.

The PAP has shown little tolerance for opposition. A regular tactic is to sue political opponents for defamation, bankrupting them so they are ineligible to stand for parliament.

The government also maintains wide powers to detain people arbitrarily and to restrict their travel, freedom of speech, and rights to free association. It has a similarly fierce attitude towards the media, intimidating local journalists into self-censorship, and keeping a close watch on the foreign media. The autocratic impulse extends into the social sphere, with laws, for example, controlling chewing gum, and fines for not flushing public toilets.

In 1990, Lee stepped down in favour of Goh Chok Tong. Then in August 2004 he in turn was replaced by Lee Kuan Yew's son, Lee Hsien Loong, usually called 'BG' because he used to be a brigadier-general.

The May 2006 election followed the usual pattern. The PAP sued the Singapore Democratic Party for allegedly impugning its honesty. The PAP gained 67% of the vote and all but two seats.

By the 2011 elections the PAP's share of the vote had dropped to 60% and this time the opposition gained six seats. The PAP's right to rule is finally being challenged.

Slovakia

The poorer part of former Czechoslovakia has been reforming rapidly

Land area: 49,000 sq. km.
Population: 5 million—urban 57%
Capital city: Bratislava, 426,000
People: Slovak 86%, Hungarian 10%, Roma 2%, other 2%
Language: Slovak, Hungarian
Religion: Roman Catholic 69%, Protestant 11%, Orthodox 4%, other 16%
Government: Republic
Life expectancy: 75 years
GDP per capita: $PPP 20,076
Currency: Euro
Major exports: Manufactured goods, machinery

Slovakia is dominated by the Carpathian Mountains. Three ranges run east to west and cover the northern half of the country. The highest point is in the Tatra Mountains—the Gerlachovský Peak at 2,655 metres. Most of the river valleys run north–south, so east–west communications are difficult. The main lowland areas are along the border with Hungary—one in the south-west including the capital Bratislava, and one in the east which includes the second city, Košice.

The majority of people are Slovaks but there are also important minorities. The largest group are ethnic Hungarians along the southern borders who since independence have been subject to increasing discrimination—especially against the use of their language. There is also a sizeable Roma, or gypsy, population, officially 2% but probably nearer 5%, many of whom are concentrated in the Tatra Mountains and live in constant fear of racist attack. Thousands have fled overseas as refugees, particularly to the UK.

Health standards and life expectancy are lower than in the more developed European countries. Though curative health services are extensive, less effort has gone into preventive health. As a result, men in particular have high death rates in middle age—probably the result of smoking, poor diet, and exposure to pollution. Education levels, however, have been high, comparable with those in Western Europe.

Industry now accounts for only one-quarter of GDP since production has fallen steeply, particularly in armaments. During the Czechoslovakia period, Slovakia also transformed imported raw materials into intermediate products such as steel and chemicals to be sent to more sophisticated manufacturing plants in the Czech lands. This pattern has continued, though at lower levels of production. More recent economic reforms have encouraged foreign investors. Volkswagen is now the largest private employer. Other major investors include US Steel, Whirlpool, and Peugeot-Citroën.

One serious consequence of the previous pattern of industrialization has been pollution: many factories have been located in valleys that trap the smoke generated by burning brown coal.

Employment in agriculture fell steeply in the 1990s and by 2007 employed only 4% of the labour force. Nevertheless, production has been maintained and the country is largely self-sufficient in food.

People who lost their jobs in industry and agriculture moved to services which account for over half of employment. Many also work in the informal sector which is thought to account for around 10% of GDP.

Slovakia's peaceful transition to a democratic free-market system was part of Czechoslovakia's 'velvet revolution' in 1990. But differences soon emerged between the two parts of the country. Slovakians thought that economic liberalization was going too fast for their less-advanced economy

Slovakia breaks with the Czechs and wanted greater autonomy. Matters came to a head after the elections of 1992.

In Slovakia the leading party was the Movement for a Democratic Slovakia (HZDS), led by a hard-line nationalist, Vladimir Meciar. This consisted predominantly of nationalists and former communists. Meciar negotiated with the Czech prime minister the terms for the 'velvet divorce'—which was not subject to a referendum since the split actually had little public support in either part of the country.

On 1 January 1993 Slovakia became an independent republic with Meciar as prime minister and another HZDS politician, Michel Kovác, as president (elected by parliament). Meciar's heavy-handed authoritarian instincts soon came to the fore as he concentrated power in his own hands, stifling opposition and curbing the press. He temporarily lost power in 1994 but was soon back and intensified his repression.

But by the elections of 1998 Meciar's grip was slipping, and he lost power to a four-party centre-right coalition led by the Slovak Democratic and Christian Union with Mikulas Dzurinda as prime minister.

In May 1999, Slovaks had their first direct election for president and chose Rudolf Schuster, who defeated Meciar.

Dzurinda set the country back on a more liberal and democratic path. He restored Hungarian language rights, for example, and started to rebuild the country's democratic institutions. In 2002 he was re-elected at the head of a slightly different centre-right coalition which enabled him to extend economic reform. For this turnaround Slovakia was rewarded with membership of both NATO and the EU, joining the latter in 2004.

In 2004, Meciar just lost the presidential election to his former right-hand man Ivan Gasparovic.

Then, in a surprising turnaround, the 2006 elections resulted in a coalition led by an anti-reform populist, Robert Fico of Smer–SD.

But after the 2011 election the outcome was a centre-right coalition with Iveta Radicova of the Christian Union-Democratic Party (SDKU-DS), as prime minister, though the coalition only has 77 of the 150 seats.

Slovenia

The richest country of ex-communist Europe, which has been president of the EU

Land area: 20,000 sq. km.
Population: 2 million—urban 48%
Capital city: Ljubljana , 0.3 million
People: Slovene 83%, Croat 2%, Bosniak 1%, other 14%
Language: Slovenian 91%, Serbo-Croatian 6%, other 3%
Religion: Roman Catholic 58%, other 42%
Government: Republic
Life expectancy: 78 years
GDP per capita: $PPP 26,753
Currency: Euro
Major exports: Manufactured goods, machinery, chemicals

Slovenia is predominantly mountainous. The highest areas are the Julian and Karavanken Alps on the north-western borders with Austria and Italy. These descend to a sub-alpine region on the edge of which is the Ljubljana basin. Slovenia has a short strip of coastline, which includes attractive beaches as well as the major port of Koper.

Slovenia's people, who are ethnically very homogenous, form an Alpine nation that has more in common with Germans, Austrians, and Italians than with the other former-Yugoslav republics to the south. Although health standards are generally good, education levels are below most EU countries.

The standard of living is higher than in other countries in Central Europe. Indeed Slovenes in general seem so contented that it is difficult to get them to move anywhere else:

foreign companies have problems persuading employees to work abroad. Even so, unemployment remains relatively high at around 9%.

Like other richer European countries, Slovenia is concerned about the ageing of the population since it offers generous pension benefits.

Slovenia only had about one-tenth of former Yugoslavia's population, but was responsible for around 20% of the output and was effectively subsidizing the poorer republics. The years immediately following independence were difficult as Slovenia lost markets in the other republics, but the economy soon revived as exporters found new buyers in the West.

Major manufacturing industries include metal products, furniture, paper, footwear, and textiles as well as pharmaceuticals and electronic appliances. Since the domestic market is very small most of the output has to be exported. About 70% of trade is now with the EU, primarily Germany and Italy. Leading companies include Lek, Krka, and Gorenje.

After 1992, following the Law on Ownership Transformation, many of the previously 'socially owned' enterprises were transferred to private hands. Some went to investment funds or were sold for cash. But many others were transferred 'internally' to workers and management in exchange for ownership certificates, or for cash at heavily discounted rates.

So far, Slovenia has made little effort to attract foreign investment, and has discouraged companies who wanted to borrow abroad.

Agriculture has become steadily less important, employing no more than 5% of the workforce, mostly on small farms engaged in dairy farming and livestock rearing. They now benefit from EU agricultural subsidies. Around half the country is covered in forests, mostly still state-owned.

Nowadays, most people work in services. Many are employed in trade and transportation: the high-tech port of Koper is a major outlet for goods from Austria. It also services trade between Eastern Europe and Asia: the

Koper serves as Austria's main port

port's largest customer is the South Korean car maker Daewoo. The capital, Ljubljana, is also at the crossroads of two of the EU's proposed new highways, from Venice to Kiev, and from Munich to Istanbul.

Another important service industry is tourism, though so far most of the two million tourists are Austrians making day trips for cheap fuel, and Italians attracted by the casinos.

In 2004, Slovenia joined the EU as the richest new entrant with a per capita GDP around 70% of the EU average and in 2008 successfully held the rotating EU presidency.

By Balkan standards, Slovenia's independence struggle was relatively painless. It had no ambitions to

take land from elsewhere, nor did it have ethnic problems. Following the declaration of independence in June 1991, the Yugoslav army made a half-hearted attempt to reassert control, but the Slovenes fended them off in a largely bloodless ten-day war.

Since independence, politics has generally been consensual, based on broad coalitions. In 1992 Slovenes opted for familiar figures. The largest single party was the centre-left Liberal Democracy of Slovenia (LDS), led by Janez Drnovsek, a member of the previous communist establishment who formed a coalition with the Slovene Christian Democrats (SKD).

In 1996 voters once again made the LDS the largest party, though with fewer seats. The government that emerged early in 1997, again headed by Drnovsek, also included the populist, centre-right Slovene People's Party and the smaller Democratic Party of Slovene Pensioners.

In 2000, after a general election when Drnovsek and LDS returned at the head of a three-party coalition. Drnovsek soon switched jobs. In 2002 he was elected president and was replaced as head of the LDS and prime minister by Anton Rop.

In the 2004 elections, however, the LDS was replaced by a coalition led by Janez Jansa of the centre-right Slovenian Democratic Party (SD). The 2007 presidential election was won by Danilo Tuerk, a left-wing independent supported by the SD.

In the 2008 general election the SD, now led by Borut Pahor, retained power, forming a centre-left coalition with the LDS, a breakaway from the LDS called Zares, and the Democratic Party of Slovenian Pensioners.

Solomon Islands

Pacific islands subject to ethnic conflict and political upheaval

The Solomon Islands consists of several hundred islands stretched over approximately 1,500 kilometres. Many have steep mountain ranges and more than 60% of the country is forest or woodland. The main islands are Guadalcanal, Malaita, New Georgia, Makira, Santa Isabel, and Choiseul.

The population is almost entirely Melanesian, mostly living in scattered rural communities. They have more than 80 languages, but most also speak Solomon Islands Pidgin. The official language is English.

Most islanders rely on subsistence agriculture. Their principal cash crops are coconuts, palm oil, and cocoa. But the main source of export income has been timber—which has provided more than 40% of government income. Logging accelerated during the 1990s and timber was soon being felled at more than twice the sustainable rate. In recent years there has been some respite since the market for timber has shrunk.

Fishing, primarily for tuna, is also an important source of income and food: the islanders have one of the world's highest per capita consumptions of fish. The largest private-sector employer is the Solomon Taiyo tuna-canning plant. The main export market is Japan.

The country's head of state is the British monarch, represented by a

Land area: 29,000 sq. km.
Population: 0.5 million—urban 19%
Capital city: Honiara, 49,000
People: Melanesian 95%, Polynesian 3%, Micronesian 1%, other 1%
Language: Solomon Islands Pidgin and many other languages
Religion: Church of Melanesia 33%, Roman Catholic 19%, other Protestant 45%
Government: Constitutional monarchy
Life expectancy: 66 years
GDP per capita: $PPP 1,725
Currency: Solomon Islands dollar
Major exports: Timber, fish, copra, palm oil

governor-general who must always be a Solomon Islander. Since 2009, this has been Frank Kabu.

Politics tends to be organized around people and issues rather than ideas. Following the 1997 elections, Bartholomew Ulufa'alu emerged as prime minister. In 1999, however, the country was shaken by ethnic violence in Guadalcanal in a struggle over land rights and jobs. In 2000 the Malaita Eagle Force seized the capital forcing Ulufa'alu to resign and replacing him with Mannasseh Sogavare of the Social Credit Party. He did not last. A general election in 2002 was won by the People's Alliance Party

By 2003, however, further ethnic violence led to chaos and an Australian-led military intervention. The country still has a 15-country 'Regional Assistance Mission'.

The elections in 2006 provoked further violence, targeting Chinese businesses in the capital. Mannasseh Sogavare was appointed prime minister but was replaced in 2007 after the formation of a new coalition led by Derek Sikua. The 2010 election, however, resulted in a new coalition led by Danny Philip of the Reform and Democratic Party.

Somalia

A 'failed state', riven by violence and now a haven for pirates

- Berbera
- Hargeisa
- ETHIOPIA
- **Mogadishu**
- KENYA
- INDIAN OCEAN
- Kismaayo

| 0 | Miles | 300 |
| 0 | Km | 480 |

Land area:	638,000 sq. km.
Population:	9 million—urban 37%
Capital city:	Mogadishu, 1.6 million
People:	Somali 85%, Bantu, Arab, and other 15%
Language:	Somali, Arabic, Italian, English
Religion:	Sunni Muslim
Government:	None
Life expectancy:	50 years
GDP per capita:	$PPP 600
Currency:	Shilling
Major exports:	Livestock, bananas (1990)

Somalia is largely semi-desert. It has rough grasslands suitable for pasture but little arable land. The highest part of the country is in the north, where rugged mountain ranges face the Gulf of Aden. To the south of these lies the Hawd plateau, beyond which there are flat sandy plains with extensive dunes along the Indian Ocean coastline.

Somalis have always been desperately poor and the incessant wars of recent decades have made them even poorer. Health standards are low and deteriorating. Around 15% of children die before their fifth birthday. Only one-third of the population have access to safe water. In most of the country the education system has virtually collapsed, and less than 10% of children enrol in primary school. There is also the ever-present threat of famine, whether from poor rains or from a collapse in the distribution system

caused by the fighting. Around 40% of the population depend on food aid. Around 1.5 million people have been displaced, many of whom are living in refugee camps in Kenya and elsewhere. Dismal conditions in the camps have also resulted in frequent outbreaks of cholera.

Through the civil war people still struggle to make a living. The most important source of income is rearing livestock—goats, sheep, and cattle—which accounts for around 40% of GDP. Two-thirds of the population are nomadic or semi-nomadic herders. Most of the others are farmers, growing sorghum and maize, chiefly in the river valleys. The most lucrative crop, and an important export, has been bananas. Indeed, some of the clan battles have been fought over control of the 120 banana plantations. Although Somalia, with its long coastline, could be a significant producer of fish, most of the catch is taken by foreign vessels.

There are parts of the country that function fairly well commercially and individual companies can work if they employ enough security guards. In 2004, Coca-Cola, for example, opened a heavily fortified bottling plant.

Since independence in 1961, Somalia has been riven by ethnic and clan divisions. Following the assassination of a newly elected president in 1969, the army took over, led by General Mohammad Siad Barre. But his rule became increasingly authoritarian.

Throughout the 1980s, several clans made an informal alliance against Siad Barre. There were three main groups: the United Somali Congress (USC—Hawiye clan); the Somali Patriotic Movement (SPM—Darod clan) in the centre and south of the country; and the Somali National Movement (SNM—Issaq clan) in the north. By January 1991, they had successfully driven Siad Barre from Somalia.

The USC claimed to have established an interim government but this was immediately rejected by the SPM, which started fighting USC forces in the south, and by the SNM, which in the north proclaimed independence for the 'Somaliland Republic'. Meanwhile, the country was stricken by a famine that killed 300,000 people.

An uncertain peace was established in 1992 with the arrival of a US-led

The state fails   UN peacekeeping force. However, this soon became bogged down and in 1995 the UN's mandate ended and its forces withdrew. By then Somalia had ceased to exist as a state.

Nevertheless efficient local administrations have emerged. Of these the most coherent is Somaliland in the north, which declared its independence in 1991 and whose president, Dhair Riyale Kahin, was elected in 2003. Its capital, Hargeisa, is relatively prosperous with wage rates higher than in some other

African countries and now has a new university. The population is around two million and most government revenue comes from customs duties. Services such as electricity and telecommunications are run by the private sector.

Another region, in the north-east, which has been autonomous since the end of 1998, though it has not declared itself as a state, is Puntland. Somaliland disputes any territorial claims by Puntland.

The rest of the country is divided between warring clans and factions. There have been numerous attempts at peace agreements. The most recent in 2004, backed by the EU, established a Transitional Federal Government (TFG) a 275-strong parliament chosen by warlords. In 2006, this established itself in Baidoa.

In May 2006, however, Islamist militias, controlled by a 'Union of Islamic Courts', started an assault on the warlords and seized Mogadishu. They were driven out with the help of troops from Ethiopia. These were supposed to be replaced by an African Union peacekeeping mission though only half the planned force has arrived, so Ethiopian troops remain.

At the end of 2007, prime minister, Ali Mohamed Ghedi of the TFG, was replaced by Nur Hassan Hussein.

Most of the centre and south of the country is controlled by two rival armed Islamic groups, al-Shabab and Hizbul Islam, probably supported by Ethiopia as well as by other global jihadist groups.

The violence now extends offshore with increasingly sophisticated acts of piracy, hijacking even the largest ships and holding the crews for ransom.

South Africa

South Africa has healed the wounds of apartheid but still suffers from serious violence

Land area: *1,221,000 sq. km.*
Population: *49 million—urban 62%*
Capital city: *Pretoria, 1.9 million*
People: *Black 79%, white 10%, coloured 9%, Indian 2%*
Language: *Afrikaans, English, Ndebele, Pedi, Sotho, Swazi, Tsonga, Tswana, Venda, Xhosa, Zulu*
Religion: *Christian 76%, Islam 2%, other 5%, unspecified 2%, none 15%*
Government: *Republic*
Life expectancy: *52 years*
GDP per capita: *$PPP 9,757*
Currency: *Rand*
Major exports: *Gold, diamonds and other minerals, metals*

South Africa has narrow coastal plains in the east, west and south. These rise to a chain of mountains that curves around the coast. The mountains are at their highest in the east where the Great Escarpment culminates in the Drakensbergs in the north. Within these mountains lies the vast plateau that makes up around two-thirds of the country.

Three-quarters of the population are black. The largest groups are the Zulu and the Xhosa, though official status has been extended to nine African languages. Of the white population, just over half speak Afrikaans, while the rest largely speak English. The 'coloured' population is of mixed African, Asian, and European descent and most of these speak Afrikaans. The majority of Asians originate from India and are English-speaking. In addition, there are thought to be up to eight million unauthorized immigrants from neighbouring countries.

The government has been redressing the injustices of apartheid. More than two million houses have been built, and many more now have piped water and electricity. But there is still a long way to go. Health standards for blacks are still similar to those in other African countries, while for whites they are comparable to those in Europe.

The government has also promoted black economic empowerment. But this remains a very unequal society: the richest 20% of the population get 62% of national income and the poorest 20% get only 4%. Unemployment is high at 24%.

Exacerbating these problems is the devastation caused by HIV and AIDS. Around 18% of the population are infected. South Africa was slow to address the epidemic because of former President Mbeki's dissident view that AIDS was not caused by the HIV virus. Now all who need anti-retroviral treatment can have it.

Another major social problem is crime. Every day, according to official records there are around 50 murders,

100 rapes, 700 burglaries, and more than 500 violent assaults.

South Africa's economic development in the past depended on its mineral wealth. It has the world's largest reserves of gold, along with many other minerals such as platinum and manganese, as well as diamonds, coal, and iron. Diamond mining is controlled by the De Beers company which has also dominated the world diamond market.

De Beers dominates the world diamond market

South Africa's manufacturing companies, which are responsible for around 20% of GDP, process these minerals, as well as agricultural produce, and make a wide range of consumer goods. Some have suffered as a result of exposure to international trade, though others including the car industry have expanded.

Agriculture now employs only around 7% of the workforce. The main crop is maize, most of which is still produced on white-owned farms. Land reform has been limited. People can now claim ancestral property seized during the apartheid years. But by the end of 2008, while most of these claims had been settled, only 4% of the land been redistributed.

Like more advanced industrial economies, South Africa has a sophisticated service sector. This includes financial services which contribute 20% of GDP and a major tourist industry that each year welcomes more than six million visitors.

South Africa's post-apartheid era started when Nelson Mandela strode from prison in 1990. In 1993 he won the Nobel Peace Prize. In the first multiracial election in 1994 his once-banned African National Congress won 63% of the votes in the National Assembly, which duly elected him president, heading a government of national unity. The main post-mortem on the apartheid years was through a Truth and Reconciliation Commission, headed by Bishop Desmond Tutu, which granted amnesties to many perpetrators of violence.

Mandela was an inspiring and conciliatory president who helped hold the country together. He retired in 1999 and his successor, Thabo Mbeki, led the ANC to even more decisive victories in 1999 and 2004.

Mbeki, was a more remote figure than Mandela, much criticized for his bizarre views on HIV and AIDS, and his unwillingness to help unseat President Mugabe in Zimbabwe. He hoped for a third term but was undermined by a power struggle with ANC vice-president Jacob Zuma who proved victorious and in 2008, despite corruption charges, was elected leader of the ANC.

In the 2009 elections, the ANC took 266 of the 400 seats in parliament and duly elected Zuma as president. More of a populist, Zuma was expected to give the government a left-wing flavour but has generally been fairly pragmatic.

The main opposition party, with 67 seats, is the Democratic Alliance, led by Helen Zille, which largely represents Afrikaaners and white liberals though now gets more support from coloureds. Another, with 30 seats, is the Congress of the People, an ANC breakaway. A third, with 18 seats, is the Inkatha Freedom party which is strongest in Kwazulu-Natal.

Spain

After making rapid strides the Spanish economy has now started to decline

Land area:	505,000 sq. km.
Population:	44 million—urban 77%
Capital city:	Madrid, 3.1 million
People:	Spanish, though with strong regional identities
Language:	Castilian Spanish 74%, Catalan 17%, Galician 7%, Basque 2%
Religion:	Roman Catholic 94%, other 6%
Government:	Constitutional monarchy
Life expectancy:	81 years
GDP per capita:	$PPP 31,560
Currency:	Euro
Major exports:	Cars, chemicals, fruit, vegetables

Spain, Western Europe's second largest country, is dominated by the Meseta, the vast and often barren central plain that covers more than two-fifths of the territory. Most of the people who live on the Meseta are in its centre, in Madrid, which was established there for political reasons. Otherwise, the most densely populated areas are those that encircle the plain.

This dispersal of population contributes to strong regional identities. Though Spain is racially fairly homogenous, it has major groups, such as the two million Basques in the north and the six million Catalans in the north-east, who are determined to defend their distinctive language and culture.

In future, however, there are likely to be fewer Spaniards of any kind. The average fertility rate is only 1.4 children per woman of childbearing age, and after around 2010 the population will start to fall.

Spain used to be a country of emigration. But now the flow is in the other direction. By 2007, Spain had attracted 5.2 million official immigrants (11% of the population) along with thousands more unauthorized immigrants. Most come from other EU countries, though the largest single group is from Morocco from where many people risk a dangerous clandestine journey across the Straits of Gibraltar.

Only 5% of the workforce are engaged in farming, but the country is a major exporter of wine, fruit, vegetables, and also of olive oil of which Spain is the world's largest producer. The rural population is shrinking fast, in some regions by 5% per year, and much of the land is turning to desert—left untended and subject to droughts and forest fires.

Rural communities on the coast also have the option of fishing—Spain's 17,000 or more boats are the largest fleet in Europe—but fishing too seems to be in decline. Of these boats, the 2,000 deep sea vessels that fish in foreign waters take the bulk of the catch.

Rather more healthy is Spanish

industry. The Basque country has long been a centre of heavy industry and machine tools. And Catalonia has thousands of small companies thriving in sectors such as textiles and shoes. But the most striking success in recent years has been the car industry, the third largest in Europe, turning out around three million cars per year of which 80% are exported. All the companies are owned by foreign transnationals.

Spain exports 80% of its car production

Agriculture and industry are now eclipsed by services, which now account for more than 60% of GDP. Of these, tourism is one of the most important. Around 50 million people arrive each year, making this the world's second most popular destination, after France. Most visitors head for the Mediterranean beaches of the Costa Brava and the Costa del Sol. However, the industry may now be in decline, faced with competition from Eastern Europe at the cheap end of the market and Asia at the top end.

Despite Spain's economic advance, major problems persist. Much of the country's progress depended on a property bubble which burst in 2008. Unemployment had been falling but by 2009 was back up to 20%.

The most difficult political problem has been in the Basque country where, in 2000, 38% of voters chose parties that favoured independence. The smallest is Herri Batisuna whose terrorist arm, ETA, has since 1969 killed over 800 people. In 2006, however, ETA declared a ceasefire and the government gave the region greater autonomy.

Spain's recent renaissance dates from the end of the dictatorship of General Francisco Franco who died in 1975. The first free elections for 40 years were held in 1977, and in 1978 a new constitution established a constitutional monarchy that also involved devolution to 17 autonomous regions. Regional governments now spend more than the central government. Those to which most power is now devolved are the Basque country, Catalonia, and Galicia.

The successful transition to democracy owes a huge debt to King Juan Carlos—especially his role in resisting a 1981 coup attempt. But one of the most significant political figures, opening the country up both economically and socially, was Felipe González, whose centre-left Partido Socialista Obrero Español (PSOE) governed from 1982 to 1996.

González and the PSOE were replaced after the 1996 election by the centre-right Partido Popular (PP), led by José Maria Aznar who was re-elected in 2000. Aznar was undermined by his support for the US-led war on Iraq and for wrongly blaming a 2004 Islamic terrorist attack in Madrid on ETA.

In the elections three days after the bombings, the voters punished the PP by returning as the largest party the PSOE, now led by José Luis Rodríguez Zapatero. He withdrew troops from Iraq and also pressed ahead with socially liberal legislation.

In the 2008 elections the PSOE and Zapatero were returned, but as a minority administration. Moreover the economic picture is now gloomy following the bursting of a property bubble. Rising government deficits are demanding unpopular austerity measures.

Sri Lanka

Sri Lanka's 20-year-old civil war has ended, but many of the underlying problems remain

Land area: 66,000 sq. km.
Population: 20 million—urban 15%
Capital city: Colombo, 2.5 million
People: Sinhalese 74%, Moor 7%, Indian Tamil 5%, Sri Lankan Tamil 4%, other 10%
Language: Sinhala 74%, Tamil 18%, English 8%
Religion: Buddhist 70%, Muslim 8%, Hindu 7%, Christian 6%, other 9%
Government: Republic
Life expectancy: 74 years
GDP per capita: $PPP 4,243
Currency: Sri Lankan rupee
Major exports: Textiles, garments, tea, gems, chemical and rubber products

One-sixth of Sri Lanka consists of the Central Highlands in the south-centre. From these highlands a series of plains spread out, though rather than being flat they are traversed by a variety of ridges and valleys. Despite its relatively small size, Sri Lanka has distinct variations in rainfall, so there is a dry zone and a wet zone. The dry zone in the north and east relies for rainfall mostly on the annual monsoons. The wet zone is in the south-west. This is the heart of the country, with year-round rainfall and most of the cultivable land, as well as the bulk of industry and two-thirds of the people.

Many live close to the coast. In December 2004, however, a tsunami devastated much of Sri Lanka's coastline and killed 31,000 people.

Three-quarters of the population are Sinhalese and Buddhist. But there are also significant minorities.

The largest of these are the Tamils, which in turn are divided into two groups: half are the original 'Sri Lankan Tamils', who are generally better educated and live in the north of the island. The other half are 'Indian Tamils', descendants of people who came later during the colonial period to work on the tea plantations. Another, smaller Tamil-speaking minority are the 'Moors', who are descendants of Arab traders. Sri Lanka's ethnic battles have primarily been between the Sinhalese and the Sri Lankan Tamils.

Despite decades of conflict Sri Lanka has been a human development success story. Education and health services are mostly free. Over 90% of the population are literate and 93% have access to basic health facilities. Notably, Sri Lanka's successes have been shared between men and women. Nevertheless, unemployment remains high and around one-fifth of the population live below the poverty line.

As a result many have migrated in search of work. Around 1.6 million Sri Lankans are employed overseas, the majority of them women domestic workers in the Middle East, who remit more than $3 billion per year.

In 1977, the government of Sri Lanka became the first in South Asia to liberalize and diversify its economy. Since then, it has seen a rapid expansion of manufacturing. This is dominated by the garments industry which now accounts for one-third of employment and produces more than half of exports. Competition is increasing from other low-cost producers but the industry has survived by concentrating on niche markets. Other important industries include footwear, food, chemicals, and rubber and petroleum products.

Agriculture remains important, employing around one-third of the labour force. Two-thirds of this is directed towards the domestic market, particularly the cultivation of rice, though one-quarter of the country's

The world's leading tea exporter

rice needs have to be imported. Sri Lanka's main agricultural exports—tea, rubber, and coconuts—are grown in the wet zone. Much of the output is now by smallholders. Sri Lanka is the world's leading exporter of tea. Production of rubber has benefited from high prices and recent investment.

A source of export income has been the mining of gems, including sapphires and rubies, as well as semi-precious stones.

The roots of Sri Lanka's civil war lay in the 1950s when the government changed the official language from English to Sinhala. By the 1980s discontent had erupted into inter-communal violence as Sri Lankan Tamils embarked on a struggle for an independent Tamil homeland in the north and east. Their armed group was the Liberation Tigers of Tamil Eelam, (LTTE), many of them children. In response, militant Sinhalese formed their own violent organization, and the security forces fought against both. The war is estimated to have killed at least 50,000 people.

To its credit, throughout the war Sri Lanka remained a democracy, in which the president holds a powerful position. Through the 1980s and early 1990s the government was in the hands of the United National Party (UNP). But the 1994 elections were won by a coalition, the People's Alliance (PA). The principal partner was the Sri Lanka Freedom Party (SLFP) whose leader Chandrika Kumaratunga was also elected president. She and the PA were re-elected at the end of 1999.

In 2001, however, her coalition fell apart and after a fresh election the UNP, led by Ranil Wickremesinghe, took over. He negotiated a ceasefire with the LTTE, but Kumaratunga accused him of conceding too much and in 2004 she dissolved parliament. The SLFP won the most seats in the subsequent election and formed a new coalition, the United People's Freedom Alliance (UPFA).

In 2005, the Supreme Court ruled that Kumaratunga could not stand again. The UPFA candidate, Mahinda Rajapakse, defeated Wickremesinghe in a presidential election.

The war ended in May 2009 after a final army onslaught against the LTTE. This also cost many thousands of civilian lives through indiscriminate shelling. Even after the war the UN accused the government of human rights abuses against Tamils interned in refugee camps.

Sudan

Sudan has now been split in two, but the new South Sudan risks becoming a failed state

Land area: 2,506,000 sq. km.	
Population: 40 million—urban 45%	
Capital city: Khartoum, 925,000	
People: Black 52%, Arab 39%, Beja 6%, other 3%	
Language: Arabic, Nubian, Ta Bedawie, many other languages	
Religion: Sunni Muslim 70%, indigenous beliefs 25%, Christian 5%	
Government: Republic	
Life expectancy: 58 years	
GDP per capita: $PPP 2,086	
Currency: Sudanese pound	
Major exports: Sesame, cotton, livestock, groundnuts	

Sudan has three main geographical regions. The north, covering around one-third of the country, consists largely of an arid, rocky plain. The centre has low mountains and sandy desert. The more tropical south has extensive swamps and rainforests. One of the country's major features is the River Nile. The Blue Nile and the White Nile flow from the south, joining at Khartoum to form the Nile itself, which flows north into Egypt.

Sudan is ethnically very diverse, with more than a dozen major groups and hundreds of subgroups. The largest, and politically the most dominant, are the Sunni Muslim Arabs who live in the north and centre. Most of the other groups are black Africans who predominate in the south and west and are either Christians or animists. The southern region has around one-quarter of the population.

Sudan has a low level of human development. Around 40% of the population are illiterate and less half the country's children enrol in primary school. Health standards too are very poor. Hospitals and clinics are primarily in the urban areas, while in the rural areas millions suffer from infectious diseases, particularly malaria and guinea worm. Poverty has been compounded by civil war which has led to several famines, driven hundreds of thousands of refugees into neighbouring countries, and displaced four million within Sudan. Since 1983, an estimated two million people have died in warfare.

Sudan is heavily dependent on agriculture, which accounts for one-third of GDP and employs two-thirds of the workforce. Sudan could be a major agricultural producer. But most people are still working at the subsistence level, growing sorghum and millet, particularly in the south and also in parts of the centre and west. However, there is also a mechanized farming sector—the result of extensive government investment. Three-quarters of this is rain-fed, particularly in the area around the Blue Nile. The remainder of the mechanized sector relies on irrigation for food and cash crops. This includes the vast Al-Gezira irrigation scheme south of Khartoum, between the White and Blue Niles, which waters the land of more than 100,000 tenant farmers. Originally this was used for cotton, but now more for sesame.

Sudanese industry is limited, consisting largely of processing agricultural crops—refining sugar, for example, and producing cotton textiles. But industry generally has been restricted by low investment and by a shortage of skilled workers, since many Sudanese have emigrated.

One of the most promising future areas is oil. Large deposits were

Prospects of oil income
discovered in the south in the early 1980s. Production was disrupted by rebel attacks, but output has now reached 500,000 barrels per day.

Sudan's north-south civil war had its roots in the rebellion of a southern army corps in 1955 who were protesting at northern domination. A military coup in 1969, headed by Colonel Jafar al-Nimeiri, appeared to offer some respite and in 1972 the war stopped. But in 1983 it flared up again. Colonel John Garang led the rebellion of the southern Sudanese People's Liberation Army (SPLA).

Nimeiri introduced Islamic shariah law, but following food riots in 1985 he was ousted in a bloodless coup.

A subsequent civilian government was replaced in 1989 by another Islamic-dominated military coup led by Brigadier Omar Hassan al-Bashir. He banned all political parties— except effectively for the National Islamic Front, now called the National

Congress (NC), which is linked with the Muslim Brotherhood. Al-Bashir ruthlessly suppressed opposition and stepped up the war against the south.

Al-Bashir was elected president in 1996. In 1998, a new constitution legalized other parties. In 1999 Al-Bashir suspended parliament and imposed a state of emergency. Then in 2000 he was overwhelmingly re-elected for a further five-year term.

Peace was achieved starting in 2002 with the 'Machakos Protocol' which included an undertaking to share power and hold a referendum on secession in the south within six years. The deal was sealed in 2005 and Garang became deputy president, only to die a few weeks later in a helicopter crash and be replaced by his deputy Salva Kiir.

Meanwhile another war had broken out in 2003 in Darfur in the west. An uprising by the black 'Sudan Liberation Army' protesting against the marginalization of the region met with a brutal response from a government-backed Arab militia, the Janjaweed. This led to horrific violence and genocide. About 300,000 people died. A series of ceasefires have been undermined by splits among the Darfur rebels. In 2008 the African Union peacekeepers were replaced by a UN force. In 2009 the International Criminal Court issued an arrest warrant against President al-Bashir for war crimes.

Following a referendum in favour, the South became independent in July 2011 and will be governed by the Sudan People's Liberation Movement. But this will be one of the world's poorest countries and tribal rivalries could make it a failed state.

Suriname

Fractious politics, corruption, and drug smuggling

Suriname has a narrow, marshy coastal plain, but most of the country consists of a vast plateau along with ranges of mountains covered with dense tropical rainforests.

Land area: 163,000 sq. km.	

Land area: 163,000 sq. km.
Population: 0.5 million—urban 76%
Capital city: Paramaribo, 243,000
People: East Indian 37%, Creole 31%, Javanese 15%, Bosneger 10%, other 7%
Language: Dutch, English, Sranang Tongo, Hindustani, Javanese
Religion: Hindu 27%, Protestant 25%, Roman Catholic 23%, Muslim 20%, indigenous beliefs 5%
Government: Republic
Life expectancy: 69 years
GDP per capita: $PPP 7,813
Currency: Suriname dollar
Major exports: Alumina, aluminium, rice

Suriname's complex ethnic mix reflects its colonial history. Today's creole population are descendants of African slaves whom the Dutch brought to work on the sugar and coffee plantations. They were subsequently replaced by indentured workers from India and from Java—whose descendants now make up the majority of the population. Others include the Bosnegers, descendants of escaped slaves, and Amerindians.

Suriname is well endowed with natural resources. The most important of these is bauxite, which is mined by US- and South African-owned companies. All of this is processed within Suriname into alumina, and some into aluminium, taking advantage of cheap hydroelectric power at Afobaka, where the dam has created a huge artificial lake.

Bauxite provides around half of export income. In addition, there are reserves of gold, nickel, and silver that have attracted the attention of Canadian companies.

Only 12% of the labour force work in agriculture, producing rice, fruit, and vegetables. But Suriname's vast tropical hardwood forests have attracted Asian logging companies, to the alarm of environmentalists.

A less official export is cocaine: Suriname is one of the main drug transit routes to the Netherlands.

Politics in Suriname has frequently been marked by violence and overshadowed by the military. In the 1996 National Assembly elections, the largest party was the pro-military National Democratische Partij, led by former military dictator Desi Bouterse. Later the National Assembly elected Jules Wijdenbosch as president.

The general election of May 2000 was won by the National Front (NF), a coalition of three ethnic parties, which used its majority to elect as president Ronald Venetiaan of the Nationale Partij Suriname.

Venetiaan discovered that the country's gold reserves had disappeared. He managed to stabilize the economy but had little impact on reducing poverty or corruption.

Following the elections in 2010 Bouterse returned as president. His Mega Combinatie party has formed a coalition with the A-Combinatie and the Volksalliantie, giving it 35 out of 51 seats in the National Assembly.

Swaziland

Africa's longest surviving absolute monarchy now stricken by HIV/AIDS

Land area: 17,000 sq. km.
Population: 1.2 million—urban 26%
Capital city: Mbabane, 61,000
People: African 97%, European 3%
Language: Siswati, English
Religion: Zionist 40%, Catholic 20%, Muslim 10%, other 30%
Government: Monarchy
Life expectancy: 45 years
GDP per capita: $PPP 4,789
Currency: Lilangeni
Major exports: Soft-drink concentrates, sugar, wood pulp, cotton yarn

Swaziland can be divided into four regions from west to east. In the west is the Highveld, which rises to 1,400 metres and covers one-third of the country. This descends first to the Middleveld, and then the Lowveld, before reaching the Lubombo Mountains, which form the eastern border.

Almost all of the people are Swazi. Heavy investment in education has resulted in an 82% literacy level. Health standards are lower, with a relatively high infant mortality rate, and they are set to fall further still as a result of HIV/AIDS: in 2008, around one-quarter of adults were carrying the virus. As a result, the population is steadily falling.

Most economic activity is based directly or indirectly on agriculture. Landholdings are of two forms. Around 60% is 'Swazi National Land', held in trust by the king and used communally for rain-fed subsistence crops and livestock as well as for cotton. The rest is 'Title Deed Land', most of which is owned by corporations that irrigate it to grow sugar cane, citrus fruits, and pineapples. These companies, like many businesses, are in the hands of trusts controlled by the royal family.

Industrial activity is centred on processing agricultural goods. In 1986, Coca-Cola moved its concentrate plant from South Africa to Swaziland, taking advantage of cheap sugar. Other similar companies have followed, and Swaziland's leading export is now soft-fruit concentrate.

Swaziland achieved independence in 1968 as a constitutional monarchy. This did not last. In 1973 King Sobhuza II suspended the constitution in what has been called the 'king's coup' and banned political activity. In 1986 his successor, King Mswati III, kept most of these restrictions in place and now rules as an absolute monarch with a lavish high-spending lifestyle.

There is also some quasi-democratic representation through the 65-member House of Assembly. The elections of 2008 were boycotted by the progressive parties so members are largely from conservative groups such as Sive Siyinqaba, which might also win an open election. Opposition comes mostly from trade unions, churches, students, and groups that form the Swaziland Democratic Alliance. In recent years there has also been some anti-government violence, including bombings.

A new constitution was approved in 2006, but offered no serious changes; indeed in some respects it gave the king even more power.

Sweden

Swedish social democracy has lost some of its gloss, but still has striking achievements

Land area: *450,000 sq. km.*
Population: *9 million—urban 85%*
Capital city: *Stockholm, 795,000*
People: *Swedish, Lapp (Sami)*
Language: *Swedish*
Religion: *Lutheran 87%, other 13%*
Government: *Constitutional monarchy*
Life expectancy: *81 years*
GDP per capita: *$PPP 36,712*
Currency: *Krona*
Major exports: *Manufactured goods, electrical machinery*

Sweden can be divided into three regions. The largest, in the north and centre, is Norrland whose mountains and forests constitute around 60% of the country. In the far south is Götaland, which includes the Småland Highlands and, on the southern tip, the plains of Skåne. Between these, in the south-centre, is Svealand, a lowland area with numerous lakes.

Sweden's people until recently were ethnically quite homogenous— the largest minorities being people of Finnish origin and small numbers of the reindeer-herding Sami in the north. From the 1970s, however, Sweden's liberal immigration and asylum policies welcomed people from further afield. In 2007 around 13% of the population were foreign born. Previously immigrants tended to come from other Nordic countries, but in the 1990s many people arrived from further afield as refugees.

Sweden's people enjoy one of the world's highest standards of living. Between the 1930s and the 1970s, Sweden created an extensive welfare state. Around 85% of healthcare expenditure, for example, comes from public sources. This has ensured that the benefits of economic growth have been equitably distributed.

Some of the credit for this must go to a strong civil society, including well-organized trade unions and a large number of voluntary organizations. Sweden has also created other institutions like the 'ombudsman', which many countries have since copied. Public-sector spending has been reduced in recent years but in 2007 still accounted for over 53% of GDP, and the government employed 33% of the workforce. Most of this expenditure is based on local taxes spent by local governments.

Sweden's early industrialization was based on raw materials like wood and iron ore, though it has neither coal nor oil and is heavily dependent on hydroelectric power. Forestry provides around 11% of exports. Minerals are also still important: Sweden produces most of the EU's iron ore. But Sweden's investment in education also ensured that the country could move rapidly to higher levels of technology.

In the internet age Sweden has the advantage that many of its people speak English well.

This small country has also produced some of the world's leading manufacturing companies including SKF, the world's largest producer of ball-bearings, and the Swiss–Swedish engineering group ABB. Some leading companies like Saab and Volvo have had their car divisions bought up by US companies, but Volvo and Scania are among the world's largest producers of trucks. And one of the most striking successes has been the telecommunications company Ericsson.

As a result, Sweden has become a major trading country. Exports are over half of GDP. Remarkably, around 40% of the shares on the Stockholm stock exchange are controlled by one family, the Wallenbergs.

Sweden is also one of the world's most generous aid donors, giving around 0.5% of GDP in official development assistance grants—one of the highest proportions in the world. It has also offered assistance to the Baltic states.

Sweden is a generous aid donor

Sweden's successful socio-economic model started to come under question in the 1970s and 1980s. Economic growth had slowed and could no longer support public expenditures. By the early 1990s the economy was deep in recession and unemployment had shot up. This forced the government to cut back on welfare spending, though by the standards of other countries, these changes have been modest. The economy has since revived. Unemployment in 2009 was 8%.

Sweden is a constitutional monarchy with King Carl Gustaf XVI as the ceremonial head of state. Since the Second World War, the country has usually been governed by the Social Democratic Party (SAP), though often in cooperation with other parties. From 1969, the SAP was headed by Olaf Palme who in 1986 fell victim to one of the world's most unexpected assassinations.

He was replaced as prime minister by Ingvar Carlsson. By the 1991 election, however, Sweden was deep in recession and the SAP lost heavily. Carlsson resigned and Carl Bildt took over as prime minister, heading a coalition led by his own Moderate Party along with the Centre, Liberal, and Christian Democrat Parties.

After the 1994 elections the SAP returned, forming a minority government, but Carlsson resigned in 1995 to be replaced by Goran Persson. In the 1998 elections, the SAP lost ground and had to make an informal alliance with the Left Party and the Greens.

This pattern continued after the 2002 elections with Persson returning as prime minister at the head of a minority administration that depends on the Left and Green parties.

Though Sweden joined the EU in 1995, support is at best lukewarm. In September 2003, despite the government's recommendation, Swedes rejected joining the euro zone.

In 2006 the voters were clearly ready for a change and gave a narrow victory to the centre-right Alliance coalition, led by Fredrik Reinfeldt of the Moderate Party. The coalition was returned after the 2010 elections but without an overall majority.

Switzerland

Switzerland has now joined the UN, but still hesitates about the European Union

Land area: *41,000 sq. km.*
Population: *7.5 million—urban 74%*
Capital city: *Bern, 123,000*
People: *German 65%, French 18%, Italian 10%, Romansch 1%, other 6%*
Language: *German 65%, French 18%, Italian 10%, Romansch 1%, other 6%*
Religion: *Roman Catholic 42%, Protestant 35%, Muslim 4%, other 19%*
Government: *Republic*
Life expectancy: *82 years*
GDP per capita: *$PPP 40,658*
Currency: *Swiss franc*
Major exports: *Machinery, precision instruments*

Switzerland is one of Europe's most mountainous countries. More than two-thirds of its territory, to the south and east, is covered by the towering peaks of the Alps. Another eighth, running along the north-west border with France, is covered by the less dramatic Jura range. Lying between these, and occupying most of the rest of the country, is a plateau around 400 metres above sea level interspersed with hills. It is within this belt of land, stretching from Lake Geneva in the south-west to the Bodensee on the north-east border with Germany, that most industrial and agricultural activity takes place, and where most people live.

Switzerland's population is formed by a conjunction of three European cultures: German, French, and Italian. A highly decentralized form of government, based on semi-autonomous states or 'cantons', has enabled these cultures to coexist and thrive. Switzerland thus has three official languages. The German-speaking majority live mostly in the north and east; the French in the south-west; and the Italian in the south-east. There is also a fourth, though not official, set of dialects— Romansch—which are also spoken in the cantons bordering on Italy.

Cultural diversity is further intensified by a large immigrant community, which makes up around 20% of the population. Around one-fifth are Italian, and another 14%, more recent arrivals, are from former Yugoslavia. These workers are very diverse: some are highly paid international business managers and bureaucrats; others are seasonal agricultural workers or hotel staff.

Poorly endowed with productive natural resources, Switzerland has relied for survival on the skills and ingenuity of its people. As with most richer countries, around two-thirds of the workforce are employed in service industries. In Switzerland, however, one of the most significant of these is banking: Swiss banks manage about a third of all private financial assets invested across borders, and financial services are about 9% of GDP and employ 3% of the workforce. Banking secrecy laws have contributed to this success. However, following the global financial crisis, which in 2009 required a bail-out for the largest bank, USB, the government was obliged by international pressure to weaken these laws. It is also negotiating double-taxation agreements with several countries.

Industry in Switzerland has been weighted towards higher levels of technology in such areas as chemicals, pharmaceuticals, and the manufacture of precision instruments and watches. Engineering and food processing are also important.

The perennial strength of the Swiss franc hampers exports, but heavy investment in research has kept Switzerland at the technological leading edge. A notable achievement was that of the Swatch watch company in fending off the digital challenge from Japan. For a small country, Switzerland is also home to many multinationals, including CIBA, ABB, and Nestlé.

Swatch repels the Japanese challenge

Agriculture, whether growing crops on the plain or rearing livestock in the mountains, employs less than 4% of the workforce but is very productive and meets around 60% of local needs. It is also highly protected, and the expensive food produced encourages many Swiss to stock up in neighbouring countries.

The beautiful Swiss countryside draws in tourists—over 11 million per year—though many are increasingly discouraged by the expense.

The most remarkable aspect of Switzerland, however, is its form of direct democracy. The central government is weak. First, because much of the power resides in the 23 cantons, each of which is responsible for education, hospitals and taxation, and has its own judiciary. Second, because most vital decisions are subject to referendums, of which there are three or four each year.

This narrows the territory for national party politics. Most parties are stronger at the cantonal level, but they do elect members to the bicameral federal assembly which since 1959 has been controlled by the same broad four-party coalition—though the populist anti-immigrant Swiss People's Party has recently become the most prominent member. The assembly in turn elects a seven-member federal council to serve as the executive.

Of these, one person is chosen to serve one year as president—a term of office so ephemeral that many Swiss are hard-pressed to name their head of state.

This decentralized system has produced stable if conservative government that has maintained Swiss neutrality and economic independence. Switzerland finally joined the UN in 2000.

In a referendum in 2001 Switzerland again rejected EU membership, though views are divided: broadly, urban dwellers and French-speakers are in favour, while rural dwellers and German-speakers are likely to be against. The country does, however, cooperate closely with the EU, for example, allowing in workers from all EU countries.

Syria

The Assad family's right to rule Syria is now threatened by widespread protests

Land area: 185,000 sq. km.
Population: 21 million—urban 55%
Capital city: Damascus, 4 million
People: Arab 90%, Kurd, Armenian, and other 10%
Language: Arabic
Religion: Sunni Muslim 74%, Alawi, Druze, and other Muslim sects 16%, Christian 10%
Government: Republic
Life expectancy: 74 years
GDP per capita: $PPP 4,511
Currency: Syrian pound
Major exports: Oil, phosphates

Syria has three main regions. One is a narrow fertile coastal strip that has year-round supplies of water. To the east of this is a mountainous zone with two ranges running parallel to the coast. The rest of the country is the Syrian Desert. Most people live in the section of land between Damascus in the south and the second city, Aleppo, near the Turkish border, though the most densely populated land is along the coast.

The population is mostly Arab and Muslim—split between the majority Sunni community and others including the Alawi, a branch of the Shia sect. Syria has had oil but is not wealthy. Around one-quarter of the workforce is either unemployed or underemployed.

The population is growing rapidly, at around 2.7% per year, and the healthcare system is struggling to keep pace. Education standards are relatively high though at higher levels the system is under strain.

Agriculture is still an important part of the economy, employing around one-third of the labour force. Two-thirds of the cultivable land is in private hands and it is evenly distributed among small farmers, thanks to extensive land reform in the 1960s. Nevertheless, the government exerts a strong influence since it controls the prices—often supporting farmers by paying more than the world price for crops.

Farmers devote around two-thirds of the land to wheat and barley for local consumption, but more significant is cotton, which generates around half of agricultural GDP and provides around one-tenth of export earnings. Some 75% of Syria's agriculture relies on the often erratic rainfall; the remaining irrigated land is mostly in the coastal strip.

Industry is also a major employer. The government has invested in heavy industries, but most activity nowadays is in lighter manufacturing such as cotton textiles, which employ one-third of the industrial workforce.

The government has tried to encourage greater private, and even foreign, investment, and the proportion of the economy in private hands has been rising and a number of new banks have started up. Foreigners are now allowed to own equity and in 2009 Syria relaunched its stock exchange.

Syria's oil will soon run out

Syria still depends to some extent on oil, which has provided around one-third of export income. Oil has been extracted in quantity only since the 1980s; nevertheless the reserves are relatively small and Syria is now a net importer.

Syria also has gas reserves which have been developed by foreign investors, Conoco and Elf Aquitaine, to supply local energy markets, hopefully releasing more oil for export. There are also plans for a major pipeline for gas exports.

The other major mineral export is phosphates. A further source of foreign exchange has been expatriate Syrian workers in the Gulf countries whose remittances are around $0.8 billion annually.

Since 1963, the government of Syria has been the exclusive preserve of the socialist Ba'ath (Resurrection) Party, though the current regime dates from a coup in 1970 when a faction representing the minority Alawi sect seized control. In 1971 the coup's leader, Hafez al-Assad, was elected president, a post he held until his death in June 2000.

Although nominally a republic with a parliament and a council of ministers, Syria is in practice a dictatorship with the president exerting control through the army and the intelligence services. Military spending is heavy—11% of GDP.

The Assad family and other Alawis control the government but most of the business community is Sunni—a division that is a source of tension. One concern about liberalization is that this could put too much power into the hands of Sunni merchants.

Syria's control has also effectively extended over Lebanon. Syria sent in troops in 1976 to help quell the civil war and subsequently retained a presence in what became a client state —though under international pressure withdrew in 2005.

Syria considers itself to be a centre of Arab nationalism and Damascus has been host to a number of dissident Palestinian groups. Syria has also been one of the most implacable opponents of Israel, with whom it has fought wars in 1967, 1973, and 1982.

Hafez Assad was succeeded by his son Bashar who was nominated by the People's Assembly and confirmed as president in a referendum in 2000. Bashar's arrival could have marked a more open era. This did happen in economic terms, making Damascus a more modern and bustling city.

But in political terms there has been little progress. The secret police have remained omnipresent, and human rights workers or dissident bloggers have been imprisoned for 'weakening the national spirit'.

It took the inspiration of the Arab spring, however, to present a real threat to Assad. The protests started in March 2011 in the southern town of Deraa, initially calling for democracy but later specifically for the end of the Assad regime. Since then the protests spread across the country. Around 1,300 people have been killed by the security forces and more than 10,000 refugees have fled to Turkey.

Taiwan

Taiwan is the most financially robust of the Asian tigers, but its relationship with China remains precarious

Land area: 35,980 sq. km.
Population: 23 million
Capital city: Taipei, 2.6 million
People: Taiwanese 84%, mainland Chinese 14%, aborigine 2%
Language: Mandarin Chinese, Taiwanese
Religion: Buddhist, Confucian, and Taoist 93%, Christian 5%, other 2%
Government: Republic
Life expectancy: 78 years
GDP per capita: $PPP 31,100
Currency: New Taiwan dollar
Major exports: Machinery, electrical and electronic products

The eastern part of the island of Taiwan, around two-thirds of the territory, consists mostly of a series of mountain ranges running north to south that descend steeply to the eastern coast. To the west, they descend more gently to a broad coastal plain that is home to most of the population. The island is geologically active: in 1999 an earthquake killed 2,448 people.

Taiwan's original inhabitants were a diverse collection of Malayo-Polynesian groups. Subsequent waves of emigration from China, however, displaced most of these groups to the mountains and they now account for only around 2% of the population.

Most people are descendants of the Chinese who had been arriving since the 17th century. A later addition, and the most powerful, however, are the descendants of the two million people who arrived following the 1949 communist revolution. These were largely the élite of the former regime: the industrialists and the professionals. More recent arrivals, but very much at the bottom of the social ladder, are the 300,000 or so immigrant workers. In recent years, however, there has also been an increase in labour emigration to China: at least 2% of Taiwanese now live and work there.

The population is highly educated and government spending on education is around 4% of GDP.

The basis for Taiwan's fairly equitable pattern of development was an extensive land reform over the period 1949–53. Most of the land on the fertile plains is still worked by small and very productive family farms that ensure the country remains self-sufficient in rice. Nowadays, however, agriculture employs only 5% of the workforce.

Taiwan's industrial development after 1949 benefited from considerable US aid. From the 1960s the USA obliged the Taiwanese to open their economy and produce goods for export. This initiated a steady upward progression: in the 1960s, light assembly work and garments; in the

1970s, heavy industry; in the 1980s, TVs; and in the 1990s, computers. Manufacturing still accounts for around one-fifth of GDP.

Taiwan is now the world's third largest producer of information technology goods, mostly subcontracting for foreign brands—90% of the circuit boards for the world's personal computers, as well as 80% of the notebook computers, are made by Taiwanese companies, many based in the Science Industrial Park at Hsinchu. However most have shifted many production operations to other countries; around two-thirds of IT hardware is now made in China.

Taiwan's industrial development is largely based on small producers—which account for around 80% of employment, and roughly half the economy. There is one company for every 18 people—the highest ratio in the world. Their effectiveness is based on extensive networking. Elsewhere, this might lead to corruption but this seems to have been less of a problem here. Taiwan tends to suffer less from financial crises partly because its financial system is based on family loans and other unofficial sources, and because Taiwan has tight capital controls. It could not in any case turn to the IMF, since it is not a member.

Taiwan has a company for every 18 people

Taiwan would have a rosy economic future were it not for the looming threat from China. Since 1949, political life has been dominated by the question of the island's status. The Nationalists (the Kuomintang, or KMT), who fled the mainland, for decades asserted that the long-term aim was reunification—assuming that they would rule the whole of China.

The USA initially backed this position, but by 1979 had opened diplomatic relations with China and started to loosen ties with Taiwan. Taiwan is not a member of the UN and has diplomatic relations with only around 30 countries.

The USA's engagement with Taiwan was also weakened by decades of an authoritarian and corrupt KMT regime. Since then, however, Taiwan has become an open democracy, with direct elections for both parliament and the presidency.

In 1996 the presidential vote saw the re-election of the KMT's Lee Teng Hui who espoused reunification. But the presidential elections in 2000 produced a dramatic change. The KMT was split between an official and unofficial candidate, opening the way for Chen Shui-ban of the Democratic Progressive Party (DPP), which espouses independence. The 2004 presidential elections also proved dramatic. Chen was shot and wounded the day before the election and won by a razor-thin margin.

Chen's presidency was to prove disappointing. This was partly because the KMT retained a parliamentary majority. But his own judgement was poor: he lost two referendums on whether Taiwan should join the UN and was caught up in corruption scandals for which he was sentenced to life imprisonment.

In 2008 the KMT regained the presidency, with a convincing victory for Ma Ying-jeou, former mayor of Taipei, who has promised to mend fences with both China and the US.

Tajikistan

Tajikistan has suffered from civil war, corruption, and increasingly autocratic rule

Land area: 143,000 sq. km.
Population: 7 million—urban 27%
Capital city: Dushanbe, 562,000
People: Tajik 80%, Uzbek 15%, Russian 1%, other 4%
Language: Tajik, Russian
Religion: Muslim 90%, other 10%
Government: Republic
Life expectancy: 65 years
GDP per capita: $PPP 1,753
Currency: Tajikistan rouble
Major exports: Aluminium, electricity, cotton

Tajikistan is very mountainous. The country is largely a collection of valleys with rapidly flowing rivers fed by melting snow and glaciers. More than half this landlocked country is 3,000 metres or more above sea level.

Tajikistan was a forced creation of the Soviet Union in 1929 that took little account of ethnic divisions. Tajiks make up four-fifths of the population but there is also a substantial Uzbek minority, in addition to Russians and numerous other smaller groups. Most Uzbeks live in the more industrialized northern region, Khujand, which is connected only tenuously to the rest of the country. The Tajiks, who are mostly Sunni Muslims, largely live in the rural areas in the foothills of the mountains and in small communities stretched out alongside the rivers and irrigation canals.

In the narrower valleys, the flat roof of one house is the yard of the house higher up—heightening vulnerability to the earthquakes to which the region is prone.

This was always one of the poorest of the Soviet republics and a civil war from 1992 further undermined human development. Infant mortality is high and more than 40% of households are below the poverty line. Around one million Tajiks migrate at least seasonally to Russia each year. Their remittances of $1.7 billion are worth around 45% of GDP.

Ironically, for a country with so many rivers, one of the scarcest commodities is clean drinking water. The country's water treatment plants are in a state of disrepair and water supplies are frequently contaminated.

Agriculture occupies around two-thirds of the workforce, though the principal crop is not wheat but cotton, which is one of the economic mainstays, taking up one-third of the land and accounting for 10% of exports. Cotton demands extensive irrigation from the rivers and canals, but run-off from heavily-used pesticides and fertilizers for cotton production has contaminated groundwater and rivers. Farmers also rear livestock and grow cereals and a variety of other crops, but the country still remains heavily dependent on food imports.

Tajikistan's dramatic terrain also contributes directly to its two other main exports. Many of its fast-flowing rivers have been harnessed for hydroelectric power. The largest dam, at Nurek, delivers 11 billion kilowatt hours per year. Much of this electricity is sold to neighbouring countries. Still, the country is generating only 5% of its total capacity.

Abundant hydroelectricity is also ideal for aluminium smelters, which absorb 40% of output—though all the aluminium oxide has to be imported. The Tursunzae smelter 65 kilometres west of the capital, Dushanbe, is one of the world's largest. Ageing equipment has contributed to a fall in output, but low world prices have discouraged reinvestment.

Though it has no bauxite, Tajikistan does have rich deposits of other minerals, which are being extracted in cooperation with foreign mining companies. These include gold and silver, along with strontium, salt, lead, zinc, and fluorspar. One of the largest companies is Zerashvan, most of which is owned by the Chinese company Zijin Mining.

Economic development is seriously hampered by corruption which the European Bank for Reconstruction and Development reckons to be the worst in Central Asia. This is exacerbated by the trafficking of opium, en route from Afghanistan to Europe.

Tajikistan's fractured geography has contributed to its complex and fractious politics which aligns along not just ideological but also religious and regional fault lines. Since independence in 1991, there has been a struggle for power between communists and Muslims and also secular democrats. Former communists remained in power following independence, but Islamic

Fractured geography; fractious politics

groups embarked on a widespread armed struggle. Many fled the country and Islamic guerrillas established a base in Afghanistan. Around 25,000 Russian troops fought on the government's side while Iran backed the Islamic force. In addition there have been attacks from the Uzbeks.

In 1994, a disputed presidential election resulted in a victory for the neo-communist Imomali Rahmonov at the head of his People's Democratic Party (PDP). In 1997 Russia and Iran supported a UN-brokered peace deal in which Rahmonov agreed to share power with the other parties, which had come together as the United Tajik Opposition (UTO), and to hold presidential and parliamentary elections. The UTO, for its part, agreed to demobilize.

Rahmonov duly rigged and won the 1998 election. In 2003 a plebiscite approved constitutional changes to allow him two more terms. In 2006 he was re-elected with almost 80% of the vote on a turnout of 90%. The 2000 legislative election was also flawed, giving the PDP two-thirds of the seats, and in the 2005 parliamentary elections the PDP claimed an unlikely 75% of the vote.

In 2007 Rahmonov changed his name to the less-Russian Rahmon. Political change is more difficult but discontent has been rising.

Tanzania

Tanzania is politically stable, but development is slow and tarnished by corruption

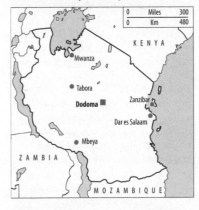

Land area: 945,000 sq. km.
Population: 41 million—urban 26%
Capital city: Dodoma, 1.7 million
People: More than 100 African groups; Arab on Zanzibar
Language: Kiswahili, English, Arabic, and many local languages
Religion: Mainland: Christian 30%, Muslim 35%, indigenous beliefs 35%; Zanzibar: Muslim
Government: Republic
Life expectancy: 55 years
GDP per capita: $PPP 1,208
Currency: Tanzanian shilling
Major exports: Coffee, manufactured goods, cashew nuts

Mainland Tanzania has a narrow flat coastal plain which rises to the vast plateau that makes up most of the country. But there is also some spectacular scenery, including in the north-east Africa's highest mountain, Kilimanjaro, at 5,895 metres. Much of the land is assigned to national parks and game reserves. The republic also includes islands in the Indian Ocean, two of the largest of which make up Zanzibar.

Tanzanians can belong to any one of 120 or more ethnic groups. None of these holds a dominant position, but effective efforts at 'nation-building' after independence, which included the promotion of Kiswahili as the national language, helped Tanzania to avoid serious ethnic conflict. The same language is also spoken in Zanzibar, though the population here—around one million—has a strong Arab component and is almost entirely Muslim.

Most Tanzanians are poor: around one-third are thought to be living below the poverty line. One-third of children are malnourished and 10% of children die before their fifth birthday. Around one-third of a million people die each year from malaria. HIV and AIDS is also now a major problem with around 6% of adults infected. The government makes anti-retroviral drugs available to all who need them. Tanzanians do however have reasonable standards of education, a legacy of earlier socialist investment.

Most people live in the rural areas from subsistence agriculture though only 8% of the land is under cultivation; large fertile zones as yet untouched. The main food crops are maize, cassava, rice, sorghum, and beans. Most smallholders also grow one or more cash crops, including coffee, cotton, cashew nuts, and tobacco. Tea and sisal are the other main cash crops, though grown mostly on estates.

Many farmers also raise livestock, notably Zebu cattle along with sheep, goats, and poultry. There is also considerable potential for developing fishing offshore as well as sustainable exploitation of freshwater Nile perch in Lake Victoria.

In Zanzibar the main export crop is cloves, but agricultural productivity is low and output is not increasing fast enough to have an impact on poverty.

Most manufacturing is based on processing agricultural commodities. Previously this was dominated by state-owned companies, though now many of these have been privatized.

Tanzania has a number of promising mineral deposits, which include diamonds as well as iron ore, nickel, and phosphates. However the main source of income is gold. There are three main gold mines, owned by Ghanaian, South African, and Australian companies, which have seen a steady rise in production.

Another area for development is tourism to the nature reserves and Africa's two largest game parks, as well as to the 'spice island' of Zanzibar. But annual tourist arrivals, around half of which come from the EU, are only 200,000—far fewer than in neighbouring Kenya.

For the first quarter of a century after independence Tanzania was governed, first as prime minister and later as president, by Julius

Julius Nyerere's influence
Nyerere. He was a hugely influential figure nationally and internationally. Nyerere's brand of African socialism had some social and political successes in Tanzania but also notable failures, including his attempts to concentrate people into villages and promote cooperative development of the land. When Nyerere retired in 1985 (he died in 1999), he was succeeded by Ali

Hassan Mwinyi.

As a consequence of Nyerere's economic failures Mwinyi from 1986 had to sign Tanzania up to an IMF-directed structural adjustment programme. Donors also pressed for an end to one-party rule by the Chama Cha Mapiduzi (CCM) party.

Multiparty presidential elections were duly held from 1992. But multiparty democracy has yet to take root. The main alternative is the Civic United Front (CUF) but this does not provide strong opposition to the CCM.

For the 1995 elections the ineffective Mwinyi was succeeded as president by Benjamin Mkapa, and he and the CCM duly won the presidential and legislative elections in 2000. In turn, he was succeeded in 2005 by Jakaya Kikwete, who vowed to take a stronger line against corruption and was reelected in 2011.

Semi-autonomous Zanzibar has representation in the mainland parliament, but also has its own president and internal administration. There has long been tension with mainland Tanzania. The CUF on Zanzibar accuses the CCM of sacrificing Zanzibar to the mainland, while the CCM says the CUF has plans to dissolve the union.

The 2000 elections in Zanzibar were violent and chaotic with the CCM presidential candidate Amani Abeid Karume declared the winner. 2001 saw a bombing campaign and an exodus of refugees. The 2005 elections were fairer, though still returned Karume.

Following the 2011 elections the CCM and CUF formed a government of national unity with Ali Mohamed Shein of the CUF as president.

Thailand

Red- and yellow-shirted political demonstrations have shaken Thailand's democracy

Land area: 513,000 sq. km.
Population: 67 million—urban 34%
Capital city: Bangkok, 5.7 million
People: Thai 75%, Chinese 14%, other 11%
Language: Thai, English. regional dialects
Religion: Buddhism 95%, Muslim 4%, other 1%
Government: Constitutional monarchy
Life expectancy: 69 years
GDP per capita: $PPP 5,840
Currency: Baht
Major exports: Food, machinery, manufactured goods

Thailand's distinctive geographical shape extends over four main regions. In the far north are mountain ranges covered by forests of tropical hardwood, while the north-eastern area is much dryer—a barren plateau that makes up one-third of the country. The southern region, which stretches down the Malay peninsula, is mountainous with narrow coastal plains. But the core of Thailand, and the focus of most activity, is the central plain—a rich agricultural area criss-crossed with rivers and canals.

Thailand's people are relatively homogenous—held together by language, religion, and a deep respect for the monarchy. The majority of people speak Thai and almost all are Buddhist—most communities have ornate temples. Nevertheless, Thailand also has distinct ethnic groups. In the northern region these include the hill tribes and in the far south Malays,

who are Muslims and have affinities with Malaysia. But the largest group of non-Thai origin, even if now mostly assimilated, are the Chinese who are the driving force in many commercial activities.

Thailand has made rapid progress in human development in recent years. Literacy is high, and secondary school enrolment, previously low, is now rising. Citizens now also have access to free basic health care.

Although incomes have increased, millions still live in poverty, particularly in the north-east, and there are many child labourers.

Thailand is a migration hub. Although many Thai workers have headed overseas, chiefly to Saudi Arabia, Brunei, and Singapore, around one million immigrants have also arrived from neighbouring countries, mainly Burma and Cambodia, to work on construction sites and sugar mills or in the 'entertainment' industry.

Thailand's population is still predominantly rural and 40% of the labour force still work in agriculture, mostly on small farms. Thailand's fertile land has enabled it to become the world's largest rice exporter.

Rapid economic growth has, however, taken its toll on the

environment—particularly on the forests. Since the mid-1960s forest cover has been halved to around 25%.

In recent years the most dramatic development has been the hectic process of modernization and urbanization. In the 1960s the government opened up the economy and invested in infrastructure. This helped to expand exports, first of agricultural products and then of manufactured goods and later moving into higher technology items.

Most of the country's industrial activity is focused on the sprawling capital Bangkok which, with its environs, now accounts for more than half the country's GDP—a rapid expansion that made the city a byword for pollution and traffic congestion. In recent years there

Above and below Bangkok's traffic

have been considerable improvements. Bangkok now boasts an overhead railway, the Skytrain, and a subway system, and pollution has fallen.

Thailand's leading source of foreign exchange is tourism. More than 15 million people visit each year, attracted to beach resorts such as Phuket and Samui, and to the often raucous nightlife, as well as the more peaceful hills around Chiang Mai.

Thailand was never colonized. It became a constitutional monarchy in 1932, and since 1946 has been ruled by King Bhumibol who is revered by most Thais. It has, however, had many autocratic or military governments and a number of coups. In 1992 the middle classes took to the streets protesting against military-backed parties, and this, along with an intervention by the king, opened the way to civilian rule.

In 1997, Thailand's economic boom came to a sudden halt triggering the Asian financial crisis and bringing down the government which was replaced by a coalition led by Chuan Leekpai of the Democratic Party (DP).

The elections in 2000 saw a victory for a new party, Thai Rak Thai (TRT) (Thais Love Thais), led by millionaire businessman Thaksin Shinawatra. Thaksin was a corrupt autocrat but had successful populist programmes, including cheap healthcare and in 2005 TRT won the election outright.

In 2006, following a murky financial deal by Thaksin's family, the army, with the king's tacit approval, carried out a bloodless coup. Thaksin fled into exile. Even so, the 2007 election was won by the People Power Party (PPP), a successor to TRT.

There then ensued an era of colour-coded politics. In 2008 yellow-shirted royalists of the People's Alliance for Democracy, blockaded the parliament and later the airport. The constitutional court then dissolved the PPP for electoral fraud opening the way for a DP-led coalition with Abhisit Vejjajiva as prime minister. This provoked violence from red-shirted Thaksin supporters.

The 2011 election pitted the DP against Pheu Thai (PT), the latest political vehicle for pro-Thaksin supporters. The campaign was transformed by the last-minute takeover of the PT leadership by Thaksin's sister Yingluck Shinawatra. An effective campaigner she helped them gain 265 of the 500 seats. Nevertheless she has drawn smaller parties into a coalition government. One critical question now is whether she will allow Thaksin to return.

Timor-Leste

A new nation that has yet to achieve political stability

Timor-Leste, formerly known as East Timor, comprises the eastern half of the Pacific island of Timor and the north-western enclave of Oecussi Ambeno. The territory consists largely of forested mountains that descend to coastal plains and mangrove swamps. Most people are Maubere, a mixture of Polynesian and Malay. Although their traditional beliefs are animist, the Catholic Church has also been important. The newly adopted official language is Portuguese—a language that few citizens speak.

Most Timorese have made their living from agriculture, growing food crops such as sweet potatoes or corn, along with cash crops—especially coffee, which has been the leading export. Farmers on the coastal plains also grow rice and plantation crops such as rubber, tobacco, and coconuts. In addition the forests yield many kinds of timber.

In the urban areas, however, there is little industry and unemployment is around 20%. The main source of national income in the future will be oil and gas since there are considerable offshore reserves, though this has led to boundary disputes with Australia.

The island of Timor had been divided between the Dutch colonists in the west and the Portuguese in the east. The Portuguese hung on

Land area: 15,850 sq. km.
Population: 1 million—urban 28%
Capital city: Dili, 60,000
People: Austronesian, Chinese, Indonesian
Language: Tetun, Portuguese, Indonesian
Religion: Roman Catholic 90%, Musllim 4%, Protestant 3%, other 3%
Government: Republic
Life expectancy: 61 years
GDP per capita: $PPP 717
Currency: US dollar
Major exports: Coffee, oil, gas

longer and East Timor declared its independence only in 1975. Indonesia promptly invaded. Within three months, 60,000 people had died. But the struggle continued—led by Xanana Gusmão's Frente Revolucionária do Timor-Leste Independente (Fretilin). Following the collapse of the Suharto regime in Indonesia, the East Timorese were offered a referendum. In 1999, despite massive Indonesian-backed violence, 80% of people voted for independence.

Elections for a constituent assembly in 2001 were won by Fretilin. Presidential elections in April 2002 were won by Xanana Gusmão. Independence was declared.

For a few years there was an uneasy peace. Matters came to a head in March 2006 when prime minister Alkatiri sacked half the army for going on strike. Soon Dili was taken over by rioters and peace was only restored by the arrival of Australian forces. Alkatiri resigned and was replaced by Nobel Peace Prize winner Jose Ramos Horta.

In 2007 Ramos Horta was elected president and he appointed José Alexandre 'Xanana' Gusmão as prime minister. In 2008 both survived assassination attempts.

Togo

Togo has yet again failed to install a democratic government

Land area: 57,000 sq. km.
Population: 6 million—urban 43%
Capital city: Lomé, 729,000
People: Ewe, Mina, Kabré
Language: French, Ewe, Mina, Dagomba
Religion: Indigenous beliefs 51%,
Christian 29%, Muslim 20%
Government: Republic
Life expectancy: 62 years
GDP per capita: $PPP 788
Currency: CFA franc
Major exports: Coffee, cotton, phosphates

Within its elongated shape, Togo has a variety of geographical regions. Inland from the thin coastal belt with its series of lagoons there is a plateau region. This extends to the Togo Mountains, a chain that crosses the country from south-west to north-east. Beyond the mountains to the north-west there is another plateau drained by the River Oti.

Togo has more than 30 ethnic groups. The largest, who live mostly in the south, are the Ewe, a group also to be found in Benin and Ghana. The population continues to grow fairly rapidly—by 2.7% per year. The greatest density is in the south.

This has put pressure on public services, particularly education. Togo has maintained higher educational standards than neighbouring Francophone countries but the system is deteriorating. Health services too are under strain and around one-

quarter of children are malnourished. Togo is also afflicted by HIV and AIDS which by 2008 had infected 3% of adults. Services are particularly poor in the rural areas.

Two-thirds of Togolese make their living from agriculture. Most are smallholders, growing such crops as cassava, millet, yams, and maize. In a normal year, Togo is self-sufficient in these crops, though it still needs to import rice. In the plateau regions and particularly in the north many people raise cattle and sheep.

Farmers can also supplement their incomes with a range of cash crops. In the mountains along the border with Ghana they grow cocoa and coffee, while to the east they grow cotton. Cotton has become particularly important and production, which takes place mostly on small farms, has more than doubled over the past decade and accounts for around one-quarter of export earnings. However, the intensity of cotton production is causing environmental damage. Efforts to further liberalize the cotton industry by selling Sotoco, the state marketing board, have been frustrated because no-one wants to buy it.

Industrial activity is dominated by phosphates, of which Togo is one of the world's largest producers. Togo's deposits are extensive and compact, and conveniently located just 40 kilometres from the capital, Lomé, which is also the main port. However output has fallen over the past few years, partly as a result of ageing equipment and inefficient management but also because Togo's deposits have a cadmium content above acceptable levels for the EU. The industry was returned to complete control by the state in 2007 and the formation of Societé nouvelle des phospates de Togo.

Togo also has a small manufacturing sector. Much of this involves processing agricultural goods such as cotton, palm oil, and coffee for export, but there are also other enterprises making cement, beer, and textiles for local consumption. Some of these companies are to be found in an export-processing zone which by 2005 had attracted 60 companies, from Europe, South Korea, and India, employing around 9,000 people, mostly in light manufacturing.

For almost 40 years Togo was ruled by Gnassingbe Eyadéma, who was appointed president following a military coup in 1967. Eyadéma, who was from the minority, northern Kabré group, headed a military government but later set up his own political party, the Rassemblement du peuple togolais (RPT). Since in 1979 the RPT was the only legal political party Eyadéma was duly elected.

By 1991, with a wave of following considerable donor pressure, Eyadéma was forced to include in the government members of opposition groups, most of whom are Ewe.

In 1992, this government came under attack from the army, provoking a nine-month general strike. In the ensuing violence more than 200,000 people were forced to flee the country. In 1993 Eyadéma was re-elected in a fraudulent election that resulted in the withdrawal of most development aid.

The 1998 election was worse. The opposition were denied access to the media and the electoral lists were tampered with. When it became clear that despite this the chief opposition candidate, Gilchrist Olympio, was winning, the paramilitary police seized the ballot boxes, the interior minister declared Eyadéma the winner, and Olympio fled into exile.

Following the election, repression was stepped up and hundreds of people were killed by the security forces. Corpses were washed up on the beaches of Togo and neighbouring Benin.

Corpses washed up on the beaches

International mediation started in 1999. Eyadéma established an electoral commission but then amended the code it produced, so the opposition refused to participate in the legislative elections of 2002. This new RPT-dominated parliament then removed the two-term restriction for presidents, allowing Eyadéma to stand again in 2003, and changed the rules so as to exclude Olympio. In another rigged election Eyadéma won again.

Eyadéma died in February 2005 and the military initially tried to install his son Faure Gnassingbé as president. Following international pressure they eventually retracted and a presidential election was held in April. As usual, this was rigged, giving Gnassingbé 60% of the vote. In the 2007 general election the RPT won easily. Gnassingbé was reelected in 2011.

Tonga

Tonga is still dominated by its king but change is in the air

Nuku'alofa

SOUTH PACIFIC OCEAN

Land area: 750 sq. km.	
Population: 0.1 million—urban 25%	
Capital city: Nuku'alofa, 22,000	
People: Polynesian	
Language: Tongan, English	
Religion: Christian	
Government: Constitutional monarchy	
Life expectancy: 72 years	
GDP per capita: $PPP 3,748	
Currency: Pa'anga (Tongan dollar)	
Major exports: Copra, squash, vanilla	

Tonga comprises an archipelago of more than 150 Pacific islands; some coral, some volcanic. Around two-thirds of the population live on the main island, Tongatapu. The islands are vulnerable to cyclones and tsunamis the most recent of which was in 2009. Tonga is also threatened by rising sea levels and a steady increase in salination.

Tongans are ethnically fairly homogenous and although they live in a lower-income country they enjoy good access to sanitation, clean water, and health care. Now they are starting to suffer from 'lifestyle diseases', including obesity and diabetes. They also have a high rate of literacy—close to 100%. Fertility remains high, but population growth has been slowed by high rates of emigration, particularly to New Zealand.

Most people still rely on agriculture, which accounts for over one-quarter of GDP. They grow food and a narrow range of cash crops—coconuts, squash pumpkins, vanilla, and melons. Half of Tonga's foreign exchange income comes from one commodity—squash pumpkins— which go primarily to Japan.

In the past, Tonga has benefited from foreign aid, but this is likely to dry up because Tonga, despite its protests of poverty, no longer qualifies as a 'least-developed country'. The other main source of foreign exchange, equivalent to around 40% of GDP, is migrant remittances from overseas workers. This too may shrink because fewer people are leaving. So further progress will be difficult unless Tonga diversifies its economy.

Tourism, with around 60,000 visitors per year, is also developing, offering the opportunity to enjoy the scenery and see hump-back whales.

Tonga has been effectively an absolute monarchy. From 1965 until his death in 2006 the country was dominated by King Taufa'ahau Tupou IV. He was succeeded by his son George Tupou V.

The 2005 election was won by the Tongan Human Rights and Democracy Movement (HRDM). Most of the seats in the parliament, however, were appointed by the king or traditional leaders. In 2006 there was a pro-democracy riot.

The 2008 elections strengthened the pro-democracy forces and a commoner, Feleti Sevele, became prime minister. At his coronation in 2008, the king announced that he would give up his political powers. The government altered the constituion so that 17 of the 26 seats were popularly elected. The Friendly Islands Democratic Party won the 2010 election but a noble, Lord Tu'ivakano, became prime minister.

Trinidad and Tobago

The oil will soon run out. Gas should fill some of the gap but more jobs are needed

Land area: 5,000 sq. km.
Population: 1 million—urban 14%
Capital city: Port of Spain, 345,000
People: Indian 40%, black 38%, mixed 21%, other 1%
Language: English, Hindi
Religion: Roman Catholic 26%, Hindu 23%, Anglican 8%, Baptist 7%, Muslim 6%, other 30%
Government: Parliamentary democracy
Life expectancy: 69 years
GDP per capita: $PPP 10,766
Currency: Trinidad and Tobago dollar
Major exports: Oil

These two islands are the southernmost of the Caribbean chain. But they can also be considered as an extension of South America: Trinidad, which has 94% of the land area, is only 12 kilometres from the coast of Venezuela. Trinidad's northern range of mountains essentially continues the coastal range of the Andes, which surfaces again as the island of Tobago. The southern quarter of Trinidad is oil-bearing land and includes the pitch lake at La Brea, which is the world's largest natural source of asphalt.

Tobago is less industrially developed. Its chief geographical feature is its main ridge which slopes down to extensive coral reefs that are popular with divers.

The country is racially very diverse—a product of its sugar-producing past. The main groups are the blacks who are descendants of the original African slave labour force, and the East Indians who are descendants of the indentured Indian workers who largely replaced blacks on the plantations after the abolition of slavery. In addition, there are smaller European and Chinese communities. The people of Tobago have long felt dominated and neglected by Trinidad.

Standards of human development are reasonably good though health and education services are uneven and young people have limited employment opportunities.

Agriculture still offers some employment but crops are now mostly for local consumption. Sugar for many years was a major cash crop and export earner and is still important for some farmers, but output has declined sharply and will fall further. There is also a small fishing industry.

Since the middle of the 20th century economic life has instead largely been restructured around oil and gas. Trinidad is a major offshore producer and also a refiner of crude oil imported from other countries in the region. In the 1970s, the sharp rise in oil prices transformed Trinidad and Tobago into one of the richest countries in the region. Some of the revenues go into a Heritage and Stabilisation Fund which in 2007 stood at $1.8 billion.

Oil production is now declining as the fields become exhausted but fortunately there are ample reserves of natural gas and production has increased steadily. This is used for local consumption and power generation as well as being liquefied for export. This should last until 2028.

Just as important, the gas now serves as feedstock for the manufacture of petrochemicals or fertilizers. The country has become the world's largest exporter of ammonia and methanol. Foreign investment in these plants and the growth of ancillary enterprises has helped reduce unemployment which in 2008 was 5%.

Two-thirds of the workforce are now employed in services. Some work in the tourist industry, though a smaller proportion than in most Caribbean countries.

Tourists head for Tobago Except when attending the annual carnival, relatively few visitors spend much time in Trinidad which now suffers from rising levels of crime, much of which is associated with drugs. The chief long-stay attraction is Tobago which has many unspoilt beaches as well as a variety of plants and wildlife in the rainforests.

With oil reserves likely to last no more than ten years, the government is anxious to diversify into other sectors, including agribusiness and marine, financial, and information technology services.

Politics in Trinidad has usually been conducted along ethnic lines. There have also been disputes between the two islands which led in 1980 to the formation of the Tobago House of Assembly. For many years the dominant party was the People's National Movement (PNM), which has generally had the support of blacks as well as the Chinese, those of mixed race, and non-Hindu Indians. The PNM saw the country through independence and held power for decades, effectively excluding the Indian community.

This pattern was broken in 1986 when the National Alliance for Reconstruction (NAR), a multiracial coalition led by Arthur Robinson, won most seats in the house of representatives. But the NAR's austerity measures proved unpopular and the coalition soon fell apart. In 1989, one of its factions formed a new party with largely Indian support: the United National Congress (UNC).

In the 1991 elections, the PNM, led by Patrick Manning, regained power, only to lose it narrowly in the 1995 elections, following which Basdeo Panday became the first person of Indian extraction to become prime minister, as the UNC formed a coalition with the NAR.

In the legislative elections in 2001 however, it took a series of ballots before the PNM, still led by Patrick Manning, finally gained a majority in 2002, consolidated in 2007.

The 2010 election resulted in a People's Partnership coalition of five parties: the UNC, the Congress of the People, the Tobago Organisation of the People and two smaller parties, with the UNC's Kamla Persad-Bissessar as prime minister. The president, elected by parliament, is George Maxwell Richards.

Tunisia

A popular revolution in Tunisia has reverberated across the region as the 'Arab spring'

Land area: 164,000 sq. km.
Population: 10 million—urban 67%
Capital city: Tunis, 728,000
People: Arab 98%, other 2%
Language: Arabic
Religion: Muslim 98%, other 2%
Government: Republic
Life expectancy: 74 years
GDP per capita: $PPP 7,520
Currency: Tunisian dinar
Major exports: Textiles, food, petroleum products

Tunisia can be divided into four main geographical regions. The northern one-third of the country consists of spurs of the Atlas Mountains but also encompasses fertile valleys and plains. These mountains descend further south to a broad plateau and further south still to a series of shallow salt lakes, known as shatts. Beyond these lies the Sahara Desert, which occupies around two-fifths of the territory.

Tunisians are almost entirely Arab, since the original Berber inhabitants have been assimilated. By African standards, they enjoy a relatively high standard of living and a low rate of population growth.

There is still some poverty, and unemployment is around 17%, but mindful of the potential for militant Islam among disaffected youth, the government has maintained food subsidies and largely free health services. Primary education is free and compulsory though literacy is still only 74%.

Tunisia's economy is broadly based. The country has been moving on from agriculture and from dependence on oil and phosphates and now has a number of manufacturing industries that employ one-fifth of the workforce. Some of the fastest growth has been in textiles, which are now the leading export, though this industry is facing increasing competition from Asia.

Tunisia has been an oil producer since 1966. There have been promising new finds but the country is now a net importer. Phosphates too now make up a much smaller proportion of exports since they are used locally as raw material for fertilizers and chemicals.

Half the workforce is employed in agriculture. Most farms are fairly small and largely worked by hand. The main crop in the fertile northern plains is wheat, along with barley and vegetables, but harvests are vulnerable to erratic rainfall.

The drier parts of the country are used to grow dates and particularly olives, of which Tunisia is one of the largest producers. The main restriction is a lack of water since the country is already using around 80% of its water for irrigation. One-quarter of the land is forest, or used for grazing animals, but livestock-raising too is vulnerable to drought.

Of the service industries, one of the most important is tourism. Tunisia's coastal resorts are favourite package-holiday destinations for Europeans. Tourism, with around six million visitors a year, accounts for 5% of GDP but revenues are fairly static. The government takes care to prevent tourists finding out about human rights abuses, and regularly bans the foreign press. The country's hotels and resorts will need considerable investment if they are to attract higher-spending visitors.

Tunisia's political system is less diverse than its economy. Since independence in 1956, it has been

Only two presidents since 1956
ruled by the same party and has had only two presidents, neither of whom have been enthusiasts for political pluralism.

Tunisia's first president was Habib Bourgiba, at the head of what was subsequently called the Parti socialiste destourien. Initially, Bourgiba was relatively progressive, promoting the rights of women, for example. But he grew increasingly autocratic and in 1975, he had himself elected president-for-life. The strongest opposition came from the underground Mouvement de la tendence Islamique (MTI). To combat the MTI, Bourgiba in 1986 appointed as interior minister General Zine el-Abidine Ben Ali. He was so successful at crushing the MTI that in 1987 Bourgiba gratefully appointed him prime minister.

This was a mistake. Ben Ali had Bourgiba declared senile, removed him from office, and in December 1987 assumed the presidency himself. At first Ben Ali seemed fairly liberal. He released hundreds of detainees and legalized opposition parties. He also renamed the party as the Rassemblement constitutionnel démocratique (RCD). In the 1989 elections, the RCD won most of the seats and Ben Ali was elected unopposed as president.

Ben Ali's enthusiasm for consensus was short lived. By the early 1990s, he was again persecuting the MTI, by then renamed Hizb al-Nahda (party of the awakening), jailing hundreds of its members and driving others into exile. He also took control of other opposition parties and restricted the press and trade unions.

In the 1994 elections, the RCD won almost every seat and elections in 1999 and 2004 produced similar results. Amnesty International reported hundreds of incidents of intimidation and torture.

The picture changed dramatically from December 17 2010 when the security forces prevented a young man Mohamed Bouazizi from selling vegetables by the roadside. In desperation he set fire to himself – triggering a wave of pro-democracy protests across the country.

Having lost the support of the army, Ben Ali was forced to flee the country. An interim government was formed with Béji Caid Essebsi as acting prime minister. An election for a National Constituent Assembly is planned for late 2011 to prepare the way for fresh elections. But many of those in the interim administration served under Ben Ali and there are fears that they may slow the process.

Turkey

Tension between Islamists and secularists threatens democracy and EU accession

Land area: 779,000 sq. km.	
Population: 73 million—urban 70%	
Capital city: Ankara, 4.1 million	
People: Turkish 70%, Kurd 18%, other 12%	
Language: Turkish, Kurdish, Arabic	
Religion: Muslim	
Government: Republic	
Life expectancy: 72 years	
GDP per capita: $PPP 12,955	
Currency: Turkish lira	
Major exports: Garments, textiles, fruit, vegetables	

Turkey is predominantly mountainous. The lowlands are mostly confined to coastal areas around the Black Sea, the Aegean, and the Mediterranean. Turkey straddles the border between Europe and Asia—though its European part is quite small. The Asian part, known as Anatolia, has at its heart the central Anatolian plateau, which is encircled by mountains. Western Anatolia consists of long mountainous ridges and deep valley floors, though the highest mountains are in the east.

The majority of the population are Turkish and almost all are Sunni Muslim. Over recent decades Turkey has been industrializing rapidly and more than two-thirds of the population live in cities. But by the standards of other industrial countries, levels of human development are low. In 2004, adult literacy, for example, was only 87%, and significantly lower for

women. And outside the cities, health services are poor. Until recently, many Turks were emigrating to the EU, particularly to Germany. The exodus has more or less ceased, but there are still some two million Turks in the EU, whose annual remittances although declining are still around $1.4 billion. However, there is also an inflow: around 2.5 million foreigners are living in Turkey, of whom one million are unauthorized.

Within Turkey, there is also a significant minority of Kurds—one-fifth of the population. The Kurds are one of the world's largest ethnic groups without a state of their own, though eastern Anatolia, where most Turkish Kurds live, together with neighbouring regions of Iran and Iraq, is referred to as Kurdistan. The government of Turkey has long tried to repress Kurdish nationalism and from 1984 to 1999 fought the guerrillas of the Kurdish Workers' Party in a war that cost more than 30,000 lives. In 1999 the PKK declared a ceasefire but resumed its campaign on a smaller scale in 2004.

With more than half the workforce employed in service industries, and one-fifth in manufacturing, Turkey has the characteristics of an advanced industrial economy. Many of the formerly state-owned enterprises, including iron, steel, and chemicals, have been sold, often to foreign investors. But most of the smaller enterprises are nationally owned. Turkey is also the world's second largest exporter of pasta.

Tourism is also a major source of employment and income, with 23 million visitors bringing in around $19 billion per year. Most of the tourists on packaged holidays come from Germany or the UK.

Turkey's diverse landscape has sustained steady growth in agriculture which still employs one-third of the workforce. Although levels of technology are lower than in Western Europe, output has kept pace with population, so Turkey is largely self-sufficient in food, and has crops such as wheat, sugar beet, cotton, and tobacco, some of which are exported.

Turkey has always had an ambivalent relationship with Europe. On the one hand, Turkey is strategically important, and has been a key member of NATO, for which it was a missile launching pad during

Ambivalent relationship with Europe

the cold war. On the other hand, it has a poor record on human rights, particularly with respect to the Kurds. Added to this has been the dispute with Greece over Cyprus.

These difficulties had made the EU nervous of inviting Turkey as a candidate for membership. Negotiations finally started in 2005, but with some EU countries vowing to put Turkish accession to a referendum —and with continuing human rights abuses—enthusiasm is waning.

Turkey's recent political history has followed an unsteady course. A series of weak coalition governments have been subject to military intervention: since 1945 there have been three military coups. The military intervened yet again after the 1995 election when the party with the largest proportion of votes—21%— was the pro-Islamic Welfare Party (RP). Eventually in 1997 they forced its leader Necmetting Erbakan to resign in a 'soft coup'.

A subsequent three-party coalition proved short lived. A further general election in 1999 resulted in another coalition led by the Democratic Left Party (DSP) with Bulent Ecevit as prime minister. Ecevit had some successes, including a new loan from the IMF and the beginning of negotiations for EU accession. But disagreements within the coalition forced another election in 2002.

This produced a surprising result: a landslide victory for a centre-right party with some Islamic origins—the Justice and Development Party (AKP)—led by Recep Tayyip Erdogan. Erdogan wants to get Turkey into the EU so as prime minister has made some gestures towards the Kurds and has tried to resolve the Cyprus question. He was re-elected in 2007 and then, in the face of stiff military opposition, Abdullah Gul of the AKP was elected president.

In the elections in 2011 Erdogan and the AKP achieved another solid victory with 326 of the 550 parliamentary seats. This just fell short of the majority needed to make constitutional changes. Second came the secularist Republic People's Party. An increased number of independents were also elected.

Turkmenistan

Turkmenistan replaces one autocratic leader with another, who has his own ideology

Land area: *488,000 sq. km.*
Population: *5 million—urban 50%*
Capital city: *Ashkhabad, 828,000*
People: *Turkmen 85%, Uzbek 5%, Russian 4%, other 6%*
Language: *Turkmen 72%, Russian 12%, Uzbek 9%, other 7%*
Religion: *Muslim 89%, Eastern Orthodox 9%, unknown 2%*
Government: *Republic*
Life expectancy: *65 years*
GDP per capita: *$PPP 4,953*
Currency: *Manat*
Major exports: *Gas*

Turkmenistan is largely a vast, sandy desert, with some low mountains to the south. The country borders on the Caspian Sea, but the main rivers, which are along the northern and southern borders, flow into neighbouring countries to the east.

Turkmenistan does, however, have many canals to provide water for drinking and irrigation, including the Garagum canal, one of the world's longest, which takes water from the Amu Darya River on the eastern border to the capital, Ashkhabad, 1,400 kilometres to the south. The scale of the diversion has, however, created major ecological problems, including the drying up of the Aral Sea into which the Amu Darya flows.

The Turkmen, who are Sunni Muslims, were formerly a nomadic people. During the Soviet era much of their culture and lifestyle was repressed, and now they are mostly settled along the rivers and canals. Following the collapse of the Soviet Union, there were considerable population movements: between 1989 and 1995, many Russians left, while over 300,000 Turkmen arrived from other republics. Turkmenistan also has a substantial Uzbek minority living along the northern border.

The population grew swiftly over this period, though more recently the birth rate has been falling. The population was fairly well educated but the quality of schooling has been slipping. Health services too have deteriorated as the budget has fallen sharply. Officially unemployment is zero, since the state guarantees work for all; in fact some observers say it could be as high as 70%. Only around 10% of the workforce are in the private sector.

Agriculture, based on irrigation, employs more than half the workforce but only 12% of GDP. The largest crop is cotton, followed by wheat. But production has been falling. The land is of poor quality and farmers have little incentive to increase output. There has not been much progress in land reform and the state, which is the monopoly buyer, pays far less than the world prices for wheat and cotton.

Production is in any case fairly inefficient: much of the irrigation water in the canals evaporates before it can be used and excessive application of fertilizers for cotton has built up toxic residues in the soil.

In the past, weaknesses in agriculture have been compensated for by production of gas. The Garagum Desert in the east holds a huge basin of natural gas—1.5% of world reserves. This is responsible for more than half of GDP and three-quarters of export earnings.

The largest gas producer is the state-owned Turkmenneftegaz which has seen sales increase following new contracts with Russia and Ukraine. These exports rely on a now ageing pipeline system which needs more investment. Sales to other countries would need new pipelines, such as the 1,500-kilometre one proposed to cross Afghanistan to Pakistan, and possibly extending to India. Originally proposed in the mid-1990s, this seems plausible now if Afghanistan becomes more stable.

Turkmenistan avoided economic liberalization and instead has been

Using gas funds to invest in industry

promoting a policy of import-substituting industrialization—using gas revenues for funds. The government has been investing in food processing and textile manufacture, as well as in a cellulose plant and steel mills. But private investors are slower to spend, deterred by massive corruption and excessive bureaucracy. The private sector accounts for only 25% of GDP.

Gas revenue has also gone into construction. Much of this is for the

gas industry itself, and also for the exploitation of oil. But there have also been a number of fairly grandiose projects, including a new international airport, a marble presidential palace, and an enormous mosque.

All this activity was directed until 2006 by dictatorial president, and prime minister, Saparmurad Niyazov. He became leader of the Turkmen Communist Party in 1986 and tried to resist economic and political liberalization. Following independence in 1991, the party re-emerged as the Democratic Party, which is still the only legal political organization. The party itself does not have a heavy workload; it meets once or twice a year to rubber-stamp presidential decisions.

Niyazov developed probably the world's most shameless personality cult. He had himself proclaimed as 'Turkenbashi'—leader of the Turkmen. One of his most striking legacies is a column topped by a revolving golden statue of himself. His heavyweight secret police, the KNB, helped suppress all forms of dissent. In 2000 his dutiful parliament requested that he declare himself president-for-life. He was happy to oblige, though in December 2006 he let them down by dying.

After some swift behind-the-scenes deals, the health minister, Gurbanguly Berdymukhamedov, was appointed president, confirmed by a dubious presidential election in 2007.

There is little sign that he will be any more liberal than his predecessor, and much of his energy goes into ensuring he stays in power. Early in 2008 he launched a new ideology, 'A State for the People'.

Uganda

Uganda now has a multiparty democracy but its president is reluctant to step aside

Land area: 236,000 sq. km.
Population: 31 million—urban 13%
Capital city: Kampala, 1.2 million
People: Baganda 17%, Ankole 8%, Basogo 8%, Teso 8%, Langi 6%, other 49%
Language: English, Luganda, Swahili, Bantu languages, Nilotic languages
Religion: Roman Catholic 42%, Protestant 42%, Muslim 12%, other 3%
Government: Republic
Life expectancy: 52 years
GDP per capita: $PPP 1,457
Currency: Ugandan shilling
Major exports: Coffee, gold, fish, cotton

Uganda lies on the equator, but because of the altitude the climate is relatively mild. The country forms part of the East African Plateau, most of which lies between 1,000 and 2,000 metres. To the west the boundary is formed both by mountain ranges and by the Western Rift Valley which incorporates lakes Edward and Albert. The eastern and north-eastern borders are also marked by mountains.

Most of the southern border runs across the world's second largest freshwater lake, Lake Victoria, which is shared with Kenya and Tanzania. Uganda's landscape shows great variety, from swampy river banks, to forests, to snowy peaks.

Uganda's population is also very diverse. It comprises dozens of major groups, but these can be divided into two main groups—the Bantu and the Nilotic. The Bantu, who live mostly in the south, make up around 70% of the population. The largest of these are the Ganda, also called the Baganda. The Nilotic group, mostly found in the north, include the Teso. While the Bantu groups have provided most of the political élite, the Nilotic people have in the past dominated the armed forces.

One-third of the population live below the poverty line. And they have one of the world's lowest life expectancies, a result partly of poor health services, but mainly of the AIDS epidemic. Uganda has, however, been a pioneer in the fight against the disease, with a pervasive and frank information campaign. This has helped reduce HIV prevalence detected in ante-natal clinics from 29% in 1992 to 5% in 2007.

Nevertheless, Uganda has already made some progress in reducing poverty and also has a donor-funded Poverty Eradication and Action Plan to boost the incomes of the poor.

More than 80% of the workforce depend on agriculture, generally working on small family farms growing basic food crops such as cassava, plantains, sweet potatoes, millet, and sorghum, mostly for their own consumption. They also grow a number of cash crops, notably coffee, which is the leading source of export income, and cotton, as well as raising cattle, sheep, and goats. In addition, a number of estates also produce tea.

With ample access to inland water Uganda is also able to catch large quantities of freshwater fish, notably Nile perch, which provide the second largest source of export income.

Manufacturing is mostly for domestic use and consists largely of processing agricultural output.

One of the most serious challenges to security comes from a rebel group in the north, the Lord's Resistance Army (LRA), led by Joseph Kony, who claims communication with the spirits. The LRA has been killing

A bizarre and violent cult

thousands of people, kidnapping children to fight as soldiers, and generally wreaking havoc. In 2005 Kony and his commanders were indicted for war crimes by the International Criminal Court. In 2008 another round of peace talks broke down and the government destroyed LRA bases in the Democratic Republic of Congo, which has made the north more secure.

Following independence in 1962, Uganda suffered decades of dictatorship and violence, much of it linked to rivalries between regions and ethnic groups. This included the reign of terror of General Idi Amin, who from 1971 killed around 300,000 people and expelled 70,000 people of Asian origin—followed by the despotic rule of Milton Obote.

State-sponsored violence only stopped with the victory of a rebel group the National Resistance Army (NRA) led by Yoweri Museveni. In

1986 he was sworn in as president, representing the NRA's political arm, the National Resistance Movement (NRM). In an effort to subdue sectarianism, Museveni banned multiparty politics. Instead he introduced a 'movement' or 'no-party' system. Members of the constituent assembly were instead to be elected as independents, though in practice almost all were NRM supporters. In 1996, Museveni was elected president with a large majority.

Museveni remained popular for a long time largely because he ushered in political stability and rebuilt Uganda's wrecked economy—removing many controls and welcoming private investment. Aid and stability produced steady economic growth.

But growth subsequently slackened and donors started to fret about defence expenditure which had risen as a result of Uganda's unsuccessful intervention in the civil war in the neighbouring Democratic Republic of the Congo. They also criticised rising levels of corruption

A referendum in July 2000 voted to retain the no-party system. But under donor pressure, the parliament in 2005 approved constitutional amendments to return the country to multiparty politics. They also, however, removed the two-term limit on the presidency, allowing Museveni to stand again.

For the 2006 elections the NRM became a party, the National Resistance Movement Organization. Musevini and the NRM were re-elected. In 2011, Museveni won two thirds of the presidential vote and the NRM again obtained a large parliamentary majority.

Ukraine

People power arrives via an 'orange revolution' but then melts away again

Land area: 604,000 sq. km.
Population: 46 million—urban 68%
Capital city: Kiev, 2.6 million
People: Ukrainian 78%, Russian 17%, other 5%
Language: Ukrainian 67%, Russian 24%, other 9%
Religion: Ukrainian Orthodox and others
Government: Republic
Life expectancy: 68 years
GDP per capita: $PPP 6,914
Currency: Hryvna
Major exports: Metals, food, chemicals

Ukraine consists almost entirely of gently rolling plains with the highest part in the west. The broad Dnieper River flows through the centre of the country, draining into the Black Sea in the south, north of the Crimean peninsula.

The majority of the population are ethnic Ukrainians, speaking a language closely related to Russian. Ukraine never existed as an independent country until the break-up of the Soviet Union, and its population distribution reflects its union with Russia. The population in the rural west is almost entirely Ukrainian while that in the more industrial east adjacent to Russia has a higher proportion of ethnic Russians. Since there are strong cultural and language links between the two communities, there is relatively little tension. Only in the Crimea, where around two-thirds of the population are ethnic Russians, are there separatist inclinations.

Since independence, Ukraine's population has shrunk by five million. Death rates rose as a result of declining standards of nutrition and health. Infectious diseases such as diphtheria, cholera, and TB surged, and health problems were compounded by alcoholism. In the early 1990s, falling numbers were offset to some extent by immigration from other former Soviet republics, particularly into the Crimea. Since then, however, the country has suffered net emigration.

Among the largest permanent departures have been of the Jews to Israel, Germany, or the USA. At the same time, there is steady labour out-migration, especially to the Czech Republic, where Ukrainians are prepared to work for desperately low wages. Around five million people are working in Europe or Russia.

This is a response to low wages and unemployment in Ukraine, where one-fifth of the population live below the poverty line. Poverty tends to be worst in the old smokestack, industrial, and coal-mining areas of the east; people living in the rural west do at least have the option of farming their family plots.

Industry has gone into steep decline but still employs more than one-fifth of the labour force. Between 1990 and 1997, industrial output fell by more than two-thirds, though in recent years there has been some growth. Certain areas of heavy industry, such as steel or chemicals, managed to sustain output, but others collapsed, particularly those related to defence and those producing low-quality consumer goods that could not compete with imports.

Ukraine has a significant mining sector and is the world's fifth largest producer of iron ore. But its coal industry is rapidly disappearing and though it has deposits of oil and gas these remain under-developed.

The majority of enterprises are now in private hands. But even in some of these the government has retained a controlling share and privatization generally has been very slow. Foreign investors have mostly stayed away, discouraged by an unpredictable legal system and corruption. Some compensation is that the black economy is thought to be equivalent to half the official GDP.

Agriculture, which also employs around one-fifth of the workforce, is in a similarly difficult situation. Ukraine is blessed with rich black

Blessed with rich black soil

soil—'chernozem'—and was considered the breadbasket of the Soviet Union. But output has fallen by half since independence—a result both of the loss of its Soviet market and a general shortage of investment and inputs. Ukraine was also slow to break up its collective farms: the necessary legislation was

only passed in 1999, though by 2000 most had disappeared.

Ukraine was one of the final Soviet republics to declare its independence. Its first president was Leonid Kravchuk. A former communist, he had little appetite for economic reform. In 1992, he appointed as prime minister the more reform-minded Leonid Kuchma. Kravchuk fell out with Kuchma and sacked him but in 1994 he conceded a presidential election—which he lost to Kuchma.

Kuchma struggled with a parliament still dominated by communists and his regime was uninspiring, notable chiefly for enriching his family and friends, and for extensive political violence. Kuchma was re-elected in 1999 only after a ruthless campaign.

The presidential elections of 2004, however, were to prove dramatic. Kuchma nominated his prime minister, Viktor Yanukovich, to run against more reform-minded opposition candidate, Viktor Yushchenko. Officials supporting Yanukovich tried to rig the ballot, but it did not work. Hundreds of thousands of orange-clad demonstrators took to the streets, eventually forcing the Supreme Court to order a repeat election in December which Yushchenko won comfortably.

But the 'orange' group of parties proved fractious. Following parliamentary elections in 2006, Yanukovich was eventually appointed prime minister, but continued squabbling led to another election in September 2007 after which Yuliya Tymoshenko headed a group of 'orange' parties. In the disputed 2010 presidential election, however, she narrowly lost to Yanukovich.

United Arab Emirates

One of the world's wealthiest countries that is a regional magnet for shoppers

Land area: *84,000 sq. km.*
Population: *4 million—urban 78%*
Capital city: *Abu Dhabi , 1.7 million*
People: *Emiri 19%, other Arab and Iranian 23%, South Asian 50%, other 8%*
Language: *Arabic*
Religion: *Muslim 96%, other 4%*
Government: *Federation of monarchies*
Life expectancy: *77 years*
GDP per capita: *$PPP 54,626*
Currency: *Emirian dirham*
Major exports: *Oil, natural gas, re-exports, fish, dates*

Apart from the Al-Hajar mountains in the Musandam peninsula along the eastern border with Oman, the territory of the United Arab Emirates, which stretches south from the Persian Gulf, is flat, barren, and low-lying—a mixture of sandy desert, gravel, and salt flats.

The UAE is a federation of seven sheikhdoms. By far the largest is Abu Dhabi, which covers three-quarters of the territory and includes the capital city. The six others lie east of Abu Dhabi. They are Dubai, Ajman, Sharjah, Umm al-Qaywayn, Ra's al-Khaymah on the Persian Gulf, and Al-Fujayrah, which faces the Gulf of Oman side.

Most people live on or near the coast. Only 17% are UAE citizens, who are mostly Sunni Muslims, though there are Shia minorities in Sharjah and Dubai. The rest of the population are immigrant workers, unskilled and professional, chiefly from the Indian sub-continent who make up 98% of the private-sector workforce. Almost all national workers are in government employment. In an effort to reduce the imbalance, the government periodically stops issuing new work permits. However the efforts seem to be fairly half-hearted. Unemployment is low and occasional crackdowns that have expelled unauthorized workers have generated labour shortages.

In the past, the people of this area made their living from subsistence agriculture and fishing, though with little good land or supplies of fresh water the agricultural prospects are limited. A combination of land reclamation and irrigation from underground aquifers allows farmers to grow dates for export and enables the UAE to be self-sufficient in fruit and vegetables. As a result of such efforts, total production increased six-fold during the 1990s. Nevertheless, most grains have to be imported.

The fortunes of the UAE changed forever in 1958 with the discovery of oil and gas in Abu Dhabi, and later in Dubai, where production started in 1969. Sharjah is the only other emirate with significant deposits. Abu Dhabi's reserves are still substantial and could last one hundred years, though those in Dubai and elsewhere are running out more rapidly. Abu Dhabi also has substantial gas reserves which could last for another 150 years.

But the UAE's wealth stems not just from current oil income but also from income from investments in foreign bonds and equity markets, particularly those of the Abu Dhabi Investment Authority. In total the foreign assets of the UAE are probably around $600 billion.

Both Abu Dhabi and Dubai have steadily been diversifying away from oil into such areas as refining and petrochemicals. Dubai has done so more extensively. It has, for example, the world's largest single-site aluminium smelter and has a major duty-free zone at Jebel. This zone is used largely for small-scale local assembly, and as a distribution centre for other parts of the region, and now hosts more than 2,500 companies.

But some of the most dramatic commercial activity was in property, controlled by the government-backed company Dubai World, which has created '7-star' hotels and artificial islands on which to build luxury

Dubai over-
reaches itself
apartments and even an indoor ski resort. This has been providing another useful source of income, since around five million people visit the UAE each year.

Most of this, however, been financed by debt, so when the property bubble burst in 2009 there were fears that Dubai itself could go bankrupt.

Abu Dhabi has been more conservative, though has made some efforts to create duty-free zones. It

probably regards Dubai's woes with some satisfaction now that it has been asked to bail out the brasher state.

The UAE is a federation. But the federation is a weak one: the federal government deals with defence, foreign policy, and immigration but only 30% of total UAE expenditures goes through the federal budget, and most of this is contributed by Abu Dhabi. The rest is spent by individual emirates which control their own natural resources and local businesses.

Abu Dhabi's economic clout gives it the greatest political power. Dubai has accepted this supremacy in exchange for economic autonomy, though it may now lose this.

The rulers of the emirates meet as the Supreme Council and elect the president. They always choose the Emir of Abu Dhabi, who is currently Sheikh Khalifa bin Zayed al-Nahyanal-Nahayan. He was appointed in 2004, following the death of his father. The Supreme Council also appoints a Council of Ministers controlled by the larger states.

Sheikh Khalifa seems to be quite popular and does not face major opposition. He can also rely on his brother, the younger and more energetic Sheikh Mohammed bin Zayed al-Nahyan. In 2006, Sheikh Mohammed bin Rashid al-Maktoum became the new ruler of Dubai and was appointed federal prime minister.

The UAE has no political parties but there does not seem to be much agitation for greater democracy, nor is there much sign of Islamic militancy. Legislative activity is handled by the 40-person Federal National Council, which is appointed and has only advisory powers.

United Kingdom

Though a leading member of the EU, the UK also tries to align itself closely with the USA

Land area: *245,000 sq. km.*
Population: *61 million—urban 90%*
Capital city: *London, 7.5 million*
People: *White 92%, black 2%, Indian 2%, Pakistani 1%, other 3%*
Language: *English, Welsh, Gaelic, Urdu, Hindi*
Religion: *Christian 72%, Muslim 3%, Hindu 1%, other or none 24%*
Government: *Constitutional monarchy*
Life expectancy: *79 years*
GDP per capita: *$PPP 35,130*
Currency: *Pound*
Major exports: *Manufactured goods, fuels, food*

Scotland, England, and Wales, together with Northern Ireland make up the state that is the United Kingdom (UK). The first three are also referred to as Great Britain. Of the four, Scotland in the north is the most mountainous, though to the west Wales too has an often rugged landscape, and there are lower mountain ranges in England, mostly in the north. The lowlands and plains that make up around half the UK's territory have rich soil and a temperate climate that allow productive agriculture.

Over 80% of the population are English. The Scottish, Welsh, and Irish, however, tend to have stronger national identities, including their own languages—though these are spoken by only a small minority. The UK has also received immigrants from former colonies—in the 1950s and 1960s from the Caribbean, and later from Africa and South Asia. Ethnic minorities make up around 8% of the population; they are exposed to racism and are poorer than average.

The British still benefit from the egalitarian welfare state created after the Second World War, though health and education services have now come under financial pressure. Meanwhile inequality is rising: the richest 1% of population owns 21% of the wealth. There are also regional contrasts: wages in the South East are 20% higher than elsewhere.

The UK was the cradle of the industrial revolution, but nowadays manufacturing industry accounts for only 14% of GDP and some industries, such as coal, have virtually disappeared. British industry was boosted in the 1960s and 1970s by the discovery of oil and gas in the North Sea though by 2005, as oil production fell, the country once again had become a net oil importer. The UK has also been an attractive base for multinational companies and has been one of the world's largest recipients of foreign investment.

Agriculture accounts for less than 1% of GDP and 1.4% of the workforce, but it is highly productive and grows around two-thirds of national needs. Outbreaks of 'mad-cow' and foot and mouth diseases and the use of genetically modified crops have, however, undermined confidence in intensive farming. The fishing industry supplies two-thirds of UK needs.

The UK has a very diverse service sector. This includes a leading global financial centre in the City of London which was hard hit by the global financial crisis. The UK is also the world's fifth most popular tourist destination. The tourist industry employs two million people, bringing in 33 million foreign visitors and $37 billion annually.

The UK is a constitutional monarchy ruled since 1952 by Queen Elizabeth II. For the past 60 years, British politics has been the preserve of two main parties: the pro-business Conservative Party and the trade union-backed Labour Party, with the party now called the Liberal Democrats some way behind.

The most radical government was the 1979–90 Conservative administration of Margaret Thatcher, which privatized many state-owned enterprises and broke the power of the trade unions. Thatcher's popularity

Thatcher's legacy peaked with victory over Argentina in the Falklands War in 1982 but she eventually alienated too many people and in 1990 was replaced by John Major, who surprisingly won the 1994 general election. In 1997, however, Major suffered a crushing electoral defeat.

The victor was Tony Blair, who had purged his party of the far left

and created a 'New Labour' party that would be neither socialist nor capitalist. One of his first challenges was Northern Ireland which for decades had been a battleground between Catholic and Protestant terrorists. In 1998, diplomacy prevailed with the Good Friday Agreement that allowed for a new directly elected assembly. In 1999, both Wales and Scotland also acquired similar assemblies.

With the Conservatives still in disarray, Blair won a further landslide election victory in 2001. For this he could also thank his chancellor (finance minister), Gordon Brown, who had fostered economic growth.

Blair had taken successful foreign policy initiatives, intervening in Bosnia and Sierra Leone, but his wholehearted support in 2003 for the US-led invasion of Iraq was far less popular. The UK also found itself in the terrorist firing line with an attack on London transport in July 2005. Blair and Labour won a third term in 2005, though with a reduced majority.

In mid-2007 Blair stepped down to be replaced as prime minister by Gordon Brown. Brown gained credit in 2008 for his decisive response to the UK banking system collapse, but subsequently lost public confidence.

Meanwhile the Conservative opposition, after a series of undistinguished leaders, settled in 2005 on David Cameron who proved adept at taking over the political centre ground. In the 2010 elections, he came close to outright victory but was only able to take power in coalition with the centre-left Liberal Democrat Party, led by Nick Clegg who became deputy prime minister.

United States

The lone superpower has one of the world's most dynamic economies, and now has its first black president

Land area: 9,364,000 sq. km.
Population: 309 million—urban 82%
Capital city: Washington DC , 5.1 million
People: White 80%, black 13%, Asian 4%, Amerindian and Alaska native 1%, others 2% (Hispanics, of any ethnic group, are 15%)
Language: English, Spanish
Religion: Protestant 52%, Roman Catholic 24%, other 14%, none 10%
Government: Republic
Life expectancy: 79 years
GDP per capita: $PPP 45,592
Currency: Dollar
Major exports: Capital goods, automobiles, industrial supplies and materials, consumer goods, food

Within its vast area, the USA has most types of landscape and climate—from tropical swamps in Florida, to the deserts of New Mexico, to the snow-capped peaks of the Rocky Mountains. The east of the country consists of the Atlantic coastal plain, which extends down from New England then broadens further south to include Florida and the Gulf coast. West of this plain are the Appalachian mountains, and then the vast areas of the prairies to the north and the Mississippi basin to the south.

West of the prairies are the Rocky Mountains and then a complex system of mountains and plateaux that stretches down to the Pacific coast. From north to south, the western landscape includes the forests of Washington state, through the beaches of California, to the semi-desert of Arizona in the south. The main land

mass has 48 states. The remaining two are Alaska, which lies beyond Canada, and Hawaii in the Pacific Ocean.

The diversity of the US landscape is complemented by the ethnic diversity of its population. Most people are white—descendants of European immigration from the 17th century onwards. Added to these were millions of blacks who came mainly to the southern states as slaves. But the USA has always been a magnet for immigrants and today 10% of the population are foreign born. Since the 1960s, following changes in legislation, immigrants come from a wider range of nationalities. In 2008, of the 1.1 million legal arrivals, the highest proportion came from Mexico, China (7.3%), India (5.7%), the Philippines (4.9%), and Cuba (4.5%). Millions more arrive illegally—there are 12 million undocumented immigrants, half from Mexico. US enterprises are only too happy to use them as cheap and exploitable labour.

The US population is generally mobile and dynamic, but it does suffer from rising inequality: between the 1970s and 2007 the proportion of national income going to the richest

10% of the population increased from 33% to 50%. In 2008, 13% of the population lived below the poverty line.

In 2009 the unemployment rate among black Americans was 16%, but for whites the figure was under 10%. Black and Hispanic families also earn around 30% less than white families. The infant mortality rate for black infants is twice the rate of whites.

Most people are educated in public schools, and many are college graduates: in 2007, 29% of the population aged over 25 had completed four years of college. Health care relies more on private provision, so the 16% of the working population who are uninsured can find themselves in difficulties—an issue that the new president has been addressing.

Americans are notable for churchgoing: 40% claim to attend a service each week. But they are also more likely than people in other countries to find themselves in jail: the prison population is more than 2.5 million.

The USA has the world's largest and most diverse economy, built on a rich endowment of natural resources that include deposits of most important minerals, including copper, gold, aluminium, and coal. It is also a major oil producer, though the reserves are depleted so the USA has to import most of its oil. Much of US life is constructed around the automobile—averaging 2.3 cars per household.

Almost two cars per household

Another vital US resource has been agricultural land. The USA produces two-fifths of the world's maize and soybeans—mostly through the agribusiness corporations that have displaced small farmers. One-quarter of agricultural output is exported—and around half in the case of cotton, wheat, and rice. This enormous output has been sustained despite shrinkages in the farm labour force and in agricultural land area. But it does rely on vast subsidies—around $5 billion per year. US cotton farmers, for example, receive half their income through subsidies.

Abundant natural resources, combined with an energetic immigrant population and a large national market, have also made US industry dynamic and innovative. Major industries include steel, automobiles, chemicals, and aerospace. And although other countries have caught up in basic production, the USA has set the pace in more advanced industries such as electronics, telecommunications, and computers.

However, the US economy, like that of all industrialized countries, is now dominated by services. In 2009 less than 1% worked in farming, forestry, and fishing, 23% in manufacturing, extraction, and transportation and the rest in different forms of services.

Since it has a huge internal market, the USA does not need to depend heavily on exports of goods and services, which account for less than 10% of GDP. Nevertheless its sheer size still makes it the world's largest exporter. There have been some efforts to open up trade, including the signing in 1994 of the North American Free Trade Agreement with Canada and Mexico. Even so, there is often strong resistance to free trade from industries such as steel that lobby for protection.

Politics in the USA is based on competition between two main parties: the Democrats and the Republicans. While there is considerable overlap in terms of support and policy, historically the Democrats have represented working people and ethnic minorities, and taken more liberal social attitudes. The Republicans have been more likely to represent big business and to adopt socially conservative positions on gay rights or abortion, and will want to minimize the size of government.

Hovering around the government are thousands of interest groups with highly professional lobbying machinery. One of the most significant recent influences has been the Christian Right.

Professional lobbyists

The US government is based on a separation of powers between the executive (the President), the judiciary (the Supreme Court), and the legislature (the Congress). Since this is a federal republic, individual states also wield considerable power through their governors and state legislatures.

The 1980s were dominated by the Republican presidency of former movie star Ronald Reagan. He turned the country sharply to the right, but a combination of vast defence expenditure and tax cuts created huge budget deficits.

Reagan was succeeded in 1988 by his vice-president George Bush. Having presided over the 1990 Iraq war, Bush lasted one term, defeated in 1992 by Democrat Bill Clinton.

Clinton had a number of successes, both in foreign and economic affairs,

and was re-elected in 1996, but came close to being forced out of office as a result of a sexual relationship with a White House intern.

The presidential election of 2000 resulted in a close and disputed Republican victory for one of Bush's sons, George W. Bush. The younger Bush started out with a right-wing agenda, making deep tax cuts for the rich and disdaining international treaties, notably the Kyoto protocol on global warming.

Bush was suddenly thrust into the international arena on September 11, 2001 by the al-Quaida terrorist attacks on the World Trade Center and the Pentagon which killed 3,000 people. He responded with a 'war on terror'. This first involved overthrowing the Taliban regime in Afghanistan. More controversially, he attacked Iraq on the mistaken assumption that its ruler Saddam Hussein was concealing weapons of mass destruction. In the 2004 presidential elections he beat Democrat John Kerry.

But with the situation in Iraq deteriorating, and a weak response in 2005 to Hurricane Katrina which devastated New Orleans, Bush and the Republicans became more unpopular.

As a result, in November 2008 the Democrats secured the election of America's first black president, Barack Obama. They also achieved majorities in both the House of Representatives and the Senate.

Obama has had to struggle with the effects of the global recession and the intractable war in Afghanistan. His options narrowed in the mid-term elections in 2010 when the Republicans gained control of the House of Representatives.

Uruguay

Uruguay had a reputation for prosperity and freedom that it is now trying to restore

0		Miles	200
0		Km	320

ARGENTINA

Salto

Paysandú

Rivera

B R A Z I L

Las Piedras

Montevideo

ATLANTIC OCEAN

Land area: 177,000 sq. km.
Population: 3 million—urban 93%
Capital city: Montevideo, 1.3 million
People: White 88%, mestizo 8%, black 4%
Language: Spanish
Religion: Roman Catholic 47%, non-Catholic Christians 11%, other 42%
Government: Republic
Life expectancy: 76 years
GDP per capita: $PPP 11,216
Currency: Peso
Major exports: Wool, textiles, beef, rice

Lodged between two much larger neighbours, Brazil and Argentina, Uruguay consists mostly of plateaux and rolling grasslands—the pampas—that are ideal for cattle raising. The highest parts of the country are in the hills of the Cuchilla Grande in the south-east. Another of Uruguay's natural advantages is its fresh, temperate climate which, combined with long sandy beaches, makes this a popular holiday destination.

Uruguayans are largely of European origin, predominantly Spanish and Italian. The small black population has immigrated from Brazil. Uruguay has long been one of the most urbanized countries of Latin America.

From the 1930s, Uruguay built up a comprehensive system of social welfare The system has been eroded in recent years, and the social security system has been partly privatized. But

Uruguayans still have high levels of health and education—even university education is free. By Latin American standards, this is also a relatively egalitarian country and one of the few where inequality has not increased in recent years. The country has a large urban middle class, while most of the poor people live in the rural areas.

Uruguay's wealth was built on the raising of cattle and sheep on the pampas. And although agriculture now accounts for only 13% of GDP, and employs only one-tenth of the workforce, it still plays a central role in the economy since meat, wool, and hides make up two-fifths of the country's exports and also supply vital raw materials for industry. Uruguayans themselves are also serious meat eaters.

Most of the land is occupied by large ranches. The herd has largely been disease-free and Uruguay has been very successful at selling meat to the USA and the EU. In recent years, it has also discovered a new market for wool in China. Most cereal output is for local consumption, but rice, which is chiefly grown in the department of Rocha on the southern coastal plain, is becoming an increasingly important export to Brazil.

Manufacturing and commerce are concentrated in Montevideo, which generates more than half the country's GDP. A high proportion of this is still in the public sector, which seems to have proved more efficient than in other countries. The public sector as a whole employs around one-quarter of the labour force. The state company ANAP refines petroleum, as well as producing alcohol and cement.

Most manufacturing is for local consumption, but recently-planted forests of eucalyptus and pine have been supplying mills making pulp and paper for export.

The country's other major foreign-currency earner is tourism, mostly to

Argentines heading for the beach

the beaches around the exclusive up-market resort of Punta del Este to the east of Montevideo. More than two million visitors arrive each year, almost all of these from Argentina.

Uruguay's trade links with its heavyweight neighbours have been further strengthened by the creation of Mercosur, a common market for the countries of the 'Southern Cone', whose secretariat is in Montevideo. There are also plans to build an $800-million giant bridge across the River Plate to Buenos Aires which would promote even greater flows of trade and tourism.

Political power in Uruguay traditionally alternated between the Partido Colorado and the Partido Nacional (also known as the 'Blancos'). This division does not correspond to any ideological divide; more a matter of personalities and history.

The pattern of control by either

the Colorados or the Blancos was broken in the 1970s. Left-wing groups, notably the Tupemaros urban guerrillas, established a strong presence and eventually provoked a coup by the military who ruled from 1973 to 1984 in a vicious fashion.

When democracy was restored, the Colorados, led by Julio María Sanguinetti, won the first presidential election in 1985. He was replaced in 1989 by Luis Albert Lacalle of the Blancos.

Throughout this period, however, a left-wing coalition, the Encuentro Progresista–Frente Amplio (EP–FA) had steadily been building support. So although Sanguinetti duly won the 1994 presidential election, the 1995 congressional elections produced a three-way split that required a Blanco and Colorado coalition.

The EP-FA candidate, Tabaré Vázquez, narrowly lost the 1999 presidential elections to the Colorado candidate Jorge Batlle and again the Colorados and the Blancos formed a coalition in congress.

In the 2004 election, however, EP–FA finally succeeded, gaining an historic congressional majority and having Vázquez elected president. The FA subsequently absorbed the EP parties. Vázquez immediately restored relations with Cuba but in economic policy proved fairly pragmatic, remaining popular while holding his coalition together.

Since presidents cannot have consecutive terms he was replaced as the FA candidate in 2009 by José Mujica, one of the founders of the Tupemaros. He defeated Lacalle of the Blancos, and the FA also gained small majorities in both houses of Congress.

Uzbekistan

Very dependent on cotton and gold, and making little progress towards democracy

Land area: 447,000 sq. km.	
Population: 27 million—urban 37%	
Capital city: Tashkent, 2.4 million	
People: Uzbek 80%, Russian 6%, Tajik 5%, Kazakh 3%, other 6%	
Language: Uzbek 74%, Russian 14%, Tajik 4%, other 8%	
Religion: Muslim 88%, Eastern Orthodox 9%, other 3%	
Government: Republic	
Life expectancy: 68 years	
GDP per capita: $PPP 1,744	
Currency: Som	
Major exports: Cotton, gold	

Uzbekistan has some mountainous terrain in the east, and a few fertile oases, but around 80% of the territory is a sandy plain that merges in the south with the Kyzylkum Desert. The country is doubly landlocked—at least two frontiers from the sea.

Most of the population are ethnic Uzbeks and live in the rural areas. Uzbeks are Sunni Muslims and tend to be more devout than those in neighbouring countries. Like other former Soviet Republics, Uzbekistan also has a substantial Russian minority. The Russians live mostly in Tashkent and other cities, though many have also left: over the period 1990–99, 845,000 people emigrated. The next largest group, the Tajiks, are to be found in older cities like Samarkand.

Standards of health are poor, and recent years of economic crisis have caused health services to deteriorate

further. Expenditure on education has also been falling, as has school enrolment. Wages too have been falling, and inequality has been rising steeply.

Housing has been privatized and more than one million people own formerly state-owned apartments.

Around one-third of the population work in private agriculture, though all farmers have to lease their land from the government. People are allocated small plots of land for lifetime use. These have helped maintain food production and account for around 75% of meat and milk production. Farmers have the advantage of fertile land and a fairly mild climate, but since rainfall is light they rely heavily on irrigation.

The most important cash crop is cotton. Uzbekistan is the world's second largest exporter of cotton, referred to locally as 'white gold'. Though most cotton producers are private or belong to cooperatives, they can sell only to the state—which pays well below the world price so they have little incentive to increase efficiency or output. With so much of the land devoted to cotton, the country does not produce sufficient grain and is a substantial importer.

Heavy and inefficient use of water for cotton production has also drawn so much water from the country's two main rivers that the Aral Sea into which they drain has shrunk dramatically—by one-third between 1974 and 1995—with severe health and environmental damage in the surrounding areas.

Over-use of irrigation is drying up the Aral Sea

Industry too has been strongly connected with agriculture, notably through the Tashkent tractor factory. Rather than opening up to the outside world, Uzbekistan has pursued a strategy of import-substitution—borrowing from abroad to invest in heavy industry such as steel and in sugar production. Foreign investment is limited. A Daewoo car plant is largely for assembly operation.

The most lucrative industrial activity is gold mining. Most of this takes place at the huge state-owned open-cast Muruntau mine. Uzbekistan is the world's ninth largest producer of gold which provides one-quarter of export revenue.

The government has declared in principle that it is moving more towards a market economy, but progress has been slow. Unemployment is officially quite low but this is probably because enterprises are reluctant to restructure and to fire excess labour, and the real rate is probably closer to 20%.

Politics in Uzbekistan is dominated by the president, Islam Karimov, who is effectively the dictator of a police state. Karimov was the communist leader before the break-up of the Soviet Union. He opposed independence, but when it came he successfully had himself elected president in a 1991 vote generally assumed to have been rigged.

Uzbekistan does have a legislative body, the Supreme Assembly, but in practice this does little more than carry out presidential decisions. The two leading opposition parties are banned and their leaders are exiled.

The only effective opposition centres around Islamic groups. Mosques that were closed during Stalin's era have been allowed to reopen, but only under strict control. The clergy are regularly harassed by the police and by the national security service. Two groups that have refused to cooperate are the non-violent Hizbut-Tahrir and an armed guerilla group, the Islamic Movement of Uzbekistan (IMU), though this is now much weaker.

For the presidential election in 2000 Karimov arranged a contest by setting up a token opponent, though he still received 92% of the vote. In 2002 a dubious referendum extended the presidential term to seven years.

The US has aligned itself with Uzbekistan in the war on terror and has an air base here, but has had little success in promoting democracy.

Despite promises of greater openness, in the parliamentary elections in 2005 no opposition groups were allowed to register as parties. In May 2005, troops opened fire on demonstrators occupying a square in the city of Andizhan, killing at least 200. Many others have been jailed.

In a constitutionally dubious presidential election in 2007, Karimov was re-elected for another seven-year term, with 90% of the vote.

Vanuatu

Vanuatu is still divided along colonial lines

Land area: 12,000 sq. km.
Population: 0.2 million—urban 26%
Capital city: Port Vila, 34,000
People: Melanesian 98%, other 2%
Language: English, French, Bislama
Religion: Christian
Government: Republic
Life expectancy: 70 years
GDP per capita: $PPP 3,666
Currency: Vatu
Major exports: Copra, beef, cocoa

Vanuatu comprises around 80 Pacific islands of either volcanic or coral origin, many of which are still covered by tropical rainforests. The country is vulnerable to natural disasters, including volcanic eruptions, earthquakes, and tidal waves.

Most of the population, who are called 'Ni Vanuatu', are of Melanesian extraction and have a pidgin-based lingua franca, Bislama, along with around 105 other languages.

Vanuatu's society remains divided by its strange colonial heritage. As the New Hebrides until 1980, the country was ruled jointly by the British and French—who had established their own schools and churches. As a result, the élite population tend to consist of either French-speaking Catholics or English-speaking Protestants, with the latter in the majority.

The population as a whole is also fairly sharply divided between the majority who live in scattered rural communities from subsistence or small-scale agriculture and the urban dwellers. The farmers grow food crops, such as taro and yams, as well as cash crops such as coconuts, cocoa, and squash for export. The cattle they raise also provide export income through sales of high-quality beef.

Tourism, mostly from Australia and New Zealand, is an important source of income. Vanuatu has also been establishing itself as an offshore tax haven, and financial service companies contribute around one-tenth of GDP. In addition there is a small manufacturing sector, mostly involved in processing agricultural and forestry products.

After independence, the political parties organized themselves along language lines—though subsequently they have splintered. The main anglophone party is the Vanua'aku Party (VP), while the main francophone party is the Union des partis modérés (UMP).

The VP won the first election but subsequently split several times to create first the Melanesian Progressive Party (MPP) and later the National United Party (NUP). In the 1980s and 1990s Vanuatu was governed by a series of unstable coalitions.

Following a snap election in July 2004, Serge Vohor of the UMP emerged as prime minister, but he was soon replaced by Ham Lini of the UMP. Following the 2008 election, in which no party gained an overall majority Natapei was back, heading a VP/NUP/UMP coalition. But he faced seven no-confidence motions and was eventually ousted in 2010. In the first half of 2011 there were four more changes of leadership, finishing with Sato Kilman.

Venezuela

Venezuela has been polarized by its populist president while poverty increases

Land area: *912,000 sq. km.*
Population: *28 million—urban 94%*
Capital city: *Caracas, 1.8 million*
People: *Spanish, Italian, Portuguese, Arab, German, African, indigenous people*
Language: *Spanish*
Religion: *Roman Catholic*
Government: *Federal Republic*
Life expectancy: *74 years*
GDP per capita: *$PPP 12,156*
Currency: *Bolívar*
Major exports: *Oil, bauxite, aluminium, steel*

The central one-third of Venezuela consists of open grasslands, the 'llanos'. To the south-east, and along the borders with Guyana and Brazil, are the Guiana Highlands, an isolated area of heavily forested plateaux and low mountains.

Despite rapid population growth—since 1958 from five million to 28 million—Venezuela remains sparsely populated. The majority of people live in the north and north-west around the northern tip of the Andes, and within this area most are now concentrated in the cities that are home to 94% of the population.

In the past Venezuela has been one of the richer South American countries, with an extensive system of social security. But poverty remains high, especially in the 'ranchos', the slums that ring Caracas. In 2007, 28% of the population lived in poverty and 8% in extreme poverty. Poverty

was heightened by the country's worst natural disaster in 1999, when mudslides killed 30,000 people.

Social services are weak. Literacy is high, but schools suffer from poor teaching and deteriorating buildings. Truancy is widespread and round one-fifth of the population have not had any formal schooling. The health system too is struggling and many hospitals have closed. However, the government has been building a parallel health programme called Barrio Adentro, based on community clinics staffed mainly by Cuban doctors.

Despite high levels of poverty, Venezuela continues to attract immigrants from Colombia to work as agricultural labourers or as illegal gold miners. There are thought to be around two million unauthorized immigrants—75% from Colombia.

Venezuela used to be an agricultural country—though only a small proportion of the land is suitable for arable farming. Nowadays, however, only 13% of the labour force work in agriculture, many of them raising cattle, and around 70% of food has to be imported. Landholding is concentrated but since 2003 the government has been redistributing both state and privately-owned land.

Agriculture was pushed into the background after 1914 by the discovery of oil near Maracaibo in the north-west. Venezuela rapidly became the world's first major oil exporter. Even today, its proven reserves are among the world's largest. Oil accounts for 90% of export income and half of central government revenue. In addition, Venezuela has significant quantities of natural gas—the world's eighth largest reserves—as well as substantial reserves of bauxite, coal, iron ore, and gold.

Oil revenues accelerated Venezuela's economic and industrial development, much of which was in the government's hands. The government nationalized the oil industry in 1975, creating PdVSA which is now the largest employer. Then it invested heavily in other sectors such as steel, cement, and petro-chemicals, but abundant oil money reduced the incentive to establish a broadly based economy.

Public expenditure and investment have usually also fluctuated along with international oil prices—rising

Government spending crashes with the oil price

when prices are high, then crashing again. The worst crash was in the early 1980s when the country was hit by falling oil prices at a time of rising interest rates and the Latin American debt crisis. But even when oil prices recovered the economy failed to recover and growth since then has been slow.

Venezuela has in the past usually been governed by one of two main political parties: the centre-left Democratic Action (AD) and the centre-right Independent Political Organization Committee (COPEI). But the political process has been deeply corrupt and presidential candidates now tend to stand under fresh colours.

Hugo Chavez, a former military officer who organized a failed coup in the early 1990s, won the 1998 presidential elections on a populist platform at the head of his Movimiento Quinta República (MVR). He then organized a referendum to establish a constituent assembly to write a new constitution.

Chavez is a return to the old-style Latin American 'caudillo', or strongman, politics, using public expenditure to win votes. He has boosted spending on programmes for the poor, who mostly support him, while the middle classes are against.

With a new constitution and a higher oil price that allowed him to increase public spending a little, in 2000 he was duly returned with 60% of the vote.

But since then the country has become increasingly polarized around efforts to unseat him. In 2002 there was a national strike followed by an unsuccessful military coup. In 2004, the opposition parties and other groups, united as Coordinadora Democrática, organized a petition for a recall referendum which Chavez won with 59% of the vote. He was re-elected in 2006 and in 2008, following a referendum, term limits on the presidency were abolished.

Chavez faces a hostile media and responds in kind. A draft law in 2009 threatens journalists with imprisonment for prejudicing the 'mental health' of the public.

Vietnam

Vietnam has partially liberalized its economy, but the Communist Party remains firmly in charge

Land area: 332,000 sq. km.
Population: 86 million—urban 29%
Capital city: Hanoi, 3.2 million
People: Vietnamese 86%, other 14%
Language: Vietnamese, Chinese, English, French, Khmer, tribal languages
Religion: Buddhist, Taoist, Christian, indigenous beliefs
Government: Communist
Life expectancy: 74 years
GDP per capita: $PPP 2,600
Currency: Dong
Major exports: Textiles, oil, rice, marine products, coffee, footwear

Around two-thirds of Vietnam consists of the Annamite mountain chain, which snakes down the length of the country from north to south. The two main lowland areas are the Red River delta in the north, which includes Hanoi, and the much larger Mekong delta in the south, which includes Ho Chi Minh City. These two are linked by a narrow coastal plain.

Most of the population are ethnic Vietnamese who live in one of the two deltas. Also to be found there are many of the Khmer minority, and more than one million people of Chinese origin, chiefly in the urban areas. In the mountain areas, the ethnic composition is more complex, with more than 50 different groups. Known collectively by the French term 'Montagnards', they have struggled to preserve their traditional lifestyles. Because of

population pressure in the deltas the government has been resettling people in the uplands. There are also many Vietnamese overseas: 750,000 people left after the end of the American war.

The Vietnamese have had good standards of health though public services are now more limited and people have to pay for most care. Education, however, has benefited from increased spending and many more children now go to school.

Poverty was reduced dramatically between 1990 and 2007, from 50% to 27%. However, there was also a rise in inequality, and a widening gap between the cities and the countryside. Around one-fifth of children are malnourished.

A major problem nowadays is corruption. Liberalization has also seen a growth in the trafficking of drugs from Laos and Myanmar.

More than half the population still work in agriculture and most people now work on their own account as households. Their principle crop, particularly in the deltas, is rice. Yields are high and Vietnam is now the world's second largest rice exporter. The main constraint is the lack of land. Much of the additional land brought under cultivation, particularly in the highlands, has been used for cash crops such as tea, cotton, coffee, rubber, and sugar. Coffee has been a notable success: yields are high and Vietnam is now the world's second largest exporter. Rubber too has done well. Vietnam also has ample fishing resources and exports seafood.

Rural communities rely on wood for 80% of their fuel but the countryside has suffered from deforestation. The government has had to close many logging companies and has banned log exports.

Since 1986, Vietnam has undertaken a process of economic renovation—'doi moi'—with greater emphasis on market forces and price incentives. This led to a rapid growth in industry, particularly in light

'Doi moi' boosts output

manufacturing areas such as garments, toys, electronics, and footwear. The footwear industry alone employs half a million people, 80% of them women.

There is also a growing tourist industry, with around four million visitors per year.

In recent years, the economy has been growing at around 8% annually and much of this expansion was driven by transnational companies, mostly from other Asian countries, which since 1988 have invested over $50 billion.

Privatization has accelerated. By 2008 around 3,800 state-owned enterprises had been partly privatized.

Vietnam has offshore oil deposits and is an oil exporter, but production is now falling. It does, however, have plentiful supplies of coal and much

hydroelectric potential.

Over the period 1959–75, Vietnam was embroiled in one of the 20th century's major wars—referred to in Vietnam as the American war. The communist forces of the north emerged victorious and since then the Communist Party has retained control.

Although the country has an elected National Assembly, which can hold government officials accountable, the country is managed by three people: the general secretary of the Communist Party, Nong Duc Manh; the prime minister, Nguyen Tan Dung; and the president, Nguyen Minh Triet.

There has however been some liberalization. The 'doi moi' policy represented some slackening of socialist economic ideology. And the constitution, as amended in 1992, establishes the government less as a socialist leader and more as a manager of the economy.

In terms of economic policy the government is divided between the conservatives and the economic liberals, but both wings would probably put the brakes on economic reform if they felt this would loosen the party's political grip.

There is little by way of opposition though there are occasional protests, as in 2001 when ethnic minorities demonstrated for freedom of religion and against the loss of their land to Vietnamese settlers. The National Assembly too is becoming somewhat more vocal, though even today 90% of representatives are party members.

Of greater official concern to the Communist Party is the internet, now used by one-quarter of the population. Critical journalists and bloggers, however, risk arrest.

Virgin Islands (US)

One of the most popular US tourist destinations

The US Virgin Islands in the Caribbean comprise three main islands: St Croix, St John, and St Thomas, and more than 50 other islets or cays. The most southerly island, St Croix, is the largest. But the most developed, with the capital, is St John.

The islands' varied scenery includes rugged mountains, dense subtropical forests, and mangrove swamps, along with white, sandy beaches and coral reefs. The climate is mild—though the islands are regularly hit by hurricanes.

The Virgin Islanders have a mixed African and European heritage. The Danes were in charge for a couple of centuries and they settled the country with African slaves to work on the sugar plantations. But with an eye to their strategic value, the US bought the islands for $25 million in 1917 and the dominant culture is now American. The islands have one of the highest standards of living in the Caribbean and have also attracted immigrants from other Caribbean islands, particularly Puerto Rico.

The main source of income is tourism, which employs around two-thirds of the workforce. More than two million visitors arrive each year, most of whom come from the USA, many on cruise ships, heading primarily for the beaches and the duty-free shopping.

Land area: 340 sq. km.
Population: 109,000—urban 95%
Capital city: Charlotte Amalie
People: Black 76%, white 13%, other 11%
Language: English, Spanish, creole
Religion: Baptist 42%, Roman Catholic 34%, Episcopalian 17%, other 7%
Government: Dependency of the USA
Life expectancy: 79 years
GDP per capita: $PPP 14,500
Currency: US dollar
Major exports: Refined petroleum products

On the agricultural front, sugar cane has given way to a wide range of fruit and vegetables grown for local consumption and to meet the demand from tourists.

The principal industrial activity is oil refining, but there are also factories making pharmaceuticals and electronic goods which take advantage of duty-free entry into the USA.

The islands are an unincorporated territory of the USA. They send one delegate to the US House of Representatives, though he or she lacks full voting rights. The islanders also elect a governor, as well as representatives who serve two-year terms in 15-seat legislature.

A Democrat, John de Jongh, was elected governor in 2007. The delegate to the US Congress, Donna Christian-Christenson, also a Democrat, was re-elected in 2006. Democrats also have a majority in the legislature—other seats being filled by a local party, the Independent Citizens' Movement and an independent member

There have at times been debates about the future of the islands, but they have not attracted much public interest and in 1993 the islanders voted in a referendum to retain the status quo.

Western Sahara

Independence now seems increasingly unlikely

Land area: 266,000 sq. km.
Population: 0.5 million—urban 81%
Capital city: L'ayoune
People: Saharawi, Moroccan
Language: Arabic
Religion: Muslim
Government: Government in exile
Life expectancy: 54 years
GDP per capita: $PPP   2,500
Currency: Moroccan dirham
Major exports: Phosphates

Western Sahara's territory on the north-west coast of Africa is largely flat and almost entirely desert. The western two-thirds of the country is occupied by Moroccan troops, while the eastern third, though with only 30,000 people, is controlled by the independence movement, Polisario.

The Saharawi are a mixture of Arab, Berber, and black African descent. Around 90,000 people, are now living in refugee camps in western Algeria. The camps are fairly well run; even so, there appear to be serious health and nutritional problems. In recent years, the country has also been flooded with around 200,000 Moroccan settlers and more than 100,000 Moroccan soldiers.

In the past, the Saharawi have survived largely as nomadic herders, with a little agriculture along the coast. But this is an area of huge potential. In 1963, large phosphate deposits were discovered at Boucraa in the north. Total reserves have been estimated at 10 billion tons. In addition, there are rich coastal fishing grounds which are currently exploited by Spanish trawlers.

Western Sahara was a Spanish colony. An armed liberation movement, Polisario, was formed in 1973 but when Spain relinquished the territory in 1976 it handed the northern two-thirds to Morocco and the southern third to Mauritania—though the International Court of Justice said that the Saharawi were entitled to independence. Years of brutal warfare ensued with Polisario working from bases in Algeria. Mauritania withdrew in 1976 and Morocco seized their portion too.

Polisario's guerrilla attacks did not shift Morocco, but their diplomatic offensive proved more successful. By the mid-1980s, more than 60 countries had recognized them as the legitimate government. The real breakthrough, however, came in 1990, when the UN managed to achieve a peace plan that would lead to a UN-supervised referendum on independence based on a 1974 census. A ceasefire was declared in 1991.

Since then there has been little progress on the referendum. Polisario, and its main supporter, Algeria, wants the electorate to be limited to anyone on the 1974 census, while Morocco, relying on its settlers, had wanted it to include anyone born there. The UN has arrived at a list of 86,000 voters but also has claims from another 130,000 people in Morocco.

Now, however, Morocco is refusing to hold a referendum at all and the process has ground to a halt. The UN now seems to consider independence as unrealistic.

Yemen

Despite recent discoveries of oil, Yemen remains poor—and could become a failed state

Land area: 528,000 sq. km.
Population: 22 million—urban 32%
Capital city: San'a, 1.7 million
People: Arab
Language: Arabic
Religion: Muslim
Government: Republic
Life expectancy: 63 years
GDP per capita: $PPP 2,335
Currency: Yemeni riyal
Major exports: Oil

Yemen has three main regions. It has narrow desert coastal plains along both the Red Sea and the Gulf of Aden. These plains rise steeply to highlands that reach 1,500 metres. Then to the north the highlands descend to a desert area that covers half the country and merges with the 'empty quarter' of Saudi Arabia.

Almost everyone is Arab but the mountainous terrain has dispersed the population so there can be differences in dialects of Arabic, heightened by historical divisions between north and south. In the 1830s, the territory had been carved up between the Ottoman empire in the north and the British empire in the south. After independence, the government in the north remained more conservative while the government in the south pursued Soviet-style socialism that diluted some aspects of Islam—particularly the restrictions it imposed on women. The two countries united in 1990 but the social and political differences remain.

As a whole, Yemen is one of the world's poorest countries. Literacy is low and around half of children are malnourished. Population growth is high at over 3% annually.

Around three-quarters of people rely on agriculture. Most of the rainfall is in the mountain areas, so farmers have had to build elaborate systems of terracing, growing basic subsistence crops such as sorghum and potatoes, as well as some cash crops, including fruit and high-quality 'mocha' coffee for export. But around one-third of irrigated land is used to grow 'qat'—a plant whose chewable leaves contain a mild amphetamine. In addition, the mountains support large numbers of sheep and goats.

With little land to spare and few other opportunities for employment, more than one million Yemenis have left to work in richer neighbouring countries. Their remittances are worth around $1.5 billion—6% of GDP.

Some respite from the country's poverty came from the discovery of oil in the 1980s. Oil now accounts for 85% of export earning and 70% of government revenue. Yemen also has natural gas reserves, and there are plans with the French company Total to expand liquefaction plants to enable more gas to be exported.

The government wants to capitalize on the country's strategic location by establishing a new Aden Free Zone. Although the first project collapsed due to corruption there have been attempts to revive the idea.

Another important source of foreign exchange has been tourism for the hardier travellers—around 80,000 of whom visit each year to explore the country's rich historical heritage. The industry was badly hit by a

Adventure holidays in Yemen
spate of kidnapping of tourists in 1999 by tribesmen wanting to draw attention to a

lack of amenities. Although tourists have rapidly been released unharmed, Yemen now has a reputation as an overly adventurous destination.

Yemen's unification in 1990 joined the Yemen Arab Republic in the north and the People's Democratic Republic of Yemen in the south. President Ali Abdullah Saleh from the north—which had three-quarters of the total population—became president of a united Yemen, and his counterpart in the south, Ali Salim al-Bidh, became vice-president.

Unfortunately, unification coincided with the first Gulf War, which forced more than 800,000 Yemeni migrant workers to return home—leading to unemployment and violent political unrest. Parliamentary elections in 1993 delivered a majority for Saleh's General People's Congress (GPC)—a mixture of military, tribal, and other groups. Al-Bidh's Yemeni Socialist Party came second—though with a majority in the south. An Islamic coalition, Islah ('gathering for

Reform'), came third. But people in the south complained of harassment and 'internal colonization' and in 1994, with support from Saudi Arabia, they tried to secede. This caused a civil war, which Saleh's forces won after six months and 10,000 deaths.

Yemen looked for international aid and entered into agreements with the IMF that required it to liberalize the economy, but low oil prices and falling remittance income forced the government to cut many services.

Even so, there was scarcely any opposition for the 1999 presidential election, largely because the Socialist Party boycotted it, and Islah nominated Saleh. Unsurprisingly, he won, with 96% of the vote. In 2001 a referendum extended the presidential term to seven years and he was duly re-elected in 2006. In the 2003 assembly elections GPC increased its already very large majority.

Meanwhile, much political power around the country remains dispersed among tribal leaders who have ample arms.

Saleh already had to deal with the five-year rebellion by the Houthi clan in the north, as well as al-Quaida groups in the east, and secessionists in the south.

But it was not until early 2011 that his government was directly threatened. Reflecting the Arab spring movements elsewhere, there ensued months of protests in favour of greater democracy. Saleh offered to step down but reneged on his promises. In June, however, he was wounded in an assassination attempt and left for Saudi Arabia for treatment. Yemen's future now seems very uncertain.

Zambia

The economy has revived with higher copper prices but corruption remains a problem

Land area: 753,000 sq. km.
Population: 12 million—urban 36%
Capital city: Lusaka, 1.4 million
People: Bemba, Tonga, and many others
Language: English, Bemba, Tonga, and about 70 other languages
Religion: Christian, Muslim, Hindu
Government: Republic
Life expectancy: 45 years
GDP per capita: $PPP 1,358
Currency: Kwacha
Major exports: Copper, cobalt, tobacco, lead

Most of Zambia consists of a plateau at around 1,200 metres above sea level. The highest point is in the Muchinga Mountains in the north-east. The territory is mostly open grasslands with occasional trees. The main river is the Zambezi, whose energies are tapped by the huge Kariba dam.

Zambians can belong to any of 70 or more ethnic groups, most of whom speak Bantu languages, though the official language is English. There are also small numbers of Europeans.

Two-thirds of Zambians live below the national poverty line and most of these would be classified as 'extremely poor'. In the past, one of the more successful areas was education: two-thirds of children are enrolled in primary schools and adult literacy is around 80%. In recent years, however, with budget cuts and low and often late pay

for teachers, and rising levels of poverty, educational standards have been falling. Zambia had also made progress in health, but here too services have deteriorated.

Zambia is starting to reverse the trends. In 2005 donors wrote off $3.9 billion of the country's foreign debt and most of the savings are going into health and education. In 2006 charges in rural clinics were scrapped. By 2007 it had increased the number of teachers by 50%.

This still leaves the spectre of HIV and AIDS: in 2007, 15% of the population were HIV-positive. As a result, life expectancy dropped by around 15 years and around half a million children have been orphaned. Still, it looks as though infection might have peaked and the government is providing free anti-retroviral therapy.

Most people rely to some extent on agriculture. Zambia could be a major food producer: only around one-tenth of the land is arable though only about half of this is currently being worked. One of the main constraints for local food availability is the pattern of landholding. Much of the best land is held by large commercial farmers, usually white. While they do also grow the main food crop, maize, low prices have encouraged large farms to devote more land to tobacco, cotton, sugar, and flowers for the export market.

Meanwhile the subsistence farmers work on lower-quality communal land growing maize, sorghum, millet, and other basic crops, and in some cases cotton. Efforts to transfer communal land to private ownership have been blocked by traditional leadership.

One of the most important sources of income and employment is copper, which provides three-quarters of export income. The 'copperbelt' is in the north-west of the country, bordering on the Democratic Republic of the Congo. This is the most industrialized part of Zambia, and also produces cobalt, coal, lead, and zinc. Mineral income dropped during the 1990s but lately prices have surged.

Privatization of the mines from 2000 was a tortuous business but one of the largest owners now is an Indian company, Sterlite, and companies from Switzerland, China, and elsewhere are also involved. Higher international prices have encouraged fresh investment and output has increased.

Tortuous privatization

After independence Zambia also invested heavily in manufacturing industry in areas such as food processing, chemicals, textiles, and tobacco. Initially most of this activity was in state hands, but many enterprises have been sold off.

Tourism is also growing rapidly, mainly to game parks and Victoria Falls. Over a million visitors arrive each year—many diverted from crisis-hit Zimbabwe.

Zambia's initial decades of independence after 1964 were dominated by its first president, Dr Kenneth Kaunda. In 1972, Kaunda outlawed political parties except for his own United National Independence Party (UNIP). Opposition had to come from the Zambian Congress of Trade Unions, led by Frederick Chiluba. In 1990, after political unrest Kaunda allowed other parties to organize.

Frederick Chiluba had by then created a new party, the Movement for Multiparty Democracy (MMD), and won the next presidential election. Chiluba and the MMD certainly liberalized the economy but had less appetite for democracy and in 1996, Chiluba introduced a new constitution that prevented Kaunda from standing for election. UNIP boycotted the elections that year, giving Chiluba and the MMD an easy victory.

Chiluba chose as his successor Patrick Levy Mwanawasa. He narrowly won the 2001 presidential election and embarked on an anti-corruption drive that included charging Chiluba with the theft of state funds. Chiluba was found guilty in 2007 in a case brought in London.

Mwanawasa scored a comfortable victory in the 2006 presidential election and the MMD won almost half the seats in the assembly. But he died in 2008 and was replaced in the ensuing presidential election by Rupiah Banda.

Banda seems less likely to tackle corruption. In 2009, after Chiluba had been surprisingly acquitted in another case brought in Zambia, Banda fired the head of the anti-corruption task force who had lodged an appeal.

Zimbabwe

A country faced with economic collapse and continuing political repression

Land area: 391,000 sq. km.
Population: 12 million—urban 38%
Capital city: Harare, 1.4 million
People: Shona 82%, Ndebele 14%, other African 2%, white 1%, other 1%
Language: English, Shona, Sindebele
Religion: Syncretic 50%, Christian 25%, indigenous 24%, other 1%
Government: Republic
Life expectancy: 43 years
GDP per capita: $PPP 200
Currency: Zimbabwe dollar
Major exports: Tobacco, food, gold

Zimbabwe's dominant topographical feature is the High Veld, a broad ridge around 1,500 metres above sea level that runs from south-west to north-east, and covers around one-quarter of the territory. The land drops on either side of this to the plateau of the Middle Veld, and then to the Low Veld, which reaches the frontiers at the Zambezi River in the north-west and the Limpopo in the south-east. Though it is in the tropics, Zimbabwe's elevation ensures a fairly mild climate.

The main population group are the Shona, who settled in the north and west of Zimbabwe more than 1,000 years ago and have primarily been farmers. The smaller Ndebele group arrived in the 19th century and lived initially as pastoralists in the south and east. With different forms of livelihood, the two coexisted fairly peaceably. More disruptive was the arrival from South Africa at the end of the 19th century of a small number of whites. Disappointed by not finding much gold to mine, instead they took the best land—the 'white highlands'— a seizure that rankles to this day.

When black majority rule was achieved in 1980, the government invested heavily in education and primary enrolment ratios soon shot up to 100%. There was similar progress in health with a steady rise in life expectancy. Since then, the country has been on a downward spiral. Economic collapse led to cuts in government expenditure, and fees were introduced for education and health services. Primary enrolment has fallen and clinics and hospitals are undermanned and badly equipped.

Health has also been badly affected by the AIDS epidemic. In 2007, 15% of adults were HIV-positive and life expectancy had fallen to 43 years.

By the standards of Sub-Saharan Africa, Zimbabwe has a fairly diverse economy. Sanctions imposed during the final years of white rule in the 1970s had forced the country to develop its own manufacturing industry, including steel production. Though output declined when price controls and protection were removed, even by 2004 manufacturing still made up 14% of employment. Zimbabwe also has an important mining industry, particularly for gold, which makes up 15% of exports.

Agriculture remains central to the economy, and accounts for 15% of employment. Before the current crisis, in years of good rainfall, Zimbabwe's farmers could even grow enough food to export, as well as producing cash crops such as sugar and cotton, and particularly tobacco, which is the major export earner.

Agriculture has always been overshadowed by questions of land distribution. After independence the government, nervous of provoking

Skewed patterns of landholding

'white flight', left many of the large landholdings intact. Whites made up less than 1% of the population but owned one-third of arable land—leaving black farmers to work the drier, lower-lying territory.

Both industry and agriculture are now in a desperate state mostly due to general mismanagement and corruption: over the period 1999–2006 the economy shrank by 5% per year. Since 2001, agricultural output has dropped by one-third. In 2008, before the economy switched to the US dollar, inflation was over 100,000%.

When Zimbabwe achieved black majority rule in 1980, Robert Mugabe became prime minister at the head of the Zimbabwe African National Union-Popular Front (ZANU-PF). In 1985, the party was re-elected and in 1987 the parliament replaced the office of prime minister with that of an executive president. Until 2008, Mugabe and ZANU had won all assembly and presidential elections.

Mugabe amended the constitution 14 more times—generally to augment his own power as Zimbabwe degenerated into a one-party state, and his corrupt and repressive regime became increasingly unpopular.

Early in 2000 a more effective opposition emerged as the Movement for Democratic Change (MDC) with the backing of trade unions, churches, and others. In 2000, Mugabe finally lost a constitutional referendum.

ZANU zealots responded by attacking and killing MDC members, and Mugabe, to gather popular support, condond the violent seizure of white commercial farms by 'veterans' of the civil war.

Despite this and widespread intimidation and violence ZANU almost lost the June 2000 parliamentary election to the MDC. In 2002 in a doubtful election Mugabe was returned as president. By 2003 the land seizures had led to famine. Mugabe stepped up the repression, frequently arresting, and sometimes torturing, opposition MPs. The 2005 parliamentary elections were duly rigged and ZANU maintained its grip.

By the 2008 elections, the economy was in ruins. The MDC led by Morgan Tsvangirai won 99 seats to ZANU-PF's 97. Tsvangirai also won the first round of the presidential election but refused to participate in the second round which he knew would be rigged. In 2009, after international pressure, a power-sharing agreement was reached, leaving Mugabe as president but with Tsvangirai as prime minister. In practice, Mugabe has been able to cling to much of his power.

Smaller countries

American Samoa

Pacific nation relying on the USA, tuna, and tourism

People: 65,000, of whom 92% are Samoans, bilingual in Samoan and English.
Life expectancy: 74 years
Government: Administered by the USA.
Capital: Pago Pago
Economy: GDP per capita: $PPP 8,000.
Main export: canned tuna

American Samoa is a US territory in the South Pacific. It consists of three main high volcanic islands and two atolls. Most people live on the main island of Tutuila, which also has a US naval base. More than 90% of the land is communally owned.

Many Samoans have emigrated to the USA. This has been offset by immigration from Asian countries such as Taiwan and South Korea, though immigration restrictions are becoming tighter.

The territory is economically highly dependent on the USA, which funds half the government budget. Around one-third of the workforce are employed by the government. Another one-third work in the two tuna canneries that produce the main export. Goods can enter the US duty-free providing the finished product has at least 30% locally originating material, so foreign investors can gain a foothold in the US market.

Although the US president is head of state, citizens elect their governor as well as members of the parliament, the Fono. In 2000 they chose a Democrat, Tauese Sunia, as governor. When he died in March 2003, he was replaced as governor by Togiola Tulafono, who was re-elected in 2008.

Andorra

Distinctive co-principality between France and Spain

People: 83,000, speaking Catalan, French, Spanish and Portuguese.
Life expectancy: 83 years
Government: Constitutional co-principality.
Capital: Andorra la Vella
Economy: GDP per capita: $PPP 41,000.
Main export: electricity

Andorra is a tiny, mountainous country lodged in the eastern Pyrenees between France and Spain. Less than one-third of the population are native Andorrans. Most are Catalan Spanish, with some French, and the majority live in and around the capital, Andorra la Vella.

The main industry is tourism, which makes up over 80% of GDP. Millions are drawn each year to the spectacular scenery and the ski resorts. Duty-free shopping is also an attraction for the Spanish and French, particularly since Andorra uses the euro as its currency. It is also distinctive in that it has neither income nor value-added tax, though there are municipal charges.

Andorra has a unique constitution. By tradition, the heads of state are two 'princes', the French president and the Bishop of Urgel in Spain. But this is largely a ceremonial arrangement. In 1993, Andorra adopted a new constitution, reducing the powers of the two princes and giving full sovereignty to the Andorrans, who elect a 28-member general council. The 2009 elections returned the Social Democrat Party, with Albert Pintat Santolària as general council president.

Anguilla

Politically lively British Caribbean dependency

People: 14,000, most of whom are black or mulatto. *Language:* English.
Life expectancy: 81 years
Government: Dependency of the UK.
Capital: The Valley
Economy: GDP per capita: $PPP 8,800.
Main export: lobsters

Anguilla is a flat coral and limestone island in the eastern Caribbean. Most of its people are black or mulatto though some East Asian workers and other have arrived to work in the construction industry.

The climate is dry and the soil is not very fertile so agriculture is limited. In the past, the main source of income has been fishing, but Anguillans nowadays rely more on remittances from migrants, luxury tourism, and financial services.

In 1969, Anguilla broke away from a self-governing federation with neighbouring St Kitts and Nevis—in a strange incident that required the dispatch of a squad of London policemen to restore order. After a decade of negotiations, in 1980, Anguilla became a British Dependent Territory with some autonomy exercised through an 11-member House of Assembly.

In 2000, the government was in crisis when the House of Assembly could not summon a quorum. A new election resulted in a victory for the conservative United Front coalition led by Osbourne Fleming, who won again in 2005. Fleming retired before the 2010 election, which the coalition, now led by Hubert Hughes, also won.

Antigua and Barbuda

Tourist paradise tinged with crime and corruption

People: 83,000, mostly black and English-speaking. *Life expectancy:* 75 years
Government: Constitutional monarchy.
Capital: Saint John's
Economy: GDP per capita: $PPP 19,400.
Main exports: petroleum products and manufactures

Antigua is the larger and more populated island in this two-island state in the eastern Caribbean. Its most significant geographical features are its white, sandy beaches. It has a warm, dry climate.

The standard of living is high and most of the income comes from tourism: over 250,000 visitors arrive each year either on cruise ships or to stay in the islands' exclusive resorts. Agriculture and manufacturing are mainly for local consumption and to supply the tourist industry.

The country has also become an offshore financial centre and has more than 70 banks which it combines with offshore internet gambling. Unfortunately, its lax regime has also made it an attractive money-laundering centre.

Antigua's head of state is the British monarch, but for more than 50 years the country was under the political sway of the Bird dynasty and the Antigua Labour Party (ALP), first Vere Bird, who died in 1999, and then his sons Lester and Vere Junior—a family often accused of corruption. Dynastic rule finally ended in March 2004, with a victory for the United Progressive Party, led by Baldwin Spencer, who won again in 2009.

Aruba

A rocky Caribbean dependency of the Netherlands

People: *103,000, ethnically diverse, speaking Dutch and Papiamento.*
Life expectancy: 75 years
Government: *Dependency of the Netherlands. Capital: Oranjestad*
Economy: *GDP per capita: $PPP 21,800. Main export: oil products*

Aruba is a flat rocky island in the Caribbean with few natural resources. Its people are a diverse ethnic mixture—with strains of Dutch, Amerindian, African, and Spanish. The majority are Roman Catholic. They have one of the highest standards of living in the Caribbean. Around one-quarter of the population are immigrants from the Netherlands, Venezuela, Colombia, and other Caribbean islands.

The chief industry is tourism which, with over 700,000 visitors per year, accounts for around 40% of GDP. But there is also a modest offshore financial services sector. In addition, the island refines some Venezuelan oil in a refinery owned by the US company El Paso.

Aruba was part of the Netherlands Antilles until 1986, when it was given 'status aparte', which involves having a Dutch governor but a locally elected assembly that deals with internal affairs. Earlier plans for political independence have now been shelved, though the Netherlands is keen to encourage financial independence. The 2009 elections gave the most seats to the Aruban People's Party led by Mike Eman, who became prime minister.

Bermuda

Atlantic island of wealth that could become independent

People: *64,000. Black 55%, white 34%, other 11%. Language: English.*
Life expectancy: 80 years
Government: *Overseas territory of the UK. Capital: Hamilton*
Economy: *GDP per capita: $PPP 70,000. Main export: pharmaceuticals*

Bermuda comprises one main coral-based island in the Atlantic. Bermudans enjoy a higher standard of living than their colonial masters in the UK. Two-thirds are black and the rest white; around one-quarter are foreign born. There have been violent racial tensions but conflict today is resolved through the political system.

Bermuda has little agriculture or manufacturing. One of the main sources of income, employing around one-third of the workforce, is the offshore financial sector, particularly insurance. Another is up-market tourism which directly or indirectly employs more than half the population. The island receives a quarter of a million visitors per year, though the industry has suffered several years of decline.

Bermuda is an internally self-governing British overseas territory. For 30 years the government was formed by the white-based United Bermuda Party. But the 1998, 2003 and 2007 elections were won by the black-based Progressive Labour Party.

Independence was rejected in a 1995 referendum, and subsequent unofficial polls suggest that most people, particularly the whites, are content with the status quo.

British Virgin Islands

A dependency of the UK that prefers the US dollar

People: 24,000. Most are black or mulatto with some whites and Asians.
Language: English. Life expectancy: 77 years
Government: Overseas territory of the UK. Capital: Road Town
Economy: GDP per capita: $PPP 39,000. Main exports: rum, fish

There are 36 British Virgin Islands in the eastern Caribbean, but only 16 are inhabited. The four main islands are Tortola, Anegada, Virgin Gorda, and Jost Van Dyke. The islands are hilly with many lagoons and coral reefs.

The majority of Virgin Islanders live on Tortola. Some work in agriculture and fishing, but their main source of income nowadays is tourism; some land-based, the others on yachts. Tourism accounts for around half of GDP.

Because of close economic links with the neighbouring US Virgin Islands, the currency is the US dollar.

Since the mid-1980s the British Virgin Islands has built up a large offshore financial services industry. By 2004 it had 544,000 companies registered locally. These include accountancy and legal firms, and banks and insurance companies, which are staffed by locals and expatriates.

As an overseas territory of the UK, the British Virgin Islands have an appointed governor and a 15-member legislative council. In the 2007 elections, the first under a new constitution, the National Virgin Islands Party took control, with Ralph Telford O'Neal as premier.

Cayman Islands

A congenial location for footloose international banks

People: 49,000. Mixed ancestry 40%, black 20%, white 20%, other 20%. Life expectancy: 80 years. Language: English
Government: Overseas territory of the UK. Capital: George Town
Economy: GDP per capita: $PPP 44,000. Main export: turtle products

These are three coral-based islands in the Caribbean, mostly flat and occasionally marshy, with white, sandy beaches and coral reefs. The islanders are on average far richer than their British colonial rulers—and pay no income tax. There are also many immigrant workers who make up around half the workforce.

Tourism is the largest employer. Around 300,000 people stay each year in this up-market destination, with many more visiting on cruise liners, the majority from the USA.

The Cayman Islands is also a major offshore financial centre and tax haven—with over 500 registered banks and trust companies. The country is a favourite location for company registration—with 68,000 enterprises claiming this as home.

This is a British overseas territory with an appointed governor, who since 2005 has been Stuart Jack. He works with an Executive Council which includes members of the Legislative Assembly.

In the 2009 election to the 15-seat Assembly the United Democratic Party, headed by McKeeva Bush, gained a majority. There is no pressure for independence, though there are plans for greater autonomy.

Christmas Island

Phosphates, crabs, and tourism in the Indian ocean

People: *1,400. Chinese 70%, European 20%, Malay 10%. Language: English.*
Life expectancy: data not available
Government: *Territory of Australia. Capital: The Settlement*
Economy: *GDP per capita not available. Main export: phosphates*

Christmas Island in the Indian Ocean was left largely uninhabited until the discovery of phosphates at the end of the 19th century. Most subsequent mining work was done by immigrant labour, primarily from the Cocos Islands, Malaya, and Singapore.

Since 1958 Christmas Island has been administered by Australia. In 1985, the island established its first local assembly and in 1994 the islanders voted overwhelmingly in a referendum against independence. Christmas Island now has a shire council as an electoral district within Australia's Northern Territory.

As the phosphate mines became exhausted, the Australian government transferred ownership to a private operator and a new lease was signed in 1998.

The government also started to invest in infrastructure in order to build up a tourist industry—including a casino, golf courses, and facilities for scuba diving. The island is well known for its 120 million red land crabs which each year between October and January migrate from the forest to the coast.

Since 2000 Australia has operated an offshore detention centre here for processing asylum seekers.

Cocos Islands

Australian dependency that aims for Australian wealth

People: *596. Cocos Malays (who are Muslim) and Europeans.*
Languages are Malay and English
Government: *Territory of Australia. Capital: West Island*
Economy: *GDP per capita not available. Main exports: copra, coconuts*

The Cocos Islands, officially called the Cocos (Keeling) Islands, comprise two main coral atolls in the eastern Indian Ocean. The only inhabited islands are Home Island and West Island.

Most of the Cocos islander community, who are Muslims, are to be found on Home Island. They are descendants of people who were brought from Malaya, East Africa, and other countries to work on the coconut plantations. Those on West Island are largely Australian government employees and their families.

Though there is some fishing and horticulture, their main source of income is still coconuts. Most of their work is organized by the Cocos Islands Cooperative Society, which supervises the production of copra and other activities.

The islands came under Australian control in 1955. In 1978, the Australian government purchased most of the land owned by the Clunies-Ross family and distributed it to the islanders. In 1984, in a UN-supervised referendum they voted to become a part of Australia. There is a non-resident Australian administrator as well as a seven-seat Shire Council.

Cook Islands

Pacific islands steadily losing people to New Zealand

People: *12,000, of whom 88% are full Polynesian. Life expectancy: 71 years*
Government: *Self-governing, in free association with New Zealand.*
Capital: Avarua
Economy: *GDP per capita: $PPP 9,000. Main exports: copra, fruit*

The Cook Islands consist of 15 islands and atolls scattered across two million square kilometres of the South Pacific Ocean.

Most Cook Islanders are of mixed Polynesian descent and speak Cook Islands Maori as well as English. More than half are on the island of Rarotonga. Many have also emigrated: there are 60,000 Cook Islanders in New Zealand and 20,000 in Australia.

Around one-fifth of the islanders depend on agriculture. There is also some industry, including the manufacture of clothing and footwear. However, most of the economy is based on tourism, with 65,000 visitors per year, and financial services, with over 1,200 registered companies, along with income from remittances.

The Cook Islands have full self-government in 'free association' with New Zealand, which deals with foreign affairs. The head of state is the British monarch. She and the New Zealand government are represented by a high commissioner, but most decisions are taken by the prime minister and the 24-member parliament. Since 2010, the prime minister has been Henry Puna of the Cook Islands Party.

Dominica

Caribbean island heavily dependent on banana exports

People: *72,000, predominantly black and English-speaking. Life expectancy: 75 years*
Government: *Republic. The president is elected by the legislature for a five-year term.*
Capital: Roseau
Economy: *GDP per capita: $PPP 7,893. Main export: bananas*

Dominica is a volcanic island in the Caribbean with rich fertile soil. Most people are of African or mixed descent and around 40% work on small farms, growing subsistence crops. They also grow bananas for export but the industry has suffered from hurricanes and from the loss of preferential access to the EU. Other crops such as mangoes, grapefruits, avocados, and oranges, are also grown but offer low returns.

Without a long-runway airport or extensive beaches, Dominica's tourist potential is limited, though there are facilities for cruise-ship visitors. A better prospect is eco-tourism, which can capitalize on the rich fauna and wildlife. Dominica is also an offshore financial sector, with 13,000 registered companies.

For most of the period following independence in 1978, the government was in the hands of Dame Eugenia Charles and the conservative Dominica Freedom Party (DFP). She retired in 1995 and the DFP lost an election to the centre-left United Workers' Party. Following the 2000 election, Roosevelt Skerrit of the Dominican Labour Party formed a coalition with the DFP but won absolute majorities in 2005 and 2010.

Faroe Islands

Fishing nation moving towards independence

> **People:** 48,000. Faroese, speaking Faroese and Danish.
> Life expectancy: 79 years
> **Government:** Dependency of Denmark.
> Capital: Tórshavn
> **Economy:** GDP per capita: $PPP 31,000.
> Main export: fish

The Faroes are a group of 17 rugged and largely treeless volcanic-based islands in the North Atlantic. Most Faroese live on the islands of Streymoy and Eysturoy.

The people are descendants of Viking settlers. The country's name means 'sheep islands' in old Norse, and raising sheep is still the main land-based activity, but the primary source of income is fishing, mainly for cod. Recently there have been some offshore oil discoveries. The islands also benefit from substantial subsidies from Denmark.

Since 1948, the Faroes have been a self-governing community within the kingdom of Denmark—though in order to protect their fishing interests they are not part of the EU. They have the oldest parliament in the world, the Løgting, founded more than 1,000 years ago, but they also have two seats in the Danish parliament.

Independence is on the agenda. The 2008 elections, as usual, produced a coalition government, and Kaj Leo Johannesen was chosen as prime minister.

Most representatives support independence, which the Danish parliament has said it will grant—if and when the Faroese ask for it.

Falkland Islands

British territory suddenly prosperous from seafood

> **People:** 3,140, of British origin, speaking English.
> **Government:** Dependency of the UK.
> Capital: Stanley. Territory also claimed by Argentina as the Islas Malvinas
> **Economy:** GDP per capita: $PPP 35,000.
> Main exports: wool, mutton

The Falkland Islands consist of around 200 islands in the South Atlantic. This British dependency is disputed with Argentina, which claims it as the Islas Malvinas. The main islands are East and West Falkland.

The population are of British extraction. They have a British-appointed governor but the islands are self-governing with their own executive and legislative councils. Many people have made their living from agriculture, predominantly sheep farming. But the economy is now booming from selling licences to foreign fishing vessels, bringing in $40 million per year. In addition there are 40,000 tourists annually.

The UK and Argentina had been trying to resolve their dispute when the Argentine military government invaded in 1982, provoking a war which ended when British troops reoccupied the islands. The UK maintains a military presence.

The islanders, now financially independent, want to maintain the status quo—a view the UK supports. Relations with Argentina improved following an agreement to cooperate, but subsequently deteriorated because of UK claims over seabed oil deposits in the South Atlantic.

Gibraltar

Spain and the UK still dispute the status of 'the rock'

> **People:** 28,000, mostly of British and southern European descent.
> **Life expectancy:** 80 years
> **Government:** British dependency, though claimed by Spain. **Capital:** Gibraltar
> **Economy:** GDP per capita: $PPP 38,000. Main exports: mostly re-exports

Gibraltar is a massive limestone and shale rock joined by a narrow isthmus to the coast of southern Spain. It holds a commanding position at the entrance to the Mediterranean.

Gibraltar's population reflects its long history as a British naval base: around two-thirds of the people are native Gibraltarians, of mixed British and southern European descent, while the rest are foreign workers.

The British military presence now accounts for less than one-tenth of Gibraltar's economy. More important are tourism, with more than five million visitors per year, and offshore financial services—the latest development being internet gambling.

Spain wants sovereignty over Gibraltar while the UK insists on respecting the wishes of the Gibraltarians who have refused to unite with Spain. Gibraltarians have self-government in all areas but defence, and are British citizens.

In July 2002 Spain and the UK announced that their preferred solution was joint sovereignty, but an unofficial referendum organized in 2002 by Gibraltar's chief minister, Peter Caruana of the Gibraltar Social Democrats, found that 99% of Gibraltarians opposed the idea.

Greenland

The world's largest island, but not very green

> **People:** 58,000. More than 85% are Greenlanders; most of the rest are Danish.
> **Life expectancy:** 70 years
> **Government:** Dependency of Denmark.
> **Capital:** Nuuk
> **Economy:** GDP per capita: $PPP 20,000.
> **Main export:** fish

Two-thirds of Greenland lies within the Arctic Circle and most of the territory is covered by an ice cap up to eight kilometres thick and holding 10% of the world's total fresh water—though some of this is now melting.

Greenlanders are mostly a mixture of the native Inuit and settlers from Scandinavia. Traditionally they have survived by hunting seals, but today, more people make a living through fishing, mainly for shrimp. There has also been some zinc mining but this has now ceased, and left Greenland even more dependent on subsidies from Denmark worth over $11,000 per person per year. Future prospects may depend on capitalizing on the melting ice for hydropower, and possibly becoming a location for global internet servers. There are also hopes of offshore oil.

Greenlanders are Danish citizens who elect two representatives to the Danish parliament. In 2009, however, Denmark ceded some of its powers, and from 2010 the local parliament took control of domestic affairs.

The 2009 elections resulted in a victory for the socialist Inuit Ataqataqiit party. Its leader Kuupik Kleist became prime minister at the head of a three-party coalition.

Grenada

A spice island of strategic importance to the USA

People: *90,000, mostly black, speaking English or a French patois.*
Life expectancy: 75 years
Government: *Constitutional monarchy.*
Capital: St George's
Economy: *GDP per capita: $PPP 7,344.*
Main exports: nutmeg, mace

Grenada is a volcanic island with a rugged mountainous interior and an attractive coastline that has many secluded bays and beaches. Most Grenadans are black, while a minority are of mixed race.

The country's fertile soil and abundant rainfall allow around one-quarter of the population to make a living as small farmers. Apart from growing food crops, they also specialize in nutmeg.

But the main source of foreign exchange nowadays is tourism. There are more than 130,000 annual stopover visitors, and a similar number arrive on cruise ships.

In recent years the largest bilateral donor to Grenada has been Taiwan, in exchange for political support. In 2004 Hurricane Ivan caused over $800 million-worth of damage.

Grenada, which achieved independence from the UK in 1974, has a history of political conflict. Turmoil in 1983 resulted in the killing of prime minister Maurice Bishop, and provoked a US invasion.

Democracy was restored, though it has been unstable. In 2008, the Democratic Congress, headed by Tillman Thomas, was re-elected with a comfortable majority.

Isle of Man

From buckets and spades to banking in the Irish Sea

People: *76,000, Manx and British.*
Life expectancy: 79 years
Government: *Dependency of the British Crown, rather than of the UK.*
Capital: Douglas
Economy: *GDP per capita: $PPP 35,000. Main exports: tweeds, kippers*

The Isle of Man lies halfway between England and Ireland in the Irish Sea. The highest point is Snaefell, at 620 metres, but most of the territory is low-lying and used for agriculture.

Its people, the Manx, are of Celtic descent. Health and welfare services are at least equal to those in the UK and incomes are now similar.

Until recently, the island relied heavily on budget tourism. But this has been eclipsed by financial services. Low tax rates have attracted more than 60 banks, and finance and related services now account for more than half the GDP. As a result, the Isle of Man's economy has been very healthy, with annual growth of about 5% combined with low inflation and unemployment under 1.5%.

The Isle of Man is not part of the UK; it is a crown dependency with a lieutenant-governor appointed by the British monarch. But it has considerable autonomy: its parliament, the Tynwald, makes its own laws and sets the tax rates. It also applies immigration controls: even British citizens need a work permit.

Politics is mostly consensual. The parliament elects a chief minister who in turn chooses a cabinet, but there are no political parties.

Kiribati

Small islands scattered over a vast area of ocean

People: 112,000, most of whom are Micronesians and Christians.
Life expectancy: 63 years
Government: Republic, with a president elected by the legislature. Capital: Tarawa
Economy: GDP per capita: $PPP 1,295. Main exports: copra, fish

Kiribati consists of 33 islands or atolls spread over three million square kilometres of the Pacific. Since most of the territory is less than a metre or two above sea level, it would be an early victim of global warming.

Its people, called the I-Kiribati, are mostly Micronesians and around one-third live on one atoll, Tarawa. Many I-Kiribati also work outside the islands, either elsewhere in the Pacific, particularly in Nauru, or as sailors on foreign vessels.

There is little fertile soil, so agriculture is limited, though farmers do grow taro as well as breadfruit, bananas, and especially coconuts to export copra.

Most people earn their living from fishing. Kiribati's control over such a huge area of the Pacific gives it considerable fishing potential, particularly for tuna, both from its own boats and by selling licences to foreign fleets. Even so, the country is heavily dependent on aid.

Kiribati, formerly the Gilbert Islands, gained independence from the UK in 1979. Politics is personality based. There is a 39-member parliament. Anote Tong was elected president in 2003, narrowly beating his brother Harry.

Liechtenstein

Commuters head for one of Europe's richest nations

People: 35,000. Alemani, though many people commute in from neighbouring countries. Life expectancy: 80 years
Government: Constitutional monarchy. Capital: Vaduz
Economy: GDP per capita: $PPP 85,382. Main export: machinery

Liechtenstein is a tiny country between Austria and Switzerland. The western one-third lies along the valley of the River Rhine. The eastern two-thirds is mountainous, formed by the foothills of the Alps.

The people are descendants of the Alemani tribe and many still speak the Alemani dialect, which is a variant of Swiss German.

Liechtenstein has some agriculture, primarily raising livestock. But more important are precision engineering factories, tourism, and the financial services industry. Attracted by Liechtenstein's low rates of corporate taxation, and secretive banks, more than 60,000 companies are registered there. These activities also employ people from neighbouring countries—every day more than one-third of the workforce commute in.

Liechtenstein is a principality, with Prince Hans Adam II as head of state. Through a 2003 referendum, the country gave even more power to the king making him close to an absolute monarch.

In 2009, elections to the single-chamber parliament gave a narrow overall majority to the centre-right 'Patriotic Union' Party, with Klaus Tschutscher as prime minister.

Marshall Islands

Reductions in US aid will have serious repercussions

People: 60,000. Micronesian, speaking English or Marshallese dialects.
Life expectancy: 71 years
Government: Republic in free association with the USA. Capital: Majuro
Economy: GDP per capita: $PPP 2,500
Main exports: fish, coconut oil

The Marshall Islands consist of 30 low-lying islands, along with hundreds of islets, scattered across two million square kilometres of the Pacific. All are threatened by global warming. They include Bikini atoll, which was used as a US nuclear test site, and the largest, Kwajelein, which is home to a US missile testing site.

One-third of the population live in the slums of Majuro. Many Marshallese still suffer from diseases linked to radiation exposure, though they are entitled to compensation from a trust fund.

On the outer islands, most people survive through subsistence farming, fishing, and remittances from people working on other islands. Those on the larger islands have benefited from considerable US aid.

Since 1986, this has been an independent country in 'free association' with the USA, which looks after defence and other matters. In 2003, the country signed a new compact with the USA, guaranteeing funding of around $800m per year.

There is a 33-member legislative assembly. Following a no-confidence motion in 2009, the president, chosen by the parliament was Jurelang Zedkaia, an island chief.

Monaco

A minuscule, and not very democratic, city-state

People: 32,000, of whom almost half are French; only 16% are Monegasque.
Life expectancy: 80 years
Government: Constitutional monarchy, ruled by the Grimaldi family. Capital: Monaco
Economy: GDP per capita: $PPP 30,000.
Main exports: services

Embedded in France's Mediterranean coast, and less than two square kilometres in surface area, Monaco is the smallest country in the UN, though landfill and reclamation from the sea have increased the area by 20% over the past fifty years. Monaco is highly dependent on France: and almost half its residents are French citizens attracted in part by the lack of income tax.

The primary industry is tourism—four-star and upwards. The casino in the Monte Carlo district is one attraction, and the largest employer; the harbour is always packed with luxury yachts. Many banks and insurance companies find this a comfortable location. Monaco also has some industries, including pharmaceuticals and electronics. More than 25,000 people commute in each day from Italy and France.

Since 2005, Monaco's hereditary head of state has been Prince Albert II —the latest in the line of succession of the Grimaldi family. There is a legislative national council, but executive power is in the hands of an unelected, and always French, minister of state whom the Prince chooses from a list presented to him by the French president.

Montserrat

Still suffering the effects of a disastrous volcanic eruption

People: *5,000. Black, with a few expatriates.*
Language: English.
Life expectancy: 73 years
Government: *Overseas territory of the UK. Capital: temporary capital is Brades*
Economy: *GDP per capita: $PPP 3,400. Heavily dependent on aid*

Montserrat is a volcanic, forest-covered island in the Eastern Caribbean. Its Soufrière Hills include a volcano that erupted in 1997 laying waste to the southern two-thirds of the island, including the capital Plymouth, the airport, and the port. The volcano is still active so almost everyone now lives in a safe zone in the north. The population includes indigenous black residents and some white expatriates.

Rehabilitation got off to a controversial start when the British government argued that Montserrat was asking for too much aid. Relations are still fractious, but aid has financed new houses, roads, and service and there will be a new capital at Little Bay. Agriculture now offers limited employment opportunities. Construction has offered an alternative. And the opening of a new airport, could eventually help revive the tourist industry.

Montserrat is a UK overseas territory, with its own executive council and a nine-member legislature. In 2007, Peter Waterworth was appointed governor. In the 2009 election, the Movement for Change and Prosperity won six of the seats and Reuben Meade became chief minister.

Nauru

An island that has been almost entirely excavated

People: *14,000. Nauruan 58%, other Pacific Islanders 26%, other 16%. Most speak Nauruan or English. Life expectancy: 64 years*
Government: *Republic, with a president elected by the parliament*
Economy: *GDP per capita: $PPP 5,000. Main export: phosphates*

Nauru is a 21-square-kilometre coral island, the centre of which had a plateau of phosphate rock allegedly formed by guano (bird droppings), leaving just a narrow coastal strip of fertile land.

Nauruans grew wealthy with the income from phosphate exports. These have in the past funded generous social services but also encouraged an unhealthy, sedentary lifestyle. Most of the mining was done by immigrants from China and from neighbouring islands.

Now that the deposits are virtually exhausted, Nauru has found itself in desperate straits. By 2002 the country was declared virtually bankrupt and was having to be propped up by Australian aid, partly in exchange for which Australia has been allowed to establish a detention centre on the island to process asylum seekers. Attempts to develop its offshore banking centre were thwarted in 2003 by a money-laundering scandal.

Independent since 1968, Nauru has a 20-member parliament that elects one member as president. The political culture is very unstable. Since 2007, following the 17th change of government in 14 years, the president has been Marcus Stephen.

Niue

The largest uplifted coral island, linked to New Zealand

People: *1,300. Polynesian with some Europeans and others.*
Government: *Self-governing, in free association with New Zealand.*
Capital: Alofi
Economy: *GDP per capita: $PPP 5,800 . Main exports: root crops, coconuts*

Though geographically part of the Cook Islands in the South Pacific, Niue is politically separate. This is the world's largest uplifted coral island.

Most people are Polynesian and speak their own Niue language. They live predominantly from agriculture, working on family plantations, growing subsistence crops such as yams and taro, as well as cash crops that include coconuts and passion fruit. The island also has a sawmill as well as some fruit processing factories.

Lacking local opportunities many people have migrated to New Zealand, where around 22,000 Niueans now live. Niue's population has halved over the past 30 years.

The main exports are root crops, coconuts, honey, and handicrafts. The island is heavily dependent on remittances and on aid from New Zealand, which at times has made up around two-thirds of GDP.

Since 1974, Niue has been a self-governing territory in free association with New Zealand, which appoints a high commissioner. The island has its own nine-member legislative assembly. Following the 2008 elections, Toke Talagi became prime minister.

Norfolk Island

Australian island settled by convicts and mutineers

People: *2,100. European and Polynesian. Around 30% were born on the Australian mainland and 23% in New Zealand*
Government: *External territory of Australia.*
Capital: Kingston
Economy: *Main exports: agricultural products, postage stamps*

Norfolk Island is a volcanic island which lies north-west of New Zealand in the South Pacific. Though most of the land has now been cleared, the island still has many of its distinctive pine trees, and most of the coastline consists of cliffs. Originally a penal colony, Norfolk Island was later settled by descendants of mutineers from *The Bounty* who migrated from Pitcairn Island. Around 80% are citizens of Australia; most of the rest are citizens of New Zealand.

There is some agriculture, but the main source of income nowadays is tourism from Australia and New Zealand, with around 40,000 visitors per year.

Relations with Australia have at times been strained and there have been several efforts to clarify the island's status. It is not a dependent territory but an 'integral part of the Commonwealth of Australia'.

The Australian governor-general appoints an administrator, but the island has its own nine-member legislative assembly and raises its own taxes. In 1991, the islanders voted against any change in their constitutional status. Only since 1992 have they been able to vote in Australian federal elections.

Northern Mariana Islands

Offers duty-free entry to the USA for garment makers

> **People:** 88,000. Chamorro, Carolinian, and Asian. Life expectancy: 77 years
> **Government:** Self-governing commonwealth in political union with the USA. Capital: Saipan
> **Economy:** GDP per capita: $PPP 12,500. Main export: garments

The country consists of 22 volcanic or coral islands in the Pacific. The largest indigenous group are the Chamorro. They and other islanders are US citizens. But more than half the population are immigrant workers, particularly from the Philippines and more recently from China.

Few people work in agriculture. Many more now work in the tourist industry, which caters to around 300,000 visitors per year. But many of these are from Japan and Korea and the numbers dropped sharply in 2005 when Japan Airlines pulled out.

The main employer has been a foreign-owned garments industry. But this is being phased out now that other countries are being granted equivalent access to US markets and the country is under pressure to comply with US labour standards.

As a commonwealth of the USA, the islanders get substantial aid. They will probably need more. The closure of garment factories and a decline in tourism has cut revenues.

The islanders elect their own governor, who since 2005 has been the Republican, Benigno R. Fitial. There is also a legislature, though this often just rubber-stamps the governor's proposals.

Palau

Probably the world's most intensively governed country

> **People:** 20,000. Palauans speaking Palauan, English, or other languages.
> Life expectancy: 70 years
> **Government:** Republic in free association with the US. Capital: Koror
> **Economy:** GDP per capita: $PPP 8,100. Main exports: shellfish, tuna

Palau consists of more than 300 volcanic or coral islands in the Pacific, most of which are enclosed by a barrier reef. Palauans are Micronesians. More than half live on the island of Koror.

Many Palauans are still involved in subsistence agriculture and fishing, but the main sources of cash income are government employment and tourism. Around one-third of the workforce are employed by the government There is also income from fishing licences sold to foreign fleets. Around 90,000 tourists visit each year.

Palau's emergence from the status of UN Trust Territory was a long and tortuous process—during which one president was assassinated. In 1994 Palau joined the UN. It has a compact of free association with the USA which until 2009 provided around $500 million annually, some of which has gone into a trust fund.

The national president, elected in 2008, is Johnson Toribion, but there is also a two-tier legislature, and a 16-member Council of Chiefs. In addition there is an elected government and legislature in each of the country's 16 component states, one of which has a population of 100.

St Helena

A remote but strategic island in the South Atlantic

People: 7,600, of mixed European, African, and Asian descent. *Language:* English
Government: Dependency of the UK. *Capital:* Jamestown. Grouped with Ascension Island and Tristan da Cunha
Economy: GDP per capita: $PPP 2,500. Main export: tuna

St Helena is a mountainous island of volcanic origin, almost 2,000 kilometres from the coast of Angola, with a mild subtropical climate. Its remote but strategic location has in the past made this an important port of call for sailors and notably for aircrews during the Second World War. The most famous visitor was Napoleon Bonaparte who was exiled here in 1815.

The island has few resources. Many people make a living from fishing, primarily for tuna which is frozen for export. Agriculture is limited, though some farmers can grow maize, potatoes, and other crops and raise some livestock. With few opportunities, many people emigrate. The islanders have requested automatic British citizenship, but this has been refused. One continuing issue is that of an airport, which the UK is reluctant to fund.

St Helena, as a British dependency, has an appointed governor, but it also has a legislative council which includes 12 elected members. There have been pressures for greater autonomy. St Helena itself has other islands as dependencies, including Ascension Island, which has a US airfield, and Tristan da Cunha.

St Kitts and Nevis

Two Caribbean islands that have stayed together—just

People: 39,000, mostly black and English-speaking. *Life expectancy:* 72 years
Government: Federation under a constitutional monarchy. *Capital:* Basseterre
Economy: GDP per capita: $PPP 14,481
Main exports: machinery, food, electronics, tobacco

St Kitts (short for St Christopher) and Nevis is a federation of two volcanic islands in the eastern Caribbean. Three-quarters of the population live on St Kitts. Standards of human development are high.

On St Kitts, many people used to work on sugar plantations but these have now been closed. As an alternative the government has encouraged small-scale manufacturing and St Kitts has now two electronics companies.

A more promising source of employment is tourism. The islands attract around 120,000 people each year, mostly fairly up-market, who occasionally stray from the beaches to visit historic sites. Nevis in particular has also become a centre for financial and business services with more than 15,000 registered companies. However the islands have also attracted drug traffickers and organized crime.

The 1997, 2000, and 2004 elections to the 11-person National Assembly were won by the St Kitts-Nevis Labour Party with Dr Denzil Douglas as prime minister. Nevis is richer, and itching to separate from St Kitts. A referendum in 1998 failed to get the requisite two-thirds majority.

St Pierre and Miquelon

French fishing islands in the North Atlantic

> **People:** 7,000. French, of Basque and Breton origin. Life expectancy: 79 years
> **Government:** Territorial collectivity of France, but self-government through a local assembly. Capital: Saint-Pierre
> **Economy:** GDP per capita: $PPP 7,000. Main exports: fish and fish products

St Pierre and Miquelon comprises two main islands off the southern coast of Newfoundland, Canada. The islands are fairly bleak and their chief importance is as a French presence in the North Atlantic, offering an opportunity to exploit the rich fishing grounds.

Most of the people are descendants of fishing communities who settled here from France. Their main activity is still fishing for cod. At times this has led to disputes with Canada, though these have now been settled with agreements on quotas. Even so, with the fishing industry in long-term decline, there are fewer opportunities and many of the islanders have been emigrating to Canada. The territory is heavily subsidized by France.

Living in a territorial collectivity of France, the islanders are entitled to send a deputy to the French National Assembly. Locally, they have their own prefect, who since 2008 has been Jean-Pierre Brecot.

They also elect their own 19-member territorial council. The political parties include some local parties and others that correspond to those in metropolitan France. After the 2006 elections the most seats went to Archipel Démain.

San Marino

One of the world's smallest and oldest republics

> **People:** 30,000. Sammarinese and Italian, speaking Italian. Life expectancy: 82 years
> **Government:** Republic. The joint heads of state are the elected captains-regent. Capital: San Marino
> **Economy:** GDP per capita: $PPP 41,900. Main exports: ceramics, textiles

Enclosed within Italy, San Marino lies near the Adriatic coast on the slopes of Mount Titano. Although an independent republic, San Marino is necessarily linked in many ways to Italy, and a high proportion of residents are Italian.

Agriculture is still an important source of income, based on cereals and livestock. But the country also has a range of industries. The main one used to be stone quarrying. Legend has it that the state of San Marino was founded by a 4th-century stone-cutter.

Now that the quarries are mostly exhausted, manufacturing industry is devoted to a range of other products including ceramics, textiles, and electronics. San Marino has growing banking and tourist industries, with three million visitors per year, and also does a good trade in postage stamps which provide around one-tenth of government revenue.

San Marino has an elected 60-member 'great and general council'. The 2008 elections resulted in a victory for a right-wing coalition, the Pact for San Marino. The council chooses two of its members to serve for six months as 'captains-regent' who are joint heads of state.

Seychelles

Tourist island where the largest employer is a fish cannery

People: 87,000. Seychellois, a blend of Asian, African, and European.
Life expectancy: 73 years.
Languages: English, French, and Creole
Government: Republic. Capital: Victoria
Economy: GDP per capita: $PPP 16.394
Main export: canned tuna

The Seychelles consists of over 100 islands in the Indian ocean, of which two-thirds are uninhabited. Most Seychellois, who for Africa have a high standard of living, live on the island of Mahé.

The mainstay of the economy is tourism, which employs around one-third of the labour force. Some 160,000 visitors arrived in 2008, mostly from Europe. The Seychelles has little agricultural land but has extensive fishing grounds and the tuna-fish cannery is the largest employer and biggest export earner—though 40% of the workforce has to be imported from China and elsewhere.

Following a coup in 1977 politics was dominated by Albert René, who with his People's Progressive Front (SPPF) ran the country as a one-party socialist state until 1993 when he reverted to multiparty democracy and switched to free-market policies.

René won a series of presidential elections and the SPPF dominated the National Assembly. René stood down in 2004, making way for James Michel who won the 2006 and 2011 elections. The main opposition party is the Seychelles National Party, led by Wavel Ramkalawan.

Tokelau

Three scattered atolls in the South Pacific

People: 1,400. Most are Polynesian, speaking Tokelauan and English.
Government: Overseas dependency of New Zealand. No capital; there are centres for administration on each island
Economy: GDP per capita: $PPP 1,000. Main exports: stamps, handicrafts

Tokelau consists of three widely separated atolls in the South Pacific—Atafu, Nukunonu, and Fakaofo. Each has a series of islets around a lagoon, which would be at risk from any rise in the sea level.

The people are of Polynesian origin. Most are Christian. They have some subsistence farming though the soil is thin and infertile. The main crops are coconuts, copra, breadfruit, papayas, and bananas. There is also some livestock-raising and fishing, and a few people work in small enterprises producing handicrafts for export.

Tokelau also gains income from issuing licences for fishing in its waters and selling postage stamps, as well as from remittances from overseas workers. But the budget has to be balanced with substantial aid from New Zealand and elsewhere.

Tokelau is a dependent territory of New Zealand and has an appointed administrator. In addition there is an assembly, the General Fono, for which elections are held every three years. However, there are now moves to change Tokelau's status to make it a self-governing territory in free association with New Zealand and a new constitution is being drafted.

Turks and Caicos

Caribbean islands mired in corruption

> **People:** 22,000. Black and European, speaking English, with unauthorized Haitian immigrants. Life expectancy: 75 years
> **Government:** Overseas territory of the UK. Capital: Grand Turk
> **Economy:** GDP per capita: $PPP 11,500. Main export: seafood

The Turks and Caicos islands are two groups of about 40 islands south of the Bahamas. Flat and sandy, the islands are surrounded by coral reefs. Only eight are inhabited.

Two-thirds of the residents are black, the 'belongers'. The rest are expatriates or unauthorized Haitian immigrants. Around 20% work in agriculture, growing vegetables and citrus fruits. Others fish for lobster and other seafood that are the main source of export earnings. Tourism is also an important source of income, mostly on the island of Providenciales which has a Club Med resort. Finally there is a financial services industry that covers more than 3,000 registered companies.

As an overseas territory of the UK, the islands have their own governor, as well as a local executive council and a 13-member legislative council. There are two political parties, the People's Democratic Movement (PDM) and the Progressive National Party (PNP). The PNP won the elections in 2003 and Michael Misick became premier. In 2009, however, as a result of widespread and systemic corruption, the governor suspended the constitution and restored direct rule from the UK.

Tuvalu

Owner of a valuable internet domain name—tv

> **People:** 12,000, mostly Polynesian. Life expectancy: 69 years
> **Government:** Constitutional monarchy. Capital: Funafuti
> **Economy:** GDP per capita: $PPP 1,600. Main export: copra. Relies heavily on remittances from emigrant workers

Though its name means 'eight together', Tuvalu actually comprises nine scattered Pacific islands and atolls. Many people make their living from fishing or from subsistence agriculture, but around half those in employment work for the government.

The government gets 40% of its revenue from fishing licences from foreign vessels. But it has also made considerable sums from sales of postage stamps and coins, and more recently from selling access to its international telephone codes, initially for phone sex. Its internet domain name, 'tv', has been leased for $2 million per year.

Tuvalu was formerly linked with Kiribati as the 'Ellice' in the Gilbert and Ellice Islands colony. Its head of state is the British monarch who is represented by a local governor-general.

The country has a 12-member directly-elected parliament. Since 2010 the prime minister has been Willy Telavi.

In 2000, Tuvalu joined the United Nations. It hopes to use this status to promote international action on climate change since global warming through a rise in sea levels threatens to submerge parts of the country.

Vatican

A sovereign state in Rome, the headquarters of the Holy See

People: 783. Italian, Swiss, and many others, speaking Italian and Latin.
Government: Monarchical-sacerdotal.
Head of state: Pope Benedict XVI
Economy: GDP per capita: data not available. Currency: Euro. Main exports: stamps, religious mementos

The Vatican is a 44-hectare enclave on the banks of the River Tiber in Rome. This is a territorial state which is the headquarters of the Holy See, the central government of the Roman Catholic Church.

The head of state is the Pope, but most of the temporal duties, including international relations, are the responsibility of the secretary of state, currently Cardinal Tarcisio Bertone. The Vatican City is administered by a governor.

Those who live and work within the Vatican become citizens. Some 3,000 other workers commute in daily from Rome. Almost everything needed within the Vatican has to be imported. But the Vatican has its own diplomatic service, currency, radio station, and daily newspaper. The Swiss Guard are responsible for internal security.

The Roman Catholic Church derives its income from members' contributions, an extensive investment portfolio, and properties all over the world. These are administered by the Institute for Religious Works. In 2007, the Vatican's expenditure was $373 million, of which around $70 million was contributed by Catholic dioceses around the world.

Wallis and Futuna

Pacific islands heavily dependent on French aid

People: 15,000. Polynesian, speaking Wallisian and French.
Government: Overseas collectivity of France. Capital: Mata-Utu
Economy: GDP per capita: $PPP 3,800. Main exports: taro and other crops. Relies heavily on French aid

In addition to the islands of Wallis and Futuna the country includes the island of Alofi. These are formerly volcanic islands in the South Pacific. The people are mostly Polynesian, with a few French. The Roman Catholic Church has a strong influence.

Most Wallisians make their living from subsistence agriculture, growing taro, yams, bananas, and other crops. The main cash crop is coconuts, grown for copra. They also fish for tuna. A few other people work for the government. The country relies for survival on French aid.

With limited work prospects, many people have emigrated. Some 20,000 are in New Caledonia—more than live at home—and their remittances are a major source of income.

The islanders send a deputy and a senator to the French National Assembly, but they also have their own territorial assembly. In addition to the equivalents of the parties in metropolitan France, there is also a local party, the Union locale populaire, though there is no pressure for independence. Wallis and Futuna also has a territorial council made up of three traditional kings and three other appointed members.

International organizations

While countries take care to protect their independence, they inevitably have to cooperate with others, either on a bilateral basis or, as is happening increasingly, through a wide range of inter-governmental—usually referred to as international—organizations. Within these, they follow a mutually agreed set of rules and in some cases cede a degree of sovereignty. The following section offers a brief review of the most significant international organizations, global and regional, to which countries are likely to belong.

United Nations

The international organization with the largest, almost universal, membership is the United Nations (UN). The UN is the successor to the League of Nations which was set up in 1920 following the First World War. The League was supposed to prevent further disputes, but it failed to do so, largely because it was unable to secure universal membership, notably that of the United States, and it proved powerless to prevent a Second World War. It was only after that conflict that the victors in 1945 established a stronger body, the United Nations, which from the outset included all the major powers.

The initial membership of the UN—51 countries—comprised largely those that were independent at that time. So most members came from Europe, or from Latin America where the majority of states had won their independence in the early 19th century. Asia at that point had just four members: China, Iran, and the Philippines, plus British India whose component parts had yet to gain independence.

Over the decades, membership would more than treble. Expansion of the UN often happened in rapid bursts, as from 1960 after decolonization when 16 new members joined from Africa, and from the early 1990s after the disintegration of the Soviet Union, when 15 new members joined from Eastern Europe. The newest arrival, South Sudan in 2011, brought the membership to 193.

In its largest body, the General Assembly, UN decisions are based on the principle of one country, one vote. But five countries hold a privileged position in the UN's powerful executive body, the Security Council, which was designed to respond quickly and decisively to international events and, if necessary, take military action against offending states—something the League had been unable to do. As with the League, however, the intention of the founders of the UN was also to implicitly freeze a post-war status quo, so the Security Council was to be dominated by the victors. It was created with five permanent members—China, France, the Soviet Union, the United Kingdom, and the United States—supplemented by a number of temporary members, now ten, which were to be elected for two years each. Crucially, each of the five permanent members of the Security Council has a right of veto over any of its decisions.

With the passage of time, this has led to the anachronistic position that France and the UK, which are now only middle-ranking powers, retain disproportionate influence. There have been some changes: in 1971 nationalist China, based in Taiwan, was finally displaced by the Republic of China, and in 1991 the Soviet Union ceded its Security Council seat to Russia. However, more fundamental reforms which would recognize the rise of global powers such as Brazil, Germany, India, or Japan have foundered on the unwillingness of France or the UK to accept any demotion.

Any independent state agreeing to the UN charter should be able to join, but it will need both a Security Council recommendation and a two-thirds majority backing in the General Assembly. During the years of the Cold War, a number of countries had their applications vetoed either by the United States or the Soviet Union. Nowadays, the most persistently contentious issue is that of Taiwan, whose membership would be vetoed by China which regards Taiwan as one of its own provinces.

The primary function of the UN has been political, aiming to settle disputes among its members peacefully through diplomacy; as the UN charter puts it: 'to save succeeding generations from the scourge of war'. The UN can, however, also itself authorize the use of armed force against aggressors, though in practice its military activities have generally been limited to peacekeeping—sending in troops to enforce an existing peace settlement. This involves member states volunteering equipment, along with troops to wear the UN's 'blue helmets'. In 2010 there were 17 such missions in operation, with more than 90,000 troops. One of the longest-running missions, since 1964, has been in Cyprus. More recent peacekeeping missions include those in Sudan, Timor-Leste, the Central African Republic/Chad, and Sudan. Member countries that provide troops are reimbursed by the UN, and for some countries such as Bangladesh which has more than 10,000 troops on duty, this is a useful source of income.

In addition to its political functions, the UN is also very active in the economic and social spheres, through a large number of agencies. Many of these predate the UN. The International Telegraph Union, for example, had been formed in 1865. And the International Labour Organization (ILO), founded in 1919, had also been affiliated to the League of Nations. But the freshly created UN would also add a number of new organizations such as the World Health Organization (WHO), the United Nations Educational, Scientific and Cultural Organization (UNESCO), and what is now called the United Nations Children's Fund (UNICEF), as well as other organizations, including the United Nations Development Programme (UNDP), which would emerge later.

These agencies report to the UN's Economic and Social Council (ECOSOC)—a body that comprises 54 members of the General Assembly. Some of the organizations, like the United Nations High Commissioner for Refugees (UNHCR), or the United Nations Conference on Trade and Development (UNCTAD), are classified as 'programmes' or 'funds'. Others, such as the Food and Agriculture Organization (FAO), or WHO, are classified as 'specialized agencies'.

Their management structures can be very different. For example, the governing body of the ILO, which is a specialized agency, has representatives not just of governments, but also of trade unions and employers' organizations. UNICEF, which is a fund, is also distinctive in having direct contact with the donor public, through 'national committees' of volunteers. The headquarters of all these bodies, are scattered around the world, though mostly in the developed countries, with quite a few in New York or Geneva. The United Nations Environment Programme, however, has its headquarters in Nairobi, Kenya.

In addition, individual agencies will have offices in the developing countries in which they work. Given the large number of agencies there is always the potential for overlap and duplication which has sometimes led to rivalries and 'turf wars' as each agency sets its own priorities—complicating life for host governments. Both UNICEF and ILO, for example, are concerned with child labour. The UN has therefore been trying to streamline activities in each country so as to present 'One UN', with the senior representative of UNDP acting as the Resident Coordinator for all UN activities.

Bretton Woods organizations

Also affiliated to the United Nations, though not accountable to it, are two other post-war organizations, created in 1944 to rebuild the global economic system—the International Monetary Fund (IMF) and the World Bank. The IMF was designed to promote international monetary cooperation. When a country's foreign exchange reserves drop to dangerously low levels it can borrow from the IMF providing it is prepared to subject itself to the IMF 'conditionalities'— the remedies it prescribes for restoring the country to good financial health. The IMF itself borrows the necessary funds from its members who receive interest—which is repaid by the borrowers, though the IMF also takes a cut to pay for its services.

While IMF conditionalities should vary from one situation to another, in practice they have tended to follow a fairly rigid free-market template— stabilize, liberalize, privatize, and deregulate. For example, borrowers normally have to remove restrictions on imports or foreign exchange, or on the flow of capital. They also usually have to cut public expenditure.

These measures often prove politically unpopular; indeed in some countries they have provoked 'IMF riots'. At the time of the Asian Financial crisis in 1997, for example, the IMF was criticized for issuing rigid, and wrong, prescriptions that only pushed countries into a further downward spiral. As a result, many countries resolved never to fall into the hands of the IMF again, and took the precaution of building up vast foreign exchange reserves. The IMF's economists subsequently recanted, conceding, for example, that abolishing capital controls during a crisis can be damaging. But by then its disillusioned client base had shrunk significantly, along with its revenue. For the IMF, the 2008–09 global economic crisis came as something of a relief, since it forced some countries to ask for funds, though this time they faced less rigid conditions.

While the IMF's job was to tide countries over temporary financial difficulties, the World Bank was established to lend funds over a longer period to finance specific development projects, such as building roads or schools. For this purpose, it offers two main channels. One, the International Bank for Reconstruction and Development (IBRD), lends funds at close to market rates. Another, the International Development Association (IDA), offers loans at concessionary rates to the poorest countries. To raise funds for the IBRD the World Bank sells its own bonds on the international financial markets, servicing the bonds with the interest it receives from borrowers. For IDA funds, however, it has to rely on periodic replenishments from donor countries.

Like the IMF, the World Bank has also been quite prescriptive, requiring countries to follow its free-market line. Indeed the two Washington-based organizations, along with their host government the United States, have been described as representing the neo-liberal 'Washington consensus'. The World Bank, like the IMF, has been losing some of its rationale, since many developing countries, such as India or China, can themselves now borrow just as cheaply on the international capital markets—which do not package their loans with advice on potential environmental destruction or the abuse of human rights.

An even more pressing concern for many developing countries is the fact that the Bretton Woods organizations are dominated by the developed countries. Unlike the UN, the voting systems in the governing bodies of these organizations are based on the size of members' contributions. In the IMF, for example, the United States has 17% of the votes, followed by Japan and Germany with 6%. The developing countries are therefore pressing for reform of the governance of both institutions, so as to reflect the shift in global economic power and influence.

World Trade Organization

The Bretton Woods institutions were originally intended as a triad, with a third body that would ensure orderly international trade. The aim was to prevent trade wars—the kind of 'beggar my neighbour' policies, chiefly consisting of import restrictions or tariffs, which were thought to have contributed to the recession of the 1930s. The trade body, however, was slower to emerge. At first, it took the form merely of a treaty, the General Agreement on Trade and Tariffs. Only in 1995 did the treaty became an institution, the World Trade Organization (WTO). Countries that join the WTO have to abide by its rules. They must, for example, treat all other members equally—by allowing every member country the same access to their markets, with no discriminatory tariffs. The WTO also serves as an international court for trade disputes, since members agree to abide by its rulings.

In addition, the WTO organizes rounds of multilateral global trade negotiations, through which all members try to agree lower tariffs simultaneously, along with other policies that might encourage global trade. The current series of talks, the Doha round, which started in 1991, has been particularly contentious since the developed countries, notably those in the EU, are reluctant to cut their massive agricultural subsidies, while also being unwilling to permit developing countries to raise tariffs to protect their poor farmers from the ensuing flood of cheap food imports. The WTO is more democratic than the Bretton Woods organizations since it is based on one member, one vote. Nevertheless, the developing countries protest that it often serves the interests of multinational corporations which lobby developed country governments to protect corporate interests.

G8 and G20

For all governments, the overall political forum is the United Nations, but the leaders of the most powerful democratic countries also meet on a more exclusive basis to set their own priorities. One of these meetings is the series of summits of the 'Group of eight' (G8) industrialized countries. The first meeting, originally of six countries, was held in 1975 in response to the oil crisis. By 1998 the membership had expanded to eight—Canada, France, Germany, Japan, United States, Russia, the United Kingdom, and the United States.

While the UN is grounded in formal international agreements, the G8 is much looser, functioning more like a private club. Its annual meetings rotate from one member country to another, with no permanent secretariat, or indeed even a fixed agenda. Instead, officials from each country, referred to as 'sherpas', meet throughout the year to prepare for the summit. In the early years, the G8 typically focused primarily on economic issues and trade, but subsequently expanded its remit to encompass anything from international terrorism to global poverty. In addition to the annual summits of leaders, there are also a number of working groups consisting of ministers from government departments. At all these meetings, however, countries arrive not at formal agreements but 'understandings.'

For the rich countries, the G8 has the advantage of allowing them to meet on a more informal and personal basis, with less of the diplomatic protocol that can slow things at the United Nations. From the perspective of those excluded, however, the G8 is a secretive and unaccountable body, and has come under increasing fire from non-governmental organizations, particularly those hostile to globalization.

The justification for the creation of the G8 was that its members had compatible governmental and economic systems—and were responsible for more than half of global output. But this rationale has been undermined from two directions. Russia, for example, which in the 1990s appeared to be on the road to democracy has steadily been sliding into autocracy. At the same time, the G8's economic significance has gradually been eroded by the emergence of new global giants, notably Brazil, China, and India.

In response, the G8 has been supplemented in various ways. One is through the creation in 1999 of a 'G20' group of finance ministers or central bank governors, whose annual meetings deal with many of the economic functions of the G8. Its additional members are Argentina, Australia, Brazil, China, Italy, Indonesia, the Republic of Korea, Mexico, Saudi Arabia, South Africa, and Turkey—and the European Union.

Regional organizations

In addition to having global links, all countries have ties with countries in the same part of the world, with whom they may consider themselves to have cultural links or common political interests. The United Nations itself also has five regional economic commissions—for Africa, Europe, Latin America and the Caribbean, Western Asia, and Asia and the Pacific.

Most other regional organizations have primarily been concerned with trade. Listed below, travelling from west to east, are some of those that aspire to broader cooperation, or even integration.

Organization of American States

The Organization of American States (OAS), whose charter was signed in 1948, has inevitably been dominated by its richest and most powerful member, the United States. Its headquarters are in Washington, DC. Canada, though now the OAS's second largest contributor, was wary

of US domination and delayed joining the organization until 1990. In 1991, with the arrival of Guyana and Belize, the OAS achieved its current full membership of 35 states.

The OAS grew out of US initiatives in the Western hemisphere, starting with a series of conferences from the 1890s. The OAS declares its four 'pillars' to be: democracy, human rights, security, and development. Initially most of the emphasis was on security. From the outset, the OAS was primarily a mutual defence pact and during the Cold War it was essentially an anti-communist alliance. For this reason, Cuba's position was always going to be contentious. Although it was an OAS founder member, Cuba was suspended in 1962. It was finally invited to return in 2009, though it seems to have rejected the invitation—spurning a club it has long derided as the 'Yankee Ministry of Colonies'.

Although it has in the past helped settle disputes between members, the OAS has often been sidelined. It played little part, for example, in the settlement of the wars in Central America in the 1970s and 1980s. In recent years, however, it seems to have had a fresh lease of life and is placing more emphasis on human rights and on the strengthening of democracy in the region. It came to the fore in 2009, for example, when, following a military coup, it suspended the membership of Honduras.

European Union

The European Union (EU) is by far the most successful inter-governmental body, in that its members have ceded a significant degree of sovereignty—particularly in trade, for which the EU has established a 'single market' with no tariffs.

The EU is the latest incarnation of what was founded in 1961 as the European Coal and Steel Community. Its current form reflects the Lisbon Treaty, ratified in 2009, which among other things established an EU presidency and a common representative for foreign policy. Following a series of expansions eastwards, the EU now has 27 members.

With its headquarters in Brussels, the EU has become a large and powerful body in its own right. It has three main decision-making institutions. At the top is the European Council which represents individual member states. This is chaired by the President, currently Herman Van Rompuy of Belgium, and consists of the heads of state or government of member countries, along with the High Representative for Foreign Affairs and Security Policy, currently Baroness Ashton of the United Kingdom, and the President of the Commission, currently José Manuel Barroso from Spain. Complementing the European Council, is the Council of Ministers which has both executive and legislative functions—though this is more a collection of different councils consisting of ministers from each country whose attendance varies according to the subject under discussion. 'Ecofin' for example, deals with economic and financial issues.

Alongside the European Council is the 736-member European Parliament, which has representatives directly elected by EU citizens every five years. Members of the European Parliament form seven groups according to the ideologies of their national political parties. Following the 2009 elections, the largest grouping was that of the centre-right European People's Party with 265 seats, which includes the Christian Democrat Union from Germany and the Union pour la République from France. Second, with 184 seats, is the centre-left Progressive Alliance of Socialists and Democrats, which includes the British Labour Party and the Spanish Socialist Workers Party. The main function of the parliament is to pass EU legislation, though the adoption of legislation is carried out in conjunction with the Council of Ministers.

The third main institution, the EU's equivalent of a national bureaucracy, is the European Commission. This body, which is independent of national governments, consists of 27 commissioners, one from each country, each of whom is responsible for one of 27 policy areas—from trade to energy to transport. The Commission makes proposals for legislation to the parliament and also monitors compliance on existing legislation such as that on the common agricultural policy. The commissioners are supported by around 23,000 civil servants.

Another prominent EU institution is the European Court of Justice, with one judge from each country, which settles disputes between members and rules on whether countries are meeting their treaty obligations. Another is the European Central Bank, which manages the monetary policies of the countries that have adopted the common currency, the euro—16 of the 27, plus five other non-EU countries.

The original impetus behind what is now the EU was implicitly to bind together the leading countries of Europe so as to prevent future conflict. It retains some of that drive by drawing former communist countries into its democratic and capitalist fold. Another of its purposes is to present a substantial counterweight to the United States and Russia and emerging powers such as China, India, and Brazil. In many international meetings, European countries are represented by the EU.

Integration brings benefits but it also has costs. The main cost is an erosion of national sovereignty as member countries, and their citizens, become subject to a growing body of EU legislation. Throughout Europe there is tension between europhiles who welcome greater integration towards something like a 'United States of Europe', and eurosceptics who want to limit the EU's influence—or escape it altogether. This spectrum is also matched to some extent at the national level. Historically, France and Germany have been considered among the more pro-European core countries who have signed up to most of what the EU has to offer. The UK, on the other hand, which claims a 'special relationship' with the USA, has tried to straddle the two centres of power. The UK only joined the EU in 1973, and subsequently stood aside from a number of EU initiatives, including the adoption of the euro, and the Schengen agreement which removes common border controls. Some of the Nordic countries, such as Denmark,

are also more reticent EU members.

Commonwealth of Independent States

The Commonwealth of Independent States (CIS) was established in 1991, with its headquarters in Minsk in Belarus, to shoulder some of the coordinating functions that had been performed by the former Soviet Union. The CIS comprises Russia and 11 of the former soviet republics; three others, Estonia, Latvia, and Lithuania refused to join the CIS. Although it was founded to coordinate economic, foreign, social, and other policies, in practice it remains a vehicle through which Russia can maintain its influence. The CIS has never been very coherent since each member chooses how to participate. It has also been undermined by inter-member disputes which have extended to military conflicts between Armenia and Azerbaijan, for example, and between Russia and Georgia. For the future it seems likely that members will be drawn to other more useful associations, or even towards the European Union.

African Union

The African Union evolved from the Organization of African Unity (OAU). The OAU had been established in 1963, aiming to create greater solidarity and cohesion between African states in the post-colonial era. It had some successes, in resolving border disputes between members, for example, in supporting independence movements, and in serving as a focal point for opposition to the apartheid-era South Africa. However, it was severely constrained by its principle of non-interference in its members' domestic affairs.

In 2002, following an initiative from Libya, the countries of Africa replaced the OAU with the African Union (AU), a more ambitious project aiming to move along lines similar to those of the European Union. Unlike the OAU the AU has the capacity to intervene in member states to support democratic governance and uphold human rights. The most practical expression of this has been the creation of AU peacekeeping forces, the first of which was sent to Burundi in 2003, consisting of troops from Ethiopia, Mozambique, and South Africa. Since then, the AU has become involved, often in conjunction with the United Nations, in policing disputes in a number of countries. In 2010 this required a presence in Burundi, Somalia, and Uganda, though these missions and others have generally been small and not very effective.

The AU has also stood up to oppose military coups. In 2008, for example, it assembled a joint military force to reverse a coup in the Comoros, and in 2009, following coups in Madagascar and Mauritania, suspended their AU memberships. In 2010 it also successfully asked the UN to impose sanctions on one of its own members, Eritrea, for supporting jihadist groups in Somalia. On the other hand, the AU has proved less decisive when addressing difficult issues involving more powerful members such as Zimbabwe.

The AU headquarters are in Addis Ababa, Ethiopia, where it has also established its administrative body, the AU Commission, along with the Pan-African Parliament, whose members are elected indirectly by national parliaments. AU heads of state meet once a year. They also aim in the longer term to establish an African court of justice and a central bank—though these seem fairly remote prospects.

Arab League

The Arab League, officially the League of Arab States, is one of the oldest regional groupings, founded in 1945 to strengthen the ties between the Arabic-speaking countries in the Middle East and North Africa, which now number 22. Since then it has given rise to numerous institutions, covenants, agreements and resolutions. Its forums include the Council of Joint Arab Defence, and the Economic and Social Council, along with specialized ministerial councils.

Some of the benefits have been cultural and linguistic. The Arab Organization for Education, Culture, and Science has developed curricula, and carried out literacy programmes and adult education. It has also encouraged the Arabization of computer systems.

But on the political and economic front the results have been less impressive. Cooperation is hampered partly because of the diversity of political systems which range from a free-wheeling democracy in Lebanon to deeply conservative autocracies as in Saudi Arabia. They also have different ideas about freedom of expression. Thus Qatar is host to al-Jazeera TV, which often pokes fun at Egypt's president Mubarak.

Saudi Arabia was unimpressed in 2003 by a Libyan plot to assassinate King Abdullah. There have also been rifts over issues such as the Iraq war, supported by some members, but vigorously opposed by others.

Efforts at trade integration have also made little progress. The Arab common market has only five members, and less than 10% of Arab trade is with other Arab countries.

Association of South-East Asian Nations

The Association of South-East Asian Nations (ASEAN), which has its headquarters in Jakarta, Indonesia, was launched at the height of the Cold War in 1967 by five countries: Indonesia, Malaysia, the Philippines, Singapore, and Thailand. This was essentially a political gesture of mutual solidarity against the encroachment of communism, though it also aimed to build stronger relations between member countries, many of which had lingering border disputes. In the years after the ending of the Vietnam war, the group eventually expanded to ten, with the arrival of Brunei, Burma (Myanmar), Cambodia, Laos, and Vietnam. This also signalled a change in direction, with less emphasis on security and more on economic cooperation.

ASEAN has generally been a loose association and its annual meetings have often been little more than talking shops since its members are determined not to interfere in each others' affairs. Even when its summits issue declarations, these are frequently worded in vague terms, or establish objectives that its members

may lack either the will or the means to implement. Nevertheless, ASEAN can claim some successes. Since its founding, apart from occasional border skirmishes, no two ASEAN members have had a large-scale war.

In the 1980s, ASEAN was also instrumental in dealing with Vietnam's intervention in Cambodia to despatch the Khmer Rouge. ASEAN also helped found in 1994 the ASEAN Regional Forum which addresses political and security issues across 27 Asian countries. In addition, it has established the 'ASEAN + 3' group to increase cooperation with China, Japan, and South Korea.

On the economic front, ASEAN has been relatively passive. There is an ASEAN Free Trade agreement but many of the member countries have achieved economic cooperation by default, as they have created integrated cross-border production systems, notably for electronic goods. They also have numerous other bilateral and multilateral agreements. Another ASEAN venture has been the Chiang Mai Initiative which has set up a system of mutual currency reserves, though this is still on a small scale.

In terms of influencing the behaviour of its member states, ASEAN has often proved feeble. One of its most contentious steps, in 1997, was to admit Burma. In so doing, it claimed it would draw the pariah regime into a regional embrace and thus into the wider international community. This promise remains unfulfilled: Burma has carried on much as before.

The ASEAN members have tried to make the organization more forceful. In 2003 they agreed three 'pillars' for the ASEAN Community—security, economic, and socio-cultural—which together would help it create a 'just, democratic and harmonious environment'. Then in 2007 they agreed an 'ASEAN Charter' to provide a stronger legal and institutional framework, which by 2008 had been ratified by all members. The charter proclaims adherence to principles of good governance, fundamental freedoms, human rights, and social justice. But it also carefully reiterates the principle of mutual non-interference, so progress is likely to be slow.

South Asian Association for Regional Cooperation

The South Asian Association for Regional Cooperation (SAARC), which has its headquarters in Kathmandu, Nepal, was established in 1985 by Bangladesh, Bhutan, Maldives, Nepal, Pakistan, India, and Sri Lanka, with Afghanistan joining in 1997. It aims to provide a platform in which members can work together in a 'spirit of friendship, trust and understanding'.

Representing a combined population of more than 1.5 billion people, SAARC should be one of the world's most significant regional groupings. Unfortunately, friendship and trust have been in short supply— particularly between the two main powers, India and Pakistan. India has been notably unenthusiastic about SAARC which it sees more as a forum where the smaller countries can air anti-Indian complaints.

The lack of mutual trust has undermined SAARC's capacity

even to promote trade. SAARC is supposedly aiming for regional free trade by 2015, but given the dense thicket of tariffs and regulations still in place this seems a remote prospect. As a result, South Asia remains the world's least integrated region. Most of its effective inter-country agreements tend to be bilateral, largely between India and other countries.

Pacific Islands Forum

Founded in 1971 as the South Pacific Forum, and in 2000 renamed the Pacific Islands Forum (PIF) this consists of 16 states in the Pacific and has its secretariat in Suva, Fiji. The largest members are Australia, New Zealand, and Papua New Guinea. Most of the rest are small island states. Initially, there were doubts about whether Australia and New Zealand should be full members given their economic clout and their dominant position as aid donors to the other countries, but eventually it was concluded that the Forum could serve as a channel through which the small countries could exert some influence on the two larger ones. Another advantage of having the larger countries is that each provides one third of the operating budget.

The PIF is still a fairly informal grouping that lacks a constitution. While this may seem a weakness, it does permit the annual meetings to react quickly to issues as they arise. Critics argue, however, that it is little more than a talking shop and that it is often difficult to get Pacific island countries to rise above national concerns and arrive at regional consensus.

In the past, members have also

been reluctant to interfere in each others' affairs. In 2000, however, following a coup in Fiji, the Forum agreed, in the Biketawa Declaration, on some guidelines for engaging with issues of 'regional security'—which in this region is largely a code phrase for military coups or national political unrest. The Declaration commits member governments to principles of equality and good governance and permits the Forum to provide mediation. In truth, effective military intervention is likely to come largely from Australia and New Zealand, though the Forum provides an umbrella of regional legitimacy. In recent years, the Declaration has enabled peacekeeping forces to be despatched to Nauru and Tonga. Following a coup in 2009, Fiji was suspended by the Forum until it holds a democratic election.

Apart from security issues, the Forum initially focused on trade and economic concerns, achieving some agreements on issues like transport and fisheries. More recently, however, it has extended its work into social areas including education and health. In 2005, the Forum agreed a Pacific Plan with a number of initiatives to promote economic growth, sustainable development, good governance, and security.

Indicator tables

Notes on indicator tables

These tables summarize economic and social data produced by the United Nations Development Programme (UNDP). Full tables are available in the annual UNDP *Human Development Report*. An entry of '..' indicates that data are unavailable or inappropriate for that country.

Table 1. Income and poverty

GDP $PPP—Gross domestic product per capita at purchasing power parity (PPP). For an explanation of GDP and PPP see pages ix–x. While this gives some indication of the relative wealth of countries, it does not take into account the distribution of income: many Indians, for example, earn more than many Americans. Nor does it necessarily offer guidance to quality of life: people in Cuba, for example, may have a low income but good health services. Dissatisfaction with GDP as an indicator led to the creation of the UNDP human development index.

HDI ranking—This is the human development index ranking, based on a composite of indicators for income, health, and educational attainment. In 2009, the ranking included 182 countries. This list is not exhaustive because it excludes non-UN member countries, such as dependencies, and also those countries for which there were insufficient data. Although this ranking was produced in 2009 it is based on data for earlier years.

Poverty—The proportion of people living on less than the international poverty line of less than $1.25 per day. These data are largely for developing countries. In industrialized countries the poverty line is usually set relative to average incomes, so is not comparable. See the note on poverty on page x. The data are for the latest available years.

GDP annual growth—For an explanation of GDP, see page ix. The average growth rate over the period is calculated from total GDP expressed in the national currency.

Distribution of income—The ratio of proportion of national income going to the highest and to the lowest 10% of income earners. Note that this refers to income; not to wealth or total assets. Contrasts in wealth can be much more marked, but more difficult to measure. The data are for the latest available years.

Gini-index—This is another measure of income distribution. It takes a value between 0 (absolute equality) and 100 (one person owns everything). Data are for the latest available years. The figure tends to be higher in developing countries, though there can be significant differences even between these countries: those in Latin America, for example, tend to be more unequal than those in Asia.

Inflation—The average percentage change in the consumer price index. It should be noted, however, that different groups in the same country can experience different inflation rates. The poor are particularly sensitive to the price of food.

Table 2. Health and population

Life expectancy at birth—How long a newborn child can expect to live, on average. World average life expectancy for 2007 was 68 years. Women have a biological advantage that should enable them to live six or seven years longer than men, so there are normally more women than men. However, where there is severe discrimination against women that affects their health the differences can be far less. In Nepal, for example, the discrimination is so great that, on average, men outlive women.

Population, natural increase—The rate of population growth, excluding the effect of migration. It is thus the birth rate minus the death rate.

Net migration rate—The total number of immigrants to a country minus the number of emigrants over a period, expressed as a percentage of the person-years lived by the population of the receiving country over that period. A negative figure indicates net emigration.

Dependency ratio—The population aged 65 years and above expressed as a percentage of the population of working age (15–64 years). A high ratio makes it more difficult to pay adequate pensions.

Safe water—The proportion of the population with reasonable access to safe drinking water, either treated surface water, or uncontaminated water from other sources such as springs or wells.

Malnutrition—The proportion of children under five years old who weigh less than they should for their age. Another indicator of malnutrition is stunting, which refers to low height for age. Children can also suffer from deficiencies in important 'micronutrients' such as Vitamin A or iodine. Children are most vulnerable to malnutrition in the first two years of life. Opportunities for growth lost during that period cannot be recovered: stunted children grow up as stunted adults.

Table 3. Education and gender

Adult literacy rate—The proportion of the population aged 15 and above who can read a short, simple statement on their everyday life. This information is collected either from national censuses or from household surveys. However some countries do not collect this information and instead substitute data on school attendance.

Educational attainment—The distribution of the population according to the highest level of education attained or completed. Developing countries have in the past put most of their efforts into ensuring that children go to primary school. Nowadays many developing countries, having achieved close to universal primary enrolment, and gender parity, and turning their attention to secondary education.

Parliament, percentage of women—The proportion of seats held by women in a lower or single house and, where relevant, in an upper house or senate. Globally, the average is 19%. Some countries have tried to force this ratio up by insisting on quotas either within the parliaments or within the list of candidates from parties. Others, such as Pakistan, have respectable looking ratios, only because the women members are serving as proxies for their husbands.

Income ratio— The ratio of the female non-agricultural wage to the male non-agricultural wage. In some countries women make up close to half the workforce, but they are still less likely to occupy senior position, and they generally earn less than men even for the same work.

Table 1. Indicators for income and poverty

	GDP $PPP per capita 2006	HDI ranking out of 182 2007	Poverty % < $1.25 2000–2007	GDP annual growth, % 1990–2007	Income distribution Ratio of richest 10% to poorest 10%	Gini-index	Inflation % 2006–2007
Afghanistan	1,054	181	..	..	..	..	17.0
Albania	7,041	70	<2	5.2	8.0	33.0	2.9
Algeria	7,740	104	6.8	1.4	9.6	35.3	3.5
Andorra	41,235	28	..	..	..	..	..
Angola	5,385	143	54.3	2.9	74.6	58.6	12.2
Antigua and Barbuda	18,691	47	..	1.8	..	..	..
Argentina	13,238	49	4.5	1.5	31.6	50.0	8.8
Armenia	5,693	84	10.6	5.8	7.9	33.8	4.4
Australia	34,923	2	..	2.4	12.5	35.2	2.3
Austria	37,370	14	..	1.8	6.9	29.1	2.2
Azerbaijan	7,851	86	<2	2.9	2.9	36.5	16.7
Bahamas	20,253	52	..	..	..	..	2.5
Bahrain	29,723	39	..	2.4	..	..	-5.5
Bangladesh	1,241	146	49.6	3.1	6.2	31.0	9.1
Barbados	17,956	37	..	..	..	..	4.0
Belarus	10,841	68	<2	3.4	6.1	27.9	8.4

Belgium	34,935	17	..	1.8	8.2	33.0	1.8
Belize	6,734	93	..	2.3		..	2.3
Benin	1,312	161	47.3	1.3	10.8	38.6	1.3
Bhutan	4,837	132	26.2	5.2	16.3	46.8	5.2
Bolivia	4,206	113	19.6	1.3	93.9	58.2	8.7
Bosnia and Herzegovina	7,764	76	<2	11.2	9.9	35.8	..
Botswana	13,604	125	31.2	4.3	40.0	61.0	7.1
Brazil	9,567	75	5.2	1.2	40.6	55.0	3.6
Brunei Darussalam	50,200	30	..	-0.3	..	..	0.1
Bulgaria	11,222	61	<2	2.3	6.9	29.2	8.4
Burkina Faso	1,124	177	56.5	2.5	10.8	39.6	-0.2
Burma (Myanmar)	904	138	..	6.8	..	..	35.0
Burundi	341	174	81.3	-2.7	6.8	33.3	8.3
Cambodia	1,802	137	40.2	6.2	11.5	40.7	5.9
Cameroon	2,128	153	32.8	0.6	15.0	44.6	0.9
Canada	35,812	4	..	2.2	9.4	32.6	2.1
Cape Verde	3,041	121	20.6	3.3	21.6	50.5	4.4
Central African Republic	713	179	62.4	-0.8	15.7	43.6	..
Chad	1,477	175	61.9	2.4	11.8	39.8	-9.0
Chile	13,880	44	<2	3.7	26.2	52.0	4.4

Table 1. Indicators for income and poverty

	GDP $PPP per capita 2006	HDI ranking out of 182 2007	Poverty % < $1.25 2000–2007	GDP annual growth, % 1990–2007	Income distribution Ratio of richest 10% to poorest 10%	Gini-index	Inflation % 2006–2007
China	5,383	92	15.9	8.9	13.2	41.5	4.8
Colombia	8,587	77	16.0	1.2	60.4	58.5	5.4
Comoros	1,143	139	46.1	-0.4	60.6	64.3	::
Congo	3,511	136	54.1	-0.2	17.8	47.3	2.7
Congo Dem. Rep.	298	176	59.2	-4.3	15.1	44.4	16.9
Costa Rica	10,842	54	2.4	2.6	23.4	47.2	9.4
Côte d'Ivoire	1,690	163	23.3	-0.7	20.2	48.4	1.9
Croatia	16,027	45	<2	3.0	6.4	29.0	2.9
Cuba	6,876	51	::	::	::	::	::
Cyprus	24,789	32	::	2.5	::	::	2.4
Czech Republic	24,144	36	<2	2.4	5.3	25.8	2.9
Denmark	36,130	16	::	1.9	8.1	24.7	1.7
Djibouti	2,061	155	18.8	-2.1	12.8	40.0	::
Dominica	7,893	73	::	1.4	::	::	3.1
Dominican Republic	6,706	90	5.0	3.8	25.3	50.0	6.1
Ecuador	7,449	80	4.7	1.2	35.2	54.4	2.3

Egypt	5,349	123	<2	2.5	7.2	32.1	9.3
El Salvador	5,804	106	11.0	1.8	38.6	49.7	4.6
Equatorial Guinea	30,627	118	:	21.1	:	:	:
Eritrea	626	165	:	-0.7	:	:	:
Estonia	20,361	40	<2	5.3	10.4	36.0	6.6
Ethiopia	779	171	39.0	1.9	6.3	29.8	17.2
Fiji	4,304	108	:	1.6	:	:	4.8
Finland	34,526	12	:	2.8	5.6	26.9	2.5
France	33,674	8	:	1.6	9.1	32.7	1.5
Gabon	15,167	103	4.8	-0.7	13.3	41.5	5.0
Gambia	1,225	168	34.3	0.3	18.9	47.3	2.1
Georgia	4,662	89	13.4	1.8	15.9	40.8	9.2
Germany	34,401	22	:	1.4	6.9	28.3	2.1
Ghana	1,334	152	30.0	2.1	16.1	42.8	10.7
Greece	28,517	25	:	2.7	10.2	34.3	2.9
Grenada	7,344	74	:	2.4	:	:	4.2
Guatemala	4,562	122	11.7	1.4	33.9	53.7	6.5
Guinea	1,140	170	70.1	1.3	14.4	43.3	:
Guinea-Bissau	477	173	48.8	-2.6	9.5	35.5	4.6
Guyana	2,782	114	7.7	2.9	25.5	44.6	12.3

Table 1. Indicators for income and poverty

	GDP $PPP per capita 2006	HDI ranking out of 182 2007	Poverty % < $1.25 2000–2007	GDP annual growth, % 1990–2007	Income distribution Ratio of richest 10% to poorest 10%	Gini-index	Inflation % 2006–2007
Haiti	1,155	149	54.9	-2.1	54.4	59.5	8.5
Honduras	3,796	112	18.2	1.5	59.4	55.3	6.9
Hungary	18,755	43	<2	3.3	6.8	30.0	7.9
Iceland	35,742	3	:	2.5	:	:	5.1
India	2,753	134	41.6	4.5	8.6	36.8	6.4
Indonesia	3,712	111	:	2.3	10.8	39.4	6.4
Iran	10,955	88	<2	2.5	11.6	38.3	17.2
Ireland	44,613	5	:	5.8	9.4	34.3	4.9
Israel	26,315	27	:	1.7	13.4	39.2	0.5
Italy	30,353	18	:	1.2	11.6	36.0	1.8
Jamaica	6,079	100	<2	0.6	17.0	45.5	9.3
Japan	33,632	10	:	1.0	4.5	24.9	0.1
Jordan	4,901	96	<2	2.0	10.2	37.7	5.4
Kazakhstan	10,863	82	3.1	3.2	8.5	33.9	10.8
Kenya	1,542	147	19.7	0.0	21.3	47.7	9.8
Korea, South	24,801	26	<2	4.5	7.8	31.6	2.5

Kuwait	47,812	31	..	1.8	..	..	5.5
Kyrgyzstan	2,006	120	21.8	-0.4	7.3	32.9	10.2
Laos	2,165	133	44.0	4.2	7.3	32.6	4.5
Latvia	16,377	48	<2	4.7	10.3	35.7	10.1
Lebanon	10,109	83	..	2.4	..	..	..
Lesotho	1,541	156	43.4	2.4	39.8	52.5	8.0
Liberia	362	169	83.7	1.9	12.8	52.6	..
Libya	14,364	55	..	..	..	..	3.4
Liechtenstein	85,382	19	..	..	..	..	..
Lithuania	17,575	46	<2	3.0	10.3	35.8	5.7
Luxembourg	79,485	11	..	3.3	6.8	30.8	2.3
Macedonia	9,096	72	<2	0.4	12.4	39.0	3.5
Madagascar	932	145	67.8	-0.4	15.9	47.2	10.3
Malawi	761	160	73.9	0.4	10.5	39.0	8.0
Malaysia	13,518	66	<2	3.4	11.0	37.9	2.0
Maldives	5,196	95	..	5.1	..	..	7.4
Mali	1,083	178	51.4	2.2	11.2	39.0	1.4
Malta	23,080	38	..	2.6	..	..	1.3
Mauritania	1,927	154	21.2	0.6	11.6	39.0	7.3
Mauritius	11,296	81	..	3.7	..	..	8.8

Table 1. Indicators for income and poverty

	GDP $PPP per capita 2006	HDI ranking out of 182 2007	Poverty % < $1.25 2000–2007	GDP annual growth, % 1990–2007	Income distribution Ratio of richest 10% to poorest 10%	Gini-index	Inflation % 2006–2007
Mexico	14,104	53	<2	1.6	21.0	48.1	4.0
Moldova	2,551	117	8.1	-1.3	9.4	35.6	12.4
Mongolia	3,236	115	22.4	2.2	8.6	33.0	9.0
Montenegro	11,699	65	..	3.8	..	..	..
Morocco	4,108	130	2.5	2.0	12.5	40.9	2.0
Mozambique	802	172	74.7	4.2	18.5	47.1	8.2
Namibia	5,155	128	49.1	1.8	106.6	74.3	6.7
Nepal	1,049	144	55.1	1.9	14.8	47.3	6.1
Netherlands	38,694	6	..	2.1	9.2	30.9	1.6
New Zealand	27,336	20	..	2.1	12.5	36.2	2.4
Nicaragua	2,570	124	15.8	1.9	31.0	52.3	11.1
Niger	627	182	65.9	-0.6	15.3	43.9	0.1
Nigeria	1,969	158	64.4	1.1	16.3	42.9	5.4
Norway	53,433	1	..	2.6	6.1	25.8	0.7
Oman	22,816	56	..	2.3	..	..	6.0
Pakistan	2,496	141	22.6	1.6	6.7	31.2	7.6

Palestine	..	110	..	..	..	..	3.5
Panama	11,391	60	9.5	2.6	49.9	54.9	4.2
Papua New Guinea	2,084	148	35.8	-0.6	21.5	50.9	0.9
Paraguay	4,433	101	6.5	-0.3	38.8	53.2	8.1
Peru	7,836	78	7.9	2.7	26.1	49.6	1.8
Philippines	3,406	105	22.6	1.7	14.1	44.0	2.8
Poland	15,987	41	<2	4.4	9.0	34.9	2.4
Portugal	22,765	34	..	1.9	15.0	38.5	2.8
Qatar	74,882	33	..	..	..	..	13.8
Romania	12,369	63	<2	2.3	7.6	31.5	4.8
Russia	14,690	71	<2	1.2	11.0	37.5	9.0
Rwanda	866	167	76.6	1.1	18.1	46.7	9.1
Samoa	4,467	94	..	2.9	..	..	5.6
Sao Tome and Principe	1,638	131	..	..	..	..	..
Saudi Arabia	22,935	59	..	0.3	..	..	4.2
Senegal	1,666	166	33.5	1.1	11.9	39.2	5.9
Serbia	10,248	67	..	0.0	..	..	6.4
Seychelles	16,394	57	..	1.4	..	..	5.3
Sierra Leone	679	180	53.4	-0.3	12.8	42.5	11.7
Singapore	49,704	23	..	3.8	17.7	42.5	2.1

Table 1. Indicators for income and poverty

	GDP $PPP per capita 2006	HDI ranking out of 182 2007	Poverty % < $1.25 2000–2007	GDP annual growth, % 1990–2007	Income distribution Ratio of richest 10% to poorest 10%	Gini-index	Inflation % 2006–2007
Slovakia	20,076	42	<2	3.4	6.8	25.8	2.8
Slovenia	26,753	29	<2	3.5	7.3	31.2	3.6
Solomon Islands	1,725	135	::	-1.5	::	::	7.7
South Africa	9,757	129	26.2	1.0	35.1	57.8	7.1
Spain	31,560	15	::	2.4	10.3	34.7	2.8
Sri Lanka	4,243	102	14.0	3.9	11.7	41.1	15.8
St Kitts and Nevis	14,481	62	::	2.8	::	::	4.4
St Lucia	9,786	69	20.9	1.3	16.2	42.6	2.5
St Vincent and the Grenadines	7,691	91	::	3.0	::	::	7.0
Sudan	2,086	150	::	3.6	::	::	8.0
Suriname	7,813	97	15.5	1.8	40.4	52.9	6.7
Swaziland	4,789	142	62.9	0.9	22.4	50.7	5.3
Sweden	36,712	7	::	2.3	6.2	25.0	2.2
Switzerland	40,658	9	::	0.8	9.0	33.7	0.7
Syria	4,511	107	::	1.5	::	::	3.9
Tajikistan	1,753	127	21.5	-2.2	8.2	33.6	13.1

Tanzania	1,208	151	88.5	1.8	8.9	34.6	7.0
Thailand	8,135	87	<2	2.9	13.1	42.5	2.2
Timor-Leste	717	162	52.9	..	10.8	39.5	10.3
Togo	788	159	38.7	-0.2	8.3	34.4	1.0
Tonga	3,748	99	..	1.7	..	..	5.9
Trinidad and Tobago	23,507	64	4.2	5.0	14.4	40.3	7.9
Tunisia	7,520	98	2.6	3.4	13.3	40.8	3.1
Turkey	12,955	79	2.7	2.2	17.4	43.2	8.8
Turkmenistan	4,953	109	24.8	..	12.9	40.8	..
Uganda	1,059	157	51.5	3.5	13.2	42.6	6.1
Ukraine	6,914	85	<2	-0.7	6.0	28.2	12.8
United Arab Emirates	54,626	35	..	-0.1	..	..	..
United Kingdom	35,130	21	..	2.4	13.8	36.0	4.3
United States	45,592	13	..	2.0	15.9	40.8	2.9
Uruguay	11,216	50	<2	1.5	20.1	46.2	8.1
Uzbekistan	2,425	119	46.3	1.2	10.3	36.7	..
Vanuatu	3,666	126	..	-0.4	..	..	4.0
Venezuela	12,156	58	3.5	-0.2	18.8	43.4	18.7
Vietnam	2,600	116	21.5	6.0	9.7	37.8	8.9
Yemen	2,335	140	17.5	1.6	10.6	37.7	10.0
Zambia	1,358	164	64.3	0.1	29.5	50.7	10.7

Table 2. Indicators for health and population

	Life expectancy years, 2007		Population natural increase, %	Migration net rate	Dependency ratio, old age	Safe water % access	Malnutrition % children under 5
	Female	Male	2005–2010	2005–2010	2010	2006	2007
Afghanistan	43.5	43.6	2.7	0.7	4.3	78	39
Albania	79.8	73.4	0.9	-0.5	14.4	3	8
Algeria	73.6	70.8	1.6	-0.1	6.8	15	4
Andorra	::	::	::	::	::	::	::
Angola	48.5	44.6	2.6	0.1	4.7	49	31
Antigua and Barbuda	::	::	::	::	::	9	10
Argentina	79.0	71.5	1.0	0.0	16.6	4	4
Armenia	76.7	70.1	0.7	-0.5	16.1	2	4
Australia	83.7	79.1	0.6	0.5	20.7	::	::
Austria	82.5	77.0	0.0	0.4	25.9	::	::
Azerbaijan	72.3	67.6	1.2	-0.1	9.5	22	7
Bahamas	76.0	70.4	1.1	0.1	10.3	3	::
Bahrain	77.4	74.2	1.6	0.5	3.1	::	9
Bangladesh	66.7	64.7	1.5	-0.1	6.1	20	48
Barbados	79.7	74.0	0.4	-0.1	14.4	::	6
Belarus	75.2	63.1	-0.5	0.0	18.6	0	1

Belgium	82.4	76.5	0.2	0.4	26.4	..	..
Belize	78.0	74.2	2.1	-0.1	6.7	9	7
Benin	62.1	59.8	3.0	0.1	6.1	35	23
Bhutan	67.6	64.0	1.4	0.3	7.5	19	19
Bolivia	67.5	63.3	2.0	-0.2	8.0	14	8
Bosnia and Herzegovina	77.7	72.4	-0.1	-0.1	19.6	1	2
Botswana	53.3	53.2	1.3	0.2	6.1	4	13
Brazil	75.9	68.6	1.0	0.0	10.2	9	6
Brunei Darussalam	79.6	74.9	1.7	0.2	4.9	..	..
Bulgaria	76.7	69.6	-0.5	-0.1	25.5	1	..
Burkina Faso	54.0	51.4	3.5	-0.1	3.9	28	37
Burma (Myanmar)	63.4	59.0	1.1	-0.2	8.1	20	32
Burundi	51.4	48.6	2.1	0.8	4.7	29	39
Cambodia	62.3	58.6	1.6	0.0	5.6	35	36
Cameroon	51.4	50.3	2.3	0.0	6.4	30	19
Canada	82.9	78.2	0.3	0.6	20.3	..	..
Cape Verde	73.5	68.2	1.9	-0.5	6.8	20	14
Central African Republic	48.2	45.1	1.9	0.0	6.9	34	29
Chad	49.9	47.3	2.9	-0.1	5.5	52	37
Chile	81.6	75.5	1.0	0.0	13.5	5	1

Table 2. Indicators for health and population

	Life expectancy years, 2007		Population natural increase, %	Migration net rate	Dependency ratio, old age	Safe water % access	Malnutrition % children under 5
	Female	Male	2005–2010	2005–2010	2010	2006	2007
China	74.7	71.3	0.7	0.0	11.4	12	7
Colombia	76.5	69.1	1.5	-0.1	8.6	7	7
Comoros	67.2	62.8	2.6	-0.3	5.2	15	25
Congo	54.4	52.5	2.2	-0.3	6.8	29	14
Congo Dem. Rep.	49.2	46.1	2.8	0.0	5.2	54	31
Costa Rica	81.3	76.4	1.3	0.1	9.5	2	5
Côte d'Ivoire	58.3	55.7	2.4	-0.1	7.0	19	20
Croatia	79.4	72.6	-0.2	0.0	25.6	1	1
Cuba	80.6	76.5	0.4	-0.3	17.5	9	4
Cyprus	81.9	77.3	0.4	0.6	19.0	::	::
Czech Republic	79.4	73.2	0.0	0.4	21.6	::	1
Denmark	80.5	75.9	0.1	0.1	25.6	::	::
Djibouti	56.5	53.7	1.8	0.0	5.4	8	29
Dominica	:	:	:	:	:	3	5
Dominican Republic	75.2	69.8	1.7	-0.3	9.8	5	5
Ecuador	78.0	72.1	1.6	-0.5	10.6	5	9

Egypt	71.7	68.2	1.9	-0.1	7.3	2	6
El Salvador	75.9	66.4	1.4	-0.9	12.0	16	10
Equatorial Guinea	51.1	48.7	2.3	0.3	5.1	57	19
Eritrea	61.4	56.8	2.9	0.2	4.5	40	40
Estonia	78.3	67.3	-0.1	0.0	25.2	..	..
Ethiopia	56.2	53.3	2.7	-0.1	6.0	58	38
Fiji	71.0	66.5	1.5	-0.8	7.7	53	8
Finland	82.8	76.0	0.2	0.2	25.9	..	..
France	84.5	77.4	0.4	0.2	26.2	..	..
Gabon	61.5	58.7	1.8	0.1	7.2	13	12
Gambia	57.3	54.1	2.6	0.2	5.2	14	20
Georgia	75.0	68.1	0.0	-1.2	20.7	1	3
Germany	82.3	77.0	-0.2	0.1	30.9	..	..
Ghana	57.4	55.6	2.1	0.0	6.3	20	18
Greece	81.3	76.9	-0.1	0.3	27.2	..	..
Grenada	76.7	73.7	1.3	-1.0	10.6	6	..
Guatemala	73.7	66.7	2.8	-0.3	8.2	4	23
Guinea	59.3	55.3	2.9	-0.6	6.1	30	26
Guinea-Bissau	49.1	46.0	2.4	-0.2	6.4	43	19
Guyana	69.6	63.7	1.0	-1.0	9.5	7	14

Table 2. Indicators for health and population

	Life expectancy years, 2007		Population natural increase, %	Migration net rate	Dependency ratio, old age	Safe water % access	Malnutrition % children under 5
	Female	Male	2005–2010	2005–2010	2010	2006	2007
Haiti	62.9	59.1	1.9	-0.3	7.3	42	22
Honduras	74.4	69.6	2.3	-0.3	7.3	16	11
Hungary	77.3	69.2	-0.4	0.1	23.8	::	2
Iceland	83.3	80.2	0.9	1.3	17.4	::	::
India	64.9	62.0	1.4	0.0	7.7	11	46
Indonesia	72.5	68.5	1.2	-0.1	9.0	20	28
Iran	72.5	69.9	1.3	-0.1	6.8	6	11
Ireland	82.0	77.3	0.9	0.9	16.7	::	::
Israel	82.7	78.5	1.5	0.2	16.4	::	::
Italy	84.0	78.1	-0.1	0.6	31.3	::	::
Jamaica	75.1	68.3	1.2	-0.7	12.2	7	4
Japan	86.2	79.0	-0.1	0.0	35.1	::	::
Jordan	74.3	70.7	2.2	0.8	5.9	2	4
Kazakhstan	71.2	59.1	0.9	-0.1	10.0	4	4
Kenya	54.0	53.2	2.7	-0.1	4.8	43	20
Korea, South	82.4	75.8	0.4	0.0	15.2	::	8.0

Kuwait	79.8	76.0	1.6	0.8	3.2	::	10.0
Kyrgyzstan	71.4	63.9	1.5	-0.3	7.7	11	3
Laos	65.9	63.2	2.1	-0.2	6.1	40	40
Latvia	77.1	67.1	-0.4	-0.1	25.4	1	::
Lebanon	74.1	69.8	0.9	-0.1	10.8	0	4
Lesotho	45.5	43.9	1.2	-0.4	8.4	22	20
Liberia	59.3	56.5	2.8	1.3	5.7	36	26
Libya	76.8	71.6	1.9	0.1	6.6	29	5
Liechtenstein	::	::	::	::	::	::	::
Lithuania	77.7	65.9	-0.4	-0.6	23.7	::	::
Luxembourg	82.0	76.5	0.3	0.8	20.5	::	::
Macedonia	76.5	71.7	0.2	-0.1	16.9	0	6
Madagascar	61.5	58.3	2.7	0.0	5.6	53	42
Malawi	53.4	51.3	2.8	0.0	6.1	24	19
Malaysia	76.6	71.9	1.6	0.1	7.3	1	8
Maldives	72.7	69.7	1.4	0.0	6.4	17	30
Mali	48.8	47.4	2.7	-0.3	4.3	40	33
Malta	81.3	77.7	0.1	0.2	21.2	::	::
Mauritania	58.5	54.7	2.3	0.1	4.6	40	32
Mauritius	75.7	68.5	0.7	0.0	10.7	0	15

Table 2. Indicators for health and population

	Life expectancy years, 2007		Population natural increase, %	Migration net rate	Dependency ratio, old age	Safe water % access	Malnutrition % children under 5
	Female	Male	2005–2010	2005–2010	2010	2006	2007
Mexico	78.5	73.6	1.4	-0.5	10.0	5	5
Moldova	72.1	64.5	-0.1	-0.9	15.4	10	4
Mongolia	69.6	63.0	1.2	-0.1	5.8	28	6
Montenegro	76.5	71.6	0.2	-0.2	18.8	2	3
Morocco	73.3	68.8	1.5	-0.3	8.1	17	10
Mozambique	48.7	46.9	2.3	0.0	6.2	58	24
Namibia	61.2	59.3	1.9	0.0	6.1	7	24
Nepal	66.9	65.6	1.9	-0.1	6.8	11	39
Netherlands	81.9	77.6	0.3	0.1	22.9	..	..
New Zealand	82.1	78.1	0.7	0.2	19.4	..	..
Nicaragua	75.9	69.8	2.0	-0.7	7.5	21	10
Niger	51.7	50.0	3.9	0.0	4.1	58	44
Nigeria	48.2	47.2	2.4	0.0	5.8	53	29
Norway	82.7	78.2	0.4	0.6	22.7	..	..
Oman	77.3	74.1	1.9	0.1	4.7	18	18
Pakistan	66.5	65.9	2.3	-0.2	6.9	10	38

Palestine	74.9	71.7	3.2	0.0	5.5	11	3
Panama	78.2	73.0	1.6	0.1	10.4	8	7
Papua New Guinea	63.0	58.7	2.4	0.0	4.3	60	35
Paraguay	73.8	69.6	1.9	-0.1	8.4	23	5
Peru	75.8	70.4	1.6	-0.4	9.3	16	8
Philippines	73.9	69.4	2.0	-0.2	6.9	7	28
Poland	79.7	71.3	0.0	-0.1	18.8	..	..
Portugal	81.8	75.3	0.0	0.4	26.7	..	..
Qatar	76.8	74.8	1.0	9.4	1.3	..	6.0
Romania	76.1	69.0	-0.2	-0.2	21.3	12	3
Russia	72.9	59.9	-0.4	0.0	17.9	3	3
Rwanda	51.4	47.9	2.6	0.0	4.5	35	23
Samoa	74.7	68.4	1.8	-1.8	8.6	12	..
Sao Tome and Principe	67.3	63.5	2.5	-0.9	6.9	14	9
Saudi Arabia	75.1	70.8	2.0	0.1	4.6	10	14
Senegal	56.9	53.9	2.8	-0.2	4.4	23	17
Serbia	76.3	71.6	0.0	0.0	21.1	1	2
Seychelles	..	..	..	..	..	13	6
Sierra Leone	48.5	46.0	2.4	0.2	3.4	47	30
Singapore	82.6	77.8	0.3	2.2	13.8	..	3.0

Table 2. Indicators for health and population

	Life expectancy years, 2007		Population natural increase, %	Migration net rate	Dependency ratio, old age	Safe water % access	Malnutrition % children under 5
	Female	Male	2005–2010	2005–2010	2010	2006	2007
Slovakia	78.5	70.7	0.0	0.1	16.9	..	..
Slovenia	81.7	74.4	0.0	0.2	23.5	..	..
Solomon Islands	66.7	64.9	2.5	0.0	5.4	30	21
South Africa	53.2	49.8	0.7	0.3	7.1	7	12
Spain	84.0	77.5	0.2	0.8	25.3	..	..
Sri Lanka	77.9	70.3	1.2	-0.3	11.4	18	29
St Kitts and Nevis	..	..	..	..	..	1	..
St Lucia	75.5	71.7	1.1	-0.1	10.1	2	14
St Vincent and the Grenadines	73.6	69.4	1.0	-0.9	10.0	..	..
Sudan	59.4	56.3	2.1	0.1	6.4	30	41
Suriname	72.5	65.3	1.2	-0.2	9.9	8	13
Swaziland	44.8	45.7	1.4	-0.1	5.9	40	10
Sweden	83.0	78.6	0.2	0.3	28.1	..	..
Switzerland	84.1	79.2	0.1	0.3	25.5	..	..
Syria	76.0	72.2	2.5	0.8	5.2	11	10
Tajikistan	69.3	63.7	2.2	-0.6	6.0	33	17

Tanzania	55.8	54.2	3.0	-0.1	6.0	45	22
Thailand	72.1	65.4	0.6	0.1	10.9	2	9
Timor-Leste	61.5	59.8	3.1	0.2	5.8	38	46
Togo	63.9	60.4	2.5	0.0	6.3	41	26
Tonga	74.6	69.0	2.2	-1.7	10.3	0	..
Trinidad and Tobago	72.8	65.6	0.7	-0.3	9.5	6	6
Tunisia	76.0	71.8	1.0	0.0	9.6	6	4
Turkey	74.2	69.4	1.2	0.0	8.8	3	4
Turkmenistan	68.8	60.6	1.4	-0.1	6.2	..	11
Uganda	52.4	51.4	3.3	-0.1	5.2	36	20
Ukraine	73.8	62.7	-0.6	0.0	22.1	3	1
United Arab Emirates	78.7	76.6	1.3	1.6	1.3	..	14
United Kingdom	81.5	77.1	0.2	0.3	25.1	..	..
United States	81.3	76.7	0.6	0.3	19.4	..	..
Uruguay	79.8	72.6	0.6	-0.3	21.8	0	5
Uzbekistan	70.9	64.5	1.4	-0.3	6.6	12	5
Vanuatu	72.0	68.1	2.5	0.0	5.7	41	20
Venezuela	76.7	70.7	1.6	0.0	8.7	10	5
Vietnam	76.1	72.3	1.2	0.0	9.3	8	25
Yemen	64.1	60.9	3.0	-0.1	4.4	34	46
Zambia	45.0	44.0	2.6	-0.1	6.0	42	20

Table 3. Indicators for education and gender

| | Adult literacy rate 1999-2007, % | | Educational attainment, % of population, 2000-2007 | | | Parliament % women | Income ratio women's to men's |
	Female	Male	Less than upper secondary	Upper secondary, non-tertiary	Tertiary	2009	1996–2007
Afghanistan	12.6	43.1	..	..	..	26	0.24
Albania	98.8	99.3	63.0	29.6	7.4	7	0.54
Algeria	66.4	84.3	92.1	7.6	..	6	0.36
Andorra	..	..	48.0	34.8	16.1	25	..
Angola	54.2	82.9	..	..	..	37	0.64
Antigua and Barbuda	99.4	98.4	..	..	..	17	..
Argentina	97.7	97.6	65.7	23.2	11.1	40	0.51
Armenia	99.3	99.7	18.4	61.2	20.4	8	0.57
Australia	..	..	..	..	..	30	0.70
Austria	..	..	26.2	57.9	15.9	27	0.40
Azerbaijan	99.2	99.8	16.5	70.2	13.3	11	0.44
Bahamas	..	..	28.9	70.2	0.3	25	..
Bahrain	86.4	90.4	50.3	38.4	11.2	14	0.51
Bangladesh	48.0	58.7	82.9	12.9	4.2	6	0.51
Barbados	..	..	75.7	23.1	1.1	14	0.65
Belarus	99.7	99.8	..	..	..	33	0.63

Belgium	..	..	42.3	31.0	26.8	36	0.64
Belize	..	..	74.2	13.6	10.9	11	0.43
Benin	27.9	53.1	85.6	12.2	2.2	11	0.52
Bhutan	38.7	65.0				14	0.39
Bolivia	86.0	96.0	61.6	23.8	14.0	15	0.61
Bosnia and Herzegovina	94.4	99.0	..			12	0.61
Botswana	82.9	82.8	..			11	0.58
Brazil	90.2	89.8	70.4	21.2	8.1	9	0.60
Brunei Darussalam	93.1	96.5	..			..	0.59
Bulgaria	97.9	98.6	40.4	41.3	18.0	22	0.68
Burkina Faso	21.6	36.7				15	0.66
Burma (Myanmar)	86.4	93.9				..	0.61
Burundi	52.2	67.3				32	0.77
Cambodia	67.7	85.8				16	0.68
Cameroon	59.8	77.0				14	0.53
Canada	..	..	23.7	38.1	38.2	25	0.65
Cape Verde	78.8	89.4	..			18	0.49
Central African Republic	33.5	64.8	..			10	0.59
Chad	20.8	43.0	..			5	0.70
Chile	96.5	96.6	..			13	0.42

Table 3. Indicators for education and gender

	Adult literacy rate 1999–2007, %		Educational attainment, % of population, 2000–2007			Parliament % women 2009	Income ratio women's to men's 1996–2007
	Female	Male	Less than upper secondary	Upper secondary, non-tertiary	Tertiary		
China	90.0	96.5	..	..	..	21	0.68
Colombia	92.8	92.4	64.7	25.4	9.7	10	0.71
Comoros	69.8	80.3	..	..	..	3	0.58
Congo	71.8	90.6	..	..	..	9	0.51
Congo Dem. Rep.	54.1	80.9	..	..	..	8	0.46
Costa Rica	96.2	95.7	64.7	18.5	15.0	37	0.46
Côte d'Ivoire	38.6	60.8	..	..	..	9	0.34
Croatia	98.0	99.5	40.2	45.4	13.9	21	0.67
Cuba	99.8	99.8	59.6	31.0	9.4	43	0.49
Cyprus	96.6	99.0	41.3	33.8	24.9	14	0.58
Czech Republic	..	..	14.5	73.0	12.5	16	0.57
Denmark	..	..	25.8	43.7	30.3	38	0.74
Djibouti	..	..	..	..	..	14	0.57
Dominica	..	..	88.8	5.7	5.0	19	..
Dominican Republic	89.5	88.8	..	..	..	17	0.59
Ecuador	89.7	92.3	..	..	..	28	0.51

415

Egypt	57.8	74.6	::	::	::	4	0.27
El Salvador	79.7	84.9	75.6	13.8	10.6	19	0.46
Equatorial Guinea	80.5	93.4	::	::	::	6	0.36
Eritrea	53.0	76.2	::	::	::	22	0.50
Estonia	99.8	99.8	27.9	42.3	27.5	21	0.65
Ethiopia	22.8	50.0	::	::	::	21	0.67
Fiji	::	::	::	::	::	::	0.38
Finland	::	::	30.9	38.8	30.3	42	0.73
France	::	::	42.6	35.9	19.8	20	0.61
Gabon	82.2	90.2	::	::	::	17	0.59
Gambia	::	::	::	::	::	9	0.63
Georgia	::	::	16.3	57.8	25.8	6	0.38
Germany	::	::	21.5	57.1	21.4	31	0.59
Ghana	58.3	71.7	::	::	::	8	0.74
Greece	96.0	98.2	51.0	25.7	23.3	15	0.51
Grenada	::	::	::	::	::	21	:
Guatemala	68.0	79.0	84.8	11.2	3.7	12	0.42
Guinea	18.1	42.6	::	::	::	::	0.68
Guinea-Bissau	54.4	75.1	::	::	::	10	0.46
Guyana	::	::	::	::	::	30	0.41

Table 3. Indicators for education and gender

	Adult literacy rate 1999–2007, %		Educational attainment, % of population, 2000–2007			Parliament % women 2009	Income ratio women's to men's 1996–2007
	Female	Male	Less than upper secondary	Upper secondary, non-tertiary	Tertiary		
Haiti	64.0	60.1	..	..	..	5	0.37
Honduras	83.5	83.7	..	..	..	23	0.34
Hungary	98.8	99.0	36.5	48.9	14.7	11	0.75
Iceland	..	..	37.4	30.3	27.6	33	0.62
India	54.5	76.9	..	..	..	9	0.32
Indonesia	88.8	95.2	..	..	..	12	0.44
Iran	77.2	87.3	..	..	..	3	0.32
Ireland	..	..	40.0	31.2	26.4	15	0.56
Israel	88.7	95.0	23.9	33.1	39.7	18	0.64
Italy	98.6	99.1	59.5	30.4	10.1	20	0.49
Jamaica	91.1	80.5	..	..	..	14	0.58
Japan	..	..	26.1	43.9	30.0	12	0.45
Jordan	87.0	95.2	..	..	..	8	0.19
Kazakhstan	99.5	99.8	29.5	56.1	14.4	12	0.68
Kenya	70.2	77.7	..	..	..	10	0.65
Korea, South	..	..	36.2	40.4	23.4	14	0.52

Kuwait	93.1	95.2	74.4	17.3	8.3	3	0.36
Kyrgyzstan	99.1	99.5	23.0	62.1	14.9	26	0.55
Laos	63.2	82.5	:	:	:	25	0.76
Latvia	99.8	99.8	19.7	60.0	20.3	20	0.67
Lebanon	86.0	93.4	:	:	:	5	0.25
Lesotho	90.3	73.7	:	:	:	26	0.73
Liberia	50.9	60.2	:	:	:	14	0.50
Libya	78.4	94.5	:	:	:	8	0.25
Liechtenstein	:	:	:	:	:	24	:
Lithuania	99.7	99.7	23.5	50.8	25.7	18	0.70
Luxembourg	:	:	39.0	39.7	21.3	23	0.57
Macedonia	95.4	98.6	52.2	35.6	12.2	28	0.49
Madagascar	65.3	76.5	:	:	:	9	0.71
Malawi	64.6	79.2	94.8	4.7	0.5	13	0.74
Malaysia	89.6	94.2	61.3	27.1	8.0	15	0.42
Maldives	97.1	97.0	:	:	:	12	0.54
Mali	18.2	34.9	:	:	:	10	0.44
Malta	93.5	91.2	77.2	12.0	10.8	9	0.45
Mauritania	48.3	63.3	:	:	:	20	0.58
Mauritius	84.7	90.2	79.2	17.7	2.6	17	0.42

Table 3. Indicators for education and gender

	Adult literacy rate 1999-2007, %		Educational attainment, % of population, 2000-2007			Parliament % women 2009	Income ratio women's to men's 1996-2007
	Female	Male	Less than upper secondary	Upper secondary, non-tertiary	Tertiary		
Mexico	91.4	94.4	69.7	15.3	14.9	22	0.42
Moldova	98.9	99.6	::	::	::	22	0.73
Mongolia	97.7	96.8	46.6	41.1	12.2	4	0.87
Montenegro	94.1	98.9	22.6	61.4	16.1	11	0.58
Morocco	43.2	68.7	::	::	::	6	0.24
Mozambique	33.0	57.2	::	::	::	35	0.90
Namibia	87.4	88.6	::	::	::	27	0.63
Nepal	43.6	70.3	::	::	::	33	0.61
Netherlands	::	::	34.8	38.6	26.0	39	0.67
New Zealand	::	::	28.7	40.1	25.9	34	0.69
Nicaragua	77.9	78.1	::	::	::	18	0.34
Niger	15.1	42.9	::	::	::	12	0.34
Nigeria	64.1	80.1	::	::	::	7	0.42
Norway	::	::	14.5	53.8	31.7	36	0.77
Oman	77.5	89.4	::	::	::	9	0.23
Pakistan	39.6	67.7	76.7	17.1	6.3	21	0.18

Palestine	90.3	97.2	68.8	12.8	18.4	..	..
Panama	92.8	94.0	66.0	23.1	10.4	17	0.58
Papua New Guinea	53.4	62.1	..	..	..	1	0.74
Paraguay	93.5	95.7	72.6	23.6	3.7	14	0.64
Peru	84.6	94.9	53.7	26.0	16.3	29	0.59
Philippines	93.7	93.1	62.6	26.4	8.4	20	0.58
Poland	99.0	99.6	..	..	..	18	0.59
Portugal	93.3	96.6	77.4	11.4	11.2	28	0.60
Qatar	90.4	93.8	59.0	20.1	20.9	0	0.28
Romania	96.9	98.3	47.3	43.6	9.0	10	0.68
Russia	99.4	99.7	..	..	..	11	0.64
Rwanda	59.8	71.4	..	..	..	51	0.79
Samoa	98.4	98.9	..	..	..	8	0.40
Sao Tome and Principe	82.7	93.4	..	..	..	7	0.47
Saudi Arabia	79.4	89.1	65.8	19.2	14.9	0	0.16
Senegal	33.0	52.3	..	..	..	29	0.55
Serbia	94.1	98.9	..	..	..	22	0.59
Seychelles	92.3	91.4	51.8	36.8	7.4	24	..
Sierra Leone	26.8	50.0	..	..	..	13	0.74
Singapore	91.6	97.3	41.2	39.2	19.6	24	0.53

Table 3. Indicators for education and gender

	Adult literacy rate 1999-2007, %		Educational attainment, % of population, 2000–2007			Parliament % women 2009	Income ratio women's to men's 1996–2007
	Female	Male	Less than upper secondary	Upper secondary, non-tertiary	Tertiary		
Slovakia	..	..	19.2	67.6	13.2	19	0.58
Slovenia	99.6	99.7	26.4	55.5	18.1	10	0.61
Solomon Islands	..	..	..	..	..	0	0.51
South Africa	87.2	88.9	73.0	18.1	8.9	34	0.60
Spain	97.3	98.6	58.6	17.8	23.6	34	0.52
Sri Lanka	89.1	92.7	..	..	..	6	0.56
St Kitts and Nevis	..	..	..	..	..	7	..
St Lucia	..	..	..	..	..	17	0.50
St Vincent and the Grenadines	..	..	..	..	..	18	0.51
Sudan	51.8	71.1	..	..	..	17	0.33
Suriname	88.1	92.7	..	..	..	25	0.44
Swaziland	78.3	80.9	..	..	..	22	0.71
Sweden	..	..	20.7	51.1	27.0	47	0.67
Switzerland	..	..	21.4	52.3	26.2	27	0.62
Syria	76.5	89.7	89.6	5.1	5.3	12	0.20
Tajikistan	99.5	99.8	21.0	68.3	10.6	20	0.65

Tanzania	65.9	79.0	98.4	0.7	0.9	30	0.74
Thailand	92.6	95.9	:	:	:	13	0.63
Timor-Leste	:	:	:	:	:	29	0.53
Togo	38.5	68.7	:	:	:	11	0.45
Tonga	99.3	99.2	25.9	66.2	7.9	3	0.57
Trinidad and Tobago	98.3	99.1	:	:	:	33	0.55
Tunisia	69.0	86.4	:	:	:	20	0.28
Turkey	81.3	96.2	76.8	14.7	8.5	9	0.26
Turkmenistan	99.3	99.7	:	:	:	:	0.65
Uganda	65.5	81.8	93.5	1.6	4.8	31	0.69
Ukraine	99.6	99.8	25.6	36.0	38.0	8	0.59
United Arab Emirates	91.5	89.5	:	:	:	23	0.27
United Kingdom	:	:	:	:	:	20	0.67
United States	:	:	14.8	49.0	36.2	17	0.62
Uruguay	98.2	97.4	75.3	15.1	9.6	12	0.55
Uzbekistan	95.8	98.0	:	:	:	16	0.64
Vanuatu	76.1	80.0	:	:	:	4	0.69
Venezuela	94.9	95.4	63.9	21.7	12.8	19	0.48
Vietnam	86.9	93.9	:	:	:	26	0.69
Yemen	40.5	77.0	:	:	:	1	0.25
Zambia	60.7	80.8	:	:	:	15	0.56

Oxford Paperback Reference

The Kings of Queens of Britain
John Cannon and Anne Hargreaves

A detailed, fully-illustrated history ranging from mythical and pre-conquest rulers to the present House of Windsor, featuring regional maps and genealogies.

A Dictionary of World History

Over 4,000 entries on everything from prehistory to recent changes in world affairs. An excellent overview of world history.

A Dictionary of British History
Edited by John Cannon

An invaluable source of information covering the history of Britain over the past two millennia. Over 3,000 entries written by more than 100 specialist contributors.

Review of the parent volume
'the range is impressive ... truly (almost) all of human life is here'
Kenneth Morgan, *Observer*

Company of Liars

'Combines the storytelling traditions of *The Canterbury Tales* with the supernatural suspense of Mosse's *Sepulchre* in this atmospheric tale of treachery and magic' **Marie Claire**

'A richly evocative page-turner' **Daily Express**

'Imaginative, hideous, irresistible' **Sunday Telegraph**

The Owl Killers

'Scarily good. Imagine *The Wicker Man* crossed with *The Birds*' **Marie Claire**

'I'm completely engrossed' **Mark Radcliffe**

The Gallows Curse

'Bawdy and brutal' **Simon Mayo**

'A gem of a story, meticulously researched and told with bloodcurdling relish' **News of the World**

The Falcons of Fire and Ice

'A compelling blend of historical grit and supernatural twists' **Daily Mail**

'Teeming, invigorating' **Guardian**

A GATHERING OF GHOSTS

KAREN MAITLAND

REVIEW

First published in 2018 by Headline Review
An imprint of HEADLINE PUBLISHING GROUP

First published in paperback in 2019 by
Headline Review

1

Cataloguing in Publication Data is available from the British Library

ISBN 978 1 4722 3591 6

Typeset in Adobe Garamond Pro by Palimpsest Book Production Ltd, Falkirk, Stirlingshire

Printed and bound by CPI Group (UK) Ltd, Croydon, CR0 4YY

HEADLINE PUBLISHING GROUP
An Hachette UK Company
Carmelite House
50 Victoria Embankment
London, EC4Y 0DZ

www.headline.co.uk
www.hachette.co.uk

I shall see a world I will not like – summer without blossom;
kine without milk; women without conscience; men without
courage; conquests without a king; woods without mast; seas
without fish; faulty judgements of old men; false precedents
of judges. Every man a betrayer, every son a destroyer . . .
An evil time.

The Prophecy of the Morrigan from
Cath Maige Tuired (The Battle of MaghTuireadh).
Irish, *circa* ninth century in its written form

Soften your tread. The Earth's surface is but bodies of the
  dead,
Walk slowly in the air, so you do not trample on the
  remains of God's servants.

Abu al-Alaa al Maarri, eleventh-century poet
and philosopher, born AD 973

Cast of Characters

Hospitallers' Priory of St Mary, Dartmoor

Prioress Johanne – sister in the Order of the Knights Hospitallers of St John of Jerusalem and elected head of the Commandery of St Mary

Sister Fina – sister of the Order of the Knights of St John and keeper of the Holy Well

Sister Clarice – bookkeeper and steward

Sister Basilia – infirmarer, in charge of the infirmary for the sick and elderly

Sister Melisene – hosteller, responsible for the hospitality for travellers and pilgrims

Goodwife Sibyl – lay servant who cooks for the sisters

Meggy – local widow and lay gatekeeper of the priory

Sebastian – disturbed patient in the infirmary

Father Guthlac – a blind patient in the infirmary

Brengy – the stable boy

Dye – a kitchen scullion and sister of Brengy

Hospitaller Brothers

Knight Brother Nicholas – warrior-monk and steward of the Knights of St John

Brother Sergeant Alban – groom and non-military serving brother of the Knights

Hob – a carter from Buckland

Commander John de Messingham – preceptor of the Knights of St John at Buckland

Lord Prior William de Tothall – head of the Knights of St John of all England, based in Clerkenwell

Knight Brother Roul – Knight Hospitaller from Clerkenwell

Villagers on Dartmoor

Morwen – young cunning woman, daughter of Kendra

Kendra – blood charmer and former keeper of the Holy Well

Ryana and **Taegan** – Morwen's elder sisters

Deacon Wybert – local parish cleric

Tinners

Sorrel – woman with a withered arm

Todde – seeking his fortune as a tinner

Master Odo – owner of the tin-streaming works

Gleedy – Master Odo's right-hand man

Eva – cook for the single tinners

Chapter 1

Hospitallers' Priory of St Mary, Dartmoor
Eve of May Day and Feast of Beltane, 1316

That night, of all nights, Sister Fina was late. If she had arrived on time to close the holy well beneath the chapel, perhaps she might have averted all that came after, but she hadn't. And it was Sister Clarice who was to blame. Never let that woman start talking if you're in a hurry.

'Could I beg a moment of your time, Sister Fina?' she'd say.

But it never was just a *moment*.

Sister Basilia, who wouldn't hear a bad word said about any soul, not even if they'd murdered every child in Widecombe, once told her fellow Hospitallers they should be thankful for Clarice's *gift* of words, as it pleased her to call it, for she said the pedlars and merchants were so battered down by them they gave her what she wanted at half the price just to get away. The other sisters had rolled their eyes, for Basilia was cheerfully determined to see God's blessing in everything, even a burned bun, which vexed them even more than Clarice's nagging.

That night, Clarice's *little word* was about the extravagant

1

use of candles, or was it beeswax polish? Probably both, Sister Fina couldn't remember. She'd long stopped listening, though that hadn't stopped Clarice talking, and by the time Fina eventually hurried across the priory courtyard it was already dark. She'd had to light a lantern to avoid tripping over the abandoned pails and pitchforks littering the cobbles. The priory cat, which was ignoring the mice and hunting for scraps of roast mutton, hissed as the sister's heavy black skirts clipped her tail. Fina giggled, for the little beast sounded just like Clarice sucking her breath through her teeth at the wanton waste of yet another candle. But it served the old steward right: if she hadn't lectured her for so long, Fina wouldn't have needed to burn one.

Even though the buildings surrounding the courtyard gave some shelter, the cold wind almost ripped the cloak from Fina's shoulders as she picked her way towards the chapel. But it sounded even louder inside, as if the devil was beating his tail against the stone walls in a violent rage, furious that he'd been shut out. Fina glanced up at the tiny stained-glass window above the altar. She was always afraid the wind would blow it in if the rain didn't smash it first. Basilia said that the casement was too narrow to come to any harm, but Fina took care not to stand too close.

Fina was the youngest of the eight sisters at the priory and taller than all of them. Her shoulders were perpetually hunched as if she was trying to make herself shorter, but her red-raw bony hands and feet looked as if they'd been intended for someone twice her height, and she'd been given them by mistake.

She hurried across the stone floor, the cold seeping up through her thin leather soles, and locked the pilgrims' door at the opposite side of the round chapel, which allowed worshippers to enter the church without going through the priory. She didn't want any villagers slipping in while her back was turned. Then she ducked beneath the low arch of the doorway that led to the well. The narrow stone staircase spiralled down into the darkness and, far below, she could hear the splashing of water in spite of the roar of the gale outside. But even before she'd taken a step, something made her draw back.

The rock walls of the staircase always glimmered for they were covered with a fur of delicate green moss that radiated a strange emerald-gold light whenever the candles were burning below, like glow-worms twinkling on a summer's night. Pilgrims gasped in awe when they first saw it. When they thought no one was watching, some scraped their finger-nails down those walls to steal what they imagined to be a strip of precious gold, but they found themselves grasping only a handful of wet mud. That gave them a fright, thinking St Lucia had turned gold into dirt to punish their thieving. Fina had been tending the well daily for a year or more and the golden light was as commonplace to her as a loaf of bread, but what she saw that night certainly wasn't. The walls were glowing with a ruby-red light that throbbed and pulsated like a beating heart. She felt as if she was staring down into the belly of the hill slashed open. The rocks were bleeding.

The sight so terrified her that she almost slammed the door and fled, but she was more afraid of her prioress's tongue. That woman's glare could freeze the sun in the sky. Fina forced herself to examine the walls again. But she could make no sense of what she saw. Was the red glow coming from a fire in the cave below? She sniffed, but there was no smell of smoke and, besides, there was nothing much to burn down there, except the St Brigid crosses left by pilgrims

3

or the rags they dipped into the holy water. And Prioress Johanne always insisted those filthy offerings were cleared away nightly.

Still clutching the lantern, Fina slowly descended the uneven stone steps, holding herself tense and ready to retreat at the first sign of flames. The holy spring gushed out of a gap between three rocks in the wall and poured into an ancient stone trough, just long and wide enough for a man to lie in, as if he was in his own coffin. Fat yellow candles burned on the spikes that had been driven into the rock on either side of the spring. The melted wax dripped down the rock face to form frozen waterfalls at the base. But when Fina reached the point on the stairs where the interior of the cave became visible, she thought she saw something red glowing at the bottom of the trough, as if a scarlet flame burned beneath the water. It was there only for the blink of an eye. Then it was gone, and soft yellow candlelight flickered across the rocky floor once more.

Ducking under a low jag of rock, she stepped down into the cave and edged towards the spring, thinking that a pilgrim must have thrown a jewel into the water, which had caught the light, but there were only the usual bent pins and silver pennies in the trough, nothing else, except a few stems of the creamy-white flowers of may blossom floating on the surface – another offering from a villager that would have brought a frown to Prioress Johanne's brow if she'd seen it.

Some village girl had probably been using the flowers to sprinkle herself with the spring water in the belief that on the eve of May Day it would turn her into a beauty. As a child, Fina had watched the servant girls in her father's manor house do such things and was almost tempted to copy them now, but her prioress's face rose in front of her, like an archangel with a flaming sword. She'd never be able to hide such a sin from her inquisitorial gaze.

Fina scooped out the dripping flowers, crushing them in her fist. A stray thorn pierced her palm and she winced, glancing guiltily up at the painted wooden statue of St Lucia above the well. The saint knew her thoughts and was punishing her.

Averting her eyes, Fina searched the cave for what her prioress called 'rubbish' – a bandage stiff with dried blood, a three-armed cross woven from rushes, and a crude doll fashioned from reeds and wrapped in a white rag. By now, Fina knew all of the little holes and crevices in the cave where the local women tried to hide such things, and it didn't take her long for the cave wasn't large. There was room for only four or five people to crowd in around the well, though mostly they came in ones and twos.

The figure of St Lucia, patron saint of the blind, stood in a niche above the spring, for the sisters had dedicated the well to her. The long wooden dagger in her hands pointed menacingly at the pilgrims as if she meant to kill any sinner who despoiled it. Johanne had had the statue installed there when she had been elected as prioress eight years ago, to remind everyone that they should pray to the saint that the waters might heal them. No one ever dared say as much to the prioress, but in truth only the sisters of the Knights of St John and a few of the pilgrims ever offered their prayers to her at this well.

Old Kendra and her tribe of daughters, who once were the keepers of the spring, called it Bryde's Well and they'd cursed the whole priory on the day it was blessed for St Lucia. Prioress Johanne had forbidden them to come near the place, but the villagers who crept down to the cave still whispered the old name and made their prayers and offerings of clooties, pins and three-armed crosses not to the saint gazing down at their spring, but to the ancient one, the stone face that stared out at them behind the spring through a veil of water.

The prioress had not brought her to the holy well. That face had been watching over the spring centuries before the first Hospitaller sisters had set their dainty feet upon the moors a mere thirty years ago. Compared to that ancient carving, the sisters were no more than blades of grass beneath an ancient oak.

Fina tried never to look at the stone carving, though the face always drew her gaze, like a viper lying coiled in the corner of a room. It was hard to make out the features beneath the cascade, especially in the flickering candlelight. Basilia said it was a woman's face, with ears of wheat sprouting from her eyes and mouth. Melisene was sure it was the face of the sun, with tongues of fire leaping from it. The prioress said it was the face of a she-devil, who now lay crushed beneath the holy feet of St Lucia.

But Fina saw a skull surmounted by a warrior's helmet, with burning spears shooting from it, and when she was alone in the cave, she saw those spears dance with flame and the skull turn to stare at her. Even though she tried to convince herself it must be a trick of the candlelight flickering over the twisting water, even though she knew the demon had been crushed, still she could not shake off the feeling that the she-devil was very far from dead.

She shivered and, taking care not to look at the stone carving, rolled up the black sleeve of her kirtle. Clenching her jaw against the cold, she plunged her arm into the icy water to scoop out the glittering silver coins from the bottom of the pool. Even the bent pins had to be collected, for whenever they had amassed a boxful, they were sold to be melted down for their silver. But it was like trying to snatch minnows with your bare hands. The pins and coins were never where they appeared to be under the water. Over the past months, Fina had learned the skill of catching them, but that night, perhaps because she was still unnerved by the red glow she had seen, her fingers were as clumsy as those of an old woman with palsy, and the ripples she made as she lunged for them only sent them drifting further away. She gave up. She was hungry for her supper. The prioress would not come down here so late. She'd try again in the morning.

She blew out the candles on either side of the spring. Shadows closed in, like a pack of wolves, and only where the feeble light from her horn lantern flickered over the walls did the moss still glow with a green-gold haze. Once the light was gone, like the water, the moss turned black. She hurried up the stairs and into the safety of the chapel, slamming the door to the staircase behind her, as if the darkness might come bounding after her.

As the door banged, there came a yelp of fear. Fina spun round. The chapel had been empty when she'd gone down, she was sure, just as she was certain she'd locked the far door leading to the outside. But now a little boy was standing by the stone altar, gripping the corner tightly in both hands, turning his head this way and that, as if trying to see what had made the sudden noise.

He looked seven or eight years old, his tangled black hair curling over the top of a brown homespun jerkin. Fina thought he must be travelling with a family who'd taken shelter in the pilgrims' hall for the night and, as children

7

do, had gone exploring and somehow found his way into the chapel through the door from the courtyard, which she'd left unlocked.

'What are you doing in here, child?' She took a few paces towards him, intending to usher him out. 'The holy well is closed for the night. You—' She broke off. He was cringing, his arm raised over his face as if he expected a blow.

She held up her hands to assure him she meant him no harm. 'We'd best get you back to your kin before they start to fear the wisht hounds have taken you.'

She'd meant it as a joke, but he seemed even more terrified.

'Come,' she said, as gently as she could. 'Supper will be served soon and you don't want to miss that. There'll be a good hot soup. Well, the soup will be hot, at least.'

Good was not a word anyone bestowed on the cook's meals in the pilgrims' hall. Even when he did flavour the pottage with herbs or a bone stock, all you could ever smell was burned beans. That man could scorch water. The sisters always gave heartfelt thanks that Goodwife Sibyl cooked for them.

Fina raised the lantern, more to let the child see that she was smiling than to study his face. Only then did she realise the boy wasn't looking at her. His head was turning from side to side, as if he couldn't understand where her voice was coming from. His eyes were as dark as the peat-black bog pools, clear and unclouded. Twin reflections of the flame in the lantern blazed in the wide, bright pupils, but he couldn't see that light. He was blind.

She touched his shoulder and he started violently. Then his fingers inched up to grasp hers. He clung to her with a hand as cold as the water in the well below, yet his touch seared too, like ice sticking to bare skin, and she had to force herself not to flinch.

It took Fina and the boy some time to reach the pilgrims' hall, though it was only across the corner of the courtyard. The child was afraid to move. He stumbled on the cobbles and kept stopping abruptly whenever he thought he might bump into something.

That evening, only five people occupied the long, narrow chamber where travellers in need of a night's shelter ate at the scrubbed table and slept on the straw pallets on the floor. Two were pedlars, the others a master cordwainer and his pregnant wife, the last an old woman who, from her torn but costly gown, looked as if she had once known better times. But none recognised the boy or remembered seeing such a lad with anyone on the road.

Leading the child out into the courtyard again, Fina pulled him into the infirmary, which stood alongside the pilgrims' hall, where the sick, the frail and those travellers in need of many days or weeks of rest were cared for. There were a dozen beds and most were occupied.

Sister Fina's gaze darted at once to the far corner, where Sebastian sat curled on a heap of sheepskins as far from the fire as he could get. He'd been there longer than any of the others, longer than most of the sisters, and though he wasn't an old man, the hair that tumbled down his back was white and his limbs thin as worms, every joint swollen and twisted at odd angles. He was staring at a crucifix in his lap, clumsily rubbing the wounded hand of Christ with the tip of a finger, as if he was trying to soothe the hurt. Fina was relieved that he appeared quiet and calm tonight. She did not want him frightening the boy, for Sebastian would sometimes cower and scream, as if he was being tormented by all the demons in Hell. Many of the servants whispered that he was possessed but, curiously, it was often Prioress Johanne who calmed him when he was seized by these evil spirits. Although Fina couldn't begin to imagine how, for in her experience the prioress was more formidable than a legion of devils and more likely to scare someone out of their wits than into them.

Sister Basilia, the infirmarer, was at the other end of the hall, apparently giving instructions to one of the female servants. There was a mulish expression on the maid's face, and she folded her arms sullenly, staring at the long table on which the remains of supper still lay – burned mutton broth by the smell of it. Basilia kept smiling as if she was quite certain the woman would do whatever she was plainly resisting. She reminded Fina of a plump, eager spaniel, always wagging her tail and jumping up, convinced that everyone she met wanted to be friends.

She broke off as she caught sight of Fina and bustled over, while the servant seized the opportunity to escape, collecting the wooden bowls from the table with ill-tempered bangs and clatters.

Basilia beamed down at the child still clutching Fina's hand. 'And who have we here?' She gave the black curls a vigorous pat. The boy shrank back. She chuckled. 'Shy little fellow, isn't he?'

'Not shy, Sister.' Fina hesitated, then guided him to an empty bed. She prised his icy fingers from her hand and pressed them to the straw mattress. 'There . . . a good, warm place to sleep. Can you feel the wall behind? You stay here. I'll be back in a moment.'

The boy stood where she'd left him, his hands dangling,

his head turning this way and that to follow the many voices and clatter of dishes, but he made no attempt to touch anything around him.

Fina returned to Basilia and drew her aside. 'I found him alone in the chapel. He's blind, but I don't believe he can have been so for long – he has not learned to use his hands to discover where he is and he can't follow sounds, as Father Guthlac can.'

She nodded towards an elderly man sitting close to the fire, his fingers and lips moving as he recited his paternosters, counting them off on his string of beads. But his mind seemed not entirely focused on his devotions, for he cocked his head, listening to the chatter around him, smiling at this, frowning at that, occasionally calling a remark mid-prayer. His sight had faded gradually over the years, but with the help of his deacon he'd still been able to perform his duties as parish priest. Like most, he had never been able to read much Latin and had always gabbled the services by rote, so his parishioners scarcely noticed when darkness had closed in upon his world.

Basilia glanced over Fina's shoulder at the boy. 'Who brought the poor mite here?'

'I don't know, but someone must have. He can't have found his own way in. He's not even able to cross a room alone. He can't tell me where he came from or who he is. He hasn't uttered a word. I don't know if his kin have abandoned him to our care, or they mean to return for him, if he can be healed.'

Basilia regarded her with sad, reproachful eyes, as if she'd betrayed their faith in thinking that St Lucia might not perform a miracle. 'Imagine leaving a child when he needs you most. What mother would do such a thing?' She puffed up her chest like an indignant hen. For a moment Fina thought she would march over to the boy and gather him up in her arms, like a baby.

They all knew that it was Basilia's greatest sorrow that she had no children. But with a litter of lusty sons to provide for, in addition to his daughters, her father had been able to offer land enough only for one of his girls to acquire a husband of suitable rank.

'Maybe he has no mother,' Fina said. 'And no one else can spare food for him. The famine is biting hard and if he can't work . . . You see how he is. He can do nothing for himself.'

'But he can learn,' Basilia said firmly. 'And there's no one better to teach him than Father Guthlac. He'd still be out tending his flock, if his poor swollen legs would bear him up.'

She lumbered over to the boy, seized his hand and dragged him towards the old priest. In her eagerness, she didn't watch him closely enough and the lad collided with the corner of the table, setting the remaining bowls and spoons on it rattling.

Father Guthlac turned his head towards the sound. 'Who's that?' he called. 'Don't know that tread. I reckon they've been supping too much mead by the way they're crashing about.' He chuckled to himself.

Basilia took the boy by the shoulders and steered him close to the old man. 'A boy brought to us, Father Guthlac. Sister Fina believes he's newly blind and no one's shown him how to get about for himself. We thought you might teach him.'

The old priest raised his hand to silence her. 'Come closer, boy.' He extended a wrinkled hand and grasped the child's sleeve. The boy tried to pull away, but Father Guthlac had dealt with a good many little sinners in his time and held him firmly by the shoulders. He lifted the boy's arm by the cloth and ran his hand down it until he found the fingers. The old man stiffened, hunching forward in the chair and sucking his breath in noisily through his teeth. His hand darted to the boy's face, running lightly over it, like a spider.

Then the old priest gave a cry of horror and jerked his hand away, as if he'd been stung. His sightless eyes flashed wide in fear. Seizing the staff beside him, he struggled to his feet, his paternoster beads slithering to the floor. He tottered backwards, crossing his breast as if the devil himself had risen up from the ground in a cloud of sulphur. Clutching the corner of the table, he brandished his staff towards where Basilia and the boy stood.

'Drive him out!' he shrieked. 'Drive him out from these halls now.'

'Father Guthlac!' the infirmarer protested, wrapping her arms protectively across the boy's chest. 'Whatever has possessed you? He's a little boy, a helpless child. Didn't you hear me tell you he's blind?'

'If you don't put him out this very hour he'll destroy us. Destroy us all! I know what you are, boy. You may fool those who can look but don't see. But I can see you, boy – see you plain as sin. Be gone, foul creature of darkness!'

Fina rushed forward to try to quieten the priest and help him back to his seat, but he was waving his staff so wildly she was forced to retreat. The servants and the patients who could move had backed away to the corners of the hall as if they were afraid the blind priest might charge towards them. In the corner, Sebastian was moaning in fear. He shrank against the wall, trying in vain to cover his head to protect himself, but he could not raise his poor twisted arms.

13

Fina tried to placate the old cleric. 'Father Guthlac, we can find the boy somewhere else to go in the morning. In any case, his kin may have returned by then. But he'll have to stay here till daylight. The gates are locked for the night.'

The old man's mouth twisted in fear and rage, a stream of grey spittle trickling from the corner of his mouth. 'You want your sisters to be alive come cockcrow, then heed me, Sister Fina. You take that demon, bind him tight, and throw him into the sucking mire. For I give you fair warning – if that boy sleeps beneath this roof this night, not one of us will be spared the curse he'll bring down upon our heads.'

Chapter 2

Sorrel

That was the day I knew I had to go, though I'd no notion where or why. I only knew I felt the urge flooding my veins, as swallows sense the icy grasp of winter stretching out to crush them, even before the first leaf has fallen, and know they must fly before it's too late. *Too late!* Yes, that was the ghost that had come to haunt me. Every dawn for months I'd woken in dread knowing that something was wrong, so very wrong, but what? *She* gave me no answer. She spoke no other word to me, as if I was talking to the wind or the sun or the moon. She said nothing but *Come*.

I was kneeling furthest downstream from the rest of the village women. The bailiff's wife always squatted highest upstream so that no dirt or lye from the others' washing could touch hers. That was her spot by divine right and no one dared usurp it, even though that day she wasn't with us. The lowest place was left for me. No one wanted water from my clothes near theirs for fear that my misfortune might flow on to them. The river carries curses from one person

15

to another, like the wind carries dust from one man's field and blows it into another's eyes.

But I was used to my neighbours drawing away from me, glad of it really, for then I need not join in with their prattling, though I could hear them bellowing to each other over the rush of the water.

'Don't know why we're bothering to wash these clouts,' one shouted. 'The rain'll never hold off long enough for them to dry.'

'To stop the menfolk bellyaching that their breeches are stinking and lousy,' her neighbour replied. 'Spend half our lives doing things just to stop them complaining.' She ferociously pounded her husband's shirt with a stone as big as her fist, as if it was his head she was battering.

'You'll never get rid of the lice,' another called, 'no matter how long you hold them under the water. Cunning little beggars. I reckon they're the only beasts left in this village that aren't starving.'

I scrambled on to a wet boulder that jutted into the water, and used my chin and my good arm to twist my shift into a rope, slapping it against the rock. It would not get as much dirt out as scrubbing, but the cloth was so threadbare it would fall apart if I rubbed too hard.

The river was running high and would run higher still, if we didn't get a dry spell soon. A year or so back, the boulder I was sitting on was so far clear of the water that on a hot day you could spread your linens on it to dry. But it had been months since we'd glimpsed even enough blue sky to make a cloak for the Holy Virgin. The village children talked about summer as if it was some fanciful tale a storyteller had invented.

A sudden surge of water smashed into the rock, which

shuddered beneath me. The river swirled and foamed around the boulder, tugging at it as if it meant to tear it free and send it sailing downstream. The sky was darkening. A chill wind wrapped itself around me, breathing ice on to my wet skin. Another squall was coming in. But upstream, the women chattered on, like the babble of water over stones.

A dark red flash in the river captured my gaze. I thought it might be a flower, though I couldn't make out what kind. The current swept it towards me. It spun on the surface, drowned and rose again, then was tossed from one side to the other, as the eddies caught it, until it bumped against the boulder on which I sat and was trapped behind it, trembling as the spray buffeted it. Now that it was close, I could see what it was – hound's-tongue. Its crimson petals are beautiful to look at, but it stinks like dog's piss and poisons sheep or cattle if they swallow it. I reached down to fish it out before it could do any harm.

But as my fingers stretched towards it, the water around it began to turn red, as if the petals were bleeding into it. The stain widened and spread till it touched the riverbanks. I snatched my hand back and jerked upright. For as far as I could see, above and below me, the whole river was blood red. But the women were still pounding their clothes in it. Red liquid ran down their arms and dripped from their fingers. Scarlet droplets glistened on their faces. The shirts they were scrubbing were stained crimson. But the women were still washing, still chattering, still laughing.

I scrambled to my feet, leaping back to the bank. 'Stop! Stop! The river— Can't you see?'

Their hands froze half in and half out of the bloody water.

17

They stared at me, then at the river and back at me, gaping as if I was making no more sense than a cawing crow.

'The water! It's full of blood! Look at it!'

They looked. Then they dropped the sodden scarlet clothes on to the bank and came hurrying towards me. I pointed down into the bloody water and they gazed slack-jawed.

Then one laughed and snatched up the dripping flower, waving it at the others. ''Tis only a sprig of hound's-tongue.'

'You want to keep that.' One chuckled. 'You put that inside your shoe on a journey and no dog'll come nigh you nor bark at you.'

The finder rolled her eyes. 'Blood, indeed. A flower, 'tis all. You blind, as well as crippled, Sorrel?'

'Hound's-tongue.' Another snorted. 'If you ask me she's been bit by a mad hound. First sign it is, being scared of water.'

'Then she'd best keep this. They say it cures that too.' The woman tossed the stinking flower at me. It caught in my hair and I felt drops of water from its sodden petals trickle down my cheek. I pulled it out, and stood there foolishly clutching it.

They were laughing, but I could see the fear in their eyes as they stared at me. I was afraid too, but I made myself glance down. Under the leaden sky, the river was as transparent and colourless as the tiny elvers that wriggled through it.

Drops of rain began to splatter on to the rock. The women hurried off to gather up their washing and hasten home. They glanced pityingly at me over their shoulders as they walked away, and I knew they were whispering about me.

The wind snarled, tearing at my skirts, dragging the bindings of my hair loose and whipping the strands around me. The branches of the willows snaked across the water towards me, as if they were trying to hook me and pull me away from the river into an air I couldn't breathe. The clouds were bruised purple in the sulphurous light. At any moment, the storm would break over me. I should run home, but I couldn't move.

Come, you must come to me.

It was the wind playing tricks on my ears, as the river had done to my eyes.

Come to where the fire burns.

It was one of my neighbours. She was coming back to urge me to hurry to the byre I called home.

But I didn't turn my head, for I knew I wouldn't see anyone walking towards me. It was not a neighbour. It was not the wind. I knew that voice. I'd heard it calling to me since I was nine summers old, but until now only in my dreams. Was I asleep? I felt the sharp sting of rain against my face. Was I running mad?

Fire and water wait for you. The time is now.

The foul reek of hound's-tongue burned my nostrils. I stared down. My right hand was balled so tightly into a fist I could no more move its fingers than I could those on the withered left. I had to will it to open. Slowly, painfully, my fingers uncurled. The mangled red petals of the flower lay in my palm. They had stained my skin scarlet.

A flash of lightning startled me into looking up. On the opposite bank a fox, a black fox, stood motionless, its head turned towards me, ears pricked. We stared at one another. Its brush streamed out in the wind, like a flame from a blazing torch. The animal reek grew stronger till I was almost choking. And I knew that, even if I let the flower drop, it

19

would not leave me. The smell was in the air, in the water, in the beast that stood watching me.

Come! Come to where the fire burns in my heart.

A growl of distant thunder rolled across the sky. The black fox had vanished, but its stench still rode upon the wind.

Chapter 3

Prioress Johanne

A sudden chill drenched my skin as if the door to my chamber had been flung open and a cold wind had barged in. The sensation was so intense that I glanced behind me, but the door remained firmly latched. I poked at the sluggish fire, trying to prod it into a blaze. Exhaustion: that was why I felt so cold. My head throbbed. I longed to beg Sister Clarice to retire for the night and discuss the accounts in the morning, but my steward was one of those women who never needed more than an hour or two of sleep and couldn't understand why others wasted so much time in their beds.

Clarice ran a crooked finger down the column of figures she'd inscribed on the parchment. The numbers stood erect in such straight lines that not even a master mason with a plumb line could have schooled them better. 'At least our calves are fetching twice what they did last year,' she announced, with grim satisfaction. 'The farmers who graze their cattle on the lower pasture lost many beasts because of the rain and floods, but we've been spared the worst up here.'

She thrust aside the parchment and pulled another towards us, impatient to move on. She was a small, compact woman, but her black kirtle was tight about her. She had not so much as a pinch of spare flesh on her frame, but saw no reason to squander cloth on folds her body would not fill. She had come late to the Sisters of the Knights Hospitallers. She and her husband had, as donats, made a generous gift of land and money to the order, but had continued to live in the outside world. Clarice had managed her husband's warehouses and properties while he travelled through Europe buying and selling merchandise for his cargo ships, but after his death from a fever in France she had made her full profession. I sometimes found myself wondering if that had been wise, for Clarice was unaccustomed to consulting with others, much less having her decisions questioned. If I had not heard her take the oath I might have been tempted to believe that, while she had vowed poverty and chastity sincerely enough, she had omitted obedience.

She was still talking and now rapped her finger against another sum on the parchment, as if she was disgusted by its indolence. 'But the offerings left at the well are lower, far lower.' She stretched out this last phrase letting it vibrate in the room, as if she was an ancient prophet announcing the destruction of a wicked city.

I moved the candle closer, peering at the column with smarting eyes. 'Lower, yes, but only by a trifle, Sister Clarice. I am sure that will make little difference when set against the income from wax and wool.'

Clarice gave an exasperated sigh. 'The point is, Prioress, that the number of people coming to the well has increased so the amount collected from the offerings should have increased too, but it has not.'

'But the famine—'

A shriek echoed across the courtyard outside. We both jerked round. Someone was shouting and seemed to be in great distress. I struggled to my feet. 'Sebastian! I should see what can be done to calm him.'

'Didn't sound like him,' Clarice said. 'But, if it was, he's quietened now.' She scuttled to the door and dragged it open, peering out across the dark, rain-drenched courtyard. The wind roared in, scattering the parchments on my table and sending billows of smoke and sparks from the hearth swirling about the room.

'Sister, have a care for the fire,' I protested, kneeling down to gather up the documents before the wind blew them into the flames. But before I could rise again, I heard another bellow as, somewhere, a door opened and slammed.

Clarice flattened herself against the wall as Sister Fina came hurrying in, tugging a small boy behind her. Clarice latched the door while Fina stood coughing and flapping ineffectually at the clouds of smoke. In spite of the weather, she had not fastened her cloak, which was trailing wet and muddy behind her. The drenched skirts of her black kirtle clung to her legs.

'What's the commotion? Sebastian again?' Clarice demanded, before I could utter a word.

'It was Father Guthlac,' Fina panted, wiping her wet face on her damp sleeve. 'The boy . . . I never expected . . . how could I? After all, he's just a little boy.'

'Calm yourself, Sister Fina,' I said firmly, in an attempt to remind both women that *I* was prioress and they were standing in *my* chamber. 'Why don't you and the boy come closer to the fire and dry yourselves, before you get the ague?'

Up to then the child had been half hidden behind Fina.

She grasped him by the shoulders and guided him gently towards the hearth. Although he was not in any danger of being burned, he shrank back as he felt the warmth, holding up his hands to shield his face as if he couldn't understand the source of the heat.

I wondered if he might be simple. Sadly, there were many such children born to villagers in those parts. They would rock back and forth, or shriek uncontrollably at the sight of something as commonplace and harmless as a feather. Changelings, the local people called them. They swore that pigseys stole human babies and left their own strange offspring in their place. But the boy's expression was not vacant like those children's. As he turned, he blundered into a stool and I realised he couldn't see, not even the flames of the fire that were dazzlingly bright in the dimly lit room.

'Who is this child? A villager?'

Fina gnawed her lip. 'I found him alone in the chapel, Prioress, when I was closing the well. He won't tell us his name and no one in the pilgrims' hall recognises him. I thought perhaps Father Guthlac might help him, being blind himself, but as soon as he touched him . . .' She twisted her long fingers as if she was attempting to plait them.

'Oh, for goodness' sake, spit it out, Sister Fina,' Clarice snapped. 'Prioress Johanne and I have a great deal to discuss.'

'The accounts can wait until the morrow, Sister Clarice,' I said firmly. 'Don't let me detain you if you have things to do.' My eyes felt as if they had been skinned. I couldn't even look at another column of figures much less make sense of them.

I was relieved as I saw Clarice scuttling towards the door,

24

but instead of opening it, she plumped herself down on a narrow bench beside it and folded her arms as if to make plain she was waiting for an explanation.

Fina glanced at the boy, who had backed as far away from the fire as he could and was crouching in the corner. She edged closer to me, lowering her voice, though in that small chamber the child could hardly fail to hear her, unless he was deaf as well as blind. 'Something about him seemed to upset Father Guthlac. He started yelling that we'd draw a curse down on us all if we let him stay. He said . . .' Fina swallowed and dropped her voice to whisper '. . . he said we should drown the boy in the mire. He's a priest!' she added, in a shocked tone, as if that fact had somehow escaped me. 'He's always so kindly and placid. What should we do, Prioress?'

'Do? I think that is plain enough. Unless you want the whole priory to be kept awake you had better keep the boy well away from Father Guthlac. I imagine his shouting has also alarmed the other patients. Is Sebastian distressed?'

She shrugged.

'Keep the boy out of the infirmary,' I said. 'We want no more disturbances for our patients.'

'But, Prioress, suppose Father Guthlac is right? They say the blind have the gift of second sight, and the boy did appear in the chapel after I'd locked the door. How could he have got in there unless by dark magic? And there's something else . . . When I opened the door to the well, the rocks . . . they were running with blood.'

'Blood? Whose blood?' I demanded. 'Has someone been hurt?'

Fina was twisting her fingers again, like an anxious child. 'I didn't mean . . . Not real blood, but the rocks were glowing blood red. In all these months I've been sister of the well, I've never seen such a thing.'

'Sister Fina, you are not an unlettered cottager! What you *think* you saw was a reflection of the candlelight on water, nothing more. You know the walls glow when the candles are lit. As for the boy being in the chapel, I dare say he was crouching in a corner when you came in as he is now, afraid to move or call out since he couldn't see who was walking about. But if you are going to take fright at shadows, perhaps I should appoint another sister to take charge of the well and you can work in the kitchens. In my experience plucking chickens and pounding dough is a sure cure for any strange fancies of youth.'

From her perch by the door Clarice gave an impatient snort. 'The boy got in because she neglected to lock the pilgrims' door properly. It wouldn't be the first time I've found it open after she swore she'd closed it. You don't attend to what you're doing, Sister. Always daydreaming. The prioress is right, some honest toil in the kitchens would soon set your head firmly back on your shoulders.'

Fina opened her mouth to protest, but I was in no mood to listen to an argument between the two. 'It matters not how the boy got into the chapel. Our duty is clear. He is a child without kin, at least until a relative claims him, and as such we are pledged to give him shelter and care. He shall stay here until we see if St Lucia, in her mercy, will intercede for him. Eventually he can be sent to one of our brothers' priories to be—'

I was interrupted for the second time that evening by a rapping at my door. How many more people were going to come charging in tonight? Could no one solve the slightest problem without running to me?

26

Clarice rose to unlatch the door. The gatekeeper, Meggy, edged in, a ragged piece of sacking grasped tightly over her head against the rain. Her florid cheeks glistened with water, and more dripped from the end of her broad, fleshy nose. Some in the priory muttered that the widow was too old for such a post, but she'd spent most of her life ploughing the family's own strip of land and helping her late husband in his blacksmith's forge, hefting iron, pumping the furnace bellows and holding the heavy carthorses. Years of such toil had left her with brawn enough to deter most unwelcome intruders. Besides, I knew that she had nowhere else to call home now.

'Prioress, Hob and his lad have come from Buckland with the supplies.'

'At this hour?' The carter usually arrived in the morning in time to unload his wagon and eat in the pilgrims' hall at noon, before setting out again. For reasons I had never understood, Hob always refused to stay overnight, preferring to sleep in the stables of inns or even on the open road rather than in our warm hall.

All our wine, with goods such as parchment and black cloth, which we could not purchase from the local markets, was delivered to us the commandery at Buckland along with any messages from the preceptor of the Knights Hospitallers there, John de Messingham. But I had noticed in these past months that letters sent out by the Lord Prior, William de Tothall at Clerkenwell, intended for all the priories of the English order were dispatched to us through Buckland, instead of straight to my hand, which annoyed and alarmed me.

The commandery at Buckland was small, the number of professed brothers seldom rising above seven, but the community of sisters in the nearby preceptory of Minchin Buckland had, of late, swollen to almost fifty professed women and dozens more lay servants as, one by one, Lord William had persuaded the priories of sisters across England to enter the cloister at Minchin Buckland and live under one roof, with the knights at Buckland tasked with the noble duty of protecting and serving them. Although, if you asked me, *guarding and gaoling them* was a more accurate phrase.

But we would *not* be joining our sisters to sit sewing and praying for the souls of our brothers, of that I was determined. We were needed here, caring for those who crossed the moors, for they were fraught with peril. Outlaws hid among the rocky tors and deep valleys watching for vulnerable travellers. Seemingly lush swards of grass concealed mires that could suck down men and horses alike. Bone-chilling mists descended without warning, sending men wandering in circles till they died of cold or exhaustion.

'Hob has more sense than to risk crossing the moor at night,' I said. 'What's brought him so late to our door?'

Meggy shrugged her broad shoulders. 'Slow journey, Hob says. Up to their hocks in mud all the way, they were. Said they spent more time digging the wagon out than riding on her.'

'Tears of Mary!' Clarice muttered. 'They could have had their throats cut on that lonely track when it's as black as this, especially if the cart got stuck. There are more men turning to robbery every day, desperate for food since this famine took hold. Why didn't he and the boy take shelter for the night?'

'I reckon he'd a mind to do just that, but the knight riding with them was determined to press on to reach here. Asking to see you, he is, Prioress. I told him you'd not be best pleased to be disturbed. "Priory's locked up for the night," I said. I'd not have let him through my gate at this hour, save that he wears the cross of your order.' She gestured with her chin towards the white *cross formée* on the shoulder of my left sleeve.

'A brother Hospitaller? Here? Did he ride with the wagon to guard it?'

'If he did,' Meggy muttered, 'it was a foolhardy thing to do. Hob knows how to slip past any trouble quietly without drawing attention to himself. Folks see his cart being escorted by a knight with his white cross flashing out in the dark for all to see and he might as well have the King's herald marching in front of him bellowing, "Here's a fellow wants robbing."'

'If Commander John de Messingham sent a brother from Buckland, he must have had his reasons for doing so,' I said sternly. It did not befit a servant to question the actions of the Knights of St John.

All the same, I confess that I was a little curious too. The brother was probably on his way to board a ship at Fowey to voyage to the Citramer, our order's heartland on Rhodes, for every Knight of St John who held office was obliged to serve a season there or on our ships in the Turk-infested Aegean. But an unsettling thought burrowed into my head. Had the knight come to collect the responsions that every priory was obliged to pay to the mother house at Clerkenwell? But that money was not yet due. My stomach lurched. Why would he have come to demand it early?

'Where have you left our brother?'

'Sister Melisene took him to the guest chamber.' The expression on Meggy's face made plain that, had it been up to her, the knight would have cooled his heels out in the rain. 'There's a groom with them. He's supping with the knight. I told Hob and his lad they could bed down in the pilgrims' hall, seeing as how it's half empty.' She grinned. 'Hob's not best pleased about spending the night here. Said he'd sooner be on his way, but I told him, I said, "There'll be no unloading that cart this side of dawn. Sister Clarice will want to check every last keg and bundle, and she'll not be of a mind to do that in the dark." Isn't that right, Sister?'

Clarice gave her an approving nod.

'Then I had better find out what brings this knight to our door,' I said, reaching for my cloak.

'But what shall I do with the boy?' Fina wailed.

I'd almost forgotten she was there. The boy had not moved out of the corner where he was crouching, though he leaned towards us as if he was listening. I grasped his little hands and pulled him to his feet. His fingers lay limply in mine, cold as a corpse's. I examined his face carefully, feeling his arms and ribs. He offered no resistance.

'He's pale and thin, but not as emaciated as most children who've taken to the road in search of food.' I straightened. 'Give him something to eat and put him to sleep in the pilgrims' hall. And ask Sister Basilia to give Father Guthlac a sleeping draught to calm him. He can recognise faces by touch and it may be that something about the boy's features reminded him of someone he once knew. If it was a person he thought dead, the shock might have caused him to utter such wild words.'

Fina began, 'Father Guthlac said—'

But I cut her off. I had no time or patience to listen to more wild talk about boys appearing by magic and rocks bleeding. 'I'm sure Father Guthlac's reason will be restored

30

by morning,' I told her firmly. 'He may even be able to tell us who the child is and where he lives.'

When I ventured out into the courtyard I walked the long way around so that I could pass beneath the casements of the infirmary. I stood for a few moments outside listening, but all seemed quiet within. The rain had eased into a fine wet mist, but the wind dashed it so hard against my skin that I was soaked before I'd reached the door of the lodgings that were reserved for highborn guests and clergy. I tucked stray strands of wet hair beneath my black veil, hoping I did not look quite as bedraggled as I felt.

Inside, two men were seated at either side of the table, fishing rabbits' legs from a dish of egurdouce, and gnawing them so voraciously that I would have sworn they had not eaten for a week. They clambered to their feet as I closed the door behind me, wiping the thick wine sauce from their fingers to their napkins and inclining their heads somewhat curtly.

'I am Knight Brother Nicholas, late of Buckland,' the older of the two said.

He had the appearance of one who had been forged of steel, not flesh, from the silver-grey of his close-cropped hair and the stubble on his chin, to the hard, angular bones of his face and rigid stance. I suspected that though his knee might bend when he prayed to God, his back never would. A single glance at his hard-muscled frame and corded neck told me he had known battle.

'This is Brother Sergeant Alban, my groom, also late of Buckland,' he added.

The groom, a bow-legged, wiry fellow, shuffled awkwardly and glanced shrewdly up and down the length of me through narrowed eyes, as if he was appraising the worth of a horse in the marketplace.

'I heard a man shouting earlier. If someone is causing trouble, I'll soon put him to rout.' Nicholas gestured towards his sword, hanging from a peg on the wall next to his cloak.

'That will not be necessary,' I said firmly. Why do men think a sword or a bow is the answer to every problem? 'It was merely Father Guthlac. Until recently he was the parish priest here, but now, in the winter of his life, he's being cared for in the infirmary. Something distressed him, but he is usually a mild-tempered man. I doubt he will disturb you again.

'But you said you were late of Buckland, Brother Nicholas? Then I assume you are journeying to join our brothers on Rhodes. God speed you both, and grant you safe passage. But it must be a matter of some import that forces you to travel by this route. Have you a message for us?'

Even as the words left my lips, I knew the question was foolish. Neither Commander John nor the Lord Prior of England would waste the services of two members of our noble order to deliver a message that could as easily have been carried by the carter. Hob had often carried them before and, since neither he nor his son could read, they could be entrusted with letters even of a delicate nature.

'We're not bound for Rhodes, Sister,' Nicholas said, 'which I deeply regret, but for which Alban here is profoundly grateful. He doesn't lust for the sea as his mistress, do you, Brother?' He jabbed at the groom's stomach with the point of a knife still dripping with red wine sauce.

The groom scowled and muttered something under his breath, but all the while his gaze kept darting back to the plate of rabbit, like a starving ferret.

32

'This is our journey's end, Sister,' Nicholas announced with a cold smile. 'The Lord Prior of England has commanded us to serve the order here in the Priory of St Mary. I have come to relieve you of the heavy responsibilities you have had to bear. You need have no more anxieties. I shall take charge now.'

Chapter 4

Hospitallers' Priory of St Mary

The fire spat petulantly in the hearth and the wind rattled the door and shutters, but even these sounds were less deafening than the chilling silence that had descended upon the guest chamber after Brother Nicholas's announcement. The speech he had been rehearsing ever since they had set off had withered on his tongue before he'd uttered a single word of it.

The summons to attend his commander at Buckland had filled him with elation. The arrival of two new brothers from Clerkenwell was a sign that two of the existing brothers were to be moved on, and Nicholas had prayed, with considerably more devotion than he had for many years, that he would be one of them.

All the men at Buckland grumbled that they had become little more than servants and bodyguards to the sisters of the Knights of St John, who complained almost daily about the brothers allowing their livestock to wander on to the sisters' land or not providing them with the provisions or wine they claimed they were entitled to either in sufficient quantity or quality. Nicholas would have been eternally grateful never to spend another moment in the company of those demanding harpies.

He was aching to return to the Citramer, to be back in the fight against the Saracens, with the blood and sweat of battle in his nostrils, hoofs thundering beneath him, and on either side his brother knights, their mouths roaring death, their eyes shining. He longed to be sitting round the fires at night reliving the glory, his belly full of good wine, the smell of roasting meats mingling with the scents of rose, amber and musk, which clung to the skin long after aching muscles had been soothed by silent, dark-eyed maids.

So certain had he been that he was to be sent back to Rhodes that Commander John had been obliged to repeat his instructions twice before Nicholas had grasped them. His fellow knight was indeed to set sail, but Nicholas had served far longer in the Citramer than most, and fighting in that heat took its toll on the body.

'Let young blood fight the Saracens,' Commander John had said. 'The order has a far more subtle and insidious enemy to defeat here in England, one that threatens our very existence. Think of it as a mission behind enemy lines,' he had added, with a smile. 'The task will be simple for a man of your talents and it will take no time at all. And if you succeed, as you surely will,' he spread his hands, 'then Lord Prior William will undoubtedly wish to send you back to Rhodes, not as a fighting man this time but as a commander. And, believe me, Brother Nicholas, you will need all your wits and cunning to uphold the English cause there among all those knights of France. Prove your talents in this matter, Brother, and you will rise.'

Nicholas's dealings with the sisters at Minchin Buckland

had led him to believe that flattery, combined with his authority as a man and a knight, was the best way of handling women, and with fifty noble, literate but bored sisters pitted against seven overworked and harassed brothers, it was his only weapon. But his commander had assured him that managing the eight women who were alone up here on these remote moors would be as easy as bridling a well-schooled mare. The sisters would be overjoyed to have two men to protect them.

Commander John had nudged him in the ribs and winked. 'Next time I see you, Brother, you'll be twice your girth. They'll be falling over each other to make sure you have the tastiest dishes and softest linen.'

His commander had been right about the food: the rabbit in egurdouce, which was even now congealing in the dish, was delicious, but the expression on the prioress's face told him she was as far from overjoyed as it was possible to be.

Prioress Johanne was a small, neat woman with a straight, narrow nose, and startlingly vivid blue eyes, rendered all the more piercing by the black veil that covered her hair and hung in damp folds at either side of her sharp cheekbones. She held her head so upright, that her neck, not covered by any wimple, seemed to elongate even as he watched. Steam was rising from her wet clothes in the heat of the small room, so that she looked like some avenging spectre rising from a tomb.

'Here!' she finally spat at him. 'And precisely what do you intend to take charge of here?'

Alban dropped back into his chair, making clear that he wanted no part in this conversation by stuffing his mouth with large sops of bread dipped into the rich sauce.

Nicholas's belly was rumbling and he had to fight hard

to drag his gaze from the dish. Couldn't the woman have given them an hour or so to eat before she had burst in? 'Sister Johanne, my orders—'

'Prioress,' she said, enunciating each syllable in a voice as brittle as ice.

He ploughed on, ignoring the interruption. He was damned if he was going to allow her to correct him like some errant schoolboy. 'I am ordered to take over the stewardship of the priory and to oversee the running of—'

'Sister Clarice is our steward. I can assure you that she is more than equal to the task, having run her husband's warehouses and estates for many years before making her profession. She has never once failed to collect the produce and rents owing to this priory, nor, as I am sure the Lord Prior would be the first to acknowledge, failed to deliver a perfect reckoning to Clerkenwell. I shall compose a letter to the Lord Prior and Commander John, which you may deliver to Buckland, thanking them for their care, but explaining we have no need to deprive the order of the valuable services of two brothers who, I am sure, are needed to serve in the fleet.'

Nicholas could contain himself no longer. He reached over to the pot in front of Alban and stabbed one of the rabbit's legs savagely with the point of his knife. He dragged it out, tearing at the flesh with his teeth and gulping it down. He took a certain satisfaction in watching a spasm of disgust and annoyance flash across Johanne's brow. *You can stamp and frown as much as you please, Mistress, but this is a battle I am going to win.* When he had finally stripped the leg of its meat, he waggled the bare bone at her. 'Your sister may have found it easy enough to collect the rents and tithes in times past, but the longer people stay hungry, the more obdurate they'll become. It'll take more than the pretty face of a woman and her sweet tongue to persuade them to hand over what they owe. Believe me, Prioress, half the commanderies across England cannot raise enough to send the responsions that are due to Rhodes.'

'If, on your own admission, our brothers in the other commanderies have not persuaded their tenants to pay their rents, whereas we have, it must be self-evident that Sister Clarice's "sweet tongue" has proved considerably more effective than those of our brother knights. Perhaps you would like Sister Clarice to give them some instruction on how it should be done.'

An angry flush spread across Nicholas's cheeks. 'And perhaps you would like to read the Lord Prior's direct orders, Prioress.'

He snapped his fingers at the groom, who reluctantly laid down the bone he was gnawing and fumbled in a leather scrip, finally thrusting a roll of parchment into the knight's hand. Nicholas slid it across the table towards Johanne. 'I take it you are familiar with Lord William's seal and signature.'

The prioress unrolled the parchment using only the tips of her fingers, as if she was handling a soiled arse-rag. As her gaze marched over the letters, Nicholas watched the muscles of her jaw clench so tightly that he fancied he could hear her teeth splintering.

'I am sure I need not remind you that you and your sisters have taken a sacred vow of obedience, as have I, and we are all subject to the Lord Prior's rule.'

Johanne's blue eyes glinted as cold as winter ice. 'And may I remind you, Brother Nicholas, that prioresses, unlike commanders, are not appointed by the Lord Prior but elected by their own sisters.'

Nicholas slammed his fist on the table, jolting Alban's elbow so that the sauce-soaked bread he was about to pop into his mouth slid half across his face. He wiped it on the napkin, cursing under his breath and darting furious glances at the knight.

Nicholas, trying to hold his temper in check, ignored him. 'Doubtless you are aware that the other communities of our sisters in England have been compelled to move to Minchin Buckland where they can be kept safely cloistered under one roof – all except this one. If the Lord Prior should decide to close this priory and move you to Buckland too, you would no longer be prioress. So, if you value your independence, *Sister* Johanne, you had better learn to bend your neck before it is broken for you.'

'Do you dare to threaten—'

'I make no threats, Prioress.' Nicholas held up both palms in the mocking gesture of a man who shows he is unarmed. 'I merely offer you my humble advice.' He stuffed a hunk of bread into his mouth with one hand, while pouring wine into two goblets with the other. He offered one to Johanne, who curtly refused it.

'You must forgive me, Prioress. It's been a long, cold journey and my tongue grows sharp when my backside is aching from the saddle. Years of riding in stinking, sweaty armour in the heat . . . Well, let us just say that it gives a man sores that nag at him worse than any scold who was ever sentenced to ducking.'

'I will ask Sister Basilia to make up an ointment for you.'

'Ask her if she'll rub it in for him too.' Alban chuckled.

'I'm sure Brother Nicholas does not need the assistance of a woman to find his own arsehole,' Prioress Johanne said sweetly.

Alban's jaw dropped, revealing a mouthful of half-chewed rabbit.

'I will leave you to your rest, Brothers.' Johanne inclined her head curtly. 'We will discuss this matter further in the morning.'

She stepped out into the darkness of the courtyard, pulling the door closed. She had barely taken a step when it opened again and Nicholas slipped out behind her.

He pressed himself close to her to be heard over the wind. 'I fear that we have not made the best of beginnings, Prioress, but I must obey my orders and you will find it a good deal less painful to help me to carry them out than attempt to fight me. Do I need to remind you that it's been but seven short years since the Knights Templar were first interrogated in England and their treasures plundered, even though the order had believed itself safe here? Indeed, when the persecutions of the Templars began nine years ago in France, many fled to England to escape the torture and burnings, believing the Inquisition would never come to these shores. But come they did. Now there is no country left on God's earth in which those heretics and sodomites may hide. I've heard men mock the Dominican Inquisitors as *Domini canes*, the hounds of God. But all good jokes are built upon a savage truth, for when hounds are in full cry they will tear to pieces any beast that crosses their path, whether or not it is their quarry.'

'And what has that to do with this priory?' Johanne said. It was too dark for Nicholas to see her expression, but her tone was as sharp as a freshly honed blade.

'Since the Pope gave so many of the Templar lands to our order, there are rumours that the greedy eyes of the kings are now turning towards us. If they attack us, they will gain the lands and property of both orders at a single stroke. Our

Lord Prior is determined to ensure they have no excuse to cast us into the same prisons where so many Templar knights lost their lives. He has eyes everywhere, searching for the smallest spark of corruption, heresy or sorcery in his order, and if he suspects so much as a glimmer of it, he'll grind the guilty beneath his boot before they can ignite a blaze that will burn us all on the heretics' pyre.'

'Are you suggesting—' Johanne began indignantly.

'I suggest nothing. I merely wish to remind you that the sisters from the other priories were not gathered together in Minchin Buckland merely for their safety but to ensure that no whisper of scandal attaches itself to them.'

'Because they are the weaker sex?' she said. 'Have you forgotten, Brother Nicholas? It was the *Knights* Templar who stood accused of immorality and heresy, not their sisters.'

He spread his hands. 'As you say, Prioress, the weaker sex. And a fortress is only as strong as its weakest wall. As every defender knows, that is the wall you must shore up first.'

'Don't you mean shut up?' the Prioress snapped.

A satisfied smile creased the corners of the knight's mouth. The arrow had struck its mark. 'Sleep well, sweet Sister,' he murmured, as he strode back into the darkness.

Chapter 5

Sorrel

I must have been six or seven days upon the road when I finally met the one she'd sent to guide me. Now that I think on it, he was a strange one to send, for he didn't even realise he was her messenger. No more did I, at first.

I'd been searching for a place to shelter for the night, for though I couldn't see the sun behind the leaden clouds, the grey light was fading and bedraggled black rooks were massing in a small coppice of trees around me. Dusk was swiftly approaching. My legs ached and I was dizzy with hunger. While I still had enough daylight to see and strength left, I had to build a fire and cook the few wild worts I had been able to glean from beside the track as I'd trudged through the rain. If I didn't, I'd simply sink down in the mud and fall asleep, no longer caring whether I ever struggled up again.

But just when I despaired of finding shelter, I emerged from the trees and saw a drovers' hut and pinfold a little way off the track. The pinfold was no more than a clearing of sodden earth and mildewed weeds, surrounded by a broad

42

bank. A stone trough stood in one corner, fed by a spring that flowed out of the bank and spilled over on to muddy stones beneath. By the look of it, no sheep or cattle had been driven this way for many a month – part of the bank had fallen away and one of the walls of the tiny wattle hut, which huddled next to it, had blown in. Still, half a shelter was better than none, and I'd little fear that any drover would be needing it. Since the famine had taken hold, few had any beasts left to sell and those that still lived were so weak they could scarcely be driven a mile to market without collapsing.

The earth floor of the hut was slimy and wet, but at least the roof gave some protection from the wind and rain. I wedged the sagging wattle wall upright, as best I could, and gathered a few stones to set a fire on. With my flint and iron and few precious wisps of dried flax, I managed to get a flame burning and tried to coax it into a blaze, but the wood I had gleaned along the way was so damp that the fire gave more smoke than flame. I set my pot of water and worts over it, but there was scarcely enough heat to warm my numb fingers, much less the iron pot. For the hundredth time since I'd set out, I asked myself why I'd left my village. I'd precious little there but at least I'd slept warm and dry.

I didn't know where I was going or even where I was. A week ago, that hadn't seemed to matter, for by the time I'd left the riverbank that morning, I'd had only one thought in my head – that I must go. It was all I could do to stop myself running straight out of the village and keep running until my legs could no longer move. Had I finally gone mad? Yet I knew I would go mad if I didn't follow that voice. I felt like a dog barking itself into a frenzy and leaping against the chain, almost choking itself in its desperation to break free and reach the one who was calling it.

43

I'd gone straight from the river to my father's cottage and bundled the few things I would need to survive into a small iron cooking pot that had once belonged to my mother. My father was, as usual, dozing by the fire. He grunted and cursed me as I tiptoed about gathering what I needed. He didn't open his eyes to see if it was me who had disturbed his rest. But for once his curses made me smile. I'd been raised on them, weaned on his scorn, you might say. It was almost comforting to hear the old goat swear at me for the last time. It was like a parting kiss or a dying blessing. For I was determined that, even if I starved or froze on the road, nothing would make me return.

My father constantly told neighbours and strangers alike that he should have smothered me when I was a babe, as soon as it became plain that my left arm, which had been wrenched from its socket when I was dragged from the womb, would never mend right. I can move it a little, use the back to help balance things, but there's no strength in it. My hand hangs withered and useless, like a rosebud that's been half broken off its stem.

When I was little, I would try to pull my arm off when no one was looking. It felt as if it didn't belong to me, as though the limb of some dead animal had been sewn on to my body. But at other times it seemed almost alive, a giant leech that had fastened on to my shoulder. It was swelling, bloated and full of blood, while the rest of my body shrank and shrivelled, like a dry leaf. I'd tear at the arm, bite it and scream till I was exhausted. But no matter how I hated it, I couldn't rid myself of it. I'd found out, earlier than most, that crying changes nothing.

And I'd learned to do with one hand what other women could do with two, including take care of my brother and the old goat, which I'd done ever since he'd driven poor Mam into her grave.

I'd turned in the doorway and watched my father scratch his belly in his sleep. He'd stir himself when my brother trudged home from the fields. They'd sit, the pair of them, either side of the hearth, drinking ale and grumbling about why their supper wasn't ready. I wondered how long they would wait before hunger drove either of them to look for me. And how many days would pass before they realised I wasn't coming back.

I wouldn't return, but where was I going? Ever since I'd left the village, the voice that had been calling to me so insistently had stayed silent, and the dreams I'd had in the few brief hours of sleep I could snatch, huddled in the wind and rain, were not of fire and flowing springs, but of Mam sitting on the threshold of our cottage in the summer sun, plucking pigeons. I was exhausted, soaked and starving, traipsing through mud and rain to reach some place I knew not where or why. What had I done?

I stirred the iron pot. The water was beginning to steam a little, but even if the wood burned long enough to soften the handful of blighted leaves and roots, I knew they wouldn't ease the hunger pains for long, probably only make them worse. And if I became too weak to walk . . . *Help me. Speak to me. Show me where to go.* I closed my eyes, trying to hear her voice. But the only sounds were the dripping of rain from the sodden thatch and the hissing of the wind. Even the rooks had fallen silent.

It was then that I heard the noise, not the voice in my head, but something outside. A distant clanging and clattering, like a ghost dragging chains. I shrank back. Mam used to tell me tales of tatter-foals that haunted lonely tracks

45

after dark, demons who took the guise of wild, shaggy horses and frightened travellers into bogs with their creaking and rattling, or else swept them on to their backs and carried them into lakes to drown them. Heart pounding, I peered out through the broken wattle and in the witch light saw a monstrous creature crawling down the track. I couldn't make out what it was at first, but then I laughed. It was nothing more than an ancient, swollen-kneed donkey ambling past. The noise came not from the poor beast, but from the shovels and picks strung from the wooden cradle on its sway-back.

Someone was leading the animal, though I couldn't see him clearly until he seemed to spot the drovers pen' and stepped off the track towards it. The donkey stubbornly resisted, until it caught sight of the water trough and bolted towards it, almost ripping the leading rope from his master's hands and rattling the tools so violently that the rooks flapped up from their night roost, cawing in alarm.

The man looked as if he'd not slept beneath a roof for weeks. His hair hung in long matted rats' tails and his clothes were so patched and dirty I reckoned even his lice had gone off to beg for alms. I'd have taken him for a beggar if it hadn't been for his donkey and all his tools.

Would he come to the hut? I pulled my knife from the sheath on my belt and held it ready to defend myself, though it was scarcely much of a weapon. My first thought was to slip out while he was occupied with watering the donkey, but the possessions I'd been carrying in the pot lay heaped in the corner, and my supper was just beginning to boil. I was ravenous. If I didn't get something hot inside me, I was afraid I'd not survive another night out in the cold and rain. I couldn't bring myself to tip it away. Besides, I had found the hut first: why should I give it up to him?

But if he did mean me harm, I daren't risk being trapped. If he chose to block the doorway, I'd never get past him. Leaving the pot simmering over the sulky fire, I crept out and crouched behind the bank.

The donkey, having drunk his fill, had turned his attention to the sparse patches of rank weeds and his master was sitting with his back to me on the side of the trough, pulling off his tattered boots. With a squeal of protest, followed by a sigh of relief, he dipped his blistered feet into the cold water. Mud and grass drifted from his filthy soles. But something else rose slowly to the surface. It was a single purple-red flower. It was crushed almost beyond recognition inside his boot, but I knew it, and it made my spine tingle. Hound's-tongue.

Come! Come to me.

He dried his feet on a bit of rag, then thrust them back into his worn boots. Without even a glance in my direction, he dragged the protesting donkey over to the more sheltered part of the pinfold and tethered it there, then edged towards the drovers' hut with a wariness that matched my own. He stood a few paces off and sniffed the air. The smell of wood smoke mingled with boiling worts gusted over the bank. His fingers slid to the hilt of the knife in his belt, as mine had done. But I sensed he was more afraid of being attacked than intending to do another harm. He glanced back at the donkey. Would he retreat and journey on?

The time is now. Follow. Follow.

'You're welcome to share my fire for the night, Master.' The words were out before I even knew I would utter them. *She* had sent him and I couldn't let him walk away.

He jerked round. His gaze darted to the knife in my hand, but instead of raising his own as most would have done, he shuffled backwards a few paces. I slid my blade back into its sheath and gestured towards the hut.

He still hesitated, as if he feared I was leading him into a trap, but finally a broad grin split his long, thin face. 'Kindly offered, Mistress, and I'll not offend you by refusing. My friends call me Todde. I'll not trouble to tell you what my enemies call me,' he added, with a chuckle, 'for a woman with such a generous nature could never be one of them.'

It was only as I led the man across the muddy pen that I realised what I had done. I hadn't enough to fill my own belly, never mind his, and so little wood that the fire would surely die away before the hour had passed, if it hadn't already. Suppose he became angry and thought I had tricked him.

As soon as he entered, Todde crouched, pressing himself as close as any man could to the miserable fire as if he meant to suck all the heat from it. He stank like an old wet dog that had rolled inside the carcass of a rotting pig. He pulled out a battered wooden bowl from somewhere inside his jerkin and I tipped some of the herb broth into it. I found myself giving him a bigger share than remained in the pot for me, for Mam had always given the menfolk the largest portion of any food to be had, and I had learned to do the same. I felt a flash of resentment towards him and anger at my own foolishness. The few mouthfuls I had of the watery broth made my stomach cry out louder than ever. What had possessed me to invite this stranger to my fireside?

But when he had devoured every last drop, Todde rubbed his belly and, grinning, offered me his thanks. I stared at him slack-jawed: I'd never heard any man thank a woman in my life, much less me.

We sat in awkward silence for a while, rubbing our wind-chapped hands over the dying embers. Times had been hard

48

for everyone since last year's harvest failed, and what precious seed had been saved for planting had rotted in the sodden earth or been gobbled by mice and birds that were as hungry as we were. Some folks from my village, the freemen, had packed up their families and left for the towns to see what work they could find. Most drifted back, saying it was even worse there. They'd been forced to sleep in graveyards and doorways – they couldn't find enough work pagging loads for merchants or scavenging bones for the lime-makers even to put food in their children's bellies, much less have any left to buy shelter for the night. They said the pilgrims' halls in the monasteries were full to bursting, and so many crowded to the alms windows when a dole of bread or broken meats was to be had that the food ran out long before those at the back could fight their way near enough to reach it. Many now tramped from village to village in the desperate search for work or to beg food. I guessed that Todde was simply another.

He scratched his armpit. 'Where are you bound, lass?'

I shrugged. It sounded foolish to say I didn't know and worse still to ask a stranger if he could tell me.

'Hunger driven you on the roads, has it? Aye, I've seen a good many along the tracks in search of food and work. But seeing as you treated me so kindly, I'll let you into a secret that'll serve you well.'

He lowered his voice, looking round with great exaggeration as if he was about to reveal where a crock of gold was hidden, though he could have stood on the middle of the pinfold and shouted it – there was no one, save me, to hear him.

'The King's decreed that any man has the right to look for tin on anyone's land as he pleases, without let or hindrance. There's not a sheriff or lord in the land can stop a tinner digging wherever he's a notion to, even if the land belongs to the Archbishop of Canterbury himself. But see now, it's pointless just searching anywhere. In most places a man could dig for a year and not find enough to make a sheath for a mouse's pizzle. But I know of a place where such quantities of tin lie just beneath the sod that a man can make his fortune as easily as scooping up coins spilled from a sack.'

'So where is this place of riches? Up there in the clouds with the angels, is it?' I said. My aching belly and light head were making me irritable. How could this help me? Had *she* really sent him or had the hound's-tongue been nothing more than a crushed flower that might be found in the shoes of a hundred pedlars, beggars and pilgrims to ward off savage dogs?

Todde grinned. 'You'd think you really *were* in Heaven if you saw the place. For with less digging than it takes to plant a row of beans, a man can become as wealthy as a lord. All you have to do is make a trench, then let water from a stream wash through it to sweep away the worthless gravel and leave the heavy tin. Then you pick it up, like you were gathering eggs from a hen's nest.'

Todde shuffled on his backside a little closer to me, and I found myself rocking away from his sour breath. 'It's where I'm bound right now,' he said. 'They call it Dertemora.' He frowned, then chuckled, striking his head with a palm. 'Whatever put that word on my tongue? Rain has rotted my wits. Dartmoor, that's the place.'

The fire smouldering on the stones burst up in a blaze, the flames clawing so high I was afraid they would set the hut ablaze, but Todde seemed not to notice. A shimmering ribbon of molten ruby gushed out from a brand and flowed down into the heart of the fire as if a spring was spouting from a rock.

Fire and water wait for you.

The voice was so clear, I thought someone was standing behind me. I whipped round, but the doorway was empty. And when I looked back at the fire it was no more than the heap of smouldering embers it had been a few moments before. But I knew. In that moment, I knew that this was the place I had to reach.

Todde was still talking: '. . . soon as word spreads every Jack and Jill in the country will be heading up to those moors to claim the richest seams. So, you want to get yourself there as fast as you can.'

'Have you not eyes to see?' I snapped. Lifting my withered arm, I let the useless hand dangle in the glow of the dying fire.

'Aye, I saw it,' he said softly, and a spasm of what might have been pain flickered over his face, as if it was his own arm that was hurt.

'So!' I demanded furiously, angry with him for making me show my arm, like some whining beggar. 'Even if I was to find a seam, how would I keep it, much less work it? With strapping men desperate to make their fortune, even a woman with two good arms couldn't hold a claim against them, especially if there's no law against any man digging where he wants.'

'Maybe you couldn't. But I reckon there must be more wants doing in a tinners' camp than digging. Same as reaping the grain. There's some that scythe and some that gather, tie and stook the sheaves, and others that keep all the hands fed. Haven't you noticed the pedlars and merchants haven't been coming to the villages these past months? They know there's few can afford their prices now. I reckon they must all be flocking to sell whatever they have on Dartmoor, for that's the only place where men have money enough to buy. They live off the fat of the land, those tinners.'

A sly smile creased the corners of his mouth. 'But, now, here's the thing. When those tinners come back after a long day's work bone-weary, their bellies roaring for food, they'll not want to go out again to find wood for the fire, fetch water or pluck a chicken. They'll want a good hot meal on the table afore their backsides touch their stools.'

Todde nudged my cooking pot with the toe of his mud-caked boot. 'If they're earning a king's ransom, I reckon there's a fair few who'd be willing to part with a good measure of it to someone who could put hot food in their bellies night and morn, and maybe fetch them wood enough for a warm fire to sit by of an evening.'

I stared into the ash of my tiny fire. Now that the heat was gone the wind, blowing through every broken slat in the walls, sliced bone-deep through my wet kirtle. I couldn't remember how it felt to wear dry clothes. The rain never stopped. If what Todde had said was true, there would be dry dwellings, blazing fires and good food where a host of men and women were tinning. Fetching and cooking for tinners couldn't be worse than tending my father and brother, and I'd be paid in coin, not curses, for my trouble. If I tried to survive on the road alone, I'd soon be too weak from hunger and cold to continue searching. If she was calling me to this place, then surely she was showing me a way to live. With food in my belly and warm shelter each night, I could continue to search for her and I would find her. I had to.

Chapter 6

Hospitallers' Priory of St Mary

Brother Nicholas was dreaming that he was struggling in the sea, trying desperately to swim to the shore, but seagulls were attacking him, diving at him, pecking at his eyes, and shrieking loud enough to make his ears bleed. He jerked awake and the birds vanished, but their cries did not. It took a few moments for his fogged brain to remember where he was and even longer to grasp that the noise was coming from the courtyard outside, and they were not the cries of gulls but of women.

He groped for his cloak, which he'd laid over the bed for extra warmth, stumbled to the door and dragged it open. He peered out cautiously. The courtyard was crowded with people standing in small groups staring at the far end, though what was arousing their interest he was unable to see without stepping on to the muddy cobbles. He realised his feet and legs were bare. He shivered.

'What hour is it?' he called, but if any heard him, no one replied.

An overexcited pedlar's dog began to herd a paddling of

squawking ducks towards him, their wings flapping wildly. Still unnerved by his dream of the gulls and not wanting the creatures to invade his chamber, he slammed the door and sank into a chair. The chamber was freezing. The fire was nothing more than ashes, and there was no sign of any servant bearing breakfast or even water in which to wash his face.

He had overslept, and the burning in his belly, the sour taste in his mouth and dull headache convinced him that the cause was too much strong wine on a stomach that had been far from lined with solid food: by the time Nicholas had returned to the guest chamber, Alban had devoured every fragment of rabbit and every morsel of bread. There wasn't even a spoonful of sauce left in the pot.

In a foul temper, Nicholas had clambered into the narrow bed with the intention of rising early to question the old priest in the infirmary before he tackled the prioress again and began the tedious process of examining the priory's ledgers. Commander John had told him that the procurator at Clerkenwell was convinced there was something wrong with the account records he was receiving from the priory, not that he could put his finger on what was amiss. Nicholas had been ordered to ferret it out, whatever *it* might be.

He already had a shrewd suspicion of what he might uncover. Farmers and millers always tried to hide the true worth of their land and stock, so they could pay less rent and hand over fewer beasts and sacks of flour. It was the same the world over – Cyprus, Rhodes, England: the local people tried to cheat the order if they could. But a seemingly casual gossip with a long-serving parish priest might enable Nicholas to discover a little more about the true value of the lands, properties and livestock in those parts and the real wealth of the priory's tenants. Old men were always willing to talk for hours if they could find someone interested enough to listen, and Nicholas was interested.

Women were easily gulled and too soft-hearted. He'd wager his own horse that the farmers were handing their coins to Sister Clarice and rubbing their hands in glee that she was fool enough to charge them a fraction of what they should have paid. Little wonder she found it easy to collect the rents. Well, the tenants' days of ease were over. A seasoned knight was in charge now, and those thieving hands were about to be chopped off.

Nicholas plucked his hose from the crumpled heap of clothes and, with a shudder, tried to pull them on. They were still wet and stinking from the journey. He badly needed a fresh pair, but he couldn't see where his clothes chest had been put. Hob had plainly not troubled to unload the cart, and Alban had lacked the wit to retrieve his personal possessions from among the provisions.

It had been the journey from Hell. It had rained without ceasing, numbing limbs and faces, while the wind, slashing through their sodden clothes, had stung like salt on a flayed back. The loaded wagon had repeatedly become mired in the twisting ribbon of calf-deep mud that Hob had called a road. Both Nicholas and Alban had been obliged to dismount to help free it and they'd no sooner have it rolling forward again than it would start sliding back down the hillside or a wheel would become wedged behind a stone buried in the ooze.

Once they were up on the moors, they'd no longer been able to make night-camp for fear that the light from the fire would bring every outlaw in those parts swarming to them, like flies to a corpse. This shire was reputed to be one of the most lawless in England. Its inhabitants thought themselves so far from the sheriff's men in Exeter that they could rob and murder as they pleased and never be caught. Even the white *cross formée* on the breast of a knight's black cloak was no protection, these days. All the Hospitallers' garb meant to any man in England now was rich pickings.

Then, finally, as they had breasted a rise, Hob had stopped the wagon and pointed through the driving rain to where three or four tiny red dots hung somewhere in the darkness, suspended between Heaven and Hell. 'That,' he had announced, with grim satisfaction, 'is your new home, Masters, and may God have mercy on your souls.' He'd crossed himself. 'I hope you're praying men, Brothers, 'cause you'll need more than a sword to protect you up there. Other side of that priory stands the most accursed hill on the whole moor. Old 'uns called it Fire Tor, but some call it Ghost Tor. You can hear the dead whispering among those rocks. Hungry ghosts, they are. There's many has heard them talking, and some even followed the voices into the caves up there. Followed them in, Brothers, but never came out, for once you cross into the deadlands, there's no coming back.'

Nicholas shivered as he dragged his damp, mud-caked shirt over his head and struggled to thrust his arms through the clinging sleeves. He was still wrestling with his jerkin when Alban sauntered in, dragging a blast of cold air with him.

'It's well you're dressed,' the groom observed.

Nicholas grunted. 'In sodden clothes! Where's my chest? Hasn't that idle carter and his frog-witted son unloaded it yet? Go and chase them, and while you're about it, tell one of those servants to fetch me some breakfast, a good one too, seeing that I didn't get any supper.' He glowered at Alban.

The groom was leaning against the door, scraping mud from under his nails with the point of his knife. He was a scrawny, ferret-faced man, indifferent to people, but there wasn't a horse he couldn't calm and control in spite of having lost, as a lad, his left forefinger and half of the middle one to a badger cub. He didn't bother to look up until Nicholas drew breath.

'Be lucky to get anything to eat this side of noon,' he said morosely. 'All the servants and sisters are outside, chattering like a flock of starlings. There'll not be a peck of work done this day by my reckoning.'

'And the prioress allows this?' Nicholas said incredulously. If this was how the priory was run, then Johanne would find herself in a cloister in Buckland before the month was out.

'I reckon she's more to fret over than the servants just now. There's been a death, one of the patients in the infirmary.'

'So? That can hardly be anything unusual. Does work cease every time someone dies?'

'Aye, well, if the gossip in the courtyard is to be believed, it's not death that's unusual but the manner of it. There's some out there whispering *sorcery*.' A sly grin crept across Alban's face. It was not often he could render the knight speechless and he was evidently relishing the moment.

In spite of his thumping headache, chilled body, raging thirst and general ill-humour, a prickle of excitement and fear coursed down Nicholas's spine. He'd been sent to uncover financial mismanagement, but now it seemed there might be a whole mess of corruption to be exposed. He grabbed his still-damp cloak and hurried out into the courtyard.

It was easy to guess which door led into the infirmary from the gaggle of people clustered around it. He forced his way through them and flung open the door. Prioress Johanne spun round, surprise and anger darkening her face as she saw who had entered.

The room was long and plain, save for a wall painting of the Blessed Virgin over the doorway, staring down with a dispassionate gaze. A fire blazed in the hearth in the centre of the floor. The reeds of the thatch above were blackened with smoke, which drifted among the rafters in sluggish grey clouds until it found its own way out. A row of narrow, but high beds were ranged down the length of one wall, each divided from its neighbour by a wooden partition to keep out the draughts. A scrubbed table and benches stood on the opposite side of the hearth, while an assortment of rough wooden stools, copper lavers and small iron-banded chests crowded together in the corners of the room. A screen woven from rushes had been set in front of a bed in the far corner. Nicholas could hear faint moans and whimpers coming from behind it. Evidently one patient was too ill to be moved, unlike the others: the rest of the beds were empty, the covers thrown back in disarray as if the occupants had been dragged out in haste.

Nicholas was puzzled. Many came to the infirmaries of St John in the hope of making a good death, and life carried on around them as they died. Indeed, it was right and fitting that it should, for it helped others prepare their souls for what must come to every man and to search their consciences for the answers to questions that would be asked of them in their final confessions. Why had these patients been removed?

Prioress Johanne and two more sisters of the order were standing close to the only other bed that appeared to be occupied, though Nicholas could not immediately see who was lying there. The plumper of the two sisters took a step towards him, as if she intended to shoo him out, but the prioress caught her sleeve and gave a slight shake of her head.

Nicholas strode towards them. 'I understand that one of your patients has died, Sisters. Brother Alban said . . .' He hesitated. Remembering the grin on Alban's face, he was suddenly afraid of sounding foolish. Suppose the groom had been joking. It wasn't something any normal man would find funny, but Alban had a twisted sense of humour.

Johanne moved, blocking Nicholas's path. 'There is no need for you to trouble yourself. As you say, the man is dead. The sisters will prepare the body for burial. Brengy, the stable boy, has been sent to the village to warn the sexton and Deacon Wybert to prepare for burial. All is in hand,' she said briskly. She gave a cold smile and took another pace forward, as if trying to ease Nicholas out of the door. But he stepped around her and, in a few strides, was at the bed.

The linen sheet had been drawn up to cover the face of the body lying beneath. After a lifetime of war, Nicholas was hardly a man to shrink from the sight of a corpse. He whisked back the sheet, exposing the gaunt face. An elderly man was lying on his back, his clawed hands raised to either side of his face, as if he had been trying desperately to push something or someone away. His milky eyes were wide open and his neck arched. His mouth gaped in a silent scream and his wrinkled face was contorted in an expression Nicholas could only describe as pure terror. He had seen such horror frozen on the faces of women and men who had been shrieking for mercy even as the fatal blow descended, from a knight's sword or a Turks' scimitar, but never on an old man who had died peacefully in his bed.

As he had done many times for his brothers slain on the battlefield, Nicholas grasped the wrists, intending to pull the arms down and cross them as a Christian should lie in death, but they were rigid, unyielding as stone, the flesh so cold it made his fingers sting to touch it. He snatched his hands away with a gasp. 'He's been dead for hours. Why was he left like this? Surely you know that a body stiffens and then it is impossible to straighten it. Who was watching these patients through the night?'

'That is the puzzle,' Johanne said softly. 'Father Guthlac has *not* been dead for hours.'

Nicholas stared at her. 'This is the parish priest? But I wanted to talk to him. You didn't tell me last evening that he was dying.'

'I had no reason to think that he was,' Johanne said. 'He was restless throughout the night, unusually so. Sister Basilia and Sister Melisene both went to comfort and calm him at different times, not least because the other patients in the hospital were distressed by his cries. Father Guthlac fell quiet in the darkest hour before dawn, and it was thought he was sleeping, but when dawn broke, he was discovered . . .'

'I found him like this, Brother,' Basilia said. Her plump hands were trembling and her eyes were red, as if she'd been crying. 'He was having terrible nightmares, but he had no fever or sickness, did he?' She turned in appeal to Melisene, who nodded silently.

'I tried to straighten his body. I made to close his mouth and cross his arms, but he was as stiff then as he is now, and cold as iron. There is no explaining it. The fire's been burning

60

all night and his is one of the closest beds to it, in consideration of his age, and he'd good thick coverings.'

'There's one explanation,' Melisene muttered.

'And one I do not wish to hear repeated, Sister Melisene,' Johanne snapped.

'What's this?' Nicholas demanded. 'If the sister knows what has caused this unnatural death, I insist on hearing it.'

'Father Guthlac was old, blind and frail. There is nothing unnatural—'

Nicholas pointedly turned his back to the prioress, raising his voice above hers. 'Come, Sister, what is it you wish to tell us?' He laid his hand on Melisene's shoulder in what he imagined to be a reassuring and encouraging gesture. She backed away in alarm, gazing in mute appeal at Johanne.

'Very well,' Johanne said. 'I dare say you will hear the rumours, anyway. Last evening a blind child was found abandoned in the chapel. When he was brought into the infirmary Father Guthlac became upset. As I said, he was an elderly man, and the old often become agitated by the slightest change around them. He was given a sleeping draught to soothe him.'

'Am I to understand he was poisoned with the sleeping draught, given too much or the wrong herb was administered?'

'I gave it to him and I've never made a mistake with my simples,' Basilia protested indignantly.

'It was the boy who did this,' Melisene blurted out. 'Father Guthlac said if we let him stay under our roof, he'd destroy us. Destroy us all! Those were his very words. And that boy was tormenting the poor old priest all night long. Punishing that sweet, holy man for speaking out God's truth.'

Nicholas stared down at the priest's corpse, the face still

contorted in terror. 'So, Alban was right,' he breathed. 'It was sorcery!'

'It was no such thing,' Johanne said firmly, glaring at Melisene. 'Brother, you are a learned man and you surely cannot believe that a frightened blind child, who cannot utter so much as a single word, much less a curse, would have the power to murder a priest when he wasn't even in the same room. The child is helpless. He cannot even find his bed unaided or feed himself without his hands being guided to the dish. Father Guthlac had become agitated, a state well known to bring about convulsions in those who are frail. Look at the way his neck is arched. That is doubt-less why the body is so rigid.'

Nicholas eyed the corpse doubtfully. 'If this priest was indeed seized by a convulsion that is yet further evidence that *maleficia* is at work. The innocent are often thrown to the ground in such fits by evil spirits conjured by a sorcerer or a witch in order to prevent them from speaking the truth. I must examine the boy for myself.'

'I forbid you to do any such thing, Brother Nicholas.' The prioress's voice reverberated through the silent hall, like a watchman's horn. 'The child has been abandoned and appears to have only recently lost his sight. He is terrified. Interrogating him over matters of which he can know nothing will only frighten him still further. Besides,' she said, lowering her voice, 'any attempt to question him would be useless. As I have told you, he cannot speak.'

She took a deep breath, as if she was trying to hold her anger in check. 'You told me last night that the Lord Prior is anxious that nothing is spread abroad which could be used by those who seek to destroy the Knights of St John. The parish priest, Father Guthlac, was a good man, but the old often lapse into strange fancies. I've known some who thought their own children were trying to murder them or that their wife was their mother. When men are dying their wits may flee their heads long before their souls depart their bodies. But I fear that if Father Guthlac's words were repeated outside this priory by one of the servants or travellers' – she glared pointedly at Melisene – 'there are some simple and unlettered folk who would believe them. The villagers still remember Father Guthlac as he used to be and they would be only too eager to see malicious witchcraft in his death. I am sure you would agree, Brother Nicholas, it would not be in the interest of any in our order to have the name of the Hospitallers coupled with that.'

She fixed him with a stare that had been known to stay the hand of even the most drunken or violent of men. And, to his surprise, Nicholas heard himself telling the prioress he would make it known that he had examined the body, and could assure the servants Father Guthlac had died in his sleep of nothing more sinister than plain old age.

Afterwards he convinced himself he had done it for the sake of the order. Neither Commander John nor the Lord Prior would thank him if tales of murder by sorcery at a Hospitallers' priory were spread abroad. A scandal such as that could be used to bring down the Knights of St John, as it had the powerful and wealthy Templars.

On the other hand, if a private report about the true manner of the old man's death were to reach the Lord Prior at Clerkenwell, it might prove the very weapon Lord William needed to force these obdurate women into Minchin Buckland and keep them safely cloistered there till they withered and died.

What better way to prove himself worthy of returning to

Rhodes as Commander Nicholas than to save the Hospitallers from disgrace and deliver the sisters to Buckland at one stroke? It would give him great pleasure to see the prioress stripped of her rank and forced to bend the knee, very great pleasure indeed.

Chapter 7

Sorrel

Todde was walking ahead of me on the narrow moorland track, dragging his donkey through the mire and deep icy puddles, while the poor beast tried repeatedly to jerk his head rope from his master's frozen fingers and snatch at a few blades of coarse muddy grass. The teetering stack of Todde's tools and Mam's cooking pot tied to his back clinked and clanked with every step.

I stumbled after them, gazing up at the iron-dark clouds swirling around the towering rocks, the black bog pools quivering under the lash of rain, the withered, stunted trees hunched and bent like aged men against the wind. I began to wonder if the sun had ever shone on these moors or ever would. Why had I been led to such desolation? Why had I ever listened to a voice in my head? I plodded on, my mind as numb as my face and feet, no longer bothering to pick my way around the deep ruts. My shoes were already so sodden and heavy with mud that another puddle would make no difference. Besides, the moor was as wet as the track – there was no avoiding the mire wherever you trod.

My flesh had turned into cold earth. My bones had become gnarled roots. I was a corpse melting back into the soil of the grave.

Somewhere overhead a buzzard screamed, like a cat in pain, but almost at once there was another cry, this time human. Three men were running down the hillside towards Todde and the donkey, long knives flashing in their hands. Before he'd a chance to draw his own blade they were upon him. One held him from behind with a knife at his throat, as the second grabbed the leading rope of the donkey, while the third straddled the path, staring towards me and tossing his knife from hand to hand.

'Take one step closer, woman, and we'll slice his throat from ear to ear. Then we'll come for you.'

Chapter 8

Morwen

'It won't do no good,' my eldest sister, Ryana, grumbled, but she might as well have saved her breath.

When Ma had been journeying in the spirit lands she'd have to do whatever the magpie or the other creatures told her, else we'd all suffer. Leastways, that was what she said. Taegan, the middle one of us three, reckoned Ma only did what she'd already made up her mind to do, and we suffered anyway. Once, when Taegan was only ten summers old, she'd dared come right out with it and say as much. Ma thrashed her for those words till she howled worse than the ghosts on Fire Tor. The beating didn't stop her thinking it, but it taught her to keep her thoughts from Ma's sharp ears, most of the time anyway.

Ma let Ryana argue with her sometimes, though never win, for she said Ryana would become the well-keeper when Ankow came to lead her own soul to the lych-ways. She said the gift always passed to the eldest daughter, but I watched Ryana whenever she said the spirits were talking to her and I reckoned she couldn't hear anything, just pretended to.

Not that it mattered: those black crows at the priory wouldn't let her or any of us near the well since they'd built the chapel over it.

'We'll go up to the Tor three nights from now,' Ma said. 'Moon will be nine days old by then, according to my reckoning.'

The leather curtain that served as door to our cottage was pulled aside to let the grey light of morning trickle in. Rain dripping from the thatch was blowing through the gap and collecting on the earth floor in a long puddle just across the threshold.

Ryana stared at it and scowled. 'How you going to see the moon, if it's pissing like this?'

'I seen what's coming,' Ma said. 'There'll be mist, but night'll be dry enough for what needs doing.'

She squatted on the edge of her low bed. The frame was larch poles lashed together, then criss-crossed with cords. It was light enough to move close to the fire on a winter's night, and when space was needed, it could be pushed back against the low wicker partition that separated the livier, where we ate and slept, from the shippon, where the goats were bedded. Ma pulled an old sheepskin from the bed and wrapped it around her shoulders. She always felt the cold after she'd been journeying.

Whenever I came back from journeying, it was as if I was seeing everything around me through a drop of water, which made it bigger and brighter. I'd see each tiny feather on a skylark's body, even if it was flying high above me, or watch a raindrop grow fatter and fatter till all the colours of the rainbow burst inside it before it ran down a leaf. I saw each shining buttercup and stone as if for the first time, looking right into them, the colours so bright they dazzled me, the grains in the pebble so sharp, they were bigger than a rock. I longed to ask Ma if she saw these things too. I wanted to share the colours and the things I'd seen with someone who understood, but I could never tell Ma that I journeyed. I wasn't the firstborn. I had no gift, no right.

'That bread done yet, Morwen?'

I started guiltily at the sound of Ma's voice and, wrapping my hand in the edge of my kirtle, dragged the flatbread off the hot stones set round the edge of the glowing peat fire. It was scorched and crumbled into half a dozen pieces before I could tip it into Ma's lap. It was mostly pounded meadow-sweet and rush roots, and they never held together like flour. But since those black crows had stolen our well, there'd been few coins pressed into Ma's withered palms, not enough to buy flour from the village mill, and what little grain we sowed turned black and rotted afore the shoots were even halfway grown.

Ryana took small, quick bites from the edges of her bread, like a sheep nibbling grass. She looked like a sheep, with thick curly hair always falling into her eyes, big yellow teeth and a long face that rarely smiled.

Ma gobbled down the bread as hungrily as if she'd been walking all night, though I knew her wizened old carcass had never left her bed. Ryana was supposed to keep watch when Ma was journeying, see that her body came to no harm, for if someone were to wake her or move her while her spirit was gone, it would never find its way back. But Ryana always fell asleep, so it was always me who watched.

Ma licked the crumbs from her fingers and frowned. 'Tae's a long time fetching that water.'

When Ma had woken, that was where I'd told her Taegan had gone, but the truth was my sister had slipped out in the night, just as soon as she thought they were both asleep. She'd be in a byre somewhere with Daveth or his brother. Tae giggled that the lads shared everything, including her. But Ma would kill her if she found out. She'd forbidden Taegan to see the brothers. There'd been a bitter feud 'twixt her granddam and theirs, and probably their granddams before that, but Ma refused to speak of the cause. Don't suppose she could even remember. Besides, Ma reckoned men were only good for one thing and that was getting a woman with child. She'd told us to keep well clear of the whole tribe and trouble of them till we wanted a babe. Ma was staring suspiciously from Ryana to me, and I knew only too well whom she'd pick on to question.

Afore Ma could speak, I leaped up and pulled a sheepskin cloak around my shoulders. 'I'll go and find Tae. Maybe she's taken a tumble, ground's so slippery.'

Go and warn her was what I really meant, else Taegan would come sauntering back, blithe as a blackbird, without the water she was supposed to be fetching. She'd get a beating, all right, but I'd get a worse one for lying to Ma.

Outside, I snatched up a couple of pails and set off towards the stream that bubbled and meandered through the high clumps of sedges and marsh grasses. Clouds rubbed around the great rocks of the tor, like fawning dogs, and swirled down between the folds of the hills. The sound of two stones being struck together echoed through the rain. But it was only a stonechat – such a great noise for such a little bird. I glanced back at our cottage.

It had been built sloping sideways on the hill, so that water and animal piss would run out through the drain hole in the lowest wall of the shippon. At the opposite end, the highest corner of the cottage was formed around a great craggy boulder that stuck out of the ground, and at the back of that, among the smaller rocks, there was a cave. You'd not see it in passing, unless you knew where to look. In good times, we kept our stores in it. But since the famine, it was mostly empty save for Ma's tools that she used for her charms – a hag-stone, a hare's foot, feathers of owls, jays and magpies, the bones and skin of a viper, the skull of a badger and many other things the beasts and birds had given her over the years.

But there were other treasures too, things only Ma and Ryana were allowed to touch, especially the ancient white stone carved with the three women, their arms and faces pressed so tightly together that when I first discovered it – I'd been no bigger than a rabbit then – I'd thought they were all one creature, which in a way they were: the women were Brigid and her two sisters, who are one. The Bryde Stone, Ma called it. One night, when they thought I was sleeping, I'd watched Ma teaching Ryana how to feed the stone with milk from the cows and with wild honey. Ryana held it just as Ma showed her, but I could see she didn't feel it stirring in her hand or see the cold blue flame that burned around it, not like I did.

I dropped into a crouch and stayed very still. Through the rain and low cloud, I'd caught a fleeting glimpse of someone moving further down the hill. At first, I thought it might be Taegan, but if Ryana was the sheep-face of the family, Taegan was the cow, with a shaggy mane of ox-blood hair, and breasts and buttocks that, no matter how scrawny her face became when food was short, never lost their undulating curves.

As for me, Ryana said I was the scroggling of the family, stubby and skinny, the shrivelled apple left on a tree that isn't worth the picking. But I'd the reddest hair of all three

71

of us. Once, I heard an old hag in the village say it was so fiery-looking that any man who touched my pelt would surely get his fingers burned. And he would too, I'd make sure of that.

A woman was toiling up the rise between clumps of heather, but as she came closer I saw she was altogether brawnier and much older than either of my sisters. A withered memory fluttered feebly inside my head. There was something familiar about her, but it died before I could revive it. She looked to be making for our cottage, for there was no other dwelling or road on this tor. She was taking the long way round to avoid the mire, but I could find a way through it, using stones and markers you'd never see, unless you knew what to look for.

I ran home like a hare and arrived before our visitor appeared above the rise.

'Woman's coming, Ma. Reckon she wants a cure.'

'How would you know, Mazy-wen?' Ryana sneered. 'She might want a curse.'

That was the name she'd tormented me with when I was small. She still used it whenever she wanted to remind me of how stupid I was.

Ma struggled up from the bed and took her place on a low stool behind the fire in the centre of the floor, poking it vigorously with a stick till the flames leaped and crackled. She liked her customers to find her there and be forced to peer at her through the smoke. She could make smoke form the shapes of birds and beasts, if she wanted.

'Fetch her in, Morwen.'

The woman was standing a little way from the door, her ruddy face glistening with sweat after her climb in spite of the cold wind. She was wearing a faded black gown, patched

72

and worn but not homespun, like the village women. She shuffled awkwardly, evidently trying to make up her mind whether to come or go. Those who'd not come to Ma before often lost their nerve at the last moment. It was my job to coax them. I reckon even Ma knew I was good at that, not that she would ever say as much.

'Kendra knew you'd come.' I smiled encouragingly. 'She's been expecting you. She's waiting for you inside.'

Still the woman hesitated.

'No need to be afeared of her. She wants to help you.'

Finally, she seemed to make up her mind. Wiping the sweat from her nose with the hem of her gown, she ducked through the door.

I slipped in quietly behind her and crouched with my back to the rough stone wall. Ma would send Ryana for her journeying things if they were needed, but if she wanted fresh herbs, she usually sent me. Ryana could barely tell larks-claw from lungwort. But one glance at Ma's face told me something was wrong.

The woman was standing in front of the fire, her meaty arms crossed defensively, but Ryana was on her feet glaring at her, and Ma, crouching low over the flames, was spitting on the back of her fingers, as if our visitor had the evil eye.

Ryana turned on me. 'What you let her in here for? Don't you know who she is?'

'Morwen was no more than seven summers when last she saw her,' Ma said. 'That child's not been to our Bryde's Well since, 'cause Meggy wouldn't let us through the gate!' She thrust her hand towards the woman's face, her fingers clawed in a curse.

The woman staggered back a few paces, but she did not run out. 'I only do what I'm bade, as well you know, Kendra.

I was birthed on this moor same as you and my ma dipped me in Bryde's Pool just like you done to your chillern. But what can I do? You're blessed, you are, got three strapping daughters to take care of you if you're ailing. My man's dead and the only boy I bore who lived to be full-grown was taken to fight the Scots and I haven't heard so much as a whisper of him these past ten years. I've not a stick or stone to shelter under that I can call home. Those sisters of St John are good to me. Give me work, a warm bed, and a good bellyful of food – even give me their old robes while there's still plenty of warmth and wear left in them.' Meggy plucked at the faded black cloth.

'Gave you *my* work, that's what they did. Mine! Took the living from my family that's been keeper of that well since ever it first sprang from the rock.'

'It's Sister Fina who looks after the well. I'm keeper of the gate, that's all, as well you know, Kendra.'

'Aye, I know it well enough. I know you won't let us through it, not even when they're all abed.'

'I've my orders,' Meggy said sullenly, staring at the earth floor as if she was suddenly ashamed.

'So, if these women been so good to you, what you come to me fer, Meggy? They thrown you out?'

The woman glanced round and picked up a low stool. She set it opposite Kendra, on the other side of the fire, then squatted awkwardly on it.

'Ankow came for Father Guthlac's soul. I could tell someone had been taken when I woke with a shiver in the night. I knew for certain Ankow had passed through our gates, for a cold wind follows him wherever he goes. And when death's bondsman enters a dwelling, he never leaves alone.'

'That what you come to tell me?' Ma sneered. 'I saw black

Ankow galloping across the moors on that skeleton of a horse, with his hounds baying at his heels. I knew he was hunting souls. The magpie told me it was the old priest he was after. Good riddance, says I.'

I stared at Ma. Had she really known? She hadn't said, but she'd rarely tell me what she'd learned from the birds.

Meggy stared into the flames. 'Thing is, Kendra, it wasn't a natural death. That much every soul in the priory knows. There was a blind boy found in the chapel after dark that night. Abandoned by his kin, they reckoned, but I never saw him come through my gates, nor anyone leave that evening who could have brought him. Father Guthlac took against him the moment he came close to him, said the lad should be drowned in the mire, so as he could do no mischief. But the prioress wouldn't have it, insisting the boy stay. But his spirit hag-rode the old priest all night till he was worn to rags. You could hear his cries of torment all over the priory. Pitiful they were, poor soul. The boy's evil spirit left Father Guthlac at cockcrow and that was when the old priest died. He'd warned us the boy would destroy us all, and he was the first to be taken.'

She spread out her wrinkled hands before the blaze and lifted her head to look at Ma. 'Who is he, Kendra? He's not human, that's for certain, 'cause I never leave those gates when I'm on duty, save to piss, and when I do those gates stay locked till I return. No mortal being could have got in, unless they used sorcery to make themselves invisible.'

I expected Ma to send Ryana to fetch her stones or feathers, but she sat staring unblinking into the fire. Then her lips parted and her finger stirred. As the column of smoke drifted up, the heart of it began to swirl dense and dark. All of us leaned forward to see what creature might be hovering above the flames. But it was neither bird nor beast that shaped itself in the smoke. It was the head of a man in profile, great beetle brows hung over deep dark eyes, craggy cheeks and a sharp angular nose. Then, like a wraith, it dissolved away, drifting up into the blackened thatch.

'You saw,' Ma said.

Meggy nodded, her face glistening with sweat from the heat of the fire. 'I saw, Kendra. And I've seen it afore on Crockern Tor, when there's a great storm a-brewing. I'd know that face like my own. But what does it mean?'

'The old man of Dertemora has protected these moors from danger since the day the tors rose into the sky and the first lark sang above them. Old Crockern senses danger, death. He's come to warn us evil has come to this place. The boy's eye is closed now. But if he opens it, malice will pour from it, like a river in flood.'

I'd seen old Crockern's face in the smoke too, plain as I saw Ryana dozing in the corner, but I'd seen something more. Someone was standing behind him, a shadow, but not of him. There was a woman in the smoke who wasn't Ma or Meggy or Ryana. I knew, like you know when a storm is rolling in, it was her that Ma should have been paying heed to. Ma should have listened to her.

'You must learn the boy's name,' Ma was telling old Meggy.

'He doesn't speak. And none knows who he is or where he comes from.'

'Bring me something of the boy's, something I can use to ask the spirits. But take care how you get it. He mustn't know, else he'll put you in the grave alongside the priest. You come to me on Fire Tor three nights from now. I'll make a charm that'll keep him from Bryde's Well. He must not touch that water else his middle eye will be opened.'

76

Meggy gnawed her swollen knuckle. 'Can you not give me summat to protect me now?'

Ma stood up and felt along one of the beams that supported the thatch, then pulled down a three-armed cross woven from reeds. I knew what she would do and, without waiting to be asked, I handed her the sack in which she kept her bits of wool. Ma said naught, but pulled out three strands, one black and one white as the sheep that had given them, and a third stained red with madder – Brigid's colours. She wove them sun-wise around the heart of the cross and held it out to Meggy over the flames of the hearth fire.

'Dip it three times three in Bryde's Well and put it under your pillow, along with a sprig of hare's beard to guard you from demons while you sleep. That's when he'll come to torment you if he learns what you're about.'

Meggy reached out, but let her hand fall without grasping the cross. 'Prioress says Christ's cross is all we need to protect us from the evil one. She don't hold with Bryde's crosses, specially this sort. Says it's pagan. If someone were to see it and tell her . . .'

'Then you'd best make sure they don't,' Ma said tartly. 'The four arms of his Christ's cross didn't protect old Guthlac, did they? And if you thought they would protect you, you'd not have come to me.' Ma thrust the woven cross bound with the three strands of wool at Meggy again and this time she took it, pushing it down inside her gown between her breasts, glancing uneasily behind her through the open door as if afraid she was being watched.

Ma held out her hand again, and Meggy fumbled in the small cloth bag dangling at her waist and drew out a silver coin, which she laid in Ma's palm.

'Three nights hence,' Ma said, as Meggy ducked out of

the door. 'And, remember, you'll not be safe till we know his name. Bring me something, do you hear? And whatever you do keep the boy away from Bryde's Well until the charm is ready. If his eye ever opens, there's not a soul on these moors will escape the darkness.'

Chapter 9

Sorrel

The long, sharp blade glinted in the sodden grey light, as the outlaw pressed his knife harder against Todde's throat. The whole world seemed to stand still, like a waterfall frozen as it crashed over the rocks. Even the wind held its breath. A single buzzard hung above the tor, its wings scarcely flapping, its clawed beak pointed down at Todde, who was hunched, motionless with fear. The bird watched as if it knew death waited below.

The outlaw glanced up at it and gave a harsh caw of laughter. 'Maybe I'll slice this little man's belly open, spill his guts, and he can have the pleasure of watching that hawk pick them over as he dies. Bird looks hungry for his supper, what do you reckon?'

From the grins on the faces of all three men, I was certain they'd have carried out their threat in a flea's breath and taken pleasure in it too. There was naught either Todde or I could do, save stare in dismay as the outlaws dragged the donkey to the top of the hill, every pick and shovel, pot and blanket that we'd owned along with it. For a moment, I saw the two men and the ass standing like grey rocks against the sky, then they vanished, as if someone had taken a giant cloth and wiped them away.

As soon as they were safely out of sight, the outlaw holding Todde whipped the blade away and kneed his backside, sending him sprawling face down in the mud. The robber took off after his companions and was gone.

Todde was bent double, shaking and retching, though his belly was empty. I thought it best to leave him be. Men don't like a woman to see them afraid, though I didn't blame him – only a mooncalf wouldn't be scared when he'd come that close to dying. He was straightening himself when I heard a shout on the track behind us. A man was trotting towards us on the back of a stocky little horse. In a flash, Todde had drawn his knife and was running back down the slope. He darted in front of me and stopped, his arms thrown wide as if he was trying to defend me. I stared at his ragged back. I couldn't imagine my own brother doing that for me.

'If you're thinking to rob us, you're too late,' Todde bellowed. 'Thieves just made off with all we had, so you'd best get on your way afore I do you some mischief.'

'Rest easy, sir,' the man said. 'I'm no outlaw, merely a hardworking tinner.'

Shaken though I was, I couldn't drag my gaze from the man's face. He had such a bad squint in one eye that it was staring into his hooked nose, and his pale, wispy beard did not conceal the absence of a chin. His mouth receded straight into his puckered neck. He looked like a plucked chicken.

He swung his leg over the saddle and slid off the horse. 'Saw them dragging your beast off, but I was too far away to help.'

'Caught me off guard, they did,' Todde muttered sullenly.

'But if you hadn't distracted me by riding at the girl, I'd have been after them in a flash and I'd not have stopped until I took back what was mine.'

The man shook his head sadly. 'It's as well you didn't catch up with them. You'd be lying out on that tor with a knife in your guts if you had. They've been known to cut a man's throat just for the fun of it. And it's not just travellers they prey on. They'll pick off any tinner who's fool enough to be working alone.'

Todde frowned and took a pace forward. 'Here, did you say you're a tinner?'

The man nodded. 'Gleedy they call me, on account of . . .' He pointed at his squint eye and laughed. 'Work for Master Odo. He owns the tinners' rights to the valley over yonder. There's dozens of men stream for him there. If you're looking for work in these parts, he might take you on.'

Todde shook his head vehemently. 'I'll work for no man. I know how it is – the master standing there supping while the men break their backs, then he takes all the tin they've sweated for and tosses them a crooked penny for their trouble. Well, he'll not be using me as his donkey. I'll stream for myself and what I earn I keep. I've got rights, you know. Can dig on any land I please that's not being already worked for tin, doesn't matter who owns it. That's the King's law,' he added, jerking his chin up as if challenging the tinner to disagree.

'I'll not argue with that,' Gleedy said. 'But how are you going to dig with no tools?'

'What's it to you?' Todde demanded.

Gleedy shrugged. 'Just being neighbourly. But I'm not one to force help on those that don't want it.' He made as if to climb back on the horse.

But I edged forward. 'Wait! You can't blame him for being chary after what's just happened.'

Gleedy lowered his foot and turned.

I grimaced at Todde, trying to make him see sense. 'Listen to him, Todde. You can't dig with your bare hands, and even if you could start streaming, it might be days or weeks before you found tin and then you've got to take it somewhere to sell. How are you going to eat till then?'

Gleedy nodded sagely. 'Master Odo doesn't need to go looking for men to hire. He's any number coming to him begging for work, 'specially now food is so scarce. Turned three men away just yesterday and they were all strapping lads. But, deep down, he's a heart as soft as a virgin's breast. If I was to tell him of your misfortune and how you've a crippled woman to feed, he might give you a chance. Worth a try, isn't it?'

I felt the heat of my fury and indignation rising to my face and had to force myself not to stalk away. Pride doesn't fill an empty belly.

Todde opened his mouth as if he would object, but Gleedy jumped back in before he could. 'I know you want to be your own master and so you shall, if you play it right. Master Odo would pay you a generous sum for each pound of tin you dug and that's a rich seam he's found. You'd be able to put food in your bellies and save a good bit too. So, by my reckoning, if you were to work for him a month or so, you'd easily earn enough in that time to buy the tools you'd need to set up on your own.'

Todde still didn't seem convinced, but clearly he couldn't think of anything better.

Gleedy told us to wait while he rode ahead to speak to the master. He knew how to handle Master Odo, he said, and it seemed he did, for a while later we saw him riding back with a grin as broad as a barrel.

'Master Odo's even got an empty cottage for you, all ready and waiting, it is. You come along with me.'

We followed him up the track, then turned off along a broad path, churned up by the hoofs of many horses. In the wettest places rocks, wood and bundles of bracken had been thrown down to make stepping-stones through the mire, but the mud oozed over the top when we trod on them. The track wound up a steep rise. I couldn't see so much as a byre, much less a cottage. Then I caught sight of a huge column of dense black smoke rising up from the other side of the hill, as if a great dragon lay coiled there.

Todde saw it too and yelled in alarm to Gleedy, 'Something afire over yonder.'

Gleedy reined in his horse and turned, waiting for us to catch up. He gave another of his wide grins, showing a mouthful of teeth so crooked and crowded, it looked as if he'd grown twice the number of any normal man.

'That smoke? It's from the blowing-house. New it is,' he added proudly. 'Every man used to have to crush the rock and find wood to burn the ore before he could take it to a stannary town. Even then it was worth only half its weight for it had to be smelted again there. Now it's all burned together and the fire's so hot it has to be done just once. That's Master Odo's doing, and it doesn't come cheap, I can tell you. When you think on it, those outlaws did you a favour. If you'd been working for yourself you'd have been put to four times the labour. Come on!'

His horse ambled on, evidently knowing the path so well that it was able to step on to tussocks of grass to avoid the worst of the mire. There were no sounds out here, save for the mournful cry of a bird and our feet squelching through the mud. I'd never known such quiet and a strange peace settled on me. I felt welcomed, wanted, as if the land had thrown a blanket of softest wool around me and gathered me up into its arms.

But as we breasted the rise, the noise that burst in my ears made me stagger backwards: iron hammers smashing granite, stones crashing into buckets, the whinnying of pack-horses, the shouts of men, the bellows of women and the yells of children. I had never heard such a violent clamour, even at St Stephen's fair. It was as if the ground had yawned wide before me and I was staring down into the pit of Hell.

Chapter 10

Prioress Johanne

Sebastian was quiet now. He had suffered another of his night terrors. I say 'night', but they came upon him just as frequently during the day. He'd shake and whimper, trying to crawl away from something only he could see, but his twisted, wasted limbs would not support him, and even the few feet he managed left him pale as death and sweating in pain. I held him in my arms, murmuring softly and stroking his long hair as he fell into an exhausted sleep. His hair and beard were as white as sea foam but soft and silky, like a baby's. He allowed the sisters to wash and gently comb them, but they could not cut them. He shrank in terror whenever he saw the glint of shears or a blade near his face, even if it was in my hands.

I laid him back on the sheepskin and watched him for a few moments to make sure he had not woken. Then I pulled the rush screen across to close off his bed and edged away.

The boy was sitting alone on a bed at the opposite end of the infirmary. I had ordered he be housed there after the death of Father Guthlac. Basilia could not keep leaving the infirmary to tend him in the pilgrims' hall, and Melisene, the priory's hosteller, said he was making the other travellers uneasy, although when I pressed her she was forced to admit he had done nothing to disturb them. And that, it seemed, was the problem – he did nothing.

I knew Melisene wanted the boy out of her hall. She was accustomed to raucous pedlars, chattering pilgrims and old men-at-arms who ate heartily and drank deeply, but mostly, unless they were trapped by sudden snow or dense fog, these travellers were away about their business during the day, leaving the hall empty, so that she and the lay sisters could sweep, straighten pallets and mend fires in peace without, as Melisene put it, 'the idle getting under my feet'. But the child never moved, and that unnerved the servants. It wasn't natural, they muttered, especially not for a boy.

I walked down the length of the hall, pausing to speak with the other patients as I passed. When I reached the corner where the boy sat, I pulled a stool close to his bed and sat down. I took care not to touch him, fearing to startle him, but he flinched away as if he sensed someone was close to him. He looked even thinner and paler than I recalled, his wild dark curls framing a face that was as white as breath on a frosty morning, the skin almost transparent. Was he ill?

'I am Prioress Johanne. Do you remember my voice?'

He turned his blank stare towards me. The dancing candle flame glittered in the pupils of his dark eyes, like the sun reflected on the surface of a deep pool. His eyes were so bright and clear that, for a moment, I was convinced he must be able to see, but when I passed my hand close to his face, he did not blink or move.

Basilia came bustling over, an anxious look on her plump

features, as if she was afraid I'd come to find fault. 'The boy's been fed, Prioress, though he still makes no attempt to spoon anything down himself or feel for a piece of bread. He'll bite and chew it when it's put against his mouth, but . . .' she faltered, staring down at him, '. . . even the changeling children we've had brought here soon learn how to feed themselves. They're so greedy for food they'd snatch from anyone's trencher.'

'Perhaps those who had the care of him fed him like a baby.'

The boy's head was still turned towards me, but his eyes seemed fixed on some point behind me in the hall. His stare was so intense that, even though I knew he was sightless, I found myself looking round to follow his gaze, but could see only one of the other patients who lay tossing wildly and whimpering in her sleep, as if she was having a terrible dream. For a moment, I almost thought the boy was . . . I reproved myself sternly. The child had done nothing. He hadn't moved. The woman was doubtless in pain, which was making her restless. That was all.

'We cannot keep calling him *the boy*, Sister Basilia. He must have a name.'

'But he still hasn't spoken, Prioress. He can't seem to tell us who he is.' Basilia lowered her voice, her gaze flicking sideways towards him, as if she was afraid to look at him directly. 'He may have been born dumb, or whatever robbed him of his sight might have taken his voice too or his memory.'

'Then until we can find out who he is, we shall call him Cosmas after the great saint and physician who protects the blind.'

I grasped the child's hand, which still felt as cold as

granite. He jerked, but did not pull it from mine. 'Cosmas,' I pronounced slowly and clearly. 'You' – I prodded his chest to make my meaning clearer – 'will be called Cosmas.'

I watched his face for any sign that he had heard or understood. The candle flame still reflected in his black pupils, but I could have sworn that it had turned blood red. Startled, I dropped his hand and stared up at the candle guttering on the wall, thinking it must be smoking or tainted, but it was still burning yellow, and when I looked back into Cosmas's eyes that was how it burned there too, like any true reflection. The flame must simply have flickered in a sudden draught. What was wrong with me? I was starting to imagine things, like Sister Fina. I had spent too many hours lying awake, fretting over the arrival of Brother Nicholas. That was the cause. I rose and drew Basilia away from the boy, trying not to betray my momentary disquiet.

'How does he respond to his eyes being bathed in the holy well?'

Basilia's cheeks and neck flushed, and her gaze shifted to the painting of the Blessed Virgin on the wall, as if she was looking for divine protection or absolution. It was hard to know which.

'He's not yet been to the well . . . I would have asked Sister Fina to take him, but she's not come into the infirmary these many days and I had to stay with the patients. Poor things, they are still so distressed over dear Father Guthlac's death. They were all so fond of him – some of the older ones had known him since they were young . . . And we've all been so busy trying to make sure Brother Nicholas would not discover anything to find fault with.'

The longer she babbled, the more excuses tumbled out, so thick and fast that I knew none was quite the truth.

'I understand, Sister Basilia, but I insist you take Cosmas to the holy well tomorrow, whatever your duties here. I am sure the servants can manage the patients for the brief time you are absent. The boy was brought to us for healing and we must have as much faith as those who entrusted him to us. Ask Sister Fina to open the well early for you and take him there before the pilgrims are admitted. If he becomes distressed when the water touches him, it will not upset them. Do it, Sister Basilia. I will hear no excuses.'

The sudden rapping at my door sent my quill sliding across the parchment, leaving a trail of black ink. I hastily dabbed at it before it could dry, cursing. I was angry with myself for jumping like a guilty child: I'd no reason to feel shame. All the same I moved swiftly to the far corner of the room, where I had dragged my narrow bed away from the wall. A section of panelling had been lifted out to reveal a gap, just wide enough to squeeze through. The tiny chamber beyond had been designed as a private chapel and, should the need ever arise, a place of safety in case of attack. But it served another purpose now: it contained a stout chest standing hard against one wall, and it was into this that I slid the parchment and small bag of coins.

The knocking came again, more urgently this time. I stepped swiftly out, closing the wooden panel and pushing my bed back against it. Satisfied that all was safely concealed I pulled back the bolt on the door of my chamber.

Sister Clarice pushed the door barely wide enough to slip through, then shut it again, leaning on it and panting a little, as if she had been hurrying. Tiny beads of sweat glistened in the furrows of her forehead in spite of the chill weather. She reached into her leather scrip and pulled out a bag, which she held towards me with a triumphant smile.

'Three more rents collected, Prioress.'

'Did any complain?'

She grimaced. 'Master Rohese almost refused, saying that it was a week early, but I told him that if he waited until the due time then a brother knight would be collecting the monies owed. They'd search every corner and cranny of his byres and barns, and record everything down to the last egg and bale of wool to find reason to demand double from him.'

'Sister! You should not have said that. We don't know that Brother Nicholas has any such intention. No knight of St John would seek to take more from any man than was due to God and to his servants.'

As prioress it was my duty to defend the reputation of the knights of my order, but I would not have wagered a clipped farthing on the truth of it. Unlike most of my sisters, I had served on Cyprus before the order's headquarters was moved to Rhodes and I had seen at first hand the cruelty and arrogance of some of my brother knights, which had given the people of that isle good cause to rise against the Hospitallers, though I would never have breathed as much to a living soul, not even to Clarice.

The sister's sharp chin jerked up at my reprimand, and there was the same flash of fury in her dark little eyes as there had been when Nicholas had demanded she surrender the priory's accounts.

'It is no sin to speak the truth, Prioress,' she retorted indignantly, 'and I said no more to Master Rohese than I believe. I was forced to sit with *Steward* Nicholas, as it pleases him to dub himself now, as he went through my figures, and I can tell you he'd barely glanced at the records before he was claiming a property here had been undervalued or an acre of plough land there should yield twice more than I had recorded. He was accusing me of not checking thoroughly, not that he said it in so many words, but he might as well have done for we both knew what he meant.'

She snorted furiously, and then the corners of her lips twitched in a faint smile. 'But Master Rohese is no man's fool. There are many hidden valleys and oak woods on the moors. I suspect he'll already be moving some of his cattle and stores to places where a stranger would never find them. Not,' she added firmly, fixing me with a baleful stare, 'that I encouraged the tenants to do any such thing, Prioress.'

I fought back an answering smile. It wouldn't do to let Clarice know how secretly delighted I was, for I shouldn't be condoning such devious behaviour. But I was all too well aware that the moment Brother Nicholas had the rents in his possession he would have financial control of the priory. We would be forced to beg him for every farthing, like a servant asking their master for coin to pay the butcher. I could picture the smug satisfaction on his face as he debated whether or not to grant us a pittance, and the delight he would take in his power to refuse. He could and, I had little doubt, *would* make our work impossible and our lives unbearable. But so long as the coins remained in our chests and the keys on my chain . . .

'Thank you for your . . . efficiency, Sister Clarice. If I can deliver the responsions to Clerkenwell in full, using our own messenger, it may convince the Lord Prior that we can manage our affairs perfectly well and should be left here to do so.'

I found my gaze straying to where my bed stood against

the wall, guilt welling in me again. Was I breaking my vow of obedience or merely being a wise and prudent servant of God? It was a question I had wrestled with ever since I had become prioress, but not one on which I could seek guidance from any in our order or even from gentle, bumbling Father Guthlac. He had been a godly man of simple faith and I doubted such dilemmas had ever troubled his sleep.

Clarice crossed to the small fire and stood with her back to it, warming herself. With the flames flickering behind her steaming skirts, I saw again, for one terrible moment, the Templars burning on the heretics' pyres, heard the echo of those screams, smelt the stench of burning flesh. Brother Nicholas might have heard of the torments that had befallen them. But I had witnessed them. I had seen with my own eyes – would to God I had not – the once noble warriors blinking in the harsh light of day as they were dragged naked from their dark dungeons, seen the crowd spit at them, hurling insults and dung, where once they had thrown flowers. Sometimes I think it is a blessing to be blind, for then you cannot see those horrors again and again as if they were burned into your own eyes.

I knew far better than Brother Nicholas how quickly kings and crowds could turn. The Templars had lent money to finance half the wars and buildings in Christendom. They had fought and defended nobles and kingdoms. They had seemed as invincible as the Holy See itself. But how swiftly the hands of sovereigns that once had proffered coins and jewels could wield axes, ropes and flaming brands.

And the spinning of the stars in the heavens had brought us into a far more treacherous time than ever the Templars had known. When the populace starves and grows desperate, kings and bishops alike look for some way to turn aside the people's anger. Tether a bear for the hounds to rip to pieces, and the brutes will not turn on their masters. I shuddered, as if the dead had walked over my own grave. How could I protect my sisters from those flames if the Hospitallers were next to fall?

Chapter 11

Sorrel

The steep-sided valley was as raw as an open wound in a man's belly. The land had been stripped bare of grass, the earth and stones that had been torn from it lying heaped in great barren wedges of black and rust-coloured rubble at either side of a broad trench. A lake had been gouged out at the head of the valley, the water held back by a great wall of earth and stone into which a sluice gate had been set. Wooden channels ran from it in zigzags down the hillside towards the ditch at the bottom in which a line of men were digging. As I watched, a man standing by a sluice gate put a ram's horn to his lips and blew a single deep, mournful note across the valley. At once the tinners threw their shovels on to the bank and clambered out of the trench, only just in time before the man dragged on a rope to heave up the gate and a great gush of water flooded down the wooden leats into the ditch, pushing a tide of mud and gravel out into an old riverbed below.

At the top of the valley, directly below the lake, there was a large, square building, with roughly hewn granite walls and a slate roof from which a fat plume of tar-black smoke was rising into the grey sky. But Gleedy was pointing towards the lower end, where a score of tiny black huts were clustered.

'There now, that's where you'll be living, cosy as a flea in a king's bed. See how close to the digging it is? There's some workings on these moors where you'd have to trudge miles before you could start, but here it's just a lamb's skip and you're there, fresh as a skylark in spring, and no time wasted.'

I followed Gleedy and Todde down the sodden track into the valley. Everything was stained with black and red mud, from the faces and clothes of the men, women and children, who toiled with their shovels filling buckets, to the packhorses that stood, heads drooping, in the mizzle. The water in the river downstream of the ditch was so thick with it that it looked as if it was running with cow-dung instead of water. The mud had even hunted down the dandelions and the few blades of grass hiding among the cracks in the rocks to suffocate them too.

The tinners turned their heads to watch us as we picked our way round the spoil heaps, sidestepping the women with yokes on their necks who were heaving buckets of stones over to the granite blocks where the men were pounding them with iron hammers. A few muttered to each other, but I couldn't catch the words. A couple offered me weary smiles in exchange for my own, but mostly their faces were blank, indifferent.

The 'cottage' Gleedy led us towards turned out to be no more than a crude bothy on the edge of a group of others like it. The one he pointed to was lowest down the hillside and I knew at once that all the rain and piss from the neighbours' huts would run straight down into it. The walls were made from grass sods laid flat atop one another. But at least the cold earth floor was sloped, which was a mercy, for all the water had run into one corner, where it lay stagnant and black, like the rest of the valley.

'There's not even a shelf or bed boards,' Todde grumbled. 'You expect us to sleep on the wet earth?'

We'd been sleeping on nothing but mud for days, but I guessed that Todde was trying to cling to the one remaining snippet of dignity that had not been stripped from him by the outlaws.

Gleedy gave a high-pitched giggle. 'What would you want with a shelf? Those outlaws took all you had to put on it.' He tried to pull his features into a serious expression. 'You spread a good thick layer of bracken on the floor and you'll sleep like a babe at the breast.' He grasped Todde's shoulder. 'But, seeing as you've had a bad time of it, I'll tell you what I'll do. Master Odo keeps some stores 'case any should need them. I'll go there now and take a squint.' He laughed again and pulled down the bottom lid of his cross-eye in case we'd missed the joke. 'See if I can't find you a couple of snug sheepskins to rest your weary bones on.'

Todde ambled down to watch the streaming while I went to hunt for bracken and kindling to build a fire. I might as well have saved my aching legs. All the bracken had been stripped bare near the camp, and I was too afraid of stumbling into a sucking mire to wander far with the light fading. I found a few twigs but I knew they'd be too wet to burn unless I could borrow the heat of a neighbour's fire to dry them.

I heard the horn sound again, two notes this time, and by the time I found my way back, the mud-soaked women were trudging up to their huts, where they knelt to blow on the embers of glowing peats and coax them into a blaze.

Their menfolk squatted by the filthy stream, splashing water on their hands and heads. Dirt and sweat ran down their faces, leaving black trails on their skin, as if the devil's slugs had crawled there.

Gleedy had returned and was pushing a bundle of moulting sheepskins into Todde's arms. He caught sight of me, and beamed, his forest of teeth glinting in the darkness of the doorway. 'I've not forgotten you, Mistress. You'll be glad of this, I'm thinking.' He held out a rusty iron cooking pot almost burned through. 'I've slipped a parcel of dried mutton in there and an onion or two. Now, don't you be thanking me. Master Odo would never forgive me if your man was too weak from hunger to work come morning. Sleep sound.'

'He is not *my* man!' I blurted indignantly, and instantly regretted it when I saw Todde flinch. I'd not meant to hurt him – he had suffered enough blows to his pride that day. But, all the same, it was as well that it had been said. I had no illusions that any man would want a creature like me in his bed, but I'd not suffer myself to be in bondage to anyone again, as I had been to my father and brother. Todde had wanted to be his own master: couldn't he understand that a woman might want that too?

Gleedy grinned, gave a little bow and ambled away through the smoke of the cooking fires, swallowed by the thick grey sludge of evening.

'You want to watch what you take from Gleedy,' someone muttered behind me.

I turned to see a tall, handsome woman leaning on the corner of a hut.

'He'll not give you so much as a cat's turd for nothing. Deeper you climb into his purse, harder it'll be to pull yourself out.'

'What do you mean?' I asked, but as she spoke, one of the tinners slouched past. She caught my eye, giving a warning shake of her head.

The tinner scowled at her. 'Supper ready, Eva? My belly's roaring louder than a rutting stag.'

She waited until he had trudged further up the rise, then moved a pace closer to me. 'Just remember what I said, but don't speak of Gleedy or the master to any folk here. Those two have ears and eyes everywhere.' She gave a grim smile. 'Cheer up. You'll soon get the lie of this place.' She reached out to give my arm a squeeze, but it was my withered arm she grasped. She jerked at the feel of it, staring down at my flopping hand, and bit her lip. 'Ah, you poor creature.'

If any other soul had said it I'd have snapped at them, like a baited dog, but the haunted expression in the woman's eyes told me she'd scars aplenty of her own.

Eva's gaze darted to the twigs I was clutching. 'That won't warm a spoonful of water.' She jerked her head towards the huts higher up the slope. 'I cook for some of the men and chillern that has no women to tend them. My fire's up there by that old tumbled wall. If you bring along some of that mutton to add to the pot, you're welcome to join us, just for tonight, mind.'

Half a dozen men, three young boys and a girl, no more than seven or eight summers old, were already gathering around a bubbling cauldron, when Todde and I slipped among them. Eva had lit the fire close to an ancient stone wall that had half tumbled down, but was still high enough to give shelter from the wind, and she'd built two short, low walls out of the fallen stones to shield the fire-pit on either side. The great iron pot was balanced on top of them so it straddled the flames, but did not burn.

After several minutes of stirring, and calls from the men to hurry, Eva doled out the pottage into old wooden bowls. Some of the men had brought their own spoons made from mutton bones, the rest lifted the bowls to their lips and slurped, scraping the shreds of meat that were caught on rough wood into their mouths with black-rimmed fingernails.

It was hard to say what the pottage had been made from, for the fragments of meat, bone and herbs in it had plainly been reboiled so many times they'd no taste left in them. Most likely the cooking pot was never emptied. Eva just added whatever the day brought her, a bit of fat bacon, a handful of herbs, a lump of dried fish or whatever bird could be snared. But I would have gobbled a bowl of boiled mice that night. My belly ached for hot food. And everyone sitting around that fire likely felt the same for no one spoke: they were too busy eating.

As soon as the bowls were empty, most of the men handed them back to Eva and silently slunk away, too bone-weary to do anything but sleep. A few others sat on, talking to Todde, asking him where he'd come from and if he'd ever been streaming before. To hear him crowing you'd think he knew as much about tin as a master potter knows about clay. A flame of excitement flared in his eyes, as he boasted of the money he'd made in times past. He seemed more cheerful now that he'd a bellyful of hot food, as if he'd forgotten he'd had a knife held to his throat just hours before. But the tinners were grinning at each other behind Todde's back. They could tell he was talking through the seat of his breeches.

I gazed up at the moors. I'd thought I knew just how dark a night could be. Back home, when the moon was hidden by clouds and there was no light to be seen, save for the glow of a stockman's fire across the fields or a sliver of candlelight beneath a shutter, I had called that darkness. But it was only dark because the sun had gone. Here in this valley it wasn't simply that the light had vanished from the sky, the darkness there was alive, a great black river of malice that swept down the hillsides and filled the valleys, a chill, choking tide that could drag you down and drown you in despair.

I lifted my head and stared up at the top of the hill. She had called me to these moors, but not to this valley surely. This was not the place. It couldn't be. But where was I to go? What was I to do?

Why have you brought me here? Speak to me, I begged.

But only the wind answered and if it had words for me I couldn't understand them.

I shivered. Pulling my cloak tighter about my shoulders, I forced myself to turn my back on that desolate valley. But as I stared into the flames and smoke of Eva's fire, I glimpsed movement: something was glittering and swaying in the old stone wall behind the fire. I screwed up my eyes, trying to make out what it was. It must be the wind stirring clumps of grass or herbs growing between the stones.

Then the breath caught in my throat. Those were not grasses. They were the heads of vipers, their eyes gleaming in the firelight, their forked tongues flickering as they tasted the air. From every crevice and hole in the wall more heads were emerging, swaying back and forth, dozens of them. I scrambled to my feet, backing away in alarm.

The men's heads jerked up and one by one they followed my gaze, trying to see what I was staring at. Then a ripple of raucous laughter ran through them, and they nudged each other in the ribs, pointing at the wall and at my face.

'They's only long-cripples,' one man chortled. 'Nest in the wall, they do, and the heat of the fire makes them poke out their heads. They'll not come out to harm thee in this weather. Too cold and wet.'

'But on warm days you need to watch where you're treading,' another added, 'especially if you go barefoot. Moors are swarming with those creatures.'

'And you wants to give your bedding a good shake too, afore you lie down, creep in under the bracken, long-cripples do, looking for a warm place. Old Will got bit right on his cods one night. They swelled up and turned black as the devil's toenails. Could hear his screams all the way to Widecombe.'

The men, still chuckling, fell to telling tales about those they had known who had been bitten by vipers, blithely ignoring the dozens of glinting black eyes watching them unblinking from the wall.

But somewhere in the distance, a hound began to bay, another joined in, then another, their calls echoing across the moors, as if a pack of dogs were running after their prey. One by one, the tinners fell silent, tense, listening, straining to peer out into the darkness, and this time, I saw fear on every face.

Chapter 12

Morwen

My sister Taegan lay curled up asleep on a heap of old skins in front of the banked-down fire. Ma gave her a sharp poke in the ribs with her stick. She groaned and rolled on to her belly. Ma raised her stick again and cracked it down across Taegan's broad backside. She let out a yelp and scrambled to her feet, rubbing her eyes.

'Up, you idle trapes,' Ma said. 'We've work to do on Fire Tor.'

Taegan hadn't been asleep for longer than a cow's tail, having wobbled in after dark, her breath stinking of sour mead. She'd collapsed on to her sleeping place without bothering to drag off her gown, and was snoring before her face hit the skins. She scowled at Ma, rubbing her bruises. 'So, what did you wake me fer? I'm not Ryana, you blind old gammer.'

''Cause I'll be needing all three of you tonight. Ryana's already out fetching my tools. Morwen, you pick all the worts I bade you fetch?'

I patted the sack that was already slung over my shoulder.

I'd been searching for two days and nights, because some of the herbs could only be picked at dawn and others by moonlight if they were to be in their full power. And I remembered to give the earth some honey to make amends when I plucked the vervain, for that's a holy herb. Ryana never did.

Ma grunted. 'Bring water too, Morwen. Tae, you carry wood for the fire, and mind you bring enough to bait it till dawn this time, else you'll find yourself walking back to fetch more. Stir yourself, girl. Sun's not going stay abed for you.'

The wind was raw and chafing, driving the clouds across a waxing moon. A *duru moon*, Ma called it, when it was more than half full, for it opened the doorway into the kingdoms of beasts and spirits. It gave us little light that night, but such things never mattered to Ma. She always counted the nights of the moon, whether she could see it or not, notching them off with an iron blade on a gnarled branch she'd cut from the twisted oaks in the valley. She said you might as well try to empty a river with an acorn cup as make a charm or a curse on a night when the moon wasn't in the right phase to give it power. Some charms needed a waxing moon, others a waning, a full or horned moon. I tried as hard as I could to learn the ways, but some secrets Ma would only whisper to Ryana, not that she ever paid any heed.

We made a slow procession, winding up the slope to the rocks at the top of the tor. By day, there was barely a track to be seen, nothing that would lead a stranger up to that place, but at night the path shone out as clear as the streams that ran down the side of the hill. Ryana walked first and most easily for Ma's tools were light. Ma followed and, for all that she dug her staff into the ground to lean on, she was as sure-footed as the sheep that grazed the moor. I walked behind her with the wort sack over one shoulder, the heavy skin of water over the other and the bundle of kindling sticks in my arms, which Taegan had thrust at me behind Ma's crooked old back. Taegan trailed behind, hefting a faggot of wood on her back, supported by a broad strap across her forehead. She'd heeded Ma's warning and was carrying as large a load as she could manage, though she whined like a chained dog all the way up.

In the last few feet before we reached the rocks the wind was so strong that I staggered under the weight of all I was carrying. I had to stop to steady myself. My eyes streamed. Here and there, far off in the darkness, blurs of red and yellow light shone like the eyes of beasts. They were the shepherds' fires burning far below in the valleys and the fires of the outlaws high on distant tors. Wiping my stinging eyes, I hurried up into the lee of the great towering rocks. Ma groped along in the darkness till she found the crevice between them and vanished inside. Ryana passed her the bundle of tools and squeezed in after her. But I hung back, listening. From deep within came a hollow knocking and a plaintive, high-pitched keening, as if many women were sitting deep inside the hill, mourning and weeping, throwing the melancholy notes from one to another, like the plovers that lived on the wind. Passing strangers and villagers, who heard the sounds, were always agasted and swore that Fire Tor was haunted by demons, or else it was the souls of the dead trapped inside, crying to be released. But I was never afeared: my skin tingled with excitement at the sound.

'Stir your arse, Mazy-wen,' Ryana called from inside. 'Ma's waiting.'

I pushed the water skin, herbs and kindling inside, then gathered up the wood Taegan had dropped and passed it through the cleft, stick by stick.

'Tell Tae to keep watch for Meggy,' Ma called, her voice echoing through the rocks.

'Fer why?' Taegan groused. 'It's so dark, I couldn't see a herd of white horses if they galloped over me. I've a good mind to go home.'

But I knew she'd not dare, just as I knew she would never set foot inside Fire Tor. When Taegan had reached her seventh birthday, Ma had tried to take her in, promising it as a great treat. But I reckoned that malicious cat, Ryana, had been whispering and pistering in little Taegan's ear, scaring the wits out of her, because later she'd tried to do the same thing to me. Ryana always sneered that when Ma had led Taegan up to Fire Tor, Tae had screamed and fought like a cornered weasel, even scratching and biting Ma's arms, rather than be pushed through that narrow gap into the dark, though she was so small she'd have passed through as easily as a mouse's whisker through a fox hole. Neither Ma's threats nor slaps had persuaded her to go in then or since, though I doubted she could have squeezed her fat udders through now, even if she'd wanted to.

I turned sideways and slid through into the space between the rocks. Inside it was so dark that my eyes ached. The blackness rippled towards me in waves. I felt for the wall of rock and slid down it, until I was squatting on the damp earth. The whispering spirits slithered around me in the darkness. They crept so close, I could feel them brushing my cheek and stirring my hair as they passed. Another voice began to sing. It was Ma keening softly with the spirits of rock and earth, begging them for the gift of fire.

There was a hollow click as she struck iron and flint. A tiny scarlet spark flashed out in the darkness, then yellow flame ran over the kindling, growing and leaping till a ruby-red glow filled the long, narrow cave, and the shadows of the spirits pranced around us.

The walls of the cave were formed of craggy rocks, with a huge flat stone over the top making a roof. At the far end, on the other side of the blazing fire, a narrow slab rose out of the earth, which was always kept covered with a long white woollen cloth.

Ryana was slumped against one wall, her bulging sheep-eyes half closed, yawning. Ma crouched before the flames, her head bowed, her grey hair tumbling loose down her back. She had spread her tools before the fire. She could do that by touch even in the dark for she knew them better than she did her own daughters – the white Bryde's Stone, a fox tooth, the hide of an unborn fawn, an ear of barley from a long-ago harvest, a raven's feather and the dried hind leg of a black dog.

Ma took a scallop shell that had been packed with dry moss and sprinkled with animal grease. She laid a burning stick from the fire to it. Sparks glittered across the moss and it began to burn with a bright yellow flame that rose steadily higher and thicker. Ma carried the shell into each corner of the cave, circling it there, muttering to the guardian spirits, before she returned to the fire.

Her wrinkled hand snaked out, and I passed her the water skin, watching as she poured water on to the earth in a circle around the treasures she had laid out. When it was complete, she hunkered down and took a small piece of dried meat from the bag that hung about her withered neck. It was the flesh of a red cat she had dried many moons ago. She chewed the dried meat, and when it was soft, she held it up in her hands before the flames, offering it to Mother Brigid, then laying it solemnly before the fire. Praying over each of her horny old palms in turn, she pressed them over her eyes and began to rock back and forth, singing with the voices in the cave, crying out to the duru moon to open the doorway and let the spirits come and go.

She tilted forward, rolling her head, her long grey hair flung out like a whip till it was inscribing a full circle in the air. The shadows on the walls were circling with her, spinning faster and faster, until it seemed that Ma was as still as a rock and the cave spun around her in a great dark vortex.

She turned her head, glaring back over her shoulder, and gave a savage snarl, like a wildcat, her yellow teeth bared, lips curled. She flung her arms wide and I saw red fur erupting through her skin, the glint of dagger claws on her outstretched fingers. Then, just as swiftly, they were gone. A long breath hissed from her open mouth, like the sound of rushing water, and she was limp again, bowed forward, her hands stretched out to the fire, and I knew that the spirit had passed through Kendra.

Ma held out her hand to Ryana for the worts I'd collected, but she'd fallen asleep in the heat of the fire. I crawled over and dragged the sack into my sister's lap, shaking her awake. While the charm was being woven, Ma would let only Ryana touch the herbs. Once I'd kicked her, my sister handed Ma the sprigs in turn, each bound with a strand of red wool. The murmurs in the cave grew louder, as if the sight and smell of the sacred herbs excited the spirits, drawing them closer, as the bloody carcass of a goat draws down the kites and ravens.

Ma lifted each wort in turn, singing its name and virtues to the mournful notes the spirits lent her.

Remember, Vervain, what you revealed,
Holy wort you are called, the enchantment breaker,
And you, Henbane, who call the ancestors to speak,
Call them to gather and call them to guard . . .

And so she continued, raising each sprig in turn before the seeing-fire – *bryony, wormwood, whortle, crow leek, corncockle, cleavers, adderwort.* Finally, she gathered them all together and bound them slowly, three times, with a strand of white wool as she sang:

Against demon's hand and dwarf's guile,
Against the elfin kingdom and the night hag's ride,
Against the fiend which flies from the west and the
*   north, from the east and the south,*
Against the eye of darkness that must not be opened.

Ma struggled stiffly to her feet and walked the bunch of worts three times around the seeing-fire, trailing them through the smoke. Then she stepped over the flames and through the smoke so that it swirled around her. Finally, she approached the cloth-covered slab at the end of the narrow cave. My stomach lurched. I knew well what lay beneath, but whenever Ma went to that rock, I felt as if I was seeing it for the first time. Ma nodded to Ryana, who lumbered to her feet and pulled the cloth from the ledge.

The corpse lay on its back, bathed in the blood-red firelight. The shadows of the spirits crawled across his chest. He was naked, his skin leathery, tanned to the colour of a thrush's wing, his hair and beard startlingly flaxen against his dark face. His lips had shrivelled back, revealing long teeth.

Beneath his chin, the skin had shrunk away from the long

gash across his throat, pulling it open as wide as a mouth. Ma had cleaned the wound in his neck and his corn-coloured hair, soaked in the blood that had poured from the skull-splintering blow to the back of his head. She had neatened his beard with her bone-comb and plaited a hag-stone into it. But she could not wash the look of shock and horror from his face.

Ryana had laid a river-polished black pebble on his fore-head and I had woven a bracelet for his right arm from the prickly twigs of the flying thorn that grew in the crevice of a rock without touching the earth. And it was to the hand of that arm that Ma now touched the bundle of worts. Then she pressed them against his bare feet, his head, mouth, breast and the withered worm of his pizzle and cods.

'Let me go, Kendra!' A voice as deep as a grave echoed around the cave. 'Bury me, burn me, release me from this torment of darkness.'

I stared up. The shadow of a man hung above me, grey, swirling, as if it was formed of smoke, except the eyes. His eyes were solid, living, burning like twin embers. The mouth opened wide as a scream and dark as a grave.

'Kendra, set me free from this misery and desolation. I beg you.'

Ma laughed. 'Not you. I'll never free you. You are Ankow. You were chosen. She gave you to me. See all the pains I've taken to keep your body safe. You should be grateful. You'll do my bidding till you come to carry me safe to the lych-ways, and then you'll serve my daughter, her daughter's daughter, till the sun and the moon vanish from the sky.'

His howl of despair and anguish made even the rocks tremble. I covered my head with my arms, feeling his wretch-edness stab deep inside me, but Ma only chuckled. She broke off a piece of the henbane and laid it below Ankow's naval. The curly golden hairs on his belly caught the fragile leaves and held them, like a spider's web. And the shadow vanished.

'Meggy's come.' Taegan's voice echoed through the cave.

Ma grunted and, taking the bundle of herbs, shuffled towards the narrow entrance, closely followed by Ryana.

It was left to me to cover the corpse again, extinguish the moss still burning in the scallop shell and collect the water skin, before finally emerging into the darkness. I shivered in the raw air after the warmth of the fire. Meggy and Kendra were already huddled in the lee of the rocks.

'. . . set the charm afire,' Ma was saying, 'and walk round the cave three times, so as smoke circles it. And mind yer walk round it with the sun.'

'Wasn't born with a goose cap on my head,' Meggy retorted indignantly. 'I know right way to walk.'

Ma snorted. 'No telling what queer notions you've picked up from those carrion crows you bide with.'

The gatekeeper bridled, but Ma ignored her and continued, 'As you walk, you must say the will worth, tell the spirits what you seek – "May this smoke keep him from the well as the flame keeps the wolf from the flock." And you've to mean it.'

Meggy shivered. 'Never fear, Kendra. There'll not be a charm worked that was ever meant more.' She glanced towards the east. 'I'd best hurry. There's not much time.'

Ma grasped her arm. 'This'll only keep the fiend at bay. If we're to destroy him, you must fetch me something I can use against him. We have to learn his name.'

Meggy nodded, tucked the bunch of herbs beneath her cloak and vanished into the darkness. Ma began to pick her way back down the hill, with Ryana and Taegan hurrying ahead, anxious to get back to their beds.

The echoes of the ghosts still whispered and sang through the cave. I turned to look back at the towering rocks. A red-orange glow danced and flickered through the crevices. The villagers said that whenever the heart of the tor was burning in the darkness it meant Ankow was riding up to the rocks on his skeleton horse, carrying the souls of those who had died. The fire would burn until dawn, and any who were awake in those parts would see the glow of those flames and tremble, afeared that, before the year was out, Ankow would drag them through that crevice into the lands of the dead.

Chapter 13

Hospitallers' Priory of St Mary

'Are you on your way to open the well, Sister Fina?' The words sprang out of the shadows, making the young woman start violently, for it was not yet dawn, and she'd thought the courtyard deserted at that hour.

Fina held up her lantern as Brother Nicholas stepped out in front of her, his black cloak billowing in the chill breeze. He was grinning, as if causing her to jump had amused him. The lantern lit his face from below, giving his hard features a wolfish, almost malevolent expression.

'You're about your duties early, Sister,' he added. He squinted up at the dark sky where as yet only a pale stain of light marked where the sun would rise.

'Sister Basilia wants to anoint a blind boy's eyes today with the water from St Lucia's well. But some of the pilgrims and villagers have heard . . .' She pressed her long fingers to her mouth afraid the words would escape if she didn't lock them in. Prioress Johanne had sternly warned all the sisters that they must be guarded in what they told the brothers about the priory, but Fina was certain the

112

prioress meant her. She was the one the prioress was warning and all the other sisters knew as much. Sister Clarice was always watching her, waiting for her to make a mistake, so she could go tattling to Johanne, trying to convince her that Fina was too young and stupid to be in charge of anything more than plucking chickens. The prioress had started to believe her. *An unlettered cottager*, Johanne had called her . . . Well, that was what she'd meant, anyway.

Nicholas had turned his head away from Fina, distracted by the sight of one of the scullions, Dye, emerging yawning from a doorway to fetch water. Her hair hung loose and tangled, and she was clad only in a torn shift that hung off one shoulder and barely covered her thighs. But the slattern seemed to hold more interest for the knight than Fina. She could sense he was about to walk away. He obviously thought she was too slug-witted to be worth talking to.

Words began tumbling from her mouth. 'Sister Basilia thinks it would be less frightening for the blind child if he was bathed in the well before the pilgrims arrive. The cave is so small, he might easily be jostled.' She offered Nicholas the small gift of information, like a tiny child might offer a feather or a daisy, knowing it was worthless, but wanting it to be accepted.

The knight took a pace or two towards her, so she was forced to look up at him. The hem of his cloak brushed her ankle as the wind gusted. His arm was so close to her breast that if she swayed forward his flesh would press into it. In spite of the breeze, she felt the hot flush on her cheeks and quickly lowered her eyes. *Why did he stand so close? He should not. He must not.* But she didn't step back.

113

'Pray excuse me, Brother. I . . . I need to unlock the door and light the candles.'

She tried to step round him but, whether by accident or design, he moved too, in the same direction and she found her way blocked again.

'Then allow me to escort you, Sister. It would not be gallant to leave you to stumble around in a dark chapel. You might fall and hurt yourself.'

'There's no need to trouble yourself, Brother. I've tended the well so often, I could find my way if I was blind.'

'Ah, but now that we brothers have arrived to protect you, you must allow us to assist you, else we shall feel we've made this tedious journey for nothing. Besides, I haven't yet had a chance to pray at this miraculous well of yours. You surely wouldn't forbid a brother knight his spiritual consolation when your sweet sisterhood exists to bring us such comforts.'

Something about the way his tongue lingered on 'comforts' made the back of her neck tingle.

'But the boy will probably resist having water splashed on his eyes. Your prayers would be disturbed by the commotion.'

'Believe me, I have offered some of my most devout prayers with a sword in my hand and men screaming all around me in the thick of battle. The squeals of one small boy will hardly disturb me. Come, I insist!'

He was not smiling now and the fingers that suddenly gripped her arm were not gentle.

They walked in silence across to the chapel door, where he released her. Frightened now, she fumbled with the bunch of keys that hung from a chain around her waist, praying that Basilia would come quickly, though she'd be no more help in getting rid of Nicholas than a puppy bent on licking

114

a robber to death. Besides, what was there to be afraid of? A brother knight of her own order merely wished to pray. That was all . . . all he'd said.

But the moment the door swung open Nicholas pushed into the chapel behind her, pressing so close she could feel his hard belly against her back. The single ruby light burning above the altar served only to make the darkness of the chapel more intense. Her hands were shaking as she lit the candles set ready on the iron spikes at either side of the door.

But Nicholas stepped away from her and curtly inclined his head before the crucifix hanging above the altar. Seeming to forget she was there, he began to pace slowly around the circular chapel in the space between the eight pillars and the outer wall, murmuring his *Aves* to honour the Blessed Virgin, as the sisters did in their private devotions. As silently as she could, Fina slipped across to the other side of the chapel and unlocked the door to the well. Now that she could see he was praying, her breathing slowed a little. What had she imagined could possibly happen?

But she had descended only two or three steps, when she heard footsteps approaching the door above her. It was customary to circle the chapel three times. Surely he could not have finished his prayers already. Panic gripped her once more.

'Wait!' she called. 'Let me light the candles below, then you will see a great marvel as you walk down.'

She hurried down as fast as was possible on the slippery, uneven steps. In the darkness below she could hear the gush and splash as the water poured into the stone pool. The spring was flowing strongly today. As she reached the bottom step, she lifted the lantern to light her way across the floor of the cave. She was about to take the last step down, but she snatched her foot back with a muffled scream. The floor of the cave was moving, heaving, as if the stones were alive and wriggling up out of the earth.

'I am still waiting for this marvel, Sister Fina,' Nicholas called, from the top of the stairs.

'I . . . cannot light the candles, Brother.'

She tried to think of some excuse that would stop him coming down. But it was too late – she could already hear his footfalls on the stairs. He'd evidently grown bored with waiting. She ran back up, taking the steps two at a time, heedless of slipping, and almost cannoned into him.

'The candles are wet, Brother,' she gabbled breathlessly, 'too wet to light, and the floor of the cave is slippery with water. The pool sometimes overflows when the spring gushes strongly. You mustn't soak your fine leather shoes. When it's safe, I will call you.'

She took a determined step up, though it meant pushing her cheek against his groin. The spiral staircase was only wide enough for one person to move safely and, as she'd hoped, he had no choice but to retreat back into the chapel to allow her to pass. As soon as she was safely through the door to the well, she locked it behind them and, on the pretext of fetching dry candles, she fled.

Her first thought was to warn Prioress Johanne, for she couldn't possibly open the chapel to pilgrims, not with . . . But as she was hurrying across the courtyard towards the prioress's chamber, she saw Basilia emerge from the infirmary, leading the boy by the hand. Fina had forgotten about them. There were only two other keys to the well. One hung in the prioress's chambers, for she had the key to every door in the priory, but the gatekeeper, Meggy, had the third, and if

116

Basilia went to the well door and found it was locked, she might fetch Meggy to unlock it. Basilia must not take the boy down there, not now, not today.

Fina picked up her skirts and ran towards the waddling infirmarer and the boy, intending to steer them away before they reached the chapel, though she hadn't any notion of what explanation she would give. But charging across a yard as slippery as a greased pig with mud and goose-droppings was bound to end in disaster. Fina had taken no more than a few paces before her foot slid from beneath her and she came crashing down. Spears of pain shot through her arm and knee as she lay in the stinking, wet ooze, wanting to clamber up, but too shaken to attempt it. Basilia dropped the boy's hand and hurried over, almost slipping herself in her haste.

'Sister! Sister, are you hurt?' She prodded and squeezed Fina, as if to discover whether she was fat enough to be slaughtered. The shriek Fina uttered when she touched her arm made her stop. Over Basilia's plump shoulder, she glimpsed Nicholas emerging from the chapel. He was the last person she wanted to see as she sprawled in the mud and filth. But he had evidently seen her and was approaching all too swiftly.

'Help me up, Sister,' she begged, dragging heavily on the infirmarer's arm. Fina succeeded in standing, but her knee gave way in a flash of pain and she almost collapsed again. 'Sister,' she whispered urgently, 'you mustn't let anyone unlock the door to that well . . . There is something down there . . . I must find the prioress . . .'

'The only thing you must do, Sister Fina, is to get that arm and leg attended to,' Basilia said firmly.

'Listen, please,' Fina begged.

'I have her, Sister Basilia,' Nicholas's voice broke in, and before either woman could protest he'd scooped Fina up in his arms and was striding across the courtyard towards the sisters' dorter.

Chapter 14

Prioress Johanne

'Whatever possessed you to gallop across the yard like a stable boy, Sister Fina, especially when it was so muddy?' I snapped, as soon as I reached the narrow bed on which she lay. 'Even a child would know to take more care.'

I hadn't intended my words to sound as harsh as they did, but my annoyance was partly born of guilt. The arrival of Brother Nicholas was occupying all my thoughts, so that I'd scarcely noticed how muddy and treacherous the courtyard had become. I should have given orders for it to be scraped and strewn with straw, but there was so little left that the servants were having to use bracken to bed the horses and could scarcely cut enough for that. The tracks were so wet that the horses pulling the sledges became mired to their hocks and it was easier for the servants to fetch what little bracken they could carry on their backs than spend hours dragging the exhausted beasts from the mud.

Fina blinked at me, then struggled to sit up, but I pushed her down. She was deathly pale, whiter even than the bleached linen sheets in which she lay. Basilia, with the help of two servants, had managed to wrestle her dislocated knee back into place. She'd also bandaged Fina's arm and, though our infirmarer believed it to be cracked, the bone had not snapped in two or pierced the skin. I guessed that she had given Fina something to dull the pain while they had dealt with her knee, for her eyes were unfocused, and when she spoke, her words were slurred.

'Unclean spirits. The devil's spirits, mustn't . . . he mustn't see.'

'Sister Fina,' I said sharply, intentionally so this time, trying to rouse her, 'are you speaking of little Cosmas again? You are a sister of the Knights of St John, not an ignorant village woman. He is just a child come to us for help, nothing more. You have been taught that our Lord himself said, "Whoever receives a child in my name, receives me." It is compassion and prayer the boy needs, not accusations of demons and sorcery. The draught Sister Basilia gave you has made you drowsy. I will leave you to sleep. We will speak again when you wake and I trust that by then your good sense will have returned.'

I hoped the syrup Basilia had given her was the cause of this nonsense and not the recurrence of the trouble that had afflicted her when first she joined us. She had come to our priory straight from her father's house. She had appeared shy and naïve, which was natural enough for any highborn young woman who had seen nothing of the world, but I also sensed a bitter resentment in her.

Not all of our noble sisters enter the order entirely by their own choice, though they must swear that they do. It is not unknown for their kin to persuade them that they have but two alternatives: to become a sister of the Knights of St John where they are comparatively free to travel and work, or to enter a nunnery to be walled up in the cloister with little to do, save pray and idle away the weary hours in games and gossip. Though, if Brother Nicholas had his way, we would not even be offered that semblance of a choice.

But to a young woman who from childhood has dreamed of marriage and being mistress of her own house, even such freedoms as the Knights of St John can offer may seem like the bars of a cage. I knew that Fina saw me as an ageing prioress who could not possibly understand the torment and longings of youth, but I did. I had not made my profession in the order willingly, though my dreams had been very different from hers. But my own father had decided that placing two of his children in different orders was politically and spiritually expedient, for then the religious of two orders would offer masses for his soul when he departed this life, and Christ must surely regard one, at least, with favour.

It was the reason I had appointed Fina keeper of the well, although Clarice had warned me she was far too young. It was true she was immature and often given to strange fancies, but I hoped that such responsibility might help her settle in the order, even grow to love it as I had done. And it had appeared to work. St Lucia's blessed water had wrought a miracle. Fina had come to treasure the well, as fiercely protective of it as old Meggy was of her gate or Sibyl her kitchen.

I tried to assure myself that her wild words now were due to the shock of the fall and anxiety about who would care for her beloved well. When she was rested, she would doubtless be herself again.

I had already turned away when I felt the tug of a hand on my skirts. 'Not the boy. The well . . . go alone to the well.'

Fina was trying to drag herself towards the ring of keys

that lay on a stool near her bed. She almost tumbled out as she tried to reach them. She was becoming so agitated that I knew she wouldn't rest until she had them, so I placed them in her hand and, for the second time, pushed her back against the pillow. 'Lie still, Sister, or you'll do yourself more harm.'

But Fina grasped my hand and pressed my fingers around an iron key so hard that the edge bit into my flesh, making me wince.

'Alone,' the young woman repeated, her face screwed up in an effort to voice the single word.

I took care picking my way across the courtyard. I certainly didn't want to find myself lying in a bed with an injured back or worse. I'd given orders that the worst of the mud and dung should be scraped up, and planks and old sacks should be laid in front of doorways for people to walk on. Two of the servants, their skirts looped up to keep them out of the filth, were already at work with Brengy, the stable boy, and his young sister, Dye. They were shovelling piles of stinking sludge into a wheelbarrow. They paused to blow on their wet, cold hands and glowered at me, but I was used to that, and merely nodded. My duty was to manage the priory for the good of all, not to curry affection.

A thick rolling mass of granite-grey clouds turned morning to twilight. Inside the chapel it was almost too dark to see the pillars. The light from the lamp barely grazed the altar over which it hung. But as my eyes adjusted, I saw that the door leading down to the well lay open. Had the pilgrims been admitted? I crossed to the door through which they would enter the chapel, but it was still bolted.

Footfalls sounded on the stone floor behind me and I

turned to see the redoubtable form of Meggy lumbering away from the well door, lantern in hand. She started as she caught sight of me in the shadows, and guilt flashed across her face.

'Should you not be at the gate, Goodwife Meggy?'

'Heard poor Sister Fina had a bad fall. Didn't know if she'd opened the well for pilgrims. Went down to see if candles had been lit.'

'You could have asked one of the sisters to do that.'

'Didn't want trouble them. 'Sides,' she added, with a mulish shrug, 'my niece is these two years married and still not with child, so I said I'd fetch some holy water from the well for her, so St Lucia will grant her a babe.'

'St Lucia is not known for such miracles. Your niece should light a candle to the Blessed Virgin and ask her for a son.'

Before I had banished the cunning woman and her tribe of daughters, barren women had come to bathe naked in the spring, not seeking a blessing from a holy saint of the Church but to ask some pagan goddess to make them fecund. Some still tried, and one had fled dripping, without stopping to pull on her gown, when the sisters had surprised her. I was fond of Meggy. She was a loyal soul and, in her own way, good and honest, but she was still a villager at heart. No matter how often I had reproved her, she still could not entirely shake off the superstitious ways of the moorland people. I had little doubt that if any of the spring water were to be given to Meggy's niece, it would be the old goddess who would be invoked, not St Lucia.

I was just on the point of reminding Meggy for the hundredth time that she was a gatekeeper in a Hospitallers' priory, and that the order did not employ those who prayed to idols, when I noticed that the candle flame in the lantern she was clutching was guttering wildly. Her hand was trembling. 'Are you ill, Meggy?'

The woman shook her head, glancing behind her at the open door leading down to the holy spring.

'Is something wrong with the well? Sister Fina seemed to want me to inspect it. Did you see anything amiss?'

She made a strangled noise as if she was about to speak, but had choked on the word. She thrust the lantern into my hand. 'Been coming to the well since I were a girl and old Kendra's ma was keeper of it. I've never seen anything like this afore. You take a look for yourself, Prioress. Don't know how we'll get them out, unless they've a mind to go.'

'Get who out?' I demanded. 'Is someone down there?' I edged towards the door and dangled the lantern inside, but all I could see were the first few steps curving down. I strained to listen for any sound of movement below, but heard only Meggy's ponderous footsteps retreating across the chapel floor behind me.

The gatekeeper paused, grasping the latch of the chapel door. 'Old Father Guthlac was right. There's a curse come to this place with that boy and it'll not lift till he goes.' With that she wrenched open the door to the courtyard and was gone.

Taking a deep breath and a firm grip of the lantern, I edged down the first of the uneven steps. The craggy walls began to twinkle with the familiar green-gold shimmer. Normally, the sight filled me with a wondrous joy and peace, but today I was too anxious about whom I might discover below to see the beauty of it.

I'd reached the point on the stone steps where I should have been able to see the glow from the candles that burned beside the spring, but the only visible light came from the lantern I carried and the glimmer from the walls. The trickle and splashing of water grew louder as I descended, but I could hear nothing else. No voices or feet rattling loose stones on the cave floor.

I paused on the second to last step. The lamplight had caught a flicker of movement. At first I thought the water in the pool was overflowing and had flooded the cave. I lowered the lantern so that the light fell on the floor. But it wasn't water rippling across the floor. It was frogs, hundreds of glistening green-gold frogs. There were so many that they were climbing over each other, eyes bulging, throats pulsing in and out. The stone trough was heaving with them, all jostling to find space. Even as I stared, more crawled out over the lip of the pool, and they'd no sooner leaped to the floor than another wave clambered out behind them.

I had seen tiny froglets in the damp cave before, but they had vanished into the cracks and crevices of the rocks as soon as light fell upon them, and there'd never been more than two or three at a time. Frogs were generated from the filth of mud and slime, and there was certainly enough mud up on the hills after all this rain to breed a plague of them. If I'd seen them out there on the moor, in that desolation of sucking mires and black pools, I would not have been surprised. All the foul creatures of the night dwelt in those bogs. But that the pure, holy water of the well should spawn such vile beasts, the very symbols of sin and wickedness, was unthinkable. Now I understood what Sister Fina had meant when she spoke of evil spirits. Words from The Apocalypse of St John pounded in my head, accusing, flaying –

de ore pseudoprophetae spiritus tres inmundos in modum ranarum sunt enim spiritus daemoniorum. From the mouth of the false prophet, three unclean spirits like frogs, for they are the spirits of devils.

I hastily made the sign of the cross. We would have to close the well to pilgrims until a way could be found to cleanse the cave of such an abomination, but word must not get out. If it became known that the very water that was meant to heal and bless had generated a swarm of evil spirits, the pilgrims would never return. But what reason could I—

I jerked around at the sound of a low whistle. Brother Nicholas was standing on the stairs behind me, staring down over my head. Swiftly I raised the lantern in front of me, trying to plunge the floor of the cave into darkness, but even as I did so, I knew it was too late. The knight had already seen all that I had wanted to conceal.

'So, this is why Sister Fina was so anxious I shouldn't visit the well this morning.'

He pulled the lantern from my grasp and, crushing me against the wet wall, squeezed past me. He descended a few paces, peering at the heaving floor and the frogs swarming out of the trough. He raised his foot and stamped on two of the little creatures that were trying to crawl up the steps. They screamed as they died. He ground their corpses beneath his boot, scraping off the gory remains on the edge of the stone step.

'At least we know they're made of flesh and blood.' He gave a mirthless laugh. 'So, what has brought this plague upon your house, Prioress Johanne? I assume the pilgrims don't usually have to share their holy water with these foul imps. Some might be tempted to think God is punishing your stubbornness, like he punished the pharaoh of Egypt. But then frogs are a sign of heresy too, are they not? If the Lord Prior learns of this . . .'

We both knew I did not need him to complete that threat.

'On the other hand, Prioress, with no pilgrims bringing offerings until our little friends have departed, dear Sister Clarice's ingenious accounts might, in the end, prove accurate. Wouldn't that be a curious justice? Perhaps that's why you have been visited with such a plague, Prioress. She claims the offerings are low, so to keep her from the sin of falsehood, Our Lord is ensuring that they are.'

Brother Nicholas thrust the lantern back at me with a triumphant grin. 'I should make sure you close the door when you leave. You won't want Satan's imps infesting the entire priory. You'll find that much harder to conceal.'

Chapter 15

Sorrel

I heaved the bucket of wet tailings into the barrow and watched as the lad dragged it up the slope of the spoil heap and tipped it over the edge. *Tailings?* A month ago I'd not even known the word existed, but in the past few days I'd had to learn fast.

The tinners thought it great sport to talk in a language only they could understand. *Carry that to the bundle*, they'd say, then crack their sides laughing as I set off in the wrong direction. I'd asked the women what I was meant to do and they'd sighed and rolled their eyes, as if I was a moon-touched bairn, but they'd shown me how to tip the gravel on to a long, sloping stone slab where the water separated it into *heads*, which contained the heavy tin, and *tailings*, the lighter waste. They'd warned me they'd only show it me once, mind, for they couldn't afford to lose time teaching me instead of earning.

But Todde, being a man, wouldn't ask. The first time, he filled the buckets with the useless stones instead of the tin gravel. The tinners winked and grinned to each other but said naught. They let him heft those buckets all the way up to the crushing stones and watched him pound the gravel till the sweat streamed off him. Only when he'd hauled the broken stone up the slope and into the blowing-house did a bellow of laughter break out from the men below, nearly as loud as the blast of the horn.

The stoker came storming out, railing at the miners. 'Spitfrogs! Nugheads! Suppose you think that's funny, do you? If I'd not seen what this codwit was up to, his dross would have ruined the firing and it'd have had to be done all over again. Fuel doesn't come cheap, you know. Master Odo would have taken the cost from all our wages.'

'He will anyhow,' one man muttered darkly.

'Can't expect us to watch Todde like a babby,' another called up. 'He said he were old hand at tinning.'

'Well, you'd better make sure he learns to be and fast too, else we'll all be going hungry.'

Give him his due, Todde swallowed his humiliation and was learning. His arms were strong, and though he'd still a way to go afore he matched the skill of the experienced tinners, he was the last to set his hammer down when the horn blew at dusk and the first to lift his shovel at crack of day. The tinners began grudgingly to nod their approval behind his back.

At night, after we had gulped down the few mouthfuls of our meagre supper, I had taken to wandering up to Eva's fireside to get warm. Most days I was lucky if I could gather enough kindling and dung to keep the cooking fire alight long enough to heat the pot. Neither Todde nor I had time to cut and dry peat on the moor, which most of the tinners' wives burned, for we had to spend what hours of light there were trying to earn money. The dried mutton Gleedy had given us was now no more than a lingering flavour in water that contained only herbs and a handful of dried peas I'd borrowed from Eva. She was a good soul and I'd often catch her slipping a piece of meat or dried fish into the hand of one of the small children who'd creep up to her hut after the tinners had gone in the hope of a bite. It was a pity she had no bairns of her own, for she'd have been the best of mothers.

She would glance up when I hesitantly approached, jerking her head towards a place close by the fire, as her way of saying I was welcome to share it. Mostly we sat in silence for, unlike the women I'd known back in the village and the tinners' wives, she never gossiped about others. Neither did she talk about her own life or worries. She spoke only when she needed to, as if words were precious coins that should not be wasted on fripperies.

But I reckon she did have troubles, for when she was passing by my hut, a day or so after we arrived, I noticed a great purple bruise on her cheek. I asked her if she had slipped in the mud, but she shrugged and trudged on up to her cooking fire by the ruined wall.

A tinner's wife watched her until she was out of earshot, then nodded towards her retreating back. 'I'll wager a man gave her that. It's not the first time I've seen her with a black eye or a cut lip. Cook their suppers and warm their beds for them, and that's the payment you get.'

I wasn't surprised by the bruise. Father had given them to Mam often enough. It was what a man did once he'd got a woman to wed him. She was his then, to use as he pleased, like a beast he'd bought at market. But all the same I was puzzled. I wondered which of the tinners who gathered around Eva's fire she'd taken as her lover. I scrolled through their faces in my head. I couldn't think of one I'd want in my bed, not that any man would want me, as my father had constantly reminded me. But I wondered why Eva put up with such treatment, if she wasn't married to the man. There were plenty of others to choose from in the camp.

While Eva never grudged me a share of her fire I felt bad that I'd nothing to offer in return. Then one night, after the tinners had gone back to their own huts, she settled down to pluck a few snipe she had snared. I picked one up, wedged the carcass between my knees, and swiftly ripped out the feathers with my good hand, neither tearing the skin nor leaving any shafts. I caught Eva watching me out of the tail of her eye. She said nothing, so I took up another bird and worked on, silently plucking and handing them to her to be gutted and dismembered ready to be dropped into her pot. Later, as I walked down to my own hut, I heard hurrying footsteps behind me. I felt something soft and cold thrust silently into my hand. It was one of the plucked snipe. I turned to thank her, but she had already vanished into the darkness.

Each night after I dragged myself away from the warmth of Eva's fire, I had to return to the misery of the cold, wet hut I shared with Todde. He was usually snoring by the time I crept in, for I tried to stay out until I thought he would be sleeping. I did not feel completely at ease sharing a hut with him. Sometimes, I'd wake in the night and listen to him breathe, watching his chest rise and fall. Once his fingers had lingered when they'd accidently brushed mine as I was handing him a bowl. I'd snatched my hand away and he'd dropped his gaze, staring at the ground as if he'd never seen it before, his face flushed. I still didn't know what to make of him.

131

Shivering, I wrapped myself in one of the damp, stinking sheepskins Gleedy had given us. My legs ached with the cramp, and my toes swelled and itched for my feet never seemed to get dry, but in spite of that, sheer exhaustion always dragged me into sleep. I dreamed of black mud and grey water.

Even trapped inside the mire of those dreams, I knew I must find the woman who had called to me. I was desperate to hear her voice again. Every night I begged her to speak to me. But she had fallen silent and I couldn't rouse her, couldn't reach her. Some nights, I was sure she was hidden just out of sight along the track. If I could reach the next bend in my dreams I would find her, but my feet were gripped fast in the mud and, as I struggled to free myself, I was jerked awake into another ash-grey dawn by the shrill blast of the ram's horn echoing across the valley.

Don't leave me here in this valley. Speak to me. Show me how to find you!

The sun was sinking behind the ridge above the tin workings and a dark tide was seeping up the valley before the ram's horn sounded the two notes to signal the end of the long day's hard toil. The bairns were the first to throw down their tools and escape. Older children began plodding up to their parents' huts, carrying the little ones, who were already falling asleep, utterly exhausted after a day of fetching and carrying, filling buckets and sorting through gravel. But all who wanted to eat had to work.

I rinsed the grime from my face and hands in the puddles left in one of the wooden leats. I'd soon discovered that they were cleaner than the river into which all the dirt from the gravel washings and the filth from the camp were emptied.

Using my teeth, I pulled off the mud-soaked rag wrapped around my fingers and dangled my hand in the icy water, sucking in my breath with a hiss as it touched skin rubbed raw and weeping from the rope handle of the bucket. I'd always thought the skin on my right hand was as tough as tanned cow's leather, but I'd not spent long days at home lugging heavy buckets full of wet earth and stones.

'You best stir yourself else it'll all be gone,' a woman called, over her shoulder, as she scurried by.

I glanced up. A straggling procession of men and women were making their way along the valley towards the blowing-house. The whole camp seemed to be on the move. I hurried to catch up with the woman, trying not to slip in my haste.

'Where they all going?'

She frowned at me, as if she thought I was mocking her. She glanced down at my withered hand, and her frown lifted. 'Course, you're the woman came with the squab. Don't you know what day it is?'

I shook my head and she gave a weary smile.

'Aye, well, can't say as I blame you. One day's same as another here. But it's Saturday, when old Gleedy doles out our wages and opens his stores so we can buy what we need for the week, seeing as how we can't get to market. But flour and beans have been real scarce lately, so you'll have to hurry if you want any.'

I scrambled after her as she toiled up the muddy track, my spirits lifting. At last I'd receive some money for my toil and, best of all, buy something to put in the pot, a good measure of beans or peas, flour to make flatbread, perhaps even a little meat. My belly ached for solid food after the watery soup I'd been eating. I remembered the excited gleam in Todde's eyes that night in the drovers' hut when he spoke of the merchants flocking to sell to the tinners. *They live off the fat of the land, those tinners do.* I'd seen precious little fat since we'd arrived, but perhaps tonight we would eat well.

The choking smoke and metallic stench from the blowing-house billowed down to meet us, wrapping itself around us in the damp air. The door was open and a heavy, scorching heat blasted out into the chill night. A fiery glow lit the inside, turning the faces and clothes of all blood-red, as if we were staring into the maw of a roaring dragon. Outside, a great waterwheel churned, pumping the giant bellows for the furnace. But the tinners and their womenfolk were already hurrying past it towards a lower stone building, hidden from the valley below by the blowing-house in front. Two huge hounds had been chained to one side of it, and were straining at their spiked collars, leaping and barking savagely at all those coming towards them, though their leashes were too short to allow them to reach anyone. I fervently hoped their tethers were strong.

I edged in through the open door and found myself gawping like the village mooncalf at the sight inside. Gleedy was standing towards the back of the long building, surrounded by more kegs, boxes and sacks than I'd ever seen, even on market day. Sheepskins were piled on shelves next to cats of salt. A tower of buckets teetered precariously next to a stack of shovels. Bundles of rope and cord swung from the hooks on the beams beside several ladders, a swathe of axes and barrels of every size.

In what little free space remained, men and women had formed a straggly line in front of a broad oak table. A hunched, pimple-faced youth was perched on a stool behind the table. A ledger, a small brass-banded wooden box, a guttering candle and a dish of ink and quills were crowded so near to his elbows it was a wonder he didn't send them all flying when he moved. As each person came forward, he consulted the ledger, then slowly counted out coins, which he laid in a row close to him, as if he feared they'd be snatched away before he'd finished. Only when the person had made their mark in the ledger on the spot where he pointed did he slide the coins towards them.

Eva was hovering near the table, and the men I'd seen eating at her fireside on the first night each handed her a few coins as they passed. She looked more cheerful than most of the men and women, who scowled as they scooped up their paltry wages and hurried towards the stores. A couple of men argued with the lad, insisting he look again at the sum written in the ledger, but he jerked his head in Gleedy's direction and shrugged. Still grousing, but under their breath, the men made their marks.

Gleedy seemed too busy to notice the dark looks he was getting from the men near the table, or maybe he was indifferent to them, for those tinners he was dealing with seemed equally disgruntled. He was emptying a measure of withered peas into a small sack held open by one of the women. The coin she proffered vanished instantly into the deep leather pouch dangling from the belt around his hips. He beckoned to another woman and two rushed forward together, shoving each other in an effort to be served first. In the time it took to say a paternoster, he'd sold half a dozen salted sprats from a barrel, a handful of iron nails, a battered cooking pot, a length of rope, some dried beans and a small measure of rye flour to different customers, without once pausing to think or search for anything. Several times, though, he shook his head, showing empty barrels and sacks, and I realised my

135

neighbour had been right. There might be plenty of nails and shovels to be had, but you can't fill a belly with those. I would have to make haste to claim what I was owed and buy whatever food I could.

But before I could join the queue in front of the table, someone began shouting and my belly lurched. I recognised the voice only too well.

'All the graffing and crushing I've done, and you're saying that's all I have to show for it!' Todde bellowed. 'Don't you think you can cheat me, boy, just 'cause I'm newly come.'

He reached across the table and seized the lad's jerkin, hauling him up off the stool so that his toes were barely grazing the floor. The boy struggled to prise the great fists off. Todde dropped him and he tumbled backwards over the stool, letting out a high-pitched yelp, like a kicked puppy.

Gleedy strode across, elbowing people aside. 'What's all this?'

Todde rounded on him. 'This lad of yours is trying to get me to make my mark in that book to show I've had all that's due to me. Means to fob me off with a tenth of what I'm owed and he'll slip the rest into his own purse. I know his sort. Just 'cause he can read, he reckons he can gull us. But I wasn't hatched in a goose's nest. Here!' He brandished a notched stick in Gleedy's face. 'I've got the tally of every bucket I've fetched to the blowing-house and I've been keeping my ears and eyes open. I know exactly what they're worth.'

Gleedy took the stick and counted, then pulled the ledger towards him. 'See for yourself. What's written in the ledger exactly matches your tally, every last bucket.'

'Then why is this little arse-wipe only giving us this?'

Todde pointed to the few coins that now lay scattered across the table.

'That's what's due to you after what you owe has been paid.'

'Owe? I owe no man.'

'Sleep out in the open on the moor, do you?' Gleedy asked.

Todde frowned in bewilderment, but I caught the looks the other men were exchanging around him, and before Gleedy had said anything more, I suddenly understood what Eva had been trying to warn me about that first night.

'You sleep in a snug cottage, isn't that right? So there's rent to be paid for it, not to mention the sum you owe for the dried mutton and onions you've been stuffing your belly with and those sheepskins keeping you warm. Then there's the hire of the buckets, picks, hammers and shovels you've been using and wearing out. And there's taxes to be paid to the King on every ingot of tin that leaves here, afore it can even go to market. Fact,' he said tapping the ledger. 'What you owe Master Odo comes to more than twice what you've earned this week. So, by rights, the lad shouldn't be giving you anything. But I told him to pay you enough to buy a bit of food for your pot. "Can't work if they're starving," I says. You can pay the rest off each week, till you're clear.'

Todde had turned scarlet with rage. 'You thieving bastard! You never told me that!'

Gleedy shrugged. 'Only a fool would think he'd could use another's man's tools to work another man's land, then keep all the profits.'

'And it takes a bigger fool to think I'd stay,' Todde retorted. 'You can tell Master Odo he can work his own land. I'll not pay him for the pleasure of doing his work.'

Gleedy grimaced in the mocking way the mummers do when they're pretending to be sad to make a crowd laugh. 'I see now I should have explained to you how a free man earns a living. You being a villein, you'll not have worked for wages afore.'

The blood drained from Todde's face and his eyes bulged, as if Gleedy was pressing his hands round his throat. 'I'm no serf,' he spluttered. 'I was born a freeman. I'm a tinner!'

'You are now and living under the King's protection. Long as you remain a tinner, your old master can't touch you. But if you was to leave here with no tools to prove your trade and no claim on any land, the lord who owns you would have you dragged back at the horse's tail. And if you are the villein they've been hunting, he'll have you hanged for a thief too, 'cause you didn't run empty-handed, did you?'

Todde was shaking his head violently, but couldn't utter a sound.

Gleedy patted his shoulder. 'But don't you go losing so much as a peck of sleep over it. Long as you stay here, you're as safe as if the King's own guards were standing watch over you. You're in good company too. Half the men and women here are runaways. But they know their old masters can't touch a hair on their heads, leastways providing they don't make trouble and get themselves thrown out.'

I felt as stunned as Todde looked. I'd realised from that first night in the camp that Todde had never seen tin-workings before, much less been a tinner. But most men boast of skills they don't have. It had never occurred to me that he was a runaway, a bound man. Men who flee their manors and masters usually run with little or nothing. He had had tools and the donkey! There would be a price on his head, and anyone in the land could claim it whether they delivered his head with or without the rest of him. My face must have betrayed my shock, for Todde looked anguished as he caught sight of me.

'I swear there's not a word of truth in it, Sorrel,' he protested. 'That bastard's tongue is as forked as a viper's. He's been lying to us from the start. It was no chance meeting we had with him on that track. He was waiting for those outlaws to rob us. He probably pays them to do it.' He pointed towards the stores. 'It's how he gets all that. I reckon if you searched back there you'd find all my tools and your pot too.'

Fury gathered in his face and he twisted round to face Gleedy again. 'I'm right, aren't I? Those outlaws, they *are* in your pay!'

Two men were edging up behind Todde. Gleedy's eyes flicked towards them. I could see what was coming.

'Todde, leave it!' I pleaded, but he ignored me.

Gleedy laughed. 'Mine? You think cut-throats like that'd answer to me?' He pulled a comic face, exaggerating his squint. 'Rob me blind they would.'

The tinners laughed far more heartily than the feeble joke deserved. Gleedy sauntered away, as if the matter was settled. But Todde hadn't finished. With a bellow of fury, he launched himself at the retreating back, but had taken no more than a pace before his arms were caught by Gleedy's two henchmen, who pinned him against the wall. One drew back his fist aiming it at Todde's belly, but a burly tinner caught his arm, shaking his head.

'He'll be no use to Gleedy if he can't work, will he?'

The two men hesitated, glancing at Gleedy, but he was already serving his next customer. Reluctantly they let Todde go. Men and women averted their eyes and began talking loudly again, determined to pretend nothing had happened. But the tinner who'd saved him laid a firm hand on Todde's shoulder.

'Like your woman says, let it go. Be grateful he left you with enough coin to buy a bite. Seven days is a long time to go begging for scraps from neighbours when you've naught in your pot, and he's been known to do that and far worse to men that cross him once too often.'

Chapter 16

Hospitallers' Priory of St Mary

Brother Nicholas lifted his gaze from the accounts ledger as the candle flame snuffed itself out, leaving only a wisp of smoke. He rubbed his dry, smarting eyes, surprised to discover it was already dark inside the chamber. He'd been working by the light of a candle all afternoon: little daylight found its way over the high wall and in through the tiny casement, still less when the day was as leaden and grey as this. It must be later than he'd realised.

He'd been chasing the capricious figures from one parchment roll to another, but it was like trying to catch a shape-shifter by the tail. He was sure the accounts were not correct but, like the procurator at Clerkenwell, he could not pin down exactly where they were wrong. Money was missing, he was almost certain of that, but if it was, then from where? How was it disappearing, and into whose purse? The sisters were not spending it on living in luxury, if the food being served was any measure of that. So how far had the corruption spread? Were they all conspiring in this theft, if indeed that was what it proved to be, or was it just that crafty old hag Sister Clarice? She'd certainly been extremely anxious that he should not examine the books.

The prioress had told him that Clarice had managed her husband's warehouses before she joined the order. And she had certainly gained the skills of bookkeeping, Nicholas had to admit. In fact one might say she had learned the art rather too well. Just what other lessons had she learned from some of the less scrupulous merchants and sea-captains? But he would uncover the truth. He refused to be made to look a fool by some withered old vecke, who should be spending her dotage sitting silently in a cloister, sewing altar cloths or whatever it was that nuns did.

The door of the priory's small guest chamber was flung open and slammed shut with equal vigour, sending a violent blast of glacial wind whirling around the room. The draught snatched up the parchment Nicholas had earlier laid to dry on the table and sent it spinning to the floor. But he was too busy trying to keep the quills from being blown into the embers of the fire to notice where his letter had landed, until he heard the ominous crunch of parchment and glanced down to see it crushed beneath Alban's muddy boot.

'You frog-witted ox! I know you're used to living in a stable, but haven't you at least learned how to enter a room without wrecking it?'

'Why did you toss it on the floor if you didn't want it walked on?'

Ignoring Nicholas's splutter of curses and insults, he lumbered over to his narrow bed and flopped down on the thin straw mattress so heavily it was a marvel that the wooden legs beneath didn't splinter or the cords snap.

Nicholas retrieved the parchment and examined it with dismay and fury. Wetted by the sodden footprint, black ink was dribbling down the page. Every word the loutish boot had stamped on was obliterated. 'Look!' he spat. 'It's taken me half the afternoon to write this and it is not even fit for an arse-wipe now.' He crumpled it and hurled it savagely into the fire. He instantly regretted it and tried to pluck it out, but it was already burning, turning his carefully crafted sentences to smoke and ashes. He cursed again. He could at least have used the fragments that were not smeared to jog his memory. God alone knew when Hob's wagon would return from Buckland, but when it did, Nicholas wanted to be sure his reports were ready to be dispatched immediately with the carter.

Alban lay on the narrow bed, watching him indifferently. 'You're in a sour temper. Fleas biting your arse, are they? It's me that should be grousing. While you've been sitting warm by a fire, scratching away like some fusty old monk, I've been up to my oxters in freezing water, trying to clean a cartload of mud off your rouncy, then pluck the burrs and thorns from his belly and legs. Where have you been riding that poor horse?'

'You'll be up to your armpits in cold water again, this time head first, if I find so much as a mote of dirt on that beast. I was out at dawn, freezing my cods off in a gale, riding these blasted moorlands and trying to prise some sense out of toothless old gammers and their drooling offspring, all of whom do nothing but grunt and stare at you till you start to wonder if you shouldn't be talking to their hogs instead.'

Alban gave a sly grin. 'And there was me thinking we're pledged to be the serfs and servants of the blessed poor, whom we should treat as if they were our lords and masters.'

'If you want to serve the poor, you'll find them out there grubbing around in the mud and rain, believing they'll make their fortune by digging for tin in other people's dung heaps. It's not the poor I've been chasing for days, but those who feign poverty, those who think they can fool me by hiding their cattle or concealing their kegs in caves. Everywhere I go it's the same story.'

He twisted his hands into the mockery of a fawning man. '"God's chollers, sir, you've come all this way and got yourself fair drownded on the road, and all for nothing, seeing as how I've already paid the holy sisters. Can't ask me to pay the rent twice, now can you, sir?"'

'They think that by paying those harpies half what is due to our order, they can cheat us out of the rest. And that's when I can find my way to their wretched little farms and watermills. Ask anyone round here for directions and they either pretend they can't understand you, or deliberately send you in the opposite direction. And if they can lead you into a mire or through an icy stream on the way, they'll do it and think it rare sport.'

He punched the lumpy straw mattress on his own bed as if it was a man's face. 'But they'll rue the day they tried to deceive me, as will that wily old embezzler Sister Clarice. There is more mischief to be uncovered here than a few missing rents. I can smell the stench of it. That boy they claim to be blind, for one. Sister Melisene was certain he sent a spirit to torment and murder the old priest. But the prioress maintains the boy is dumb and can't be examined. She's deliberately keeping him from me.'

Alban grunted. 'You want to be careful if you go chasing after little boys. There's some might start saying you have a fancy for them. I keep my ears cocked in the taverns and marketplaces. There's talk that the knights of the white cross are no better than the knights of the red, buggers to a man.'

In two strides Nicholas was at the bed. He grabbed the front of Alban's shirt, hauling the groom up till his face was inches from Nicholas's own. 'I'm no sodomite and no Templar.'

'Hold hard!' Alban protested, trying to wrench himself out of Nicholas's fists. 'I don't take kindly to those insults any more than you do. The last man I heard say as much, I knocked straight through a wall. I'm just telling you what's rumoured.'

'I know full well what's rumoured!' Nicholas bellowed, thrusting Alban back hard on to his bed. 'Ever since the Templars confessed to kissing each other's arses and pissing on the cross, half of Christendom has been whispering that the Knights of St John indulge in the same foul practices.' His fists were clenched so hard that his knuckles had turned white. All the honour, all the glory so hard won throughout the centuries by noble men who had pledged their lives and souls for Christendom, all that gave meaning and purpose to his own life had been turned to dust and dung in the eyes of men and God because of what the Templars had done. They had soiled the very title 'knight' and all it stood for.

He slumped back on to his bed. 'Those fools in France didn't act swiftly enough. They burned the Templars they caught, but they let half of them slip through their fingers and escape to foul our nest in England, and our king won't supply enough men to hunt them down. Says his coffers are bare and we all know why – he's been stuffing the pockets of his own little catamites. Even when they do arrest the Templars hiding here, they're not put to the rack properly, because thanks to King Edward's lily liver, there's no one in England competent enough to do it without killing them

145

before they've uttered a word. They have to smoke out every last one of the Templars and burn them for the filthy vermin they are. Until they do, we will all be tainted.'

'Aye, but till that happens, you want to be careful not to set tongues wagging here 'cause women will use anything they can against a man if it serves them. More cunning than a skulk of vixens, they are.'

'They can try any trick they dream up,' Nicholas said. 'But I swear that before I'm done I'll uncover every festering sore that the prioress is trying to hide from me and she will pay dearly for them all.'

Chapter 17

Sorrel

As soon as I plodded up to Eva's fire I could see that something was amiss. Usually by this time she'd be threading her way between the men and children, collecting the bowls that had been scraped clean and hovering impatiently over those who were trying to make the pottage last by savouring every mouthful. But that evening she was crouching on a rock beside the fire, hunched over the flames as if she couldn't soak up enough warmth, caged inside her own thoughts.

On most nights the menfolk, though always weary, brightened as the hot food slid into their bellies, cheered by the knowledge that a few precious hours of sleep lay between them and another round of muscle-tearing work. They'd pick through the bones of the day, teasing each other, bickering or grumbling, and when there was nothing left to chew over, they'd dice, or play a few rounds of 'cross and pile', flipping a coin and betting on which side up it would land. The tinners gambled on anything and everything, even how far they could spit or piss. I reckoned they were in as much debt to each other as to Gleedy.

But Eva's mood had settled over the men, like a winter's fog lingering over a pool. She didn't glance up as, one by one, the tinners dropped their wooden bowls in a heap next to her and, with a wary glance, ambled off back to their own huts. As the last vanished, I slid on to a stone close to the fire and held out my hand to catch its warmth, rubbing the heat, like an unguent, over my aching knees. From below came the sounds of families settling in for the night: the babbling of voices, children yelling, iron pots clanging, and the rasp of whetstones as men sharpened tools ready for the coming day. A shower of scarlet sparks rose, like a flock of birds, into the darkening, smoke-filled sky as someone tossed another peat on to the embers of their fire. Still Eva didn't stir. Even in the orange-red glow of her own fire she looked pale, her eyes sunk into dark hollows.

'You sick?' I asked softly.

Her head jerked up, as if she hadn't realised I was there. By way of answer, she struggled up and bent to collect the bowls the tinners had discarded. I scrambled to my feet and went to help her, taking a pile from her hand. She stooped again to gather up the rest, but as she reached out, a cry escaped her and she remained doubled over, clutching her belly. I dropped the bowls I was holding and guided her back to the rock. It was only as I tried to ease her down that I saw the dark stain on the grey granite, glistening in the fire-light. The back of her skirt was soaked with blood. She stared vacantly, as if she couldn't remember what she was doing.

I laid a hand on her arm to try to get her attention. 'Your moon time come? You want me to find you some rags?'

She frowned, and for a moment I thought she would shrug and say nothing as she usually did. But even from that brief touch, I'd felt the cold clamminess of her skin. She was

148

shivering too. She glanced around her, as if to make quite sure there was no other within hearing.

'I was with child. But three days ago he . . .' Her mouth worked silently as if she could not drag the words out from the mire of her thoughts. She gave a deep sigh. 'I fell. Soon after, the bleeding started. The babe swam away on the red tide. It was no bigger than a pea pod. But the blood won't stop. It grows worse.'

I stared at her helplessly. Cobwebs and puffballs wouldn't work on bleeding like this, even if I knew where to find them on these sodden moors. I'd never seen anyone staunch this kind of flow. I was the last of my mam's children, and no other women in our village would have wanted me near them when they were birthing their babes. Neighbours, who knew they were with child, would cover their eyes as I passed and shrink back between the cottages, for fear my shadow would touch them and their unborn babe would be maimed in the womb.

'Is there a woman in the camp who helps with childbirth? They might have the knowledge . . .' I began.

'Any woman who had such skills wouldn't need to dig in the mud for stones,' Eva said. She winced, pressing her clenched fists into her belly, as she braced herself against the pain. 'But before you came, there was a woman here living with one of the tinners. She'd been born on the moor. Her man . . . he made enemies. One night, he was found with a knife wound. Threatened they'd do worse to him next time. Pair of them wanted to leave that night, but the gash kept bleeding each time he moved.'

Eva paused, her breath coming in short shallow pants. I could see it was costing her dear to speak so much, but I knew she'd not be telling me unless the tale was important.

'The woman . . . daren't leave her man, afeared they might do him more injury while he lay helpless. But she told me there was a blood charmer, lived out on the moors. She begged me to take a clootie to her, get her to charm it. I did as she bade me, fetched it back and she bound it round the wound. I'd not have believed it, but . . . the bleeding stopped.'

Eva drew her knife from her belt, and cut a strip of cloth from her skirt. Struggling to her feet, she soaked it in the bloody patch on the rock. Then she thrust the wet rag into my hand. 'You'll take it? Take it to the blood charmer. There's none else who'd do me such a kindness here.'

Chapter 18

Morwen

Ma wiped her fingers round the inside of the bowl and sucked them to be sure she hadn't missed a drop of the rabbit stew. Ryana scraped the burned fragments from the bottom of the cooking pot, nibbling them with her long yellow teeth. We'd managed to make the rabbit stretch among the four of us for two days, adding more herbs, stale breadcrumbs and water each time we heated it, till even the remembrance of flavour had been boiled out of the bare bones. The rabbit had been payment from one of the villagers – he'd wanted to know where to find his horse – and had likely been poached from one of the warrens, but Ma never asked where her payments came from.

The clouds had been lying low all day on the moor, wrapping our cottage in a wet grey mist, like a soiled bandage. In the shippon, on the other side of the wicker partition, the three goats had settled down to sleep as if they thought it was already nightfall.

'Have to go foraging tomorrow, the lot of ye,' Ma said. 'And I mean proper foraging, Tae. Don't think you can disappear all day and saunter back with nothing to show for it but a bunch of thyme, else you'll not be getting a share of anything in the pot.'

Ryana glanced sullenly at Taegan, muttering low so Ma wouldn't hear, 'She knows where to get fed and it's not from our pot.'

Ryana and I both suspected that Daveth and his brother stole food for our sister from their own ma, but Taegan never brought it home to share after her nights in the byre with them.

There was a clatter as the goats suddenly scrambled to their feet.

Ma raised her head. 'Someone's coming.'

'It'll be old Meggy back with summat she's lifted from the boy,' Ryana said, looking interested for once.

Meggy had come to our cottage just the day before. It was plain to see the old woman was in a fair twitter, which I could tell pleased Ma. She'd not refuse to help Meggy, especially if it meant thwarting the black crows – besides, she couldn't afford to turn away any chance of payment – but she'd not forgiven her, and if we could give her a fright, Ma reckoned that was no more than Meggy deserved.

'Birds tell me they've closed Bryde's Well,' Ma had announced, as soon as Meggy had seated herself on the stool.

The villager who had brought Ma the rabbit had told her that.

Meggy had nodded. 'I burned the charm and said the words like you told me, Kendra.'

'Never fails if it's done right,' Ma had told her.

'But you didn't warn me what that charm would do,' Meggy had protested indignantly. 'Made my skin crawl when I saw it. Frogs, thousands of them, swarming all over the floor and walls. Water was so thick with them it looked like a mess of pease pottage. That kept the boy from the well, right enough, kept them all from it. But for how long Kendra?'

Ma had ignored the question. 'Yer fetch me anything from the boy?'

Meggy had shaken her head. 'They've got him in the infirmary. Don't even let him out in the yard by himself to piss, case he bumps into summat and hurts himself.'

'You could give him something,' I'd said. 'That'd be a reason to get close.'

Ma had glared at me and Ryana had cackled with laughter. 'What's she to give him, Mazy-wen, a slap?'

But Meggy wasn't laughing. 'Maybe I could . . . a toy or some such.'

'Have a care, Meggy,' Ma had cautioned. 'If it's summat of yours, he'll use it against ye.'

Meggy had looked more affrighted when she'd left than when she'd arrived, and Ma had had a good chuckle over that. All the same, she'd meant the warning to be heeded. Ma never ignored what the spirits had shown in her smoke. She might not admit it, but deep down I could tell she was as worried as Meggy was about what trouble the boy might bring to the moors.

Ma jabbed me in the ribs with her stick. 'Move yer arse, Morwen. Get yourself out there and see who's coming.'

'It'll be Meggy, I'm telling you,' Ryana said, but Ma ignored her.

I slipped through the door and crouched behind the great rock that formed the corner of our house where I could watch without being seen. The clouds were lifting as a chill wind gathered strength. A flock of starlings took flight, weaving and twisting, like a giant puff of smoke. I gazed up at them, trying to read the shapes they made – a spiral with three twists, a hound's head black against the darkening sky, and then they were snaking down towards the valley. A black hound, a wisht hound, in the sky over the Fire Tor! Should I warn Ma? I could already hear Ryana sneering, *Mazy-wen can't even read the signs of rain when it's falling.* I felt myself shrinking. Say naught. That was safest.

Someone was coming, but it wasn't Meggy. The dregs of light had almost gone and I couldn't see her clearly, but I knew the old woman's shape and the way she moved. This one seemed younger, a thin whip of a creature. She was trying to cross the stream using the stepping-stones, but she was slipping and wobbling, like a duck on ice, as if she was unused to balancing on such things.

I darted back inside to warn Ma. 'It's not Meggy,' I said, grinning at the sulky scowl on Ryana's face. 'Must be a blow-in, I reckon. She's not from the moor for she doesn't know where to put her feet.'

I darted outside again. The stranger was standing afore our threshold, staring at the cottage, her face half hidden under a bit of old cloth she'd wrapped about her head against the wind and wet.

'Is it Kendra you're seeking?' I asked, gently as I could.

'Is she the blood charmer?'

'Best on the whole moor,' I said, though for all I knew, Ma was the only blood charmer – I'd not heard tell of another.

Kendra had already taken up her place on the low stool behind the fire and had veiled herself in the smoke by the time I coaxed the stranger in.

The woman huddled just inside the doorway. Her eyes, blue as speedwell flowers, darted round as she stared at each of us in turn. She looked as frightened as a mouse dropped into a box of cats. Her face was gaunt and her homespun clothes worn and caked with mud. They were heaped in layers over her thin body, as if she was wearing everything she owned, though if she was, how would she warm herself come winter?

Ma studied her curiously, then beckoned to her with a crooked finger. 'Come closer, Mistress. Old Kendra'll do you no harm, less ye mean harm to her.'

The stranger took a pace or two towards the fire. She ignored the stool and squatted on her haunches. As she moved, I saw that her left arm was dangling, like a broken wing, the skin on the limp hand mottled and waxy.

Ma noticed it too. She gestured to it. 'Can't do nothing about that. If you'd been brought to me as a babe, maybe . . . but it's too late now.'

The woman started to say something, but she spoke so low I couldn't make it out.

Ma leaned forward. 'Speak bold, Mistress. There's none here to hear your secrets save us and the spirits.'

The woman's eyes widened and she glanced around as if she thought she might see imps hanging from the beams or peeping at her from under Ma's bed.

'I'm living at the tinners' camp. A woman there . . . a friend. She took a fall and lost the bairn she was carrying, three days since. The bleeding won't stop. She said you once charmed a clootie for a tinner at the camp who'd been stabbed and his wound stopped bleeding.'

I could see Ma preening herself. It always put her in a good humour to hear her charms had worked.

'Eva said you might . . . I don't know what to do, how to help her. I'm so afraid she might die.'

155

'And what do they call you, Mistress? Have a name, do you? Everything must have a name.'

'Sorrel.'

Ma licked her lips, as if she was tasting the sound of it. 'A healing wort, sorrel is, and one old Brigid marked for her own. Well, now, Sorrel, you brought a clootie from this friend of yours?'

Silently the woman held out a rag rusty with drying blood. Her fingers were smeared with crimson. She passed the clootie over the flames of the hearth fire to Ma.

Ma held it in both hands, inscribing a circle with it in the smoke of the fire. Her eyes half closed, she began to chant:

> *'Stone woman sits over a spring and water stands*
> *   still as ice.*
> *Earth dries up, skies dry up.*
> *Veins dry up and all that is full of blood dries up.'*

She raised the rag. 'It shall be so, till milk flows from a rock.' She held the clootie out to Sorrel. 'Go home now. Tell her to press this to the flow of blood and it'll stop.'

But I knew it wouldn't. Something had gone amiss. There was no power in the charm – I could feel it.

Sorrel rose, holding the bloodstained rag carefully, as if it was as fragile as a blown egg. She looked at me, afraid to speak directly to Ma. 'I've no coin, not till . . .' She trailed off, her face racked with misery.

'Can't pay in coin for a blood charming,' Ma said. 'Silver works against it, see. But food I'll take,' she added hopefully.

Sorrel gave a helpless shrug. 'I've none of that either.' She flinched as if she thought Ma might curse her or snatch the rag away.

156

Ma hesitated, but we could all see the woman was not lying. She'd known more hunger than we had, by the look of her.

Finally, Ma grunted. 'See you bring me summat when you have it, else my curse'll follow.' She fixed the woman with a gimlet eye. 'But don't you go telling anyone, else half the moor folk'll be looking to me to give them charms for naught.'

Sorrel nodded and gave her a grateful smile, then ducked out of the doorway and into the night. I followed, and as soon as we were a few paces from the cottage, I hurried to catch up with her, but she had already broken into an ungainly, lopsided run, plainly anxious to take the charm back to her ailing friend. But it wouldn't work, I knew it, though I couldn't tell her.

'You're newly come to Dertemora?' I called after her. It was a foolish question and had tumbled from me before I even knew what I was saying.

But she stopped. No, more than that. She jerked as if something had struck her violently from behind. She walked back to me. Her eyes searching my face as if I was someone she had once known, but had only just recognised.

'What did you say? What did you call this place?'

'Dertemora. That's the name for these parts.'

The clootie fell from her hand. She stared around her into the darkness, as if she'd been asleep and had just woken and was trying to remember where she was.

The wind bowled the rag across the ground. I ran after it and snatched it up. I cupped it between my palms, silently repeating the charm. I saw blood running down flesh and willed it to stop. I watched it dry and felt the cloth grow hot, then icy cold in my fingers. Eva's flow would stop now, I was certain, though I could never tell Ma what I'd done.

But when I pressed the clootie back into Sorrel's hand, I almost dropped it again as a great flash of heat and pain coursed up my arm. For a moment, as we both held the rag, it was as if the moors had vanished and the two of us were alone inside the darkness of the tor cave. I could see Ankow's corpse lying on the slab, could hear the dead prowling around us, their high-pitched keening circling me and this woman, binding us together with a rope of fire. But Sorrel was not alone: a great dark shadow was hovering behind her, a giant bird, its wings spread wide, as if they were hers.

The rag was jerked from my fingers and the two of us were standing once more on the wild moor. Sorrel's eyes flashed wide in wonder, staring into mine. 'You,' she breathed, so softly I half thought it was the sigh of the wind. 'You hear her voice too. But who is she?'

'I know you . . .' I began. 'I saw you before in the smoke.'

She half opened her mouth to speak, but a shout rang out from the doorway behind us.

'Morwen, get you back in here. Ma wants some worts fetching.'

I turned to Sorrel, but she was already hurrying away, only glancing round once to look at me. But once was enough. I knew she had felt the heat flash between us too, though plainly she had not understood it. Had she seen what I had? No . . . No, she had sensed something, but she hadn't seen it. She couldn't. Not yet. She didn't know how to see. There was a power in her that was not yet awakened and she didn't know it. But I did. And in that moment, I realised I'd been waiting all my life for her to come and find me.

Chapter 19

Prioress Johanne

The knock on my door was timid and Sister Fina's entrance into my chamber more so. She edged through the door, still limping, and stood more hunched than usual, cradling her bandaged arm. She stared at the corner of the table, but said nothing. She was always ungainly, but that morning she seemed more awkward than ever, like a broken statue, repaired with parts from a different figure. I urged her to sit, not from kindness but irritation at her shuffling.

Fina winced as she perched on the edge of a stool.

'I'm glad that you are recovered enough to leave your bed,' I said. 'Are your injuries mending?'

Fina nodded unconvincingly, but still didn't look at me.

'Sister Basilia says she urged you to rest in bed for another day or two at least for you've eaten almost nothing. Are you sure you are well enough to be up? You look pale.'

It would have been nearer the truth to say she looked hag-ridden, with circles under her eyes and a dark stain on her lip where it appeared she had been biting the skin.

'I . . . The day is long when I am alone in the dorter and all the other sisters are working. I'd rather be busy.'

'Reading, contemplation and prayer are our work too, Sister Fina, and even the most infirm of sisters can do that from her bed.' I felt a pang of unease as I said it. If my sisters were forced into the cloister at Minchin Buckland, reading and contemplation would be all the work they were permitted to do. I shuddered as I imagined the endless years stretching ahead, filled with nothing but attending services and praying for the souls of our brother knights, one day the same as the next, each year identical to the one before and the one to come. Yet if I could not convince the Lord Prior to let us remain, I might well find myself trying to persuade my sisters of the glories of just such an existence. Could I really bring myself to urge them to submit to that? I'd have no choice. The Pope would never release us from our vows, and even if His Holiness could be persuaded to allow us to join another order, there were none we could enter that would allow us to escape the prison of the cloister.

But if Brother Nicholas uncovered the truth, I would not be granted even the small mercy of the cloister. Maybe none of us would live to see Minchin Buckland. What had he said the night he came? *The Lord Prior has eyes everywhere, searching for the smallest spark of corruption or sorcery in his order, and if he suspects so much as a glimmer of it, he'll grind the guilty beneath his boot.*

And if that boot descended, what would become of those we cared for? Where would they draw their last terrified breath?

Sister Fina shifted on the stool. Even that small movement made her wince.

'Sister Basilia also tells me you have refused to take the syrup of poppy she prepared to ease your pain.'

'It makes me drowsy . . .'

'So, you'd rather bear the pain. That is to your credit, Sister Fina. Fortitude is a virtue we should all endeavour to acquire, and when you reach my great age, your back aches and your bones creak, believe me, you will have need of it.'

I laughed, expecting Fina to smile at the joke or assure me that I was not old. But Fina did neither. She probably saw a decrepit old woman when she looked at me, and on days like this, I felt it.

'Even if you were fit to work, Sister Fina, we have been obliged to close the holy well, as you must know, and I cannot tell you when we will reopen it to the pilgrims. That matter rests in God's hands.'

For the first time since I'd entered the room Fina's head snapped up and she looked at me directly. 'The frogs are still there?'

'They were this morning when I inspected it.'

'It's the boy . . . The boy has brought a curse on the well, hasn't he?' Fina demanded fiercely. 'Father Guthlac said he would destroy us.'

'The boy has not been near the well, Sister Fina. How could a mere child conjure such a plague when he cannot even dress himself?'

'Then why have they come?' she wailed.

God knew I had prayed for an answer to that question for many hours, but I had received none, unless I had closed my ears to His voice. Was this a punishment for my disobedience?

Fina was watching me, her thin fingers plucking restlessly at the folds of her black gown.

'If it is any comfort, Sister Fina, I observed several dead frogs among the living this morning. It may be that the rest will also die, since there can be little for them to eat down there. But, for now, all we can do is wait and pray.'

Clarice, in whom I had confided, had suggested several remedies, each more drastic than the last. She seemed to imagine we were defending our priory from the Saracens, for boiling oil and fire had been among those she most enthusiastically advocated, even capturing a few herons and letting them loose down there to eat the frogs. But most of her suggestions seemed likely to inflict as much damage on the holy spring as the animals. If we'd been suffering a plague of mice, we could have laid poison, but who knew what might poison a frog? I suspected few people had ever had cause to wonder.

'But you have to drive the frogs out,' Fina insisted. 'You can't just wait.'

I was losing my patience with the young woman. Only that she appeared to be in so much pain prevented me from taking her to task as thoroughly as she deserved. 'Sister Fina, kindly remember that I am the prioress, and it is not for you to question me! But I will endeavour to believe that your intemperate words arise from a deep concern for the pilgrims who have undertaken long journeys to reach us. Therefore, to ease your mind, I will tell you that Sister Clarice has been kind enough to sit in the chapel and receive the villagers and pilgrims. If they should ask why the cave is closed, she tells them rocks have fallen and are being cleared. Thanks to the protection of the blessed St Lucia, not a soul was hurt. If they have brought an offering or some token to leave at the well, she takes it and assures them that it will be placed by the well as soon as it is safe to do so. In the meantime, she offers them ampullae of holy water, if they wish to buy them. So, you see, they are not sent away disappointed.'

'But how can she get to the spring to fill them? The frogs . . .'

'She does not need to go down there. They have already been filled,' I assured her.

'But how?'

I sighed in exasperation. 'The water in them has been blessed in the chapel, which is directly above the well. St Lucia will hear the prayers of the pilgrims if they are made in faith, so we must ensure they have no cause to doubt, Sister Fina.' I stressed 'doubt', trying to make her understand that the precise origin of the water need not be discussed with outsiders.

'So, you have made Sister Clarice the well-keeper now!' Her hands were balled in her lap so tightly the knuckles were white.

Given her wild outbursts since she'd entered my chamber, I was sorely tempted to say yes. But I reminded myself of why I had given the duty to her in the first place. I did not want her to sink back into the strange melancholy and wildness that had afflicted her before. The well had healed her then. At least standing guard over it might help her again.

'Sister Clarice is the priory's steward.' I was determined that whatever orders he had received, Brother Nicholas would claim that title only when I was in my grave. 'As such, she has work enough for three sisters and needs no more added to her burden. Just as soon as you are recovered, I will ask you to resume your duties as keeper of the well. For you are sorely needed.'

'I am recovered now,' she blurted out eagerly, and her eyes suddenly looked alive again. 'I can sit in the chapel and receive the pilgrims and villagers much better than she can.

163

Many of the villagers recognise me now and they know they can trust me with their offerings.'

'Are you sure you are strong enough?'

She nodded earnestly, in the manner of an over-eager child trying to convince their mother they can be trusted with an errand. It only increased my doubts.

'You will remember what to tell the pilgrims and villagers. Mention only the fall of rocks. No talk of the frogs or curses must escape the priory.'

She was already limping to the door, moving hastily but so clumsily I was afraid she'd take another fall.

'Sister Fina.'

She turned and regarded me warily as if she thought I was about to snatch her treasure from her.

'I would remind you again to be guarded in what you say to the two brothers. If Brother Nicholas finds you alone in the chapel, he may wish to speak to you.'

She flushed. Then her chin jerked up defiantly. 'I kept him from seeing the frogs. I wouldn't let him down there and I locked the door.' She gnawed the raw patch on her lip. 'But I don't understand why the brothers are here. And why shouldn't we talk to them? We've nothing to hide. We've done nothing wrong.'

'In spite of your good efforts, Brother Nicholas has discovered our troubles with the well.' I did not tell her that I had neglected to bolt the door behind me. 'But it is . . .' I hesitated. How could I make her understand the danger we were in?

'The Lord Prior has sent the brothers here because he is anxious about all of his priories in England. I am sure you will not have ever encountered any men or women of the Order of the Knights Templar, for it has been seven years since any dared wear the red cross in England, and you cannot have been more than ten years old then and even younger when the order was attacked in France.'

'When I was little my older brothers played being Templar knights. They used to make the servant boys be Saracens so they could fight them. My brothers had to win because they were knights.' Her mouth softened into a smile, but it vanished almost at once and she frowned. 'But there are no more Templars, so why should the Lord Prior care about them?'

'Because there are many similarities between their order and ours. Over the past two centuries, the Templars gained much wealth and land across Europe, and they were answerable only to the Pope, no matter whose domain their land and castles were in. Their financial and military power made many uneasy, King Philip of France for one. Nine years ago, without warning, orders were given that all the French Templar knights were to be arrested on charges of . . .'

I swallowed the words I was about to utter. There was no need to shock this sheltered young woman by telling her the foul things they had been accused of. And what did it matter now? In Paris, the knights' Grand Master, Jacques de Molay, who had confessed under torture, retracted his confession and was burned alive as a relapsed heretic. He died promising that woe would fall on those who had condemned them, and before the year was out both Pope Clement and King Philip of France were dead. But their deaths came too late to save the lives of the hundreds of Templar knights they had destroyed.

'Prioress Johanne?'

I dragged my attention back to Sister Fina, standing

impatiently by the door. 'The knights were arrested on charges of sorcery and heresy, charges the Pope was forced to investigate since they had been levelled by a Christian king.'

She shrugged, as if all I was saying was of no consequence to her, the rambling tale of some old woman about things that had happened long ago. But it had been just two years since Jacques de Molay was so cruelly executed. The wind still carried his ashes.

I fought to hold back my temper. 'Can you not understand the danger, Sister Fina? Our order, too, has wealth and lands that others covet, and we, too, are answerable to none but the Pope. The Lord Prior fears we might suffer the same fate.'

'But they were heretics. We're not guilty of any crime.'

I wanted to scream at her, *Neither were they*. Their only crime was their arrogant belief in their own survival and their blindness to the jealousy of powerful men. But I dared not say any of this. It was heresy to defend anyone who stood accused of being a heretic, and even worse to proclaim innocent those who had been found guilty of such an unforgivable sin.

'Sister Fina, you must try to understand that the Lord Prior wants to ensure that no one can falsely accuse our order of wrongdoing. There are many, both outside our order and within it, who believe that women who are neither married nor under the rule and government of the cloister may be led into wantonness and heresy because our sex is more easily tempted, as the serpent tempted Eve. The Lord Prior believes that to silence any who might point the finger, all the women in our order should be cloistered so that none may accuse us of any sin.'

Fina's eyes flew wide with alarm and fury, as if I was

responsible. 'But my father swore I would not be forced to take the veil. I only agreed to my vows, because—'

I held up my hand to stem the flow. I was in no mood to listen to any more of her angry outbursts. Did she really imagine I would not resent this even more than she did?

'That is why I urge you to consider carefully anything you say to the brothers, Sister Fina. We must all do everything we can to convince Brother Nicholas that we are above suspicion. Such wild thoughts as you uttered just now, about the curses and the boy, will only fuel our brother's fire. We must guard our tongues, chew every word twice before we utter it to ensure we give him nothing he can use to drag us to Minchin Buckland. If he does, you will not be the keeper of anything, much less your beloved well.'

Chapter 20

Morwen

'Told you, Ma, see? They've lit a fire inside the stone circle and they've penned their horses in there too.'

Even Ma could not deny the evidence of her own eyes, though she'd not believed me when I'd told her. I was hurrying home after setting snares when I'd spotted the smoke in the distance, rising thin as a sapling into the evening sky. I was always careful to set my traps well away from any cottage or track for all the land belonged to the King and abbots, so they said, not that I'd ever seen an abbot tending a cow or a king planting a bean.

I'd known something was wrong. No one lived near that stone circle and the villagers only ever went to it on the eve of Beltane or Samhain and it wasn't the night for either of those bonfires to be lit. No one went inside that circle, save for those feasts, or let their beasts stray in there. If they did, it would call down a curse on their heads.

Ma rose from where she'd been crouching in the darkness and picked her way towards the stream. She didn't bother to search for the stepping-stones, but hitched her skirts up to her scrawny thighs and splashed across. She'd gone barefoot all her life, and her soles were as horny as the trotters of sheep. She was as surefooted as them, too. I followed and we edged around the knoll on which the stone circle stood, then crept up the slope on the far side so we could look down into it. Ma squatted, still as a grey heron watching for fish.

Three men, wrapped in sheepskin cloaks, sat on low stones around the fire pit, which they had dug in front of the queen stone, the widest and tallest of all the standing stones in the circle. One man was sharpening his long knife while another turned a wooden spit on which a couple of skinned hares were impaled. The wind was gaining strength again and I could smell the roasting flesh in the smoke gusting towards us. Ma would never kill or eat a hare: some cunning women could turn themselves into hares and you might slay a woman instead of a beast. But, though I'd never taken a bite of one, I couldn't stop my empty belly growling. The meat smelt as sweet as rabbit.

A skinny hound, which looked as ravenous as I was, prowled around the three men occasionally making little dashes forward as if it meant to snatch a carcass off the spit in spite of the heat of the fire. It retreated whenever one of the men aimed a kick at it.

The flames sent shadows and lights writhing across the circle, and by its light I saw that the spaces between the waist-high stones had been stuffed with prickly furze bushes and old wicker hurdles to form a pen, in which a half-dozen squat little packhorses were corralled alongside the men, with a big-boned, muscular beast that was obviously meant for hard riding, though few in those parts could have afforded such a valuable animal.

'He's got his fat backside on the cup stone,' Ma growled in outrage.

A man was squatting on a broad stone that lay horizontally in front of the queen stone. It had in it a round hollow, as if an apple had been carved into the top, where milk and sometimes honey were offered to Brigid. No villager would dare touch the stone, unless they were making an offering, much less offend the spirits by setting their arse against it as if it were a midden heap to be shat on.

The hound must have caught Ma's fierce whisper for he wheeled round in our direction and came bounding up between the stones, barking and leaping, though he couldn't jump over the thorns. In an instant two men were on their feet, their knives drawn. The third, still seated on the cup stone, peered warily into the darkness. He reached down and felt for a bow and an arrow, sliding both on to his lap. The horses, alarmed by the dog, bunched together, pricking their ears and circling restlessly.

'Who's there? . . . If that's you playing the fool, Hann, I'll give you the drubbing of your life.'

One of the men thrust a torch into the fire and, when the end was burning, hurried across to where the hound was barking, holding the brand out over the stone. But the wildly gusting light fell on nothing but grass and rowan whipping in the breeze.

'Is there something in that bush over yonder? I heard there was all manner of beasts on these moors, hellhounds and wild cats.'

'Wild kitten, more like, if it can be hidden in a bush so puny and wizened. It's outlaws you want to be fretting about, not cats.' He glanced back at the man seated by the fire.

'You reckon we should send old Whiteblaze out, see if he can flush 'em?'

The man by the fire rose to his feet, the bow gripped between his fingers, though not yet drawn.

'Ma!' I whispered. 'If they let the hound loose and he shows them where we are . . .' Kendra knew charms that could make even strange dogs lie down, but they wouldn't work against arrows. 'Come away quick, Ma, afore they get that hurdle pulled aside.'

Keeping low, I wove down the slope, expecting Ma to follow, but instead I heard her voice riding the wind, like a hawk.

'That's a sacred circle, that is. You get yourselves and your beasts out of there else I'll make you curse the day your ma whelped you.'

One man let out a strangled squawk, like a hen that had been sat on. But the other thrust the burning torch in the direction of Ma's voice. Ma rose out of the grass, brandishing her staff, her wild grey hair fanning around her in the wind, her arms flung high and wide.

For a moment, the men just gaped. Then they burst out laughing.

'That's your hellhound, Jago, a mad old vecke.'

'It's you who reckoned it were murdering outlaws,' the other retorted. He put his hands in front of his face, rocking from side to side in mock terror, and sang out in a high-pitched voice, 'Have mercy on us, old woman. Don't hurt us!'

Ma yelled her threats again, but I could barely hear them over the raucous laughter and insults of the men, as they pranced around in mocking imitation of her.

I ran back to her, trying to pull her away.

'Look,' Jago yelled. 'There's another. You're right, there's a whole gang of those outlaws need taming. I'll wrestle with the maid and you can have the old hag.'

'Not if I get to the maid first.'

They tugged at the furze bushes jammed between the stones, cursing as they scratched and pricked themselves.

I shoved Kendra down the slope. 'Run, Ma, run.'

She hesitated, but even Ma could see that we couldn't fight these men, leastways not like that. We ran. Ma bounded down the hillside towards the stream, though I knew she'd not make for the cottage. She'd not want to lead the men there, but there were plenty of hollows and rocks she could lie low in. Ma knew the moors better than the faces of her own daughters. She could still run like a hare, but she couldn't keep it up for long, not like she used to. I raced off in the opposite direction, trying to draw the men away from Ma. I knew it was the maid they'd chase after, not the old hag.

It was hard going, tearing over the tussocks of coarse grass and heather, but I was more surefooted than the men lumbering behind me. They'd not see me in the dark. Then I heard the dog fall silent. It must have bounded free from the pinfold and was searching for my scent. Almost at once I heard the deep baying of a hound that had picked up a trail. But was it mine or Ma's? Maybe, if we were lucky, it was a deer's.

I ran on, trying to keep to the low ground, so the men wouldn't see me against the sky. I was making for a place upstream where the water had cut a hollow deep beneath the bank at the point where the river curved around a stand of wizened oaks. I knew if I dropped down into that and crouched beneath the overhang, the men could search all night and they'd never find me. But that hiding place wouldn't fool a hound and I could hear it behind me now, its baying growing louder, more excited. I risked a glance over my shoulder. The men were following their hound, stumbling over the uneven ground, the flames and the smoke of the torch streaming behind them in the night's sky.

'Find her, Whiteblaze! Harbour the little witch.'

The hound was running at full chase now. My back tensed. At any moment, I expected to feel the beast's hot breath on my legs, its sharp fangs sinking into my flesh. If Ma had been with me, we could have driven it off with her stave, but I had nothing save the small knife in my belt and dogs don't back off at the sight of a knife. Only if it sprang on me would the blade be of any use, and then only if I turned to face it afore it leaped. But it was the men who followed the hound I feared more.

A pain tore at my side. I was gasping for breath. My legs felt as shaky as if I had the ague. The hound was gaining. I could hear it ripping through the heather and the bushes behind me. A boulder jutted up in front of me and the rush of the water crashed suddenly on my ears. In my panic, I'd almost run past my hiding place. I slithered to the edge of the river and slid down. But exhaustion and fear had made me clumsy. My bare foot slipped on the wet rocks, and I plunged under the icy water.

Gasping and choking, I struggled to stand, but the rain-swollen current dragged me over the water-polished, slippery stones. I threw myself forward and managed to half crawl, half stagger into the hollow beneath the overhanging bank. In hot summers, when the streams were low, a gravel ledge was clear of the water, but now the river filled the whole bed, though it was shallower and a little calmer at the edge of the curve. I crouched in darkness, clenching my teeth to stop them chattering as the freezing water swirled and frothed around my thighs.

Over the thunder of the water and the rumble of stones, I heard the hound scrabbling directly above me. It was so close, I could hear it panting. Then it did what it was trained to do: it began to howl, its cry carrying right across the moor. It had cornered the quarry, trapped the prey and now it was calling its masters to deliver the kill.

I could no more break out than a vixen could escape from its den when the dogs had found it. I knew I wouldn't be able to scramble out of the river on the other side: I'd been to that spot enough times to know that the bank was too high and slippery. I'd have to wade far upstream against the swift current and the hound would keep pace with me all the way.

I pressed my shoulder into the cold wet earth and tried to think of all the charms and curses I'd learned when I'd listened to Ma teaching Ryana. But I could remember nothing that would silence the barking of a dog. And even if I'd known where to run to, I wasn't sure I could move now – my feet and legs had grown so numb that I couldn't feel them, much less move them. Over the wind and water, I could hear the voices of the men, out of breath, but urging the hound to hold. I was shaking and, though I tried to tell myself it was just the cold, I knew it wasn't.

Then, suddenly, there came a whistle so high and sharp it seemed to pierce the wind, like an arrow through flesh. The dog's barking ceased and it began to whimper. The whistle sounded again, even more piercing than before, and this time the dog yelped as if it was in pain and I heard it crashing away across the dark moor, as if a pack of wolves were after it.

The men were cursing and bellowing for it. 'I'll thrash that brute when I get hold of it,' one yelled.

'Leave the wretched beast. Look for the girl. Whiteblaze was standing on the riverbank. He must have chased her in. She probably thought to make him lose the scent by taking to the water.'

I buried my face in my knees and tried to cover my arms with my wet hair, so he'd see no gleam of pale skin. I shrank back against the earth. They were walking up the bank. The black water turned to foaming blood, as the red flames of their torch passed over it.

Mother Brigid keeps the men from the river, as the flame keeps wolf from goat. But I had no herbs to burn, no charm to weave, only the will worth, only that.

They were almost overhead. If they stepped too close, too heavily, and the sodden earth gave way, they'd crash down on top of me.

'What's that?' one called.

My heart had risen so far up my throat, I thought it would choke me.

'There, look . . . something moving . . . A black beast.'

'It'll be Whiteblaze come crawling back.'

'No, no! Over there, by that stunted tree. It's twice his size. It can't—'

A shriek split the night, a terrible sound, as if all the force of the wind had been balled in a giant fist and hurled towards them. The men echoed it with their own fainter cries of fear as they fled back towards the safety of the stone circle and the fire that burned there. I was shaking with cold and fright, but I daren't leave my hiding place.

Above me, I heard something moving again. Had the men returned, the hound?

175

'Morwen?' A woman's voice, but it was not Ma's or either of my sisters'. 'The men have gone. But you best get home quickly, case that dog comes nosing back.'

I could barely stand and fell several times as I splashed along the edge of the river, battling against the slippery stones and current to find a place upstream where I could pull myself out. As I struggled to clamber up, I felt a hand grip my arm. I stared up. A woman was standing on the bank above me. It was too dark to see her face, but as she hauled at my arm, a tingle shot through me, brief and brilliant as the flash of a kingfisher on the river. It was the woman who had come to our cottage, the woman I had seen in the smoke.

I clambered on to the bank, and stood there dripping, my jaw so stiff I could barely unclench my teeth enough to speak.

'Sorrel, what . . . are you doing out on the moors?'

'Went to your cottage to leave some dried mutton for Kendra. Only a mouthful, it is, saved from my portion, but I'll bring more when I can.' She gave a brittle laugh. 'Don't want her curse to follow me. Eva stopped bleeding, like she said. She's on the mend now. I meant to thank Kendra . . . to thank you. It was you, wasn't it, made the charm work? I could feel when I took it from you . . . I can't explain . . . But I could feel it had . . . *changed*.' Sorrel slowly folded her fingers, staring down at her hand as if it still held the blood charm.

'Your sister said you and your mam had gone to the stone circle. But I'd heard a couple of tinners talking earlier. They said they were going to pen a horse they'd found in a circle of stones that was well away from their valley. They didn't want it to be seen by the master's henchman. I reckoned there might be trouble if you ran into them, so I started

176

after you. Wanted to warn you. Saw you run off and the men send that hound of theirs after you.'

'Someone whistled, sent it flitting.'

Sorrel chuckled softly. 'Learned that trick when I was a bairn. I was always afraid of the village dogs. Boys would set them on me to make me run. Thought it funny, the way my arm would flap about and throw me off balance so I couldn't go straight. An old man in the village, who'd been lame since he was a lad, saw the dogs leaping at me one day and sent them howling off by whistling. Taught me how.'

'And the shriek that frightened the wits from those men? You did that too?'

Sorrel shook her head. 'Not me . . . I mean, I don't think . . . No, I couldn't do that. I was so angry I wanted to, but how could I have done it? But then . . . who?'

We stared out into the vast hollow dome of darkness. In the distance, we could see the tiny glow from the tinners' fire in the stone circle. The wind rattled the bushes and shook the grasses. Clouds tumbled over each other as they charged across the inky sky, but nothing else was stirring – at least, nothing that was of this world.

Chapter 21

Hospitallers' Priory of St Mary

For the third time that morning, Deacon Wybert paused and glanced uneasily at his congregation in the priory's chapel, catching sight of their frowns and furtive grimaces. All through the mass his voice had grown ever more strident, the words gushing out of him like liquid shit from a man with the flux. Now he hesitated, staring wildly around.

Nicholas flicked his fingers impatiently, urging him to continue. The knight's evident annoyance only served to unnerve the village deacon even more. He stammered, lost his place in his head and could only recover by returning to the beginning of a lengthy prayer he had already said and reciting it again, for he'd learned the service by rote, listening to old Father Guthlac.

Nicholas gave a deep-throated growl, alarming several of the elderly village women, who edged away from him, as if he might drop to all fours and start biting. Prioress Johanne glowered at him. But he ignored her. He was impatient to see the mass ended, so that he could grab the deacon before he escaped again.

Nicholas suspected the man had been deliberately avoiding him: each time he'd called at his cottage, his housekeeper had sworn Wybert was in another village. Having been thwarted by Father Guthlac's untimely death, he was determined to learn what the deacon could tell him, but from the way the gibbering fool was muddling his way through the mass, Nicholas was beginning to think that wasn't going to be much. On the other hand, he was obviously easily intimidated so he might be frightened into letting something slip.

Something else was annoying Nicholas even more than the deacon's babbling. He glanced around. What was that infernal noise? It was also alarming the handful of elderly villagers and pilgrims who leaned wearily on the pillars at the back of the circular nave, exhausted from having trudged miles across the moors to reach the priory. Now they, too, seemed anxious for the service to be over and not just to claim the bread and meat that would be doled out afterwards.

Up at the altar, the deacon's hand shook as he made the sign of the cross over the silver chalice containing the wine, now transformed by his gabbled words into the Holy Blood of Christ. His prayer ended, he sank heavily to his knees on the cold stone, raising the sacred cup to his lips. But the noise he had been trying to drown grew louder in the silence, a dull but skin-crawling buzzing. Several villagers peered nervously up into the thick shadows of the thatched roof, as if they feared a swarm of bees or wasps might be hanging there.

Wybert lowered the chalice and glanced fearfully towards the door that led down to the holy well, whence the droning seemed to come. Was some evil spirit trapped there? The whole village knew Kendra and her daughters had cursed the sisters the day they'd turned them from the well. Could this be their revenge? Now that Father Guthlac was dead, would he be called upon to vanquish whatever was down there? He shuddered.

'Deacon Wybert, you must finish the mass,' Prioress Johanne prompted.

Startled by the sound of her voice, he almost spilled the consecrated wine. The buzzing was invading every crevice and corner of his skull, driving out all other thoughts. He raised the chalice again and took a gulp of wine, but panic made him gasp for air at exactly that moment. The burning liquid was sucked into his lungs, and he choked violently, coughing and flailing for breath. The precious Blood of Christ spewed from his mouth and ran in red streams from his nose as he fought for air. He fell forward on to his hands and knees. The chalice clattered on to the stone and rolled away, leaving an arc of crimson wine spreading over the flags.

Sister Basilia reached him first, pounding on his back with her broad hands so hard it felt as if she had broken his ribs. But, gasping and vomiting, he was too weak even to crawl away from her ministrations. As the other sisters pressed around him, the prioress retrieved the chalice from among the feet and set it safely on the altar.

The villagers and pilgrims crowded forward, their excited chatter obliterating the sound that, only moments before, had so perplexed them. This spectacle was far more enthralling than any strange noise, for if this man had been struck down by God at the very instant the Blood of Christ had touched his lips, he must have committed some terrible sin. At least, that was what the pilgrims were telling each other. The moor folk, though, were glancing in awe at the locked door to the well. Perhaps it wasn't God he'd angered but old Brigid herself, for hadn't he been standing on the very spot where her spring gushed out below his feet? But God or goddess, whoever had struck him down, they were determined to enjoy his gruesome demise.

The prioress was equally determined to disappoint them. She tried to clear them from the chapel – though at first not even her authority could prise them away from the entertainment.

Sister Melisene knew her customers better. She hurried to the door, flung it open and bellowed above the din that she was off to distribute the dole of food. The villagers hesitated, but empty bellies triumphed, and they limped and shuffled after her. The pilgrims were harder to disperse, but by the time the prioress had succeeded in closing the door behind them, the deacon was hunched miserably on the stone floor, his chest heaving painfully and his face still scarlet. Otherwise it appeared he would live. As silence fell on the group of sisters, the buzzing grew louder till it filled the chapel, blotting out all the sounds of life outside.

Nicholas, ignoring Wybert, was prowling around the chapel trying to determine where the noise was coming from. He soon realised it was loudest by the well door. But it certainly wasn't frogs croaking. He marched over to Fina and thrust out his hand. 'Give me your keys.'

She turned wordlessly to the prioress.

'I think you may find it more convenient to ask Sister Fina if she would be kind enough to unlock the door,' Johanne said evenly. 'The wooden bolt often swells because of the damp, but she has a way of coaxing it.'

Nicholas was in no mood to ask any of the women anything. He snatched the ring of keys from Fina's hand and advanced on the door with one thrust in front of him like a lance. He regretted his impulse almost at once, for no amount of wriggling would make the prongs connect with the slots in the bolt on the other side of the door and he had no way of knowing if he had thrust the right key into the hole or if it was merely as stubborn and obdurate as the women who guarded it.

The sisters watched in silence, though he could sense their supercilious glances to each other behind his back. His temper and frustration were reaching boiling point. But even he realised he'd been beaten.

'Like everything else in the priory, it would seem that you have arranged it so that this bolt will yield its secrets only to a woman. Sister Fina, will you please open this door?'

His humiliation was complete as the bolt slid back for her as smoothly as ale slips down a thirsty man's throat.

But as it swung open, both stench and noise charged out, smashing against Nicholas like a battering ram – the sickening reek of hundreds of rotting frogs and the buzzing of the thousands of blow-flies that swarmed over them. The iridescent green vermin crawled so densely over the steps and walls that the very rock itself seemed to be undulating. It was as if he was staring down into an open grave. For a moment, he saw – he *thought* he saw – the putrefied remains of a man lying on the stairs beneath the pall of flies, as if a corpse had tried to claw its way out of a tomb.

Clamping his sleeve to his mouth to stop himself retching, Nicholas backed away from the stairs, stumbling over his own feet in his haste to escape. But though he kept telling himself it was only the shadows cast by the candlelight and the heaving mass of greenbottles, he could not shake the ghastly image from his head. All the sisters had taken an involuntary step backwards, clamping hands over mouths and noses, their eyes wide with shock and disgust. Mercifully, the flies were too cold and lethargic from the chill of the cave to fly out in any number, though a few were escaping into the chapel.

Johanne recovered first and slammed the door shut. Clarice snatched up the first thing she could seize, which happened to be a white linen manuterge with which Deacon Wybert had dried his hands during the unfinished mass, and vigorously swatted the few flies creeping out under the door. Wybert gave a feeble squeak of protest at the desecration of a holy cloth but no one paid him any heed. He began furiously batting and brushing at his clothes and tonsure, as if he could feel the tiny creatures crawling over his skin.

'There must be thousands of them,' Basilia mumbled, through the hand she still pressed across her nose and mouth, for the flies might have been contained behind the closed door, but the stench had escaped to fill every corner of the chapel, making even the strongest stomach heave.

'That is hardly to be wondered at,' the prioress said sharply. 'Flies are born of corpses. Judging by that smell, it would seem all the frogs have died and the flies have sprung from their remains. We must—'

She broke off. Sister Fina's moans were becoming ever louder, as though she was about to start screaming. She was staring fixedly at the bottom of the door, where more flies were emerging from the gap between wood and stone. Johanne seized her arm, and turned her, dragging her a few paces towards the courtyard door.

'Sister Fina, go at once and fetch wet cloths to stuff around the door. Otherwise the chapel will be filled with flies.'

Fina stumbled towards the courtyard, darting horrified

glances back at the well door, as if, at any moment, the whole swarm would burst through it.

Nicholas strode after her and stood in the open doorway, gulping air. For once the smell of stable dung and burned beans from the pilgrims' kitchen seemed almost as fragrant as a summer meadow.

'I can't be expected to celebrate mass with that noise and loathsome stench,' Deacon Wybert said, clambering shakily to his feet. 'It's not seemly.' He stumbled past Nicholas and almost hurled himself into the courtyard in his haste to leave.

Nicholas tried to grab his arm. 'I want a word—'

The deacon gagged, then vomited, barely missing Nicholas's boots. He staggered out of the gate looking as if another bout might overtake him. Nicholas, an expression of disgust contorting his face, decided there was little point in trying to detain him. 'Lily-livered fool,' he muttered. 'A spell in the order would do him good. Our priest brothers sing mass standing knee deep in blood, with the screams of dying men and horses as their choir, and they don't stumble over a word.' He glared at Prioress Johanne. 'But your gutless deacon has a point. No one will attend services here while the place is swarming with more flies than a dunghill, and all the time the door to that chapel remains shut the order is losing valuable income. How do you propose to rid it of that vermin?'

'For the present, there is nothing that can be done,' Johanne said briskly, 'except ensure the well door is sealed as best we can.'

'Hare's gall in milk will kill them,' Basilia said. 'I always leave some dishes of that in the casements of the infirmary when the weather is warm.'

'I fear we would not be able to catch enough hares to dispatch as many flies as we appear to have,' Johanne said.

184

Basilia looked crestfallen, but instantly brightened. 'I'm sure I've read in one of my herbals that burning fleabane and willow herb together will drive them away. But I can't remember if it was dried or fresh. We use fresh fleabane to mix with the rushes, but that's to keep down the fleas, so perhaps dried—'

Her prioress stemmed the flow by laying a hand on her arm. 'Sister, it is a good thought, but I fear any attempts to drive away the flies will merely send them pouring into the chapel and then they'll be crawling over the kitchens and the infirmary. We must be patient.'

'Patient!' To Nicholas, the buzzing seemed to be growing louder. He was sure he could feel the stone beneath his feet vibrating as if the creatures were dashing themselves at the roof of the cave. 'How many plagues are we to endure? It appears you can't maintain a simple holy well, much less a priory and all its lands and tenements. May I remind you, *Sister* Johanne, that this priory belongs to the order and exists solely to carry out our ministry to the poor and sick, and to collect the revenues the Citramer so desperately needs. And if the one appointed to have the care of it is found wanting, she, with all those sisters who support her, will swiftly be removed, by force should that prove necessary.'

Chapter 22

Sorrel

That evening I did not go up to Eva's fireside, though I told Todde that was where I was going. Eva was growing stronger. Since she shared the food she cooked for the tinners, she ate better than most of the women in the camp and, unlike them, she'd time to snare birds and animals for the pot too. But though the bleeding had stopped even before I'd returned with the charm – I reckon it must have happened at the moment Morwen held the cloth – there was still something draining out of Eva day by day, as if her spirit was shedding invisible drops of blood or weeping tears that could not be seen or heard. Even after all we'd shared that night, she talked no more to me than she had done before. She never spoke of her lost bairn, even when I asked her, but I could see in her eyes the edges of her soul freezing over, like the ice creeping towards the centre of a pond. And I climbed up to her fire of an evening as much for her sake as to seek warmth for myself. But not that night.

Why did I lie to Todde? Why would he care where I went? I was nothing to him. There were plenty of women with two good arms in the camp. Why would he or any man look twice at me, save in disgust? But, all the same, I knew he'd try to dissuade me from going out on the moor at night. He'd not ventured out there himself since we'd arrived in the camp. Few of the tinners would leave the valley alone after dark, for fear of the hounds we heard howling across the moors. But I had to go, just as I'd had to leave my village and set out on this journey. Morwen held the answer I'd been searching for. She knew whose the voice was and why she was calling me. I was growing more certain of that each day.

It was already dark by the time I neared Kendra's cottage. A chill wind blew through the sedges and rustled every furze bush. I couldn't stop myself constantly turning my head, certain that some beast was creeping through the long grasses behind me. I should not have come. When I'd ventured to this place before, I had been fretting so about Eva that a herd of dragons could have roared past me and I'd not have noticed. But now I heard every scurry among the heather, heard the cry of all the birds winging towards their roosts.

Kendra's cottage squatted like a black toad in a cold puddle of moonlight. A flickering tongue of gold-red light darted out beneath the leather curtain that hung in the doorway as if it was searching for grubs. I stopped. What reason could I give for coming? I'd no food to spare. I didn't even know what I would say to Morwen if I could speak to her alone. It was madness.

As quietly as I could, I began to back away. The slippery wet stepping-stones in the stream glinted in the moonlight. The bubble and rush of the water seemed louder than when I'd crossed a few moments ago and I stared down, trying to balance myself. But as I stepped out on the other side I collided with something in the darkness that was both soft and hard. I almost pitched backwards into the stream, but a hand grabbed my arm and steadied me.

'I knew you'd come.' Morwen's eyes glittered so brightly, it was as if a candle burned behind them. She put a finger to her lips, nodding towards the cottage, then beckoned to me to follow. We climbed up the side of the stream. When the clouds hid the moon, Morwen didn't falter. She slipped around every stone and bush, every mire and mound, as if, like a cat, she could see as well at night as in the day. She darted ahead and I lumbered behind, slipping and tripping, until suddenly she vanished. I called out, terrified she'd been swallowed by one of the quaking bogs the tinners feared.

I breathed again, as I heard her voice, but I couldn't make out where it was coming from. Then the moon sidled out from the clouds and I saw that I was standing on the rim of a hollow on the side of a hill, as if a giant ladle had been plunged into the earth and scooped it out. A small pool, black and glistening, lay in its heart. I could just make out the figure of Morwen crouching close to the water. I stumbled down after her, squelching through patches of oozing mud that lay hidden beneath sodden grass.

She grinned, her teeth flashing white in the moonlight. 'I knew I could fetch you back to the cottage.'

'Fetch me? No, you didn't. I came looking for you,' I said indignantly. I'd had a lifetime of my father and brother ordering me to do whatever they pleased. I knew that was the way of it with men. But since I had walked away from them, I was determined to be commanded by no one.

'I can teach you how,' Morwen chirruped, oblivious to my annoyance.

She was as excited as a bairn wanting to show off some

new-found treasure. Before I could stop her, she was yanking at the cloth twisted about my head. My hair whirled out in the breeze.

Morwen seized my good hand and thrust a slender stick into it. 'Elder,' she announced. 'Ma uses it to summon spirits too.' She guided my hand, scratching three circles in the sodden earth, a small one, then a larger one around it and the third around that. 'You must offer the spirits something.'

'Ow!' I flinched as she tugged a few strands of hair from my scalp, and bound them rapidly around some worts. Even in the dark I could recognise one by its scent: rosemary. Mam had planted a bush of that near the door of our cottage, but it died the year she did.

'This is what I used to fetch you,' Morwen said happily. 'Rosemary to bind us, yarrow to call you, and rowan to guard against wicked spirits that might harm you on the journey or might appear in your guise to trick me.' She leaned over and dipped the sprigs in the black pool, sending ripples running outwards across its surface. She touched the dripping herbs to the east, south, west and north of each circle in turn, letting the water shower on to the earth, though if you asked me, it was so sodden, it scarcely needed that blessing.

'See, that's what you do, but all the time, as you do it, you must say the will worth, say it and want it more than anything else. See the person in your head walking towards you, like I saw you. And you came.'

She sat back on the wet grass and, though I couldn't see her expression, I could hear in her voice that her smile had faded. 'Ma says only the eldest daughter has the gift, been that way since first our granddams had the keeping of Bryde's well afore the black crows stole it from us. Ma was the eldest, and her ma too afore that. But I know Ryana can't journey or speak to spirits. She fools Ma into thinking she can 'cause Ma is so sure it's her that has the gift, not me. Ma won't let me speak of it. She says there's ways and secrets that can only be passed to the eldest, else the spirits will grow angry and take revenge. But you can feel them, hear them, like me. I knew it from the moment we both held that clootie. I can talk about these things to you. I can show you.'

'But before I came for the blood charm, you didn't know me to call me. It wasn't your voice I heard in my dreams. Whose was it? Why did she bring me here?'

Morwen didn't answer. Frustration and disappointment fermented inside me. I'd been so sure.

'I thought you'd know,' I burst out. 'When you said "Dertemora". I thought that meant you knew. It was a sign.'

'It's just the old name for these moors,' Morwen said. 'Everyone calls it that . . . all the villagers. Only the blow-ins, like the black crows and the pedlars, call it summat different.'

'The tinners call it Dartmoor.'

'Aye, that's it, but no one birthed in these parts would say it. It's not respectful. 'Sides, Old Crockern, the spirit who guards these moors with his wisht hounds, he'd not know it by any name save Dertemora. How could any ask him to protect it, if they don't know its rightful name? But . . .' she shuffled closer to me '. . . this voice of yours, what did it say?'

It was a night for whispering secrets so I told her what I had shared with no one. I sensed that she alone would not mock me or think I'd run mad. I told her about the river turning to blood, Todde and the hound's-tongue, and how, since I was nine summers old, I'd heard a woman's voice in my dreams. That day by the river I'd heard her again, but this time when I was awake. I told how I'd walked away from my village, my father, my life.

She listened in silence.

'Who is it that calls to me?' I finished.

For a long moment, she said nothing, then finally she murmured, 'Maybe Ma could see her face in the smoke . . . Maybe she could show you.'

I reached out to take her hand. 'I reckon you've more skill in your fingers than Kendra and all your kin before her. I know the answer lies with you. You can show me. I know you can.'

Chapter 23

Prioress Johanne

Even with the door to my chamber firmly closed, I could hear Sister Melisene shouting from the other side of the courtyard. 'Now you stop that at once,' Melisene bellowed, 'or I'll feed all the meat to the swine. They have better manners.'

I flung open my door and immediately regretted doing so, for the cacophony emanating from the other side of the priory gate was worse than that of a mob of drunken revellers after a Christmas feast.

'You heard the sister,' Goodwife Meggy shouted, through the small grid in the huge door. 'Kennel your tongues and stand quiet, else this gate stays shut.'

She slammed the shutter and took a step back, murmuring something to Melisene and the servant with her, who were both balancing baskets on their hips. But the clamour outside, far from abating, was growing louder. There were even thumps on the stout wooden door, as if people were hammering on it with their staves. I dragged my cloak about my shoulders and hastened across the courtyard, trying to avoid the worst of the puddles. 'What is happening out there?'

The gatekeeper folded her meaty arms, glowering at me as if I was responsible for the disruption at her gate.

'Tinners' womenfolk and their brats, that's what. They've the gall to come here begging for food. Claim there's none to be had in the villages round about. Got more sense than to sell it to them, that's why. Hid it where those thieving tinners won't find it. They'll need every bite they can find to fill the bellies of their own families, if this harvest is as bad as the last, which it looks fair set to be. Now those tinners have come here demanding alms, shoving our old folks and cripples to the back of the queue. It's not right. You ought to send them packing, Sister Melisene.'

The hosteller gnawed her lip. 'I don't like them elbowing the frail aside, but I can't just turn them away. Some of those children are so thin they look as if their arms would snap if you touched them, and the mothers are nearly as gaunt. But last time I took the meats out to them, the stronger children and some of the mothers just snatched it straight from the basket before I had a chance to share it among them.'

'Then it is up to us to see that it is distributed fairly,' I said. 'Fetch four more baskets and divide the food equally between the six. Meanwhile I will go out and speak to them.'

'That rabble?' Meggy said. 'Don't you turn your back on them else they'll have the clothes ripped off it.'

Behind me I caught the piteous cries of Sebastian through the casement of the infirmary. The noise must be carrying in to him as loudly as it had into my chamber. I had to fight the impulse to go to comfort him, but I had to trust one of the other sisters to do that. Better for him and the other patients that I dealt with this disturbance.

193

If Brother Nicholas heard this . . . Sweat drenched my body, and I felt as if I was standing in front of the great fire in the kitchen instead of out in the courtyard in the damp, chill breeze. I took a deep breath. Thanks be to the Holy Virgin, our troublesome brother was one problem I did not have to deal with at this hour. Meggy had told me he had ridden out on his black rouncy early that morning. I had cursed the news then, worried about where he might have gone and what he might discover, but now I was relieved. After the plagues of frogs and flies, I did not need him accusing me of being unable to carry out the most basic of the Hospitallers' duties – the dispensing of alms. If Nicholas sent word to Clerkenwell that we couldn't even deal with a few beggars, they would have me removed before the ink on the report was dry. And the thought of what they might discover once I was no longer there made me shiver. I gave myself a little shake. This was foolishness. They would discover nothing, and I would not relinquish my duty as prioress until the day they laid me in my grave. All the more reason to ensure peace was restored at our gates before Nicholas returned.

I tried to focus once more on the commotion outside, which, though it seemed impossible, was growing ever louder. I seized Meggy's stave, then told her to open the wicket gate in the great door and bolt it behind me as soon as I had passed through. The gatekeeper regarded me dubiously, as if I was intending to walk out into a pack of ravening wolves. 'They're just women, children and crippled old men,' I assured her.

'Savages is what they are!' But Meggy did as she was bade and opened the wicket gate, though barely wide enough for me to squeeze through, then slammed it shut again.

For a moment, I found myself almost agreeing with Meggy, for a crowd surged towards me, jostling me so closely, I could barely breathe. I felt small hands burrowing under my cloak, stealthy as those of professional cut-purses.

But almost at once a cry went up. 'That's not her. That's not the one as brings us meats.'

The rabble drew back a little, staring at me sullenly. I recognised a few faces, old women and a lame man who came regularly to mass. But as Meggy had predicted, a group of emaciated but belligerent women and children had pushed them roughly to the back and were keeping them there. Some of the children were pawing me again, stretching out filthy, spindly arms and crying out in the high-pitched wheedling voices that hardened beggars use to solicit alms.

'You will all be given something. But I will not tolerate the scenes of yesterday.' I seized one of the more persistent urchins by the wrist, dragging his hand from my skirt. 'You, boy, sit over there. You, and you, join him. I said, "Sit!" Get right down on the grass. No one will have a bite until everyone is sitting, and if anyone gets up again before they have been given their meats, they will get nothing.'

I sorted them into six groups. At first some of the boys remained standing defiantly, staring at me with mutinous eyes, some even jeering, but eventually their mothers pulled them to the ground, and finally those children who had come alone reluctantly followed. It took much heaving and groaning before some of the elderly women were able to lower themselves on to the sodden grass. I was sorry to force them to it and felt every twinge of the pain in those aching backs and swollen knees, but I could see no other way.

When all were seated, I called to Sister Melisene and to the servants who had gathered at the window in the gate to watch, ordering them to bring out the baskets. As they passed out the food, several of the tinners' women and children began to demand two, three, even four portions for ailing children and old folks back at their camp, but I shook my head firmly when the servants looked doubtfully at me.

A boy sprang to his feet and ran towards an old woman. He snatched the bread and mutton bone from her hand as he passed, racing off down the slope with it, the old lady's wails following hard on his heels. Three other children leaped up, and before any of us could stop them, they'd grabbed the food from those who had already received their share and run off. Seeing what was happening, the other villagers quickly hid their portions in their clothes or the sacks tied to their waists.

When they were finally convinced that the baskets were empty and no more was forthcoming, the tinners' families clumsily picked their way down the slope, plainly unused to walking over the boggy grass. We helped the old and infirm to their feet, and Melisene went to see what she could find in the kitchens for those who had been robbed of their alms.

An elderly woman, who came often to mass, hobbled up to me. 'You'll not have any meats to give soon, Prioress, if those tinners aren't stopped.'

'Perhaps next time we will have to take the alms they need to a place nearer their camp so that the villagers are not pushed aside here.'

The old woman shook her head till the wrinkled skin of her neck wobbled like a cock's wattle. 'You'll not be needing to take it closer to them for they're coming closer to you. Seen it with my own two eyes, I have. It's you who'll be begging for alms soon and you'll not get any from them.'

The rumble of hoofbeats distracted me before I could ask

what she meant. Brother Nicholas was cantering up the rise on his black rouncy. Its coat was glistening with sweat, and wisps of pinkish white foam stained the corners of its mouth. He had ridden the horse hard. The villager was still talking as he reined in dangerously close beside us, forcing me to pull her out of his path.

The old woman cocked her head, watching him slide from the saddle. 'Take more than him to drive them off. Kendra's curses have done no good. And if she can't banish them, no one can.'

'Drive who off?' I demanded in, I confess, a somewhat irritated tone. I was preoccupied with wondering which properties my brother knight had been sniffing around this time. But it was Nicholas who answered me.

'I rather think the goodwife is referring to the tinners. I found your cowherd trying to round up the cattle. It seems he was watering them at the stream when a gang of men and their dogs charged them, scattering the beasts in all directions. They claimed the cattle were trampling their bounding markers.'

'Bounding?'

Nicholas snorted, sounding not unlike his horse, which was pawing the ground and tossing its head, impatient to be taken to the stable. 'When the tinners want to commandeer a new site, they mark the corners of their boundaries by cutting turves and flipping them over, then lay stones at the edges of their claim. All they have to do then, it seems, is to inform their so-called tinners' court that they have placed their bounding markers, and provided another tinner hasn't already laid claim to that spot, they're free to start tearing up the land, digging for tin.'

'But those are our grazing rights,' I said indignantly.

'Were!' Nicholas said. 'I've seen their tin works in the far valley. Once they start digging here, there won't be a blade of grass left fit for grazing or water a beast could drink without poisoning itself.'

'That's what I've been trying to tell you,' the old woman cried triumphantly. 'They'll be doing to you what they done to us. Tore off my door to use as firewood, they did, and their dogs killed my last hen, while they stood and laughed. You mark my words, they'll be cooking their suppers on your threshold, if you don't stop them, and it'll be your cattle that they'll be stewing in their pots.'

Her rheumy eyes spotted Melisene returning with a small basket of food for those souls who'd been robbed, and she hobbled away to claim her share, though I did not recall seeing anyone snatch food from her hand. But perhaps I was mistaken.

Nicholas took a tighter grip on his restless horse's bridle, trying to hold the powerful beast in check. 'I wouldn't generally wager a dog's turd on any prediction made by some old village crone, but I'd gamble my own rouncy that she'll be proved right about this. Those tinners have the King's law on their side and it's the most badly worded statute that was ever drawn up since Nebuchadnezzar was crowned. Tinners could dig up this priory stone by stone, if they chose to claim there was tin beneath it. And claim it is all they need to do. According to stannary law, *their* law, no landowner may "vex or trouble them", which means, in effect, whatever they want it to mean. If a farmer so much as waves his fist at them or sets a dog on them to try to stop them digging up his crops, they say he has *troubled them* and can have him fined or worse. There's no court in this land that can curb their rights, unless King Edward chooses to change the

law, and he won't do that while they're making the tin he needs for his wars, and lining his coffers with the tax they're paying on every ingot. He doesn't care who in his kingdom suffers, so long as he has enough money to lavish gifts on those pretty lads he has tumbling into his bed.'

'That is dangerously close to treason, Brother Nicholas,' I warned.

His mouth curled in contempt. 'So, now I am obliged to listen to a woman try to school me in my duty to the King. Perhaps, my sweet prioress, you'll not be so quick to defend your sovereign when those tinners have taken your livestock, your water and your land, and you discover that our illustrious king is too busy fondling his friends to spare a moment to restrain these marauders. We will have to deal with this ourselves. The Lord Prior will not want to cause trouble, but Commander John at Buckland itches for a fight. He'll not be so squeamish. I'll send word to him. A few well-armed knights and sergeants-at-arms riding down on those tinners without warning will soon put them to flight.'

The tinners might be wolves, I thought, but Nicholas was a fox of the most cunning breed. Crying *help* to Buckland to come to the aid of defenceless women unable to protect themselves all alone on the moors – the brothers would love that. It would be just the excuse they needed to herd us safely into the fold of the cloister, leaving Brother Nicholas free to dig as deep as he pleased into the affairs of the priory. He would not be content with our removal. Nicholas was an ambitious man, itching for command. He wanted to discover something, anything, to earn the gratitude and respect of the Lord Prior. And when a man is so determined to expose evil, he can take even the miracles of a saint and present them as the work of the devil.

I met his gaze levelly. 'If a farmer's raised fist can be counted a vexation to the tinners, Brother Nicholas, I can only imagine what offence they might take if a party of armed knights came charging down on unarmed men, not to mention their wives and children.'

Nicholas laughed. 'They'd have a tough time serving an appeal against us to bring us to their tinners' court, much less imposing a fine. You must name a man to charge him. Hard to do that when he wears no coat of arms.'

I tapped the white cross on my cloak. 'A Knight of St John may not bear his own arms, but he does bear the arms of Christ, and God's knights do not ride down famished women and children, who are the very ones we are all pledged to serve. I have seen those tinners' families today with my own eyes. What they do, they do because they are desperate.'

'As you will be if they are allowed to invade our order's land unchecked. You women are too easily deceived by a tearful beggar's brat. You do realise that their mothers pinch them to make them cry, don't you? The tinners I saw in the valley were not hungry, they were greedy for wealth, and if their women and children go without food, it is because their fellow streamers would see them starve rather than share what they have. They're a pack of dogs gone wild.'

'Perhaps so, but when the knights have ridden back to safety at Buckland, where do you think the tinners will come to vent their fury? If you bring men-at-arms from Buckland to this place, all you will succeed in doing is starting a war between this priory and the tinners, and if they are the wild dogs you claim them to be, our cross and our veil will be no protection. You will not send to Buckland, Brother Nicholas, not as long as I remain prioress here.'

200

'And how long will you remain prioress, Mistress, with no livestock and no lands to support you? You had best ensure that no more plagues beset your holy well, for you will need every bent pin those pilgrims throw into it.'

Chapter 24

Sorrel

'Listen!' Morwen said, tilting her head towards the rocks towering above her on the top of the tor.

But I was panting so hard the only thing I could hear was the river of blood thundering deep within my ears. I was used to working all day on an empty belly, bred to it, you might say, like the donkey which walks round and round, day after day, turning the grinding stone. I'd spent my life hauling water from the village well, hefting firewood, pummelling the washing and wrestling with our own small patch of stony land, but though that makes your back and belly as strong as a blacksmith's, it doesn't prepare your aching legs for climbing up a steep tor with a spiteful wind snatching the air from your mouth before you can even take gulp of it.

It was all very well for Morwen. Like all those who were born on the moors, every step she'd taken in her life must have been either up a hill or down, and she could bound up a steep slope like a cat up the thatch on a roof. But by the time I'd staggered to where Morwen was crouched at the top of the tor, sweat was crawling down my face, and I'd such a burning pain beneath my ribs all I could do was flop down on the wet grass and lie there. Great grey clouds rose up, one behind another, like walls of stone, but a beam of dazzling sunlight, thin and straight as a golden arrow, slipped between them, striking the twisted branches of a thorn bush that grew out of a crack in the rock above my head, catching the raindrops that clung to the leaves and turning them into a shoal of sparkling rainbow fish.

I pointed. 'Beautiful,' I gasped.

Morwen lay down beside me, her head nearly touching mine as we stared up at the rainbow fish darting among the waterweed of thorn.

'Some nights,' Morwen murmured, 'the moon turns all the rocks to silver. The sky's as dark and soft as a mole's pelt, full of great frosted stars. When the summer is dying, the hills burst into yellow and orange flames. The bracken burns red as a fox, and rowan berries glow like hot embers. And when Brigid brings the sun back in the spring, the black moor pools are so still and calm you can see the clouds in them as if the skies were below your feet 'stead of above your head. When I was little, I'd watch the birds in the sky drifting in the pool, as if they were flying through water. I thought if I jumped in I'd find a whole new world down there. Tried it once, but all I found was mud.' She sat up, frowning. ''Tis all mud now. The moor is hurting. Nothing's right.'

She rolled over and tugged at me like a fretful bairn till I sat up. 'But listen,' she repeated.

I couldn't hear anything on that high peak, save the roar of the wind as it flattened the tawny grasses, but as my heart stopped thudding from the climb, I began to hear another sound burrowing out beneath the wind's shriek, a hollow knocking, as if a corpse was beating against the stone walls of his tomb with his own bones. And voices too, but they were murmuring to each other so low that I couldn't make out the words, or maybe they were speaking in a strange tongue, for I could no more catch the meaning than grasp that shaft of sunlight in my fist.

Morwen was watching me intently, her flame-red hair gusting in the breeze, her great green-grey eyes hungry as those of a stray cat. I felt like a beast with a plump bird clamped between my jaws and she was judging how best to snatch it. I knew she wanted me to tell her something, but I didn't know what.

She shook her head impatiently. 'You hear them?'

'Who's in there? Your mam and sisters?'

Morwen grinned, showing a missing front tooth. 'Ma only comes here at night when there's a certain moon she needs for her charms. That's why I brought you here in daylight, 'cause I knew it'd be safe then.'

The voices inside the cave grew louder, though their song was so mournful it hurt like a fist reaching into your chest and twisting your heart. I ached with misery at the sound, as if I was watching someone I loved weeping at a graveside. I wanted to take away their pain, but I couldn't.

'I knew it. I knew they'd speak to you.' Morwen breathed the words softly, as if she feared to disturb them. 'Come on.'

She edged sideways through the crevice, though she was so slender I reckoned she could have walked in face on. Then she stuck her hand back out between the rocks, groping for mine. Our fingers interlaced and she threaded me through the gap till we stood hand in hand, listening to the voices swirling in the darkness. Our breathing slowed and, without meaning to, I found I was drawing in breath with her and letting it go as she did, so that it felt as if the damp, earthy air was entering me through Morwen's body. We had melded into one creature.

A cold, hoary light trickled through the crevice, like dawn breaking after a winter's night, and I saw we were standing in an oblong cave. The floor sloped down towards the entrance and was bare but for the ashes of a fire set about by blackened stones and a few sticks of scorched wood.

Morwen let go of my hand, crouched and carefully heaped fresh tinder on the little nest of feather-soft ashes. She struck a flint and iron together several times until a bright flash of sparks caught the pile of twigs and a fragile flame guttered along its edge. She blew down a hollow reed, until the tinder was ablaze, then added some of the charred sticks.

'Ma'll not see the glow in daylight, but we'd best keep it small.'

The firelight flickered over the far end of the cave, which before had been in darkness. A slab of stone was covered with a white woollen cloth, but it wasn't lying flat. It concealed something beneath.

'What's under that?' I was curious, itching to raise it and look, but I didn't want to offend Morwen by prying.

As she glanced up, I jerked my chin towards the slab. That gesture would have made my old mam give one of her fond, sad smiles – she said I started pointing with my chin long before I could walk: if my good hand was grasping a crust of sucked bread or a shiny pebble, I had no other hand to point with. Mam said it broke her heart to watch me, for she could see even then that I longed to seize the world with both hands. But I reckoned one hand is more than big enough to catch life by the tail, if you can make that fist strong.

Morwen's gaze sidled towards the cloth, but she didn't

look at me when she answered. 'I'll show you one day, but not yet.'

There was something under there that bothered her. I could sense it. I glanced back at the cloth and shivered, as if Morwen's unease had jumped across the space between us.

She made me sit facing her on the other side of the fire, the way I'd faced Kendra across the smoke. 'Look,' she commanded. 'Look into the fire, then you'll see. And you must say the will worth, like I told you down at the pool. You must ask the woman who called you in your dreams to speak again and show you her face. You must want it so much that you feel like you'll break.'

While my mam still lived, I used to stare into the flames every evening, but it seemed now that that was centuries ago, far back in the embers of my childhood. When had there last been time to sit still of an evening without falling asleep from exhaustion? But now I felt myself shrink down into that little bairn again, when I'd watched fiery boars thunder through charcoal-black forests of towering trees, and golden salmon leap over ruby waterfalls, and great black and amber birds hover over the bloody corpses of the slain. Those things were as familiar as my own cottage back in my village. But what I saw in the flames in that cave were not those creatures. This was an alien fire.

Gnarled and twisted oaks, wizened and shrunken, old as the tors and their spirits as dark. But they are not trees at all. They're wrinkles and veins, nostrils and eyebrows. They're the face of a man, an ancient man, with great cavernous sockets for eyes. A black hound leaps from one of the sockets, his coat aflame and crackling with bright ruby sparks. A second dog bounds out and . . .

I turn my head. A woman is sitting before the fire. I see only

206

her back, only her outline dark against the leaping flames. She is weaving cloth on a loom of silver birch poles. She has spun the warp threads from soft rushes, green as spring, but as she picks the weft thread in and out of the warp, it shimmers and dances, crimson as rowan berries, golden as sunbeams. And I see it is not thread at all but living flames. She is weaving a cloak of fire.

Something darts across the cave. I catch it on the very edge of my vision. But I know if I turn to look at it, it will dissolve. I glimpse it again. It shines as though it has been cast from moonlight, yet it has a shadow at its core, a deep blackness.

I feel a tingle on my back. The skin on my shoulder-blades stretches and bubbles, as if maggots are burrowing out. They burst the skin, emerge quivering. They're attached to my body like my wizened arm, yet while that lies limp, they're uncrumpling, unfolding, blood is pumping through them, my blood. They're swelling, fluttering, and I can no more make them lie still than I can make the fingers of my left hand move. Shadows are gathering around me, drifting closer, like shoals of curious fishes. If I turn my head, they dart away, only to swarm close again when I stare back into the flames.

'Fire and water wait for you. The time is now.'

It's her voice. The one that called to me in my dreams and from the river.

'Come deeper into the fire, come. It will not harm you. It is cold fire, a fire of ice. It is my fire, come.'

Something lies beneath those twisted oaks, deep at the heart of that fiery core, a fox, a black fox. It doesn't look at me, yet I know if I walk towards it it will rise and run before me, deeper.

'Deeper.'

The whispers around me grow deafening, as if I'm being

sucked into the roar and crackle of the fire. The creatures of the shadows are flying at me now. Their damp grave-breath drifts against my face, their dead icy hands stroke my hair, and I know that if I let go I will float away with them through that sea of cold black flames.

'Let go. Come deeper. Come to me!'

My head jerks up and, with every last grain of will I possess, I push a single word into my throat and force it out between my lips. 'NO!'

The shadows rise, swirling around me, like a clamour of rooks. The whispers slither back through the cracks in the rock.

'What did you do that fer?' Morwen's voice held all the bitter frustration of old Kendra's.

But the green eyes that stared at me through the veil of smoke were filled with betrayal, like those of a child who'd had a juicy plum snatched from her hand before she'd taken a bite.

'Did you see . . .' I began, but I'd no words for what I'd seen or heard, so the question drifted down to the trampled earth.

Morwen glowered at me. 'What? What did you see? Tell me!'

I tried, but it was like trying to make a shattered pot whole again. The more I spoke, the less sense it made. Nothing was joined together, nothing I had seen had an ending. I was sinking down in nameless dread, knowing only that if I had taken one more step I would never have been able to return. It was as if I was gazing down into my grave – no, worse than that: I was staring through it into whatever living darkness lay beyond.

All the time I was trying to tell her what I'd seen and heard, Morwen sat with her knees drawn up to her chin, her arms wrapped round them, fists clenched. Her gaze was fixed upon the fire. She didn't glance at me once.

'Brigid. That's who's been calling you,' she said dully, after a long silence.

'Saint Brigid?'

'No!' she snapped. '*Our* Brigid. Her whose well the black crows have stolen. Mother Brigid of the fires and the sacred springs. That's who you saw. She made these moors with her own hands, sang the wells out of the earth and called the rivers to run to the pools. It was her you saw at her loom. She was weaving her mantle. She returns at Imbolc and spreads her mantle over the earth to protect the land and drive away the last snows of winter.'

Brigid, the old goddess. Some memory fluttered to life deep inside me. Mam used to spill a little milk for old Brigid sometimes, if the butter was stubborn and wouldn't come in the churn. The parish priest used to rail against such things in his sermons, but Mam said the old mother belonged to the women and was no concern of any man.

I'd not thought of it since I was a bairn, but now a picture came into my head of Mam tying a strip of cloth on a bush outside our cottage at sunset. It must have been winter then, for I remember the puddles were frozen and Mam's fingers were so blue with cold she had to keep blowing on them as she tied the knot. I'd thought it a strange thing to be doing – Mam was always careful to gather in before nightfall any clothes she'd put out to dry for fear they'd blow away or be ruined by beasts.

''Tis Brigid's mantle,' Mam told me, though I could see it wasn't even big enough to make a cloak for my wooden doll. 'This night Brigid rides through the village to bless the mantles of those who do her honour, and it'll bring good fortune to our home if she blesses it.'

Mam promised that next year she'd show me how to hang the cloth, but she never did, for by then she was dead.

Morwen's head whipped round, fury blazing in her eyes. 'You don't even know who Brigid is, do you? So why did she call *you*? I have the gift. Why not me? Why can't I see her?'

I could feel her hurt, but I didn't know what to say. I wished it had been her. I didn't want this. 'What does she need from me?' I asked, hoping that Morwen's pride would be soothed a little if she could teach me again as she had down by the black pool.

But my question seemed only to bait her anger. 'Why ask me? You should have asked *her*. She wanted to tell you. Why did you refuse to listen? I would have gone to her when she called. I wouldn't have run away like a – a – a sheep with a tick up its arse. You have to look into the fire again. And this time you must do what I told you. You have to want her to answer.'

'I'll not be told what I *must* do,' I said, as furious as she was now.

I scrambled to my feet and stalked towards the narrow cleft that led out of the cave. 'And I'll not conjure those creatures again.' I shuddered, still fearing their grave-cold fingers against my neck. 'If you've the gift, like you say, then you look into that fire. And while you're about it, you can tell this Brigid of yours to leave me be!'

Chapter 25

Hospitallers' Priory of St Mary

Brother Nicholas, striding across the courtyard, caught sight of the chapel door and stopped abruptly. Smoke! The chapel was on fire! He started to shout a warning, but realised even as the word escaped his lips that it wasn't smoke but steam. The sun had broken through the clouds, and a shaft of light shone full upon the door of the chapel. The wet wood steamed in the unexpected warmth. Nicholas tried to ignore the curious stares of the servants who'd been startled by his bellow and were watching his progress with undisguised curiosity, as if they thought he might start capering like a court jester.

It was hardly any wonder he couldn't recognise sunlight when he saw it, Nicholas thought morosely. It was as rare as cocks' eggs on this blasted moor. In Rhodes, the heat would be shimmering off the stones and the golden sun sparkling on a clear azure sea. Grapes would burst sweet on his tongue, and a girl whose silken hair smelt of damask roses and jasmine would be smiling at him as she poured his wine.

Mud squelched out beneath the sacking that had been laid in front of the chapel threshold, covering his boots with a stinking ooze of stagnant water and goose dung. He cursed vehemently, startling the servants a second time. The Lord Prior *must* send him back to the Citramer. He'd rot from the feet up if he was forced to spend another winter in this miserable realm.

But Lord William would send him nowhere if he couldn't even prove that the accounts drawn up by some aged crone were flawed, either through ignorance or, as he was beginning to suspect, by deliberate manipulation. But every time that suspicion crept back into his head, he found himself stamping on it. It was one thing to believe that women were fools and easily gulled, but that they would be clever enough to cheat the order was quite another, especially to do it so skilfully that neither he nor the procurator at Clerkenwell was able to uncover it. That was simply not possible.

Nicholas grasped the iron ring on the chapel door, hesitating before he turned it. Frogs and flies, bolts that would yield only to a woman's touch, what new plagues was that well about to spew forth? He was starting to think it was possessed of a malice all of its own. He pushed the door. The bright pool of sunlight outside made the chapel seem darker than usual and, for a moment, it appeared deserted. Nicholas was annoyed. He'd been sure he'd find her in here and, equally importantly, find her alone. He was about to stalk out again, when he glimpsed a movement in the dark shadows.

Sister Fina was emerging from the well door on the far side of the chapel. She halted when she caught sight of him and he thought she might dart back down, like the lizards on Rhodes, which scuttled under rocks if they caught sight

212

of a man approaching. But instead she stood in the archway, as if, once again, she intended to prevent him or indeed anyone from entering.

'The . . . the well is open again, Brother Nicholas. The pilgrims will come today.' Her tone was wooden, rigid as her body.

'I was informed that it was to be reopened.'

Actually, he hadn't been told anything. But Alban had seen lights moving in the chapel after dark and had alerted him. From the shadows, he'd watched the prioress and her sisters spend half the night carrying a stinking soup of rotten frogs, maggots, dead flies and slime in relays of buckets up the steps to a pit outside the priory walls, there to be buried with lime, and all under cover of darkness. Darkness covers many sins.

They had sluiced the floor and steps with lye and lime, but he was sure he could still smell traces of the foul putre-faction creeping from the open door. Like the stench of heresy, it could not be washed away.

Nicholas closed the chapel door and strolled towards Fina. He was half amused to see the panic on her face as she retreated sideways until she collided with a pillar. She stood with her back hard against it, as if she was trying to melt into it.

'No need for such modesty, Sister. We're brother and sister in the same order, both sworn to chastity. It's no less seemly for you to be alone with me than with one of your father's sons.'

God's blood, surely she didn't think he was going to ravish her. She might be the youngest sister, but she was no beauty, though now he thought about it, he couldn't exactly decide why. It wasn't that any of her features was ugly, he decided, more that everything about her was mismatched. A nose too narrow for her face, eyes too pale for her dark hair, and breasts too small for her ungainly height. But he was prepared to convince her he found her more ravishing than the Queen of Sheba, if he had to. It wouldn't be the first time he'd bedded a woman to discover what he needed to know.

It had been pounded into him since the hour he first grasped a sword that the first and overriding commandment was to ensure the survival of the order. Nothing transcended that. If a knight must disguise himself as a Saracen to discover where an attack was planned or lie with a woman to learn what God's enemies were plotting; it was his duty to God and to the Hospitallers to do it. Such knowledge was as nobly won as any fight on the battlefield.

Nicholas glanced towards the closed door, then back to Fina. 'Sister, you collect the offerings left by the pilgrims at the well, don't you? I know what pilgrims leave as gifts at the shrines of saints, but I confess I've always been curious about what they bring to a well like this. Tell me, what's the largest sum you've collected in one day?' He laughed. 'I imagine they are quite generous, having toiled all the way up here.'

'Bent pins. That's what they bring. They drop them into the water and I collect them. Prioress Johanne says . . .' She pressed a hand to her mouth, like a guilty child, as if she had said something she shouldn't and was afraid she might incriminate herself.

Nicholas studied her carefully. *Now, just what is it that our sainted prioress says that you don't want to tell me?*

He stepped towards Fina, reaching out his hand. He fingered a stray lock of chestnut hair that had crept out from beneath her black veil.

'You have hair like silk. I noticed it before, and I should know. I've handled the finest silks in the markets of the Citramer.' He tucked the strand back beneath her veil, his fingers gliding over her cheek. 'Don't blush, Sister Fina. There's no sin in a knight giving praise to God for the lovely thing He has created. Indeed, it would be a sin to ignore it.'

Fina tried to move away, but he leaned across her, supporting himself on the pillar with one hand, which rested a breath from her face.

'You were telling me about the pins. It must be a tedious chore to collect those wretched little things day in, day out. And I dare say you have to clear up all the mess those pilgrims leave too, their bits of flowers and stinking rags covered with blood and pus. I've known warrior knights who wouldn't have the stomach for that. I only hope your prioress appreciates all you do for her. I doubt she'd bend her proud neck to lift a rag or fish around in freezing water for something as small as a pin.'

She nodded eagerly and seemed on the verge of speaking. That was the trouble with women: if you feigned the slightest interest in any mundane task they performed they'd insist on telling you about it in such tedious detail that you'd be begging for the mercy of the executioner's axe before they were done. And he hadn't come in here to listen to her babble about pins.

He stepped away from her, fixing his gaze on the bloodied head of the crucified Christ that hung above the altar. 'After dealing with all those villagers and beggars traipsing through here, it must be a relief for you to meet with the merchants and their wives who cross these treacherous moors on their way to and from the ports. Doubtless their offerings of coin and jewels are worth the trouble of collecting. If they've

215

endured the perils of the sea, they must be overwhelmed with relief to have been brought safely to shore, and those about to venture on board ship must be praying they don't perish. They're fearful going one way and grateful coming the other. I imagine it's hard to say which pays better.'

He laughed. 'Brother Alban reckons it's the merchants returning from the ports that pay more, thankful to be on solid land. I say it would be those about to set sail, for they're praying twice over – first, that they don't drown and, second, that if they must, their days in Purgatory will be short. But you know what a surly fellow Alban is, always insisting he's right. That weasel had the nerve to bet me that you collect as much as twenty shillings a day at this well when the merchants are returning. I said it was twice that sum when they're going out. I beg you to settle the wager for us, Sister Fina. And I pray you'll tell me my reckoning is nearest to the mark, for if I should lose the wager to Alban I'll be obliged to—'

He spun round as the chapel door opened and sunlight burst in, scattering shadows. But the light was blotted out moments later by the great bulk of Sister Basilia waddling through the doorway, gripping a young boy by one shoulder. She propelled him towards the edge of the stone altar, lifted his arm as if he was a doll and pressed his fingers on to the edge. His hand clasped the corner, like the claws of a little wren perched on a rock. The boy stared at Nicholas, with an unblinking gaze that the knight found both impudent and unnerving.

'Why have you dragged this village brat into the sanctity of the chapel, Sister Basilia?' Nicholas demanded. 'There's no service today and I was about to make my private devotions. I don't want to be disturbed.'

Basilia looked flustered. She glanced uneasily at the child, who hadn't moved from the altar or given any sign that he knew they were talking about him. 'Forgive me, Brother Nicholas. I didn't mean to disturb you. I shouldn't have dreamed of fetching him in here if I'd known you were at prayer. I hadn't expected you . . . That is, you don't often . . . But Prioress Johanne instructed me to bring the child here and bathe his eyes and tongue in the holy well, now that the flies . . .' She faltered, her gaze darting to the well door as if she feared a new and more terrible plague was even now gathering below.

She took a steadying breath. 'If St Lucia intercedes for the child, the holy well will cure him of his blindness and his dumbness too.' Her tone suggested she intended to be quite firm with the saint and leave her in no doubt as to the nature of the miracle that was required of her. 'Though how I'm to get him and a lantern down those steps without him falling and sending us both crashing to the bottom, I'm sure I don't know. I can barely manage to squeeze down that staircase myself. I can't see where I'm putting my feet and those steps are so worn and slippery.'

She stared pensively over her great belly. Her feet were remarkably tiny for a woman of her size. 'I know I shouldn't say it, but I'm always afraid I'll get stuck and won't be able to turn round. It's foolishness, I know, but those rocks seem to close in soon as I start on the steps.' She gave a nervous high-pitched giggle. 'If I'm to take the boy down, I fear I'll have to trouble you to help me, Sister Fina.'

Fina's eyes flashed wide in alarm and she took a pace back, as if Basilia had asked her to cradle a snake.

'Blind, you said.' Nicholas took a pace forward. 'Is this the boy who caused the death of the old priest?' He addressed the question to Fina. But she made no answer, though he hardly needed one for her panicked expression spoke louder than a town crier. 'So, this is the little sorcerer.'

'Prioress Johanne says he is just a helpless child,' Basilia retorted, lifting her head defiantly, so that her many chins wobbled.

Nicholas studied the corpse-pale lad. He was as thin and frail as a prisoner chained for months in a dungeon. The prioress may have protested the boy's innocence, but she didn't appear to have given orders that he be treated well. He looked half starved.

Nicholas did not trouble to lower his voice. 'The devil may work through any creature. His imps take the form of frogs, foxes and goats, why not a boy? We have no way of knowing if the child has ever been baptised and had the devil cast out of him.'

Cupping his fingers under the lad's chin, the knight tilted the child's head towards the light from the door, passing his other hand several times across the boy's eyes. Then, without warning, he slapped the child's face. The boy yelped, trying to protect his head with his arms.

'Interesting,' Nicholas said. 'He can make sounds.'

He dragged the child's arm away from his cheek and wrenched open his jaw, tilting his head back until it seemed he might snap his neck. He grasped the boy's arm, pulling him round to face Sister Basilia.

'He has a tongue and it's not tied to the bottom of his mouth, like some I have seen. He does indeed appear to be blind, but I can see no reason why he shouldn't speak, except for obstinacy. I wager you could cure his dumbness far more swiftly and surely than by pouring holy water into his mouth. Arm yourself with a good switch and use it hard. Tell him

218

you'll stop only when he begs you to. That will encourage him to words. I'll gladly do it for you, if you've not the stomach—'

He broke off and stared down at the stick-thin limb that he still grasped in his great fist. The boy's arm was moving, but not because he was trying to struggle out of the knight's grip. He was standing motionless. Only his arm wriggled. It was undulating in Nicholas's hand as if the long, solid bones inside were now many tiny bones, as supple as a spine, as if the soft skin had hardened into scales, as if Nicholas was holding a writhing serpent in his hand instead of the arm of a boy. The boy's hand began to open, but it wasn't a hand, it was a head, with jaws that were stretching wide to expose two long fangs.

With a cry of horror, Nicholas staggered backwards, crashing into the pillar behind him. He stared at the boy's arm. But it was just an ordinary limb made of human flesh, no different from its twin – anyone could plainly see that. Yet he knew what he had felt.

The child was staring up at him, as if he could read every wild thought that was passing through the knight's head, as if he could see the shock and fear on Nicholas's face. A shaft of bright sunlight from the open door fell across the boy, the golden sparkle reflected in the twin pupils of those great, unblinking eyes. But even as Nicholas stared, the light that bathed the child turned to thick crimson gore. Nicholas pressed his hands to his eyes, convinced that he must have struck his head on the pillar when he stumbled and blood was running down his face, blinding him. He drew his hands away and examined his fingers, expecting to find them stained scarlet, but they were clean and the light that flooded through the door was as yellow as the sun itself.

'Whatever ails you, Brother Nicholas?' Basilia cried, waddling towards him. 'You've gone as pale as milk. Do you feel faint? You should sit down.' She tried to take his arm, but he shoved her away.

She was still clucking around him, when out in the courtyard a bell began to toll for the midday meal. She gave an audible sigh of relief. 'That's the noon bell. Cook will be putting the pottage on the table.' She waddled back towards the boy. 'No time for bathing now, young man. We'd best get you back to the infirmary else you'll miss your dinner. We can't have that, can we? You're already as skinny as a mouse's tail.'

She seized the boy by his hand and dragged him out behind her to the courtyard. Nicholas stumbled to the doorway, but a sound behind him made him turn his head. Fina was standing in the centre of the chapel, making the sign of the cross over and over again, as if to protect herself against something she feared. Had she seen that serpent too? Nicholas was certain that was one question he did not want her to answer.

Chapter 26

Sorrel

I hugged the threadbare cloak to my chest and paused to draw breath, staring down at the tinners' valley. The slopes of Fire Tor had been jewelled with rosy-purple heather and butter-yellow furze, but here on the hills every bush had been torn up for fuel or bedding, leaving weeping sores in the scalp of the earth from which the mud oozed in deep rivulets. The snide wind carried a fine mizzle, which, though you could scarcely see the drops, had already soaked through my hair and skin into the bones beneath. After the warmth of the fire in the tor, I was chilled to the marrow and my belly griped with hunger. Even the roots of my hair ached with cold. But though my body craved heat, I cringed at the thought of returning to the clatter and swarm of that valley. I wanted time to think about what had happened in the cave. I needed peace. I needed to be alone, but there was no chance of that down there.

Misery had wrapped around me, like a wet mantle, and spurts of anger kept rising inside me, not for Morwen but myself. Truth be told, I was as frustrated with me as she was, more so. Why had I pulled back? Why had I not followed that voice, followed Brigid to the place she was trying to lead me when I'd had the chance? I'd been so close.

I'd left my home, my village, all I had ever known, and trudged to this midden, this hell, to follow that voice. All these weeks I'd been begging her to tell me who she was and what she wanted of me. All through the long, cold days and freezing nights I'd been trying to reach the place she was calling me to, and now, when every question was about to be answered, when I would finally see her, finally understand, I had fled!

Why can't we call that hour back, unsay what should not have been said, turn right instead of left, stay instead of running away? But time will not turn for us. Each moment melts, like a snowflake, and will not come again. And everything has changed because of it. The flight of an arrow is quicker than a breath, yet it takes a man from this world for eternity.

If I could have run back to that cave, I would have done it in a heartbeat. This time, I would have found the courage to enter the fire. Even knowing I could never return, I would have followed that voice. But it was too late. Morwen would have extinguished the flames. She would have walked away, and who could blame her?

Would Brigid simply abandon me here in this desolation? The thought terrified me, but I knew I deserved no better. Worse than that, I had driven away the only person who could help me find my way. Morwen would not summon me again, of that I was certain.

I jumped as the mournful echo of the ram's horn drifted towards me on the wind. It couldn't be that late, could it? The tinners would already be trudging back to their huts.

Would this be my life from now on? But where else could I go? Where else would I find work when so many were tramping the roads in search of it? Listless, I began to pick my way down the muddy track, weaving from one side to the other, trying to find a stone or a clump of rotting grass to step on, every muscle tense, fearful of slipping in the glutinous mud. The fires outside the huts had already been poked into life, their gusting flames huddled beneath the tiny shelters.

It was hard to see clearly through the smoke and mist, but there was no welcoming glow in front of my own hut. Couldn't Todde at least have mended the fire? After all, he practically sat on it half the night. The only time I ever got near its warmth was when I was stirring the pot.

I picked my way along the bank of the soup-thick, muddy stream. A figure was limping towards me. His pace quickened as he caught sight of me.

'Where have you been?' Todde snapped. He was annoyed, that was plain, and a wave of guilt overcame my irritation. Hunger makes everyone waspish and he'd spent the day shovelling and breaking stones, knowing he'd have next to nothing to show for it come Saturday. We were so deep in debt we couldn't afford to lose a single minute, and I had thrown away whole hours up in that cave, and for what? I couldn't have felt more wretched if I'd dropped a loaf of fresh bread into the mire.

He broke off, tilting his head as though he was trying to look at someone behind me. 'Where's Eva?'

His tone had changed so suddenly that it took me several moments to understand what he was asking.

'She'll be up at her fire by now, doling out the men's supper,' I said, feeling all the more guilty that I wasn't at my own hearth stirring our pot.

'Aye, well, that's just it, she isn't. Seeing as how neither of you were at your fires and no one had clapped eyes on either of you all afternoon, we thought you'd gone off together foraging, setting snares or some such.'

There was no reproach in his voice and I was grateful that he didn't add 'when you should have been streaming'. Not that I answered to him: he wasn't my kin. But, all the same, he'd a right to expect that I'd help pay back what we both owed.

'Did she tell you where she was going?' Todde asked. 'Her cooking fire's dead and her pot cold. Beans in there still hard as nails. Only thing blazing is the men and lads who want their food. They'll be spitting like a nest of weasels if she doesn't come soon.'

'She's a lover, one of the men she cooks for,' I told him. 'Maybe she's with him.'

Todde gave a snort of laughter. 'I reckon it'd have to be *all* the men she cooks for to keep her rolling in the hay so long. There's not a tinner in this camp who'd have the strength to keep a woman pleasured for so much as an hour after a day's graffing.' His grin vanished and anxiety furrowed his forehead. 'No, I reckon if the two of you weren't out together, then something's amiss. Maybe she's had a fall and done herself a mischief. Some of the tinners she cooks for were muttering about starting a search. I best tell them she's gone off on her own. It'll be dark soon.'

I glanced up at the sky. The sun had long sunk behind the tor, and the shadows were creeping down the valley. Had Eva collapsed somewhere? Maybe the bleeding had started again. Was that possible? If Morwen had stopped it with a charm, did she have the power to start it again with a curse – revenge for what I'd said and done?

Todde squeezed past me on the sodden track, and as he did so, he put his hands on my shoulders. 'At least you're back without harm. I was starting to fret. You ought not to go wandering on those moors alone, not with the pack of wild dogs that keeps howling and those outlaws. It's not safe for a lass like you, not safe for anyone.' He hurried away.

I didn't know whether to be glad he was watching out for me or annoyed that he thought I couldn't take care of myself. I realised he hadn't asked where I'd been or what I'd been doing, not that he had the right but that wouldn't stop most men. Todde was a strange one, all right.

I heard the commotion before I had even clambered up to Eva's hut. Most of the unmarried tinners and the children had wandered off to try to cadge food at neighbours' huts or join in the search for Eva. But the few men who still lingered were shouting and arguing with each other, as they stared balefully at the big iron pot of cold pottage. Exasperated, I elbowed them out of the way, got a blaze going, then hefted the pot on top to start cooking.

It took time to heat, for it was a large cauldron with enough in it to feed a score of men and children. I stirred the pot and blew on the fire, trying to make it boil quicker. All the while men and children kept wandering up, demanding to know if they could eat. I reckon some would have devoured it half cold, dried beans and all, if I hadn't slapped them away with the ladle.

I kept glancing up, expecting to see Eva hurrying up the rise. I couldn't understand what had happened. Even if she'd gone to fetch more peats for the fire or to check her snares, she would have left the pot simmering. She always began to cook long before the horn sounded, for she knew the men would be wanting to eat as soon as they had climbed up the hill. If she had set out on some errand, it must have been much earlier in the day and plainly she had expected to return by mid-afternoon to make ready.

Darkness filled the valley. It had stopped raining, but sullen clouds still covered the moon and stars, threatening to tip more water upon the sodden earth at any time they pleased. Voices drifted up from below still calling for Eva.

Was she lying on the moors with a broken leg, having slipped in the treacherous mud? If she'd gone foraging for herbs she could be anywhere in the vastness of that wilderness. She might have stumbled into one of the sucking mires or fallen from the towering rocks, or be lying unconscious with Morwen's curse upon her, the life-blood seeping out of her. I shivered in the cold wind. If she wasn't in the valley they'd never find her tonight, not in the dark, and by morning . . .

Holy Virgin, keep her safe. Let them find her before it's too late.

Chapter 27

Hospitallers' Priory of St Mary

This time it was not Sister Fina who discovered it but the pilgrims or, rather, an old woman and her daughter with a tiny, wizened child. They had shuffled into the chapel, and had grudgingly parted with a coin to visit the well. Meggy had recognised the old woman at once. They'd grown up together in the village. As children, they'd played in the woods and streams. As mothers, they'd gossiped as they'd pounded their clothes in the washing pool, while their infants clung to their skirts. As widows to men who had died long before their three score years and ten, they'd helped each other to lay out their husbands' corpses and lent a comforting hand as the bodies were laid in the earth. But that was a lifetime ago, another age, another place.

Now the old woman's cow was ailing, she said, and its milk had dried, as that of so many down in the boggy valleys. She knew she should have driven the beast up to the common pasture on the high moor for the summer, but too many of her neighbours had lost their animals up there these past months, stolen and butchered by the outlaws or driven into such terror by the tinners' noise and their dogs that they'd run headlong into the mire or broken their legs tumbling over rocks. Her little granddaughter was sick too, wheezing so she could scarcely draw breath. But the old woman was sure Brigid would heal both child and cow, for when times had been better hadn't she always left a drop of milk or honey by her hearthside for Brigid before she'd blown out the rush lights at night? And she swore to do it again, if Brigid would only draw down the cow's milk.

Meggy had wanted to keep her old neighbour talking, learn all the news and gossip from the village, but the old woman had been chary, ill-at-ease, eyeing Meggy's black gown as if her friend had been replaced by a changeling. She wouldn't even meet Meggy's gaze. When the gatekeeper talked of the times when they had first been kissed at the harvest feast or when the boys had stolen their clothes as they bathed naked in the stream, she had shaken her head and muttered that she didn't remember.

Meggy, wounded, had reluctantly sent them into the chapel, but had warned them to tell Sister Fina they'd come to ask the blessing of St Lucia for the ailing child. They had done as they were bade, reciting the words as carefully as a charm, if unconvincingly. But they could have told Fina they were coming to buy a curse from St Lucifer, for their words fell like sand thrown at a closed door.

Fina had caught a flash of movement outside in the courtyard and had already turned from the two women before they'd finished speaking, fearful that Sister Basilia was bringing the boy back. She had not made another attempt to bathe his eyes, but Fina had hardly dared to set foot outside the chapel, not even to relieve herself, for fear they'd slip in. She had started to close the well earlier each day, relieved when she could put the key through the wood and feel the bolt slide into place, knowing it was safe again until morning.

Like a fly trapped in a jar, the image of the boy staring at Nicholas as if he was cursing him buzzed ceaselessly round her head, Nicholas backing away from him, the horror unfolding in his face. What kind of power did that child possess to terrify a battle-hardened knight? Sister Basilia must not bring Cosmas back to the well. If the boy ever set foot in the holy cave, he would destroy it, just as the old priest had prophesied.

It was all she could do to stop herself running to the chapel door and slamming it shut. Bolt it. Keep him out. Keep them all out, Brother Nicholas too, especially Brother Nicholas. For days, she had been picking over his questions and her answers, telling herself what she should have said, what he would have replied, until she could no longer distinguish memory from imagination. He had been trying to catch her out, trick her, like Prioress Johanne had warned her he would, but what had she told him? He should not have been standing so close. He should not have touched her. She couldn't think, couldn't remember. It wasn't her fault.

Their offerings of coin and jewels are worth the trouble of collecting. That was what he'd said. *Jewels and coin . . . jewels and . . .* He was accusing her of stealing. The merchants were *paying twice over.* He'd said that too. Once to the priory and again to her: that was what he'd meant. He was calling her *thief.* But she'd never seen any jewels. So, someone was sneaking in here, to *her* well, and stealing.

A shriek made Fina jerk upright. Then she heard the sound of worn shoes slip-slapping on stone. The young woman with the grizzling child on her hip rushed up the steps from the well. The older woman followed more slowly, panting hard. Her daughter flung open the pilgrims' door and hurried out of the chapel into the cold, watery light.

The older woman took a few paces, then staggered back, leaning heavily against the chapel wall, her hand clutched to her chest. 'Spring's running blood – blood!'

Fina shook her head, trying to clear her thoughts. 'No . . . No. I've seen the walls turn red once before, but it's the light. Prioress Johanne said it was a trick of the light. If you go right down to the bottom it vanishes. There's nothing red in the spring.'

'I have been right down and I tell you it's blood, not water, in that spring.' The old woman peeled herself from the wall and tottered towards Fina. 'Give me back my coin,' she demanded, holding out a cupped hand.

Fina gaped at her. 'But it was an offering to God. You can't take it back.'

Many villagers and pilgrims grumbled at being asked for an offering, but none had ever asked for its return. Even if the cure they sought did not seem forthcoming, they left their coin in hope, in faith.

'What can't this woman take back?' a voice thundered behind them. Brother Nicholas strode into the chapel.

Chapter 28

Hospitallers' Priory of St Mary

Startled by the sudden appearance of the black-robed knight, the old crone who was arguing with Sister Fina staggered a few paces backwards until she collided with one of the pillars. She gripped it hard on either side as if she intended to uproot it and hurl it over her head towards Nicholas if he came any closer.

He glanced from her to the equally alarmed Fina, trying to decide which of the two he was likely to get more sense from, not that sense was something he expected from any female.

Fina found her voice first and an ugly flush, like a rash, spread from her neck to her face as she stammered some kind of explanation, but it made no sense to Nicholas. The old crone, sensing her chance to retrieve her money slipping away, loudly repeated her demand. Nicholas's jaw tightened. He wasted no words on either of them. Seizing the old woman by the arm, he thrust her out of the pilgrims' door, slamming it so hard behind her that the saint in the stained-glass window trembled.

Fina was now standing in front of the doorway to the stairs, with her back to them as if she intended to prevent him from going down. That made Nicholas all the more determined to do so. Evidently, there was something down there that the sisters were again trying to conceal. He gripped Fina by the shoulders and dragged her aside. With a silent prayer that it wasn't another swarm of flies or some equally foul creature, he began to edge down the stairs.

Above him he could hear Fina babbling about a trick of the light and a red glow, but as far as he could see the walls glistened with their customary luminous greenish-gold. It was unnatural and a little unnerving, but he supposed it was all to the good that pilgrims should be awestruck upon entering the place, without the need for artful displays of silver and jewels that other shrines had to install in order to strike wonder into their visitors.

Nicholas ran his hand down the oozing wet moss on the wall, pressing against the stone in an attempt to keep his balance on the uneven steps. The stairs seemed even narrower than he remembered. It felt as if the walls were squeezing together, crushing his broad shoulders. Some might say it was to his credit that he ventured down at all, for he still expected to hear the dreadful buzzing of those flies. His skin crawled at the thought of them and several times he scrubbed at his face, sure he could feel them alighting on him. But there were no sounds except for those of his boots grinding grit against the stone steps and the water splashing far below.

As he emerged in the cave at the bottom, he found himself sweating in spite of the chill, damp air. And then he saw it. A thick, deep-red liquid was running out between the three rocks and splashing into a scarlet pool.

Each time Nicholas watched a brother priest say mass, the

cleric would raise the goblet of wine and proclaim that it had turned to blood, the sweet, precious Blood of Christ. Some priests in the order seemed convinced that what they drank *was* His Blood. Some saints had even declared they had knelt at Christ's feet and suckled it as it gushed from his side. The idea revolted Nicholas. He could only assume that those men had never seen battle, had never seen a man hacked to death. He had watched gore pour from the terrible wounds of friends and enemies alike, tried to hold men's guts inside bellies slashed wide open, and pinched tight the wounds in men's gurgling throats as the blood scalded his hand. Nicholas could attend mass only because what was in the chalice looked and smelt like wine and, though he could never admit as much even to his confessor, he was certain it remained wine.

But when holy water in a well turned to blood surely not even the most pious priest in their order would regard it as a blessing.

Take thy rod; and stretch forth thy hand upon the waters of Egypt, and upon their rivers, and streams and pools, and all the ponds of waters, that they may be turned into blood: and let blood be in all the land of Egypt, both in vessels of wood and of stone.

For a moment, Nicholas felt a tingle of exaltation, the kind that an inquisitor feels when he sees the first flame running up the robe of the heretic on the pyre. The wicked are punished. These obstinate women are chastised. God has vindicated the righteous.

But the full import of the words of Holy Scripture, which had bubbled into his head at the sight of the bloody spring, suddenly punched him hard. *All* the ponds, *all* the streams, *all* the vessels of water! It was one thing for God to have

233

cursed this well as a sign to the women, but suppose He had indeed cursed all the water in the priory, all the water on the moor. Suppose, as in Egypt, it had all turned to blood. When a man has known the desperate agony of thirst, while roasting in heavy armour in the heat of a battle as the sun beats mercilessly down, he might be forgiven for fearing that particular curse more than most.

Nicholas almost fled straight back up the stairs in his desperation to find out how far the scourge had spread. But something penetrated his brain and stopped him. He stared at the sluggish stream of blood gurgling out of the cave wall. Another image floated into his head, of crossing the courtyard at Buckland when the servants were slaughtering the pigs, sheep and geese for the great Christmas feast . . . the sharp frost sparkling on the thatch in the winter's sunshine, the white breath of the sweating men, the squeals and shrieks of the beasts, steam rising from the eviscerated carcasses of the pigs as they were hauled up on chains to drip from the beams, then the rivers of blood running between the cobbles to freeze in great scarlet puddles. And with that sight came the stench of dung, guts, singed bristles and, above all, blood, the unmistakable sweet-metallic smell of fresh blood.

This cave, like any battlefield, should be reeking of it, but it wasn't. The only thing he could smell was damp and the tang of a creek when the tide has gone out. He took a pace towards the narrow coffin-like trough, and scooped up a handful of the red liquid. He sniffed it. Not blood, but mud. Stinking red mud.

Chapter 29

Sorrel

I shivered, trying to burrow deeper into my cloak though it offered precious little warmth against the clinging drizzle. Below me, in the darkness of the tinners' valley, a single line of guttering torches flickered past the ditch and climbed towards our huts. The shadows of the men splintered into tiny fragments against the hillside, so that a swarm of rats seemed to scurry alongside the tinners.

One of the boys had been dispatched to the camp with the news that a body had been found. You could see he was not best pleased at being sent home before he'd had a chance to take a squint at the corpse, but he cheered a little when he found himself the centre of a crowd who were, for once, desperate to hear what he had to say. All the womenfolk had gathered around Eva's fire to wait for the men to return. And waiting was all we could do.

'Looks like they've got a body slung over that pack beast,' someone murmured.

'Holy Virgin, let it not be Eva,' I said. 'The boy said it

was so dark out there they couldn't be certain till they turned the corpse.'

'Could be a beggar starved to death out there. More dying every day.'

'They'd not be troubling to fetch a dead'un back here, if it wasn't one of us,' another said grimly.

One by one the women who had been squatting on the ground clambered to their feet, standing still and silent as owls, watching the procession of scarlet flames crawling towards us. They stepped aside to make a space as the man leading the horse drew level and tethered the beast to a post near the old wall. Those men who followed him said nothing and looked at no one, not even their own wives. They blew on numbed fingers with lips that were pinched and almost rigid from cold. Strands of hair snaked down beneath hoods and wriggled across foreheads glistening with rain and black sweat. Crouching at the fire, they stretched out their hands towards the flames, begging for warmth. Todde was hunkered down among the other men, his shoulders hunched, his hands clenched under his armpits. I could see he was exhausted – they all were.

The bundle dangling over the horse's back was shrouded in the patched cloaks of several men, and bound with straw ropes, but no one made a move to cut it loose and lay it down. Whoever it was, we couldn't leave them hanging there, like a sack of grain. If no one else would do it, I'd have to. I drew my knife and began to saw at the rope. But a hand on my shoulder tugged me back.

'Leave it be. That's no sight for women or chillern,' a gravelly voice told me. 'It's Eva, as far as we can tell. No sense in taking her down, when we'll only have to lift her back up for the horse to carry her to her grave.'

My heart seemed to shrivel in my chest. Eva – dead! I fought against the tears that threatened to choke me.

'But she should be washed . . . shrouded, made decent,' I protested. 'She doesn't deserve to be carted about slung over a horse, like carrion.'

'Carrion's about the right word for what's left of her. Foxes and birds have been at her. Nothing left of her belly. Chewed off her hands and feet too, and . . .' He swallowed hard and didn't finish.

'But how could she be dead? Foxes couldn't have killed her.'

The old tinner flapped his hand at me as if I was a squawking hen. 'Hush, woman. When the fire's out, there's no point asking if it was wind or rain that did for it. Eva's gone and there's no more to be said.'

'Good deal to be said, if you ask me,' another man retorted. 'She's been ripped to pieces, bones gnawed like she was a roasted sheep.' He stared over his shoulder at the impenetrable black tide that lapped around the camp. 'Whoever did this is still out there. Could take any of us next.'

'Hounds gone wild, I reckon,' a woman told him. 'Masters have died or driven them out to fend for themselves 'cause they've not got scraps enough to feed them. Dogs start roaming in packs and they get so crazed with hunger they'll hunt down anything that moves. I've heard them howling many a night.'

'And there's men gone wild too,' another tinner said, raising his head as he crouched by the fire. 'I heard tell of a family, ten of them, maybe more, used to lie in wait for lone travellers, picked them off one at a time, dragged them to their cave and killed them. Whole family would feast on the corpses. Didn't even bother to cook them, liked their meat raw. When the sheriff's men hunted them down, they found a heap of human bones and skulls in the back of the cave. Hanged the lot of them, they did, even the children who were barely old enough to walk because they'd have grown up to do the self-same thing. No cure, there isn't, not once they get a taste for human flesh. It's like when a dog's taken to savaging sheep.'

His gaze crept towards the corpse slung over the horse, then darted back to the safety of the fire. 'I reckon that's what we got ourselves up on the moor, a nest of those corpse-eaters. I saw the marks on that body and I'd swear on the devil's arse and the Holy Virgin both that they weren't dog bites.'

There was a rumble of voices, some agreeing, others telling him he was talking out of his backside.

'Either way, the tinners' court can't handle this,' the stoker from the blowing-house said, raising his voice above the rest. 'Even they've not got the powers to deal with murder. Coroner will have to be fetched, and if he reckons there's men in these parts feeding off human flesh, he'll call in the sheriff and roust men-at-arms to smoke them out.'

'No cause to go involving them.'

Heads jerked round as Gleedy slithered out from the shadows beside a hut. As if he'd dragged it with him, a vicious gust of wind tore across the fires, flattening the flames and biting deep into wet skin. The tinners shuffled uneasily. How long had he been standing there?

'That's Eva we found,' the stoker said, getting to his feet. 'And if someone's murdered the poor mare, we want to see them hanged for it.'

One of Gleedy's eyes flicked towards the horse and its burden. His face was drawn and haggard in the firelight, as if he'd not slept. 'So long as they hang the right man, but those coroners aren't bothered about that. Fat and lazy as hogs in a mudbath, they are. Just want to collect as many fines as they can for the King's coffers. The more they collect, the more they can cream off.'

'That's the pot calling the pan burned-arse, if ever I heard it,' a woman whispered in my ear.

Gleedy gazed around at the faces, gaunt as skulls in the firelight. 'And if I go sending for the coroner, it'll be you and your families that'll suffer, and I'll not let that happen. Eva cooked for you and she'd never see any man starve even when he had naught. She'd not want to bring down trouble on your heads. We all know who it was, who did this. I warned Eva time and again only to use food from my store, but she would insist on foraging. You all know she was a bondswoman. I warned her that there's hirelings combing every inch of these moors with their hounds for runaway serfs and villeins to drag them back to their masters for a heavy purse. I reckon she got caught by one.'

Gleedy jerked his head at the woman who'd spoken of the wild dogs. 'Those are the hounds you hear baying. Those serf-hunters are watching this valley night and day for the chance to snatch back any runaway who strays beyond it. And they think it great sport to set their hounds on any poor wretch whose only crime is to want to live free. They enjoy watching them ripped to pieces, just like their masters love to see a hind's throat torn out by their dogs.'

I stared at the limp bundle hanging from the horse. So, Eva had been on the run from her master, just like Todde and, like him, she'd dared not leave the protection of the tinners. So why had she gone out there yesterday?

I glanced at Todde. He was staring fearfully at the moor,

doubtless wondering if a hunter out there was lying in wait to run him back tied to a horse's tail, battered and bruised, then deposit him, more dead than alive, at his lord's feet. Had Eva, too, stolen from her master? Was that why they'd hunted her down, or had the lord of the manor merely wanted her returned to make an example of her to deter others from following?

'All the same,' the stoker said, 'bondswoman or not, if murder's been done the sheriff ought to be summoned. Even a serf can't be done to death without a fair trial.'

Gleedy shook his head gravely. 'I want poor Eva's killer brought to justice as much as any man here. If I could lay my hands on the bastard who did this to that good woman, I'd string him up myself by his cods. But you all know, same as I do, that the sheriff will not risk losing his office by pointing the finger at any lord or his hirelings. Those coroners and sheriffs all have good parcels of land, and you know how jealous the landowners are of the tinners' rights. They'll not go looking for anyone else, my friends, 'cause a chance like this is what they've all been waiting for. Eva was working here, they'll say, so it must have been a tinner that killed her and dumped her body on the moor, hoping the birds would peck it clean. The sheriff will arrest any man in this valley who can't prove he was in plain sight of others every moment of yesterday afternoon – anyone who slipped away to shit, anyone going to fetch a tool or a swallow of ale. I know how they reason. He'll hang the tinner, drive his woman and children on to the moors to starve, and announce justice has been done.'

The men were glancing anxiously at each other. There must have been at least a few moments that day when they'd vanished from sight behind a spoil heap or were crouching down in the ditch scraping up the last handful of gravel.

Gleedy stared pointedly round us all. 'And there's a few among us I reckon wouldn't relish being examined too closely by any coroner or sheriff. Isn't that right, Toddy?'

Todde's eyes blazed in the firelight. 'Don't you go accusing me. I was here streaming same as everyone. Isn't that right, Sorrel?'

All the faces around the fire turned to me. I tried to speak, but the word wouldn't come. I stared at Todde and saw fear flash across his face as he suddenly remembered I hadn't been in the valley when Eva had gone missing.

Gleedy grinned, his swarm of teeth glistening, like maggots, in the firelight. 'Think how the justices will see it. Most tinners in the camp have their womenfolk with them to warm their cods. But Sorrel told me, the first day she arrived, that you weren't her man, quite adamant she was, as I recall. But the justices, they'll look at you and say, "Here's a fellow has needs, just like us, and if the woman he's with won't satisfy those needs, well, he's bound to get roused up." They'll say you forced yourself on Eva and killed her to stop her telling. There's plenty here will swear to it that they saw you attack me in the warehouse for no good reason. You've a hot temper on you.'

'I never laid a hand on her, you dogshit!' Todde launched himself towards Gleedy, fists flailing. But Gleedy had already stepped swiftly back behind the fire. Three of the tinners grabbed Todde, wrestling him away.

Gleedy held up his hands. 'Course you didn't,' he cooed, his tone as soothing and slippery as butter. 'We all know that. It was the men hunting runaways, like I said. But I'm just telling you how a sheriff would reason it.'

241

I saw looks passing between some of the women and glanced at Todde, uneasy now. I shook my head, trying to rid myself of the chilling thought.

He was still struggling in the clutches of the men who held him, but the fight had drained out of him and he looked as if he might collapse if they let him go. The men sensed it too, for they lowered him to the ground, where he sat, head in hands, rocking and groaning.

The men and women had all fallen silent. They sat round the fires hunched, withdrawn into their own thoughts, while the shadow of the corpse on the horse's back moved restlessly over them as the beast shifted.

In the wall behind Eva's fire, vipers' heads were appearing, swaying back and forth, their eyes glittering, their forked tongues slithering in and out between their fangs, tasting the chill night air. Then, far out in the great lake of darkness, a hunting horn sounded, vibrating across the hills, and at once came the howls of a great pack of hounds as they took up the cry to seek and kill.

Chapter 30

Prioress Johanne

'Tinners.' Sister Clarice flung the charge across our supper table. 'That's who's to blame for our well turning red. They're damming every river and stream and building their leats right across this moor. Not to mention pouring the filth from their workings straight into the water that the villagers use downstream. It's a wonder they haven't poisoned every man and beast for miles. Some of those streams coming from their workings are so thick with mud and silt you could walk dry-shod across them.'

'For once, I find myself obliged to agree with you, Sister Clarice,' Nicholas said. 'I've seen it myself. The water spewing from those workings is much the same colour as that oozing from the spring.'

He laid aside the beef bone he'd been digging into with his spoon, finally forced to concede there was not a shred of marrow left inside it, and was reaching for the last bone on the platter, which had been placed between him and Alban. But Alban's fingers were nimbler, even though he had lost the two on his right hand, and he snatched it up. Nicholas glowered at him.

'But it's a holy well,' Sister Basilia protested. 'A healing well. The water is miraculous, pure. It gushes from the rock. It doesn't come from any of those rivers on the moor.'

Nicholas snorted, staring pointedly at the marrow bone Alban was burrowing into. I suspected the sergeant was well aware of the knight's hostile glances, and was taking a childish and somewhat malicious pleasure in having beaten his superior to the prize. Nicholas dragged his attention back to Basilia.

'Your holy water comes from the moors. Even you can't be so naïve as to think it flows from solid rock.'

Basilia flushed and stared down at her trencher. I knew I should have intervened, but in truth I was relieved that at least this discussion seemed to be silencing Sister Fina. These past few days, I never knew what wild words would suddenly burst from her. It wasn't so bad when we sisters were alone, but whenever Brother Nicholas was within earshot, I felt as if I was trying to stamp out sparks blowing across a field of dry corn, knowing that he was listening for the smallest thing he could use to destroy us. But for the moment, happily, he had launched into one of his many gory tales of the Citramer to prove his point to poor Basilia.

'Some years back, I found myself part of an army laying siege to a Turkish citadel. An old man had attached himself to us and sold us information, for he knew the area well. He'd been captured in his youth and forced to become a Mussulman upon pain of death, but he'd always borne a hatred for them. He was willing to work for us, though he was crafty enough to make sure he was paid first. Anyway, this citadel of theirs had plenty of wells inside and well-stocked food stores. And they taunted us each day that we'd run out of supplies before they did, even waved their meats and bread at us from the battlements.

'But this wily old serpent sidled up to our commander and told him to send men out into the countryside to find some pigs, slaughter them and leave the carcasses to bloat in the sun for a couple of days. The soldiers cursed him with every foul plague they could think of. They craved fresh meat and thought it a wicked waste.'

Nicholas glared at Alban again, who was chewing a lump of beef, apparently paying no heed to his lecture.

'The commander even had to set a guard over the pigs to stop his men sneaking in and stealing a leg or two, and the guards were even more incensed when the carcasses began to rot. They were having to stand next to the putrid stench. But when the pigs were good and ripe, the old man told us to wedge them in a particular river where the water would wash over them. None of us could see what good that would do, except add to the stink and flies, but a day or so later everyone in the citadel was vomiting and sweating in agony. The poisoned water from the stream had somehow flowed into their wells.' He raised his hand as Basilia seemed on the point of interrupting again. 'You can call that another miracle if you want – the Turks probably called it a curse – but years ago, someone told me that all the streams, springs and pools in the world rise up from one vast lake deep beneath the earth.'

Fina raised her head, glowering at Nicholas as if he had grossly offended her. 'Hell lies beneath the earth, deserts of fire and vast howling whirlwinds where thieves spin for ever in the darkness,' she said, like a child reciting a lesson.

Nicholas frowned, evidently wondering if we'd missed the

point of his story. But the word *thieves* sent a shiver of danger down my spine. Why had she selected that particular sin?

I rapped sharply on the table, trying to divert her. 'The water in the courtyard well still runs pure and clear,' I said, with as much cheer as I could muster. 'And one blessing of this rain is that it is even higher than usual. So, we shall certainly not want for water. We must give thanks for that.'

Basilia beamed at everyone, nodding so vigorously that her chins wobbled like a calf's foot jelly.

But Nicholas glared at me. 'And if this filth seeps into the well in the courtyard?'

'We have cisterns and barrels that are filled only with rainwater. They cannot be polluted.'

Nicholas let out a snort of derision. 'Even on this accursed moor, it cannot rain for ever. You must bend your neck, Prioress, and send word to the Lord Prior to ask for the knights to ride to our aid. It is only a matter of time before our cattle and sheep are poisoned and those of our tenants. We have a duty to protect them and, besides, it is in our interest to do so. How will they pay their rents if they have starved to death?'

The eyes of every sister were fixed on me and they all seemed to be holding their breath. Even Alban had stopped eating.

I laid down my knife and fastened my gaze on Nicholas. 'I assure you that I am fully aware of my duties both to my tenants and to my order. I trust no one will ever have cause to say I have neglected either. I thank you for your concern, Brother Nicholas, but if you imagine that the tinners will be frightened off by a band of knights, you are much mistaken. They are tough men, not armed with swords, I grant you, but more than capable of cutting down a horse and rider with such implements as they can weld. They will fight and they have the law on their side.'

A candle, guttering on a spike on the wall behind Nicholas, threw his face into deep shadow, but I did not need to read his expression to know the fury that was written there.

'As you yourself have already so eloquently pointed out to me, the tinners are answerable only to King Edward. They live and work under his protection. The Lord Prior is not an imprudent man. And, as you have also told me, Brother Nicholas, he is anxious not to give the King any reason to act against our order. I am sure Lord William is wise enough to realise that, should the tinners send word to the King that a group of heavily armed Hospitaller knights had ridden down upon a camp of defenceless men, women and children lawfully engaged in the King's business, injuring, if not killing, many, he would take that as an act of war against the Crown.'

'Defenceless!' Nicholas exploded. 'They are terrorising the villagers.'

'Even a king may be caught in a snare of words, if they are well twisted, especially if they are delivered by a limping old man and a child with a bandaged head, which you may be quite certain they would be. Besides, the King needs tin and he needs the taxes from it. He does not need the Knights of St John.'

My chair scraped back on the stone floor as I rose. The sisters hastily scrambled to their feet and it gave me not a little satisfaction to observe that Nicholas and Alban, who shared their benches, were obliged to stand too, as they were pushed back.

'There are those among us who feared that the frogs and the flies meant the end of our holy well, but those plagues passed. The spring will soon cease to flow red, of that I am certain. Have faith, sisters, and pray. This is but another test of our courage.'

I guessed that Clarice would follow me to my chamber, but I was in no mood for her talk of falling revenues. I hurried across the darkened courtyard, pausing only to listen outside the casement of the infirmary for any sounds that might tell me Sebastian was distressed. Mercifully all seemed quiet. I unlocked the chapel door and slipped inside, pulling the wooden bolt into place behind me. I did not light the candles. I did not want anyone to know I was there. I craved just a few moments of peace and solitude. I knelt before the altar, bunching my robe beneath my knees to cushion them against the cold, hard flagstones. The bowed head of the crucified Christ, hanging above me, glowed beneath a halo of soft light shining down from the oil lamp above.

How long could I contain Nicholas? I truly believed what I had told him that declaring war on the tinners would bring more danger and call down the King's wrath. That was always the trouble with men. They thought every problem and vexation in the world could simply be put to flight by the point of a sword.

But was Nicholas really so foolish? He knew we were deceiving Clerkenwell, although I was certain he had no real proof yet. Otherwise he would have acted. He was arrogant, and he was no bookkeeper, that much was evident, but he wasn't stupid. The more we thwarted him, the deeper he would dig. As the proverb says, 'Suspicion has double eyes.' But if he started a feud between the priory and the tinners, he would not need to prove his case to finish us. I had to prevent any report of his from leaving the priory.

I stared at the open hand above me nailed to the cross. The painted blood around the wound in his palm shone fresh, as if Christ was hurling it down upon us. 'Why, Lord? Why do you punish us with yet another plague on your holy well?'

Guilt burned in me. Was He punishing me for deceiving our holy order? But if we failed, if I failed and the priory closed, what would become of all the frail creatures we sheltered – our servants, Meggy and Sebastian? Who would care for Sebastian? He would not survive without us. It was my duty as prioress to protect those I had taken a sacred oath to serve.

But an insidious voice nagged in my head and it was one I couldn't smother. It whispered that I had once knelt before another altar, placed my hands in the lap of another prioress and, before my brothers and sisters, had vowed faithfully to serve the order of the Knights of St John of Jerusalem with my life. To honour one vow meant breaking another. What if there could be no right way, no sinless path, only a greater or lesser wrong? And which was the greater wrong? Who would tell me, when God refused to answer me?

If any of my sisters had come to me, I would have told her that her duty was obedience to the order, yet as their leader, as the one responsible for so many lives, was it not them I should put before all else? We were pledged to be the serfs and servants of the blessed poor. Should I say, 'I cannot protect you, because I have vowed to obey the order. My oath of obedience is more important than your survival. My soul is worth more than your life'?

I stared down at the cold flags, painfully aware that I was kneeling directly above the holy well. I must have faith! I had assured my sisters that we had survived the plagues of

249

frogs and flies, which had seemed to spell doom for the well. We had prayed and they had vanished as if they had never been. Like the woman with the issue of blood who reached out in faith to touch Christ's robe, I had only to reach out my hand in faith and the spring would cease to flow red. I had only to believe it.

Chapter 31

Hospitallers' Priory of St Mary

Meggy sat on her stool, slumped against the wall, a snore escaping her from time to time as she drifted comfortably between dozing and sleep. It was not yet time to damp down the fire and huddle beneath the blankets, and she was enjoying the blessed warmth from the flames. It was the only time her feet ever seemed to be warm these days. Although a stone was heating on the edge of the hearth, ready to be wrapped in sacking and slid under the covers, it always cooled long before morning. But she counted herself lucky. The hut, though barely long enough to lie down in, was dry, the door solid, and her belly was full of hot pottage. As she knew only too well, there was many a widow, like her, who that night would be huddling hungry in a doorway or coughing her lungs out with a dozen others in a wet, lice-infested byre.

Meggy was dreaming of her son, not the stocky lad who'd marched off to fight the Scots, but years before when he was little, small enough to run beneath a horse's belly. They were in her husband's forge and her son was giggling as he tipped a box of horseshoe nails on to the ground. She was scolding the lad and scrambling to pick them up before his father returned, but no sooner had she put some back in the box than the little lad scattered more. The nails were growing smaller and smaller, harder to gather, and there were so many. Now the boy was banging the tongs against the anvil, making it ring through the smithy. The noise would bring his father striding in, and the child would be in trouble.

Meggy jerked awake and, for a moment, she thought the sound she could hear was the echo of her dream, but it was coming from the gate. Someone was singing, not the plain-song of the chapel or the bawdy chorus of a drunken pedlar, but an eerie high-pitched keening, like the ghosts on Fire Tor.

She dragged her cloak about her shoulders and hurried out to the small gridded window. The rain had stopped for now, but thick clouds crowded against the moon, smothering the light. Meggy cursed herself. Roused suddenly, she had neglected to bring the lantern. She was on the point of returning for it when she caught a wisp of singing again, like a feather spun towards her on the wind. It was a woman's voice, she was sure of that, and it was coming from outside the priory. But who would be out there on the moors at this hour? A soul needing help? The sound certainly tore at your heart, like a woman grieving over a new grave.

She almost found herself drawing back the beam that braced the gate. But she knew the tricks the outlaws used. She slid back the shutter covering the gridded window and peered out. Torches were left burning on either side of the main gate to guide latecomers and the lost, at least until the midnight hour, by which time they had usually burned away. The insipid orange light clawed at the darkness, as the wind twisted and flattened the flames. But she could see no one standing in front of the door, no grieving mother or shivering child.

A chilling thought gripped her. Was it the voices of the dead that had awoken her, the restless spirits that roamed the moor trying to find their way back to the villages of the living? The corpses of murderers, self-murderers and madmen had been thrown into those black, sucking mires for centuries, for they weren't welcome in hallowed ground. The souls of the innocent, too, wandered the lych-ways: the babies born maimed, left out to perish; the beggars dying alone, unshriven, their corpses lying out for the birds to pick their bones. Their ghosts would never lie quiet for they had no graves to rest in.

Meggy crossed herself, and reached for the shutter, struggling to slide it back over the grid, but the wood had swollen in the months of rain. As she fought with it, the wind gusted the flames of the torches, and as the tongues of light snaked out, for a moment they illuminated a tall figure standing motionless in the darkness, unmoved by the buffeting of the gale. Meggy glimpsed a bone-pale face, a tangle of long hair darting out like lightning bolts about the skull. She couldn't tell if it was a ghost or human, only that it was staring at her, as if it was her that it had come for. The wind gusted once more and both torches were extinguished, as if that spirit had snuffed them out. All the world was plunged into darkness. But the singing rose again, riding the wind, like a shrieking hawk.

Her hand trembling, Meggy caught the edge of the wooden shutter and heaved with strength born of fear. The wood slid home with a crash, and she turned, leaning her back against the great oak gate, trying to recover her breath. Then a cry escaped her.

The boy was standing behind her in the dark, empty courtyard, his blind eyes suddenly glowing ghost-green, like a wolf's in moonlight. His arms were stretched out towards the gate, as if he was trying to grasp the wild notes that were rising like a flock of birds. He turned his head slowly, until he seemed to be staring right at the door that led into the chapel. A sudden silence filled the courtyard – even the wind held its breath. Meggy had never seen such deep blackness, known such dark silence. She could no longer hear the low rumble of water in the cave echoing up through the stones or the wind crying on the moors. It was as if she was lying in her own grave, beneath the earth. No sight, no sound, only darkness.

And she could not know that in that moment the prioress's prayers had been answered – the spring had ceased to flow red. In fact, it had ceased to flow at all.

Chapter 32

Morwen

'But what does it mean, Kendra?' the old villager wailed.

Her tiny granddaughter lying at her feet gave a faint mew and tried to turn herself over, but the effort was too great and she flopped back, grizzling fretfully. There wasn't a peck of flesh on her and the floor of our livier was digging into her sharpened bones. I winced for her, knowing the pain of sleeping on the hard earth, but Ma and her visitor glanced at the small bundle indifferently, as if it was a kitten they might toss into a pool to drown. Neither moved to lift her.

'Whole moor is turning widdershins,' the old woman continued, ignoring her grandchild. 'I've never known such rain. I've not a wort or a pea that's not rotted or been gobbled by birds. There's not a single chick hatched by my hens this year that didn't die of the gapes even afore it got its proper feathers, and now I've only one old bird left, and I daren't let her out of my cottage for fear a fox'll snatch her or the tinners.'

At the mention of tinners, Ma spat into the fire. 'There's your answer. You want to know why all's gone amiss, it's those tinners penning their beasts in the sacred stone circle. Using cup stone as a seat for their filthy arses and pissing on the queen stone.'

I shifted on my haunches by the door. I'd not seen the tinners actually pissing on the great stone, but they probably did, for they treated the sacred circle worse than a midden.

'Tearing up the land, they are,' Kendra said. 'Dragging the streams from their natural beds.' She wagged her finger at the old woman. 'But Old Crockern'll have his revenge on any man who tries to harm Dertemora. He always does in the end. You dare to scratch Crockern's hide, the old man will break you, body and soul. You mark my words, he'll let loose his black hounds from their kennels in the woods and he'll hunt those tinners to the highest peak of Dewerstone and drive them over the edge. Smash on to the rocks below, they will, every man jack of them. And wisht hounds will be waiting to feast on them. There's many have fallen to their deaths there and their corpses never found.' Ma's eyes gleamed with satisfaction.

The old woman shuddered. 'Dewerstone's an accursed place. Still, I reckon it's no more than those tinners deserve.'

I shivered too, thinking of Sorrel. I'd been so vexed with her that I stormed and railed around Fire Tor loud enough that it was wonder Ma hadn't heard me down in the cottage. But after, I could have cut out my tongue. I'd thought she might come back to the cave, when her temper cooled, but by the time I went out she was long gone. It was my fault. I'd been journeying afore I even had words to explain it, thought it was natural, that everyone did it. But they don't, not even my own sisters. Too late, I realised that Brigid had brought Sorrel to the moor for me to show her how, but I'd driven her away.

I had to find her. I'd summoned her to come. But she was strong that one: if she felt it, she could resist it. I closed my eyes, silently calling her, pleading with her. But suppose she came back to the cave and I wasn't there. I had to get up to it. If only the old woman would leave, Ma and Ryana would fall asleep and I could slip out.

I peered around the old woman at the naked child lying between her and Ma's fire. They seemed to have forgotten the little mite was the reason her granddam had come. The girl whimpered. She was barely three summers old, but looked like a wizened crone. Her ma had run off in the night with the last sack of dried beans they'd had, or so the old woman said. 'Left me and the little 'un to starve to save her own skin,' she grumbled, spitting on the floor.

The little girl stared up from the deep pools of her eyes, her ribs fluttering in and out, like the breast of a trapped bird. I could tell she was finding it hard to breathe, lying flat on her back, but she was too weak to sit up. I had to stop myself darting over and lifting her, but Ma's staff lay within easy reach of her crooked fingers. She didn't take kindly to any interference when she was working.

Over in the corner, sheep-face Ryana was supposed to be making a powder to heal the sick child, pounding a roasted mouse in an ancient stone mortar, with dried lungwort plucked from beneath an oak tree. But Ryana's hands were idle and her mouth had fallen slack, as she listened to the two old besoms.

'Between the tinners tearing the lights and liver out of the land and the black crows nesting over Bryde's Well, is it any wonder Brigid is angry?' Ma said.

The old woman clutched at a small bag around her neck, fashioned from a scrap of grey rag, one of Ma's charms. She lowered her voice as if Mother Brigid herself might be eavesdropping at the door. 'I swear on my husband's grave I saw the water in her well turn to blood right before my eyes. Now I hear tell it's run dry. That true, is it, Kendra?'

'It is!' Ma sounded as if she'd stopped up the well herself. 'Blood gushed out, as if someone stabbed Old Crockern in the heart. Three days and three nights it ran with blood, filled the whole cave. Then the gore turned to dust, like burned bones on a pyre, and a great wind filled the cave and blew it away. Not a drop of water has flowed from that spring since.'

I studied Ma's face. That hadn't been quite like Meggy had told the tale when she'd come to her, angry and afraid. But Ma said the birds had already shown her when she was journeying – leastways that was what she said to Meggy.

The old woman fingered the amulet around her neck. 'My grandfather used to tell stories round the hearth come winter, and he said the well ran dry once before. Year of the long drought, it was, when all the streams and rivers dried. Must have been when your granddam, no, your great-granddam was keeper of the well. She told the villagers the streams wouldn't flow again till the well did. Terrible cruel, they say it was, whole flocks of birds lying dead, like black snow, on the ground and they reckon you could hear the moaning of the cattle all night as they died. People, they died too, but silently, like the birds.' Wonderingly, she shook her grey head. 'Terrible cruel,' she repeated to the fire.

Ma nodded. 'My great-granddam, she sat outside the well fasting and praying to the goddess.' She reached over and poked Ryana in the ribs with her stick. 'You paying heed to this, girl? See that you do, but keep grinding while you listen. Your ears are stuck on your head, not your hands, so there's no cause to stop moving your fingers.'

258

Ma stared into the burning heart of the peat fire. 'My great-granddam didn't move from the spot for days. Her lips were so parched they cracked wide open and her skin sank down to her bones till folks thought she'd already died and it was a skeleton sitting there. Then old Brigid spoke to her and told her what she must do. Told her to catch herself a live long-cripple, and cut its head off on the cup stone, so blood would run into the cup, then anoint the queen stone and the well stones with the viper blood. Then she was to hang the body of the long-cripple in the branches of a twisted oak tree in the wood, high as she could reach. She did exactly as Brigid had bade her, and afore dawn the first cloud appeared, by midday rain was falling and by nightfall the well was flowing again, streams too.'

The old woman glanced towards the hide hanging in the doorway to keep the rain from blowing in. Drops of icy water were dripping from the bottom edge into the stagnant green puddle that oozed over the threshold. 'Isn't rain we need this time, that's for certain. Rivers are brimming over, land is sodden, but still the spring is dry.' Her voice cracked, brittle, fearful.

'Brigid is angry.' Ma spread her fingers in the smoke of the hearth fire. She shut her eyes, muttering to herself as she slowly closed her hand. The smoke vanished into her clenched fist, as if it had been sucked up. Kendra unclenched her fingers and the smoke slid out again, slithering up towards the blackened thatch above. 'Brigid's closed the water in her fist. Only when she opens her hand will the spring flow again.'

By the time I had managed to escape from Ma's cottage and run up the hillside, I was afeared that if Sorrel had come she would be long gone again, thinking she'd been mistaken and I'd not summoned her.

Ma had blown the pounded mouse and lungwort through a cow's horn into the little girl's throat, making her cough, but she was too weak to struggle. It would ease her chest, Ma said, help her to breathe. But still the old woman was in no hurry to leave. Why waste peats on your own fire when you could share the warmth of someone else's? She'd glanced pointedly towards Ma's iron cooking pot, evidently hoping she could share whatever might be there too, but in that she'd be disappointed. Unless Taegan returned with a bite of food from Daveth and his brother, the pot was likely to remain empty, save for a spider that had foolishly scuttled in.

Halfway up the hill, I was forced to stop to draw breath, though usually I could run all the way to the tor. The rocks above me were hidden beneath a dense fleece of clouds, and the grey mist rolled down towards me, clinging soft as lamb's wool to my skin, cutting off all the sounds from below of running water and the cackle of the ravens. I shivered. Hunger always made me cold. I took a gulp of air to call, and the grey fret slid like syrup down my throat.

'Sorrel?' *Please let her be there.*

She was sitting by the entrance, legs drawn up, her head resting on her arms. She scrambled to her feet, relief and misery mingled in her eyes. We stared at one another, both opening our mouths to speak, both stopping to let the other go first. In the end, nothing was said – nothing needed to be said. But something had changed and it wasn't our quarrel. There was a deep hole in her, a wound, like when a green branch has been torn from a tree.

I grasped Sorrel's cold, ragged fingers and pulled her gently through the crevice into the cave. Only when the gold and ruby firelight was lapping across the walls did either of us speak.

Sorrel kept her head bowed before the crackling wood, while the voices of the cave nudged each other and whispered at her back. In a dull voice, she told me that the woman I'd charmed the clootie for was dead – not from the bleeding, she added quickly. They'd found Eva on the moors, her body mauled by dogs. I remembered Ma's tale of Dewerstone and the wisht hounds, but it was hounds of men that had killed her, Sorrel said. Hirelings hunting for bondsmen or women who had taken off without leave.

'Maybe if I'd talked to her more. Maybe if . . . Why did she go wandering out on to the moors? She knew there were men searching for runaways.'

I saw the grimace of pain on Sorrel's face and leaned forward. 'The seven sisters spin the thread of life, that's what Ma says. Spin it and cut it. You and me, we don't hold the blade. No mortal does.'

Sorrel stared deep into my eyes. 'But I have to know why. Why did Brigid bring me here? I've dragged myself all this way to sleep in a hovel that's not fit for goats. I break my back for a mouthful of food and I'm so far in debt to the master of the tin works that I may as well be bound like Todde, for I can no more leave than he can, all because I thought I heard a voice calling to me. All because a flower fell into a stream. But I did hear her. I know I did. What does she want of me? I ran away before. I was afraid to let go, afraid I'd never be able to return. And I'm sorry, so sorry. But now, after Eva, I need more than ever to find the answer. You can tell, can't you? . . . You can see?' The words hovered in the air between us, commanding and pleading.

I gnawed my lip, hesitating. 'Not the fire this time. It's not the way for you. I shouldn't have tried to make . . . You're afraid of it. And you'll not find your way to Brigid through the flames, not alone, because fear'll pull you back. But you've a friend who's travelling the lych-way now and that opens another path for you. And there's one who journeys between life and death, between the place of the living and the place of the spirits. He walks in the shadow. But if you want his answer, you must walk with him.'

Chapter 33

Sorrel

Morwen did not explain what she would do – perhaps there weren't names for such things. I watched her groping along a dark, rocky ledge. Her hands closed around a lump on the craggy shelf, as though she was gathering it out of the rock and pressing it in her fingers, like a child moulds a snowball from a drift of snow. As she lifted it, I saw that it was a roughly hewn bowl stuffed with fat-soaked dry moss, fashioned from the same grey stone as the rock on which it had stood.

She set it down near the cloth-covered ledge at the far end of the cave that I'd noticed when she'd first brought me into Fire Tor. She pulled a burning stick from the fire and lowered it towards the bowl. Tiny scarlet sparks flashed across the moss, like a flock of coloured birds taking flight from the moors, then a flame shot upward growing in strength and height until it was the silky yellow of a buttercup.

Morwen turned, holding out her hand to me. I stumbled to her side across the uneven floor, sick with apprehension. She pulled me down so I was kneeling beside her. Then, grasping the edge of the white woollen cloth, she tugged it aside. I had to stifle a cry. The light from the bowl of burning moss flickered across the corpse of a man stretched out on the slab, but it could not penetrate the black hole in his throat. In those first few moments, I barely noticed his nakedness, or the blondness of his hair, or the withered lips drawn back over yellowed teeth. It was the wound that held my gaze. It seemed to yawn wider and wider as I stared at it, a gaping mouth stretching to devour me.

Had the body been lying there last time Morwen had brought me into the cave? Had I been sitting next to a corpse and not known it? I'd realised that something lay beneath the cloth, but I'd never thought . . .

There was no stench of decay, only of herbs and a bitter-sweet smell lingering about his hair. He was like the dead cat Mam dried in the smoke when I was a bairn and hung from the rafters to keep us all safe from sickness.

'Who is he?' I whispered.

Morwen lightly touched a bracelet of thorns that had been bound about his wrist. The nails on his hand were long and smooth, almost as black and shiny as the pebble that lay on his forehead. 'Ankow – he collects the souls of the dead.'

I knew what Ankow was condemned to do. But I'd never dreamed I'd see his face, not till he came to take me. Mam used to say his head could turn right round, like an owl's, so no soul could hide from him. But she said only the dying ever saw his face when he threw back his hood and reached out his hand to seize them.

Morwen stared down at him, a strange expression in her eyes that was almost pity. 'Brigid chose him for Death's bondsman.'

'She killed him?' My voice sounded tight and high. I could barely squeeze out the words.

Morwen laughed softly. 'Not her. Another killed him. She marked him.'

I must have looked as bewildered as I felt, for she smiled.

'It was two, three summers back. I was late coming home from foraging, near owl-light it was. The sun had already dipped behind the tor. I saw a woman standing near the track, behind a rowan tree. She'd her back to me and was watching the path, so I knew she was waiting for someone, but she was peeping out so cautiously that, at first, I thought she might be afeared someone was following her and that was why she was hiding. So I hid behind a furze bush. Then I saw him leading his horse along the track. The beast was lame, hobbling. The man was hunched against the wind, not paying heed to anything save the track in front of his feet, like some folks do when they've fallen into a mire-mood. He passed the woman without seeing her. Then, as soon as his back was to her, she called out. He must have heard her, for he pulled the horse up and started to turn. That was when she hit him a good hard blow to the head with her stave. The crack was so loud a cloud of starlings flew up from a tree close by.

'He lost his hold on the horse and collapsed on all fours. He was trying to get up again, but he was too dazed to stand. She grabbed his hair then, dragged his head back and cut his throat. He was on the ground, gurgling and grasping his neck. Blood was spurting out through the fingers of his gloves. I ran for Ma, but by the time we got back he was dead and there was no sign of the woman. But the corpse wasn't alone: there was a white cow with a red blaze between its horns shaped like a snake. It was standing over the body, guarding it.

'Ma knew it was Brigid's cow, for there's no cattle with

those markings in these parts. Brigid had chosen the man as Ankow, so we brought him up here and the cow followed and lay in front of the cave for three days and three nights while Ma prepared the body with honey and herbs, so she could smoke it.'

'You told no one about the murder? Did no one come searching for him?'

Morwen's brow creased in puzzlement, as if the idea had never occurred to her. 'Who would we tell? Brigid gave us Ankow to care for. That's Ma's business, that is, like her worts and charms.'

She reached out and grasped my good hand. 'Makes no odds who killed him. He's Ankow now. As long as his corpse stays whole, his spirit can come and go between this world and the deadlands, so he can help the newly dead find their way to the lych-ways across the moors that'll lead them safe to Blessed Isles. But if his body rots and vanishes back into the earth, he can't return here. That's why Ma has to keep it safe.'

Before I realised what she was doing, she'd pressed my hand on to the corpse's chest. It was cold and leathery. I tried not to pull away. I didn't want to see the scorn in Morwen's eyes if I flinched from a dead man. This time I had to follow. There might not be another chance. A corpse could do me no harm.

'He's Ankow,' she repeated. 'Walk with him.'

I tried not to look at the gaping black hole and the bared mustard teeth. 'How can I?'

I saw a flash as the yellow flame in the bowl glinted on a blade that had suddenly appeared in Morwen's hand. I almost sprang to my feet, but I knew, though I had no words to explain it, that she would never hurt me. I let her take my good hand and turn it palm up. Morwen lightly pierced the tip of my forefinger with the point of the knife, but it was done so swiftly I felt nothing, or maybe she could charm away pain. As if I was a doll made of rags, my hand lay limp in hers as she held my finger over the corpse and squeezed so that three drops of my blood fell into its mouth. They ran down one of its teeth outlining it in crimson.

'He must taste the blood that your spirit lives in,' Morwen said. 'Then he will know and remember.'

Afterwards, in my dreams, I would see a blackened tongue slide out between those teeth and lick them, savouring my blood.

Again Morwen pressed my hand to the cold chest. 'Now say the will worth, like I taught you when we spoke to the spirits. Ask Ankow what Brigid wants of you. You must want to hear the answer. But don't ask her, unless you swear to do what she commands.'

'But how do I know if I can until I know what she wants of me?'

'You listened to her voice before. You came 'cause you couldn't find peace until you did. You can go, if you want, leave now and never ask, though you'll never rest easy. But don't ask unless you swear to do her bidding,' Morwen repeated, 'else she'll take your life and your soul.'

I knew she was right. I had come all this way, drawn by a voice that dragged my spirit behind it as if it was in irons. She would chain me to her echo for ever and without mercy, until I stood and faced her. I closed my eyes and spoke the words, trying not to flinch from the cold, dead flesh beneath my fingers. I waited. I felt . . . I thought I felt . . . a throbbing against my palm, as if the dead man's heart was flickering

267

into life, beating faintly then thumping hard, drumming against my fingers, but perhaps it was only the blood pounding through my own body.

My hand is sinking through that cold leathery skin, pulling me down behind it. I am dragged through white bone and scarlet flesh, where blood runs like purple fire through great pulsating veins, pulled through the heart's fleshy walls, thick as granite rocks, into a great dark cavern thronged with people, but they do not look at me. They do not even lift their eyes to look at each other. Each is the only creature walled inside their misery. They walk in single file through the cave, coming in from the darkness behind me, going out into the darkness in front of me; their faces swim at me like pale moons in a black lake. Some are as gaunt as skulls, others battered and bruised. There are ancient ones, wrinkled as the bark of trees, and those so young they have not known even a single day of life, a single hour. On and on they come and vanish, a great army of the dead.

One face looms out of the rest, a face I know and do not know, for her face is like a mask, a carving of a face I once knew. Only the eyes are alive, glittering like white flames.

Eva? Eva!

The twin eyes flicker towards me, the hands stretch out, as if begging me to catch hold of her and pull her from the line. As she lifts her head, I see her neck, see the marks.

Eva?

I stretch towards her, trying to touch her fingers, trying to snatch her back. But the eyes blink as if they know they are seeing something that isn't really there, a shade, a wraith, a ghost that might have been me. The face turns away and she is gone.

A voice cries behind me in the darkness, torn and rent by anguish, till it is but the ragged tatter of a human sound.

Let me go! Release me from this torment.

I half turn, though I am mortally afraid of what I will see. A grey shadow hangs in the darkness, the shape of a man, but constantly swirling and re-forming, as if it is a creature so racked with pain it cannot keep still. Its flesh is despair, and only its eyes burn. The mouth opens, dark and deep, like a wound in a throat.

I beg you, set me free!

I felt two arms tighten around me and realised I was leaning back against Morwen and tears were running down my cheeks. I pulled away and scrubbed at them with my sleeve, trying to control my trembling.

Morwen was watching me, shaking her head in bewilderment. 'You didn't walk. You didn't find her, did you? But you were crying like a babe.'

'Eva, I saw Eva, she looked so lost, her face, I thought . . . I thought she'd be whole again after . . .'

'In the Blessed Isles, she will be, but she's not reached them yet. She still walks the paths of the newly dead. But you knew she was gone. There's summat else,' Morwen said.

'When she turned, I saw her neck, the marks on her throat. She'd not been hunted down by the hounds. She'd been strangled. She was dead long before the dogs got to her. The serf-hunters wouldn't have throttled her. If they'd caught her, they'd want to take her alive to claim the reward.'

Morwen shrugged. I could see the death of a tinner's woman meant no more to her than the death of this corpse in the tor. She was not cruel: she merely accepted that that was the way of things. Extinguishing the bowl of blazing moss, she sprinkled a pinch of the charred fronds on to the man's chest. As I stared at the gaping black hole in his throat, I knew it was him who had pleaded with me.

I grasped Morwen's arm. 'He's in pain, in torment. He begged for release. How can we help him? We must.'

Her eyes softened, turning to the colour of fresh spring grass in the firelight. 'I know,' she said. 'But Ma'll not let him go. Brigid chose him, he must serve until—' She turned her head away from me, as if someone else was talking to her. 'The spirits are gathering, waiting for us. Brigid is calling them. Soon, it will be soon . . . I can feel it.'

'But what do the spirits say?' I demanded, almost shouting at her in frustration. 'Do they know why I was brought here?'

'Do the spirits know, Mazy-wen?'

I jerked around. Ryana was standing just inside the entrance, a look of malice and triumph on her long sheep-face.

'So, our little Mazy-wen has been telling you she can speak to the spirits, has she? She couldn't even charm a wart away. The gift only passes to the eldest and that's me.' Her mouth twisted into a sneer. 'Mazy-wen, Mazy-wen, doesn't have the wits of a squawking hen.'

Morwen stood up, facing her sister, her fists clenched. 'And you haven't even the wits to invent a new rhyme. You made that up when you had milk-teeth, and you still couldn't think it up without Taegan helping you.'

Ryana's grin vanished, and she slowly paced forward, pushing her face close to Morwen's. 'You won't think your-self so clever when I tell Ma you brought a blow-in to Fire Tor, one that's a tinner's bitch 'n' all. You wait, Ma'll thrash you till she breaks every bone in your body and she'll bring down such a curse on your head, you'll be tormented night

270

and day until you drown yourself in the mire just to escape her. You're a traitor, Mazy-wen. You've betrayed Ma and all our kin. The spirits will never forgive you for what you've done. Never!'

Chapter 34

Hospitallers' Priory of St Mary

Meggy was jerked from her doze by the fire as something struck her on the knee and slithered to the floor. Claws scrambled over her head in the thatch and before she was even fully awake, she was already batting at her skirts. Mice again! There were always a few scurrying about, but these past days they'd been swarming in from the moors and fields, driven from their flooded holes and starving. Gnawing at everything, they were, from leather to candles. The cooks were at their wit's end trying to keep them from spoiling every keg and sack in their stores.

The rush lights in Meggy's little hut had burned away and the glow of the embers of her hearth fire in the centre of the floor barely illuminated the stones that encircled it. Meggy winced at the pain of her stiff legs as she hauled herself to her feet to light some more rushes, not that they did much to keep the mice from running over her, cheeky little beggars. She took a pace and almost fell headlong as she stumbled over something lying on the floor. The room was so dark she couldn't make out what it was. She stooped awkwardly to pick it up. It crackled beneath her fingers. She lit a rush candle in the embers of the fire and held the flame close to examine what she held.

'Bless my soul, a brideog.'

She sank back in her chair, cradling the little object. It was a doll made of rushes, which had been soft and green when it was fashioned but were now dry and brittle. It was clad in a crudely made gown, stitched from a scrap of white woollen cloth, now almost black with dust, smoke and cobwebs. A white hag-stone was strung about the doll's neck to prevent thieves and evil spirits from entering the home. The head had been gnawed. The mouse must have knocked it off the beam above her. She'd forgotten it was even up there. Dimly now, she recalled hiding the doll somewhere in the shadows of the thatch when she'd first moved into the gatekeeper's hut, along with a crumbling sprig of rosemary from her wedding day, an eel skin to ward off stiff joints, and even a torn scrap inscribed with words she couldn't read, but words once written could charm or curse and, whatever they were, it was bad luck to destroy them.

She knew the prioress didn't hold with such things, but she'd always kept a brideog in her cottage, as her ma and granddam had done before her. Brigid blessed the hearth and kept the family safe, the goats in milk and the chickens in lay.

As she fingered the little doll, a smile softened her lips, though she didn't know it. It had been on Bryde's Day when her husband, Arthur, had asked her to be his wife. She'd been no more than fourteen or fifteen summers then. All the unwed girls had gathered, as they did every Brigid's Eve, to sit with the new brideog doll they'd made until the fire died down. Then they'd raked out the ashes and laid their shifts in front of it. The next morning, they'd raced to see whose shift Brigid had walked over, leaving her ashy footprints on the bleached linen. Meggy had been delighted when she'd found her shift marked. She had been wed within the year, just as Brigid had foretold. It wasn't till long after her son was born that Arthur confessed he might have given old Brigid a helping hand: he'd crept into the livier that night and smeared ash on Meggy's shift so she would be persuaded to accept him.

Meggy chuckled to herself. Arthur had been well named for he truly was a bear of a man but soft as butter. She still missed him and she missed her lad. If only she knew what had become of her son. How could you mourn for him, how could you let him go, if you didn't know whether he was alive or dead? Hope was a cruel tormentor.

The old gatekeeper stared up at the low rafters, from which dangled sacks, lengths of cord, rusty cooking pots, a pair of deer antlers, a couple of broken snares. It was all that she'd managed to salvage from the house she'd once shared with Arthur. His blacksmith's tools and their few sticks of furniture had been sold long ago. She would have sold these bits, too, except they were so worn out no one would buy them. She scarcely knew why she'd kept them. Always banging her head on them, she was. But she couldn't bring herself to toss them on the midden.

Meggy had never been one to throw anything away, however worn or broken, for she knew only too well that tomorrow it might be all you had to call your own. It wasn't much to show for a lifetime of toil, but she'd a roof to shelter her, a fire to comfort her aching bones and food to put in her belly. At this moment, that was worth a king's ransom to her. There were many who didn't have a mite of what she had, the poor villager who'd tottered past her gate earlier that day for one.

For the hundredth time, Meggy cursed herself. She should have made the woman come in, dragged her in. Maybe in the morning after a warm night's sleep and a full belly, she might have seen another way. But Meggy knew better than most the cruelty of having the pride ripped from you when that was all you had left.

Dusk had already been circling the priory when Meggy had slipped outside to light the torches on the priory wall. Heavy amber clouds were pressing down upon the hills, covering the tops of the tors, and the wet wind that howled up the hill sliced through her, like a butcher's knife. She'd no mind to linger out there and was hurrying back inside when she'd caught sight of a hunched figure limping up the hill. Meggy's eyes were not what they'd been in her youth and she squinted into the fading light, trying to decide if it might be a pilgrim seeking shelter or, more likely, from the painfully slow and unsteady progress, a beggar. Best wait and see before she went to the trouble of bolting the gate, for it was getting harder and harder to ram that rain-swollen beam back into place, or maybe her old arms were weakening – she angrily dismissed that thought.

Meggy had hovered just inside the open gate watching the figure stagger closer. The pitiful creature looked so frail that Meggy feared the next gust might bowl her straight back to the bottom of the valley. As she drew level with the gate, Meggy realised she was probably no older than she herself was, maybe younger, but the woman had no more flesh on her bones than that of a corpse left a year in the gibbet. The only sign of life in her was in her eyes, sunk deep into the hollows of her face.

Meggy stretched out a welcoming hand. 'Here, come you

275

in and rest. Sisters'll find you a bite of food and a place at the fire.'

The woman had stopped, her hand clutching her heaving chest. She stared up at the walls of the priory, as if she'd thought it a trick of pigseys to deceive her, a castle of mist and smoke. Meggy had repeated her offer of food and warmth, but the woman had flapped her hand, as if shooing Meggy's words away.

'I'll not trouble you,' she wheezed. 'Never begged in my life. Least I can say that much . . . Reckon I'd vomit anything I tried to eat now, anyhow. Waste it'd be. You got bread to spare, you give it to the chillern. It's the young 'uns who need food now.'

She stared out at the desolation of grey, rain-soaked moors. 'When we were girls dancing so carefree, we never thought it would end like this. There's a curse come upon Dertemora, a terrible curse, and I'm not sorry to be leaving it now.'

'Leaving? Where're you bound?' Meggy had asked. The woman didn't look strong enough to reach the next valley, much less find her way off the moor.

But she didn't reply. Leaning heavily on her stave, she had turned and hobbled away. For a while Meggy watched her, puzzled and uneasy. The distant figure paused and raised her head, staring up at Fire Tor, invisible and cloistered behind a wall of cloud. Then the stave fell from her hand and she stretched out her arms towards the tor, like a child begging to be lifted up. Suddenly Meggy understood what she intended. *It's the young 'uns who need food now.*

Grief and anger blazed up in the old gatekeeper. Her own lifetime, all her memories, all her past happiness seemed to be hobbling away from her in that brave, frail figure. And it wasn't right – it wasn't fair that any soul should come to such a bitter and lonely end. Yet fear also laid its cold hand on Meggy's back, for she knew that if it wasn't for the priory she'd be taking that same walk and there'd be no one left even to remember her name. But as she watched, the cloud rolled down the hillside, like a great wave. The woman was gathered up into the mist and vanished.

A great bone weariness sank over Meggy. Her limbs felt suddenly twice as heavy as they ever had before. *When we were girls dancing so carefree, we never thought it would end like this.* No, she should have been sitting side by side on a bench with her neighbours in the summer sun, stretching out steadying hands as grandchildren pulled on their skirts, dragging themselves up on to their chubby legs while the parents toiled in the fields. That was how it should have been. That was how it had been for their mothers and grandmothers for generations before that. Why had it all vanished into the mist? Whole lifetimes washed away in the rain. *No, we never thought, did we, never dreamed, it would end like this.*

Tears gathered in Meggy's eyes. How many months was it since she'd last set foot in the village? She didn't belong there any more. Friends she'd known all her life stared at her whenever she returned, glancing sidewise at her black cloak as if she was collecting taxes or bringing bad news. They didn't meet her eyes when she spoke to them, and their little children stared unsmiling at her, hiding behind their ma's skirts. What reason had she to go back? Yet if the priory was closed by those knights or destroyed by that boy, where could she go now? Would she, too, end her days out there starving and alone on the tor?

Anger and fear grow unseen in the darkness like the wind. At first you only notice the shutters rattling or a branch creaking, but as the night wears on it gathers strength till it has the power to rip the thatch from a roof or burst a door wide open. So it was with Meggy for the dregs of that night. She lay in her narrow bed, tossing this way and that, till anger had grown to rage and fear to terror. Even when she finally slept she was assaulted by dreams of the brideog, which had grown a long snout with sharp teeth and scurried about the blacksmith's cottage, gobbling every bite of food in the place, before scuttling up the wall to devour the side of pork hanging from the rafters. But Meggy suddenly realised it wasn't a joint of meat at all, but her own little son dangling there, and woke with a shriek of horror.

She lay sweating and trembling. Was the brideog falling on her a sign of anger from the old goddess? Was she being punished for neglecting her, turning her back on the old ways?

For the first time in many months, the fire in her hearth had died and no amount of fresh kindling, poking and blowing revived the flames. It was another bad omen. A body always feels more chilled after a restless night, especially one that has survived as many winters as old Meggy. So, grumbling to herself, she slipped out of the door of her hut, fire pot in hand, to fetch some burning embers from the kitchen.

A few servants were straggling in and out of doors bearing fresh peats and ewers of water. One of the scullions, Dye, was bent sideways trying to lug a heavy pail of piss and night soil across the cobbles. She had already spilled some over the hem of her skirt and on the rags that bound her feet and more of the foul liquid was sloshing out as she staggered forward. The girl was going to be soaked long before she reached the midden. The scullion gave a furtive glance around her, then tipped the contents into the shallow gully that ran down the middle of the courtyard. She jumped guiltily as she caught sight of Meggy watching her, but though the gatekeeper wagged a finger at her, Dye knew she wouldn't tell, and flashed her a cheeky grin as she scurried back inside.

It was only as Meggy was halfway across the yard that she saw the boy standing in the corner, deep in the shadow of the wall. Had he moved even a hand Meggy might have noticed him sooner, but he was as still as a grotesque on the church wall, with his cheek and ear pressed against the granite stones, as if he was trying to hear what they were whispering. Like Dye, Meggy glanced guiltily about her. All of the servants were about their work inside and the courtyard was deserted now, though she could hear Brengy, the stable lad, whistling as he tended the horses.

Meggy stepped closer to the blind child. 'What you doing out here, lad?'

When he gave no sign that he knew she was there, she made to touch his shoulder, but even as she stretched out her hand towards him, fury bubbled in her belly and clawed up into her throat. Why did he live, when somewhere out on the rain-drenched moor a good woman lay cold and dead?

He'll destroy you! He'll destroy you all.

He had already murdered the saintly old priest, and now he had destroyed the well. Cursed it with frogs, flies and blood, then stopped it flowing at all. Ice slid down Meggy's stiff old spine. She saw the boy again standing in the court-yard in the dead of night, his hands stretched out, summoning the spirit to her gate; she heard the unearthly keening rever-berating through the black empty sky; she saw his face turning towards the well as the water ceased to flow. Brigid's spring had kept the villagers safe and healed them since the first dawn had broken over the tors. That water had given Meggy a son. And this demon had destroyed it and he would destroy the priory too, for without the well the pilgrims wouldn't come.

Bring me something of the boy's, something I can use to ask the spirits.

Only Kendra could stop him and she must. Meggy had tarried far too long. The falling brideog was a sign. Brigid had sent it, and now Brigid had led the boy to Meggy. She must do it before it was too late. She might never get a second chance.

Meggy reached for the knife dangling from the sheath at her waist, not to stab the boy, no – as much as she loathed him, fear of what he could do to her would not let her hurt him – but simply to cut a lock of that curly hair or a snippet of cloth.

Take care how you get it. He mustn't know, else he'll put you in the grave alongside Father Guthlac. Gently now, quietly, before he senses anything amiss. He doesn't know you're there.

The old woman's grimed fingers trembled as they reached towards the child. What should she take?

Hair, take hair. It weakens a sorcerer's power. But will he feel the strength leave him?

The boy lifted his head from the stones in the wall, as if they'd whispered to him, warned him of danger. His arm banged against Meggy's hand. His eyelids jerked wide at the touch and, with an animal mew of fear, he started forward, his shoulder striking the rough stone wall as he staggered too close to it. The blow knocked him off balance and he toppled over, banging his knee against the shards of stones that were hammered sidewise into the courtyard floor.

The boy twisted himself into a sitting position, bending

over his injured knee, which was drawn up under his chin. His arms covered his head and face. He whimpered as if he expected blows to fall on him. The sight of a child cringing away from her twisted Meggy's guts. In that moment, as he was on the ground, he suddenly looked too helpless, too vulnerable to be feared. Maybe there was something about that curly head or that wordless cry that reminded her of another boy, before he was marched away and vanished into the first icy mist of autumn with a hundred other men and boys who never returned.

The courtyard was still empty and the tiny whimpers of the boy were no louder than a nestling's cheeping. No one else seemed to have heard him. Meggy hurried back inside her hut and found some clean rags and a jar of unguent Sister Basilia had given her last winter to soothe chilblains and chapped hands. She dipped the rags in a pail of cold water and returned to the boy. He hadn't moved. He just hunched, shivering, as if afraid to stir for fear he might fall off the world. Crouching, she wiped the blood from the bruised and grazed knee, then gently massaged the ointment over the torn skin. Her fingers were rough, the skin jagged. He winced but did not pull away, as if he understood she meant no hurt to him. Finally, she bound his knee with the dry rags.

She was helping him to his feet when a servant ambled out of the kitchens. 'Whatever has he been up to now?' She hurried across, holding her skirts clear of the mud. 'I only left him for the blink of a hen's eye. Set him against the wall where he'd be out of harm's way.'

'Took a tumble,' Meggy said. 'Grazed his knee is all.'

The servant shook her head, her expression flitting between annoyance and anxiety. 'Boys – always charging about! I tell you a herd of rampaging goats is easier to mind than any lad. And as for that one, he's . . .' She shuddered. 'Keep the butter from coming and make the cows run dry, he would.'

She seized the boy's sleeve, seeming unwilling to touch his flesh, but as she gave him a tug towards the infirmary, he pulled away from her and fumbled for Meggy's hand. He clutched it in his icy fingers, lifted it to his mouth and pressed a soft kiss in her horny old palm, before allowing himself to be hauled away.

Meggy stared down at her hand, still feeling the warmth and softness of the child's lips in it, as if she was holding a sun-warmed rose petal. Her throat felt uncommonly tight.

Something caught her eye, flopping back and forth in the breeze. It was stuck in the mud between the stones of the courtyard. She picked it up. A damp rag, the one she'd used to wipe the boy's knee. It was stained scarlet with his blood.

Bring me something of the boy's.

He had given it and didn't realise it had been taken. Just minutes before, Meggy would have been elated. But now . . . She stared down at her palm again. Maybe she should just rinse it clean. No need to go bothering old Kendra. Then somewhere far off, carried towards her by the breeze, she heard the tolling of the village bell, deep and melancholy, a death knell. Had they found that poor woman, or was it another soul taken from this world? Stuffing the bloody rag into the pouch at her belt, Meggy hurried towards the gate.

Chapter 35

Sorrel

I am standing on top of a towering tor. The great boulders beneath my bare feet are drenched in the crimson and purple light of a huge blood-red sun. Across a ravine there is another stack of rocks, even higher, and flat as a vast table. The sun is drifting down behind those stones, and they shimmer gold in its dying rays. Far, far below, the moor stretches in every direction, but it is already in darkness. Tiny fires, glittering like shoals of little fish in a black pool, mark the cottages and those who dwell in them. As I lift my gaze, I see that I am no longer alone. A little boy I do not recognise stands on the rock opposite me. He does not move. He does not look at me. I think he cannot see me. Maybe I am a wraith, a ghost.

As I watch, dark clouds are gathering in the sky, but as they roll towards the tor, I see that they are not clouds but birds, hundreds of them, thousands. All carry twigs and branches in their beaks. They drop them at the feet of the little boy. More fly towards us, black against the scarlet sun. The pile around the boy grows wider, higher, till he is standing on top of a great mound of wood. It glows gold in the dying light.

But it isn't glowing. It is burning. Tiny flames are shooting up through the pyre, like seedlings. They are taking hold, clawing upwards. The boy still stands motionless in the centre. The flames are writhing towards him. I try to cry out, to warn him, but no sound comes. Dark shapes leap from the wood, like mice from a burning hayrick, foul creatures, their bodies long and sinuous as weasels, snouts full of sharp teeth, eyes black as spiders. They are streaming over the rocks and running down the hillside, a great tide of them, pouring down on to the people below huddled around their fires. I hear the screams. I see the fires going out one after another as the wave of beasts spreads out from the tor and still the boy does not move.

I have to reach him. I have to stop this. The moor is writhing, shrieking in agony as the creatures swarm over it.

Then I hear it, hear her voice. Brigid's voice.

Protect the child.

I cannot see her. I cannot see. The sun has vanished. The flames die. All is darkness now. But the beasts are still swarming down the hillside. I hear their claws slithering over the rocks, smell the reek of death on their breath.

Protect the boy.

Who is he? Tell me! Tell me how to find him!

But she does not answer.

She never answers.

Chapter 36

Prioress Johanne

An explosion of screeches, yells and crashing metal jerked me from my morning prayers. I flung open the door of my chamber, then leaped back, covering my head with my arms as a dozen wings and claws flew straight at my face. For a moment, I could see nothing except a shrieking black snow-storm of flapping feathers and furious eyes. A great vortex of ravens, rooks and crows were swooping and rising in the courtyard, as if they'd been trapped there by some invisible net and were beating their wings trying desperately to escape.

The servants, brandishing brooms, ladles and frying pans, waded among the birds, trying to drive them off, but that only added to the creatures' panic and fury, and for a few moments, it seemed that they might succeed in driving the humans back into the buildings. Then above their cries came the clang of the chapel bell. The birds rose again in alarm, but as the bell swung back and forth, its great echoing gong proved too much even for them. They rose into the grey-pink dawn and flapped towards a distant tor, in dense dark cloud.

I leaned against the doorpost and muttered a prayer of thanks, trying to steady my breathing, but as I opened my eyes I noticed one of my sisters standing motionless in the courtyard, her arms stretched out in a cruciform, her head bowed. Black feathers, still drifting down in the breeze, brushed her face and caught on her veil and in the folds of her black kirtle. Around her lay the bodies of maybe half a dozen birds, some dead, others still twitching feebly, smashed out of the air by the servants' flailing brooms and pans.

Taking care not to slip on the fresh bird dung that now covered the courtyard, I hastened towards her. 'Sister Fina! Whatever . . . Are you hurt?'

Of course she was, I could see that. Scarlet blood trickled down her white face from several deep and livid scratches, and there were more on her long bony fingers, yet it was not those wounds that concerned me for I could feel the burning sting on my own neck where a bird's claw had raked me. No, it was that Fina hadn't moved. She still held her arms stretched out, like stiff wings. Her eyes were closed and she was so still that had she not been standing I would have assumed she had fainted.

'Sister Fina, look at me!'

The young woman slowly raised her head. Her eyelids, almost blue and transparent, opened suddenly, like shutters on a casement being flung ajar, but the tawny eyes beneath were as sightless as the painted eyes on a wooden saint, the pupils so small that they were mere dots. I caught hold of one of her wrists and pulled the arm down to its natural position at her side, giving its owner a little shake. Fina's other arm flopped down with the first, as if the two were attached by strings.

'The place for contemplation, Sister Fina, is in the chapel, not in the middle of the courtyard. We are not nuns who make a show of public penance and humility.'

Sibyl, one of our cooks, clutching a heavy iron frying pan splattered with feathers and blood, shuffled towards us. She was scarlet in the face and sweating after her efforts, and the twist of cloth about her head was pulled rakishly askew, almost covering one eye. She glanced up at Fina, who remained statue-still.

'I reckon that tumble she had a while back must have scrambled her wits. Why didn't she get herself indoors when the birds started swooping?'

'Perhaps it happened so suddenly she was too frightened to move,' I said. It was the only explanation I could think of, but even that made no sense. Fear would surely have made someone cover their head with their arms, or cower in a corner, not stand there with their body and face exposed, like a holy martyr welcoming death. 'What on earth caused the birds to mob the yard like that?' I glanced anxiously up at three carrion crows, which had once more alighted on the roof.

Sibyl attempted to stuff the greasy grey locks of her hair back beneath her spattered head-cloth. 'That nuggin, Dye, caused it, that's who. You wait till I get my hands on her. I told her to take the scraps from last night's supper to Sister Melisene for the poor basket. I said, "There'll be beggars pushing and shoving round that alms window before alms bell has finished tolling, and the sooner we're rid of them, the safer we shall all be." So, I told Dye, I said, "You take the meats to Sister Melisene, so they'll be ready to dole out straight way, afore those vagabonds get themselves settled in front of our gate."

'But that girl's not got the brains of a dried bean. She goes ambling out of the kitchen, mindless as a headless herring, without even thinking to cover the tray. A bird swoops down from the roof to snatch at the bread, gives her a fright and she drops the whole mess of it, scatters bread and meats all over the yard. And before you can say "pickled pork", there's a whole flock of those vicious creatures swooping down. Dye came tearing in as if the wisht hounds was after her, as did everyone else in the yard. We all grabbed up something to drive them off, but her' – Sibyl jerked her head towards Fina – 'she just stood there, like she was St Cuthbert talking to the beasts. It's a wonder they didn't peck her eyes out, mood they were in. I've seen them do it to lambs as can't protect themselves.'

As if her words had indeed become flesh, her gaze fastened on a corner of the courtyard behind Fina and she gave a little cry. She scuttled over and returned to my side, the limp and bloody bodies of two chickens dangling by their feet from her fist. 'Look! See what those flying imps have done, murdered two of my finest laying hens.'

'Murdered.' Fina repeated the word clearly and slowly, as if it was one she had never heard before. 'He'll do it again. I have to stop him.'

We stared at her. Her eyes were alive now, bright, almost luminous in the leaden morning light.

Sibyl took a pace backwards, holding the iron frying pan like a shield across her chest. 'Blessed Virgin and all the saints . . . is she saying someone's been murdered?'

I felt a chill hand clutching at my heart, but I shook my head emphatically. 'Of course not! She is simply upset by the birds.'

I glanced back at Fina's pale face, her strange eyes. Could Sister Basilia have given her more dwale? But there was no reason to do so.

'The boy murdered Father Guthlac,' Fina gabbled. 'Sibyl says, he's calling up his flying imps to destroy us. You saw them. I had to pray against them. I had to drive them away.'

I glared at both of them in exasperation. 'Goodwife Sibyl knows they were just birds, not imps. She merely meant they were causing mischief, didn't you?' I said firmly, and ploughed on before she could argue. 'And, for the last time, Father Guthlac was not murdered by anyone! Cosmas never left the pilgrims' hall that night. You know as well as I do that he is blind. He can't find his way out of any chamber unless someone leads him. The child was asleep. How could he possibly have harmed anyone?'

In truth, I was not at all sure Cosmas had been asleep for, now that I thought about it, I'd never seen him sleeping, no matter what hour of the day or night I had visited the infirmary, but that was hardly a sign of evil.

Sibyl frowned. 'There's folk do say that when we dream our spirits leave our bodies and they travel to distant places—'

'But they don't kill anyone there,' I snapped.

Sibyl eyed me doubtfully. 'My uncle went to his bed hale and hearty, and he was found dead as boiled beef come morning, face all purple and swollen like he'd been strangled. Only a mouse could have got into his livier. And they reckon that's how the spirit came in, as a mouse, then murdered him. It was his brother that did it. Two of them had been butting horns like rutting stags for years and his brother wasn't the least surprised when he was told the news. Bold as a magpie, he said he knew already because he'd dreamed his brother was dead that very same night. That's as good as admitting it was his spirit who'd killed him.'

'But Cosmas has not killed anyone.'

'And you have proof of that, have you, *Sister* Johanne?'

Brother Nicholas peeled himself away from the cart he had been leaning against and covered the short distance between us in a few strides. I cursed silently. Just how long had he been standing there and what part of Fina's wild accusations had he heard?

My brother knight did not wait for me to answer, as if he had already dismissed anything I might say. Instead he took a pace closer to Sister Fina, almost pushing his bulk between us and deliberately turning his back to me. 'Was the boy out here in the courtyard, Sister Fina? Did you see him summoning those birds?'

Fina was staring at him stricken, obviously remembering too late that I had warned her not to speak of the boy or of sorcery in Nicholas's hearing.

'Brother Nicholas,' I said coldly, before she could reply, 'as I have already reminded Sister Fina, Cosmas cannot leave the infirmary unaided, and if one of the servants had brought him out here, they would hardly stand around idling to watch the child cast spells. The birds flocked here because, as Goodwife Sibyl told us, a scullion was clumsy enough to drop a tray of meats and scatter them all over the yard. There is no more mystery to the birds swooping down than—'

'Prioress!' Sister Basilia was standing in the doorway of the infirmary beckoning frantically. 'You'd best come. It's Sebastian. The commotion in the courtyard frightened him and nothing'll calm him. He's begging for you.'

I hesitated, torn between Fina and Sebastian. It was dangerous to leave my sister alone with Nicholas. He would not rest until he had bullied and cajoled her into giving him the answers he wanted, no matter if they were true or false.

But as Basilia called again, Nicholas turned towards her, his curiosity evidently aroused. I could not have him questioning

Sebastian. I walked swiftly towards the infirmary, hurling orders over my shoulder, which I hoped would give Fina and Sibyl ample excuse to scurry away to the noisy, bustling kitchen, where Nicholas would find it impossible to trap and question them.

The rush screen was pulled beside Sebastian's bed, but I could hear his cries before I reached him. Not the shrieks and screams that at times could carry right across the courtyard, but the quiet sobs of misery and despair that were far worse, for they drove knives through my soul.

He was lying curled on his side, trying in vain to hold his hands over his head, but lifting his arms was agony for him and his limbs trembled with effort. That he even tried was testimony to a pain that was worse than any his flesh endured. I could not even hold him without causing him hurt. So I stroked his soft white hair, now matted with sweat, as he thrashed to escape the noise from the courtyard.

'Let them die!' he was sobbing. 'Sweet Jesu, spare them this agony, let them die now! Let it be over. Make them stop! Stop!'

'Hush now. You're safe in the priory. Nothing can harm you. It was only the shrieks of birds you heard through the casement, fighting for crumbs in the courtyard.'

'Swords! Can't you hear them? It's begun. You must come away now. Quickly, quickly, before it's too late! The tunnel . . . we must reach the tunnel. Run!'

'Sebastian, try to listen. All is quiet. The banging and shrieking is over.'

But the courtyard was never really quiet except at night. Always someone was dragging a rake across the stones, clanking a chain against a pail, or banging burned fragments from an iron pot. I had grown used to it, but for a man like

Sebastian, lying helpless in his bed, tormented by his dreams, every clang, scrape and squeak must have felt as if a blacksmith was driving nails through his skull. How could he understand what the most innocent of sounds outside the wall of this infirmary signified when in his nightmares even a pail of water might be an instrument of pain?

'No one is attacking us, Sebastian. It was only Sibyl and the servants banging pans to drive off the crows. No one has come with swords into this priory. You are safe here.'

I prayed with all my soul that he was. But I had lied when I said none had brought a sword within our walls. For one man had, and he could summon a whole army against us. And there was no tunnel in this priory. No hiding place for us.

Sebastian's shoulders would no longer take the strain of holding his hands above his head. His arms collapsed in despair as if he had given up trying to defend himself.

Still murmuring promises of safety to him that I did not know I could keep, I smoothed the sheet on which he lay, gently lifting swollen joints and settling them again on folds of sheepskin to take the strain from them. I knew better than any of the other sisters at which angle each wasted limb should lie to give him more ease, for I had been tending him the longest.

But as I cupped my hand under his wrist to move his arm, his fingers suddenly seized mine, pulling me close to him. His eyes, sunk deep into their dark sockets, searched my face.

'Please, I beg you do what I have asked.' His breath was sour and cold, like an old man's, though he wasn't old.

'Lie still, Sebastian, try to sleep. I'll remind the servants to be careful and quiet. I know noise distresses you.'

'No!' He spat the word at me, tugging on my hand with all the strength he could muster. But his grip was so feeble and clumsy that a small child could easily have drawn their hand from his, but I did not.

'You must listen to me . . .' he croaked. 'I can't do it . . . You . . . It must be you.' But fear and exhaustion had already taken their toll. I saw that he had sunk into sleep.

Chapter 37

Dertemora

Meggy stepped out of the narrow wicket gate and dragged it shut behind her. The moment she left the safety of the courtyard walls, the wind leaped on her back, snarling and roaring, trying to claw her cloak from her. It took several moments of wrestling with it before she managed to untwist it and free her arms. The torches on either side of the gate were guttering so wildly that she had to feel for the hole where she could push the long iron key through the wood, and wriggle the prongs to slide the bolt into place. She'd persuaded Dye to listen for the bell. The girl owed her a favour and she knew it.

Meggy turned to face the wind, and though she kept her head down, her eyes watered and her nose dripped from the smart of its bite. But she'd been born and raised in these parts and to her, like all villagers, the wind was merely the great moor breathing. If it ever stopped blowing, they'd have thought the moor had died. But even Meggy was forced to admit it was panting hard and angry tonight. There was malice prowling out in the darkness.

Her fingers strayed to the rag stuffed into the pouch

hanging from her belt. It was stiff now with the boy's dried blood. Should she take it to Kendra? She hesitated again, half turning back. She'd watched them bring that woman's body down from the tor, slung over the back of a shaggy little packhorse, like a sack of grain, but the poor soul had been lighter than grain, just a sack of dried chicken bones, good for nothing, save dumping in the ground. Meggy had stood in the gateway, watching a man and a little boy leading the horse down the narrow sheep track, listless, stumbling, their heads bowed, but not in grief. What little energy they had left could not be wasted on useless feelings.

You got bread to spare, you give it to the chillern. It's the young 'uns who need food now.

Meggy had run inside and grabbed the first thing she could find in the kitchen, then ignoring Sibyl's outraged bellows, hurried out, calling to the man to wait. The boy's eyes had popped at the sight of the dead chicken, its feathers matted and crusted with rusty blood where the birds had mobbed it. He snatched at it with such a wild, ravenous expression, that Meggy was afraid he'd sink his teeth into the raw flesh and devour it like a fox. But his father pulled the bird from his son's hand and thrust it into a sack next to the sodden corpse. He said not a word to Meggy, as if she had struck him a final blow, the last of so many that he could no longer feel the pain. But his face turned briefly towards her, angry, bitter, ashamed.

If the priory closed, she would be driven out on to the moors, for she had nowhere else to call home, no one who would take her in. And when she died out there from cold and hunger, there wasn't a single person from the village who'd waste their time searching for her corpse. No one would bring her back to lie in the earth beside her husband, her mother and grandmother. The villagers would leave the ravens to pick her clean, like a dead sheep. She had only the sisters at the priory now. They were all that stood between her and death.

Worse than death, for what happened to those souls who had no priest to give them a Christian burial, no family or friends to set salt on the corpse to protect it from the torment of evil spirits? There was no rest for such souls, no Heaven, no Blessed Isles. They wandered for ever on the desolate lych-ways, haunting the mires and lonely tracks, trying to find their way back to the living, back to homes long vanished. She'd heard them scrabbling at doors in the night. She'd seen their white shades flitting over the marshes, no longer human, twisted into the foul creatures of darkness, neither living nor dead.

She'd seen her future, felt it when the boy had turned his face to Bryde's Well and destroyed it, destroyed the life blood of her mothers and grandmothers, which had poured out for them ever since the old goddess had first struck the rock and called the spring forth. Meggy had seen what that demon could do. She had stared out into the black desolation, listened to that dark silence. She knew what was coming for her.

It must be done. He must be defeated before it was too late. His blood had been delivered into Meggy's hand. Brigid herself had put it there.

Meggy dug her staff into the muddy track and turned her face towards old Kendra's cottage. Behind her, the burning torches beside the gate fought and lost their battle against the darkness that rolled over the priory, enveloping it like a sorcerer's cloak.

Meggy paused on the track, digging her stave into the soft ground to steady herself against the violence of the wind.

She strained to listen, peering through the dark. Now that her eyes were accustomed to it, she could just make out the outlines of the furze bushes and thorn trees closest to her as they shuddered in the wind. It didn't bend them, like it did the birches – they were too stiff and low to the ground – but tonight it was giving them a fair old thrashing.

She heard the sound again. It was more than wind in the grass and it wasn't the whistling of the ghosts up on the tor. It sounded like the barking of a hound, maybe more than one. Meggy took a firmer grip on her stave. The wind was tossing the sound back and forth across the hills, like a juggler, first, it seemed, on the right of her, then behind, next in front. She peered down the valley. She could see shapes moving. Were they just bushes blown by the wind, or were creatures running out there, quartering the ground? Her eyes were watering, blurring everything.

The sound came again, a deep baying of hounds on the scent of their quarry, but from behind her now. She jerked round, her heart banging in her chest. How close was she to Kendra's home? She'd thought she could walk this track blindfold, but she'd lost all sense of how far she'd come. She didn't know whether to turn back for the safety of the priory or make for the cottage. She took a step down the track, but at once changed her mind and started back the way she'd come.

The howling was closer now, she was sure of it. She broke into a limping run, but she was struggling up the slope and the path was as slippery as raw egg. She knew she couldn't keep it up for more than a few paces.

She slowed. Her breath tore at her throat. It was only the farmers' dogs howling their protest about being chained up, she told herself firmly. No reason for her to go tearing across the moor and risk breaking a leg. She should know better at her age. But still she hesitated. Perhaps, after all, it was wisest to return to the priory. It was a foul night and an old besom like her might easily slip in the mud or on a loose stone, and then where would she be? Lying out there in the cold and wind, that's where, and it might be hours before any soul found her, for that feckless scullion would saunter off to her bed without thinking to tell anyone that the gatekeeper had not returned.

The wind shrieked about her, driving thick clouds across the sky, like ships in a storm. Meggy dug her stave into the hillside and pulled herself up the track. Her legs were as heavy and stiff as fallen tree trunks now. A black shape moved on the hummock ahead of her. A tree? There wasn't one, not there, unless she was being pigsey-led. They did that, the pigseys, imitating bushes and rocks to fool you into thinking a track was familiar so you'd follow it and become so hopelessly lost you'd wander in circles till you died of exhaustion.

Meggy wiped her streaming eyes and squinted, trying to see if it was shadow or . . . She spun round as the baying sounded again, closer, much closer. A cloud peeled back from the moon. A thin shaft of snow-cold light fell on the earth and the shadow on the hummock reared up. The black horse and black-clad rider seemed to be staring straight at her. She couldn't see the man's face – she wasn't even sure he had a face. There was nothing beneath the hood of his cloak, save the shadow of the grave.

'Ankow,' Meggy whispered, but even as she breathed the name, the moon was swallowed again by the cloud and the figure melted back into the night. The pulsing notes of a hunting horn throbbed on the wind and at once the baying of the hounds excitedly took up the cry. Now she could see them, black streaks streaming towards her, bounding over rocks and crashing through bushes. She turned and ran down the hill, her throat afire, her breath ripping at her side. But even terror could not lend her speed enough. Four huge paws crashed against her back as the lead hound leaped towards her. The old woman was sent sprawling on the ground, rocks grazing her face and hands, though she scarcely recognised the smart of it for the weight of the creature on her back, its scalding breath against her neck, the globs of spittle falling from its jaws and running down her face. In an instant, the others were upon her, their claws raking her skin as they scrambled over her arms and legs, barking and yelping their excitement.

She tried to drag her arms towards her to cover her face but the hounds trampling back and forth over her kept her pinned down in the stones and mud. She was rigid with fear, expecting any moment to feel their huge teeth tearing at her flesh.

Holy Mother Mary and sweet Brigid, help me!

Dimly she knew that she should be making confession of her many sins in these last moments, but all she could pray was that she would not know the agony of being eaten alive. They were tearing at her side, scrambling against her ribs, snarling at each other as they tried to push their great heads beneath her. She tried to press herself hard into the ground to protect her belly.

The belt around her waist tightened as one of the hounds seized the pouch that dangled from it and shuffled backwards with it clamped between its jaws. Another sank his teeth into it and they pulled against the belt, dragging Meggy sideways over the sharp stones until she thought she would be cut in two. She screamed against the pain, but as a third beast tried to seize the little bag, it snapped from the belt and the three huge dogs tumbled over and over, yelping and growling in a tangle of teeth and fur.

An ear-splitting, high-pitched whistle stabbed through the barking of the hounds. The dogs stiffened, shaking their heads, but after a moment they returned to sniffing around her and trampling over her helpless body. The whistle came again, even more piercing this time, slicing through the darkness, like an arrow. The hounds lifted their heads and whimpered, and with a final kick against Meggy's thighs, they ran off.

Everything inside Meggy was urging her to scramble to her feet and flee before they returned, but she couldn't. She lay there, her arms and legs sprawled where the hounds had left them, as if their weight was still pinning her down. She couldn't feel anything. She couldn't seem to remember how to move even her hand, as if the beasts had torn her into tiny pieces and scattered them. Then, as she gradually became aware of the wind's roar, pain flowed back into her aged body: the smart and sting of the grazes from the stones and the scratches from the hounds' claws, the bruise on her side where her belt had been dragged into her flesh, the aching and throbbing of her chest and legs from where she had been hurled to the ground. Slowly, stiffly, she pushed herself to her hands and knees, and twisted herself round, until she was sitting on the sodden track.

It was only then that she saw a figure standing on the rock above her, like a giant bird, feathers fluttering ragged and wild in the wind. It was staring down at her as if it was about to swoop and snatch her up in its talons. Meggy gave a muffled cry, pressing her hand to her mouth, and tried to struggle to her feet, but her legs and arms were trembling. She was as weak as a nestling. The figure watched her impassively as she dug the stave into the ground and dragged herself upright. Then it turned and vanished into the night.

Chapter 38

Sorrel

I crouched by Eva's old fire pit, stirring the broth in her great iron cauldron. In truth, I needn't have bothered for the broth was so watery that there was nothing to burn. The handful of lamb's cress had turned to pulp in the water. One of the tinners' wives had told me that the white shaggy-mane mushrooms I'd found on the edge of the track were good for eating, but they had dyed the green broth almost black. Apart from some bulrush roots and withered thyme there was little else floating in the pot, save for a few strips of dried salt fish. It was not nearly filling enough to stave off hunger, but what else could I offer the menfolk?

The dried beans and meat that remained in Gleedy's store now cost more than a man could earn in a week. I had tried to lay snares as poor Eva had done, but I didn't know her secret places and, so far, mine had remained empty, except for one which had caught a fox, and that had been torn to pieces by beasts and birds, leaving only a bloody skull and a few bones.

Every night I walked up to Eva's hut to cook the supper

for the single tinners, as she had done, still half expecting to find her there, adding peats to her fire or worts to her pot. And every time I felt a stab of grief on seeing that the stone she used to sit on was empty, save for the dark stain of blood, which had never washed off, even in all that rain. Who had strangled her? Why? I couldn't imagine that anyone had held a grudge against her, not Eva. I'd lost the only two friends I'd ever had, Eva and now Morwen, for I doubted old Kendra would ever let her near me again. I'd lived friendless all my life, since Mam died. I should have been accustomed to it and I was, except that once you have found a treasure and had it taken from you, it leaves a gaping hole in you where there was none before.

A snatch of ripe meat drifted down the hill on the wind. All around, neighbours lifted their heads and sniffed. Rumour was one of the women had driven some ravens off the maggoty carcass of a hare. It was rank, exceedingly so, but she wouldn't share it. Some said sourly they'd not touch it anyway – might fill your belly and satisfy your cravings for an hour or so, but just wait till the cramps and flux began. You'd be wishing you'd stayed hungry. But, given half a chance, most would have snatched it from the woman's pot.

Although the shelter Eva had built kept the rain from the fire, the wind still forced its way through the smallest crack, stirring and swirling the flames, as if they were its supper. It was almost nightfall and the horn to end the day's work would soon be echoing over the valley, though it had seemed to be blown later and later these past few days. The men grumbled that they could scarcely see the spades in their hands by the time Gleedy gave orders to sound the signal. Tonight I was thankful for that, though, for by the time the tinners had trudged up here, it would be dark enough to mask the colour of the broth and at least it smelt a deal better than it looked.

I stared down the valley towards the hell-red glow of the distant blowing-house furnace. Beyond that there was nothing but a cavern of blackness, as if the land had fallen away into the void and everything – workings, moor and men – had simply tumbled into a dark abyss. Was Morwen out there somewhere on the moor, by the black pool, or sitting by her own hearth in her cottage? I was desperate to see her. My dreams had become so vivid I was afraid to sleep.

Who was the boy I'd seen? Did he live in this world? I'd heard Brigid's voice telling me to protect him. Was that what she'd called me here to do? But how could I, if I couldn't find him? Only Morwen seemed to know what my dreams meant, but I daren't seek her out for fear of bringing more trouble to her. I'd little doubt that Ryana had carried out her threat to tell the old hag.

The night before, I had stared into Eva's hearth fire, trying to journey as Morwen had taught me, though I was terrified that, without her, I would not be able to get back. I walked into the flames, trying to reach the heart of the fire where I was sure Brigid would be waiting for me. But at the centre of the fire was the moor, and I was standing on a rock in darkness, staring down at a pack of hounds streaking across the hillside. At first I couldn't see what they were hunting and then I saw it was a little black cat. I was so afraid those huge dogs would tear the defenceless creature to pieces that I whistled to drive them off. But instantly the moor vanished, like a candle flame blown out by the wind, and I was sitting by Eva's hut again. What did it mean? What did any of it mean? Without Morwen, nothing made sense.

I spooned some of the broth into my own bowl, clutching

the warmed wood in my good hand, so that the heat could seep into my skin. But I was too ravenous to hold it for long. I drank it swiftly and, for a few minutes at least, the hot liquid fooled my belly into thinking it was food and the gripping ache eased a little.

Something brushed my shoulder, making me jump. Gleedy was standing just behind me. The firelight gleaming up from below melted away his face, leaving only a forest of crooked teeth and the whites of his eyes hanging in the shadows. 'You swallowed that faster than my hounds gulp their suppers. Hungry, are you?' He dipped the ladle into the great iron pot and raised it to his lips, taking a noisy slurp, then letting the remainder trickle back into pot. 'Weak as whey. That wouldn't satisfy a babe, much less men who've been digging all day. The tinners will not be paying you for a belly full of water.'

'They know what you're charging for a handful of mouldy peas,' I muttered. I wanted to say more, a great deal more, but I had to bite the words back. I couldn't afford to anger him. There was no food to be had except what was in his stores, and he could charge whatever he pleased for it. If I couldn't even buy a little dried fish on Saturday, the men really would be eating water.

His fingers touched my bare neck. It felt as if a huge spider was creeping over my skin. I slapped his hand away and scrambled to my feet, stepping away from him. He studied me with a curiously amused expression. I couldn't be sure, because of his squint, but his gaze seemed to be wandering over every inch of my body. It made my skin crawl as if his fingers were still touching me.

'I envy the tinners on nights like this,' he said, 'with their women to cuddle up to, keeping them snug and warm of an evening. There's me, all alone, at the other end of the valley. A pot stuffed full of tasty meats and a flagon of wine is well and good, but you can't really savour them when you've no one to share them with. Same as a bed, thick blankets, soft place to rest your head, but somehow you can't seem to get warm when there's no one lying beside you. I really miss my Eva on nights like this, grieve for her something terrible, I do. My bed's a lonely place without her in it.'

'*Your* Eva? What? *You* were her lover? I don't believe it.' I shuddered at the mere thought. Why would a woman as handsome as Eva even look at a creature like Gleedy? Of all the men in the camp she might have sought a little comfort with, it would surely never have been this slug. But . . . there was the bruise I'd seen on her face, the food she got from his stores to cook. What had Gleedy said to Todde? *Half the men and women here are runaways. But they know their old masters can't touch them, providing they don't make trouble and get themselves thrown out.* If Eva had been a bondswoman, she'd probably had no choice but to let Gleedy use her or be sold to the serf-hunters. Her babe! Was he the father? I felt sick.

From the other side of the valley came the long mournful wail of the ram's horn. In front of the huts, fires were blazing up as the women poked and blew on sleepy embers. The tinners would be climbing the hill hungry for their supper.

Gleedy scooped another ladle of broth from the pot, and held it high in the air, watching the liquid splash back down into the cauldron. 'I'm thinking a good wedge of fat ham would cook up a treat in that pot, maybe with some white peas. Can't you just smell them?'

'If you could cook wishes, beggars would eat like kings,'

I snapped. 'The tinners can't pay me enough to buy ham at your prices, and I'll not be digging myself more debt. Besides, I heard you say there wasn't so much as a strip of dried mutton left in the stores. And dwelling on what you can't have only makes the belly crave it more.'

Furtively glancing down towards the tinners' huts, Gleedy shuffled close to me again, catching my arm as I tried to move away. 'I always keep a little something aside for those who know how to ask nicely,' he whispered.

'Then you'd best offer it to someone else. I told you, I'll not be owing you any more.'

I tried to wrench myself from his grasp. I was desperately willing the tinners to arrive and come to my defence. Gleedy had hold of my good arm, and I hadn't the strength in my withered one to push him away. The wind gusted, shoving me hard towards him, as if it was his henchman.

'Did I say anything about money?' Gleedy said, with an injured expression, as if I'd grossly wounded him. 'Ask nicely, was all I said.'

At that moment, two of the younger tinners plodded up to the fire, juddering to a halt as they caught sight of me struggling in Gleedy's grasp. I was never more relieved to see anyone, but it was short-lived.

The two lads grinned at each other. 'Got yourself a new woman, Gleedy? Didn't take you long. Don't mind us, we'll just help ourselves.'

Gleedy ignored them and bent his face close to my ear. 'You be nice to me and I'll see there's always something hot and tasty in your pot . . . and not just this one.' He was still holding me tightly by my arm. But his other hand snaked down and he grabbed me briefly between my legs, winking at the two men, who roared with laughter.

I jerked my arm from Gleedy's grasp, almost smashing him in the face with my elbow. I was only sorry that I missed. I didn't know who I was more outraged by – Gleedy or the lads who were watching us, as if I was a dancing bear being goaded for their amusement.

'You might have been able to frighten Eva into your bed. But I'm no bondswoman. You can't threaten me.'

Gleedy spread his hands and gave a high-pitched giggle. 'You know me, I couldn't threaten a mouse. No, Eva and I had an arrangement. I helped her to set up her business. Sold her food cheap from the stores, so she could make a nice little profit. I took care of her. She never went hungry. And she was more than eager to show me her gratitude, as you will be as soon as you see all I can do for you.'

'Go on, you warm his cods for him,' one of the lads jeered. 'Keep him in a good humour and we'll all be the better for it.'

They glanced round as another man came lumbering towards us. He halted abruptly, frowning as he stared at the two of us in the firelight. But before I could yell out, he'd hastily retreated back into the darkness. It was then I realised no one was going to come to my aid. Why would they? There wasn't a man or woman in the camp who could afford to be on the wrong side of Gleedy. They'd not risk their hides to defend a woman who was no kin to them.

Gleedy watched the tinner back away, his lips gleaming wetly in the firelight as he grinned at me. 'Like you said, food's dwindling fast in the stores. I may have to start keeping what little I have for the tinners who've worked here longest and those with families. Only fair, that. You wouldn't want the little ones to go hungry.' He leaned over me. 'Course, you might be thinking, That makes no odds to a woman

308

like me 'cause I can leave. Find work somewhere else, even though I'm a cripple and there's hundreds of strong men begging to be taken on. And you may be right. You may find some tender-hearted person who'll take pity on you, same as I did. But what you got to consider is those who can't leave, like your friend, Toddy.'

He gestured towards the dark expanse of the moors. 'Serf-hunters are lying in wait out there and they've got fat purses. And, these days, a poor man like me has to scrape a few coins together wherever he can, *sell* whatever he can, even though he hates to do it. Those hunters spend many a cold night out on the moor, so when they do finally capture their quarry, they think to have a bit of sport with him, pay him back for all the trouble he's given them. Well, I don't need to tell you that, do I? You saw what happened to poor Eva. But then maybe old Toddy means nothing to you. Why should he? In which case, you just forget all about it and sleep easy.'

Gleedy yanked me forward as we trudged up the rise towards the storehouse. His pace, far from slowing as we climbed, had quickened, like that of a small boy who could barely contain his impatience for a promised treat. His fingers dug painfully into my good arm, while my other dangled, help-less, by my side.

There had been times when I was growing up that I'd wept that no man would ever want to marry me. My father and the village boys jeered often enough at the useless lump of flesh hanging from my shoulder to convince me of it. I knew all men would be as repulsed by it as they were, and even supposing there was a man who was not, he'd never wed a woman with one arm, any more than he'd buy a horse with three legs. A wife was for work, like a pig was for pork. But Gleedy was not looking for a wife and I sensed that the very thing that might repulse other men aroused him. He was excited by the thought of a woman who could not push him away, a woman who could be easily overpowered even by a runt like him.

All the while he'd been dragging me down the hill, he'd been prattling about how grateful I'd be when I saw the food and other little comforts he could provide if I was especially nice to him. But now he fell silent, taut with anticipation, panting noisily, like a dog.

A thousand thoughts jostled and fought in my head. Would he try to make me flirt with him and coax him, like I'd seen tavern girls do, or would he force himself on me at once? I tried to picture his storehouse in my head, remember where things were. What could I snatch up and use as a weapon? Should I grab something as he dragged me through the door or pretend to submit to get him off guard and hope that I could find something to use while he slept? But what if I tried to hit him and missed? What would he do to me? And if I succeeded in knocking him senseless, what then? What would happen when he recovered? And if I hit him too hard and he died . . . I'd hang!

He was dragging me past the blowing-house. The heat from the glowing furnace rushed at us through the open door. Inside I glimpsed two men, their bare backs running with glistening black sweat as they ladled the molten tin from the stone trough beneath the furnace and poured it, shining like moonlight, into the moulds. I wanted to cry out to them, beg them to help me, but I knew it was useless.

We turned around the corner of the blowing-house into the bitter wind. The two hounds had been chained so that their tethers would now reach the storehouse door. They sprang to their feet, barking and growling, making me jump back so violently that Gleedy, his fingers slippery with sweat, lost his grip on me. Impatiently he snapped at the dogs to lie down. But they took little notice until he reached into the pouch hanging from his belt and tossed them each a sheep's hoof, which they caught in mid-air and flopped down at once to gnaw.

Free from his grip I stared frantically around, desperate to run, but where could I go? I couldn't return to our hut. I'd have to make for the moor. I tried to move silently, but I was afraid to turn my back on him. If he let the dogs loose, they'd seize me long before I could scramble up the steep, slippery side of the valley.

Gleedy was dragging the growling brutes away from the storehouse door. I turned to run, but before I had taken more than a pace, a hand was clapped over my mouth from behind and a strong arm wrapped about my waist. I was dragged, struggling and kicking, away from the glow of the blowing-house and into the shadows. Hands forced me face down to the ground. A man's thighs straddled my back. I tried to wriggle free, but his weight was too great. Stinking, sweaty fingers still pressed across my mouth, half blocking my nose. I couldn't breathe. That became the only thing that mattered, trying to draw air into my lungs. My good hand was pinned under me. I managed to pull it free and grasp the hairy arm that was suffocating me. But strength was fast flowing out of me. Tiny golden sparks were bursting in my head. My body was drifting away from me. I lost my grip on his arm and lay limp.

I felt his hot breath on my ear and dimly heard a man's voice whisper to me to lie still and not make a sound. I didn't need to be told that: I couldn't move. Was Gleedy going to drag me inside or force himself on me out there in the mud?

The hand slid from my mouth, and I took a great gasp of cold air, though his fingers still rested on my chin, and I knew they'd shoot up again if I attempted to call out. The thick legs uncoiled from my back. The crushing weight lifted, but an arm tightened about my waist again, holding me down. Every muscle in my body seemed to have turned to bone.

'Playing games, are we, Sorrel? Hide and seek? And what'll be my prize when I find you?'

It took a moment to realise that Gleedy's voice was not coming from beside me. He was tramping around in the darkness, calling out first in amusement, then with growing impatience. Boots crunched over the stones just yards from me, and lifting my head, I caught a glimpse of his outline lit by the glow of the blowing-house, before a hand pressed me back into the mud.

'Keep down!'

I strained to hear if Gleedy's footfalls were coming closer, but it was impossible over the hissing and wheeze of the huge blowing-house bellows, the creaking of the waterwheel that drove them, and the clatter of the men inside.

The arm lifted from my waist and hauled me to my feet. 'Come on, this way and keep low.'

I found myself running and stumbling over stones and the stumps of bushes, with the sound of rushing water growing ever louder, until I was dragged behind a wedge-shaped heap of gravel spoil, where my captor finally released me and flopped down beside me in the mud. There was not a glimmer of light and I could see nothing of the man sitting beside me. I could hear him panting hard, though whether from the effort of dragging me or from excitement I couldn't tell. My own breath was tearing painfully in my chest. My legs felt as if my bones had melted away and I was sure I'd never be able to run again. But, all the same, I levered myself up until I was crouching, ready to spring away if the man made any move to touch me.

'What do you . . . want from me?' I groped behind me with my good hand, ignoring the scratches from the sharp gravel and thorns, as I swept my palm over the ground until my fingers closed around a weighty stone. Slowly, so he wouldn't detect the movement, I brought it to my side, ready to strike.

'Guessed he'd fetch you to his storehouse. Takes everything he buys or steals there.' The man was keeping his voice low, but I'd heard it many times before these past weeks.

'Todde? Is that you?'

'Who else?'

I could hear the cocky grin in his voice.

'Ought to be a special room in Hell for rats like him,' Todde muttered. 'I'd skin him alive, then cut out every twisted little bone in his body, one at a time, starting with his thieving fingers.'

I knew the smile had faded from his face for his tone was as bitter as ox gall. But I felt more angry than grateful. If Todde hadn't stopped me, I'd be safely up on the moor by now. Then his words penetrated my fear and anger.

'You guessed he'd take me to his storehouse? You knew what he was planning and you didn't trouble to warn me?' I tightened my grip on the stone.

'Been watching you and him, just in case, though I didn't really believe it. Even I thought a vile little weasel like Gleedy

313

wouldn't go that far. And bairns imagine all sorts, don't they, always making up wild stories, especially the lads?'

'What stories?' I whispered.

'After Gleedy accused me of killing Eva, I did some asking around quietly, case they did fetch the sheriff. Wanted to be ready to defend myself. Nobody would say anything, though I reckon a few knew more than they were letting on. But I noticed one little lad, seemed to be acting a bit queer, especially when Gleedy was around. I got him on his own and bribed him with a strip of dried meat that I'd . . . found.'

Stolen, he meant, but that hardly mattered now.

'Anyway, the lad told me the day Eva went missing he'd sneaked away when he was meant to be picking over stones and gone up to Eva's hut. He said Eva used to give him a bite of food now and again. She was soft like that. But when he got up there he heard Gleedy's voice and hid 'cause he was afeared he'd be in trouble for skipping off. He saw Gleedy punch Eva, leastways that was what the boy claimed. She tried to get away, but that weasel put his hands round her throat and throttled her. He slung her over that horse of his and led it up the track. Lad didn't know if she was dead or alive then, but it was the last he saw of her. Gleedy must have dumped her corpse on the moor. Probably hoped it would be picked clean afore she was found and recognised.'

I saw again, with a shudder, Eva's face as her spirit walked the lych-way, the dark marks on her neck, the look of terror in her eyes, and I knew it was the truth.

'Course, the lad said nowt, not even when her body was brought back, 'cause he knew he shouldn't have gone up there when he was meant to be streaming, reckoned his father would give him a whipping if he found out. He was terrified of what Gleedy might do to him too, and with good reason.'

Eva's face hung in the darkness in front of me. Fury raged through me. Gleedy had used her and killed her. Todde knew and had said nothing to me, to anyone. At best, that made him a coward. At worst . . .

'I suppose you thought that if you snatched me from Gleedy, I'd be nice to you instead. Was that it? Or maybe you thought if I'd warm his bed, I'd warm yours after. Was that what you were hanging around the store for, to beg his leavings? That's what you've been after from the beginning, isn't it?'

I was only out there because of what Gleedy had threatened to do to Todde. I should have let Gleedy sell him. He deserved it. Instead, I had almost sold myself to protect this . . . this little toad. I heard the gravel grating under Todde's backside as he moved and nearly swung at him with the stone, but behind the spoil heap it was too dark even to see his head. But I was so filled with rage, I would have pounded it to a pulp.

'Won't deny I took a fancy to you the first day I clapped eyes on you. Hoped, maybe, in time . . . But I'd never force myself on you or any lass. Seen too many masters have their way with maids against their will. Makes me mad as a lynx in a trap to see a man do that to a woman that can't refuse him. My own brother was forced to send his new bride to the priest that had married them. Things weren't never the same between them after that, and when their first babe was dragged into the world . . . Well, my brother was never sure. I'd catch him looking at the child with this cold stare, and when the little lad would try to take his hand or cuddle up against him, he'd shrug him off as if the boy were a stray dog.'

There was silence between us, filled only by the wind and the water tumbling unseen.

'You can't go back to the hut tonight. The weasel's bound to go looking for you there.'

'You think it'll be any different tomorrow? I'll have to make for the moors.'

'You can't do that! It's not safe.' Todde lunged at me.

I suppose he might have been trying to seize my hand in the darkness, but it was my thigh his fingers closed around.

'It'll be safer than here! You're as foul as he is.' I pushed myself on to my feet.

Todde grabbed the hem of my skirt. 'I swear you'll come to no harm from me. You'll starve out there, if you don't die of cold first. There's tinners and outlaws crawling all over the moor. If they find a lass on her own, they'll do things to you that would make even Gleedy sick to think on. At least in the camp you'll not starve and I'll not let that weasel lay a paw—'

But I refused to listen to another of Todde's wild promises. 'Aye, I'll not starve if I let Gleedy have his way, at least until he loses his temper and I end up as hound's meat on the moor.'

Todde was no better than those tinners. I'd cooked for them and they had just stood by and let Gleedy drag me off. Worse, they'd egged him on, just to make their lives easier. *Warm his cods for him. Keep him in a good humour and we'll all be the better for it.* Was that how Todde reasoned too?

'You don't like to see a man force himself on a woman?' I snarled at him. 'No, I wager you don't unless there's something in it for you, and that's a belly full of food. Was that your game, hold me here, then deliver me to Gleedy yourself, so you'd make quite sure to get your share.'

I caught the tail of my name being shouted into the wind. Gleedy was coming back.

Todde heard it too, he already had hold of my skirt, now his other hand fastened itself around my ankle. I couldn't pull away from him. I daren't kick out, knowing if I did, he could easily pull me over.

Fingers grabbed the back of my neck. Another arm locked about my waist, so that I couldn't turn. Gleedy's body pressed into my back. I could feel his prick hardening against my thigh, his mouth hot against my ear.

'Seems I misjudged you, Toddy. You've done me a service catching this little cat for me. You'll not be sorry. I'll see you well rewarded. Always pay my debts I do, unlike some. But you'll have to wait till morning. I fancy I'll be a mite occupied tonight.'

I struggled, but I couldn't break free from him. His fingers had slid around my neck and were pressing into my throat, not hard enough to make me choke, but as if he wanted me to know he could throttle me just as easily as he had Eva, more so, for I had only one hand to fight with.

I was still clutching the stone in my fist. I smashed down as hard as I could against the knuckles of his hand on my waist, and felt the stone strike flesh and bone. Gleedy yelped and his fingers released their grip. I brought my elbow back hard against his belly, heard him grunt and felt him sag. But as I tried to step away, I slipped in the mud and fell to my knees. Gleedy didn't hesitate: before I could struggle up, he'd grabbed my hair and was yanking my head back so far I thought my neck would snap.

Todde was on his feet beside me. 'Let her go, you bastard!'

I felt rather than saw his fist fly past my face and heard the crack as it crunched into Gleedy's jaw. Gleedy crashed down, almost dragging me with him. Todde seized my arm and dragged me upright.

317

'Run!' he yelled. 'Run!'

I picked up my skirts and scrambled up the slope, slipping and stumbling. Behind me I could hear the two men shouting. There was a long-drawn-out scream, then nothing but silence and darkness.

Chapter 39

Prioress Johanne

'Brother Nicholas, I assure you that your letters have been safely dispatched to Buckland.'

My hands clenched in exasperation inside my sleeves, where I'd tucked them, not for modesty but to warm my numb fingers, since he and Brother Alban were, as usual, blocking what little heat rose from the meagre fire.

'Any reply there may be for you will be sent with Hob when he brings the next wagon of supplies, and we cannot expect him now until the first frosts harden the ground. The servants can't even use the sledges on the moor, for fear that the horses will get stuck in the deep mud and die of exhaustion. It is the same every year in the autumn. I grant you, this year the season has come upon us unnaturally early. But there is nothing to be done, except pray that rain will soon give way to ice. Though that will bring its own horrors,' I told them grimly.

As if to prove the truth of my words, a particularly vicious blast of wind rattled the shutters so violently that I began to fear the thatched roof would be lifted off.

Nicholas shuffled forward to the edge of the wooden chair and spread his fingers over the glowing peats, trying to snatch at the warmth. 'If my letters had reached Commander John, I am certain he would have dispatched a messenger with a reply at once, not trusted to carters and lumbering wagons.' He drained his goblet and stared pointedly at the flagon of wine that had somehow found its way to Alban's side. He gestured for it to be passed back, but his sergeant studiously ignored him.

'He can't handle the eighty hinds he already has in his herd, never mind chasing after the handful up here,' Alban muttered. 'You want your letters to be answered, Brother Nicholas, you should be sending them to Clerkenwell. They've men and good horses to spare as messengers. Don't have to rely on an ancient old gammer and his mooncalf son.'

Nicholas stalked over to seize the flagon. 'I have told Commander John quite plainly that unless he sends men-at-arms, this priory will be starved out long before winter sets in, let alone spring, because there won't be a single head of cattle or sheep left on our lands that the outlaws, tinners or thieving villagers haven't butchered.'

'And I told you, Brother Nicholas, that bringing armed knights here will simply start a war we cannot win,' I thundered.

'I trust even you will not presume to consider yourself more able than a knight commander of St John to decide how best to protect one of our order's priors.'

'I am prioress here. The safety of this priory is my responsibility—'

'Can't see as it makes any odds either way,' Alban cut in. 'It's not knights we need. This place is losing more kine to the rain and the packs of starving beasts than to tinners and outlaws. Kept awake half the night, I am, by packs of wild dogs howling out on the moor, and young Brengy in the stables tells me the foxes are growing so bold they attack the sheep in daylight. You want to tell the Lord Prior it's hunters and trappers are wanted here.'

A look of disgust settled on Nicholas's face as he tipped the captured flagon over his goblet, to see nothing but the gritty dregs slithering out. 'I would indeed have informed the Lord Prior of all that has befallen this cursed priory, Brother Alban, if the prioress hadn't sent the only able-bodied man who was capable of finding his way to Clerkenwell on some spurious errand to Exeter from which he still has not returned. And *if* he ever returns,' Nicholas said savagely, 'he is likely to find only skeletons sitting round the refectory table, since we will all have long since died of hunger and thirst.' He flung the lees of the wine over the fire. 'All, that is, except a certain brother sergeant, who will have died of a burst gut having devoured every last morsel of food in the priory and our corpses as well, I imagine.'

Nicholas picked up his cloak and swirled it around to settle on his shoulders as he strode to the door. 'I will bid you goodnight, Prioress, and retire to my chamber where, if I hurry and manage to get there before my brother, I might still find a goblet of wine left in the ambry.'

The door opened and slammed shut, letting in a blast of wind, which sent a dense cloud of smoke, peat ash and scarlet sparks from the fire swirling about Alban's head. He coughed violently and overturned the chair in his haste to back away, beating out the burning fragments that had fastened them-selves on his jerkin and were smouldering in his beard. Eyes

321

and nose streaming, he stumbled to the door and out into the cold, damp air.

I waited until the second billowing of smoke had settled, then sank back on the chair close to the fire, the seat still disconcertingly warm from Nicholas's backside. I should have returned to my own chamber and my devotions, but a great leaden weariness had settled on me and I couldn't summon the energy to take even that short walk across the dark courtyard.

I knew my chamber would be nearly as cold as the open moor for I'd given instructions that the fire should not be lit in there, in spite of the season. From now on, we had to save most of the fuel for the kitchens. We had not had a single full day of sun, much less the weeks of good weather needed, to dry newly dug peats. The only way of getting peat or wood to burn was to bring it in wet and try to dry what little we could beside those fires that were still alight. Food was scarce enough, but if we couldn't cook it we would surely starve.

Somehow the presence of the two brothers at meals always left me exhausted. What was the word Alban had used so contemptuously about our sisters at Buckland? Hinds? Yes, and the two men were like rutting bucks, always roaring and charging at one another. It should have been easier when my sisters were present, but I found myself in a constant state of tension in case one prattled some indiscretion, which Nicholas would pounce on. I found myself trying to listen to half a dozen different conversations at once, so I could step in and deflect the speaker if they were drifting into perilous waters.

Fina was the worst for, like a small child or a moon-crazed beggar, whatever thought came into her head burst from her lips. Without warning she would hurl some bizarre remark across the refectory table or toss it into the courtyard. She seemed to have convinced herself that all that had befallen the well of late was somehow an attack on her, because she was its keeper. Though I had succeeded in convincing her that it was dangerous to speak of Cosmas in front of the brothers, I knew I had been unable to persuade her that the little boy was not cursing the holy spring.

Clearly the responsibility for the well had become too heavy a burden for her and I was annoyed with myself for not realising it before. Several times I had been on the verge of informing her that, when the spring started to flow again and the well reopened, I would give the charge of it to another of the sisters. But that seemed to be what she feared most, that the boy was trying to drive her away from it, steal it from her. I feared that telling her I was going to give the task to another, far from easing her mind, would only increase her unreasoning hatred of the boy and, indeed, of me. Sometimes I caught her glaring at me with such malice, it seemed she thought I was as much to blame as the child for the spring drying up. Perhaps I was. Perhaps God was punishing me for my deception.

With every day that passed, word would spread like a ripple on the pond that the well was dry, and who knew if the pilgrims and their offerings would return even when the waters flowed once more? That would give Nicholas yet more to write about in his letters. Not that any of his reports had so far reached Buckland, for Sister Clarice had paid off the messengers before they set out and retrieved the parchments. She had even recruited the ever-loyal Meggy to keep watch for signs of any servant preparing for a journey in case our brother might try to dispatch them with one of his letters.

Nicholas could wait until his beard grew as white and long as winter, Buckland would not be sending any knights. And so long as I remained prioress, I would do all in my power to ensure they never did, if I had to burn a thousand reports or bribe a hundred messengers.

But even if our brother's reports never reached Buckland or Clerkenwell, how long could we keep going? The infirmary was filling with the frail and those made sick from starvation or eating grass to stave off the agonies of hunger. The old were being left at our doorstep by families who could no longer feed them, and the crowd begging for alms at the window grew larger and more pitiful by the day, but we had less and less to give them. I had talked glibly about Hob not returning until the ground hardened, but what if he never came back? The famine was not confined to the moor: the whole of England was suffering. The time would come, if it hadn't already, when Buckland had no food to spare for us.

If we had to survive alone and feed all those so desperately in need who came to us, the money Clarice and I had hidden would be our priory's only hope. I would have to send her out with a couple of strong servants to Exeter or one of the ports to see what she could buy from any ships that docked. The merchants were already exploiting the famine and charging a king's ransom for a sack of grain, but if anyone could drive them down it was Clarice. She knew how hard won every coin we possessed had been and she would not waste a single farthing.

But first we had to rid ourselves of our two brothers. Even if she managed to slip away without Nicholas or Alban noticing, Nicholas would demand a reckoning when she returned with supplies, and not even Clarice could disguise the fact that she had spent more than we had claimed to have. Nicholas had made it his business to learn the price that was charged for every pot, fish and nail in these parts and I knew he was watching every bale and basket brought into the priory, adding up what we were spending. He was determined to prove we had money we had not accounted for. That man could turn a saint to murder, and with every day that Nicholas was here, the guilt I felt at deceiving the order was slowly dissolving, like salt left in the rain.

My head jerked up as a blast of cold air struck my cheek. Had I been dozing? For the third time that evening, smoke and ash swirled about me as the door opened and slammed shut.

'I've been searching all over for you. You'd best come quick.'

Beating the smoke away, I struggled up to see Sister Clarice, her hand already on the latch, as if she was about to dash away again. My heart quickened.

'What is it? Don't tell me we've another plague of creatures down in the holy well.'

She shook her head. 'Sebastian.'

I didn't wait for more. We hurried out into the cold night air. Before we even reached the door of the infirmary, I heard the shrieks and crashes coming from inside, but the shouts were not the usual cries of Sebastian's night terrors. I ran the last few paces and burst in. I couldn't see him at first, but I could tell from the terrified glances of the other patients where he must be. Those who could walk were huddled together at one end of the room, those still in bed had shrunk beneath their blankets. Cosmas alone sat motionless on the end of his bed, as if he was deaf as well as blind. Basilia was standing in the centre of the hall, gnawing her lip, as she

325

turned to me, her expression wavering between relief and fear. Tears welled in her eyes.

'We never thought to keep it from him.' She gestured helplessly towards the far end of the infirmary. 'Didn't think he could use it.'

I still couldn't see Sebastian, but hurried down to where he slept.

'Careful!' Basilia called behind me. 'He's not himself. If he doesn't know you, there's no knowing what he might . . .' Her words trailed off as I reached the furthest wall.

Sebastian sat curled on the floor, hidden by his bed, his head tilted sideways and his eyes squeezed shut. I stifled a gasp. His white woollen nightshirt was streaked with blood, and more oozed from beneath his fingers. At first, I thought he must have fallen and grazed himself, but there was too much blood for that. His shrieks had subsided into long, body-shuddering moans.

'Sebastian? It's Prioress Johanne. You know my voice, don't you? Open your eyes. Look at me.'

His eyelids fluttered open, but he wouldn't look up. 'Go! Go . . . away. I can finish this . . . want to finish this.'

'Finish what?' I laid my hand on his twisted leg. He was chilled to the marrow and little wonder, lying on the cold flags. 'Come, let's get you back into a warm bed. I'm sure Sister Basilia will have an unguent to soothe those fingers.'

I tried to take one of his hands in mine so that I could examine his palm. 'What have you been doing to them now?' I could hear myself speaking in the tone a mother might use to a little child. Sometimes when he was in the grip of a night terror it calmed him, but this time, he jerked furiously away from me.

'Leave me! Get away. I want this to end. In God's mercy,

let me finish it, if you will not.' He lashed out, flailing his arm, though the pain of doing so made him shriek. His elbow struck my cheek. He was too weak and the limb too wasted to do me any serious injury, but the bone was sharp and I jerked back.

The violent movement of his arm had unbalanced him and he rolled sideways away from the bed. It was only then that I saw what his thin chest had been covering. A long, sharp knife had been wedged into the gap between the bed and the wall, so that the blade pointed out into the room.

'Who left that . . .'

But even as I spoke, I realised what Basilia had been trying to tell me. Sebastian had wedged the knife there himself. He'd cut his fingers on the blade as he'd tried to thrust the handle hard into the gap. I could only imagine the pain and effort it had cost him. His joints were so crippled that on most days he could barely grip a piece of bread and raise it to his mouth, much less spoon broth from a bowl or comb his hair.

Sebastian began throwing himself from side to side, and at once I realised what he intended to do. He was trying to use his momentum to impale himself on the blade, for he had not strength or control to stab himself using his hands. I seized his shoulders, trying to drag him away from the knife, but for all that he was weak, he was in the grip of such a terrible agony of the mind that it lent him a strength even I could not overcome. I came close to being flung against the point of the blade myself in the struggle.

'Help me hold him!' I yelled.

I heard feet running down the flags behind me. Two hands seized Sebastian by the shoulders and dragged him roughly across the floor. I made a lunge for the knife, wresting it out

327

and sending it skidding to land at the feet of Basilia. Startled, I turned to see who had pulled Sebastian aside and found myself staring up into the face of Brother Nicholas. He was the last person on earth I'd wanted to witness this.

Sebastian lay trembling and helpless on the floor. Nicholas lifted him, sobbing and shaking, into his bed. Trying to suppress the tremor in my voice, I thanked the knight curtly, told him we could manage and need detain him no longer.

'You appeared not to be managing just now, but if you insist.' He stared down at the huddled figure, studying him carefully for a long time, before he finally turned away.

He strode towards the door, but as he passed the bed where Cosmas sat, his steps faltered. He glanced sideways at the boy, as if afraid to look at him directly, then swiftly and covertly, he extended the two fingers of his right hand towards him, making the horn gesture, which I had seen others use to ward off the evil eye. Then he hurried to the door. Surely Nicholas didn't still believe the boy to be a sorcerer, capable of conjuring evil spirits as he had accused him of doing that first day.

It took a long time to settle Sebastian. I tried to calm him while Basilia attempted to anoint and dress the deep cuts to his hands. At first, he resisted us both, but finally what little strength he had summoned was exhausted and he lay limp, staring blankly up into the dark shadows beneath the thatch, allowing us to move his limbs as if he had already passed from life to death. I winced for him as Basilia tried to close the wounds. How he could have had the willpower to keep pushing the knife handle into the crack while the blade sliced so deeply into his palms and fingers, I do not know, but he had lived constantly in such agony that perhaps he had barely registered this fresh pain.

Finally, Basilia forced a hollow horn between his lips, and tipped a sleeping draught into his mouth. He swallowed it. He no longer had the will to resist. Tears ran down the creases at the corners of his eyes, but they were not tears born of pain.

When the patients had finally settled quietly in their beds, I walked softly from the hall. Cosmas had been helped to bed and lay exactly where he had been put, but he was not asleep. He, too, lay staring up into the cavern of the roof above his bed. Firelight danced in his crow-black eyes. Did the child ever sleep?

It was only as I reached the door that the shock and horror of what Sebastian had tried to do fully engulfed me, and with it came such an utter weariness that I wasn't sure if my legs would bear me long enough to reach my chamber. For eight long years I had prayed that Sebastian would be healed, if not his body then at least that the torment of his poor mind would ease and he would stop reliving the horrors. But his despair was greater than ever. I felt as if a knife was being twisted inside my own chest knowing that he was in such agony of spirit he'd been driven to attempt such a terrible sin, one that would send him straight to the torments of Hell. He was so desperate to escape his own self, his own life, and I was powerless to ease that pain. Bile burned my throat, as I realised what he had been begging me to do the day the birds had mobbed the priory. He'd wanted me to help him end it then, but I had not listened, not heard him. I should have warned Basilia what he might try to do. I should have kept him safe.

I shivered as I shuffled outside. In my haste, I had not stopped to collect my cloak, and the wind cut wet and sharp through my kirtle. The courtyard was almost in darkness now. Most of the torches had burned away, and I had to tread carefully for fear of tripping. My foot sank into a deep, icy puddle I hadn't seen, and I cursed beneath my breath.

I was watching the ground so carefully I didn't notice the figure emerge from the doorway of the guest chamber until he was almost upon me.

'I trust the man is sleeping now?'

'Brother Nicholas! I thought you had retired long since. Yes, he is quiet.'

'But for how long? Our infirmaries, Prioress, are intended for those who are sick to have a place of rest and solace until they are well enough to leave. They are not fitting places for the mad or possessed. Next time it might be a child or even one of your sisters who is stabbed. Even cripples can summon surprising strength when they are seized by a fit of madness or by a demon, as you yourself discovered this night. Why has he not been moved to a monastery where there are monks with the strength and skill to deal with such unfortunates?'

'Sebastian is content here. He knows us now. It would distress him to live among strangers. We can care for him as well as any monastery,' I added firmly. I tried to step around Nicholas. 'You will excuse me, Brother, I am weary and in need of my bed.'

But the knight moved swiftly to block my way.

'Evidently Sebastian is neither content and nor can you care for him, if he was attempting to commit the unforgivable sin of self-murder to get away from this place.'

His words struck me like a blow to the stomach for they cut too near the truth. I had failed Sebastian.

I swallowed hard, grateful that it was too dark for Nicholas to see my face. 'Sebastian is tormented by night terrors. Sometimes the dark shadow of them pursues him even into his waking hours and he cannot shake off his melancholy. But, in future, we will ensure no knives are left within his reach.'

'Next time it might be poison, or hanging. There are a hundred ways a man might kill himself if he is not restrained. But there are those monks who devote themselves to caring for the possessed. They can bind him or shackle him in a cell where he can do no harm to himself or others. They can use whips and all manner of mortifications of the flesh to drive the demons out. Who knows? They may succeed in restoring him.'

It was all I could do to stop myself slapping Nicholas. 'He is not possessed! And he will be cared for *here*!'

Sebastian already suffered agony night and day, and this knight of our compassionate order was suggesting that more should be inflicted on him! I knew what they did to those they believed possessed. They suspended them in baskets over their dining tables to torment them with the sight and smell of food while denying them even a crust. They ducked them repeatedly in icy water until they almost drowned, and flogged them mercilessly so that the demons would flee their poor bodies. But they would have to hack their way through my body if they wanted to take Sebastian and subject him to that.

Nicholas pressed closer to me in the darkened courtyard, peering down into my face. 'But I am curious, Prioress. Why do you insist on keeping this particular man here in the order's infirmary? Why is it that when he becomes agitated it is you they fetch to calm him? I think, for once, you spoke the truth when you said he is not possessed. I have seen those like him before, broken in mind and in body. I know what the Holy Inquisition can do to a man.'

I'd been cold standing out in that courtyard, but the icy chill that enveloped me now came not from the bitter wind but from fear. 'You are mistaken,' I protested, trying to keep my voice low and even. 'The poor man has been crippled from birth. You've seen for yourself how wasted and twisted his limbs are. He cannot walk.'

'Few can, after oil and flames have been applied to the soles of their feet. I saw his scars. But Sebastian can consider himself fortunate. I hear there have been those whose bones have dropped out of their feet, when the torture has been applied too enthusiastically. As for his limbs – I suspect they were damaged on the rack or perhaps the strappado, if weights were also employed.'

Nicholas's tone was as cold and dispassionate as if he was talking about carving up the carcass of a hog instead of the agonies inflicted by the Holy Church on living men.

'The Inquisition serves Christ to root out heretics, Prioress Johanne. If Sebastian has indeed enjoyed their hospitality and survived, it can only mean that he has confessed to heresy. The guilt of his sin is what troubles his conscience and makes him cry out in his sleep, and the abject shame of his betrayal of Our Lord is what has driven him to try to take his own life, as Judas did.'

Nicholas glanced at the infirmary casement, then returned his gaze to me. His eyes glittered like wet granite in the light of the one torch that still burned.

'I believe you know exactly what manner of man you are harbouring in there. That's why you don't want him moved, for fear others will discover his guilt. You are sheltering a heretic, Prioress Johanne. Now, why would you do such a thing unless, of course, you yourself share those same vile beliefs? It seems those frogs were a sign from God, after all.

They were sent to expose the foul corruption and sin that lies in the heart of this priory. And I will not ignore God's warning. No sister, servant or patient will be spared if they are infected by this evil. Mark me well, *Sister* Johanne. I will see that every putrid canker in this priory is cut out and burned.'

Chapter 40

Sorrel

My first thought, as I crossed the rise at the top of the tinners' valley, had been to make for Fire Tor. It was a good shelter and I could make a fire, if Morwen had left wood enough, for my flint and steel were always in the purse at my belt. But the closer I got, the more I began to fear Kendra might have put a spell on the cave to stop me going in or that she might even be lying in wait for me inside.

The face of the corpse who lay inside Fire Tor kept rearing up in the darkness, the gaping wound in his throat stretching wider and deeper. I didn't fear to be alone with it, but I could not forget the terrible anguish in the spirit's voice, his pitiful pleading to be set free. Kendra had preserved his body and kept his soul prisoner. The old woman might reckon to make another corpse to lie with the first and trap another spirit in that cave to be her servant, especially if it was someone she wanted to punish. The thought of what she could do to torture a soul she held in her power made the torments of Purgatory sound like Paradise.

I crept instead towards Morwen's cottage, but all was in

darkness. Still, I was careful to keep my distance for fear that Kendra or Ryana might be watching for me, if the old woman had seen my coming in her smoke. But if Morwen had seen me in the flames, she would try to slip outside alone and come to me, as she had before. She'd surely know of a place I could take shelter, where no one, not Gleedy, Kendra or any outlaw, would find me. Morwen knew every rock and bush in these parts and who else could I turn to? I could never go back to the tinners' valley, that much was certain.

I huddled beneath a rowan tree to keep watch. But though I stared at the cottage for what seemed hours, no one slipped out through the leather curtain. The clouds surged across the moon. A wave of darkness broke over the cottage, then receded, exposing the tiny dwelling again, like a silver-grey rock in an inky sea.

I was exhausted. Cold burrowed deep into my bones. Every muscle stiffened. How could I tell Morwen that I was close by? We hadn't thought to invent a signal.

Will worth! Will her to come out. Charm her to come.

I'd no worts I could use, or any fire, but I plucked up grass in front of me until I'd cleared a small patch of earth. Then I drew three circles in it with a twig, as Morwen had shown me by the black pool. I unbound my head clumsily with my one hand, which in the cold had become almost as numb and dead as the other. My hair whirled about in the wind, as if each lock was a tongue darting out to taste the air. I tore a few strands from my scalp and, clamping a small stone between my knees, wound the hairs about it. I held up the stone to the darkness.

'Morwen, hear me. Morwen, come to me.'

Over and over I said the words, trying to pour all of my body and mind into them. But my strength was failing. I could scarcely keep upright. All I wanted to do was sleep, and still no one stirred from the cottage. I forced myself to sit up and try once more, daring this time to call upon Brigid to help me, as Morwen had done.

I'm kneeling by the edge of a deep lake. A horned moon is reflected on the water, the stars flash in it, so that for a moment I think I am looking up into the sky. A line of rocks, made silver in the moonlight, stretches out into the water, flat and just wide enough to walk upon. And I do walk on them, towards the centre of the lake. The water on either side of this causeway is getting deeper. If I slip, if I fall, I will plunge into its depths. But now the causeway begins to slope down into the lake. The rocks on which I'm treading lie just below the surface. My feet splash through water, soft as raw wool against my skin. The path slopes deeper. The water splashes my thighs and I stop. My feet will not stay down on the stone. I'm floating. I'll be swept off it.

I start to turn, trying to retrace my steps. But now I see where the causeway is leading. A narrow rock juts out of the lake. I think it is all rock, but now I see that a living woman crouches on the pinnacle. She's bound hand and foot. I can't see her face, but I know it's Morwen. If she moves, if she falls, she will sink, she will drown.

Kendra has put her there. Kendra will keep her there as she keeps Ankow bound to the paths of the dead.

Mother Brigid, help her! Help her to fight. Help her to break free.

The throb of a horn carried towards me on the wind. It sounded again, like the one that was blown in the tinners' valley, except this was coming from somewhere much closer than the camp. The tinners' horn wouldn't carry so far and it wouldn't be blown at this hour. I dropped the stone, ran towards a small clump of willow scrub and crouched there, straining to peer through the darkness. I could see shapes moving, but it was impossible to tell if they were wind-whipped bushes or men. Then I saw the flames of three torches, coming over the rise across the valley, heard the excited yelping of hounds. Gleedy and his two henchmen were tracking me with the pair of guard dogs he kept chained by the storehouse. The dogs had already caught my scent as Gleedy dragged me there – they'd easily find my trail.

Terror flooded through my body, and my legs, which had been aching with cold and weariness, suddenly found a new heat and strength. I fled, stumbling and tripping over the tussocks of grass. Ignoring the bushes that grabbed at my skirts and tore my skin, I ran without any idea of where I was going.

The clouds peeled away from the moon, and all at once the moor was bathed in a cold white glow. I glanced back over my shoulder. I could see the men now, the two dogs straining and panting on long leashes in front. They were making straight for the rowan tree where I had been watching the cottage. But, once there, they would quickly pick up my scent leading away. With the moon shining so brightly, they had only to look in my direction and they'd see me as plainly as I could see them. If the dogs had been off the leash and far away from the men, I could have whistled to drive them off, but now it would draw Gleedy and his men straight to me.

I dropped to my knees and found myself staring into two glowing green eyes. The creature was standing sideways, gazing right at me. For a moment, I thought it was one of the hounds, and then I saw it was a fox, a black fox, just like the one I had seen watching me back in the village when the river had turned to blood.

The beast ran a few yards away from me, then stopped again, turning its elegant head to look at me. I knew it was waiting for me. I scrambled to my feet and followed. It quickened its pace and I ran. Behind me, I could hear the voices of the men swooping towards me on the wind. The moon vanished behind the clouds. Dertemora was plunged into darkness. I couldn't see the fox and, for a moment, I thought it had gone, but it turned its head and the shining green eyes shone like lanterns in the night, drawing me on.

The fox was running towards a dark smudge of trees, tucked down in the fold of the hill. I tried to keep pace, but several times I slipped on the mud or caught my foot in the tangled stems of heather. But fear always dragged me up again, until finally I reached the trees and stumbled into the grove.

Twisted roots snaked over and around waist-high boulders, each covered with a shaggy coat of moss. It was impossible to run inside the wood, almost impossible to walk. The rocks were so closely packed together on the forest floor that I couldn't step between them, but had to clamber over. My feet constantly slipped off the wet moss into the deep, treacherous gaps between the boulders, which threatened to trap my ankles or break my toes. All the time I was scrambling deeper among the trees, the fox was bounding ahead. Every time I thought I had lost sight of it, the creature stopped and turned its blazing eyes upon me. Then, as suddenly as it had appeared, it vanished. I frantically searched this way and that, trying to glimpse a movement or a flash of shining green. But it was gone.

My leg plunged into a deep hollow between two rocks

and I realised that the space was just wide enough for me to squeeze in and sink down between them. I hunkered there, listening. From somewhere close by came a gushing and gurgling of water as it rumbled the stones on the bed of a stream. I could still hear men shouting and the excited barking of the dogs, but unless they were muffled by the trees, the sounds seemed to be growing more distant, as if the hounds had turned and were following another scent. The scent of a fox?

Chapter 41

Morwen

'You awake, Mazy-wen?' Taegan whispered.

'Course I am. Could you sleep in here?' I dragged myself up, groaning as my back scraped against the rock. It was as black as a bog pool outside, and I guessed that if Taegan had crept home, the night was half over.

Ma and Ryana had tethered me in the stone dog pen. My granddam had kept a watchdog in there years ago, afore it was killed by the King's men for they said it was a hunting hound. Ma had chained me with the old hound's iron collar round my neck, like a beast, so I couldn't lie down properly or stand up. The chain went through a hole in the rock and ended in a bar on the other side, which caught against the stone if you pulled it. Only if someone twisted the bar flat to the chain and pushed it back through the hole could you get free, but Ma was in no hurry to do that. And I couldn't reach it.

Taegan glanced towards the cottage, then squeezed her fat udders through the opening and dragged her arse in behind them. She squatted on the wet earth and pulled out

340

something wrapped in a bit of old sacking from between her dugs. 'Here, Daveth give me this. I reckon you must be fair starved by now.' She shoved a small hunk of what felt like cold meat into my hand. I was dumbstruck. Taegan never shared what Daveth and his brother stole for her. I sniffed it, sure she must be tormenting me with something putrid, but it didn't smell rotten.

'You going to eat it?' she whispered.

'What is it?'

'What do you care? It'll fill your belly, won't it? Ma'll not give you so much as a bulrush root, till you cry and beg for it.'

My belly was so empty that I'd been pressing stones against it as hard as I could to stop the pains, but I snorted. 'I'll not be doing any begging if that's what she's waiting for.'

I could only make out the whites of Taegan's eyes, but I heard her laugh.

'I reckoned as much,' she said. 'Twin yolks in an egg you are, both as contrary as each other.'

I split the meat in two, wrapped one piece in the sacking and shoved it down the front of my own kirtle, then tore a strip off the other with my teeth. It was tough and stringy, horse probably, stolen definitely. Daveth and his brother could pinch an egg from under the backside of a goose without it giving so much as a squawk. But food was, mostly, all they thieved.

'Thanks,' I mumbled, as I chewed the last mouthful.

'Aye, well, you've covered for me plenty of times with Ma.'

'That she has!' a voice growled from outside the pen. Before Taegan had time to turn, a hand grabbed a fistful of her thick hair and jerked her backwards out of the entrance.

There came a slap loud enough to scatter crows, and a yelp of pain from Taegan.

'Seems to me you need a taste of what she's getting,' Ma yelled. 'Only got one loyal daughter and that's my Ryana. She told me you'd sneaked out again. Planning to let the little traitor loose, were you? Is that what you were doing skulking round here?' From Taegan's squeals, I guessed Ma was yanking her hair or twisting her arm. 'Answer me, you idle trapes.'

I heard the sound of wood hitting flesh and Taegan's squeals became shrieks.

'Stop that!' I yelled. 'You let her be! I won't let you beat her.'

Somewhere outside Ryana laughed. 'And how are you going to stop her, Mazy-wen? You're all chained up like the mangy little cur you are.'

I might have guessed Sheep-face wouldn't be far away, taking a malicious delight in any punishment the old hag meted out to Taegan or me.

Rage boiled up in me. I yanked violently at the iron chain, trying to smash it through the granite so I could get to her. I wanted Kendra's bones to turn to ice, her skin to itch without mercy as if a thousand gnats were biting her, her guts to sear in agony as if foxes were gnawing at her belly with their sharp fangs. I had never wanted anything so badly.

Brigid, Mother Brigid, help me!

Someone was shrieking outside, but it wasn't Taegan now. Ma was howling louder than the wind in a winter storm. 'Be gone! Be gone! Stop them, Ryana, help me. Drive them off!'

'What, Ma? What is it? There's nothing there, Ma.'

Ma screamed again and tottered in front of the opening where I was chained. She was lashing out wildly with her staff and flailing, all the while trying to cover her head with

342

her free arm. Then I saw them. They were slithering out of the mire towards her, swooping down from the night sky, creatures with sharp leather wings and clawed talons flashing in the moonlight. They were fastening on her scraggy face, her scrawny arms, her wrinkled neck, her balding scalp. As soon as she dashed one away, it would haul itself back up her body, its claws clamped in her flesh, squealing in rage.

'It's her! It's Morwen! She's calling them!' Ma yelled.

She was howling every hex she could at me, but in that moment, I knew I was stronger.

'Stop her, Ryana, make her stop.'

But Ryana was staring fearfully at Ma, as if she thought she'd run mad. She couldn't see those creatures. She'd never been able to see the spirits or demons. I knew that. I'd always known that. And I laughed.

I felt the chain rattle and slither down my back. Taegan must have scrambled to the other side of the rock and released the bar. I tore it from my neck and crawled out. My legs were so cold and stiff I could barely stand, but I knew I had to run. Ryana was scuttling back to the safety of the cottage. Taegan had wisely vanished, leaving Ma leaping wildly about, as if she was dancing in red-hot shoes, beating and clawing at her own skin, and screeching every curse she knew at me. A great bubble of triumph shot up inside me. I had beaten her. Ma knew now that the gift was mine, not Ryana's. It had never been Ryana's. And it was stronger in me than it had ever been in Kendra. I was the keeper now.

Chapter 42

Prioress Johanne

Sister Clarice squinted anxiously down the hill towards the pinfold, but there was still no sign of our herd of goats the shepherd lads were supposed to be driving into it. The two of us had taken a walk outside the priory gates ostensibly to select those beasts to be slaughtered and salted for winter. Only outside the priory walls could I be sure that we could safely talk without being overheard.

I glanced at my steward. 'Brother Nicholas is no fool, Sister Clarice. I take it he has examined the accounts relating to our own beasts and the offerings left at the well?'

'He's raked every column, like a virgin maid combing her hair for fleas – only the copies of the ones we sent to Clerkenwell, of course. But he's seen our bee skeps and our cattle, so he knows – or our little ferret *thinks* he does – that they're not a true record. I've had the man pestering me from morn to night, insisting that there must be other ledgers. Sister Melisene says she saw him slipping out of our dorter when we were all at supper, and I'm sure my chest and bed have been searched, though I can't prove as much. But just let me catch him with his hand in my boxes again and I'll chop it off, knight or no knight,' she added, in a tone of such grim determination that I was convinced she would carry out her threat if he provoked her much more.

I was always careful to lock the door of my own chamber, but I reminded myself that I must make doubly sure to do so, even if I was called away in a hurry to tend Sebastian.

I had never intended to deceive the Lord Prior over the money we obtained from our livestock and the well. When I had become prioress eight years ago, the income from these sources had indeed been meagre, and for the first few years, Clarice had sent faithful and accurate accounts to Clerkenwell, delivering a third of all we received to the Lord Prior as every priory was commanded to do.

But these last few years, as the farmers' cattle and bees in the lower valleys had sickened and died in the perpetual wet and cold, ours had fared somewhat better up on the high pasture. As meat, honey and wax had grown scarce, so the price they fetched in the markets and fairs of the towns had risen, which was a blessing, at least for those who had such goods to sell. And since I had dedicated the holy spring to St Lucia, the fame of the healing well had spread far beyond the moor, carried by those making their way to and from the ports. The pilgrims had come in ever-increasing numbers, and while few left the gold and jewels that the holy relics of the famous saints could attract, nevertheless their offerings amounted to substantially more than they had been in the beginning.

But I had begun to wonder if it was judicious to record the full amount of the priory's improved income in the accounts we dispatched to Clerkenwell. The Lord Prior and his stewards would see only that the priory's wealth was increasing and the responsions would likewise rise, which would naturally gratify them at a time when most other priories were failing to deliver: Rhodes was like a dog with worms, voraciously gobbling ever increasing sums of money to support the growing fleet.

But if the seasons changed again or cattle murrain struck, our remote priory would not be able to spare the third of what little income we might receive. The Lord Prior was not a man to take such misfortunes into account. I doubted he even knew, much less cared, that bees must have flowers or that cattle sickened. He would see only that we had failed – that *I* had failed. So, like the wise virgins, whose example Christ commanded His disciples to follow, Clarice and I had, so to speak, been setting aside a little of the oil that might be needed for our lamps in the future, so that, should disaster strike, the priory could survive while the sum we had always delivered could continue to be sent to Clerkenwell.

I had not kept back the money like some crooked tax collector or coroner, creaming off coins owed to my master so that I might live in luxury. It was simply prudence, good housekeeping. I had husbanded our money, so that we could shield ourselves and those defenceless ones we cared for. But I knew those at Clerkenwell would never understand that. They would regard it as nothing less than theft, and for such sums as were involved, Clarice and I could expect no mercy. The Lord Prior would make certain that we were stripped of our right to be tried in an ecclesiastical court, and if we were tried before the King's courts we would not escape with our lives. I might deserve such punishment, but not Clarice. I couldn't bear to see such a loyal friend end her days on the gallows in front of a jeering mob and know that I had brought her to it. Brother Nicholas had to be stopped.

As if she had read my thoughts, Clarice briskly patted my arm. 'There's one blessing, Commander John's tame ferret can't send any more of his letters out until Hob returns after the frosts have set in. And I reckon we'll see the back of our two brothers once and for all then.'

'But the moment Hob returns Nicholas is bound to ask why Commander John has sent no answer to earlier reports,' I said.

'If he brings no answer, which he won't since they never received them, it's my reckoning the brothers will decide to ride back themselves, so that Nicholas can speak to the commander in person. That'll buy us time, and if he's no real proof that any sums have gone missing, what with the famine raging, Clerkenwell will more than likely decide there's no point in digging further, for they'll have greater dragons to slay than us.'

I hoped she was right, except that it wasn't simply the missing money. There was Sebastian and what Nicholas knew, or thought he knew. It hardly mattered if he couldn't prove that the accounts we submitted were false, he had only to speak the word 'heresy' and all other crimes real or imaginary would be proved. Sebastian was all the evidence of that they needed. And if Nicholas realised exactly who Sebastian was . . .

There was much I had confided over the years to Sister Clarice. She probably knew more about me than my own confessor, but not this. No one knew of it. There are some wounds that can never bear the touching and some secrets too dangerous to share with even a trusted friend.

I gazed out disconsolately over the valley. The sodden grass and bracken, beaten down by the rain, were decaying where they lay and the slopes of the hills were the colour of dung.

Even the leaves on the trees sheltering in the folds of the land had turned from green straight to dull brown with none of the gold, orange and ruby I recalled from the autumns of my childhood. The rowan berries, usually a bright splash of scarlet, had, like the purple bramble fruit and blue whortleberries, shrivelled and fallen before they had ripened. Everything inside the priory – clothes, leather-bound books, walls and hangings – was covered with mould and stank of damp and mildew. Clumps of fungus, the colour of old bones, had sprouted around the walls of the courtyard and even sprung up overnight in the chapel. The whole world was rotting away.

Clarice took a few paces forward, peering anxiously down the hill. 'There's one of the boys coming now. Where's the other with the rest of the herd? If they don't make haste, the sun will be setting again before we have even the first beast butchered.'

We started down the hill towards the stone pinfold. A solitary lad was trudging along a sheep track below us, his bare feet so caked in mud, it looked at first glance as if he was wearing boots. Three goats hobbled in front, and in spite of the whacks he kept delivering to their scrawny back-quarters with a long, thin birch switch, they did not quicken their pace. At the sight of us bearing down on him, he stopped and half turned, as if his first impulse was to run away. But he seemed to think better of it. With a last push, he herded the listless goats inside the stone wall enclosure and slid the wooden board into place.

I would be the first to admit I knew little about the care of goats. I merely gave the orders that they should be cared for, but even I could see something was badly wrong with those pitiful creatures. Their hair was matted. They were painfully thin and limping from the hoof rot. They had patches of bald skin on their backs and necks, which in some places was scabbed, or raw and oozing, as if they had been burned.

Clarice reached the pinfold first and dragged out the wooden board. Without saying a word, she clamped the goatherd's ear between her thumb and forefinger, making him yelp, and dragged him inside the stone wall. She let him go, only to seize one of the goats by its curly horns and examine the sores on its neck.

'Rain scald!' she barked, releasing the animal and turning furiously on the lad. 'You were charged with the care of these beasts. Didn't you see the signs? If you'd got them under shelter and treated the scabs straightway you could have stopped this. You've tended goats since you were in clouts, didn't the older boys teach you how to cure it?'

The boy rubbed his ear, looking mutinous. 'Goose grass, devil's leaf and great bur, that's what you use. But great bur only grows where it's dry and the devil's leaf had all gone mouldy. Could only find a bit of goose grass. Not enough for all of them.'

'And just where are the rest of the herd and the other lad who's supposed to be minding them?' Clarice demanded.

Fear suddenly flooded the boy's face. He darted a glance at the entrance to the pinfold, but I was blocking his escape.

'Had your tongue cut out, have you?' Clarice snapped. 'Answer me, boy.'

He stared at the ground, tracing a triangle in the mud with his toe. 'Tinners and outlaws stole a few, but then last night . . . We penned the goats up, we did, in the fold. Me and Kitto sat outside the gate to guard it, same as always. We had a fire lit, but only a small one, mind, 'cause we were trying to dry the rest of the wood. It was Kitto's turn to keep watch, but something woke me . . . a horn, then a pack of hounds baying answer. But no one goes hunting in the dark of the night. Kitto was nearly shitting himself. He reckoned it was Ankow riding cross moor looking to collect dead souls. He said if you heard the wisht hounds, it meant you'd be next to die. They were coming right down the hill towards us. Kitto, he opened the gate to the pinfold. I reckoned he was going to hide behind the wall, but afore he could get inside, the hounds were on us. Some went in through the gate, others they leaped clear over the top.

'Goats were bleating and the hounds were chasing them round. Me and Kitto, we only had our slings and staves. Couldn't fight that many dogs – besides, no one can fight the wisht hounds. So, we ran and hid. After a bit, we heard the horn again, closer this time, like it was calling the hounds, and they ran out and vanished.

'Kitto took off then. I reckon he hared it back to the village. I daren't move, case the hounds were still out there. Didn't close my eyes all night. First light there was this great screeching and cackling. I crept close to the pinfold. Couldn't hardly see the goats for this huge flock of crows, ravens, kites and buzzards all squabbling and fighting over the carcasses, pulling off great strips of meat. I fired my sling and yelled, but there was too many of them. They scarcely stirred.'

Helplessly he gestured towards the three goats. 'Found them wandering when I was on my way here to tell you – must have slipped out, when the hounds attacked.'

'More likely they were never in the fold and you'd lost them days ago,' Clarice said.

The goatherd looked as if he was about to protest, but thought better of it and simply shrugged.

Clarice looked as grim as I had ever seen her. 'I'd best

take some of the servants and see what can be salvaged. I'll wait here, Prioress, while you go to the kitchens and tell them to send the men down here. Ask them to bring moor sledges, so we can drag the meat back. The lad can show us where the carcasses are.'

The boy shuffled his feet. 'There'll not be much left after birds have been at them. And the goats have their throats torn out. It'll make you sick to your stomach to look at them. Best not go.'

'I can assure you, boy, my stomach is stronger than most and I've seen a good many beasts after dogs have mauled them.' Clarice eyed the goatherd suspiciously. 'Why do you not want me to see them? Will I find fewer carcasses than I expect?'

'Told you,' the boy said sullenly, 'the tinners, they stole some.'

'And if Kitto spreads the news of what's happened, the villagers will spirit away the remains of the rest.' She clamped a hand on his shoulder. 'On second thoughts, you and I had best go to find those goats now, boy. Prioress, can you send the servants down after us? Tell them to come with good stout staves and maybe a bow or two.'

I nodded. 'Perhaps our brother knights could help to guard those carcasses. Nicholas has been itching for an excuse to draw his sword. I'm sure I can see to it that the task keeps them safely occupied for the day.'

The goatherd stared from one to the other of us, evidently bewildered by the grim smiles that suddenly flashed between us. But I had underestimated Brother Nicholas.

By the time the servants and knights arrived at the pinfold where the goats had been killed, the villagers were already at work plundering the spoils, ignoring Clarice's orders to stop. As soon as they glimpsed Nicholas and Alban bearing down on them, they ran off down the slope, nimble as the goats whose remains were now slung over their shoulders. Apparently, Nicholas had charged after them and managed to give one of them a slash across the back with his sword, sending him tumbling head over heels. But the haunch of goat the villager was carrying had landed in a bog pool where it sank into the black ooze.

I had suggested to the servants that they might butcher the first sledge-load before going back for the next, pointing out that the cooks and scullions could start the boiling and preserving while they dragged back the next load. That way, Nicholas and Alban would be standing guard for the best part of the day in the rain and cold, for Clarice had left them keeping watch with only young Brengy, the stable boy, for company.

But late that afternoon, a clatter in the yard made me whirl round. Alban was leading his horse across the yard, muttering to himself. He faltered as he caught sight of me, but led the horse purposefully forward, staring straight ahead as if he had not seen me, though I was sure he had.

'I understood that you were doing us a great service by keeping watch on the pinfold, Brother Alban. Surely you're not deserting your post.'

He hesitated, clearly wondering if he could pretend not to have heard me, but reluctantly stopped. 'Only a few carcasses left down there now. Villagers have seen Brother Nicholas's handiwork with a sword. They'll not be fool-hardy enough to tackle a knight of his skill. Not worth getting their heads lopped off for a scrawny haunch of meat.'

'Men whose families are starving might well think it worth the risk,' I said, but Alban merely grunted.

I stepped closer. Blankets and packs were tied high around the saddle to help keep him firmly seated on the muddy track. Plainly he wasn't riding out to inspect some local mill or tenant's farm. 'It is late to be riding out, Brother Alban. And you look set for a long journey. Where are you bound?'

'Brother Nicholas's orders,' he said dourly. 'Reckons there to be a good two hours of riding light left.' His voice lowered as he continued grumbling, but he seemed to be addressing himself. 'Could have waited till morning, get a full day in. What difference will an hour or two make? He's not the one'll have to sleep out.'

'Sleep out?' My stomach lurched. 'Is Brother Nicholas sending you to Buckland?'

I caught a twitch of Alban's mouth at my mention of the name. For a moment, I thought his shifty expression was because I had guessed correctly, but then the real reason dawned on me.

'You're not going to Buckland, are you? Where are you bound?'

Alban glanced behind him towards the guest lodging. He seemed to be considering whether or not to answer, then finally he shrugged. 'Clerkenwell, that's where.'

'But it is madness in this foul weather. Even if the horse doesn't get mired, half the bridges have been swept away and the rivers will be far too swift and swollen to ford.' I caught hold of the reins, trying to think of some way to keep him there. 'At least stay until morning. You'll need a full day's riding even to reach a road that is well travelled. Go blundering about the moor in the dark and you'll break your neck.'

He stiffened and I knew at once that I'd said the one thing guaranteed to make a man like Alban go. 'I've yet to put my leg over a horse that can throw me, Mistress. Flood, snow and storm, I've ridden in them all and there's never been one that's stopped me going wherever I've a mind to.'

He jerked the reins from my hand and led the horse around me towards the gate. Meggy was already struggling with the beam that braced it. The old gatekeeper seemed to be having much more trouble than usual in releasing it. Everything was swelling and sticking in the wet. Alban watched Meggy for a moment or two, then impatiently nudged her aside and thrust the reins into her hand, leaving her to pat and soothe the horse, as he wrestled with the beam, but it took even him an age to drag open the heavy gate.

I watched him lead the horse out on to the moor, sick with dread. There was nothing I could do to stop him. Our time was running out far more quickly even than I had feared.

Chapter 43

Sorrel

Todde lies in the mud behind the spoil heap. Blood pools around his thighs and the earth is too sodden to soak it up. It vomits it back. It will drink no more. Todde's belly is mangled, slashed by Gleedy's knife, which stabbed and twisted and stabbed again. Gleedy's boots are stained with scarlet as he viciously kicks and stamps on the dying man.

'Think to stop me following her, did you? I'll track her down and make her watch the dogs chew you up and shit you out. She'll not be weeping over your grave, Toddy. There'll be nothing left for her to bury.'

But I do weep. I scream at Gleedy to stop. I thrash at him with my fists. But though he hears my shrieks, to him they're only the cry of the owl, and though he feels my fists, it is only the buffeting of the wind.

A fox, a black fox, suddenly appears above Todde's body on the ridge. It stands motionless, its head turned towards me, ears pricked. We stare at one another, its brush streaming out in the wind like a flame from a blazing torch. I smell its animal scent. It grows stronger. It envelops me. And I know that the boot

355

which slams into Todde's ribs and smashes his nose can do no more hurt to him now.

A tiny shriek jerked me back into another place. I couldn't grasp where I was, and I was too afraid to move. Giant flakes of snow were falling and swirling about me. One drifted down on to my face, but it wasn't cold. It was dry and scratchy. Leaves, that's what they were, not snowflakes. I was lying in a small copse, my body twisted between mossy rocks and snaking tree roots. Through the trees, I could just make out the smudge of a distant tor, black and jagged against a pale ghost light, which marked where the sun would shortly rise.

A weasel, slender and sinuous as an eel, bounded on to a moss-covered rock and stood on its hind legs, warily sniffing the air. My alien human scent hung between its hunting ground and the den it was making for in the roots of the twisted oak.

A mouse scurried through the undergrowth, trying to reach its nest before dawn, just one more rustle among the thousand whispers of the dried leaves. Only the weasel could smell it. The hunter was on top of its prey in a single leap, wrapping the muscles of its body around the mouse like the coils of a snake, squeezing the frantically beating heart. The mouse screamed. The weasel sank its fangs into the neck of its prey. The mouse shuddered, and lay still. The hunter bounded off towards another of its dens, breakfast dangling, like a hanged felon, between its jaws. They say our souls leave our bodies as mice when we dream. It was well for me the weasel hadn't noticed mine scurrying back.

I dragged myself upright in the gap between the mossy boulders, rubbing my stiff neck. As the breeze hit my back,

I shivered. My kirtle was soaked through from the sodden ground, but I was well used to waking to that.

I could hear the burble of running water somewhere close by. My throat felt as if it had been stuffed with old tree bark. I tensed, listening for any sounds of Gleedy's dogs. But they were probably back at the camp by now and he had more than likely set them on to poor Todde's corpse to obliterate the stab wounds. I gagged and tried not to think . . . tried not to think that Todde was dead because he'd protected me. I wanted Gleedy to pay and, somehow, I *would* find a way to make him pay for Eva, for Todde and for me too. I wanted to tear him apart. I wanted him to die pleading for his life and I wanted to hear the fear and pain in his voice, see them in his twisted soul.

Rage only increased my thirst. I clambered towards the sound of running water, though my legs were so numb I could barely stand. My feet kept sliding off the sodden moss, and my legs slipped down the gaps so that I was continually banging my ankles and knees on the rocks, but it was still so dark beneath the trees, I was more afraid of stumbling straight into the river.

As the first pink shaft of light penetrated the small gap in the trees, I saw a twisted rope of pearl-white mist threaded through the trunks and caught the glint of water racing beneath, spilling over the clawed roots of the nearest trees. But above the rush and gurgle of the stream, I was sure I could hear other sounds – voices, a high-pitched murmuring as if people were hidden behind the ribbon of fog, or even in the water itself. They were like the sounds on Fire Tor, but these were fainter, higher, like a host of soft-spoken women. I stopped, balancing precariously on a boulder. Were spirits watching me through the mist?

My tongue was glued to my teeth and there was a foul taste in my mouth, like the stench of wet dog. My belly ached for food. Maybe if I drank deep it would fill my stomach and stave off those hunger pangs, but I was afraid to go any closer. When I was a child, Mam had often warned me about the water sprites who lured people to the riverbanks with their singing, then reached out their long, icy arms and dragged their victims beneath the water to devour them in their slimy beds.

I peered through the billowing white mist. It seemed to glow with light of its own under the dark, twisted branches. Great beards of silver lichen hung from the twigs, stirring faintly in the breeze. As the wind swirled the edge of the fog, I thought I could see a dark shape moving, the outline of a head turning towards me. I jerked backwards. My foot slipped on the moss and I crashed down against the trunk of the tree, dashing the breath from my body.

I was fighting so desperately for air that, though I was dimly aware of something or someone moving silently towards me through the gloom, there was naught I could do to protect myself, save raise my arm to try to fend it off.

'You bang your head?'

I knew that voice.

Morwen vaulted like a cat over the boulder from which I'd clumsily fallen and crouched beside me, holding me up till I could drag air back into my lungs.

'Thirsty,' I gasped. I needed water so badly now that I couldn't think of anything else. Morwen hauled me to my feet and steadied me, as if I was a babe learning to stand, as we climbed over the rocks and roots to the river. The sky was lightening all the time, but the mist clung chill and damp to my face. We both knelt by the stream. Morwen cupped her hands in the icy water, tipping them towards my mouth, but she dribbled much of it down the front of my kirtle. I pushed her hands aside and bent forward cupping the water to my mouth, with my good hand, while Morwen grasped the back of my gown to stop me falling in. There wasn't enough water in the whole river to extinguish the fire of my thirst, much less fill my belly, but finally I dragged myself away and sat up.

Morwen reached into the front of her own ragged kirtle and pulled out a small piece of stringy meat, which she placed in my lap. I didn't stop to ask what it was, I sank my teeth into it. My belly was so empty, I couldn't stop myself gulping it down in lumps, which stuck in my raw throat for I'd barely chewed it in my haste.

I nodded at her, still sucking the last fragments from my fingertips. 'Can't remember when I last ate anything solid.'

''Bain't much. Taegan gave it me. Ma's been keeping me hungry after Ryana told her I'd taken you into Fire Tor.'

Shame and guilt burned my cheeks. 'I thought you'd saved . . . I'm sorry.'

I was aghast at my selfishness. No one had food to spare unless they took it from their own mouths. But the truth was, I'd been so eaten up by hunger I hadn't thought about where it had come from.

Morwen grinned. 'I like watching you eat. You came last night, close by the cottage.'

'How did you know?'

'Saw you when I was journeying. Found your mark when I got free.' She dug into her kirtle once more and held up the stone still bound with my hair. 'It fetched me here.'

'Got free?'

Morwen grunted. 'Ma and Ryana tethered me in the pen.

359

Ma used to tie us up in there when we were chillern, if we'd vexed her. Tell us the wisht hounds would come for us. She's not done it for a good long time – doesn't have the strength now unless that cat Ryana helps her.'

Morwen massaged her neck and, in the growing light, I could see the red ring about her throat. She'd been chained up like a dog. I glimpsed other marks too, a purple welt just showing above the top of her kirtle, and from the way she flinched as she moved, I guessed there were more on her back. Kendra and Ryana had done more than tie her up in a kennel.

'It's my fault, I shouldn't—'

She laughed. 'Never met such a one as you. I reckon if the stars came tumbling to earth, you'd be saying it was your fault for bumping into them.'

She listened without saying a word, while I told her all that happened in the tinners' valley. 'And I can't ever go back,' I finished.

She looked at me as if I was mad. 'Why would you want to? Brigid called you to her. She brought you here. She's been biding her time, waiting till you learned what you needed.'

'She spoke to me again.'

Resentment flashed across Morwen's face, though she struggled to push it away. 'What did she say?'

'Protect the boy. That's what she said. But I don't understand. What boy? You always seem to know what these things mean.'

A slow smile lifted the corners of Morwen's mouth, as if something she had been fretting over suddenly made sense to her.

'You know, don't you?' I said. 'You know who the boy is.'

Still giving me that strange look, she shook her head. 'Not yet,' she said softly, 'but I reckon I will. Tell me what Brigid showed you, what you saw when she told you to protect the boy.'

She listened patiently to the tale of the child atop the great pyre on the tor and her smile widened into a gap-toothed grin. 'Ma was wrong. I was sure she'd not looked at that smoke right, like I was sure that blood charm she gave you had no power. I know who he is now. I know his name!' Grabbing my hands, she dragged me to my feet, dancing around me in delight.

'I reckon it's almost time. The three sisters are come into their power.' She glanced up as a skein of geese flew honking towards the rose-pink dawn. 'But we'll have to find a place to hide till the hour comes. That Gleedy might decide to have another go at tracking you in daylight.'

'I want to kill him,' I said. My jaw clenched so hard I could barely get the words out.

Morwen laughed. 'Brigid will take revenge on him, never you fear. When Brigid's cubs are harmed, the great mother's more savage than a she-wolf. When her anger grows, there's nothing and no one can stand against her. You wait and see. When I went journeying, I saw—'

She broke off and tilted her head as if she was trying to see around something that was blocking her vision. 'The river's crying, can you hear it?'

'I thought it was a spirit, like in Fire Tor.'

'No, that's her, that's the river,' Morwen said urgently. 'It's her warning. When she cries like that it means death. It's beginning.'

361

Chapter 44

Hospitallers' Priory of St Mary

Although darkness had descended over the priory, fires were still blazing in the courtyard and the steam from boiling goats' heads mingled unpleasantly with the smoke from the wood and peats. The flames beneath the great iron pots and the torches burning on the walls lit up the yard with a demonic red glow, and the figures of servants and sisters emerged and vanished again into the eye-stinging fog like wraiths on All Hallows' Eve.

Brother Nicholas picked his way across the yard, slippery with mud, blood and the evil-smelling green sludge from the goats' stomachs. Although he was coughing violently, he was almost grateful for the dense smoke: it hid the stack of skinned goats' heads that had been heaped by the kitchen door, their sightless black eyes staring out at their kin already drowning in the bubbling grey water of the cauldrons.

For once, the smell of cooking meat was not making Nicholas hungry, quite the reverse, for the stench of blood and dung was equally powerful and he could not stop seeing those beasts lying scattered across the pinfold, their heads almost ripped from their bodies and their guts spilling out into the mud.

An image of another pile of heads flashed unbidden into his mind, human heads, chopped from their bodies while their owners were still screaming for mercy, or muttering fervent prayers, which were severed long before they were finished. The ransom demanded had not been paid. Saracen prisoners were executed in the full sight of their friends, wives and children as the spotless sun blazed down and the azure sea sparkled behind them. Whatever men might boast by the firesides in England, there were some orders they were forced to carry out in war that were neither honourable nor glorious. Nicholas chided himself. He was growing soft here in this cold, grey land, or maybe it was old age creeping up on him. But he didn't want to admit that.

Through the smoke, he caught flashes of steel turned ruby in the firelight, heard the rapid chopping of an axe blade smashing through bone. Three or four servants were working on a trestle table erected in the courtyard, skinning and dismembering the carcasses, cutting away the chewed flesh the dogs had mauled and tossing what could be saved into the wooden tubs behind them. They would most likely be chopping, boiling, pressing and salting meat till dawn. With the flesh so mutilated by bird and hound, it had to be preserved within hours else it would turn foul.

At least with the coming of darkness, the birds had finally departed. Not even the smoke and fires had discouraged them. Every roof ridge and post in the priory had been black with squabbling ravens and crows, while dozens of kites circled and screamed overhead, constantly swooping to snatch meat from the table or peck at the pile of heads. The birds even clawed and stabbed the servants' hands and arms, trying to steal the raw meat they were holding. They had spent more time shouting and lashing out to drive the feathered thieves away than they had skinning and cutting. But even as a maid fought off one bird, three more would be stealing from the barrels behind her. Now cooks, scullions and even the stable lad sweated, chopped and sliced at a feverish pace to make up for the time they'd lost.

Nicholas reached his chamber and barged in, slamming the door quickly before the smoke and stench could follow him. He didn't bother to light a candle before dragging off his boots. He was so tired he knew he'd probably fall asleep before he could snuff it out. From the tail of his eye he caught a glimpse of movement on Alban's bed. That idle bastard was already slumbering, no doubt with a well-stuffed belly.

'I'm not stirring from this chamber till noon tomorrow,' Nicholas announced. 'My bones are colder than a witch's dugs and my back's as stiff as the devil's prick. Christ curse this foul land and all those thieving villagers. I'd hang every last one of them by their heels till their eyeballs bled.'

It was only as he flung himself down on his own protesting bed that he remembered Alban should not be there. He was supposed to be well on his way to Clerkenwell by now. Had that witch somehow stopped him leaving?

'Alban? What in the name of Beelzebub and all his flies are you doing back here?'

Nicholas flung his boot towards the dim outline of Alban's bed. There was an explosion of squeaks and scrabbling claws as a horde of mice leaped off and bounded across the floor and up the walls. A plague of frogs, then flies, now these vermin! What next? Locusts? he thought savagely.

He'd told the prioress that the frogs were a curse sent by

364

God, a sign of heresy, and now that he had discovered the marks of the Inquisition on that man's body, he felt vindicated. Alban was even now on his way to deliver a report informing the Lord Prior of England that Johanne was harbouring a heretic. But just who was the man? A Cathar, a Waldensian? Or was it possible the pathetic creature who lay twisted in that infirmary was the remnant of a Templar knight, a foul sodomite, whose brethren had caused men to turn against not only the Templars but the Hospitallers as well. Nicholas had urged Lord William to send a group of men-at-arms without delay to seize Johanne and Sebastian and drag them both to Clerkenwell for questioning. They would soon discover the man's identity and the heinous sins to which he had confessed.

If Sebastian proved to be a fugitive Templar, Lord William would be aghast at the thought that he had been given shelter in one of the Hospitallers' own priories and profoundly grateful that Nicholas had uncovered the secret before anyone outside the order had stumbled upon him. It would certainly divert the Lord Prior and Commander John from Nicholas's failure to produce any definite proof that money was going missing.

Yet a persistent thought was buzzing at the back of Nicholas's head. One he had not even wanted to contemplate. There was another to whom Prioress Johanne had stubbornly given shelter, one accused of killing a priest. Ice water ran through Nicholas's guts as he again felt the boy's arm writhing, like a serpent, in his hand, the skin turning to scales beneath his fingers.

He had refused even to think about what had happened in that chapel, pushing it fiercely from his mind each time it edged towards him. He told himself that the boy had somehow witched his mind. He'd seen ebony-skinned magicians on Rhodes cast their staves upon the ground and watched the rods transform into serpents and wriggle away before the eyes of an awestruck crowd. Those men could make your very eyes deceive you, as if you had drunk far too much strong wine.

But seeing is one thing, touching is another, and what he had felt in his hand was no trick of the eye. It was nothing less than a demon, its foul form concealed beneath a child's soft skin, like a cankerous worm lying coiled in the heart of a peach. But he had not confronted that fiend, not sent it howling back to Hell. Instead, he had run away from the creature, like a stable boy fleeing the field of battle.

Nicholas knew he could not make any mention of that encounter in the report that Alban was even now carrying to the Lord Prior of England, not if he hoped to be returned to the Citramer again, as Commander of the English Tongue there. A holy knight of St John, who admitted he could not even confront evil in the form of a small child, would be fortunate to find himself in charge of a midden heap.

From beyond the walls of the guest chamber he heard a yelp of pain and a stream of oaths from one man, answered by raucous laughter from the others. Someone had probably slipped on their backside in the filth of the courtyard.

But something else troubled him now. The night that Sebastian had tried to stab himself, when the heretic had shrieked and raged, sending all the other patients and sisters backing away in alarm, the boy had sat motionless on his bed, like a puppet that had been laid aside and would not move again until his master took him up. Had the boy been hag-riding Sebastian, driving him into fits of terror as he had the old priest? But why would a demon torment a

366

heretic? They were both the devil's servants. And there could be no doubt that Sebastian was a proven heretic: the Inquisition did not permit anyone to live who had not fully admitted their guilt. But the prioress herself had told him that Sebastian had been tormented by these fits of madness long before the boy had arrived. So, if the boy was not controlling the man, then . . .

Was the prioress not only harbouring a heretic but also a sorcerer? She was seen to have private, whispered conversations with Sebastian, to go to him when he was seized by the visions of the torments of Hell that awaited him or the evil spirits that surrounded him. Had Johanne forced Sebastian to conjure a demon to serve her in the form of the child? Was that why she had given the man shelter? Was this the payment she was demanding from him?

The boy had appeared as if by magic, the very night he and Alban had ridden in, and he had tormented the blind priest to death before Nicholas had had a chance to question him. It was the hosteller, Sister Melisene, who had first accused the boy of murdering the old priest, and she seemed a practical creature, well used to dealing with all manner of men. Was it possible that Prioress Johanne herself had commanded this demon child to kill the old priest to stop Nicholas discovering the truth?

Words and phrases swept through his head, carried on a swelling tide of fear, but the only one he could snatch out of the flood was *malum* – evil. Evil! Put a sword in his hand and he would gladly have fought a dozen armed Saracens singlehanded. He had been trained all his life to face such enemies without flinching, but demons, sorcerers, he had no weapons to fight those. And that terrified him.

The light from the fires outside filled his chamber with a

dull red glow and shadows slithered across the wall so that it seemed no longer to be solid, but rippling, dissolving in front of him. Smoke mingled with blood, and the sickly smell of boiling flesh pushed its way in through every crack and crevice. The clatter of bones being hurled into the pails, the chatter of the servants and chopping of knives and axes grew ever louder until he could hear nothing else. He could see the servants transforming into creatures with monstrous birds' heads and lizards' tails, the great open mouths in their bellies snapping at the mangled goats. And they were laughing . . . laughing at him.

Chapter 45

Prioress Johanne

'Vinegar for pickling has almost gone, Prioress,' Sibyl said, rolling her eyes balefully in the direction of Sister Clarice, leaving no doubt as to whom she blamed. 'And the salt's finished too. Ground up the last cat of it an hour since, and that was one I rescued from the floor of the pigeon cote. Had to scrape the bird shit and feathers off it afore I could crush it. Couldn't salvage any more than a pound or so fit for use. I suppose it's a mercy the cote is almost empty and birds won't be needing salt. But what am I to do with these? I can't even set them on the drying racks in this weather.'

She gestured towards several wooden tubs that stood against the courtyard wall. They'd been covered with planks that had had to be weighted with rocks to stop the ravens and kites, which were already massing in the early-morning light, finding a way in. Our cook's hair was straggling down beneath the cloth that bound it. Her face, hands and sacking apron were streaked with dung and rusted with dried blood. She, like the other servants, looked exhausted, but she was still chivvying and scolding any scullion she caught snatching a few minutes' rest. They had worked until dawn, and now that the last goat had been butchered, the yawning servants were scrubbing the blood from the trestle tables and tossing gory axes and knives into a tub ready to be cleaned and sharpened, just as soon as they'd filled their bellies.

'And those,' Sibyl pointed through the dark doorway into the cavernous kitchen where skinned goats' legs lay piled on the long table, 'I can build smoking stacks in the courtyard if we can get enough dry wood or peats, but what use is smoking the meat if I've no salt? And before you say honey, Sister Clarice, there's not enough of that left after the pickling to sweeten the temper of a toad. We should've had more salt in the stores.' She glared reproachfully at Clarice.

Seeing my steward's eyes flash dangerously, I hurriedly stepped between them. 'None of us could have known we'd have to salt down almost our entire herd of goats, Goodwife Sibyl.'

'And those salt cats don't sprout out of the ground like turnips,' Clarice said tartly. 'Have you any idea of the price those thieving merchants are asking for them? The panners can't dry salt in the rain any more than you can dry meat. You'll just have to—' She broke off, giving me a sharp jab in the ribs. 'You might want to make yourself scarce. Brother Nicholas is coming this way and he doesn't look happy!'

The warning came too late. I tried to step swiftly into the kitchen, but my direct route was blocked by a table in the yard and I found myself trapped.

I had asked a servant to deliver small ale and a platter of boiled goat's meat to Nicholas's chamber well before the usual breakfast hour, and instructed her to inform the knight that the refectory was being used by the cook to press the head meats for brawn, there being no space left in the kitchens.

She was to tell him that his meals would be served in his chamber and the sisters would eat in their dorter until the refectory could be cleared. I had hoped in that way I might avoid him for a few hours at least, but it seemed my prayer had not been answered.

Brother Nicholas's voice rang out imperiously even before he'd reached me. 'Where is the boy, Prioress Johanne?'

'If Brengy is not in the stables, he will probably be in the kitchen with his sister, Dye, helping with the goat carcasses. Why? Did you want your horse saddled? Are you planning to leave us like Brother Alban?'

I was aware of Clarice's gaze upon me, and found myself torn between hoping we were about to be rid of our brother and dread that she had been right: he would try to ride to Buckland and deliver his report in person to Commander John. It was too much to hope that neither Alban nor he would get through to Clerkenwell or Buckland.

'I'm not looking for the stable lad,' Nicholas snapped, 'I visited the infirmary last night, but I didn't see the blind boy there.'

My stomach churned. Why had he taken it into his head to visit the infirmary? Was it to question Sebastian?

'It was most thoughtful of you to visit the sick, Brother Nicholas. I am sure that they were much comforted.'

All the time I was talking, I was trying to edge away, though it was difficult: that corner of the yard was not only greasy with blood, dung and the contents of the goats' stomachs, it was also littered with kegs, pails and the other tools of last night's butchery.

'But the boy was not in the infirmary and I wish to question him,' he repeated firmly.

'And as I have explained many times, Brother Nicholas,

the boy is not able to speak. It is likely he has been dumb from birth, and since he is also blind, we cannot even make gestures to ask him if he is hungry, much less ask him questions.' This time I pointedly turned away from him and, lifting a pail, moved it out of my path, deliberately setting it between myself and my brother knight. But as I took a pace towards the kitchen, I was startled to feel a hand grasp my arm and heard a gasp of outrage from Clarice. The knight's fingers dug into my flesh, but I refused to flinch. I stared down at my arm, then up at Nicholas, meeting his gaze full on, determined to say nothing until he had removed his hand. For a long moment, we glared at each other, neither stirring. Then he grudgingly released me, his face flushed, though from his furious expression, it was not with embarrassment.

'I demand to see the boy, Sister Johanne.' His voice rang round the courtyard, bouncing from the walls.

'*Prioress* Johanne,' I said. 'And you may demand all you wish, Brother Nicholas. I will not allow you near that child.'

'I know who that creature is and I know you have tried to keep me from him since the day he arrived, afraid no doubt that I would discover his real nature. Is that heretic you harbour also a sorcerer? Did he conjure that demon boy? Is that why you whisper with him?'

I tried once more to walk away, fearing that any answer I gave might endanger Sebastian still further, for Nicholas would twist whatever I said. As I turned, I glimpsed several of the sisters and servants gazing at us curiously from doorways and casements. Nicholas was not troubling to keep his voice low – indeed, he seemed determined that the whole priory should hear him.

'That devil's spring lies at the heart of all of this,' he

thundered. 'That's why you and your coven of sisters seek to stay here, because the well is the source of all the evil and corruption that infest this priory. Is that why the heretic was drawn here, so that he could use its malevolent power for his sorcery? I will see that well of yours destroyed stone by stone and filled with rubble, so that no one can ever enter that place again. And as for the sorcerer who lies in your infirmary, I will finish what the Inquisition started. Do you know the penalty for a pardoned heretic who relapses into his old sin? He is burned on the pyre, Sister Johanne, burned alive.'

Nicholas took a step towards me, which would have been far more menacing had he not tripped over the pail I had placed there and stumbled sideways, having to grab at my skirts, like a child in clouts, to avoid crashing down on to the filthy stones.

Flushing still deeper, he drew himself upright. 'You cannot hope to hide that demon boy in a place such as this, *Sister* Johanne. I will seek him out and I will send him screaming back to the devil, his master. And, mark this, if you dare to stand in my way, I will have you bound hand and foot to that heretic and dragged to the pyre with him. I will see the two of you burn in this life and in the fires of Hell to come.'

Though I had known from the moment he arrived that he was determined to have me removed as prioress, even I was shocked by the malice that contorted his face. He was a knight of the holy order of St John, a monk who had devoted his life to the service of God: how could he be filled with such hatred? I found myself shaking, unable even to summon the words to begin to defend myself. How could you reason with a man who thought that a blind child was a creature summoned from Hell? I was beginning to think it was Nicholas who was blind. The boy was pure innocence, entrusted to my care by St Lucia herself. The more Nicholas sought to harm him, the more I knew I must protect him. If I could guard this child, keep him safe, then St Lucia and the Holy Virgin would surely defend us.

'Prioress! Prioress!' Meggy was hurrying across the courtyard.

A man shuffled awkwardly behind her, seemingly unable to make up his mind whether to come towards us or to retreat. His gaze slid to the bloodstained table and the tub of axes and knives that were steeping in the pool of gore that had dripped from them. Flies were crawling over the tub and more were buzzing over the puddles of dung and blood in the courtyard.

The man's eyes darted briefly to Clarice, Sibyl and me, but he addressed himself firmly to Brother Nicholas, as if certain the knight must be in charge.

'Thing is, there's a body been found out on moor. Reckon it to be one of yours seeing as how he's wearing the white cross. You want him brought in here?'

Nicholas frowned. 'A knight of our order? Dead?'

'As salt pork,' the man said. 'No question of it. Dogs and foxes have been at him, birds too, I reckon.'

'If he is of our order, you may lay the body in our chapel,' I said, gesturing towards the door. 'But has the coroner given leave for it to be moved and brought here?'

The man eyed me warily, as if I was some strange talking bird, and when he did speak, he addressed himself to Nicholas again. 'No call to be fetching any coroner. Found the body alongside the river. Water flows fast and deep there and with all this rain . . .' He spread his hands, which were

374

almost black with ingrained dirt. 'Reckon he must have got lost in the dark, and stumbled into the water. Managed to drag himself out, but was too exhausted to move. He'd have lain there and died of the cold, I shouldn't wonder. Seen it afore a heap of times, I have, 'specially when men have had a drop or two more cider than is good for 'em. Hound belonging to one of our lads was sniffing around and found the body behind a great heap of tailings. We're not dumping waste there any more, see, 'cause we're working further along the valley.'

'We?'

'Tinners. That's where we found him, on tinners' land. So, you see, there's no call to be dragging coroner all the way out here. Natural death, it was, in as much as any can be called natural. Nothing for the coroner to be fretting himself over. But we thought it only right he be given a decent burial, if he was one of yours. We don't want any man accusing us of trying to hide a corpse. So, we fetched him here. Lost a day's work over it, we have.'

'No knight would go walking over the moor,' Nicholas said. 'If he was on a journey for our order he would have been riding. Where is his horse?'

'There you are, mystery solved,' the man announced happily. 'His horse must have thrown him off into the river. No crime's been committed, save by the horse, that is. Beast will be miles away by now. Still, if we should happen to catch it, I dare say there'll be a reward.' He looked Nicholas up and down. 'There's none in your order rides anything but the best blood. Worth more than most men earn in a lifetime, those horses of yours. And, like I say, times being what they are, a man can't afford to lose even an hour's work. For it's our poor wives and chillern that'll go hungry 'cause of it, though they didn't ask for any knight to go wandering into the river,' he added.

'When you have fetched the corpse in, go to the alms window and I shall instruct Sister Melisene to give you food enough for your family's supper tonight,' I told him.

'Food?' The tinner could not have looked more affronted if he'd been promised a flock of sheep and received only a skein of wool.

'Think yourselves fortunate to be getting anything,' Clarice said, eyeing him shrewdly. 'For I dare say you've already helped yourself to your own reward.'

A spasm of what might have been guilt flashed across the man's face under Clarice's steely glare, but it vanished rapidly and he turned back to Nicholas. 'Could have left him where we found him, 'stead of hefting him all the way up here,' he said plaintively.

But his efforts to appeal man to man were wasted, for Nicholas was loudly commanding us to fetch candles and a bier on which to place the body, as if he was in charge. Since we had no bier, I instructed the man to wait until the blood had been scrubbed from the trestle table and it had been carried into the chapel, then set before the altar ready to receive the body. When that had been done, two tinners lugged the tightly wrapped cadaver inside and dumped it, none too gently, on the rough wood. They slouched off towards the alms window, muttering that they should have buried the carrion under a spoil heap and spared themselves the effort.

Most of the sisters, who'd been listening to Nicholas shouting in the courtyard, now crowded into the chapel behind us, crossing themselves and staring at the bundle. Sister Fina hastened to light the candles on the altar, then pressed herself back against one of the pillars, holding it on either side as if she could not stand without support. I was on the verge of sending her out: I feared what the sight of a corpse might do to her. I knew she'd resent it, though, so I allowed her to stay.

I decided it might be wise for once to allow Brother Nicholas to have his way and take charge of the deceased. It was, after all, a brother knight who lay dead and Nicholas might recognise him. At the very least, he would be diverted from thoughts of demons and sorcery.

The head and shoulders of the dead man were encased in a filthy sack and lengths of soiled cloth had been wrapped around the body, bound tightly in place with lengths of straw rope, trussing it from neck to ankle. The tinners were plainly not prepared to waste a serviceable cloak or blanket on a corpse.

'I must discover who this brother is,' Nicholas said pompously, the moment the door closed behind the tinners. 'His preceptor will have to be informed without delay. He will want to send knights to collect the body for burial in his own priory.'

He motioned us back with his knife as if we were the fluttering wives of noblemen, who must be kept from unpleasant sights in case we swooned. Clarice raised her eyebrows at me, but I mouthed at her to let him get on with it. At least it was keeping him from the infirmary.

Nicholas set to work with his knife, sawing at the straw rope. From the tail of my eye, I saw Fina flinch as each band broke and the body beneath sagged, as if she feared that, freed from its bonds, it might leap up and attack her. Nicholas cut the final rope around the ankles and a pair of feet flopped out from beneath the cloth, revealing blue-white mottled skin and horny toenails.

'A brother of our order riding barefoot?' I said.

Clarice snorted. 'I dare say he was well shod when he was found, but those tinners wouldn't let a pair of good boots or hose go to waste. I wouldn't be surprised if we find half his clothes have accidentally fallen off in the river too.'

Nicholas took hold of the ends of the sack that covered the man's head and tugged, but the sack would not come off.

'Stuck to the skin,' he said. He tried again, almost jerking the corpse on to the floor in his impatience.

Only with Clarice's help, and a great deal of pulling, did the sack finally slide free. It gave way so suddenly that the man's head thumped back on to the wood. But it was not the crash that made those standing around cry out.

Nicholas's jaw tightened so hard that it was a wonder he didn't splinter his teeth. I realised I was doing the same. Most of the sisters were trying to disguise their hastily averted gazes by crossing themselves and murmuring prayers.

The face of death is rarely known for its beauty, but this face, or what was left of it, was enough to make the bile of even the battle-hardened rise in their throats. The nose, lips, tongue and part of one cheek had been chewed off, leaving a cavernous black hole and a mocking grin of teeth stained rust red by the man's own blood. The eyes had been pecked out and the skin of the forehead, which had been stuck to the sacking by dried mucus and blood, had peeled off the flesh beneath as the cloth had been torn away.

No one spoke. A long moment passed before even I could unclench my jaw to force out the words, 'Ch-Christ have mercy on his soul. This must be the work of wild dogs or foxes. For pity's sake cover his face, Brother Nicholas. Let us offer our brother some dignity in death.'

378

Nicholas picked up the stained, stinking sack and tried to drape it over the ravaged face. But Clarice tugged it out of his hand, letting it fall to the flags. She trotted to the chest containing the altar cloths, took out a small white linen manuterge and laid it over the head. She threw a defiant glance at Nicholas, as if daring him to protest at this misuse of a sacred object, but for once he was silent.

'We'll have to examine the rest of the body,' I said grimly. 'I fear even his closest friends would find it impossible to identify our brother from his face and it will be harder as each day passes. In my experience, decay is more rapid wherever there is a wound. Perhaps some item of his clothes or his scrip will help.'

Nicholas did not acknowledge that I had spoken, but he began silently removing the pieces of soiled cloth from the feet up, as if even he was reluctant to approach the head again.

'It would appear that for once you have been overly harsh in your judgement of the tinners, Sister Clarice. His jerkin appears to be still in place, though—' Nicholas broke off, as he lifted the sack covering the corpse's belly.

We could see at once why the tinners had been reluctant to strip anything more than the boots from the corpse, for even had they removed the black jerkin, they could not have sold it or given it away even to the most desperate of beggars. It was so badly shredded and mauled, it looked as if it had been thrown into a pack of hounds for them to fight over. And the dogs, if they had done this, had not merely savaged the cloth, but also the soft belly beneath it. The corpse had been eviscerated, the stomach, liver and guts torn out, leaving a gaping hole, black with dried blood.

'Tears of Mary!' Clarice muttered grimly. 'It seems those

379

tinners were right not to bother sending for the coroner. Only this man's ghost could tell you if he met his death by accident or foul murder. He could have taken a dozen stab wounds to the face or belly and there'd be no knowing it now. If it was the same pack of hounds that attacked our goats, there's someone hunting out there whom I'd not want to run into after dark.'

Several flies, which had found their way in through the holes in the chapel shutters, came buzzing towards the corpse and immediately crawled inside the belly. Nicholas flapped at them and quickly drew the filthy cloth back over the wound.

'We must get the corpse into the stone drying coffin until his preceptor can arrange to collect him for burial,' he said curtly. 'Otherwise this body, too, will begin breeding flies and all manner of flesh worms to infest the chapel again. And I mean to put an end to these plagues once and for all.' He glowered at me.

'But how are we to know which priory to send word to?' Now that the wounds were covered, Basilia had found her voice. 'He may have recently arrived by ship or have been making his way to a port to board one. In either case, his priory will not yet realise he is missing and be searching for him.' She gnawed her lip. 'If he had a scrip . . . or a purse.'

'If he had a purse you can be sure that's long gone,' Clarice said.

'Look!' Fina was still standing with her back pressed hard against the pillar. She gestured wildly towards the corpse. 'His hand . . . the right one, see!' Her words ended in a high-pitched shriek.

I glanced down. The man's arm had slipped, so that it was dangling off the table. I took a step closer and, pinching the cloth of his sleeve, pulled the arm upwards. The skin was dark with pooled blood and smeared with dried mud, but that did not disguise the fact that the hand was missing the forefinger and half of the middle finger.

Nicholas gaped at it. 'Alban! But it can't be. He should be on his way to . . .' He lifted the linen manuterge covering the corpse's face and stared briefly at the mauled flesh beneath, closing his eyes and breathing hard as he dropped it back. Gingerly he lifted the edge of the cloth covering the gaping hole in Alban's belly.

'His belt, the scrip . . . they're gone! Those tinners, are they still here?'

Without waiting for an answer, which none of us in the chapel could have given him, he ran towards the door.

'Aren't you going to stay with us to pray for him, Brother Nicholas?' I called, but the only reply I received was the slamming of the chapel door.

I stared again at the corpse, trying to picture Alban as I had last seen him alive, wrestling with the beam on the great door. I realised he had not been wearing the large leather scrip. On such a rough track, it would have bumped against him constantly as he rode, irritating both rider and horse. Instead it had been strapped to his saddle along with his blankets and provisions.

If outlaws or tinners had taken his horse, they had all that Alban was carrying, including Nicholas's report. It was unlikely that any man who had stolen it could read. They would simply have burned it, with anything else that might incriminate them. And if the horse was found wandering by an honest man on the moor, he would recognise the Hospitallers' seal and would surely bring both beast and scrip to us. Either way, that report would not now reach Clerkenwell. Relief washed over me, swiftly followed by guilt that my deliverance had been at the cost of a man's life. I sank to my knees and tried to pray for Alban's soul.

Chapter 46

Hospitallers' Priory of St Mary

Brother Nicholas eyed the lump of bread with disgust. Even prisoners received larger portions than this. On the other hand, did he really want to eat any more of it? God alone knew what it had been made from, ground goat bone judging by its dryness. But he knew from experience that a diet of meat alone turned your excrement to rocks. Years of hard riding in sweaty armour had caused him enough problems with his backside without the pain that came from straining for hours over a draughty shithole.

Nicholas spooned some of the goats' liver and heart pottage into a bowl and crumbled the bread into the sour gravy. He told himself he must have eaten worse in the service of the order, though not much, but as his growling belly now reminded him, he'd swallowed nothing since that early breakfast of goat's meat. No one had had much appetite after they'd seen Alban's body, but now his head was throbbing and he felt dizzy with hunger. He'd have to shovel this pigswill down.

He stabbed his knife into another piece of what, from its

texture, seemed to be a lump of lung. God's teeth, it was rank. Not even the strong sauce masked it. Were they trying to poison him? The thought took hold and he felt the whole mess rising in his gorge.

He dropped to his knees and flung open the lid of his wooden chest, rummaging feverishly until his fingers grasped an object the size and shape of a walnut enclosed in a pierced silver case, which dangled from a chain. He had bought the bezoar years before from an Arab trader and had always carried it on his belt, but since he'd been living in Buckland, he'd not felt the need to arm himself with such protection until now. Holding it by the chain, he staggered to his feet and dipped the silver case in the remaining wine, then drained the goblet in a single gulp. He breathed more easily as the goat pottage settled itself in his stomach. If it had indeed been poisoned, the bezoar would render it harmless. But he'd certainly lost his appetite again.

Alban was dead. Would he be next? A brother of his order was even now lying mutilated in the chapel on the same table they'd used to hack those goats into bloody pieces. And he knew exactly who was responsible. That boy! That devil! That demon of the witch Johanne! She had commanded him to kill Alban just as surely as he had murdered the old priest. What other explanation could there be?

In all his years in the order, Nicholas had met few riders more skilful than Alban. On the journey to the priory, when a deep mud hole had brought the sergeant's rouncy crashing to its knees or when its hind legs had slipped from under it on a treacherous slope, Alban had not only kept his seat, as if his arse was nailed to the saddle, he had got the horse back on its feet and calmed it, before the beast had realised what had happened. Nothing less than a demon from Hell could have so spooked both horse and rider to cause a man like Alban to be thrown.

Nicholas knew only too well what demons could do. In the Holy Land, he'd seen tiny swirls of dust swell into mighty whirling djinn that flogged all in their path with whips of scorpion stings. They could pluck up a rider and his horse together, carry them through the air and fling them down a dozen miles away. He had seen a soft night sky fill with a thousand shrieking black imps that had flown at bands of men, fastening on to the eyes and flesh of horses and knights, sending the horses rearing and bolting in panic while their blinded riders lay crushed and trampled beneath the plunging hoofs. An innocent form could disguise fiends of such power and evil, that even the strongest warrior was powerless against them.

Nicholas wrenched open the door of the guest chamber, admitting a blast of icy night air, and peered out into the courtyard. Unlike the night before, the yard was unusually deserted and dark. Both sisters and servants, exhausted by the hours spent butchering and preserving the meat, had fallen into their beds almost before the last rays of the sun had vanished, some barely able to keep their eyes open long enough to pick at the boiled goat on their trenchers.

A few candles burned in casements, set ready in case anyone should need to rise in the night. Their faint glow flickered around the edges of the shutters. Nicholas's gaze methodically quartered every wall in turn, like a hound trying to pick up a scent. Earlier that day he had convinced himself that the boy had been hidden in the refectory, which was why the sisters were trying to keep him out. So, while they were in the chapel praying for Alban's soul, he had searched it but, to his disgust, had found nothing but dishes of goat meat being pressed into brawn, just as Johanne had said.

He'd waited until Basilia and the lay servants were occupied in preparing Alban's body, then slipped behind the screen in the infirmary, determined to question Sebastian, the conjuror of Johanne's demon. Nicholas would have been the first to admit he lacked the skill and finesse of the Grand Inquisitors. They understood that it was not enough merely to inflict pain. Men can grow inured to it. You have to break their minds and spirits, drag them down into the sucking mire of despair and hopelessness, so that they no longer even imagine there can be a Heaven. You must convince them that only you possess the truth, that everything they once held dear is a lie, that they are a lie. You must make them believe they are not even men. And an art like that is not mastered overnight.

But, still, Nicholas was only too aware that the memory of a battle could be more terrifying than the battle itself. Simply pressing his weight on an already twisted, dislocated joint or forcing open clawed fingers could reawaken the horror in Sebastian, remind him of what he had suffered and what awaited him again if he were to be delivered back into the tender mercies of the Inquisition. That might be sufficient to loosen his tongue.

But as soon as Nicholas had approached his bed, he'd known it was useless. Sebastian lay insensible, drugged to the point that Nicholas could have smashed his leg with a hammer, eliciting no more than a groan, much less a single word that made sense. Johanne's instructions, Nicholas had had no doubt. But she couldn't keep him asleep for ever, not without killing him, and the prioress needed her sorcerer alive.

Nicholas closed the door of the guest chamber and shuffled back across to the smouldering fire. He poked savagely at the sulky embers trying to stir them into a blaze, though at this hour, the fire should have been safely damped down for the night. He glimpsed a tiny movement on the floor and brought the poker thudding down with a great clang. He felt a satisfying crunch under the iron.

'Got you, you little imp.' He crouched, then straightened again, swinging the mangled remains of a mouse by the tail. He let it dangle for a moment, revelling in the proof that the swiftness and accuracy of his sword arm had not diminished, then dropped the tiny corpse on to the fire. The impudent creature had gnawed holes in his boots during the night. Well, the vermin had paid for its crime now. But, as if they were thumbing their noses at him, he heard the scurry of a dozen more in the rafters.

Nicholas threw himself down on to his bed, which groaned beneath his weight. It was the boy. He was drawing these plagues here. All the filth and evil in the land were crawling out of the ground towards him. Those foul creatures could smell the stench of the devil in the priory. Beelzebub, Lord of the Dunghill, was calling to his minions. Even the birds of ill omen flocked over the place where the boy was, hundreds gathering on the roofs and walls, waiting and watching over their master. That demon child was the curse that infected this place, a curse summoned by the witch Johanne and her sorcerer.

And she had surrounded herself with a treacherous sisterhood who thought nothing of cheating any man who crossed their path, like every woman Nicholas had ever met. They had continually lied to the villagers about the well. They had even tried to keep Nicholas, a brother in their own order, from discovering the truth of what lay down there. Lies came so easily to their lips that they didn't even hesitate or blush with shame. They—

The well . . . That was it! That was where they'd hidden the boy. Nicholas struck his head with his hand. How could he not have known it at once? Where had the frogs and flies swarmed to their demon king? Down in that cave! That was where the fiend would assume its true form. Why were the sisters trying so hard to keep himself and all the pilgrims from the spring, even though it was losing them money? Because they knew a demon had taken up residence down there. But no more! He would send that monster howling back to Hell whence he had come. And once he had defeated that devil, the sisters would have no weapon to send out against him, as they had against Alban and the priest. Then he would see to it that the whole coven of hags was dragged to Clerkenwell in irons.

Meggy stood in the doorway of her small hut, huddled in a blanket, blinking up at Brother Nicholas. She never removed her clothes when she was abed. In fact, often she did not trouble to lie down, preferring to doze in her chair close to the warmth of the fire. At her age, sleep came quickly, but did not linger long, so she was more than a little vexed to be jerked awake by a thumping on her door. She eased it open a crack, grumbling all the while that such shocks were not good for a body at her age.

'What is it? If you've a mind to go out, the gates are locked for the night and no one enters or leaves till the Prime bell.'

'I have no wish to depart,' Nicholas said. 'I have come for the key to the chapel and the well. And before you protest, old woman, many souls in this priory may perish if I do not succeed in my work tonight.'

Meggy took a step outside and peered across the courtyard towards the prioress's casement. It was in darkness.

Nicholas caught her arm and stepped in front of her, so that she was forced to pay attention to him. 'The prioress has retired for the night. It is I who demand the keys. As a brother of this order I have every right to enter the chapel and the well at any hour of the day or night I see fit. And as a knight of St John, I can, as easily as the prioress, have you dismissed and banished from this place. You would do well to remember that.'

Meggy pulled the blanket tighter about her shoulders against the cruel wind. 'If you want to spend the night freezing in the chapel that's naught to do with me. But you might have fetched the keys afore decent folks retired to their beds, 'stead of disturbing them in the middle of the night. Up all last night I was, 'cause of those goats. Not a whisker of sleep did I have, and here you are banging on my door, dragging me from my fireside, when I've only just managed to close my eyes. It's not Christian. Call yourself a holy knight? You ought to have more consideration.'

She bustled inside, emerging moments later with a ring of keys, which she thrust into his hand. 'You make sure you lock up after you and don't you go banging on my door again tonight. You can give me the keys back at a decent hour, when the sun's at least had time to wash his face.'

Nicholas edged into the chapel and closed the door quietly behind him. He did not want to warn the demon that he was coming. He lifted the lantern and peered around.

Five fat candles had been lit around Alban's body, which still lay on the rough wooden trestle beneath a long white cloth. Raw wool had been packed beneath it to absorb the liquids dripping from it. But even though thyme and other dried herbs had been bound into the winding sheet, the stench of putrefaction was beginning to fill the chapel.

An image flashed into Nicholas's head. Bloated bodies lying strewn across a sun-scorched earth. The sky almost bronze above the shimmering heat haze. The deathly silence, save only for the head-splitting buzzing of the flies that covered the corpses in a dense black crust, and the flapping of the vultures' wings as they lumbered to the earth to stab at the swollen bellies. And that smell, the sweet, gagging stench of rotting flesh: it clung to your clothes, your hair, your skin; it slid down your throat and filled your nostrils and mouth till you could taste and smell nothing else for days, maybe for a lifetime.

Nicholas glanced up, trying to wipe the memory from his mind. He stared at the bowed head of the crucified Christ on the altar haloed by the soft light shining down from the oil lamp hanging above. Kneeling on the hard, cold stones, he bowed his own head and tried to gather his faltering strength for the battle that would take place in the depths below. The fiend, the beast of Satan, was squatting directly beneath the body of the man he had already killed. Nicholas could see it. Its great frog eyes bulging, unblinking. Its long snake tongue coiled, ready to lash out. Could it sense the presence of the knight above it? Was it preparing to take another victim? His hand slid towards the hilt of the sword hanging from his belt. Could this demon be defeated by steel? If he had been facing a man, ten men even, excitement would have borne him up, but he found himself gripped by something he had rarely known before a battle: a cold tide of abject fear.

He prayed fervently. He found himself repeating the same words over and over again, and that chilled him the more, for he could feel the demon crouching beneath him, sucking his breath from his body so that he could barely think, much less remember how to pray.

In the end, it was the pain from kneeling on the merciless stone that drove Nicholas, tottering, to his feet. He took up the lantern and advanced towards the well door. It was only as he struggled to fit the prongs of the key into the holes in the wooden bolts on the other side that he remembered he had not bolted the chapel door. He hesitated, then realised he would feel safer if he left it unlocked. At least if he was forced to flee, he would not be left grappling to open his only escape route.

He paused at the top of the steps, holding the lantern low so the light was cast as far down as it could reach, which was only the first few steps. Beyond that, the darkness rose up, reaching out to drag him down. He listened, but heard only the sound of his own heart thumping and his rapid breaths. No water splashed or even dripped. Although he knew the spring had stopped flowing, the silence was unnerving, ominous. Should he call out a challenge? Even as he opened his mouth, he knew the words wouldn't come. With one hand pressed against the wall, he edged down the stairs, but though he tried to tread lightly, the gritty mud on the soles of his shoes ground against the stone with each step he took, echoing through the granite till it sounded, even to his ears, as if a great scaly creature was clawing its way down.

As he descended, the walls began to shimmer with that green-gold glow. But it served only to remind Nicholas of what lay at the bottom: the monstrous poison-green beast. The glistening walls no longer seemed miraculous, but as if the bloated creature had slimed the walls of its lair as it

391

passed up and down, like a slug leaves a shining trail wherever it crawls. He snatched his hand away from the wall, rubbing icy wetness on his breeches. Why had he not realised from the first that this unnatural light was a sign of evil? But the steps, though he knew they were solid, seemed to undulate beneath his feet, twisting in the eerie, flickering light, until he was forced to press his hand against the wall once more, just to keep from plunging down into the bottomless mire of darkness.

He paused on the steps as the lantern illuminated the entrance to the cave. Something was squatting in the deep shadow just beyond the glow of the candlelight. He froze, heart thudding, staring until his eyes hurt, watching for the slightest twitch of movement.

He grasped the hilt of his sword, but the stairs were too narrow to draw it. He should have had it ready in his hand. He was a knight of St John. It was second nature to him. The demon had bewitched him, was fuddling his mind, making him forget the most basic lesson that had been beaten into him long before his voice had broken.

He thrust the lantern forward so that the light pierced deep into the cave. Where the creature had been squatting, he now saw only the side of the coffin-like trough, covered with green moss. Where had it gone? He swept the lantern about, but the cave was empty. Had he been staring at the basin all along, or had the demon transformed itself and was hiding? But unless it had made itself as small as a spider, there was no crevice large enough to conceal it. Unless the beast was crouching inside the trough . . .

He raised the light high, trying to peer inside the long stone basin without leaving the comparative safety of the last step. The yellow candlelight showed only a rim of red mud, and above it, on the far side, a face. He had never seen the carving without its veil of water, but now that it was naked, grinning out at him, he realised it was a skull, with vipers writhing from the dark eye sockets and darting out from between its grimacing jaws. As he moved, so the snakes in the shadows moved too. The jaws of the skull opened wider and more vipers wriggled out, their tongues tasting the air. In his terror, every word of defence he knew against the forces of darkness fled from his brain. Raising his arm to shield his face from the devilish sight, he turned to flee.

It was then that the guttering lantern light caught something black behind him on the stairs. He saw the flash of steel and instinct made him jerk back even before his mind had registered that it was a blade. His foot missed the step below him and he crashed down. His head cracked against the wall, as the knife slashed towards his face.

Chapter 47

Prioress Johanne

I eased the door of the chapel open and stood on the threshold listening. When Meggy had come to my chamber grumbling that Brother Nicholas had roused her from her bed to demand the keys, I had sent her back to her gatehouse and hurried over at once.

I had told Sister Clarice that I had managed to persuade Brother Nicholas that Alban's body could not be taken to Buckland for burial until the first frosts.

'Being as wet as it is, I reckon that if it turns colder we're more likely to have snow than frost up here,' she had said. 'That means there'll be no moving him off the moor till spring.'

'The point is, we must stop the body being taken anywhere, now or then. If Commander John or the Lord Prior should learn that a Hospitaller was murdered here, we shall have the whole order riding in to investigate. I think even Brother Nicholas understands that he couldn't get a cart further than our gates without it becoming hopelessly bogged down. But he is demanding to take the heart to Buckland for burial at least, which he says he can do on horseback, whatever the weather. And I suspect he could, given his skill as a rider and his determination. Once there, he is bound to be questioned about how the man died. Commander John will not let Brother Alban's death go unavenged.'

'Aye, and that low-bellied viper will waste no time in tattling to Commander John all he thinks he's discovered that's amiss here, including my accounts, not that he's got a mote of proof,' Clarice had added indignantly.

'Then we had better see to it that Brother Nicholas does not leave this priory,' I had said. Missing money was the least of the crimes Nicholas would report, if he had the chance. Half the priory had heard what he'd threatened and those who hadn't had certainly learned of it by now.

I cursed myself. I had already given orders that a grave be dug and that Alban should be buried at first light. With the ground so sodden and the corpse so mutilated, it would decay rapidly in the earth, and if we could keep Brother Nicholas out of the way until the grave was filled, even he might balk at digging it up, especially if we could convince him the heart would have rotted. I had thought to let the body lie before the altar until dawn, for by rights a corpse should lie for three days as Christ's had done. The soul must be given time to move on, before the earthly remains are laid in the ground. Alban was a Hospitaller and he deserved that much at least.

But now I knew that I should have had Alban interred straight away. Nicholas was probably even now cutting out the heart. Why else would he have demanded the keys to the chapel at this hour? Which meant he had determined to set out for Buckland tonight. I had to prevent that, whatever the cost.

But as soon as I entered I saw that the chapel was empty, at least of the living. Had Nicholas already removed the heart? I crossed swiftly to Alban's body and drew back the cloth that covered it. The winding sheet was still in place, the cord still tightly bound around the neck, thighs and ankles. Just to reassure myself, I laid my hand upon the chest, feeling for any signs that the ribs had been broken open. Those, at least, seemed to be intact. Relief flooded through me, swiftly followed by guilt. I had thought so little of Nicholas that it had not occurred to me that he might simply have wanted to keep vigil and offer his prayers for the soul of our brother. Though even as that struck me, I confess I was shocked that his prayers should have been quite so perfunctory, for he must have left the chapel before Meggy had even finished telling me he was there. Otherwise I would have seen him crossing the courtyard. Most knights of our order would have remained kneeling in vigil till dawn.

I knelt too, to beg mercy on Alban's soul, though I had already spent some hours in prayer for him. But my prayers now were also brief for the hour was late and I would have to rise before dawn to see that the body was laid to rest as deep as possible in the earth. I was on the point of scrambling to my feet when I heard something that sounded like slow footfalls.

I rose swiftly, fearing that someone was behind me, but I could see no one. Stone amplifies and distorts sound, so I couldn't be sure of the direction, but as I strained to listen again, I realised it was coming from below my feet. The cave! I glanced towards the well door. The pool of light cast by the trembling candle flames that stood sentinel around Alban's corpse did not reach as far as the wall and the door lay in deep shadow. But as I edged towards it, I saw it was partly open.

If that was Brother Nicholas moving about, then what on earth was he doing down there? Certainly not collecting holy water, that was for certain, for there was no sound of splashing to indicate the spring was flowing again. A thought struck me. Was it possible *he* had caused the water to run red and had blamed the tinners for it? And had he somehow managed deliberately to block the flow? He'd boasted of his knowledge of how to poison water supplies. Was that what he was doing down there now, poisoning the holy water, so that the pilgrims would fall sick and our priory would be closed? Only that morning he had threatened to destroy the well, fill it with rubble so that it could never be used again. Was he even now carrying out his threat?

I had no weapons except the knife at my belt and I knew I was no match for a trained knight. But I could not stand aside and let him do this wicked thing. At the very least, I intended to catch him in the act and confront him.

Chapter 48

Hospitallers' Priory of St Mary

Brother Nicholas felt as if he was lying at the bottom of a deep, dark, water-filled pit, staring up into a small circle of sunlight. He couldn't move his limbs. He could hear voices, but the sounds were far off, muffled. Two figures in black hovered above him, locked together, struggling, fighting. He caught the flash of a blade. Then the small patch of sunlight shrivelled into a tiny ball and vanished. All was silence and darkness.

'Brother Nicholas, open your eyes.'

Something was crawling down his face. Flies? He tried to bat them away, but his arm had been turned to stone. He could scarcely lift it. A cold ring closed about his neck. Water! Water was trickling over his skin. Was he drowning? He forced his eyes open and pain seared through his head. Only as it subsided into a dull throb did his mind begin to grasp where he was.

He was lying on his bed in the now too familiar guest chamber. Sister Basilia's massive breasts almost slapped his chest as she bent over him to peel a sodden cloth from his forehead, immediately replacing it with another dripping icy water. He winced as she smoothed it across his brow, pressing it against the tender lump on the side of his head, which seemed to be the source of his pain.

'How are you feeling, Brother Nicholas?' Prioress Johanne was standing at the foot of his bed, her face illuminated from beneath by the candle set on the table. She looked like a talking skull.

Skull, where had he seen . . .? Vipers pouring out from between the jaws . . . Someone on the steps behind him . . . someone with a knife.

He struggled to sit up, but Sister Basilia pushed down on his shoulder with her not inconsiderable weight. 'You must lie still, Brother. You struck your head, and if you get up too soon, you'll feel dizzy and sick. We don't want you falling again and giving yourself another nasty little bump, now do we?' she cajoled, in the jolly tone that a mother would use when trying to divert a small child from crying over a grazed knee.

She was right about feeling sick, though. Nicholas was afraid to sit up again for fear that he would disgrace himself and vomit. He resolved to lie still, at least until the bed stopped rolling, like the deck of ship.

'What happened? I was down in the cave . . . How did I get here?'

'You slipped,' Johanne said briskly, 'and banged your head on the rock. Meggy helped me to drag you up into the chapel. It wasn't easy getting you up those stairs. If you hadn't been found, you might have lain unconscious in the well all night, probably for days. When I saw the chapel was empty, I was about to relock it. Only by the grace and blessed intervention of St Lucia was I moved to check the well first.

We must give thanks to her for that miracle. You could easily have died of the cold. What were you doing down there in the middle of the night, Brother Nicholas? The spring is dry. And you, above all people, must be very well aware of that.' Her tone had grown suddenly sharp and suspicious.

But Nicholas ignored the question. He had one or two of his own.

'I slipped because someone crept down behind me on the stairs. Someone who tried to stab me.'

'Stab?' Johanne gave a slight smile. 'It would appear you are still a little confused from the blow to your head. I am told our recollection of events is often jumbled, when we have been rendered unconscious. Isn't that so, Sister Basilia?'

The infirmarer nodded enthusiastically. 'Some men can't remember anything that happened on the day they were injured. One I tended thought it was still Lent when it was just a week from Michaelmas. And there was a woman who couldn't even remember her own name. Swore she'd never been married when her poor husband was standing right there with the half-dozen children she'd borne him clinging to his breeches. When she clapped eyes on him she—'

'I know exactly who I am,' Nicholas bellowed, though he instantly regretted it as a spear of pain shot through his skull. He tried to keep his tone level. 'I can remember with perfect clarity what happened down in that cave. Someone was standing behind me with an upraised knife. If I'd not slipped, that blade would, at this moment, be sticking out of my back, or I'd be lying in that well with my throat sliced open. I don't know who it was because her face was hidden in the shadow on the stairs, but I do know it was a sister in this priory. I saw her black kirtle.'

Sister Basilia's hand shot to her mouth and she gave a

strangled gasp, as if she was the one who was going to faint. 'Why would anyone wish to harm you, Brother? No sister in our order would ever do such a wicked thing to any living soul, much less to a brother knight. It's that bang on your head that's making you imagine such terrible things.'

But Nicholas was not listening to the infirmarer: his gaze was fixed on the prioress's face.

'Sister Basilia is right. I fear you misunderstood what you saw, Brother Nicholas, which is hardly surprising in the confusion. You may indeed have seen someone on the stairs carrying a knife but, I assure you, no one was attempting to stab you. It seems that while Meggy was in my chamber, explaining that you had borrowed her keys, Sister Fina was crossing the courtyard to the garderobe, when she caught sight of a light moving behind the chapel window. Knowing that the chapel should have been locked at that hour, she was naturally concerned that thieves had broken in and went, rather foolishly, to investigate, instead of rousing the lay-servants. She is keeper of the well and has come to believe that it is her responsibility to protect it. She says she saw the well door open and could hear someone walking down the steps, so she followed them, and pulled out her knife to defend herself, should she have need. As soon as you turned, she saw who you were and quickly withdrew so as not to disturb you. You must have slipped as she was climbing back up the steps. The floor of the cave is treacherous.'

Sister Basilia beamed with evident relief. 'There, you see? I knew there was no harm intended. Hardly a wonder Sister Fina should have been afraid that thieves and outlaws had broken in after what had happened to dear Brother Alban, God save his soul. It's my belief it was outlaws who set upon that poor man and killed him. Since this famine came upon us, they've grown so bold and reckless that it's not safe to cross our own courtyard after dark.'

She crossed herself, gazing anxiously at the closed shutter of the chamber, as if she feared that a band of bloodthirsty robbers might even at that moment be massing on the other side of the flimsy wood, preparing to attack.

The infirmarer might have been convinced by Johanne's story, but Nicholas was as far from reassured as it was possible to be. He knew now he had seen two sisters on the stairs, as he'd briefly regained consciousness, but neither had been rushing to his aid. One had been trying to hold back the other, he was sure. But what he didn't know was whose hand had been holding the knife.

Chapter 49

Sorrel

'Tonight, it'll begin tonight,' Morwen said. 'We must summon the spirits to help us.'

My whole body tingled. Though I'd no notion what would happen, I knew this was why Brigid had called me. I could feel a power growing inside me, like a flame that runs silent and unseen, burning beneath the thatch till suddenly the whole village is ablaze.

'Are we going to Fire Tor?' I asked her.

'Can't go there. Ma'll be waiting. She'll have set her curses on the tor to keep us out. She'll try to fight us and we've no time to waste dealing with her. It's Brigid we must serve now. We must go to the place where the spirits walked long afore they were driven into the shadows. They were strong in life and are even stronger in death. You'll see.'

I took her hand, which was warm in mine, and we clambered over the ridge in the dawn light, following the path of a surging river that ran along the bottom of the steep-sided valley. Streams and rivulets ran down the sides of the gorge into the river from tar-black bog pools and glistening rocks.

A brown buzzard watched us from its perch in a dead tree, jagged as a broken tooth against the swelling granite-grey clouds.

Morwen stared up. 'It'll be a duru moon tonight – that's when the door to the other worlds opens. Ma'll be at work too, but she'll not be able to stop us.'

As we came around the curve of the hill, Morwen gestured towards a long strip of dense woodland that hung above the river. 'That's where the spirits gather. My granddam told me that, hundreds of years ago, the tribes who lived in these parts brought their cunning women and men here to be buried. The oaks are sacred. They keep watch over the dead. No one ever cuts wood from these trees, or comes here unless they know the charms and the gift to bring, 'cause Crockern keeps his wisht hounds here and they guard it.'

She tugged me back towards the riverbank. 'Afore we go into the woods, we must wash. The old 'uns will be offended else.'

We stripped off our kirtles and left them by a small rowan tree just before the first oak. Morwen scrambled naked into the river. She arched her back and shivered in the shock of freezing water, but as soon as she found a place to balance herself against the strong current, she held out her hand to me. I slipped off the bank and down into the water. I gasped. It was so cold it sucked the breath from my chest. The current knocked my feet from beneath me, and I shot under the water, but Morwen clung to me, dragging me up. I was gasping and coughing, but we were both giggling. Our teeth were chattering like magpies.

Morwen ducked three times beneath the water and I copied her. We were rigid, our lips blue, our feet stabbed with knives of ice. The wind chilled our dripping heads, but we grinned at each other, though we were so numb we could scarcely move our lips. We hauled each other back on to the bank, squeezed the water from our hair and rubbed each other's limbs to warm them.

Morwen pulled some green reeds from the bank and I watched her tie them into the rough form of a doll. A brideog, she called it. She started to make one for me, as if I couldn't manage it. I snorted, snatched it from her hand and finished it myself by wedging it between my knees as I bound a strip of reed around the neck to make a head. I glanced up and saw her smiling, but it wasn't in mockery of my efforts.

She pulled out a few strands of her hair and bound it around her brideog, then tucked it into the fork of the tree that stood at the very edge of the wood, begging the oak of the sun to give us leave to enter. I did the same.

Morwen stood for a long time, inclining her head as if she was listening to a host of people. I tried to make out what she was straining to hear.

'The trees are talking,' she said. 'Don't need words, but you can hear them right enough, hear whether they're angry or content. Granddam used to say that if the trees are silent, that's when they're at their most dangerous. You must never enter the vert then. "Take heed of the silence," she always said, "for then the trees'll strike without warning" – a great branch crashing down on someone who's vexed them, a long-cripple striking out from its nest deep between their roots. The long-cripples in this wood are deadliest of all the vipers in England. No man or beast can survive their sting. The old oaks have their own ways of exacting revenge.'

The trees in this valley were barely taller than me, gnarled, stunted, twisted, like ancient dwarfs. They were almost bare of leaves now, but hung with long beards of silver-grey lichen.

405

Their roots slithered over thigh-high boulders that covered the ground so closely that in places you couldn't lower your foot between them. Rocks and roots alike were wrapped in a dense pelt of emerald-green moss, sodden and weeping.

We picked our way through the tangle of stone and wood, clambering over the mossy boulders and clinging to the branches to balance on them where the spaces were too narrow to squeeze between them. I pointed with my chin towards some flat stones laid on top of vertical slabs of rock. 'They look like little tables.'

'Tables where the dead eat,' Morwen said. 'Bones of the dead rest inside those. Long-cripples curl round them to guard them. Seen whole nests of them, big and little 'uns, slithering through the ribs and skulls.'

We'd reached a steeply sloped clearing, ringed by oaks standing so close together their branches had grown through and around each other, frozen in woody knots that could never be untangled. A huge rock jutted from the ground in the centre, broad at the base and narrowing to a point at the top, like a spear, reaching up as if it would impale the sky.

'That's the queen stone, that is,' Morwen said, padding towards it. I followed.

The boulders were smaller there, humps beneath the skin of shaggy moss that covered the ground, like the lumps on a toad. My skin tingled with cold, but for the first time in my life I welcomed its bite. The air inside the wood was chill, damp, but quite still, as if the wind was forbidden to enter this place, but the trees were still talking.

Morwen turned this way and that, as if she was trying to catch a whisper being tossed back and forth between these ancient souls, yet somehow I understood, as well as she now, that the language they spoke was deeper than any word or thought. The two of us squatted between the small boulders, gazing up at the stone. Towering ramparts of purple cloud were bubbling up over the moors, sucking the light and colour from the earth below.

She began to sing. The notes were soft, like the distant rushing of the water running unseen in the river below. I caught the tiny whisper of movement. A wren had fluttered down. It perched on the twisted, moss-covered root of a tree and regarded us. Lifting up its tail as if it was trying to imitate the huge rock, it suddenly burst into a short piercing trill of a song, its whole body trembling with the effort, then hopped towards one of the little stone tombs and vanished, as if it had walked through the solid rock.

'There,' Morwen breathed. 'That's the tomb of the one who will lead us tonight. Bran's bird lives with the dead, she knows. Cracky-wren tells all the secrets of the living to spirits and they whisper theirs to her. Now we wait for the moon. We wait for the door to open.'

Chapter 50

Hospitallers' Priory of St Mary

The clash of metal, and the screams of men and horses smash into Brother Nicholas's ears. Sweat pours down his face and chest. His hands are slippery with it. The Saracen is running towards him, a great curved scimitar raised above his head, glinting in the burning sun. He reaches for his own sword, but as he raises it, he sees he isn't holding a sword at all but the severed head of Brother Alban. The eyes open wide and the lips move. The swollen face is pleading with him, but he can't make out the words above the howl of battle. The deadly scimitar is slashing closer and closer. The Saracen is whirling it high in the dust-choked air. At any moment it will descend on Nicholas's skull, cleaving it in two. But he can't defend himself: his arms are full of bloody heads. More and more are being heaped on top. He mustn't let any of them drop. He has to carry them to safety.

Nicholas's yells woke him as he fought to climb up out of the pit of sleep, but it was some time before he could force his eyelids open and even longer before he could make his limbs move. They felt heavy, swollen, though from what he could see of them they looked normal enough. The bedclothes were sodden and cold, drenched in his sweat. His mouth and throat were as dry as a desert and his teeth covered with a sickly, sticky film. He groaned. Just how much had he drunk last night? He struggled to remember, but fragments of images and words were scattered haphazardly in his skull, as if marauders had broken in and smashed all that was inside. And, from the throbbing in his head, they were still in there, rampaging around with war hammers.

He staggered out of bed and over to the table, reaching for the goblet that was still half full of wine. He took a deep swig, rolling it around his parched mouth to moisten it. The wine was even more foul than usual, with a bitter-sweet taste, but maybe that was because his furred tongue was like something dug out from the bottom of a midden.

But he'd drunk less than usual, that much he did remember now. Besides, he'd been known to consume a whole flagon of wine and still manage to cut the heads off a row of straw dummies at the gallop, without missing a single one. He pressed his hand to his forehead as the grisly images from the night terror surfaced again in front of his eyes, and flopped down into a chair.

It was Sister Basilia who had brought the wine to his chamber last night. She'd poured it herself and insisted on warming it before handing him the goblet. Why had *she* brought it and not one of the scullions? The servants were abed – that was what she'd said. But . . . He turned to stare at the empty bed. Alban was dead, murdered by that demon boy. For a moment, he thought it had been another nightmare, but he knew it wasn't. And last night, down in that cave, one of the holy sisters in this priory had tried to kill him too.

Nicholas was suddenly aware of how chill the chamber felt. The fire was out, the ashes cold. How long had he slept? He began to pull on his clothes, clumsy in his feverish haste. His limbs still felt leaden, his back ached and his fingers kept losing their grip. He stumbled to the door, half afraid that it might be locked, but it opened easily. He shut it again and stood for a moment, staring at the wood, trying to calm himself. Of course they hadn't locked the door. What in the name of Lucifer had made him think they could or would? He was a knight, a warrior who'd faced the enemy in battle more times than he could remember. He'd been afraid, yes – any knight who said he wasn't was either an untried fool or a liar – but he'd never panicked. He'd always been able to think coldly and clearly, even in an ambush. What was happening to him? Was that foul demon even now witching his mind?

Nicholas turned back into the room and began stuffing what he could into a small leather scrip. He'd have to abandon his travelling chest and most of what he had brought with him, except his sword, of course. It was only as that thought struck him that he glanced over at the corner where it usually hung next to his cloak. It wasn't there. He snatched down his cloak, but it wasn't hanging under it. He pivoted on his heel, staring round the small chamber. Although he knew he would not have put the sword anywhere else that did not stop him searching for it, even in places that could not possibly contain it. He was finally forced to admit it was gone, and he felt as if someone had cut off his arm. Never had he felt so defenceless.

Up to that moment, he had still been resolved to cut out and carry Brother Alban's heart to Buckland so that it could be buried with the honour due to any in the Order of St John.

He hadn't particularly liked the idle, greedy swine and, truth be told, he was sure Alban had despised him, but every knight was pledged to ensure that if the body of a brother could not be returned for burial to their own priory the heart must be laid to rest there. But now Nicholas had only one thought in his head: to escape this vipers' nest while he still could.

He peered out into the courtyard. A couple of servants were crossing to the kitchen, looking as if they were performing some kind of strange and jerky dance, as they kicked out and flapped their arms to drive off the horde of kites and rooks that were pecking at the mud and filth on the ground. Keeping close to the wall, Nicholas made his unsteady way to the stables. Once safely inside the dim interior, he breathed a little easier, drinking in the familiar, comforting smell of horse sweat and dung. He moved towards the far end where he always made sure his horse was tethered, furthest away from any driving rain and wind. But the space was empty.

'You looking for something, Master?' Brengy, the young stable boy, emerged from behind the dun-coloured palfrey he was brushing.

'My horse! Where is it? Has someone taken it out?'

Getting no reply from the boy, Nicholas grabbed him by the back of the neck and dragged him the few paces to the empty iron ring on the wall to which he always tied his mount. 'The black rouncy, where is it?'

Brengy attempted to wriggle free, but the knight's grip was too strong.

'Blacksmith took him back to his forge to shoe him. Can't do it here, can he? How do you think he'd fetch a thundering great furnace and his anvil up here? Ow!' He squealed as Nicholas cuffed his head

'There was nothing wrong with his shoes when I last rode him.'

'Knocked one loose in the night. That ugly brute's always kicking at the walls. Keeps me awake half the night.'

'I'll kick you till you can't sit down for a month if you don't mend your manners. When's the blacksmith bringing my horse back?'

'When he's finished, I reckon,' Brengy said, earning himself another resounding slap around the head.

'I'll fetch him myself—'

'Young Dye said she saw you coming in here,' a voice called from the door. 'Now what are you doing loitering in a draughty stable when there's a good hot supper waiting for you?'

Nicholas turned to see Sibyl at the entrance, her hands on her broad hips. Brengy, taking advantage of his distraction, wriggled free and ran out of the stables as if the devil himself was flying after him.

Sibyl bustled forward. 'Now you come and see what I've cooked for you. A nice mess of Saracen.'

For one crazed moment, Nicholas thought she'd meant it. He was beginning to believe these women were capable of any horror. Then he realised she was babbling about sarcenes, a sauce dyed cherry red. She was still talking as she edged him back towards his chamber, as she might herd a goose towards the butcher's knife.

She opened the door for him. 'I know how much you like your rich sauces, Brother Nicholas, but don't you be telling the sisters I made this for you. There's scarcely any spices left in the store. But I thought you needed a good supper inside you after that nasty business. And you've not eaten for nigh on two days. You must keep up your strength, else you'll end up like poor Brother Alban, God rest his soul.'

'I dined last night.'

The cook shook her head. 'Night before last, it was. Been sleeping like a bear in winter, you have. And a good thing too, if you ask me. No better physic for body and soul than a nice long sleep. But your belly must think you've forgotten where your mouth is. So, you make sure you eat every scrap. I've no bread to give you for sops, but I've put plenty of goat's meat in there you can use to mop up the sauce. Mind, it'll not taste quite the same as usual for I've had to use honey and herbs 'stead of sugar and cloves, and there's no flour. But I'm sure you've had worse.'

As soon as the door had closed behind her, Nicholas crossed to the fire, which, in the brief time that he'd been in the stables, had been rekindled. A pipkin of the dark red stew was keeping warm by the blaze.

Someone, that scullion probably, had plainly been charged with keeping watch on his door. He dared not risk trying again to leave until all in the priory were sleeping. If he could slide the brace back on the gate without the old gate-keeper hearing, so much the better, but if he couldn't, he'd threaten to kill her – he *would* kill her – if that was the only way of keeping her silent.

But he should put something warm in his stomach before setting out on the moor – he didn't intend to stop until he had put himself well beyond the reach of the priory. He pulled his spoon from his scrip and fished around in the sarcenes sauce. As the cook had said: there was a generous quantity of goat's meat in it. He scooped out a chunk.

Now that he'd smelt it, he found he was indeed ravenous. Hardly a wonder, if Sibyl was right and he had slept for

413

nearly two days and nights. Was that possible? He'd never done it before in his life, not even after battles that had raged from dawn to nightfall. But the rest certainly hadn't refreshed him. Even now he was finding it hard to think clearly, as if his mind was drowning in a mire. But a thought was slowly unfurling and taking shape in his brain. No man sleeps for two days without waking at least once. That had been no natural sleep. Only dwale or some such potion could make a man slumber so deeply, or for so long.

He dashed the piece of meat on his spoon into the fire. Did they really imagine he was foolish enough to fall for that trick again? He seized the pipkin of stew and was on the point of opening the door and hurling both contents and pot across the courtyard in his fury, but stopped, smiling grimly to himself. Very well then, let them think he had eaten it. If they checked and found him deeply asleep they would reason he could be left for hours, and wouldn't bother watching his door or setting a guard on the gate.

He couldn't risk emptying the pot on to the fire. They'd smell the stench of burning meat. He searched for somewhere to dispose of it. His eyes lighted on Alban's chest, still standing in the corner. Pulling out the spare clothes, he tipped the contents of the pipkin into the bottom. His mouth watered and his belly pleaded as the aroma enveloped his nose. Quickly he shoved the clothes on top of the mess, arranged himself on the bed and lay in the gathering darkness, waiting for the door to creak open. Even as he lay there, he could feel the women standing on the other side of the door, a silent ring of black cats, watching their prey, just waiting for it to walk into their savage jaws.

Chapter 51

Sorrel

The clouds slid over one another, parting to reveal the rising moon then veiling her again. Waves of bone-white light and darkness lapped over the twisted trees. Snakes of mist slithered around their trunks and writhed through branches shaggy with lichen. Morwen and I sat in silence till the first ray of moonlight touched the stone in the centre of the clearing. Then Morwen stood up. I sensed that whatever she was going to do she must do alone. I didn't move as she took a tiny piece of dried meat from a small bag about her neck and chewed till it softened, then pushed with her tongue till it fell into her hand. She placed the morsel at the foot of the rock as carefully and solemnly as our village priest used to lay the Host on the paten.

'Flesh from a red cat, Mother.'

She blew on each of her palms and laid them over her eyes. I kept still and silent, watching. The moonlight washed over Morwen, turning her hair to silver, her bare shoulders to marble, petrifying her, until she was one more stone among the many. For a long time, she stood quite still. Then she drew her hands from her face, reached out to grasp my withered paw and pulled me towards the queen stone. We pressed our free hands to the rough surface. We were a circle of three now, three women, three rocks. And Morwen began to sing a chant that made my skin prickle. I understood without being told that I must want the spirit to rise, think it, will it. I knew if I let it, my hand could sink into the stone, like hot iron into snow, and I would be drawn into the very heart of the rock, as I had been pulled down into the heart of Ankow, but that was not where I had to go, not tonight.

I stared towards the stone slab where the wren had vanished. The tomb seemed to hang suspended, its mantle of wet moss glittering in the moonlight. The oak crouched over it and a coil of mist wrapped itself around it, as if they were determined to protect it. Then the moonlight vanished and the tomb became a deep, dark hole in the forest of trunks. And still Morwen sang.

A thread of mist trickled from the hole, like the wisp of smoke when a candle is snuffed out. The mist thickened, rearing up, whirling, though there was not a breath of wind inside that wood. It threaded up through the crooked branches, skull-white at first, then slowly glowing poison-green.

Chapter 52

Hospitallers' Priory of St Mary

The latch on the door of Nicholas's chamber lifted softly and he heard the soft pad of leather approaching his bed. He lay still, eyes closed, trying to slow his breathing to the steady rhythm of a man deeply asleep, but it was not easy – his body was as taut as a drawn bowstring. Was it Johanne, Clarice? It would be the easiest thing in the world for any of the sisters to smother him with sheepskin as he slept or plunge a dagger into his chest, if they thought him helpless with dwale. It might be weeks, even months, before the cart from Buckland was able to return and anyone discovered the two brothers were missing. Even when they did, if the sisters disposed of his body, they could easily claim he and Alban had set out on horseback to Buckland and must have been attacked by outlaws or swept away in a fast-flowing river when they tried to cross.

The footfalls came closer. Skirts swished against the leg of the bed. Suppose the sarcenes sauce had contained not a sleeping draught but poison? Would they think him already dead? Someone was standing over him now. He could hear her breathing – short and shallow. Excited? Tense? Someone who was about to raise a dagger? It took all his willpower not to open his eyes.

Just as he thought he could hold himself in check no longer, the person softly padded away from the bed to the fire. He heard her raking and banking it down. At least they weren't planning to burn him to death, or more likely they just didn't want their precious priory to catch fire. Then came the clink of the empty pipkin as she picked it up. Finally, the soft rustle of skirts and the slap of shoe on stone as she crossed towards the door. He heard it open and felt the rush of cold air as she paused, watching him again till finally it closed behind her. Still he dared not relax, trying to maintain that steady, slow rhythm in his breathing just in case they were standing outside, listening.

Nicholas lay motionless on the bed, straining to catch the sounds outside – chatter in the yard, the distant clattering of pots, the creaking of the well rope. He was afraid of falling asleep, of lying there as vulnerable and helpless as a trussed chicken if they crept back into his chamber. He had to admit, grudgingly, one blessing of being so hungry: the ache in his belly was sharp enough to goad him into wakefulness.

How many hours had passed? Now he could hear nothing except the wind outside. Were they all safely asleep? Nicholas slid off the bed, trying not to let it creak, and tiptoed to the casement. He could see a flicker of lights in the crack beneath the shutter, which he hoped were the flames from the night torches on the walls. He eased the door open an inch and squinted out. The courtyard was deserted, silent.

He fastened his cloak about his shoulders and, keeping close to the walls, slipped around the yard towards the stables. He was certain that his own rouncy would still be missing, but he would not get far on foot across the moor. The horses pricked up their ears, stamping and snickering softly, as he entered. The only light came from the torches in the courtyard and he could barely distinguish the beasts from the blackness around them. He needed to get closer to find a mount that would bear his weight over such a distance, but only a fool would risk getting too close to the hindquarters of a strange horse for fear they might kick out.

Something creaked above him in the hay loft. Nicholas froze. The ladder cracked and swayed, as the stable lad descended. He leaped down the last two rungs and bounded towards the courtyard, clearly intending to raise the alarm. But Nicholas's training had not deserted him. He seized Brengy from behind in mid-stride, clamping one hand across his mouth and locking his neck tightly in the crook of his other arm.

He bent his head so that his lips almost touched the lad's ear. 'Listen carefully, boy. I have killed many men in my time and I can break your scrawny neck with a twitch of my arm.' He squeezed Brengy's neck harder to ensure his words were believed, slackening his hold only slightly when he began to choke.

'Now, when I let you go, you're going to saddle your fittest and strongest horse for me. And you'll do it quickly and silently. If you utter any sound or make any attempt to run, you'll be dead before you reach the courtyard. Do you understand?'

He loosened his grip a little more and the lad nodded as vigorously as anyone could, with the bulging muscles of a knight's arm beneath their chin.

Brengy, when released, worked swiftly, though he was shaking, and as soon as it was done, he thrust the reins of the horse towards Nicholas, plainly anxious to have his attacker gone. But Nicholas had not finished with him yet. Seizing the reins in one hand, he grabbed the lad by the scruff of the neck and dragged both horse and boy into the courtyard. The mud and filth in the yard mercifully helped to muffle the sound of iron shoes striking stone until they reached the gates.

'Hold the horse, boy. You move or utter a sound, I'll cut your throat.'

Nicholas had pulled aside the beam that secured the door, and was about to drag the horse and boy through, when Meggy burst out of the gatehouse. Nicholas's arm locked round Brengy's throat again.

'Get back into your kennel, woman, and stay there. If you call for help, I swear I'll snap this lad's neck.'

He dragged the boy through and, in a swift movement, swept him up and dumped him face down in front of the saddle before swinging himself up. Brengy lay across the horse, his head and feet dangling helplessly. Nicholas looked down at Meggy in the gateway, both hands pressed tightly to her mouth, as if she couldn't trust herself not to cry out.

'I'm taking the lad with me, till I'm sure that no one is following. Then I'll let him go. But if I'm followed, you'll be fishing his corpse out of the mire. And you can tell those hellcats they needn't worry. I'll be back, but when I return I'll have half the Lord Prior's knights with me and I'll see to it that they put every sister and servant in this accursed place in chains and flog the truth out of them if they have to.'

The moon was baiting Brother Nicholas or perhaps it was the clouds that were bent on mischief. He was attempting to follow the trail along which Hob had led them on the

420

night they had first laid eyes on the malignance they called the Priory of St Mary, a night that he now cursed with all his heart. Even when the moon was shining he could barely distinguish the boggy track from the mire around it, but it seemed that the clouds were watching him, waiting until he came to a bend or a place where the path was washed away. Then they flung themselves across the moon, plunging him instantly into a terrible darkness, so that even the horse faltered and kept trying to turn back.

'I hope you're praying men, Brothers,' Hob had said, ''cause you'll need more than a sword to protect you up there.' Had Hob known more than he'd told them that night? 'Once you cross into the deadlands,' he'd said, 'there's no coming back.' The carter had been talking about the tor, but he could as easily have meant the priory.

Brengy, lying face down over the horse in front of Nicholas, kept whimpering that he was going to be sick and that his ribs were hurting.

'Then I'll have to give you something to take your mind off it,' Nicholas said, bringing his riding whip down hard across the lad's backside with a savagery that was born as much from his own fear and frustration as a desire to silence him.

Brengy gave a yelp of pain and started to cough and retch. Nicholas was in no mood to have him vomit down his legs. The lad had served his purpose, keeping the gatekeeper quiet at least long enough for him to get out of sight of the priory. Nicholas's confidence was surging back now that he had escaped. None of the sisters would follow him out here. And if they did, what danger could a woman possibly pose to him out on the moor? Even without his sword, they'd be no match for a trained knight.

He seized Brengy by the back of the shirt, dragged him backwards off the horse and flung him down. He heard a cry as the boy hit the ground and derived a vicious pleasure from it. He wanted to smash the priory and everyone in it. Besides, the stable boy deserved it. He should be hanged for stealing a knight's horse and Nicholas would see that he was, just as soon as he returned.

He stared out over the black sea of the moor. The wind was gaining strength, peeling his cloak back, creeping in under it, like a woman seeking warmth. The horse had stopped again and was trying to turn, as if it was determined to go back to its stable. It was much smaller and lighter than his own rouncy and, like all the horses in the priory, was suffering from the meagre diet of mouldy hay it had been forced to endure these past months. Nicholas could feel it wheezing and heaving beneath him. He wasn't convinced it would have the stamina to carry him as far as Buckland, but as long as the beast got him off the moor and safely to an inn or monastery, where he could eat and sleep without fear of a knife being plunged into his back, that was all he asked of it. He could find a stronger mount to carry him on from there. But even the fleeting thought of food reminded him of how ravenous he was. He tugged on the reins and wrenched the horse round, kicking its sides to urge it on, though he couldn't even see the track.

But they had barely moved a few paces before the horse stopped again, lifting its head, ears laid back. Nicholas caught the faint sound of iron shoes hitting stone, the squelch of hoofs plodding through mud. Another rider was behind him. He *was* being followed! His right hand went unbidden to his side, the fingers expecting to close over the familiar hilt of his sword, as they had done a thousand times before whenever danger threatened. But they touched only the cloth of his cloak. After several moments of shocked disbelief, he remembered that his sword had been stolen from his chamber and whoever was riding down on him might well be grasping his own deadly blade in their fist.

Nicholas was as much unnerved by his own reaction as by the danger behind him. He had faced hails of flaming arrows, men brandishing glittering scimitars, howling for his blood. His heart had pounded in those battles, but his mind had, if anything, become sharper and clearer: he had been able to decide instantly where to hack through a line of warriors or to spot in a flash a man who had left himself unguarded and thrust death home. But now he was gripped with blind panic.

He slashed his whip against the horse's flank, trying to urge it forward. But the beast wasn't accustomed to such treatment and, already unsettled by the darkness and slippery track, reared and tossed its head, trying once more to turn towards the safety of its stables. Nicholas fought to bring it around. Not daring to risk the whip again, he kicked with his heels urging the horse on.

The sound of a hunting horn reverberated in the air. Short notes answered by the deep-throated baying of hounds. In daylight and in another place, Nicholas's blood would have pumped hot and full to the thrill of a chase, but in the darkness, he felt beads of cold sweat chilling his back. He hesitated, craning round to determine where the sound was coming from, but howls were bouncing from hill to hill until they seemed to be all around him. He tried to tell himself it was only a nobleman out for a night's sport. He could confront the man, even seek his aid. But the image of Alban's corpse reared up before his eyes. The face torn off by merciless teeth, the guts and liver ripped from the belly by slavering jaws that could crack open a cow's thigh bone. A pack of hounds could rip a man apart like a hare, and if these were the same beasts, they had already tasted human blood.

As the horn sounded once more, Nicholas's horse seemed to make up its mind that the dogs were behind it on the track. It veered sideways and scrambled the foot or so up on to the bank that edged the path. Before Nicholas could force it back on to the track, it was plunging down the hillside and across the open moor.

Nicholas had always prided himself on being able to master any warhorse, even the mighty destriers, capable of carrying a man in full armour, but he was used to horses that responded to commands of the legs, for a knight must have his hands free to kill and defend himself. Instinctively, he found himself using his thighs and knees to try to steady and turn the frightened creature, but that seemed only to madden and confuse it.

It careered across the dark hillside, splashing through pools and streams, stumbling as its shoes caught in heather and lurching violently as its hind hoofs slipped on mud. One moment it seemed as if it would plunge forward and snap its neck, the next fall sideways and break a leg. Nothing Nicholas could do calmed it, and he could only pray that exhaustion would bring it to a halt.

Somewhere the horn sounded again, a dark, deep throbbing that boomed back at him from every hill, and the baying of the dogs grew more excited and ferocious. Even as the horse ploughed on, Nicholas could not resist glancing behind him. A dense black tide was rolling down the slope, moving swifter than the horse. At first he thought a torrent of water had burst from some lake or dam, but as one, then another leaped high and bounded over a rock or mound of grass, he realised they were hounds streaming down the hillside, their glowing red eyes flashing out of the darkness, like a forest fire running straight towards him.

Chapter 53

Sorrel

A glowing green mist spiralled out of the rock tomb and wove through the branches of the oak above, where long fronds of grey lichen hung like the rags of a winding sheet. Morwen's song was drawing the mist upwards. It was taking shape, like the birds and beasts in the fire in the tor, but this was no animal. It was a woman, hooded, draped head to toe in robes that billowed out from her as if they were caught in a wind I couldn't feel. The creature lifted her head. Her face was dead, a mask, a skull without flesh, as solid as moonlight. But peering out through holes in the cavernous eye sockets a pair of living eyes were trapped behind the bone, as if she was a prisoner, staring out through the bars of a cage. The eyes shone with a fierce, angry knowledge, a burning fury that the dust of a thousand years could not smother.

She turned those blazing eyes and looked at us, but though I knew her power, I was not afraid. She did not mean us harm.

Call them. Call them.

The words flew through the trees between us, but I couldn't tell if I had willed it, or Morwen or the spirit, for the words were living now, darting around the grove like a flock of starlings, and we all willed them. I could feel the heat from Morwen's hand pulsing through the rock, as if her blood was pounding through the veins of the stone.

Summon them. They will obey.

Somewhere beyond the woods, the wind was gathering strength, wailing across the sucking bog pools, shrieking above the thundering river, howling through the crevices in the tors, but inside the forest of oaks all was still. A cold white mist hung between the distant trees, shrouding the clearing, shielding the living and the dead.

Call them!

I didn't turn my head, but I could feel the others drawing near in their long robes, their eyes peering out from inside their bones. They inhabited their skulls like the weasels staring out from their dark tunnels, watching, willing, waiting. The spirits drew close, tangled in the gnarled branches, their hair the silver-grey locks of lichen, their fingers the clawed twigs. Others slithered between the arching roots, and crawled towards us over the mossy rocks. And still I was not afraid.

A shudder ran through the wood, a rumbling of stones and trembling of trees. There was a great fluttering of tiny wings as, from holes and crevices too small even to admit a mouse, a huge flock of wrens flew up into the oaks, melting once more into craggy branches in their stillness.

Someone has entered the wood. Someone has violated the sacred place. He is unclean.

Chapter 54

Dertemora

The horse was tiring badly now, and it might even have allowed Brother Nicholas to bring it to a halt, if he had tried, but with the hounds baying ever closer behind him, he urged the sweating beast on, willing it to keep moving. They were teetering down the side of a V-shaped valley, and Nicholas's only thought was to reach the bottom where surely there must be a track or at least a level way, so that he could spur the horse on and find rocks or trees in which he could take shelter. If he could just climb up into the branches or scramble on to a steep-sided rock, armed with his knife, he might at least be able to defend himself from the hounds.

But, as if it had been hiding, waiting to pounce, the babble and thrashing of water burst without warning on Nicholas's ears. He heard grinding and scraping as the swollen river dragged gravel over stones, and stones over rocks. Like a drunkard weaving down a street, water crashed against one boulder, only to fall back on to another and trip over the next. The horse veered away. Herded by the roar of the river on one side and the yelping of the hounds on the other, it struck out along the side of the steep hill struggling to keep its footing in the mud and sodden grass.

The poor creature's breathing was so laboured it seemed its lungs would burst with the effort, but fear drove the horse on, slipping and stumbling, long after its strength had been exhausted. Even in his own panic, Nicholas knew it was a miracle the horse hadn't fallen already, and if it did and rolled on him . . . The image of himself lying there helpless with a crushed leg and that pack of dogs snarling over him finally goaded him into action. If he tried to rein the beast to a halt and the animal resisted, he'd surely cause it to come crashing down. He couldn't take that chance. In the dark, he couldn't distinguish anything on the ground, but he knew these hillsides were littered with rocks and boulders lying in wait to crack a man's skull, and found himself praying that he'd land in soft, thick mud . . . But what good would prayer do now? Nicholas swung his leg over the horse's back and hurled himself into the darkness.

He landed momentarily on his feet, but the slope was too steep and slippery for him to keep his balance and he found himself tumbling backwards, rolling and bouncing towards the torrent of water. He flung out his arms wildly, trying to grab anything that might stop him plunging into the river, but it was the very rocks he'd feared that saved him. His back smashed into a boulder, and he lay still, fighting to suck the air back into his lungs. But as he gasped, a savage pain stabbed through his side. He'd broken ribs often enough in the joust and on the battlefield to know exactly what had happened. It would be agony to move, but at least he could walk. He tried to tell himself he'd been lucky – better his ribs than a shattered leg or spine.

Water was roaring inches from his head, drowning all

other sounds, but he knew the hounds were still out there somewhere. Would they follow the horse or pick up his own scent? Though his body pleaded not to move, he could not indulge it. He was defenceless, lying there in the open. He dragged himself up against the boulder and stood for a moment, listening, until the wave of pain ebbed a little. Somewhere above the thrashing of the water came the throb of a hunting horn, but it was not calling the dogs back, it was driving them on to search and kill.

He rocked forward, trying to hear where the sound was coming from. He was sure the track he needed lay somewhere to his right, but the baying seemed to be on that side of him too. He turned away from it, and with his arm wrapped tightly about his ribs, he stumbled on, following the course of the river. It was in full spate and he knew he'd be swept off his feet in an instant if he tried to cross it, but if the hounds did attack, it might be his only chance.

The clouds peeled away from the moon and for a breath the valley was lit by a ghost light. Frost-white foam spun across the seething river and ahead lay a dark mass of grotesquely twisted trees, their branches black against the moonlight, writhing like the Gorgon's hair from the scalp of the earth. The sight made him shudder, but from the tail of his eye, he caught a glimpse of movement. Silhouetted on the rise above him stood a great black hound staring down fixedly into the valley. Three or four others breasted the hill and paused, scenting the air. Almost before his mind could comprehend what he was staring at, darkness had flooded over him again. He stumbled forward. He had to reach those trees. That was his only hope.

But the whole valley seemed to be conspiring to keep him from reaching them, as if a charm had been put on the wood to stop anyone approaching it. His legs were sucked back by the deep mud. Brambles sank their claws into his cloak and tried to fetter him. Rocks reared up in front of him, barring his way, and stones rolled treacherously from beneath his feet, bringing him crashing to his knees and driving waves of white-hot pain through his ribs. The roar of the water grew ever louder but each time he paused he heard – he thought he heard – the baying of the hounds. Was it simply the wind? He no longer knew. Pain and hunger were making him dizzy, but he forced himself on, clinging to the one thought that if he could reach those trees he would be able to defend himself.

As he struggled towards them, he could see the smudges of the stunted trunks that stood sentinel on the edge. Just a few more yards, he urged himself. He turned his head as he caught the howling of the hounds again. They were closer now, much closer.

A harsh *kaah* made him whip round. Something was bobbing on one of the upper branches of the closest tree. A rook, a crow? He'd disturbed a roosting bird – that was all. He ducked under the branches, and heard another loud *kaah*, this time from the tree beyond. Both birds were screeching, irritated at being roused.

The sudden flapping of wings made him glance up and he fell heavily over a moss-covered rock, banging his cheek on another. The moss cushioned the blow to his face, but he yelped from the agonising jarring of his chest. He struggled to stand upright, unable to find a place to set his feet. The forest floor around him was covered with massive boulders, so tightly rolled together or heaped on top of each other that there was hardly a gap between them, except where the trees had forced their way through.

Kaah! Kaah! More birds were appearing in the trees around him. He could hear the furious flapping of their wings. The branches were so low that their thick beaks were clattering inches from his face. This wood must be their roost. But the yelping of dogs dragged his attention from the birds. They were close now, so close he could hear the snapping of gorse and the rattle of stones as they streamed towards him. He scrambled desperately over the boulders, cursing as his feet repeatedly slipped into the deep narrow cracks between them, constantly hitting stone with knees or elbows, each time jerking his broken ribs till he could hardly bear to breathe. It was madness to try to move quickly – he could trap a leg or even snap it.

The clouds drifted from the moon, filling the wood with cold white light and sharp, twisted silhouettes. It touched the forest floor, and the shadows ran lightly across it, so that the great mossy humps seemed to move, like waves rearing up in a stormy sea. Tendrils of glowing mist slithered between the knotted roots and through the branches, not hanging as they should but creeping and writhing as if they were alive. Yet, the air was unnaturally still.

The whirling of bird wings forced him to look up. Crows, jackdaws, rooks and ravens were flying down and alighting on every branch around him, their beaks glistening silver in the moonlight, their eyes glittering inches from his own. He ducked away from them, instinctively trying to raise his arm to shield his face, but he couldn't lift it above his shoulder for the sickening jolts of pain shooting upwards from his chest. He was drenched in cold sweat and struggling to breathe. Then he heard it: the long vibrating note of the hunting horn. Except it wasn't coming from outside the wood but from inside its very heart.

Darkness flowed down over the trees again. Nicholas couldn't move. How could the huntsman have reached the wood ahead of him, and if he was here, where were the hounds? He tried to calm himself. The sound was being distorted – these rocks, these hills flung it around, so you couldn't tell where it was coming from.

Something flew straight into his face. Feathers covered his mouth and nose. Talons raked his skin. Though every blow sent pain shooting through his body, he beat it off and scrambled across the boulders, clawing up the steep slope towards the centre of the wood. He could no longer remember why he must get there, only that he could think of nothing else to do.

'Holy Virgin, make the clouds roll back! Give me light, any light,' he begged.

With abject relief, Nicholas realised that his fervent prayer was being answered. The clouds were sliding away from the moon, as if the Blessed Virgin herself had stretched forth her hand and was pulling aside the curtain. He was in a small clearing. The trees there formed a circle around a massive spear-shaped rock. But he registered the rock only dimly, merely as a backdrop to the two women who were standing in front of it.

The cold white light fell on their naked bodies, running down the curves of their breasts, trickling over their ribs, stark as the timbers of a wrecked ship, dripping over their flat bellies and pooling in the hollows of their thighs. Snakes of mist, glowing green as the walls of the well, coiled about their ankles, slithering around their legs. Nicholas gaped in disbelief. Were these women humans or sprites?

In spite of his pain and exhaustion, he could not wrench his eyes from their bodies, from where the fingers of mist were reaching up, probing, penetrating. His hands twitched as if they would reach out to touch them too. If these were sprites, he would willingly surrender to their enchantment.

'You shouldn't have come here.' The voice was all too human.

He looked at their faces, their wild hair hanging down in wet strands. He saw now that they were beggars, nothing more than common vagabonds or street whores, probably both. As the one who'd spoken turned a little, he glimpsed the livid welts on her hip and shoulder. She had probably been stripped and whipped out of the nearest town, though she'd not been flogged nearly hard enough if she was still flaunting herself.

Fear was ebbing away from him and anger bubbling up in its place, as he remembered exactly who had driven him out on to this accursed moor in the middle of the night. He stared at the women coldly, his desire for them instantly extinguished.

'Now, just what manner of game have we caught here?' he said. 'It would seem that someone is hunting these hills with a pack of hounds. Is it you they're seeking? Thieves, are you? Runaway serfs? Don't invent some lie. What you are doesn't interest me. But I will strike a bargain with you. Show me the way to an inn or somewhere I can shelter for the night and I'll see what I can do to call off your pursuers. I dare say a well-filled purse will convince them to lose you for good.'

He had expected the women to look alarmed that he had discovered their hiding place or grateful that he had offered them a means of escape. He was even prepared for them to try to wheedle some coins for themselves, but though it was hard to see their expressions, he could have sworn they were laughing. Perhaps they were simple or mad.

'Come now, my dears, you've nothing to fear from me.' He braced his arm against his ribs, trying not to gasp with pain as he talked. 'I'm a knight of St John. Not even the sheriff's men would dare seize you, if you were under my protection.'

But instead of begging for his help, the girl who had spoken threw back her head and began to sing. At least, it was a kind of singing, he supposed. The unearthly chant made his skin crawl, as if a thousand flies were creeping over it.

If the Virgin Mary had been holding up the curtain of cloud, she now let it drop and the grove was plunged into darkness again. The trees were crowding in, shuffling towards him, like evil old men, carrying their birds nearer and nearer. The raucous *kaah*, *kaah* was all around him, hundreds of wings flapping furiously, beaks snapping. Blindly, he tried to fight his way out, but could find no space wide enough to squeeze between the trunks. How had he come in? There must be a gap – there had to be! Where was it? Then he heard it. The blast of the hunting horn so close, so piercing, he knew the sound was coming from someone standing in the centre of the glade.

All at once he saw them, the great black beasts with their burning red eyes, standing in a circle around him, just beyond the ring of the trees. He saw the blue lights crackling from their coats, saw the long, pointed teeth flash white as they threw back their heads and gave that deep sonorous howl. They were slowly padding towards him, a merciless black tide, and he could no more turn them back than a man can stop a monstrous wave rolling towards a stricken ship. The pain in his chest was almost numbed by his abject fear, for even as he drew his knife, he knew deep down that not even his sword or a hail of flaming arrows had the power to save him from this foe.

Brother Nicholas's legs gave way and he sank abjectly to his knees, crawling frantically away from the women, away from the hounds. But instead of colliding with a tree trunk or stone as he had expected, his leg vanished into the gap between two rocks. He wriggled backwards, trying to ease his battered body into the shelter of a hollow. It was a small space, no bigger than a kennel, but just wide and high enough for him to squeeze himself inside.

He crouched there on all fours, the blessed solidity of rock covering his head, his back, his sides, like steel armour. Relief surged in his belly, and he gripped the knife with returning confidence and the growing excitement of a seasoned knight facing an enemy he knows he can defeat. He was protected on all sides, save the front, where he could freely wield his weapon. That was the only place the dogs could touch him, and the entrance was so narrow that they could attack only one at a time. Kill the first and second, and the hounds' own bodies would shield him from the rest of the pack. And once the beasts smelt blood, they'd more than likely turn on their own dead and wounded. He could fight these fiends and he could win.

He waited, every muscle in his aching body tense. 'Come on, you brutes, try to take me now. Who will be first to have their throat slashed?'

Through the demonic cackle of the birds in the trees above him, through the baying of the hounds that shook the pillars of Hell beneath, he heard the women laughing.

And then his back arched in agony as a dozen white-hot

stings stabbed into his legs, his sides, his neck. He stared down in disbelief as the sinuous bodies wriggled over his hands. Vipers! They were crawling over his shoulders, dropping into his hair, slithering down his face. He twisted and shrieked as their fangs struck and struck again, begging the Holy Virgin for mercy, but mercy was not granted.

When the first brilliant rays of the rising sun touched the tops of the hills, all that remained of the noble knight of St John of Jerusalem was dismembered, bloodstained bones. The birds had feasted well. Before anyone discovered them, the bones would be covered with moss, indistinguishable from the roots of the twisted trees and the rocks among which they lay scattered. Those who stumbled over the skull would think it another mossy boulder. Ferns would grow through those broken ribs, liverworts plant themselves in the eye sockets. The birds would rear their young, safe and warm, in nests lined with strands of grey hair.

Pick nothing up in that wood. Take nothing from it, for who knows whose remains you will hold in your hand? Do not disturb the resting place of the dead, for the hounds that guard it guard it well.

Chapter 55

Hospitallers' Priory of St Mary

Meggy drew her chair closer to her fire, prodding it to send the flames blazing upwards. Outside, the wind wailed like a soul in torment and monstrous shadows flickered across the courtyard walls. All the days she had lived at the priory, the old gatekeeper had prided herself that she had loyally obeyed whatever orders the prioress had given, even if she thought them foolish. She'd have laid down her life as willingly as any knight of St John, not for the order – she had little patience with that – but for Johanne and this priory. But though the prioress had given instructions that fuel was to be kept for the kitchens and the infirmary, tonight Meggy knew it was wise to disobey: the prioress might know all about the running of a priory but she was a babe in clouts when it came to understanding the ways of Dertemora. It wasn't the chill wind that drove Meggy to this small act of defiance – she was well used to feeling the cold and expected no less at her time of life. It was the hounds.

She'd heard the beasts baying across the moors and felt again their great rasping claws on her back and their fangs snapping inches from her cheek. Her hand strayed unbidden to where the little bag had once hung at her waist before they had ripped it from her, almost cutting her in two. Save for that bloodstained rag, it hadn't contained much – her son's milk tooth and an iron nail her husband had fashioned into a little cross to keep her safe. It was all she'd got left of her menfolk . . . all she'd had, for those precious objects were now scattered somewhere among the heather and grasses of the moor.

She knew the hounds had chased to stop her taking that rag to old Kendra, though she'd no idea why. Still, she counted herself blessed. There were not many souls who'd come face to face with the wisht hounds and lived to tell the tale. Those spectral hounds were hunting their prey again tonight and Meggy was in no hurry to chance her luck twice. If Brother Nicholas was still out there on those moors then God help him, for Old Crockern wouldn't and the knight deserved all that was coming to him.

Poor young Brengy had still not returned. Dye had tearfully begged the sisters to search for her brother, though she was too afraid to go out on the moor in the dark herself.

But Sister Clarice had said firmly there was nothing to be done until cockcrow. 'If we start wandering around out there at night, one of us will end up in a bog or worse, and with the wind as strong as this, even with torches we might walk within a foot of where he's lying and not hear him even if he's able to cry out.'

Prioress Johanne had agreed. 'There is little point in starting any search tonight. Sleep will serve us better. The servants and sisters can set out at first light tomorrow. We've no idea where Brother Nicholas set Brengy down, if indeed he did. He might have decided to take the lad with him all the way to Buckland.'

The prioress had looked more drawn and anxious than Meggy had ever seen her, but she knew the poor soul had more on her mind than the missing stable lad.

Meggy had coaxed Dye out of her wailing by asking her to help dip new torches and set them blazing on the walls outside, telling her that their lights could be seen as far off as Exeter. All Brengy had to do was walk towards them from wherever he found himself. The gatekeeper promised to sit up and listen for the bell all night till he came home. Besides, she told herself, she'd never be able to close her eyes with Old Crockern's hounds howling out there on the moor. Privately, she wondered if poor Brengy would ever return. She wouldn't put it past that devil Nicholas to throw the boy into the mire, like he'd threatened, just from spite, though she'd never have said as much to Dye. No sense in setting her off again.

The sisters had long retired to their beds, but Meggy found she couldn't rest until she'd checked and rechecked that the gates to the priory were locked and braced. She'd pulled a charred stick from the fire and drawn crosses on the threshold and casement of her hut, and binding knot signs around the fire to prevent evil from entering. Alban's ghost, and maybe poor little Brengy's too now, would be wandering the moor, searching for the lych-ways, and it was well known that such lost souls who'd been snatched violently from life were drawn to the living. They would try to enter their cottages and even creep into the mouths of those who lay sleeping inside, so desperate were they to remain in this world.

But in spite of the wind, the distant baying of the hounds and her own fear, the old gatekeeper was exhausted, and the heat from the fire made her eyelids grow heavy, though she tried to resist sleep. She floated in that twilight between sleep and waking, noises from this world mingling with those from some place far off in her dreams, so that when the door of her hut creaked open, it seemed to her that the sound belonged in another realm.

Soft footfalls rustled through the bracken strewn on the beaten-earth floor. The flames bowed low in the breeze from the figure passing, then leaped up wildly. A small branch popped, and cracked open as the sap oozed out with a long, hissing sigh. Old Meggy jerked upright, almost over-toppling in her chair. The boy was standing not a foot away from her, his face half in darkness and half bathed in the ruby firelight. His head moved slightly from side to side, like a snake's, as if he was trying to sense where she was.

'What do you want this time of night? How did you get here?' Meggy demanded.

Even to herself she sounded angry, but the boy didn't flinch.

He tilted his face up, but his sightless gaze was fixed somewhere over her shoulder. Meggy half turned her head, fearing that something or someone was behind her, though she knew even as she did so that it was something blind folks did. They always said old Father Guthlac was looking at angels, but she was sure it wasn't any creature from Heaven the boy saw.

He took a pace forward, shuffling, feeling his way. He was learning how to move. Then his hand followed. It groped towards her and she wanted to draw back so he wouldn't touch her, but it seemed a cruel thing to do to a child. Instead, against her will, she found herself stretching out her own gnarled hand, touching the icy fingers. The boy immediately grasped it and began to tug her towards the door.

'Hold hard! Where do you think you're going this time of night?'

She couldn't imagine what the child was doing so far from the infirmary. Had he slipped out to use the garderobe and been unable to find his way back? But no one used the garderobe at night – that was what the pisspots in the chambers were for. If the child heard her question, he gave no sign. Although Meggy's arms were far stronger than those of most women her age, from years of toiling alongside her husband in the forge, she found she could not free herself from his grip.

The boy dragged the door open and the wind barged in, clanging together the worn iron pots dangling from the rafters, tumbling ropes and nets to the floor and sending the ash and smoke from the flames swirling, like a whirlpool, about the small hut. Meggy gave a cry of alarm. The wind often made mischief when the door was opened, but the gate lodge was sheltered behind the high wall and it had never ransacked her room like this before. It was as if her home had been plucked up and set down on the very pinnacle of Fire Tor.

'We'll be knocked off our feet, if we venture out there,' she bellowed, above the roar. 'You'd best spend the rest of the night in here.'

She struggled to slam the door, but either the wind or the boy was too strong – it must be the wind, surely. The grip on her hand tightened and, though she tried to resist it, she felt herself being pulled out into the darkened courtyard, as if her wrist was bound to a plough horse. She was punched and shoved by the gale and would have come crashing down several times, had the boy not clung to her, dragging her towards the great oak gates. Shutters whimpered against the flailing wind. Pails were thrown across the courtyard. Broken twigs and chips of goat bone were dashed against Meggy's face, and now it had begun to rain again, great cold stinging drops that lashed her skin.

She fought to turn back for the shelter of her little hut, but the boy pushed her hand up against the oak beam that braced the gates. She did not have to hear him speak to know he wanted her to wrest it back and open it.

'You can't go out there, lad! There's a storm building and death hounds are hunting. I'll not open those gates tonight, even if the Good Lord himself comes a-knocking.'

He took her hand again, pushing it over and over against the brace. Still, Meggy tried to drag him away, but it was as if he suddenly weighed as much as a full-grown bull and she couldn't even lift him to move him aside. Then, above the roar of the wind, she heard the sound of the bell tolling. Someone was outside, begging for admittance. Brengy! Was that what the child had been trying to tell her? Had he heard the bell over the noise of the wind when her old ears could not?

She was already tugging at the beam, when caution stayed her hand. She had never opened the gate without checking first, and maybe it was only the fierce wind that was rocking the bell. She opened the shutter over the metal grille and peered out.

Two figures stood there, huddled against the driving rain. Neither was the stable boy – that much she could tell. But the torches outside had long been extinguished and the devil himself could have been standing before her gate and she'd not have known him from the archangel Gabriel. The pair came closer.

443

'Let us in, Meggy. Brigid needs us.'

It was a woman's voice, one she thought she knew, but it was muffled by the storm. The woman came closer, pressing her face to the iron bars. Her hair whipped up in wet strands about her head.

'It's me, Meggy, Morwen. Let us in. The wisht hounds are running. Brigid needs us. The boy needs us.'

Meggy hesitated. The prioress had forbidden Kendra and her tribe of daughters to set foot in the place, and Meggy had always carried out those orders faithfully.

As if Morwen heard her thoughts, she shouted, 'Ma's not with us. She fights against us, against the boy. Help us, Meggy.'

Meggy felt the child's cold fingers on her hand, so icy they burned. She glanced down. The boy was staring at the door, as if he could see right through it. Suddenly he pressed his face against her thigh, sinking, like a child weak with hunger, as if all the strength was draining from him and he could no longer stand. He staggered and she caught him in her arms. She lowered the boy gently to the ground, smoothing the hair from his cold little brow. Then she clambered to her feet, took a deep breath and heaved on the swollen brace beam, dragging it back.

Chapter 56

Morwen

Glancing fearfully around her, Meggy hurried us into the chapel and bolted the door behind us. She set the lantern on the floor, so the light wouldn't be seen through the casements, while she wriggled a great iron key into the door on the far wall. Sorrel edged closer to me and the boy, staring up at the statue of the bloody man hanging on the cross, as if she thought he might raise his carved head and shout for the black crows to chase us out of his house. I was afeared in that place too but not of the crows.

When I was little, and Ma was trading in the village, I'd sometimes slip away from her and peer in through the door of the church. I liked to look at the paintings of men and beasts on the walls, the colours as bright as yellow furze and the scarlet rowans on the hillside. But this chapel wasn't like that. It was a clearing in a dark wood with great pillars, like tree trunks, stretching up all round. I was never scared of the spirits of Fire Tor, or of the ghosts of the moor, but there were shadows slithering round these walls, dark, malevolent creatures who'd never lived in this world but meant

445

to bring harm to any who did, if ever they could cross into this realm.

Old Meggy wrenched open the door and pointed down into the darkness. 'Bryde's Spring is at the bottom of those steps,' she whispered, though even if she'd shouted no one outside, not even a screech owl, would have heard her over the storm. But she didn't move aside to let us pass.

'Kendra said the boy must be kept from the well. She said if his eye opens . . .' Meggy trailed off.

'Aye, Ma spoke the truth of what she saw, but she doesn't understand the meaning of it.'

Still Meggy hesitated. Sorrel took a step towards her. 'Brigid sent the wisht hounds to stop you taking the rag to Kendra. They turned you back. She wants to protect the boy.'

The old woman's head jerked up. 'How do you know that? I didn't tell a soul . . .' She stared at Sorrel, tucking her thumbs beneath her fingers, as if she thought Sorrel might witch her. 'You got the gift, haven't you, like old Kendra?'

Slowly Meggy dragged her gaze from Sorrel. 'But even if I let you take him down there, it won't do you a mite of good, for the well's dry. Spring's stopped running. Some say it's the tinners have done it, but Sister Fina reckons it's . . .' She jerked her chin towards the boy, who clung to my hand.

I shrugged. In truth, I'd no notion what we were to do down there, only that I knew Brigid wanted the boy to be taken to her well. Like Ma said, if you asked what Brigid wanted of you, you must swear to do it, else you must not ask.

Meggy looked from me to Sorrel, then seemed to make up her mind. She stepped aside from the well door and,

446

picking up the lantern, held it out towards me. She said she'd stay up in the chapel and keep watch. But I think she was afeared to come down.

Sorrel and I crept down the stone steps, with the boy between us, me first, stepping sideways so I could keep hold of the boy's hand and guide him down. His legs kept buckling beneath him, but not 'cause he couldn't see. He was as weak as if he'd been starved for days.

The walls shimmered like glow worms on a summer's night, pigsey gold, Ma called it. Suddenly the boy crumpled on to the stone step, his head lolling against his chest.

I pushed the lantern up at Sorrel and knelt on the stairs below the boy, so that I could heft him over my shoulder. He made no sound. I didn't know if he'd fainted, but even through my kirtle I could feel the frog-cold of his body. Was he dying, dead?

I was feeling for the steps with my bare toes, trying to balance the boy and walk into the darkness. But then I could no longer feel a step below and as Sorrel came down behind me with the lantern, I saw that we were standing in the cave, just like I remembered from when I was small, save that there was no water. The stone trough lay empty and raw red, like the belly of a rabbit when its guts have been torn out. On a ledge above the well stood a painted statue of a woman grasping a long dagger, pointed down at the ancient stone carving of Brigid, as if she meant to gouge out the old mother's eyes.

I laid the boy as gently as I could on the rocky floor, his eyes were open, staring straight up, unblinking, but he didn't move, though I could see his little chest rising and falling in fluttering breaths, like a wild bird when you hold it in your hand.

Sorrel crouched beside me. 'There's not even a drop of water . . . What can we do?'

I closed my eyes. I was in Ma's cottage. A little girl lay on the floor and her granddam sat facing Ma across the hearth fire. I watched Ma clenching her fingers about the smoke, saw it sucked up into her fist.

'Brigid's closed the water in her hand. Only she can release it.'

There was no need for me to tell Sorrel what to do now. She knew better than me. She settled herself on the floor, and grew still. Sorrel didn't need to sing to open the door 'twixt the living and dead. But I sang, not to summon the spirits but to hold back those creatures of shadow that prowled above our heads. They were trying to enter, trying to cross into this world, and though the moon was hidden from mortal sight, it was a duru moon. They no more needed to see its light than I needed to see an open doorway to walk through it in the dark.

I could feel them massing, trying to find the cracks to slither through, trying to claw down the walls and ooze down the steps until they filled this cave with a darkness that would smother everything. I forced the song against them, trying to make it fill every hole, block every crevice in the cave, as I begged Brigid to hear us, to open her hand and let her spring flow. My knees were pressing hard against the boy's body. A shudder ran through him, like the death throes of a mouse when its neck is snapped. The child was dying, but Brigid wouldn't open her hand. She wouldn't let the waters run.

Sorrel scrambled to her feet. She clambered into the stone trough, then on top of the rim nearest the wall of the cave. Her feet planted on either side, she balanced precariously, steadying herself with her good hand against the rock. For a moment she swayed there, then lifted her hand and reached up to the painted statue. But she had no means of holding on. Before I could get to my feet, she was toppling over. She grabbed the wooden dagger in the statue's hand and dragged it with her as she fell. Both crashed on to the rocky floor of the cave. The head of the painted woman snapped off and bounced away, slamming into the lantern, sending it, too, spinning across the floor. The candle flame guttered wildly and blew out. We were in darkness, save for the green-gold glow of the walls that lingered for a few moments, then slowly began to fade.

'Sorrel, are you hurt?' I scrambled over to her.

She was lying on the floor. I felt her head, trying to see if it was bleeding or broken, but she pushed my hand away. 'Listen!'

Water! Water splashing on stone. The spring was running again! The statue was smashed. The well was Brigid's once more. In the darkness, we lifted the boy and carried him to the trough.

'We need to wash his eyes,' Sorrel murmured.

'No, you need to. Brigid called you to Dertemora for this.'

She didn't argue. The sound of the water changed as she cupped her hand beneath the trickling spring. Then I felt the cold drops fall on my hand as she poured the water over the boy's face. Once, twice, three times three.

> *As Brigid of Imbolc brings back the sun*
> *As the Sun gives sight to earth*
> *As the earth opens to the green mist*
> *So may his eyes open to see*
> *So may his eyes open to see*
> *So may his eyes open to see.*

The spring was gushing from the rock now, the sound almost deafening as it thundered down into the stone trough. I was drenched in a fine mist. Water was creeping up the side and starting to spill on to the floor of the cave. The moss on the walls began to glow, though no candle had been lit, softly at first like the first light of a winter's dawn, then brighter and brighter, like the summer's sun glinting on a pool. As I gaped at the walls, I felt the boy drawing out from my arms. I caught sight of Sorrel. She was staring over my shoulder towards where the boy stood behind me, her mouth open as if she had cried out, but I could hear nothing above the roaring of the water. I turned my head to look at the boy, but he had vanished.

Chapter 57

Dertemora

A crackling blue lightning flash split the dark sky above the tinners' valley. The thunderclap, when it came, rampaged around the hills, hurling itself from one side to the other, like a caged beast. Rain thrashed the bare sides of the valley, streaming down in torrents, dragging the mud and gravel with it.

Inside the storehouse, a woman pushed Gleedy's sweaty head from her breasts and struggled to slide out from beneath his slippery carcass. She froze, holding her breath, as Gleedy grunted and rolled over. In the dim lamplight, she watched a silver snail's track of drool trickle down his chinless face and soak into the straw pallet. Gleedy had drunk the best part of the flagon of wine but, even so, she was amazed he could sleep through this storm.

The woman pulled down her skirts and tiptoed across the floor towards the stores. There were stacks of tools, ladders and buckets to be had, but she knew that most of the boxes, barrels and sacks that had once contained dried beans, salted meat or flour were now as empty as a staved wine cask. The

451

last time Gleedy had dragged her to his bed, she had seen him take her payment of food out of the boxes hidden beneath the empty barrels, but she dared not risk waking him by shifting them.

A dazzling white flash of lightning pierced the gaps around the shutters and door, followed a few moments later by a long rumble of thunder. But, as if it was determined to have the last word, the wind roared back even louder and the rain seemed to redouble its efforts to drill through the roof. The woman had to get to her own hut. Her young son would be terrified there alone.

Gleedy's hounds, tethered near the door, were howling and barking loud enough to summon the devil from Hell. Usually a single growl or yap would be enough to send Gleedy scurrying to them, fearful that someone was trying to break in. She'd have to find something to toss to them to keep them back from the door so that she could slip out. He usually kept a sack of dried sheep's trotters close to his bed. She edged back, watching Gleedy for any sign that he might be faking sleep, then found the sack and took out two hoofs. Without planning to, she grabbed three more, stuffing them down the front of her kirtle to add to her cooking pot.

With another anxious glance at the slumbering figure, she hurried to the door and tried to open it quietly. But she had reckoned without the wind. Before she had pulled it more than a hand's width open, it snatched the door and hurled it back against the wall, ripping off one of its leather hinges. The wind charged in, sending stacks of tools crashing to the ground and empty sacks flapping through the air, like flocks of geese. Not even a drunkard could sleep through that.

Gleedy was on his feet with a bellow before his head realised what his body was doing. The hounds were leaping and pulling so hard on their chains that they were choking themselves. The woman threw the two trotters at them, but for once they were neither interested in food nor in any intruders. They were simply desperate to get into the shelter of the storehouse and out of the storm. Gleedy gave a shout of rage, but the woman didn't wait. She fled into the rain-drenched darkness.

Gleedy was halfway across the chamber in pursuit of her when a blast of icy air on his naked cods reminded him that he was clad only in a short shirt. He was torn between his need to force the door shut and his fear of anyone catching sight of him in the lantern light. A pile of iron pots crashed to the floor, and he ran to the door, trying to wrestle it shut, but it was hanging sideways by the one remaining hinge that was also on the point of ripping from the wood. Against the force of this wind, it would take at least two men even to get it upright and force it back into the doorway. Icy rain dashed against his face as if a man was tossing buckets of water through the hole where the door should have been. Though it did not entirely sober Gleedy, it shook him fully awake.

He abandoned his efforts and hastily pulled on his breeches and a cloak. Not even attempting to lace his hose, he thrust his bare feet into the boots and ran out into the storm past the howling dogs, heading towards the blowing-house. He intended to order whoever was working there to return with him and force the door back into place, before the wind wreaked any more havoc or those thieving tinners started looting his stores. And someone had better be there manning that furnace, or he'd see to it that by morning they were tramping the moor without so much as a shirt on their backs.

But, once outside, he found he could barely stand against

the gale, much less see where he was going. Rain, grit and shards of stone were flung into his watering eyes as if they were fired at him by a hundred bowmen. Water poured off the hillside, turning paths into streams and washing away great chunks of mud and stone, sending them slithering to the valley. The river roared down its diverted course, dragging rocks the size of a man's head with it, and even though the sluice gate, which dammed the lake, was closed, the ditch below it was rapidly filling with rain and the muddy water pouring off the land above.

Just in time Gleedy glimpsed a chunk of wood hurtling towards him. He managed to sidestep it, but slipped in the mud and crashed down on his back, landing in the water that was racing down the track. Not that the wetting made much difference, for he was already soaked, but the heavy fall shook him, while the wine and his recent exertions had, for the moment, robbed him of the strength to right himself.

Lightning sizzled down, like a blazing spear thrown to earth, causing Gleedy to squint up. Two figures stood on the edge of the lake above the blowing-house staring down into the valley. He saw them only for as long as a white flash lit the moor, but his mind held on to the image as darkness closed in again. He was sure he had seen two women up there, their hair and skirts whirling about them, their arms open, held up to the sky as if they were calling upon the heavens to throw down all their arrows, daring a thunderbolt to strike them. Had they run completely mad? The wind alone could knock them into the lake or send them plummeting down into the valley. Even a village idiot would have enough wit to seek shelter. He certainly wouldn't have ventured out even this far, had that bitch not let the wind break his door. She'd pay dearly for that when he caught up with her.

He dragged himself upright, cursing as he slipped again in the streaming mud. He wiped the water from his eyes and glanced back to where he'd seen the women. It was hard to distinguish anything through the rain, but he was certain they were still there. Was it even possible to stand up there against this? Maybe they weren't flesh and blood at all.

Eva! Could her ghost have risen? The other had only one arm raised – he was sure that was what he'd seen in the lightning flash. One arm – Sorrel? Had she, too, perished out on that moor? That was what he'd wished, what she deserved. But had she, too, returned from the dead to haunt him?

A cold snake of fear slithered down Gleedy's spine and curled itself inside his guts. Whatever was up there he had no intention of staying out here alone. He ran for the blowing-house, slipping and sliding all the way. Crashing to his knees, he struggled up again, ignoring the stinging and the warm trickle of blood running down his leg. He reached the door. It was shut! It was never shut! He pounded on the wood, though no one inside could have heard him above the thunder rolling around the hills. Eventually his fuddled, panic-stricken mind grasped at the memory that even this door had a latch. He groped over the sodden wood, like a blind man searching for lost treasure. Several times his fingers came within inches of it, but he couldn't remember whether it was high or low, much less see it. By the time his numbed hand struck it, his eyes were streaming, not just from the driving rain but with tears of fear and frustration. He dragged on the latch, leaning with all his weight against the wood. As the door swung open, the wind shoved it so violently that he fell head first into the building, almost dashing his brains out on one of the granite ingot moulds inside.

As he dragged himself to his knees, scrubbing the water from his eyes, two men rushed past him and fought to push the door shut against the storm. Gleedy felt a surge of relief and annoyance, relief to be safely inside and in the company of half a dozen men, and annoyed by the grins on the tinners' faces at the sight of him sprawling at their feet. Suddenly aware of the rain dripping from his hair and the mud clinging to him, his irritation was not assuaged by the realisation that he must look like a drowned cat fished from a well. The men silently returned to the furnace, warming their hands and glancing warily at one another.

Now that he was safely inside, in the light and heat of the blowing-house, the very notion that he had seen two vengeful ghosts seemed risible and he could hardly believe that he had allowed it to take such a hold of him. His anger at his own foolishness turned in a flash on those around him, as he gradually became aware that inside the blowing-house one sound should have been louder even than that of the raging storm, but it wasn't. The great waterwheel was silent, as were the huge bellows it pumped.

'Who gave orders for that wheel to be shut down?' he demanded.

The men glanced at each other. Then the stoker raised his head. 'Reckon it were me. Water's coming down at such a lick that if we hadn't disconnected the wheel it would've broken its shaft and been smashed to pieces.'

Gleedy wasn't stupid. He knew the stoker had done the right thing. Truth be told, he'd probably saved Gleedy's hide for, as he'd had cause to remind the tinners often enough, Master Odo had paid a fortune to build the blowing-house. If the wheel had been destroyed it would not only have cost a tidy sum to replace, but smelting would have been halted for weeks. Nevertheless, Gleedy was not in a mood to show gratitude, not when he'd seen them laughing at him.

'So, you think that if the wheel isn't running, you can sit on your backsides for days till the furnace heats up again? And I suppose you imagine Master Odo is going to pay you for doing nothing. There's an assay meeting in Tavistock in less than a month and if we don't have a full load for the assay master to test, then Master Odo'll not be a happy man, not if he's made to look a fool in front of the other tin owners. If he doesn't get his money, you'll not get yours. That I can promise you. So, if you can't use the wheel, you'd best pump the bellows by your own muscle, for if that smelting is ruined by the furnace cooling, it'll be your wives and young 'uns that'll go hungry and all those men out there too. And I reckon that if the tinners learn you've lost all they sweated to dig up, you'll find yourselves caged in that wheel, being forced to turn it by crawling.'

An angry muttering broke out. One man faced Gleedy, his face contorted in fury and his great fists clenched. 'Takes two men to pump those bellows. How are we supposed to stoke the fuel, load the heads and pour out the tin too? I've not hands enough, nor strength neither.'

Gleedy flicked his fingers towards the other four men in the blowing-house. 'No sense in them sitting idle watching.'

'We've been breaking our backs all day graffing. You can't expect us to work all night too. We're meant to be sleeping. Only came in here to get warm and dry 'cause our huts are so deep in water only the fish could sleep easy in there. Rain's flooding down that hillside. I reckon most of them will have collapsed by morning.'

As if to prove his words, another long growl of thunder rumbled round the valley and the rain redoubled its efforts to pound through the roof, drowning the tinner's words, so that he was forced to bellow. 'Every tinner and his brat in this camp will be making their way up here afore long. There'll not be room to hang a man then, much less keep that furnace burning. We'll have to bed down in the storehouse. That and this blowing-house are the only dry places left.'

Gleedy made a noise somewhere between a strangled croak and a screech, as he was suddenly brought back to the reason he had made his way to the blowing-house. The very idea of the tinners and their families wandering unchecked around his storehouse in the middle of the night was enough to cause an iron band to squeeze so tightly around his chest that he could barely breathe. His jaw throbbed with pain. He wouldn't have a bean or a pot left to cook it in by morning, if that thieving rabble got in.

'You and you.' He pointed at two of the more burly and skilful men. 'I need you to fix the hinges of the storehouse door. They've been torn away and it won't close. If a strong blast of wind gets in and under that roof, it'll lift it right off.'

Gleedy was infuriated to catch the half-concealed grins that darted between the tinners. Anyone would think they were amused by the thought of the storehouse being destroyed. They wouldn't be laughing if there was no food for them to buy next week, or a new pick head, when their old one snapped or was stolen. See if they found starving to death a cause for mirth. But even he knew that that argument wouldn't persuade them to help him tonight.

'It breaks my heart to think of those poor little children and their mothers, soaked to the very marrow, toiling up here only to find that there's nowhere to shelter from the storm,' Gleedy said dolefully. 'It'll be the death of some of them, I shouldn't wonder, if they should find themselves out in this bitter wind and rain all night.'

Morwen and Sorrel steadied each other against the buffeting of the wind and gazed down into the valley, watching the glimmers of silvery white that marked the lines of the foaming river and the ditch. Streams of water ran down the raw hillsides, like blood from a flayed back. Waves whipped up by the wind and the tumbling waterfalls raced across the black lake at their feet, crashing against boulders and splashing over the dam wall and sluice gate.

The feeble red and yellow pinpricks from the tinners' fires at the far end of the valley had been snuffed out by wind and rain. All the valley had been plunged into darkness, except for the plume of smoke that gusted from the blowing-house, lit from below by an unearthly red glow, as if it was rising from the fires of Hell.

It is time!

The heads of the two women turned as one, strands of wet hair writhing into the dark sky. Rain ran from their clothes, their skin, their fingers, their bare feet, gushing like twin springs into the lake. On the other side of the water, the blind boy stood so still that it seemed even the wind was afraid to come near him. Morwen began to sing, her voice rising into the storm above the roar. Other voices, unseen, joined hers, as if a thousand curlews had taken wing.

The boy crouched on the bare, bleeding hillside. He bowed his head and Sorrel held out her arm across the raging water. Slowly, like a new leaf unfurling, he rose, but he was growing taller, broader, his fingers thick and strong, his muscles swelling, his legs lengthening. Great beetle brows hung over

459

dark eyes, craggy cheeks and a sharp, angular nose. Like an old tree awakening from the depths of a winter's sleep, the boy, an old man now, an ancient man, aged as the tor on which he stood, stretched his gnarled fingers across the lake. His voice boomed, like a thunderclap.

My land, this is my land!

Old Crockern's eyes were open and the lightning that flashed blue from them shot like an arrow across the water to the very edge of the lake. It struck the wooden sluice gate, shattering it into a thousand burning fragments. With a roar almost louder than Crockern's, the water of the lake surged over the edge, sweeping rocks and boulders and all before it, down, down into the deep valley below.

Gleedy watched the two sullen tinners struggling to hold the door of the blowing-house. He barged out in front of them, leaving them to follow. But the moment the three were outside, the wind charged into them, knocking them hard against the wooden door, as if it was trying to punch them back in through the cracks. The men stood for a moment in the lee of the wall trying to regain their balance. They wiped the lashing rain from their eyes and braced themselves to step out into the full force of the storm.

Gleedy, his arms wrapped tightly but uselessly around himself, shuffled from foot to foot, yelling at them to hurry. The tinners scowled and swore, not troubling to lower their voices, knowing the weasel would never hear them over the wind.

'When we get in there, I'll keep him occupied. You grab any food you find. Split it with you after.'

Gleedy was staring down the valley. He was sure he could hear men shouting, struggling to make themselves heard above the pounding rain and water surging through the wooden leats. He thought he could glimpse people moving along the valley below. If he didn't get that door back in place before that rabble got up here . . .

All three men glanced up as a great boom echoed across the lake above, and they were still staring as the lightning bolt sizzled through the blackness and the sluice gate burst apart. Burning fragments of wood arced scarlet through the black sky. The tinners dived for the shelter of the wall, covering their heads as burning wood showered down. But Gleedy, standing further out, seemed rooted to the spot. He stared as a plank somersaulted in the air, showering red and yellow sparks as it fell straight towards him. Only at the very last moment did his legs stir into life and he tried to run, but it was too late. The fiery wood struck him on the back, knocking him face down into the mud. Gleedy lay pinned beneath the burning plank. The flames blazed up into the darkness.

He screamed as fire danced up his back and down his thighs. Had his shirt and breeches not been sodden, his whole body would have been instantly engulfed in flames, but even so the heat was so fierce that his skin was bubbling, blistering, like that of a roasting piglet. But even as he struggled in vain to free himself, his cries were lost beneath a deafening roar.

A flood of water and boulders crashed down upon the blowing-house and the storehouse. It swept around the stone walls. The blowing-house stood its ground for a few moments, but even it could not withstand the force bearing down. Water, rocks and now the entire contents of the storehouse were hurled against its walls, with all the force of a siege engine, and the blowing-house, though sturdy, was no castle.

A corner shuddered and collapsed. The flood poured in and hit the red-hot furnace, which burst wide open, spewing steam, burning wood and molten tin into the maelstrom. The remainder of the blowing-house toppled sideways and caved in, severing the cries of the men trapped inside.

Below in the valley, the tinners stared up at the great wall of water sweeping down towards them out of the darkness, dragging men and rocks with it. Their minds were unable to grasp what was roaring towards them until the moment that the freezing tide hit them. Then the shrieks and screams of the men, women and children were drowned in the skull-splitting howl of that flood.

Gleedy knew only a moment of relief from pain as the icy water doused the burning beam, before he, too, was swept gasping and choking into the valley below, his helpless body battered and thrashed by the barrels, ladders, spades and iron pots all churning in the raging surge.

Chapter 58

Prioress Johanne

I hear the door of the chapel open and close, but I do not look round. I know who it is and why they've come. It is the darkest hour of the night, when all of the candles and torches are burned away and when the embers of the fires are buried so deep beneath their snowfall of ashes you think they can never be revived. Only the oil lamp hanging in the sanctuary still flickers, spilling a pool of blood-red light on the altar to trickle down on to the floor.

The priory has been at peace for more than three months, the kind of peace that steals over a dying man in the hour before his death, a calmness and resignation, when the fight to stay alive has finally been surrendered. And we, too, have surrendered to death.

In the cave below where I kneel, the waters have finally subsided. The night of the storm, when the well overflowed, the spring ran, like blood pumping from a wounded heart. Water rose up the steps and streamed out beneath the door, flowing across the chapel floor. We didn't discover it until the evening of the next day, for that night it rained so hard that the courtyard itself was turned into a lake, water pouring in under the doors of all the chambers, cascading down the steps into the kitchen and extinguishing every fire.

It was a month, maybe more, before the waters in the cave finally seeped away and the holy spring retreated to its bounds within the trough, flowing gently as it had before the boy came. It was only when the waters had retreated that I found the statue of St Lucia, which I had placed above the well to sanctify it and cleanse it of the old goddess. It must have been washed off the shelf in the flood. It lay in two pieces on the floor of the cave, the head snapped off as if the ancient goddess herself had hurled it down in a fit of rage. Blind and dumb now, St Lucia's painted eyes and lips had been washed away by the battering and rolling of the water. Her dagger was smashed.

The ancient stone face that gazes out beneath the veil of the spring is unmarked, unchanged, but our saint is broken, defeated, destroyed. The old goddess has won and we are forced to retreat, for the flood has weakened the foundations of the chapel. Great cracks have appeared in the walls. It is only a matter of time before it collapses. It will not survive another winter. *We* cannot survive another winter.

And now the day has come, the Day of Judgment, of reckoning. I have spent these past nights in prayer and repentance, but I do not know if I shall ever be forgiven. For I have stolen a man's life, killed him. Snatched a heretic from the Church before he could be brought to repentance and absolution. Sentenced him to everlasting Hell.

Know this, that it was my hand who stole the dwale from Sister Basilia's stores. It was my hand that poured the syrup into the Holy Blood of Christ taken from the tabernacle high in the wall of the chapel, for it was the only wine left in the priory that by some miracle the flood had not spoiled. It must serve as his viaticum to strengthen his soul for his journey into the next world, for there was no priest to anoint him with chrism or pronounce the final absolution. It was my hand that held the cup of wine mingled with dwale, not water, to his lips. My arms that cradled him and my lips that murmured soothing words to him as he slipped gently into the sleep from which he would never wake.

I cradled him against my breast as his heartbeat retreated into the far distance, as his breathing faded to the whisper of a babe. Held him as once, long ago, I had hugged him when the night terrors had woken him as a child. Had the boy glimpsed the torments that lay in wait for the man? Had he, even then, seen visions of the terrors the future held for him?

I know what foul things they had said of the Templars. Every vile act that the holy and righteous men of the Inquisition could imagine they had accused those knights of committing. Every foul demon the Inquisitors could name they had charged the knights with worshipping in ceremonies so depraved that even the devil himself would have blushed to witness them. They had tried to force Sebastian to confess that he had worshipped the head of a devil or a golden image of a dog-headed god. They had tortured him until he could not stand, then tormented him still more until his mind was smashed into shards that could never be repaired, leaving only the terrible memories that flayed him afresh every time he closed his eyes.

I had watched the Inquisition burn those Templars who would not confess. Every day I wake with the burning stench of their flesh in my nostrils and their shrieks in my ears. They say the men died bravely. They say they died as courageously as the noble knights they were, singing praises to God. But no mortal man, no man made of flesh and blood and pain, no man condemned to those cruel flames dies like a knight in battle. They perish writhing and screaming until their throats are seared and their shrieks are silenced, but even that does not bring the mercy of death.

Sebastian did not burn. He confessed and was released, if you can call it release to send him out caged in such a twisted carcass of pain. But the confession was false. I knew my brother could never be guilty of any of those foul deeds. I was his sister, his twin. I had known him and loved him since the day our mother had birthed us. I did not need him to tell me he was innocent.

Our father had played a game of cross and pile, flipped a coin, one of us to be given to the Templars, the other to the Hospitallers. If the coin had fallen the other way, as I'd prayed as a child it would, Sebastian would have been safe. And I thought God had answered his prayers. I thought God had chosen my brother, blessed him, not me. A coin, just a coin on which our whole lives were to spin for ever.

Sebastian never knew an hour's peace after the Inquisitors let him go. He wept daily because he had confessed. He cursed himself for his weakness, because braver men had gone gladly to their deaths singing psalms. I tried to tell him that the crackle of the flames had not been drowned by their singing, as the legends say. I knew the truth of it. I'd tried to convince him that he had endured far more than many who had died. What good would his denial have done? The Inquisition, the kings and the bishops would never have believed him. They did not want to believe him. What would his death have accomplished?

But he knew only guilt that his brother knights were dead

466

and he still lived. *Could that pitiful existence be called life?* And I knew that next time they questioned him he would recant that confession. He would be presented as a gift from the Knights of St John to their masters, a burnt-offering, proof of their loyalty to Pope and Crown. But I have watched a thousand men burn nightly in my dreams, each with my brother's face. I could not let them take him. Can you understand that? Can you forgive me? Will he?

Holy Virgin, let me take his place in Hell. Let me suffer while he is finally at peace. And if Heaven will not open its gates to him, then let him be granted sweet oblivion.

Chapter 59

Hospitallers' Priory of St Mary

The knights from Clerkenwell arrived just as Brother Nicholas and Brother Alban had done, after nightfall, muddy, cold and hungry. Six of them. No carter this time, for though the frosts came and went, the ice had never lingered long enough to freeze the deep mud, and more rain had fallen, widening the already flooded rivers and deepening the puddles.

It was only when they were slumped in the refectory, devouring slices of cold goat's brawn, that their leader, Brother Roul, had insisted Brother Nicholas and Brother Alban should be asked to join them.

'Brother Nicholas left us three months ago,' Prioress Johanne told them. 'Our stable boy informed me that he set out in the middle of the night, taking one of our horses, since his own had cast a shoe. I imagine you will find him at Buckland.'

Brother Roul leaned forward, frowning. 'Three months? Even with the roads as foul as this and stopping nightly for shelter he should have reached Buckland in days, a week or two at the most. But Commander John de Messingham arrived in Clerkenwell less than a month ago, and he told the Lord Prior he had neither seen Brother Nicholas nor received any of the reports he had expected from him since he was dispatched here. Both Commander John and the Lord Prior are anxious to hear from him. They've become concerned that some disaster or deadly contagion has ravaged the priory, for there's been no word from anyone. It's why we were ordered here, to discover the cause. Has Brother Alban had word from Brother Nicholas since he left?'

'He's dead,' Sister Melisene blurted out. She paused in the throes of directing a couple of servants who were carrying in the straw-filled pallets that were needed for those brothers who would have to sleep in the refectory. She hastily busied herself arranging blankets, as if the words had not come from her.

'You believe Brother Nicholas has perished?' Roul asked sharply.

'My sister is referring to the passing of Brother Alban,' the prioress said, before Melisene had a chance to jump in again. 'He tragically died shortly before Brother Nicholas left. Indeed, that is why we assumed Brother Nicholas rode off in such haste, to convey his heart to Buckland for burial, since the state of the roads made it impossible to take his body there by wagon.'

In spite of their hunger, all of the brothers had stopped eating, their knives suspended in mid-air, as if they had turned to granite.

'But what manner of death was it?' Brother Roul demanded. 'Did he die of a fever?'

The other knights around the table now began to regard each other with alarm.

'I went to the infirmary,' one muttered. 'There was an old crone in the corner so yellow she looked as if she'd been dipped in saffron. Even the whites of her eyes were like mustard.'

The door opened and Dye edged in. She began to collect the empty trenchers. Roul heaved himself from the prioress's chair, which he had been occupying, and strode over, intercepting the scullion before she had a chance to reach the door. She backed against the wall, alarmed.

'You, girl, I warrant you know all of the gossip in this place. Can you remember back three months ago, when the two knights, Brother Nicholas and Brother Alban, were here? Did either of them fall sick?'

Dye's gaze darted towards her prioress, but the knight shifted his position slightly, deliberately blocking her view with his broad shoulders.

'Not them, Master. Always tell when a man or a pig is sick 'cause they can't stomach their meats, but the brothers must have been the fittest men alive – they ate enough for a dozen men and a dozen pigs too.'

The knights around the table roared with laughter and visibly relaxed a little.

'So how did Brother Alban die, girl?'

'Wisht hounds, that's what they say, Master. He was hunted down by the terrible black hounds who haunt these moors.'

Brother Roul swilled the sour, watered wine around in his mouth and spat it into the fire, but nothing seemed to remove the foul taste of the grave that clung to his tongue and crawled up his nostrils. He'd been outraged to discover that Brother Alban's corpse was not lying in the stone drying coffin but had been hastily buried in the sodden ground where it lay rotting in several inches of water. It confirmed his suspicions that there was much about the brother sergeant's death that Prioress Johanne and the other sisters were anxious to conceal.

He had learned that neither coroner nor sheriff had been summoned when Alban's corpse had been discovered, though plainly his death had been sudden and violent, and he did not for one moment believe the servants' tales of spectral hounds. To learn that one of the two Hospitallers, who'd been dispatched on the order of the Lord Prior to investigate the financial affairs of a priory, had met with an unnatural end, and to discover that the corpse had been hastily dumped in the mire without Buckland or Clerkenwell even being informed of his demise would have aroused the suspicions of a saint. And if that was not bad enough, it seemed that within hours of the corpse being discovered, a noble knight of the order had vanished in the middle of the night without even waiting for his own horse to be shod.

Roul had searched the knight's chambers himself and, with growing alarm, had discovered that not only had Nicholas abandoned all his clothes and possessions but had left his sword hanging on a peg near his bed. A sword was as dear to a knight as his right arm. He would no more set foot outside without it than he would lop off a limb and leave that hanging on the wall. Here was certain proof that Nicholas had not willingly embarked on any journey, not even as far as the nearby village, much less to Buckland or Clerkenwell. If his brother knight had left the priory at all, it must have been as a prisoner or corpse.

Roul had ordered Alban's grave opened the very next day, much against the protests of the village deacon, who warned of all manner of ills that would follow such a wicked desecration of the resting place of a man in holy orders. The sisters had withdrawn, leaving the grim-faced knights watching while Brengy, the stable boy, and an elderly manservant dug into the squelching soil. It was almost impossible to distinguish mud from the soaked and soiled winding sheet, and they only realised the spade had cut into the corpse's shoulder when a stench as foul as Satan's farts rolled up towards them. Brengy gagged and fled, but only managed to get a few yards away before he doubled over, vomiting so violently it seemed he might retch up his own guts. The knights, though they had encountered plenty of fresh corpses in their time, found themselves stepping backwards and gritting their teeth to avoid humiliating themselves by emulating the boy.

When some of the festering grave gases had dispersed, and with a cloth clamped to his mouth and nose, Roul, with a knight who had worked as a physician on the ships, examined the corpse. It was impossible to say what had finally killed their brother, but in spite of the putrefaction, one thing was evident: his heart had not been cut from his chest. Was that evidence that Nicholas had had a hand in his brother's death? Did he fear the corpse would bleed in the presence of its murderer and proclaim his guilt? Or was Nicholas also lying somewhere, foully slain, like his brother?

Brother Roul drew himself upright in the high-backed chair and studied the woman seated in the centre of the refectory. His five brother knights were ranged along the hard benches that had been pushed against the walls so that they might have something to rest their stiff backs on during the long hours of questioning that had been in progress since Prime that morning. The winter sun had long since vanished behind the high walls, and the fiery glow from its dying rays skimmed through the top of the casement, casting a blood-red puddle of light at the woman's feet.

Roul had sent for each of the sisters in turn, with those few servants he considered might have at least some wits, but neither coaxing nor threats had produced a tale from any of them that made sense. The sisters talked of plagues of frogs, flies and water turning to blood. The servants told wild tales of hellhounds, of a blind boy who was a powerful sorcerer, and even creatures called pigseys who, though they were invisible and barely taller than a man's thigh, were apparently able to spirit away strapping knights who were powerless to resist them. And between these fanciful tales, both sisters and servants had babbled about rampaging tinners and murderous outlaws, as if Dartmoor was to be found on the edge of the world among the isles of the dog-headed men rather than in the civilised realm of fair England.

Roul glanced at his brothers. Some were beginning to look glazed, others shifting on the creaking benches, trying to ease their numbed buttocks. He knew they were willing him to call a halt for the day, but he was determined to finish it. The Lord Prior had made abundantly clear what was expected of his emissary when he'd walked with Roul alone in the private walled garden. It was the only place Lord William could be sure they would not be overheard, for he knew well the maze of listening tubes, squint holes and concealed passages that riddled Clerkenwell, like worm holes in a ship's timbers.

'No scandal, no trial, no outsiders,' he'd commanded. 'Not a breath, not a whisper of any wrong-doing must leak out. Deal with whatever you find, Brother Roul. Deal with it decisively, but discreetly.' And as the knight had knelt for his blessing, he'd added, 'If the end is in God's cause, then the means will always be sanctified.'

Roul had bowed his head obediently. There had been many times when, as a young knight, his conscience had smote him over those means, when he had been jerked from his dreams by the cries of terrified boys or tortured old men, cries that years later still pursued him through the dark and twisting labyrinths of sleep into his waking hours. And as he'd lain in his bed trying to calm his breathing, he'd wondered if the living Christ could gaze down on the agony of His enemies with such solemn-eyed indifference as He did in his painted image on the walls of the churches as He watched the damned being clawed into Hell.

But later, when Roul had seen for himself what his enemies did in the name of their god, he had come to believe that it was not for a man like him to reason the cruelties of Heaven. And there was no question of not doing his duty now. If the order was disbanded and the Knights of St John arrested as the Templars had been, the women would be lost anyway. Better that they be sacrificed than thousands tortured. Besides, if they were guilty of the deaths of two of their brothers, not to mention stealing money from their holy order when it was most needed to defend Christendom from the Turks, then Christ Himself demanded their punishment.

'Sister Basilia,' Roul said.

The plump woman was staring up at the fragment of jewelled light still gleaming at the top of the narrow casement. She smiled at him in spite of the anxiety that was evident on her face, like a child who was eager to please but afraid she would not be able to do what was asked of her.

'We have been informed that on the night Brother Alban's

corpse was brought to the priory, Brother Nicholas went into the chapel, apparently to remove Brother Alban's heart. But it would appear he never had the chance to do so because he received a blow to the head. And you were summoned to the chapel to attend to his wound.'

'It wasn't a blow, Brother Roul.'

'It wasn't? You mean he had no head injury when you examined him?'

'His head was hurt, poor lamb. He had quite a nasty bruise here.' She tapped her temple to indicate the place, then frowned. 'Oh dear, perhaps it was this side,' she said, changing hands. 'It was over three months ago and I can't quite be sure. Now let me see. He was lying—'

Conscious of the audible groan from one of his brothers behind him, Roul hastily cut her off. 'You admit his head was bruised, Sister, yet you have just said he didn't suffer a blow.'

Basilia, thrown by the interruption, shook her head, apparently trying to clear her thoughts, making her chins waggle. 'Oh dear, I'm sorry, Brother. What I meant was that a *blow* makes it sound as if someone had struck him, which they hadn't, of course. He slipped . . . down in the well. The floor's always wet. I didn't see it myself, but that's what the Prioress Johanne told Brother Nicholas. She explained to him that he'd slipped and hit his head on the rocky wall. There are lots of rocks sticking out and—'

'The prioress told him? He didn't remember?'

'He was confused. People often are after they've knocked themselves out. Like I told him, I once had to physic a woman . . .' This time Basilia caught the exasperated expression on Roul's face and managed to stop herself. She took a deep breath. 'Poor Brother Nicholas saw Sister Fina behind him on the steps holding a knife. He told the prioress he was afraid our sister was going to stab him but, of course, she'd never dream of doing such a thing. She'd seen a light moving in the chapel. Sister Fina thought it was thieves who had broken in, but as soon as she realised it was only Brother Nicholas, she came back up the stairs. And that was when he slipped. Poor Brother Nicholas thought someone was trying to kill him, but it was all a mistake. Prioress Johanne was the one who found him. She and Meggy brought him up because it would have been no use me going down, not with my girth.'

She gave a nervous giggle, but Roul ignored her. In all the fog of these wild tales, he felt he had finally glimpsed the end of a thread, which, if he could grasp it, might lead him to the truth.

He leaned forward. 'So, let us be quite clear, Sister Basilia. Brother Nicholas swore that someone tried to stab him while he was in the cave, and three other people went down to the well that night besides him – Goodwife Meggy, Sister Fina and Prioress Johanne. None else.'

Sister Basilia's eyes flew wide. 'But he was mistaken, I told you.'

'Brother Nicholas claimed an attempt was made to kill him and days later he mysteriously vanishes during the night, leaving everything behind, even his sword. Perhaps our brother wasn't as mistaken as you have been led to believe, Sister.'

Even Brother Roul did not know quite what he hoped to find when he gave orders that the sisters' dorter, the prioress's chambers and the gatehouse should be searched. A bloody knife perhaps, a bloodstained kirtle, concealed chests of money and jewels that those at Clerkenwell suspected had been withheld from the responsions. But despite Lord William's absolution, even Roul realised he could not, in all conscience, convict any of these three women merely on the garbled account given by the infirmarer. He couldn't even swear that a crime had been committed, much less that one or all three of them were guilty of it.

Alban had been buried with indecent haste and without either the sheriff or coroner being summoned, which could, if the coroner pressed the matter, result in a heavy fine, but it did not mean the women had killed him or, indeed, that the sergeant had died by any human hand. As for Nicholas, Roul couldn't prove he was even dead, though he would have wagered his own life on it. No, he needed something more before he could act. And, as if God had heard his prayers and had sent an angel to deliver it straight into his hands, Brother Roul was to discover far more than even he could ever have dared to hope.

Chapter 60

Prioress Johanne

This will be my last night on earth and I will go to my death unshriven. I will receive no mercy, but Death is not merciful to anyone. He spares neither the good nor the wicked, the innocent nor the guilty. Why then should we expect more from men?

There are many things in my past I could confess to – deeds and denials, doubts and deceptions. They all line up, shouting and whispering their accusations. And they are right to shame me for I am guilty of them all, but not of what I have done to protect this priory, to protect my brother. If I had not done those things, I would not have sinned, and yet I would have caused the innocent to suffer. If you do what God and man declare wrong, but do it for good, is that a sin? *Thou shalt not steal. Thou shalt not lie. Thou shalt not kill.* They are crimes that condemn the body to the gallows and the soul to Hell. Yet if you steal to feed the hungry, kill to spare the innocent? Does God condemn me for that?

I will not say, 'I am guilty.' To kill is not the same as being *guilty* of killing. To kill is not the same as murder. Yes, I did it – I lied, I stole, I killed – but I am not *guilty* of doing these things, for guilt means shame and regret. And I feel no shame, no regret, no remorse.

It was old Meggy who was the cause of my betrayal in the end. Forgive me, *betrayal* is too cruel a word. Poor soul, she meant only to protect us – that was all she'd ever tried to do. But in the end, it was her very loyalty to me that gave Brother Roul all the proof he needed to condemn me. They found nothing when they searched my chamber. They did not discover the little room behind the panel in which I'd concealed the chest of money, the true ledgers and even, in those last few days, had hidden Cosmas from Brother Nicholas. That room remains sealed. Only Sister Clarice knows of its existence.

They found nothing among Sister Fina's possessions. There was no evidence that she had tried to kill Nicholas that night in the cave, though she would have done it, poor creature, had I not wrestled the knife from her hand. She'd heard his threats to destroy her beloved well, and I think she would have killed a dozen knights to save it.

Finally they searched Meggy's hut. It took them a long time and they were losing patience. Both Clarice and Melisene had to hold the old woman back, while the knights threw her broken pots, torn blankets, rags and ropes into the courtyard in a jumbled heap. Her treasured possessions were only rubbish to them. It was almost dusk, when one of the knights, poking about with a stick beneath the thatch, caught something pale as it fell from the top of one of the rafters. When he carried it out into the dwindling light, we all saw what it was: a roll of parchment fastened with the order's seal, a report written in Nicholas's hand. The mice had nibbled the wax seal but sadly not much else. And I did

479

not have to read those spider-black words to know of what I stood accused. Theft! Sorcery! Harbouring a heretic!

Meggy had stolen the parchment from Alban's scrip when he had left her holding his horse as he struggled with the gate. There had not been time enough for her to warn Sister Clarice of his departure, as she'd been instructed, and besides, she knew Clarice could never have bribed Alban. Had Meggy deliberately fumbled with the beam so that she could search his scrip?

But why had she kept it? Was she afraid to confess to the theft after Alban was killed or afraid to burn something she could not read? Or was it simply that our old gatekeeper could not bring herself to part with anything? I almost smiled. Fate had played a cruel jest on us all, for if poor Meggy had not been so diligent, that parchment would long since have been dust and ashes from an outlaw's fire blowing in the wind.

I can hear them outside the door, the rustle of their robes, their whispers so low I cannot make out the words, their breathing as they lay their ears to the wood to hear if I am sleeping. My heart begins to race and my stomach churns, even though I have knelt here for hours praying, trying to convince myself that I am not afraid to die – *should not* be afraid to die. We are taught that the saints even blessed the instruments of their torture, became protectors for the very things that brought about their death because they released them from the travail of this sinful world into the bliss of Heaven.

But did they, though? In those last few hours as they waited for their execution, did their flesh not shrink from the pain of it, their minds scream in terror of what death might really hold? Christ did. Had He less courage, less faith than they, or did they not understand what death was?

Maybe I have, after all these years, finally lost my faith. I know that I am no saint. But I have made my confession as truthfully as I can, if not to Brother Roul or to a priest, then at least to the image of Christ, who hangs on the cross above the altar. Up there His wounds perpetually bleed afresh in the flickering red glow of the lamp. Blood pours from His pale body, a spring, a river, a flood, but it does not reach me. It does not cleanse me or absolve me.

I kneel outside His light and His indifferent stare is fastened on something behind me. On the door? On the place that my executioners will enter? Will they meet His eyes and hesitate just for a moment?

My body stiffens as I hear the click of the latch, the creak of wood, feel the blast of cold air enter then swirl about me, like the vortex into which the whores and harlots are cast in Purgatory. In the past few days, I have lived this moment a thousand times. I've asked myself if, when they come, I should rise from my knees and turn to face them with courage. Make them face me, make them look into my eyes as they do this. And I will let them do it. To resist would be undignified, useless.

Shall I remain on my knees in prayer, so that the last thing I shall see is the painted figure of the one I have died for hanging in His death? Shall I keep my brothers from seeing the accusation in my eyes, spare them the pain of watching the life drain from my face? Can I do it? Can I simply make myself wait and not struggle, not plead?

I hear three sets of footfalls, soft and stealthy on the stones behind me, the faint jingle of the three sword belts, three breaths quick, shallow. I know my death will be at Brother

481

Roul's hand. He would not ask another to do what he could not. Two of them stop . . . wait. Do they pray? Pray for my soul or for their brother to give him courage? The other pair of boots advances, and I can feel him standing behind me, hear the pounding of his heart . . . or is it mine? The rustle of his sleeve as he hastily crosses himself. The mutter of a hastily spoken prayer, uttered so low I cannot tell if he is praying for his strength or my forgiveness. But Christ hanging above does not bend His gaze to look at either of us. Is this what He demands? Is this why He was born and was crucified? Is it? If I could be sure . . . Do it, do it now, quickly, before my courage fails me!

No, no, please, wait! Another hour, another minute, please. I am not ready. Save me! Blessed Virgin, save me. I am afraid to die.

His arm brushes against my veil as he drops the cord over my neck. It tightens against my throat, and he twists swiftly, he twists hard.

They buried my body out on the moor, under cover of darkness. If any shred of life still lingered in it, that act alone would have snuffed it out for they dropped me into a water-filled hole, smothered me in oozing mud. I would have drowned, had I still breath in my lungs. I think my corpse will rot quickly unless, of course, the peaty water preserves my remains as it has the corpses of those who were bound and cast alive with nooses about their necks to drown in the bog pools before Christ was ever carried to this isle.

They say I will burn in Hell. That's what they say, but the devil has not come for me yet, and neither have the angels. Only you came, only you spoke to me, summoned me and I am content to sit in your company.

Chapter 61

Morwen

Sorrel and me, we built the pyre between the great rocks on Fire Tor. Took many days to dry the wood around the fire we kept burning inside the cave. But there was smashed wood enough to find in the tinners' valley, scattered down the length of it as the waters sank back to the river, where they belonged.

Those tinners who'd escaped the flood had taken whatever they could carry away from the turf huts and fled even before the flood had retreated. But when the water was gone, there were good pickings buried under the mud and gravel for the villagers to scavenge, or so they reckoned. Soon as word got out, they were swarming over the valley, like ravens squabbling and flapping over a dead sheep, trying to snatch up all the treasures they could find to use or sell – iron pots, picks, kegs and nets. Most were buried in the mud and they had to hunt for them, trying to spot the tip of a spade or the spar of a ladder they could drag up from beneath the stones. There were curses aplenty when they went to the trouble of digging them out only to find them smashed, but

483

they took them anyway: a new handle could be fixed to a spade or a staved-in keg could be burned on the hearth fire. What they really wanted was food, but the river had taken every bite as her toll.

They even stripped the corpses of the tinners and their chillern, stripping the filthy rags from them, cutting off a knife that still hung from someone's waist or wrenching off the sodden boots. Not the amulets or crucifixes, though – no one would rob the dead of those for fear their spirits would come looking for them. The bodies were left where they lay. The birds and beasts had already started picking over them for they were starving. It wasn't many days before there was nothing but hair and bones. When the water drained away, some corpses were left buried under mud and the waste gravel from the spoil heaps that had been washed down with them. Sometimes it was only when you trod on them that you felt a leg or face under your bare feet, *his* face.

Soon as we'd finished building the pyre on Fire Tor, we carried the corpse of Ankow out and laid it on it. Ma wouldn't dare come up here again, I knew that, not now she'd seen what we could do. We laid him facing the Blessed Isles, and I combed his flaxen hair over the hole in his skull, though there was no disguising the long gash across his throat. I took the hag-stone from his beard, which Ma had placed there, and the black river stone Ryana had laid on his fore-head, and the bracelet of the flying thorn I had woven for his wrist. All those I would place on the new Ankow. This one must be set free to make his last journey.

We heaped more wood and dried peats over him, for the fire must be hot to eat through flesh, and as darkness fell, we set it ablaze. The scarlet flames darted in and out of the wood beneath, and then as the night wind gathered strength, they rose up through the pyre, crackling and roaring into the night sky. The glow turned the twisting smoke red and orange and blue as it spiralled upwards. We sat side by side, Sorrel and me, as the spirits drew close about us, crawling through the stones and slithering over the grass, the breath of their wings brushing our skin, the smell and taste of wood smoke and burning flesh filling our mouths and nostrils.

Below the tor we heard the soft pad of claws on stone, the panting and whining as the hounds gathered, their coats crackling blue as lightning, their red eyes glowing out of the darkness. Old Crockern's hounds were circling the tor. They were waiting for Ankow, waiting to take him home.

A shape was rising out of the smoke, gathering itself dense and dark as a shadow on a summer's day. Only its eyes lived, glowing like the hounds' eyes in the dark. It stood, head bowed.

'I loved her.' His voice was like the wind moaning through the rocks in the cave. 'I knew it was sin, for I had already a wife, but my bride wasn't of my choosing. It was her I loved and when her father sent her away, I had to follow. I was burning up for her. She was my life, my breath, my being. But when I found her, she was changed. She would not touch me. I had come so far, risked everything, and she would not even look at me. I only meant to take a kiss. I thought if she felt my lips upon hers, her body would remember our passion and I'd see the tenderness flow back into her eyes. But once I kissed her, I could not stop myself.

'I do not blame her for what she did, for the blow or the knife or for my death. I deserved to die. But I have been punished for my love beyond all imagining. I have been the slave of Death, and I have walked the lych-ways over this moor without rest or mercy, guiding the souls in misery and

485

torment to their rest, but never reaching the place of peace. I can bear no more. Release me, I beg you, have mercy, release me and let me rest.'

'Go,' I told him. 'Another has come to take your place. You are Death's bondsman no more.'

There was a sound like a great flock of birds taking wing and at once the figure's head lifted. He looked straight at us, then his words came like the voice of an ancient door closing. 'Thank you.'

The figure of smoke grew fainter and greyer, the edges dissolving into the sky, and all at once the hounds began to howl, a great mournful cry that rolled across the moor. One by one the pairs of burning eyes vanished. The pyre suddenly collapsed in on itself, sending a shower of golden sparks high into the sky as if stars were rising back to their places.

I took Sorrel's hand and we slipped into the cave. A corpse already lay on the wicker frame above a fire that emitted only smoke. It had been well prepared with herbs and honey. It would take several weeks of smoking before I could be sure it would not decay. But Brigid had guided us to it quickly, before it had started to rot. He was her choice and we would tend his corpse carefully, Sorrel and I.

His ghost, whimpering, pleading, terrified, was already cowering among the shadows of the cave, as the leather wings and claws of the spirits slithered over him, curious to see what we had brought them to play with.

'Let me go,' he begged. 'I was always nice to you, Sorrel, isn't that right? You're a good woman. I always said so. You'd not see a poor soul tormented. There's dozens died in that valley. You could take any of them. They deserve to suffer, thieves the lot of them, murderers too. Take them, let me go . . . What do you want with me?' he wailed.

'You are Ankow, the serf and bondsman of Death,' I said. 'And you'll do our bidding, till you come to carry us safe to the lych-ways and that will be many years from now.'

Gleedy gave a great howl of despair, but he was in our storeroom now. And there would be no release for him, not while we lived.

Epilogue

Dertemora

The firelight licks about the slabs of stone inside the tor. Its scarlet fingers probe the crevices, and all the secret places that only the phantasms know. Morwen sits facing Sorrel across the fire. The light is gentle on Morwen's ancient wrinkled skin and burnishes her grey hair to the tawny red it once was, the colour of an autumn leaf. And the face that smiles at Morwen through the smoke and flames seems as young as the one she first saw at the door of Ma's cottage, though many seasons have passed for them both since then.

The white cloth on the slab at the far end of the cave is black with the dust and dirt of decades. It billows and slides from the mounds, crumpling on to the floor as if a wind has stirred it, but though the wind is baying around the rocks outside it dare not enter here. Three figures lie side by side on the shelf, a holy family, a dead family – Gleedy, old Kendra and Ryana.

Kendra perished from old age and hunger the winter after the flood, cursing Ankow when he came for her and spitting venom to the end, poisonous as any long-cripple whose head sways in the stone wall. Ryana died from making herself pottage of hemlock having mistaken it for cow parsley. Morwen always said she couldn't tell mouse-ear from mugwort. She'd warned her not to eat it, but her sister would not take telling from Mazy-wen. But Morwen doesn't call upon their spirits to aid her. She has no need. She and Sorrel have all the power they want between them and it grows. Besides, she'd had enough of their malice in life, and death will not have made them kinder.

Taegan does not lie here. She ran off to live with Daveth and his brother. She has a brood of boisterous, red-headed offspring, grandchildren too, though the Black Death took some and famine took more, but Taegan and the brothers survived. And as to which brother fathered which of Taegan's sons and daughters, they neither know nor care. For they are brothers, and they have always shared the good things that come their way.

The cross on Father Guthlac's grave has been swallowed by the earth, as have his bones. He and the old gods had long been enemies, and he recognised his foe in that child. Old Crockern felled him, like a tree in a storm. The wind has no malice, but it will destroy anything that lies in its path, when it begins to blow.

Only the ruins of the priory stand now, and they are being eaten away by storm and rain, snow and frost, and by the villagers who, year after year, carry away more of the stone to mend their cottages and build their pinfolds. The ravens and buzzards nest safe on the broken walls, and foxes burrow beneath them. The villagers have reclaimed their well, and bring their children for Morwen and Sorrel to dip in its waters, for they are the true keepers of it now.

In the tinners' valley, the sores on the land are healing.

Grass and heather, sedges and gorse have crept over the wounded backs of the hills. The river lies quietly in her old bed, but scars will remain long after those who made them are forgotten, and deep welts will for ever mark Old Crockern's hide.

Forgotten, did I say? Yes, their names are forgotten, but they are not unseen. For if you are foolish enough to find yourself upon the lych-ways when night falls, stumbling between the sucking mires and black bog pools, straining to peer into the darkness; if you turn in fear at the whispering in the reeds and the voices in the rocks, you will glimpse flickering lights moving slowly ahead of you. And if you hold your breath and watch, you will see the silent host of men, women and children who follow those bone-white lights. Some are dressed in the white robes of monks, others in the black robes of sisters, some in beggars' rags, others in tinners' mud. You will see old village priests and women skull-gaunt with hunger, men pale with gory wounds, crones as wrinkled as time itself, or babes so young they have not lived a single day. But if you see them, do not speak to them. Do not follow them. They are the shadows of the dead. And once you cross into the deadlands, there's no coming back, Brothers, there is no coming back.

Historical Notes

From 1315 to 1317, the Great Famine ravaged Europe. It was caused by extreme wet, cold weather, which affected the whole of northern Europe from Russia down to southern Italy. Crops failed and livestock died, causing widespread starvation. Even King Edward II, passing through St Albans with his court, went hungry because there was no bread. Bands of people were on the move across Europe, hoping to find food or better conditions elsewhere. Children, babies and the infirm were abandoned on the steps of churches and monasteries, the elderly deliberately wandering off to die so that the young ones could live. There was a surge in violent crime and highway robbery as people became increasingly desperate. Frightening reports circulated, claiming that some were turning to cannibalism, but no one knew if they were true or merely rumours.

The Knights Templar

The Knights Templar were originally founded in the twelfth century to protect travellers making pilgrimages to Jerusalem. But over the following two hundred years they became the international financiers of Europe, lending money and amassing huge resources of wealth and land. They were answerable only to the Pope, and this, combined with their wealth and power, posed a grave threat to the kings of Europe.

Philip IV of France particularly feared their increasing influence, so much so that in 1307 he turned on the Templars in France. He arrested the knights and their commanders on the grounds of immorality, sorcery and heresy, charges the Pope was reluctantly forced to investigate. Over the next few years, many knights across Europe were tortured to induce them to confess to these and other crimes, including homosexuality, immorality with women and alleged bizarre secret rituals, including worshipping a 'head'. Those Templars who refused to confess or who subsequently recanted confessions made under torture were burned alive.

In England, King Edward II at first flatly refused to believe the accusations levelled against the Templars, or perhaps, given the accusations of homosexuality that had been made against him, was anxious to defend the knights. He resisted Philip's demands to bring the Templars to trial, and it wasn't until a papal bull arrived in December 1308, commanding him to take action, that he gave orders that the knights were to be arrested. Even so, it was only when members of the Inquisition landed on English soil in September 1309 to begin the ecclesiastical trials in London, York and Lincoln, that the real round-up of Templars began. This was two years after the first arrests in France, and by then a number of leading Templars had fled from Europe to England.

The Pope authorised the use of torture against the Templars arrested in England, because they were accused of heresy, though torture was technically unlawful under English law. But no one could be found who could do it effectively. They either accidentally killed the victim before he'd talked or were not practised in the art of breaking him down mentally as well as physically, so few confessions were extracted in that way. Most evidence presented at the trials in England was obtained from a Templar who had previously been arrested and escaped, and was believed to have been a spy carrying information to the Grand Master in France.

The Templar order was disbanded throughout Europe by the Pope, and its wealth, when it could be found, was declared forfeit to the Pope or Crown. Templar property was mainly handed over to the Templars' rivals, the Knights of St John.

Throughout their history, the Templars and Hospitallers were rivals for power and territory. They frequently clashed in the Holy Land, the Templars supporting the baronial side of any political disputes, while the Hospitallers supported the monarchy. However, some of the noble families of Europe placed sons and daughters in both orders as a way of gaining the maximum political influence and financial advantage. It also ensured they always had one of their offspring on the winning side. Having children in both orders was a way of hedging temporal and spiritual bets.

Knights Hospitaller

The order of the Knights of St John of Jerusalem, also known as Knights Hospitaller, originated in the founding of a hospital in Jerusalem in 1080, on the site where, according to legend, the angel Gabriel foretold the birth of St John the Baptist. It was established to provide care for poor, sick or injured pilgrims in the Holy Land. Hospitals in those days were not primarily places where the sick and injured were treated, although physicians and surgeons did work in them: they were principally intended to offer hospitality, a place where the elderly, the infirm, orphans, pilgrims and travellers could rest and be provided with safe shelter, good food and spiritual comfort.

In 1099, when Jerusalem was captured by the Christians during the First Crusade, the Hospitallers became a religious and military order under its own charter, charged with caring for pilgrims and the poor, and defending the Holy Land. The establishment of the order was confirmed by a bull of Pope Paschal II in 1113, and the Hospitallers' founder, Gérard, rapidly acquired territory and revenues for his order throughout Jerusalem, which became known as the Citramer, the heartland of the order. His successor, Raymond du Puy, established a Hospitaller infirmary near the Church of the Holy Sepulchre in Jerusalem and the care offered to pilgrims expanded into providing them with an armed escort, which soon grew into a serious fighting force.

By the mid-twelfth century the order was divided into military knights and those who worked with the poor and sick. But it was still recognised by the Church as primarily a religious order and was exempt from tithes and from obedience to all secular and religious authorities, except the Pope. At the height of their power in the Holy Land, the order held seven major forts and 140 other estates in the area, the largest two being their base in Jerusalem and lands in Antioch.

But by 1289, the Muslim forces were seizing more and more territory in the Holy Land, and in May 1291, the last remaining Christian stronghold of Acre fell, after a terrible month-long siege. The Hospitaller Knights evacuated to Cyprus, where they set up their base alongside the Templars with whom they had fought at Acre. But they became embroiled in the bitter politics of Cyprus, and Grand Master Guillaume de Villaret turned his sights on

Rhodes as their new homeland, which strategically was much better situated.

In 1306, Vignolo dei Vignoli entered into a pact with the Order of St John, offering the knights Rhodes, Kos and Leros in return for their help in securing his own lands in Rhodes. The Hospitallers made several attempts to take the island, eventually laying siege to Rhodes town, and in August 1309, it surrendered to them. Under their new grand master, Fulkes de Villaret, Rhodes became the new Citramer, where the knights had to adapt from being primarily a land-based fighting force to a navy, using their fleet of ships to fight the Turks. The order's lands outside Rhodes, known collectively as the Outremer, were organised into Langues, or 'Tongues', with bases in Auvergne, Spain, England, France, Germany, Italy and Provence, each administered by a prior. But the number of Tongues and the location of their administrative centres in Europe changed frequently throughout the Middle Ages as political boundaries and alliances shifted.

Since their foundation by Gérard, the members of the order had formed a fraternity of brothers and also sisters of the order. Both took the three monastic vows of poverty, chastity and obedience. All the members of the order, both male and female, wore black with a white cross until Pope Innocent IV ordered that the battle dress of the knights should be a red coat with a white cross. Anyone wishing to become a knight or sister had to be of noble or 'gentle' birth. Those wishing to join the order who were not of that rank became serving members. In addition, like all monasteries and nunneries, the priories of the order employed a number of lay servants, hired to do the hard manual work and wait on the knights and sisters.

As had happened with the Knights Templar, the vast estates

and wealth of the Knights Hospitaller gave rise to accusations that they were living in luxury rather than in obedience to their vow of poverty. Since the Pope was the protector of the order, and the otherwise powerful bishops had no authority over it, this resulted in many bitter conflicts between the order and the bishops of the various sees in which the Hospitallers had lands and priories. This being an age in which bishops maintained their own fighting men, both clergy and kings feared the wealth and military power of what they perceived to be an alien army in their midst, whose members owed no allegiance to anyone but the Pope, and when the Pope was in conflict with a particular bishop or king, the local authorities became distinctly nervous. Little wonder that some sought to crush the Hospitallers as they had the Templars.

The order lost Rhodes in 1522, but in March 1530, Emperor Charles V of the Holy Roman Empire and Spain gave them as their base the barren rock they were to turn into the fertile island of Malta. For this they paid an annual rent of a falcon each year – the Maltese Falcon. From this base, they continued to harry the Turkish Ottoman Empire, so much so that in 1566, the sultan launched the Great Siege of Malta, which the knights broke after four months, founding a new capital Valetta, named for their grand master, Jean de la Valette. They remained there until 1798 when they were ousted by Napoleon. In 1834, they founded a new headquarters in Rome, but continued to be known as the Knights of Malta.

Today, this Catholic order currently has around 13,500 members and 100,000 staff and volunteers, who work in hospitals all over the world. Thirty per cent of the members are now women, who are given the title 'Dame'. The order of the Knights of Malta is really a state without a country. It has formal diplomatic relations with 106 countries and states, and they produce their own passports, licence plates and stamps.

In 1540, during the Reformation in England, King Henry VIII confiscated the Hospitaller priories and land, including its headquarters in Clerkenwell. An attempt was made to revive the order under his daughter, the Catholic Queen Mary I, but it was lost again under the Protestant Queen Elizabeth I. But in 1858, during the reign of Queen Victoria, an order of the Knights Hospitaller, independent of Rome, was founded, which was known as the Venerable Order of the Hospital of St John of Jerusalem in the British Realm. A few decades later it was to oversee three charities – the very famous and familiar St John Ambulance Association (established 1878), St John Ambulance of Uniformed Men and Women (1888) and, returning at last to where the order first began, the Ophthalmic Hospital in Jerusalem (1882).

The Goddess Brigid

Brigid, Brigit or Brig, the exalted one, was a triple Celtic deity. She was goddess of poetry, spring, fertility, cattle and sacred wells, and she had two sisters, Brigid the blacksmith and Brigid the healer. Her festival was Imbolc, celebrated around 31 January/1 February, midway between the winter solstice and the spring equinox. The festival marked the transition between those months ruled by the crone of winter and those by the maiden of spring. It was celebrated by the lighting of sacred fires and the pouring of water. The goddess was the guardian of high places, such as tors, mountains and hill-forts. She was also strongly identified with cows, ewes and milk, which were of vital importance in Celtic society. In mythology, she is the creator of 'keening', which is first heard when she mourns her son slain in battle.

Brideog – pronounced *bree-jog* – means *little Brigid* or *young Brigid*. At the Celtic feast of Imbolc, long lengths of rushes or straw were twisted into the rough image of a doll, which was dressed in white and decorated with leaves, stones or shells to represent the goddess Brigid. The brideog would be sprinkled with water taken from a Brigid's Well, Bryde's Well or Bride's Well, and carried by unmarried girls in procession to a great feast where the goddess would be offered food, drink and fire, in gratitude for bringing them safely through winter, and asked to bring new life.

With the coming of Christianity, the Celtic feast of Imbolc became the Eve of St Brigid's Day, 31 January. On that night, unmarried women and girls would make a crib for the brideog and sit up with her. Young men would call and be offered food and drink, but had to treat the brideog as if she was a saint and the women with great respect. Before the girls went to bed, the ash in the hearth would be raked smooth and each girl would lay out a cloth in front of it. In the morning, they would examine the ash and cloths for any sign that St Brigid had walked over them. If she had, it was a great blessing, and any girl whose cloth she had marked would be married within the year. On St Brigid's Day itself, the brideog was carried from house to house by unmarried girls, who presented the head of each household with a St Brigid's cross.

Brigid's Cross – a woven cross traditionally made on St Brigid's Eve, 31 January, but originally associated with the Celtic goddess Brigid. It was woven from freshly pulled rushes. The newly made cross would be green, symbolising the coming of spring. The cross was woven in a sun-wise direction, and it is thought that in Celtic times it was intended to represent the sun and the hope that light and warmth would return after the cold and darkness of winter.

Today there are many elaborate versions of this cross, but probably the oldest style was the three-armed cross, or *triskele*, which was hung in byres and cowsheds to protect the animals, for in Celtic society wealth was measured in cattle, not land. There are also six-band interlaced patterns, square patterns, and a cross in a circle. A Brigid cross made with a 'binding knot', which keeps out evil spirits, was usually woven and hung in the home on All Hallows' Eve (Halloween).

But after the coming of Christianity, when many of the legends about the goddess Brigid were transferred to the hagiography of the sixth-century Irish St Brigid, Abbess of Kildare, the abbess became credited with having woven the first Brigid's Cross at the bedside of a dying chieftain. St Brigid's four-armed version of the woven cross gradually became the one most commonly found hanging over doors or hearths in cottages, having first been taken to the church to be blessed by the priest. The 'pagan' *triskele* was not brought to be blessed by the Church, but it continues to be used in byres in some parts of Europe to this day.

Brigid's Mantle – There is a traditional Irish blessing, *Faoi bhrat Bhríde sinn*, 'May you be covered by Bride's mantle,' and the old Gaelic title for Brigid was 'Brigid of the Tribe of the Green Mantles'. The goddess was thought to weave a mantle of green on her loom and spread it over the earth, banishing winter, bringing spring, and protecting all those whose hearths it covered.

When the goddess was transformed by the Church into a saint, the symbol of her mantle gave rise to a legend that St Brigid decided to build her monastery in Kildare, Leinster. She asked the King of Leinster to give her the land, but he refused. Brigid prayed, then begged him to grant her the amount of land that her mantle would cover. The King found her request so amusing that he foolishly agreed. Four virgin followers of St Brigid each grasped a corner of the cloak and walked away from each other. The cloak stretched as they walked until it covered the whole area St Brigid needed for the monastery with all its outbuildings. Awestruck, the King gave her the land. Brigid built her church near an ancient oak tree, sacred to the Druids, and close to a holy well, which was then dedicated to the saint.

The cathedral of St Sauveur in Bruges, Belgium, today houses a holy relic in the form of a piece of woollen cloth, which is said to be part of St Brigid's mantle and which is venerated on her saint's day, 1 February. Measuring 21 by 25 inches, it is a dark crimson-violet. Tests carried out in 1936 showed that it was dyed with iron oxide. Cloth of a similar weave and dye has been found in early Bronze Age burials in Denmark and Ireland. However, similar home-spun weaves and dyes were still in use up to the sixteenth century, so this fragment may have been made much later than the sixth century.

The earliest known record of this relic was written in 1347 and comes from the cathedral of St Donaas in Flanders. According to tradition, after the death of the Saxon King Harold at Hastings in 1066, his sister Gunhild fled to Flanders and presented this relic and her jewels to the Church there in gratitude for sheltering her. The piece of St Brigid's mantle was removed from St Donaas before its destruction during the French Revolution.

Dartmoor

In 1182, the name recorded in official documents is *Dertemora*, meaning *Moor in the Dart Valley*. *Moor* comes from the Old English *mor*, meaning bog or swamp, and *dart* from the Celtic, meaning *the river where oak trees grow*. Since the whole of the river is unlikely to have been lined with oak trees, even in Celtic times, it suggests that the oak groves that grew along stretches of it were considered special or sacred. Groves of ancient dwarf oaks, such as the mysterious Wistman's Wood, still survive on Dartmoor, their twisted branches hung with shaggy lichen, their gnarled roots growing over great moss-covered boulders.

Glowing Moss

Dartmoor is home to an increasingly rare luminous moss known locally as Goblin's Gold, and in other parts of England as Elfin Gold, because people passing abandoned rabbit holes, caves and ancient stone huts would glimpse something shining like gold in the dark interior; when they reached in to grab it, they found themselves clutching only a handful of wet dirt.

What they had seen was in fact a tiny, fragile, frond-like moss, *Schistostega osmundacea*. It forms dense mats covering the walls of caves or damp burrows, but grows to only around half an inch in height. It is found growing in caves, tunnels or in half-buried ruins where other plants can't survive because only a very faint light penetrates. Seen against a dark background, the moss shines with a beautiful green-gold luminosity and under flash photography can appear a vivid electric blue.

The luminous effect is due entirely to reflected light. The protonema of the moss have lens-shaped cells, which focus any available light on the chlorophyll granules. These absorb only the wavelengths of the light needed to photosynthesise, reflecting the rest of the light back towards its source, so the light seems to be radiating from the moss itself, which appears to glow. These 'lens' cells can move to within 45 degrees, to adjust to a shifting light, so that at times the glow from the moss appears to pulsate.

One location where it could still be found at the time of writing was at Yellowmead, on Sheepstor, Dartmoor, in a man-made cave, known locally as a potato-cave, but which was probably once an old tinners' tool store or stone bee-hive hut.

Land Measurement and Acres

By the fourteenth century, landholdings were recorded in *acres* and *half-acres*, not *hides*, as they had been in the Domesday Book. The word *acre* comes from Old English æcer meaning *open field*. An acre was roughly the amount of land a *yoke*, or two oxen, could plough in a day. But it was an estimated size used mainly for taxation and land sales rather than a measured area. In 1195, the monks of Thame exchanged two and a half acres of land for three and a half, but recorded that the second area was no bigger than the first.

By 1250, acres were increasingly being measured using *perches* (long rods). An acre was deemed to be four perches by forty perches. But this didn't clarify things much, because the perches themselves were not a standard length, varying from around 16.5 feet to 25.5 feet in different shires. This was a problem for anyone coming from another part of the country who was trying to assess the exact size and value of a landholding. A further complication was that, even in the same place, the length of perch used varied with the type of land being measured, so poor pasture or woodland would be measured with a longer perch than prime arable land, making an acre of woodland larger than an acre of arable land, so *acre* still really referred to the taxable value of the land rather than its actual size.

The River's Cry

If you are walking along a lonely riverbank out on the moor on Dartmoor it can be quite unnerving suddenly to hear a wordless singing coming from it, which rises above the general noise of the rushing, gurgling water. On several occasions, I've thought there must be a radio playing somewhere even though there are plainly no cars or people about. The phenomenon is due to the granite rocks and pebbles in the water, which amplify and distort sound.

When the River Dart is in full spate and this coincides with a north-westerly wind, a loud booming sound is sometimes heard, known as the 'river's cry'. It was a long-held belief that the River Dart claimed a human heart in tribute every year and the 'Dart's Cry' was the warning that one of those hearing it would shortly drown or die. Down through the centuries, there have been many tales of people hearing the cry, only to hurry home to news of an unexpected and tragic death in the family. But the river's cry has also saved lives when those hearing the booming scrambled out of the river or off the rocks in fright, only to see a huge wall of water sweep past them, which would have carried them to their deaths.

Pigseys

Pigsey or *pigsy* is an old dialect word for a *pixy*, also called *piskies, pisgies* or *pysgies*. The earliest published version of *The Three Little Pigs* in 1853 comes from Dartmoor, and in this version the heroes are not three little pigs but three little *pigseys* or *pixies*. Throughout this novel, I have used the older dialect form, *pigsey*, to help separate the medieval concept of these dark, supernatural creatures from the jolly models of grinning pixies sitting on toadstools that we find in souvenir shops today.

Pigseys were mythological little people, mostly invisible to humans, who lived hidden but parallel lives. In pre-Christian times, they were simply regarded as another race with magical powers, but with the coming of Christianity it came to be believed that they were the spirits of babies who had died before baptism and therefore could not enter Heaven, but nor could they go to Hell because they were too innocent. They were transformed into pigseys to live as close to humans as they could, though they could never speak to their families or show themselves. But in this form they could punish their neglectful parents by causing mischief and would also jealously torment their living siblings by pinching them black and blue at night.

In medieval times, pigseys were considered to be frightening and often malicious creatures, vengeful and quick to anger. They were thought to inflict painful illnesses on humans, and bring disasters and chaos to farms, families and livestock. Pigseys would disguise themselves as bundles of clothes or lumpy sacks to lure children close, then snatch them and carry them off into a *pixy-house*, from which either the children never returned or were found wandering, lost and bewildered, weeks or even years later, believing they had only been away for a few hours. A *pixy-house* is the local name for a deep crevice or hidden natural cave on the moors.

Pigseys delighted in leading travellers into bogs and mires, known as *pixy-beds*, or sending them stumbling around, hopelessly lost, in the mist or snow until they died of exhaustion. Until the twentieth century, *pixy-led* was a term that conjured fear for it meant being completely lost and helpless. If you thought you were being pixy-led you could counter this by removing your coat or cloak and putting it on inside out to reverse the spell.

There are many places associated with pigseys on Dartmoor, especially Sheepstor, which has a cave among the rocks known as Piskie Cave, Piskie Grott, or Elford's Cave, so named because a member of the Elford family is said to have hidden from Cromwell's troops in this *pixy-house*. It is believed that the cave was once much larger, but rock falls have destroyed or sealed part of it and made the existing entrance even narrower, though it is still possible to squeeze in if you are slender.

Fire Tor

There are a number of *knocking caves* or tors on Dartmoor, from which the sounds of tapping, whispering human voices, wailing, music and singing can be heard. So much so that people through the centuries have believed these tors either to be the abodes of ghosts or pigseys or the entrance to Purgatory. Many have thought there were people inside and were so convinced they went in to investigate, but found the caves empty. The sounds are probably made by the wind whistling through the many crevices, and from water dripping or running unseen among the rocks, which echoes through the granite stone and hollows.

Holy Wells

There are many healing or holy springs and pools all over Devon and Cornwall. Some are known as Bryde's Well or Bride's Well, and were once dedicated to the goddess Brigid. Since the medieval Church was powerless to stop local people continuing to make offerings at these 'pagan' sites for good fortune or for healing, many were renamed and dedicated to Christian saints, and stories were constructed to explain how the saint had caused the well to bubble up. These legends often incorporate elements taken from Celtic and other pre-Christian mythologies.

The stone carving of the head or skull described in this novel, which lies behind the holy spring, was inspired by a carving that can be found on Sheepstor Church, Dartmoor; it has evidently been moved to the church from another site.

Frogs

In the Middle Ages, frogs were one of the creatures believed to be produced by spontaneous generation. This was thought true of a number of animals and insects whose juvenile form did not resemble the adult, such as flies, which were thought to be generated by corpses. Some people even advocated beating the corpses of calves to bring out the 'blood maggots', which would eventually turn into bees, while others said scornfully that this was foolish as it was well known bees were born from oxen, hornets from horses and wasps from the corpses of donkeys. Most were in agreement that frogs and toads were born from mud, not surprisingly because they would be seen in great numbers after rivers flooded in the spring.

In many ancient cults, they were associated with goddesses and venerated as symbols of regeneration and fertility. There are a number of holy wells in Devon and Cornwall, such as the one in Bovey Tracey on Dartmoor, where legend has it that little golden frogs were seen swimming in the clear water. These golden frogs were a sign that the well had been blessed or become a healing well, thanks to an act of kindness by a villager towards a beggar woman they had encountered. Originally, she was the goddess Brigid in disguise, in later centuries the Virgin Mary. But, like owls which were once sacred to ancient goddesses, when Christianity spread and the goddesses were demonised, owls became symbols of death, and frogs became symbols of evil because they sprang from mud and filth.

The passage quoted in the novel is from the New Testament, the Book of Revelation 16:13 – *And I saw from the mouth of the dragon and from the mouth of the beast and from the mouth of the false prophet, three unclean spirits like frogs. For they are the spirits of devils.* This verse also made frogs a symbol of heresy in the Middle Ages because they come 'from the mouth of the false prophet'.

In the later Middle Ages and during the witch trials of the sixteenth and seventeenth centuries, frogs were one of the animals that witches and sorcerers were most frequently accused of having as their *familiars* or *imps of Satan* to aid them in their spells and to spy on their neighbours. Lucifer and three of his imps are said to dwell in the form of frogs in Frog Well, Acton Burnell. While the imps are often seen mocking those who look into the water, the devil, true to his wicked nature, cunningly conceals himself, waiting to do mischief.

In the days of damp houses and wells, it must have been only too easy to discover a frog hiding somewhere in the home or garden of the alleged witch, and frogs would certainly have been seen in the sodden dungeons into which the accused were thrown. What more proof did you need that the 'witch' had been whispering to one of the devil's imps?

Black Hounds

The legends about the packs of wisht hounds that hunt the moors are unlike those of early accounts of the solitary 'Black Dog' or 'Black Shuck', which often appeared on isolated roads turning travellers back from danger and may have a different origin.

The talbot, or Norman hound, was brought to Britain with the Norman Conquest. It was a hunting dog bred for speed, strength and stamina, and was probably the size and build of the modern bloodhound. The Norman hound was a crossbreed, and one of the strains used in the cross was the Hubert, a huge black hound bred in the monastery of St Hubert in the Ardennes. They were so sought after as hunting dogs that they were presented as gifts to royalty. Huberts had 'red' patches over the eyes, and if some of these dogs, or strains bred from them, escaped and turned feral, hunting in packs, they might have contributed to the legends of the terrifying hellhounds or wisht hounds of Dartmoor, with their black coats and 'red' eyes. *See also Glossary – Wisht Hounds.*

Glossary

Agasted – a medieval word meaning *afraid* or *terrified*. It is still used in Devonshire dialect today.

Ankow, Ankou or Ankuo – was a legendary medieval figure found in many areas that had once been Celtic lands; belief in him in some parts continued right up to the last century. He was the servant or bondsman of Death, and was thought to be the last person to die in the parish in the previous year. Other legends say he was a prince who had foolishly challenged Death to a contest to catch a magical stag and lost, or that he was Cain who had slain his brother Abel and was doomed to collect souls as a punishment for bringing murder into the world.

Ankow was responsible for guarding the graveyard and collecting up the souls of all who died in that parish throughout the following year. He was sometimes thought to ride a black or skeleton horse, or drive a black cart. He guided the souls of the newly dead and helped them find the spirit paths, or lych-ways, along which they must travel. Ankow announced his arrival by a mysterious knocking sound or by an eerie wail or the call of an owl. Once he entered a cottage, he never left alone.

In times of plagues and deadly fevers, when people often concealed sick loved ones for fear they would be taken away, Ankow was said to ride through the villages and towns at night, marking the doors of infected houses with a red sign.

This could lead to families being walled up alive by frightened neighbours or even killed. It would be only too easy for someone to sneak out at night and mark the houses of those they suspected of being sick or against whom they held a grudge, and claim Ankow had done it.

Bait – dialect word meaning to feed a fire and also to be in a fiery mood or bad temper.

Blow-in – a derogatory name for an outsider, a newcomer, someone who was not born in the area or village.

Brideog – a crude doll made from twisted rushes or straw, representing the goddess Brigid. *For further details see Historical Notes – Brigid.*

Buddle – a rectangular surface or shallow trough with a sloping base used in tin streaming. Crushed gravel would be tipped in and washed with water. The heavy tin ore, the *heads*, would settle near the top of the slope. The lighter worthless material or gangue, known as the *tailings*, would settle at the bottom, or be washed off the buddle, and the mixed stone, the *middles*, would settle between. *Heads* produced the best quality tin, *middles* could also be used, but would need to be smelted twice, and the *tailings* were discarded. The skill lay in spotting where to draw the line between each.

Changeling children – with many medieval villagers living in small isolated communities, constant intermarriage between the same families for generations could result in children born with genetic problems. If these were serious enough to be spotted at birth, the newborn might be exposed out on the moors and left to die. But when a child who seemed healthy at birth later began to look, behave or develop in ways that were different from the other village children, it was said that the faerie folk had stolen the human baby and left their own offspring in its place. Parents were advised that if they beat or mistreated the changeling or even pretended to throw the child on the fire, the faeries would snatch it back and return their human baby. Other, kinder, remedies were for the parent to do something ridiculous, like boil water in an egg shell, which would startle the changeling baby into speaking and betraying its real parents, who would then be forced to return for it.

Chollers – Devonshire dialect word meaning *cheeks*.

Citramer – the Hospitallers' origin and spiritual heart lay in the Holy Land, and when that was lost, in Rhodes. So, no matter which nation they were from or where they served, they referred to the Holy Land or Rhodes as being *Citramer* or *home*, whereas their priories in England and mainland Europe were *Outremer* meaning *overseas*.

Clooties (also spelt *cloutie* or *cloughtie*) – These were rags or strips of cloth torn from the garments of people who were sick or seeking good luck that were dipped in the holy wells, then tied to a nearby bush or tree, especially a thorn or oak. The affected part of the body might first be washed using the wet rag, before the clootie was hung in the tree. If someone couldn't visit the holy well themselves, a relative might bring a clootie for them. Some people regarded the clootie as an offering to the goddess or spirit of the water, in which case the rag would be torn from a garment they valued.

In other cases, the intention was to transfer the sickness or bad luck to the clootie and leave the ill-fortune at the well, in which case the rag might be torn from a garment they hated or associated with the start of their illness or misfortune. As the rag disintegrated or eventually blew away, the ailment would vanish. If anyone removed a clootie belonging to someone else they would risk that person's illness or ill-fortune passing to them.

Even when a holy well was rededicated to a Christian saint the practice continued and it still does today, along with the custom of throwing coins (silver) into wells or fountains as offerings to the gods or spirits for good luck or to make a wish come true.

Cracky-wren or **Crackety** – are Devonshire dialect names for a *wren*. Some claim that they refer to the bird's tiny size, others that they derive from another of its dialect names, *Crackadee*, which imitates the sound of a wren's alarm call. The wrens' song is said to contain more notes than any other birdsong. This, together with the wrens' habit of nesting in caves and tombs, was the reason people believed the bird carried messages between our world and the world of the spirits or the dead.

Crockern – *See **Wisht Hounds**.*

Cross and Pile – a popular gambling game played by all classes, which today is known as 'heads or tails'. Two players each chose a side of a coin and bet on their chosen side landing uppermost when the coin was flipped in the air. In the Middle Ages, the game was known as 'cross and pile' because one side of a medieval coin was stamped with a cross or a Christian symbol.

Hammered coins were produced by placing a metal disc of a certain weight between two patterned dies, then striking the upper die with a hammer to stamp the pattern on the coin. The bottom die was called a pile, and by extension the reverse side of the coin, which became imprinted with the bottom die's pattern, also became known as the pile or pyl.

King Edward II was so fond of cross and pile that he was said to have borrowed money from courtiers and even his servants to keep playing.

Cross formée – a Greek cross with four slightly fluted arms or the thicker version – a *cross pattée* – appears the most likely to be the version of the cross that the Knights of St John had emblazoned on the front of their surcoats and at the top of their left sleeves during the thirteenth and four-teenth centuries, since this is the type of cross most often depicted on their seals and in carvings on their buildings from the thirteenth century. The original cross used prior to this period was probably the patriarchal cross, which has two cross-bars. The now familiar eight-pointed Maltese Cross, still used today by the St John Ambulance, was probably not adopted until the sixteenth century. But in the early history of this order there were often variations in design between countries.

Donats – wealthy lay men and women could become donats of the order of the Knights of St John, by making a single generous donation to the Hospitallers of money, land or property in exchange for spiritual benefits, while continuing to live as laity. Donats could be buried in Hospitaller grave-yards and hoped their act of piety would shorten their time in Purgatory, particularly as the Knights of St John and their priests would remember them in their prayers and says masses for their souls.

Donats had to pledge that if they ever decided to enter holy orders, they would join the Hospitallers. A number did so in old age, which enabled them to be cared for when they were frail. But some waited until they were on their deathbeds to become fully professed members of the order. Often husbands and wives would both become donats and mutually promise that if one died, the surviving spouse would make a full profession and enter the order rather than remarry, usually to ensure their estate was maintained intact for chil-dren of that marriage. A number of wills survive from this period which contain such provisions. But it was also accepted that many donats would never want to become fully professed.

Duru – Old English word meaning *door*. A waxing gibbous moon was believed to be the doorway into other realms. It was a place between darkness and light, a time of beginnings. Different curses and charms were thought to be most effec-tive when created at certain phases of a moon – waxing, waning, full or new – depending on what they were intended to do.

Egurdouce – it meant *sweet and sour*. It was sauce used with a variety of meat, including hare, mutton and beef, and also fish. Meat such as rabbits' legs and saddles were deep fried in lard, then laid in a dish and covered with fried onions and fried currants. A sauce of melted lard, red wine, vinegar, sugar, pepper, ginger, cinnamon and salt was made and seethed until it was thick, or thickened with egg yolks or breadcrumbs, then poured over the meat and served.

Garderobe – this was a feature of religious houses or wealthy manors and castles. It was a tiny room enclosing a lavatory, often built on an upper storey projecting from an outer wall, so that waste would fall through the hole beneath the seat into a pit, moat or river. Its height made it harder for thieves or invaders to climb up through the hole. If it was built in the middle of a complex on the ground floor in a monastery, a stream of water would usually be diverted from a nearby river to run beneath it and carry off the waste. Clothes were often hung in garderobes as the stench of urine and excrement was said to keep away moths.

Gleedy – the character's nickname is derived from *gleed*, an old word meaning a *squint*.

Graffing – ancient dialect word meaning *digging*.

Hag-stones – small holed stones or pebbles, which were thought to guard against evil. Hung on the back of a door, they prevented the entry of evil spirits or witches. Hung over a bed, they guarded against illnesses and nightmares. If a key was attached to them, the combination of the stone and iron was thought to be a powerful amulet against bad luck and protected the house from thieves. Many holed stones can still be found on doors in houses and barns today, and are still hung on key rings.

Hare's beard – one of the many old country names for mullein (*verbascum*). It was called this because the plant is covered with white hairs. When dried, it could be used as kindling and as candle wicks or tapers, hence other names such as *hag-taper* and *Our Lady's candle*. Witches were said to use mullein in their spells and it may be one of the nine herbs referred to in the Anglo-Saxon Nine Herbs Charm, but the plant was also thought to be very effective in warding off demons and night terrors. The juice was believed to remove warts and the leaves and flowers were infused to make a cough, cold and bronchitis remedy.

Horn gesture – this was made by tucking your two middle fingers under your thumb, while extending your index and little finger, like two horns, towards the person or object you feared. It was used to ward off evil, especially to protect against sorcery, or defend you from those you thought might be 'overlooking' you with the evil eye.

Imbolc – is Brigid's Day, celebrated around 31 January to 2 February, approximately midway between the winter solstice and the spring equinox. The origin of the word *Imbolc* is disputed. Some authorities claim it comes from the Old Irish meaning *in the belly*, and refers to the pregnancy of ewes. Other sources state that it meant to *ritually wash* or *cleanse* and some link the origin of the word to *milk*, particularly ewe's milk. During the Middle Ages, many of the traditions, images and beliefs associated with Imbolc became merged with the Christian festival of Candlemas on 2 February, which celebrated the ritual purification or cleansing of the Virgin Mary, forty days after the birth of Christ. *See also Historical Notes – Brideog and Brigid's Cross.*

Larks-claw – *Delphinium consolida*, also known as *larkspur*, *larks-toe* and *larks-heel*. It was used to pack wounds and treat the stings of scorpions. Oil from the seeds was extracted to kill lice. If tossed in front of any venomous beast, it was widely believed the creature would not be able to move until the herb was removed.

Livier – the living space for people. Dartmoor long houses often consisted of just two rooms built sideways on a slope. The livier, at the higher end of the slope, was where the family ate, slept and worked, and the room on the lower end was known as the *shippon* and that was where the live-stock were housed, especially during winter. The livier would have a central peat fire, but no chimney or windows. An open drain ran down through the middle of the shippon and out through the wall at the lowest end of the building to carry away animal waste. Both beasts and humans used the same door, and the partition between livier and shippon was often only a half-wall, so that the heat from the animals helped to keep the humans warm and the beasts, especially young or ailing ones, benefited from the warmth of the hearth fire in the livier.

Long-cripples – Devon dialect word for snakes, usually adders, but it can also mean dragonflies or lizards. Some leech or healing wells were given the name *long-cripple*, either because they cured snakebites or because they cured the same ailments as adder skins were thought to do, such as headaches and rheumatism. The groves of dwarf oaks on Dartmoor, especially Wistman's Wood, are home to hundreds of adders, which take shelter among the rocks and are said to be the most venomous adders in the British Isles.

519

Lungwort – *Pulmonaria officinalis.* Folk names include *Bloody Butcher*, perhaps because the flowers change from pink to blue like a butchered meat. It was also known as *Adam and Eve, Mary and Joseph, Mary's Tears, Spotted Dog, Beggar's Basket* and *Our Lady's Milk-sile*, because the white spots on the leaves were said to be the stains or 'sile' made by the drops of milk that fell from the Virgin Mary's breasts when she was feeding the infant Jesus on their flight to Egypt. The leaves resemble the shape of a lung therefore it was believed they cured lung conditions in both humans and livestock. Ointment made from the leaves was used to treat ulcers on the sexual organs. The plant was also said to banish sorrow and depression, and to 'comfort the heart'.

But to confuse matters, the green lichen, *Lobaria pulmonaria,* which grows on tree trunks and rocks, is also known as *lungwort* because it, too, is said to resemble lungs and was also used to treat ailments of the lungs, especially asthma.

Manuterge – from the Latin *manus*, meaning hand, and *tergēre*, to wipe. A small white linen towel used by the priest during mass for drying the hands after washing them in the *lavabo* (ewer and basin). Today it is often called the *lavabo towel.*

Mazy – local dialect word meaning *stupid* or *mad. Mazy-jack* was often used to refer to someone who was considered to be the village simpleton.

Pigseys – old dialect name for *pixies. See also Historical Notes – Pigseys.*

Pinfold – an enclosure, often circular, in which animals were penned for the night. Most medieval drovers' roads had pinfolds built along them at intervals where animals being driven to and from market could be kept overnight to prevent them straying and keep them safe from predators and thieves while the drovers slept. In Devon, the enclosure usually consisted of broad banks surmounted by a thick hedge, or wide dry-stone walls that had tunnel-like 'kennels' built into the hollow between the stones, so that dogs could stay dry and warm while keeping watch. On Dartmoor, there were also many Bronze Age stone circles and the remains of large Bronze Age circular stone-huts; farmers and drovers often turned them into pinfolds to save building one from scratch.

Responsions – each priory of the Knights or Sisters of St John was required to send a third of all their income from produce, rents and financial dealings to the central or mother house in their country. In England, this was Clerkenwell, north of London. Knights or paid agents then took the money to the procurator-general in Avignon from where it was sent to the treasurer at Rhodes. However, records show that during the fourteenth century the responsions from the English Tongue were not always dispatched to Rhodes, because the English Hospitallers were heavily in debt. In times of war between France and England, the King refused to allow them to be delivered, since this was giving money to England's enemies, not least because French-speaking knights were in the majority on Rhodes. At such times, money from the English order of the Hospitallers was given to the King to help in the defence of England.

Rouncy – (also spelt *rouncey* or *rounsey*) – an all-purpose horse, used for riding and battle. The huge *destrier*, capable of carrying a man in full armour with weapons was the most highly prized war- and jousting-horse of the Middle Ages, but the least common, and the destrier was not a good riding horse over long distances. The agile *coursers* were often preferred for hard battles fought at close quarters, but only wealthy knights could afford either of these. A poorer knight or man-at-arms would use a rouncy for fighting and distance riding. Often in the Middle Ages the nature of the expected battle would dictate the type of horse required, so if a swift pursuit was anticipated, the knights would be advised to bring rouncies rather than destriers. None of these horses were specific breeds; rather, the size, build and training of the individual horse determined what it was called.

Sarcenes – the full name of the dish was *bruet sarcenes*. A bruet was stew containing meat. A strongly flavoured meat, such as venison or goat, was boiled and drained, and the water it had been cooked in mixed with ground nuts to produce almond milk. A sauce for the meat was made from the almond milk, flour, cloves and powdered spices. It was boiled until thickened, then wine, sugar and salt were added, with a deep red dye made from alkanet root.

'Almond milk' was made from any kind of ground nuts that were available. The powder was kept dry, then mixed with a liquid, such as wine or stock, just before cooking to create a 'milk'. It was used in many sauces and puddings where today we would use dairy milk: animal milk soured quickly and wasn't available all year round, so it was mainly reserved to make butter and cheese that could be stored.

Scroggling – an old word for a tiny, shrivelled apple, which no one bothers to harvest because it is worthless; by extension, a person who is useless, good for nothing.

Shaggy mane – the edible fungus, *Coprinus comatus,* also known as Shaggy Ink Cap, because it makes good black ink. It often grows where animals have dropped dung, probably generated from spores the beast had ingested. If it is cooked after it has started to open it can dye all other ingredients black. Not to be confused with Common Ink Cap, *Coprinopsis atramentaria*, which is poisonous if alcohol is drunk with it or after eating it.

Simples – any herbs with a medicinal property.

Sledges – on Dartmoor, wooden sledges were not principally intended for use in snow, but were used to drag fodder, kindling or other supplies across the open moorland all year round. With their broad metal runners, they could be pulled over heather, rough grass, stones and mud much more easily than a wheeled cart. Old horseshoes were hammered into rough circles and nailed to the sides of the sledge, through which ropes or poles could be attached, allowing the sledge to be dragged by horses or people.

Strappado – a method of torture favoured by the Inquisition in which a victim's hands would be bound behind their back. Their wrists would be tied to a rope slung over a high beam and they would be hauled upwards, thereby dislocating their shoulders. Victims were often also repeatedly and violently dropped several feet in the air, but jerked to a stop by the rope before they hit the floor. This generally broke their arm and shoulder bones and caused serious damage to the spine. Weights might be attached to other parts of the body, such as feet and genitals, to inflict even greater agony and damage.

Todde – Old English, meaning *fox*. By the twelfth century it had become a nickname for someone who resembled a fox in some way, perhaps in hair colour, or was foxy, cunning or crafty. As with many early nicknames, it was eventually adopted by some families as their surname, usually spelt *Todd*.

Trapes – old dialect word meaning a *slattern*, *slut*, or *slovenly woman*.

Veckes and gammers – derogatory medieval terms meaning *old women* and *old men*.

Viaticum – the eucharist (bread and wine) given by a priest to the dying. It could be offered with or without extreme unction, which is the anointing of the sick with chrism (holy oil). Receiving viaticum formed part of the last rites in the medieval Church. Viaticum means 'provision for a journey', *via* meaning *way*. The journey is the one the dying person will make from this world to the life after death.

Whortleberry – *Vaccinium myrtillus*, commonly called a bilberry. It grows on low, wiry bushes. The fruit is blue with a strong sweet scent and flavour. The flesh and juice are red. They were known as *whimberries* in Wales and *whortleberries* in Devon. But many villagers on Dartmoor called them *hurts* or *urts*.

Will worth – cunning women believed a charm or curse would only work if the charmer firmly declared exactly what they wanted, really meant it and imagined it happening. In that sense, the physical charm or curse, whether it was a bunch of herbs or a curse written on lead, had no power in itself but became the focus of the cunning woman's willpower or wishes. *Will worth* meant to state your intention or your will aloud as you directed the charm or curse towards whatever you wanted it to affect.

Wisht Hounds – mythical beasts, otherwise known as *hellhounds*. *Wisht* originally meant to *bewitch* or *invoke evil*. Wisht hounds are a pack of huge black dogs with savage fangs and red eyes that hunt across Dartmoor at night, preying on lost souls and unwary travellers. After the coming of Christianity, they were also thought to snatch unbaptised babies and children from their beds and devour them, or drop their bloody corpses at the feet of their neglectful parents. To protect the baby from such a terrible fate until the infant could be brought for christening, which in winter or in remote areas might be some months, a piece of consecrated bread (the Host) was placed beneath the child's pillow.

The hounds' kennels are said to be in an ancient grove of twisted oaks, which still stands today, and is known as Wistman's Wood. Many of the stones in the wood are balanced such that they resemble dog kennels, and foxes, badgers and rabbits make homes in them.

As they run across the moor, the ferocious wisht hounds are followed by a lone huntsman swathed in black and riding a huge black or skeleton horse. He carries the hounds' 'kill' in a sack. Some say the huntsman is the devil himself, others that he is Old Crockern, an ancient god or guardian spirit of Dartmoor, whose face can be seen in profile on the rocks of *Crockern Tor*. It is he who releases the hounds from their kennels in Wistman's Wood whenever someone threatens the moor. *See also Historical Notes – Black Hounds*.

A GATHERING OF GHOSTS

Reading Group Guide

- *'You can stamp and frown as much as you please, Mistress, but this is a battle I am going to win.'* Prioress Johanne rules the priory with a firm hand, but her authority is challenged with the arrival of Knight Brother Nicholas. To what extent is this book about power?

- The well sits at the heart of life in the priory – and is central to the mystery of the story. What did you make of the plagues? Were you surprised by the identity of the blind boy? Can we find an earthly explanation for the strange happenings?

- *Great grey clouds rose up, one behind another, like walls of stone, but a beam of dazzling sunlight, thin and straight as a golden arrow, slipped between them.* What did you make of the wild and remote Dartmoor setting? How does the myth-laden landscape frame the story?

- Compare the three first-person narratives – Sorrel, Johanne and Morwen – with the chapters that take place at the priory. Does this affect how we perceive the three different women? And what impression do we get of Nicholas?

- Discuss the theme of survival in the novel, and how it shapes the actions of the characters.

- *I saw black Ankow galloping across the moors on that skeleton of a horse, with his hounds baying at his heels. I knew he was hunting souls.* How can we understand the tensions between the different models of faith and tradition in the book – the conflict between pagan and Christian beliefs, magic, wisdom and ancient lore?

- How is the role of family presented in the novel? Think about Kendra and her daughters, the home Sorrel leaves, and the bond between Johanne and Sebastian. Is family something we're born into, or something we choose?

- Why do you think Todde wants to help Sorrel? How does fear influence the way people relate to each other?

- *Not all of our noble sisters enter the order entirely by their own choice, though they must swear that they do.* Discuss the sisters' different reasons for 'choosing' a life of servitude.

- *A Gathering of Ghosts* is set against the backdrop of a terrible famine which caused widespread poverty, desperation and displacement of thousands of ordinary people as they were forced to travel across Europe in search of food or better conditions elsewhere. What parallels can we draw with our world today?

THE PLAGUE CHARMER

1361. An unlucky thirteen years after the Black Death,
plague returns to England.

When the sickness spreads from city to village,
who stands to lose the most? And who will seize
this moment for their own dark ends?

The dwarf who talks in riddles?
The mother who fears for her children?
The wild woman from the sea?
Or two lost boys, far away from home?

Pestilence is in the air. But something
much darker lurks in the depths.

Available from

REVIEW

THE RAVEN'S HEAD

1224. Langley Manor, Norfolk. Lord Sylvain has been practicing alchemy in hiding for years and now only the Apothecary's niece can help him with final preparations to forge the Philosopher's Stone.

Alchemy calls for symbols – and victims – and when a man in possession of an intricately carved raven's head arrives at the Manor in a clumsy attempt at blackmail, Sylvain has both symbol and victim within easy reach.

But the White Canons in nearby Langley Abbey are concealing a crucial, missing ingredient . . . Regulus, a small boy with a large destiny.

Available from

REVIEW

THE VANISHING WITCH

Lincoln, 1380. A raven-haired widow is newly arrived in John of Gaunt's city, with her two unnaturally beautiful children in tow.

The widow Catlin seems kind, helping wool merchant Robert of Bassingham care for his ill wife. Surely it makes sense for Catlin and her family to move into Robert's home?

But when first Robert's wife – and then others – start dying unnatural deaths, the whispers turn to witchcraft. The reign of Richard II brings bloody revolution, but does it also give shelter to the black arts?

And which is more deadly for the innocents of Lincoln?

Available from

REVIEW